ENGINEERING DESIGN

ENGINEERING DESIGN

JOHN STEPHENSON
Senior Lecturer in Mechanical Engineering
The University of Auckland

R. A. CALLANDER
Associate Professor of Civil Engineering
The University of Auckland

JOHN WILEY & SONS AUSTRALASIA PTY LTD
SYDNEY New York London Toronto

Library of Congress Catalog Card Number: 73-5277

Photocomposed by Hartland & Hyde Typesetting Pty Limited
Printed by The Macarthur Press Pty Ltd, Sydney Australia

10 9 8 7 6 5 4 3 2 1

National Library of Australia Cataloguing in Publication data:

Stephenson, John
Engineering design/ [by] John Stephenson [and] R. A. Callander.—Sydney: Wiley, 1974.
Index.
ISBN 0 471 82210 8.

1. Engineering design. I. Callander, R. A., joint author. II. Title.

620.0042

Contents

PART II

PART III

APPENDIXES

Acknowledgements

We would like to thank the people who have helped us while we have been writing this book. Our colleagues at the University of Auckland have provided information and ideas, and our approach to teaching design has been influenced by the many discussions we have had with teachers, engineers and students.

A full tribute to the many busy secretaries who worked on the typing would require a long list. We are grateful to them all. We would thank, in particular, Gail Clark, Julie Jackson, Glynis Margetts and Teresa Marinovich.

Preparation of the worked examples and of the diagrams has called for an unusually large amount of artwork. This was done by JASMaD, an Auckland firm of consultants, and we wish to acknowledge the cooperation we have had from them. It is also a pleasure to acknowledge our continuously cordial relations with the staff of our publisher.

Finally, we acknowledge with gratitude the help and encouragement we have had from our families.

Preface

This book is intended as an introduction to engineering design for students in all branches of engineering. All of the material is suitable for use in the second and third years of a four-year course and, with suitable selection of topics, one-year or two-semester courses can be served.

The book deals both with mechanical and with structural design, and it is hoped that teachers using it will not restrict their selection to one or other of these. We believe that all engineering students should have an insight into the different approaches and methods used in design, and that specialisation should not occur early in their course. These objectives can only be achieved if teaching for all the students is based on a selection made from all parts of the book.

Certainly some parts lend themselves better to this approach than others. Part I and Part III, which are described in Chapter 1 as touching upon the philosophy and the art of design, can be regarded as being of interest and importance to all disciplines. In Part II, which deals largely with the science of design, the chapters on "Fatigue Design" (7), "Shaft Design" (8), "Ties and Struts" (12), "Connections" (13), and "Steel Beams" (15), are sufficiently non-specialised to be suitable for a broad-based course of the type we advocate.

Design is a creative activity and students should be given every encouragement to develop their ideas and talents in a design course; at the same time a sound basis for the application of their knowledge of mechanics and materials and an awareness of the capabilities of existing technology must be established. All these demands need not necessarily conflict, nor will an emphasis on sound practice stifle the imaginative use of existing components or hinder ideas for new ones. Yet a teacher's hopes of nurturing creativity and inventiveness can be swamped by the volume of material that claims attention. An attempt is made in this book to deal with this material and to support the teacher who directs some of his attention to the wider implications of design.

The difficulty for the student at this stage of his education often lies not so much in the volume of material to be assimilated, as in the transition he must make from the predominantly analytical approach he has hitherto encountered to the design approach. In the future he must learn to blend synthesis with rule of thumb, analysis with empiricism, inspiration with codified regulations. Both the teacher and the textbook must help to minimise the difficulties in making this transition.

There are those who argue that engineering design should not be taught in an undergraduate course. We feel that students benefit from the utilisation in design of several of the more fundamental topics that are being taught concurrently and from the reinforcement it gives to this work.

Engineers are concerned not just with formulas and hardware but also with the standards of life. A definition often quoted is that engineering is "the art of directing the great sources of power in Nature for the use and convenience of Man". The success and standing of professional engineers must depend not just on their success in harnessing these forces but also on their concern for the effects of their technology on the natural surroundings. The challenge is there. In design, perhaps more than in any other subject, there exists the opportunity to come to grips with it.

Auckland, New Zealand
August, 1974

J. STEPHENSON
R. A. CALLANDER

Units and Symbols

We have used SI units throughout the book. For example, the *pascal* (Pa) is the fundamental unit of stress and the *megapascal* (MPa), a convenient multiple, has often been used in the book for calculations involving stresses. Accepted practice in dimensioning is to use the *millimetre* as a unit of length; with forces measured in *newtons* and many significant areas in square millimetres, a unit of stress, N/mm^2, arises very naturally. It is a convenient unit in practical design and allowable stresses are given in N/mm^2 in several British design codes. Although many will find offence in such a practice, designers will find it useful to bear in mind that 1 MPa equals 1 N/mm^2 and to compare the relative values of convenience and pedantic correctness.

Much of the information that we have collected in tables, charts and equations was in Imperial units in the original references. A good deal of conversion has been necessary, work that has not been without its difficulties. There has often been doubt about the range of values that will be selected when SI is fully implemented and about the extent and direction of rounding-off for converted values. The opportunities for errors to occur in a book of this nature have, no doubt, been compounded by the difficulties just described. We have taken as much care as we can to avoid these and other errors and will be grateful to readers who inform us of those that have escaped us.

Because most of the topics dealt with have their specialised notation, it has not been possible to use a consistent set of symbols throughout the book. We have adopted in each case the symbols best suited to our purpose, and have provided a list of symbols with each chapter.

PART I

1 The Basis of Design

1.1. INTRODUCTION

Engineering and design involve such a wide range of activities that it is difficult to give a definition of them that is succinct and descriptive. The definition quoted in the Preface delineates the boundaries of engineering, but it is useful at this stage to consider the activities in more detail.

Engineering is concerned with providing mankind with transport and communication services and with the facilities to produce and distribute goods of all sorts. The engineer supplies the scientific knowledge and technical experience needed to plan, design, construct and manage these services. His work at a professional level is not routine and his skills are to be used to solve new problems or provide new solutions to old ones. Several disciplines may work and cooperate in one field – for example, in the production of power, civil, mechanical and electrical engineers will be involved in the design and construction of plant.

Design is a central part of engineering and involves the specification and detailing of the machines, structures and systems which are to be used in engineering. It requires a wide range of knowledge and skills and is, at its best, a creative activity. As well as calling on past experience, the designer must make the best use of available research and development data if his work is to progress and prosper. The ideas and designs that are developed in response to a demand must be communicated to those responsible for production, and it is part of the designer's task to see that this is done accurately, whether it be by means of drawings, specifications or a computer programme.

In this chapter the general approach to design is considered and some of the skills and techniques involved are discussed.

1.2. THE DESIGN APPROACH

The approach to the solution of a design problem is likely to take a similar form, whatever the magnitude of the problem. Briefly, the steps involved are as follows:

(a) statement of the problem;
(b) collection of relevant data;
(c) consideration of possible courses of action;
(d) selection of preferred solution;
(e) detailed development of design;
(f) construction of prototype, development work;
(g) production.

In practice, the steps will not be as clear-cut as indicated here, and there may be feedback of information from one stage to another. However, even if the designer does not consciously recognise it, he is likely to progress through some such routine in working towards a solution; a formalisation of these steps can help in planning and executing the efforts that constitute design and students should ensure that each step is given adequate consideration.

Although seemingly a simple procedure, the formulation of the problem itself is an important step which can affect the success of the whole programme. The freedom of action of the designer may be restricted if he is asked, not to provide a solution to a particular demand, but to design a certain type of machine to fulfil a specific need. He must ensure that the problem presented to him is in fact the correct one and that his brief has not been worded in such a way as to restrict unnecessarily his freedom of action. There are, for example, many ways in which a test vehicle can be accelerated. The energy used can come from the elastic energy of solids and gases, the kinetic energy of a fly-wheel, from chemical reactions, rocket motors or linear induction motors – all these may have possibilities depending on the mass and the acceleration required. To pose a problem of this type as "the design of a compressed air system to accelerate a test vehicle along a horizontal track" will restrict the freedom of action of the designer, but may mean that alternative methods have been assessed and found wanting in some way. The designer must establish that this is in fact so before he can correctly state the problem. Note that the mention of a horizontal track will inhibit ideas of using gravity by dropping the vehicle from a height.

The collection of the data to be used in solving a problem may be the task of the designer, or he may be presented with the required information in a specification. The loading spectrum on a bridge will be needed if the structure is to be designed for fatigue in the most economical manner; such data may be available in the literature or may have to be measured or estimated from a study of existing bridges.

When an aircraft is to be designed to compete on certain routes, a great deal of information must be collected before an outline of the various possible solutions can be shaped. Some of this information will be factual, some of a statistical nature – perhaps extrapolated into the future – and some will be based on opinions and judgements. In the first category come such quantities as route lengths, airfield altitudes and allowable runway loadings; in the second, records of passenger numbers and meteorological observations. In the last category come the expressions of hope and expectation from the people who may use the equipment, and facts and fictions that make up consumer preference. The need for large windows – or no windows at all – the demand for front- or rearward-facing seats, the preference for particular styles and colours of furnishings – these are some of the imponderables that will call for compromise and will complicate decision-making.

The point has been made that some information may be presented to the designer in the form of a design brief or specification. It will become clear when the writing of engineering specifications is discussed in more detail (see Chapter 24) that the process may carry right through the first four stages now being described. Thus the writing of a specification can be considered an important part of design and can spell out success or failure for the product.

At the conclusion of stage (b), the designer may be able to visualise a number of solutions, some perhaps clear-cut, others just flashes of what might be attainable. These must be sketched out and criticised, torn apart and re-arranged, refined or discarded, until a number of possible procedures can be outlined, all of which have some chance of carrying through as a final solution. Here it may be necessary to discard some of the favoured ideas which at first sight looked full of promise, but which will not stand up to careful analysis, or for which background data are lacking; they can perhaps be stored for future reference.

Although some of this work can be a mental exercise, it is important that the engineer should be able to record his ideas and decisions, or pass them on to others, clearly and unambiguously. Engineering sketches are a vital aid for this purpose, and the availability of such sketches, drawn to scale, will greatly enhance the value of informal discussions at this stage. The sketches are also required to check the feasibility of different arrangements and to provide tentative dimensions for the computations which will confirm or modify sizes already adopted. The sketches and computations may not be very detailed, but they must be taken far enough to reduce to a minimum the hidden difficulties which can appear when the detailed design is done.

On reaching step (d), the selection of a preferred solution may be made by one man or by a committee. In some cases, more than one design will be selected for detailed development – the final selection being left until a clearer picture of their merits emerges. Here again, the ability of the engineer to communicate his ideas to others will be put to test. He will often be required to submit to management a design report in which the merits of the selected scheme are argued and perhaps compared with alternatives. It may be convenient to terminate student projects at this stage, and a description of the form of such a report is given in Chapter 24. The fate of the design report may rest in the hands of non-technical people, and the clarity of writing and drawing will count for as much as the accuracy of computations and estimates.

The design process proper ends with the completion of the detailed design, but the designer will still be expected to keep in touch with the problems and the successes that come during construction or production and even later when the project is in service. The design department cannot operate in isolation from other divisions in an engineering organisation and design work must fit into the overall pattern of activity. This is indicated in Fig. 1.1 where the makeup of a typical engineering department is shown, along with the production, sales, maintenance and research groups.

This diagram, which is focused mainly on engineering services, illustrates the way in which the design functions fit in with the other activities. The heavy rectangles illustrate positions of authority and the solid lines indicate lines of authority and communication. The dotted lines illustrate informal channels of communication between various departments or activities enclosed in light rectangles. The ideas and requirements that originate in other departments will be passed on to the design group through these channels, which should be clearly defined so as to foster easy communications. The detailed organisation will depend on many factors, such as the size of the company and type of product.

As the products of our technology become more complex and the design process more demanding, increased interest is being taken in the details of the

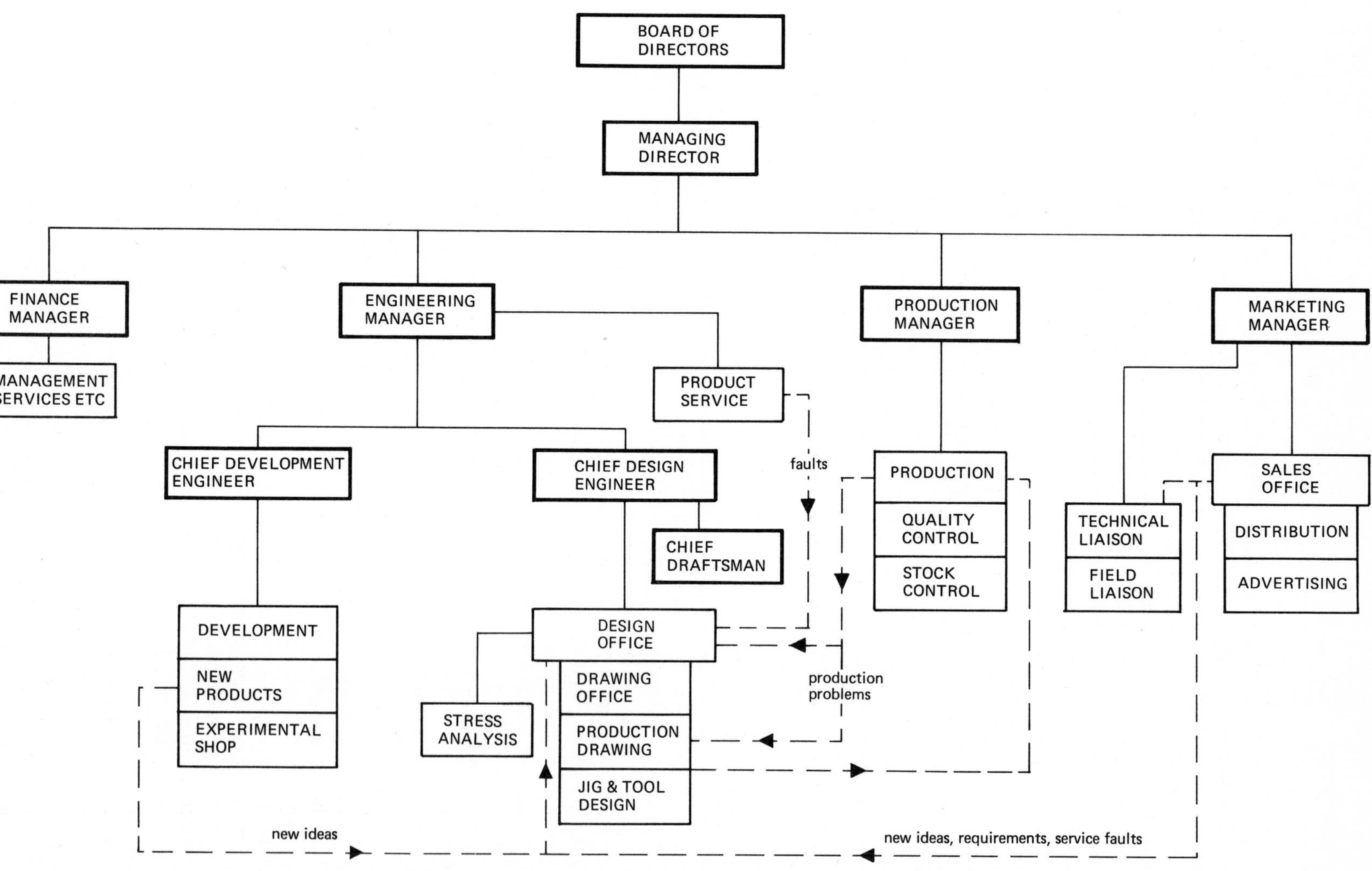

Fig. 1.1. *The place of the design office in an engineering organisation.*

way in which individuals set about design. A variety of methods and techniques has been investigated and formalised, and procedures have been developed to assist and control the search for ideas and answers. A description of many of these methods has been given by Jones[1]. In classifying the various design methods he identifies three stages in the design process – divergence, transformation and convergence.

The collection of new data and the building up of possible solutions constitute, at the outset of the design process, the divergent stage. In the second stage, the manipulation of data that have been collected may be done at the linguistic or at the mathematical level, or the transformation may take place in some "creative leap" which can be stimulated by a variety of methods but cannot be controlled. In the final, convergent stage, the detailed shape of the solution is reached as various sub-problems are eliminated. These stages may be related to the more formalised steps described at the beginning of this section.

Jones points out that one objective of the transformation must be to pattern the problem in such a way that none of the sub-problems left to the convergent stage are critical ones that may require a repetition of the cycle. In this book we are largely concerned with providing the student with the background that is required to ensure that he understands and can manipulate these sub-problems. Without this background the success of the convergent stage cannot be assured.

1.3. ANALYSIS AND SYNTHESIS

There are two fundamentally different approaches that can be taken in steps (c) and (d) described in Section 1.2.

If a conventional solution is adopted, it is possible, on the basis of experience and preliminary calculations, to reach quickly a stage where the design may be analysed in detail to determine the extent to which it meets the specification. From these analyses of strength, rigidity, performance and so on may spring a variety of new ideas and requirements which can be fed back to modify the initial solution; but the basic approach is a "cut-and-try" one in which a solution is adopted and analysed. The first solution in this analytical approach need not necessarily be conventional; invention and innovation can be the basis for a solution which is then subjected to analysis and further development. However, there is a tendency for this approach to be used unimaginatively and to handicap advance in design.

As the name implies, synthesis is a process in which the solution is built up, each element being selected so that the synthesised whole will best meet the various design requirements. An idealised view of the process of synthesis is one in which the specifications and requirements are formulated in such a way – perhaps in the form of equations and boundary conditions – that their manipulation will lead to some optimum solution. Since these requirements may consist of spatial relationships, economic and manufacturing requirements, material and reliability specifications, and even aesthetic qualities, it is clear that a direct mathematical assault on a major problem using this approach in its fullest sense would require enormous resources and information that are not at present available. However, there are situations where these methods can be used to produce optimum results; also in many cases they are used along with an analytical approach, in a step-by-step process, in such a way that it becomes hard to separate one from the other.

The methods of synthesis promise to provide the ideal solution, but only to the limit set by the information that is fed into the system and processed, perhaps by computer. Under these circumstances, the designer has less control over the compromises and modifications that are needed to give a balanced design: his art has been formalised and a balance sought by the application of numerical figures of merit. The complete synthetic design approach is a complicated one, but the objectives can be borne in mind and can serve as a guide to the collection of data and development of skills and techniques that will be required to make increasing use of it. Until the resources to apply it are available, the skill and art of the designer will count for much in the success of a design.

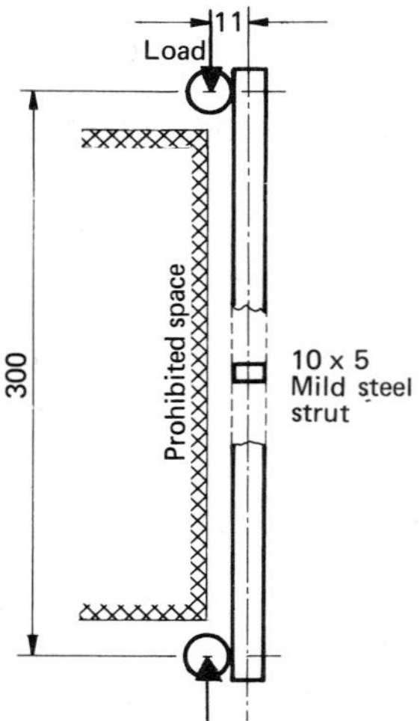

Fig. 1.2. *Eccentrically loaded strut.*

The difference between the two approaches is illustrated by a problem involving both analysis and synthesis. This was tackled by a group of engineering students who were first asked to analyse the failure load and mode for a strut eccentrically loaded in compression (Fig. 1.2); the offset loading was determined by a restricted region in space into which the structure could not encroach. Several nominally identical prototype specimens were then tested; the second part of the exercise was to design and build a minimum-weight structure which, subject to the same space restrictions, would have the same failure load as the prototype.

In this case, the weight of the prototype structure was 1.72 N; the failure load, defined as the first maximum in the load deflection curve, was 1110 N. Further details of this exercise are set out in Problem 1.1.

In analysing the failure load for the prototype, it was first necessary to consider the limiting conditions and to determine the critical one. When this was done, it was found that the column would fail in buckling in the plane of the eccentric loading. That this analysis was straightforward, was confirmed by the good agreement between the forty estimates (see Fig. 1.3). The main discrepancies arose from the yield strength assumed: some students were satisfied to take a typical value for mild steel, while others tested a sample of the material used for construction. This analytical procedure is little different from an exercise in applied mechanics.

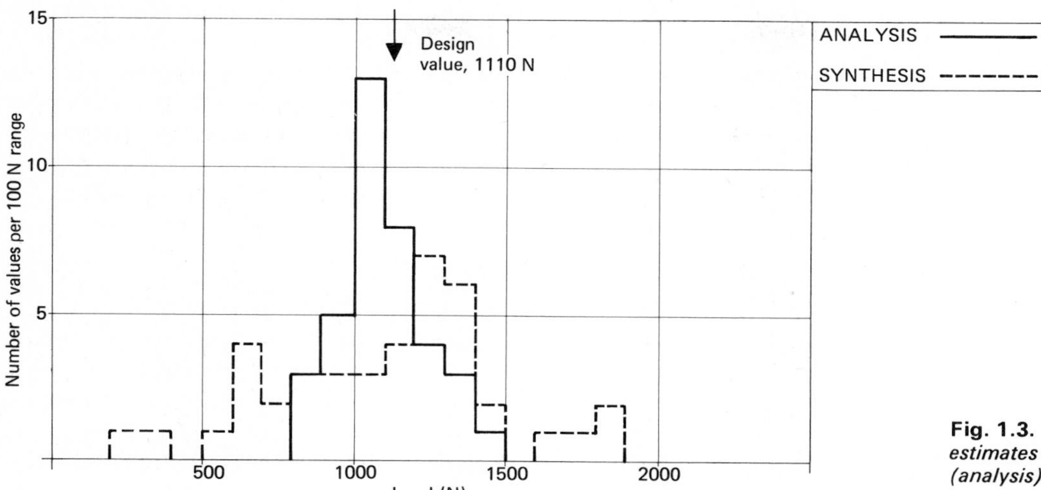

Fig. 1.3. *Frequency distribution for estimates of failure loads for strut (analysis) and for test values on struts designed and constructed (synthesis).*

When attempting to synthesise a suitable solution for a minimum-weight structure, decisions must be made as to the basic layout, material to be used and joining methods. Once the basic layout has been fixed, the strength requirements at the critical sections can be used to determine suitable dimensions; these sections can then be connected up to complete the design.

In the exercise now being considered, this procedure led to a number of different layouts and to a range of failure loads considerably wider than that encountered in the analysis (see Fig. 1.3). In the process of synthesis with some design objective other than strength and stiffness – in this case low weight – there are more uncertainties and more locations at which, because of attempts to cut back weight, failure may take place. This factor was exemplified by the number of the samples that failed, not in the main buckling mode, but at some apparently minor joint. The lightest strut to meet the specification weighed 0.40 N.

1.4. DESIGN SKILLS

The student can sometimes bring to bear on his design problems a lively imagination that is unfettered by a knowledge of the traditional ways of doing things. This freedom may help to offset his initial inexperience in practical engineering and production methods; however, his inexperience will show up in the stages of the design where a knowledge of what can be achieved and what has in the past proved impractical is helpful. His lack of knowledge of existing technology will be a handicap in step (e) – the detailed development of a design. This highlights the call for different skills at different stages in design; an imaginative approach is useful in the initial conceptual stages and practical design and manufacturing experience are essential when the detailed design is taken up.

Some individuals have a natural talent for design that is an expression of special creative and inventive powers. Many of the skills that engineers can acquire during their education and training – for example, a capacity to think and visualise in three dimensions – will always be more highly developed in such individuals. It is worth now considering briefly some of the other qualities, innate and developed, that determine the design competence of an engineer.

As he nears completion of his academic education, the young engineer will usually be competent in certain special fields. Unless his training has specifically been designed to give him detailed and fairly long-term industrial experience, his deficiencies, as far as the ability to undertake extensive design work is concerned, will also be clear-cut. His capacities and deficiencies in education and experience are summarised below.

(a) He will have a good knowledge of the basic sciences and should have the ability to relate this to specific design and engineering problems.
(b) His mathematical knowledge will be of adequate standard for most design problems, and will include experience in the use of computers. In some courses he will have had experience in applying mathematical techniques of considerable sophistication to analytical problems in engineering specialties, such as structural analysis, systems analysis and heat transfer.
(c) His competence in drafting will usually be very limited (although he can, if required to, pick this up rapidly).
(d) He will have a general, but rather superficial, knowledge of the structure of business and industry and of typical industrial manufacturing operations.
(e) He will have a broad knowledge of the properties of generic groups of engineering materials but is unlikely to be familiar with the properties of specific materials.
(f) He will not have much knowledge of the dimensions, availability and range of application of standard engineering components and hardware.
(g) He will have little knowledge of the cost of materials, components and machining operations, and will have had little experience in working to a cost ceiling or to a tight time schedule for absolute finality in a task.
(h) He will have little training in the concepts of industrial design.
(i) He will probably be aware of some of the social consequences of a developing technology but will, like many others, be uncertain how to handle the problems that arise.

This summary is, of course, a considerable generalisation. And the deficiencies that are mentioned are usually expected and sometimes accepted by employers, who are prepared to make opportunities available to young graduates to build up their experience.

1.5. THE DESIGN OF THE BOOK

It would be reasonable to suppose that this text should be designed with an eye to the skills that have just been discussed, so that the expected skills will be developed and the deficiencies that have been observed will be in some measure rectified.

The next chapters deal with some of the arguments that must be resolved to provide a basis for the detailed analysis or synthesis of a design. The loading assumptions and failure criteria, some simple formulas and typical material properties—these are dealt with as preliminaries because the satisfactory planning and assessment of a design depend on them. These chapters are grouped as Part I of the book which could be described as touching upon the philosophy of design.

Part II consists of chapters in which some of the most commonly used methods of analysis are described: these constitute the scientific basis of design. At this stage considerable call is made on the methods of applied mechanics and strength of materials. Adequate coverage of these is usually available to students in their standard texts and, where possible, lengthy derivations have been avoided. Emphasis is placed on the assumptions, simplifications and limitations of the various methods, and on the way in which they may form the basis for a standard or a code of practice.

Part III deals with the art of engineering and with areas in which the work of the designer overlaps and requires an understanding of the skills, methods and capabilities of other trades and professions. Here an attempt is made to illuminate, if not to eliminate, some of the deficiencies observed in young engineers when they first engage in design work in industry. For example, a brief discussion of the way in which an engineer and industrial designer may cooperate is given in Chapter 21.

A selection of design information is set out in the Appendixes. No attempt has been made to provide an authoritative listing of the properties of currently available materials, but rather we have sought to describe the behaviour of a representative range with sufficient accuracy for use in design studies and with sufficient breadth to make the book self-contained for a good deal of this work.

REFERENCES

[1] J. C. JONES, *Design Methods,* Wiley Interscience, London, 1970.

BIBLIOGRAPHY

The design process is examined in some of the books and publications listed below, and the others provide a useful background to the way in which designers go about their work. They are relevant to the work of those who do design as well as those who manage or teach it.

ASIMOW, M., *Introduction to Design,* Prentice-Hall, Englewood Cliffs, 1962.

BEAKLEY, G. C. and LEACH, H. W., *Engineering, an Introduction to a Creative Profession,* Macmillan, New York, 1972.

CALLANDER, R. A. and STEPHENSON, J., "Teaching Engineering Design", *New Zealand Engineering,* **27**, 9, September 1972.

CIT, *Proceedings of the First Conference on Engineering Design Education,* Case Institute of Technology, Cleveland, 1960.

CLEGG, G. L., *The Design of Design,* Cambridge University Press, Cambridge, 1969.

FIELDEN REPORT, *Engineering Design,* DSIR–HMSO, London, 1963.

FRENCH, M. J., *Engineering Design: The Conceptual Stage,* Heinemann, London, 1971.

GREGORY, S. A. (ed.), *The Design Method,* Butterworths, London, 1966.

IED, *Conference on the Teaching of Engineering Design,* The Institution of Engineering Designers, London, 1964 and 1966.

JOHNSON, R. C., *Mechanical Design Synthesis,* Van Nostrand Reinhold, New York, 1971.

KRICK, E. V., *An Introduction to Engineering and Engineering Design,* 2nd ed., John Wiley, New York, 1969.

RODWELL, C., "Engineering Design Management", *The Chartered Mechanical Engineer,* July 1971. (Gives 38 more references.)

PROBLEMS

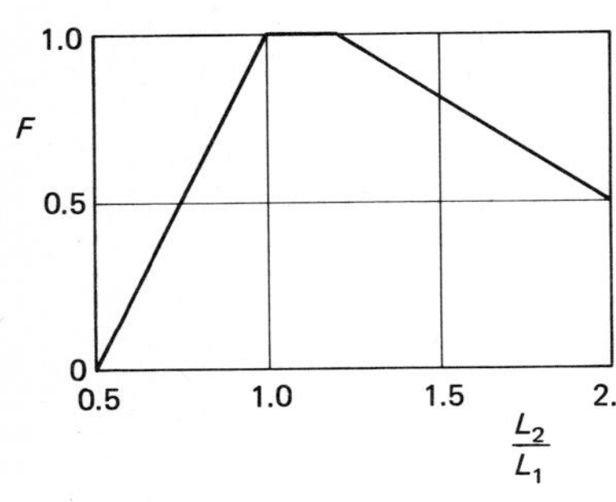

1.1. Construct the strut shown in Fig. 1.2 and test it in compression. Record its failure load L_1, and weight W_1. Design and build a minimum weight structure to withstand the load L_1, using any suitable design methods you are familiar with. Test it, and record the failure load L_2, and weight W_2. Work out a score in accordance with the formula

$$\text{Score} = 100\,F\;\frac{1.2\,W_1 - W_2}{W_1}$$

taking a value for F from the sketch. Scores over 90 rate well.

1.2. Devise other structures and loading systems and proceed as in Problem 1.1. How do the formula for the score and the values of the factor F influence the designer?

1.3. Although in design practice the engineer must balance the cost of finding out against the cost of not knowing, the student is faced with a situation where great importance is placed on finding out, regardless of cost. Comment on this statement.

1.4. In the Fielden Report (see "Bibliography"), Professor R. E. D. Bishop is quoted as saying that "a man who has spent years designing machine tools would probably design a bicycle which looked like a milling machine". What measure of truth do you think there is in this statement?

1.5. To what extent can you accept the old adage: "If a design looks right, it is right" as a basis for judging the merits of a design?

1.6. The local authority of the city or town in which you live is to offer a prize for the best preliminary design for a plaza or city centre. Suggest a framework for the design brief that will be presented to competitors. Show how you would encourage the inclusion in the design of any features that you regard as important. Describe these features. Does their inclusion restrict the creativity of the designers?

1.7. You have decided to enter the competition described in Problem 1.6. What would you regard as the ideal team? How would you set about generating ideas for the project?

1.8. A declining mining industry is the cause of unemployment in a relatively unspoiled part of the country. To provide employment some industries—including a paper mill and a plastics factory—are to be developed. What undesirable effects can these industries have on the environment? What steps can be taken to minimise these effects?

1.9. Select a product with which you are familiar and which you consider requires redesign. Describe how you would go about finding the best design solution. To what extent will you have to rely on past experience?

1.10. Definitions of design quoted by Jones[1] range from "Decision-making, in the face of uncertainty, with high penalties for error" to "The performing of a very complicated act of faith". Comment on these and any other definitions you can find, and make up your own that is relevant to engineering design.

1.11. There is a move towards bringing increasing numbers of people into the design act, particularly when the environment is affected. How do you think this trend will change the way the engineering designer works?

1.12. A tank is to carry 250 litres of hydraulic fluid for a drive on a mobile drilling rig. Sketch a suitable tank showing all the fittings necessary for attachment, filling and any other requirements you can visualise.

2 Load Analysis

2.1. INTRODUCTION

In some design work the engineer is given very complete guidance in the determination of loads by a code which may be based partly on theory and partly on experience; in other situations it will be left to him to establish suitable values. It is intended not to describe here any particular codes in detail, but rather to indicate in a general manner how the problem of covering a variety of loading cases is dealt with in practice.

Once a decision has been made as to what loads must be considered, they can be analysed and displayed in the way that best demonstrates the effect they will have on the structure and best suits the stress analysis that is to follow. This display will usually take the form of some graphical loading diagrams, and a description of these is given in the last part of the chapter.

2.2. TYPES OF LOADING

Before any detailed design work for a machine part or structure can be undertaken, the loading under various conditions must be determined. In some cases this operation will be quite straightforward; in others, considerable investigation may be required and the designer can still be left in the position of having to a large extent to use his judgement.

Frequently, the problem lies not so much in evaluating the loads as in deciding on what is a reasonable criterion. Given a formula for calculating loads due to wind, it remains to select a design wind velocity: a change from 30 m/s to 42 m/s will double the load.

When a designer encounters an unusual or unfamiliar loading situation he will often look to existing codes and standards to see if similar cases are treated there. This can be a fruitful approach, provided that he recognises that some of this information may be empirical, and hidden limitations to its use may exist.

A good discussion of loading for one particular case is given in the British Standard Specification for Steel Girder Bridges.[1] Clause 2 of the Standard sets out the forces to be taken into account.

Clause 2

For the purpose of computing stresses the following items shall, where applicable, be taken into account:

a. Dead load.
b. Live load.
c. Impact effect.
d. Lurching effect.
e. Nosing effect.
f. Centrifugal force.

g. Longitudinal force.
h. Wind pressure effect.
j. Temperature effect.
k. Frictional resistance of expansion bearings.
l. Forces on parapets.
m. Erection forces and effects.
n. Forces and effects due to earthquakes, ice packs, subsidence and other similar causes.

Subject to the provisions of other clauses, all forces shall be considered as applied and all loaded lengths chosen in such a way that the most adverse effect is caused on the member under consideration.

(Extract from BS 153, *Steel Girder Bridges*, is reproduced by permission of the British Standards Institution, 2 Park Street, London W1A 2BS.)

This Standard goes on to define the *dead load* as the weight of the structure and any permanent loads fixed thereon. The *live load* is defined as the weight of traffic on the bridge, and a series of standard highway and railway loading patterns are set out in detail.

Note that the loading comprises dead load, live load and a group of dynamic loads. In addition, the designer must take into account forces arising from temperature effects and those that may occur during erection of the structure. He is reminded finally of other forces and effects that may occur which might be labelled "forces of nature". To take these into account in a design will clearly require many statistical records of the phenomena concerned.

Of the dynamic loads, impact and centrifugal forces are encountered in many machines and structures. The former are often catered for by increasing the live load or the safety factor by a percentage determined by the severity of the case; the latter can usually be computed with some accuracy. The lurching and nosing effects that are listed are confined to loads set up on railways. The longitudinal force (item g.) arises from vehicles braking or accelerating.

The wind pressure effect (item h.) is dealt with in some detail in the Standard. A wind pressure value of 1400 Pa (approximately corresponding to 40 m/s) is specified, although the engineer may, under special conditions, have to consider a pressure calculated from the formula $p = 0.88\,v^2$ where p is the pressure in Pa and v is the velocity in m/s. The pressure is assumed to act horizontally and normal to the side of the bridge and the areas involved are the net areas seen in elevation for the windward part of the structure. Forces on the leeward girders are also taken into account, with some alleviation for the sheltering effect from the windward part of the structure.

Having determined the various forces involved, the designer must decide what combination of them should be used for the design. Here the Standard can be quoted again.

Clause 16

The following combinations of forces shall be considered:

a. The worst possible combination of dead load with live load, impact, lurching and centrifugal force.

When a member whose primary function is to resist longitudinal and nosing forces due to live load is under consideration, the term live load shall include these forces.

b. The worst possible combination of any or all of the forces listed under a. to l. inclusive in Clause 2.
c. The worst possible combination of forces during erection.

d. The worst possible combination of any or all of the forces listed in Clause 2, at the discretion of the engineer.

(Extract from BS 153, *Steel Girder Bridges*, is reproduced by permission of the British Standards Institution, 2 Park Street, London W1A 2BS.)

Note that it is not always the case that the greatest load sets the design condition. For example, when considering the overturning effect of wind acting on the live load on the bridge, the code stipulates that the wagons or vehicles shall be considered to be unladen.

The approach illustrated by BS 153 has been described at some length since it can be taken as a model for other design problems. Even if a code is very detailed, it is clear that there is room for the designer to discover new and perhaps critical loading patterns. It is possible for a structure that has been designed for all likely operational loads to be loaded to failure while being transported to the assembly site. Students should lose no opportunities to study any codes that are available; they should seek, where possible, the reason or theory behind the rulings given.

When no code is applicable, the loads to be taken into account should be listed methodically and broken down in the manner best suited to the way in which the stress calculations are to be made. This breakdown must depend to some extent on the behaviour of the materials under the load patterns that are expected. If there is, for example, a possibility of brittle fracture of the material (say, due to operation at low ambient temperature) there will be a need to pay increased attention to the estimation – and perhaps the reduction – of shock loads. In the same way, when varying loads are present, there is the possibility of fatigue failure; the load analysis should be approached with the object of preparing information that can readily be applied to fatigue design.

In mechanical engineering such a wide variety of machines and structures is encountered that few of them are the subject of a standard. However, where a public risk is involved (as, for example, with cranes and boilers) the designer is guided by a code.

In some smaller machinery, the dead loads are small when compared with the live and dynamic loads from belt and gear drives, inertia, centrifugal and out-of-balance forces, and vibration; in larger machines and structures the dead loads become increasingly important. Many of the dynamic loads and forces can be determined using the theory and methods of dynamics, though in some cases experimental measurements must be resorted to.

The designer can seldom build up a picture of the loading pattern without becoming aware of ways in which the initial layout can be improved. Any attempt to proceed at this stage by calculations carried out in isolation from drawings and sketches of the structure will mean that opportunities for improvements are lost. The need for carefully made scale sketches throughout the formative stages of a design must be strongly emphasised.

2.3. LOAD ANALYSIS

When a complicated loading on a beam or shaft is encountered, it is usually not possible to calculate directly the stresses set up; the first step is to break down the loading into its simplest components and to calculate any unknown forces or reactions. It is then most convenient to resolve the forces and moments into their components acting along or about three axes. These axes are so selected that one force component is taken axially along the shaft or beam; the other two force components are resolved into two orthogonal planes which contain this axis. These are termed the *axial force* and *shear forces* respectively. The moments are taken so that one component is a couple about the longitudinal axis: this is termed the *torque*; the other moments are resolved perpendicular to the torque axis: these are the *bending moments*. These forces and moments are usually displayed on diagrams which show clearly the complete loading picture.

With experience, it is possible for the designer to gain a feel for those forces and moments which will be significant. For example, in many mechanical engineering applications the axial loads do not cause high stresses, especially where the shafts are short and there is no possibility of buckling. In the same way, the shearing stresses in such cases are often small and usually reach a maximum in that part of the shaft cross-section where the bending stresses are zero. This does not mean that the designer can automatically neglect such forces; in a short, deep beam the shear stresses are likely to be of importance. With experience, the designer will learn to concentrate his effort on those locations where time and care are best repaid in terms of an efficient structure.

There are reasonably well standardised sign conventions for forces and moments and the ones described below will be used in this text.

Axial force (AF)	Tension force $+ve$ Compressive force $-ve$
Shear force (SF)	Positive when all external forces acting on the part to the left of a given section, when summed, tend to move that part upwards.
Bending moment (BM)	A force to the left of a given portion produces a positive bending moment when it tends to rotate the portion on which it acts in a clockwise direction. A simpler statement to visualise and remember is that a sagging or positive bending moment tends to produce a concave upward curvature, and conversely, a hogging or negative moment produces a convex upward curvature.
Torsion moment (TM) or torque	A sign is not often needed, but the turning moment can be taken as positive when it would tend to turn a right hand thread in the positive direction of the coordinate axis.

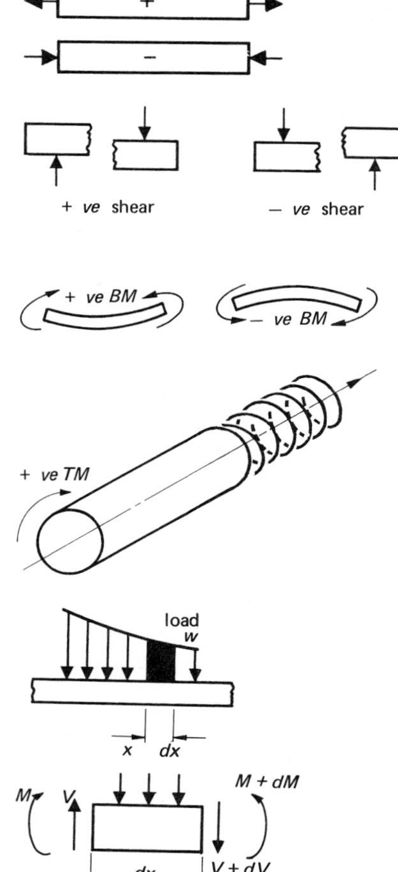

The relationship between loading, shear force and bending moment can be expressed in general terms and used as an aid to drawing and checking the diagrams.

Consider a small element dx along the length of a beam where the load per unit length is w ($+ve$ downwards). The bending moment, M, and shear force, V, are shown in their positive sense. The change in shear, dV, in length dx is wdx, and for equilibrium of the element sketched

$$V - wdx - (V + dV) = 0$$

$$\therefore dV = -wdx \qquad (2.1)$$

Now, taking moments about the left-hand end of the element, we have

$$M - (M + dM) + (V + dV)\,dx + wdx\,\frac{dx}{2} = 0$$

and neglecting second-order terms,

$$-dM + Vdx = 0$$

or

$$\frac{dM}{dx} = V. \tag{2.2}$$

If the load is concentrated at a point, there will be a step change in the shear force diagram; from Equation (2.2), the gradient of the bending-moment diagram must then be indeterminate. Several useful points follow from these equations. For example:

(1) when $w = 0$, $dV/dx = 0$, i.e., V is constant;
(2) when w is constant, the gradient of the shear-force diagram is constant, positive or negative depending on whether w acts up or down;
(3) when V is positive (or negative) the slope of the bending-moment diagram is positive (or negative);
(4) when V is zero, $dM/dx = 0$, i.e., there is a maximum or minimum in the bending-moment diagram;
(5) the area under the load diagram is equal to the change in shear – this follows from Equation (2.1), $dV = -wdx$, $V = -\int wdx$;
(6) the area under the shear-force diagram is equal to the change in bending-moment – see Equation (2.2).

These and other uses of Equations (2.1) and (2.2) are illustrated in the examples that follow.

Example 2.1

We consider the case of a simply supported beam (the reactions of which supply forces, but no couples) loaded at its centre. The shear force (SF) and bending-moment (BM) diagrams are shown in Fig. 2.1.

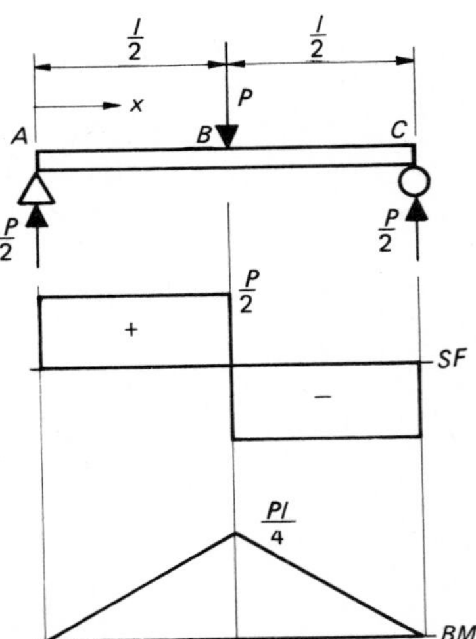

Fig. 2.1. *Simply supported beam with central load.*

Here it is clear that the reactions at A and C are $P/2$; the maximum BM is $Pl/4$.

Although in this case the shape of the diagrams can readily be visualised, the constant shear force between A and B is confirmed by Equation (2.1). In this region, w, the distributed load, is zero, i.e., $dV/dx = 0$, x being measured to the right from A, and V = constant = $P/2$. In this region also $dM/dx = V$ from Equation (2.2), so that $dM = \int V dx$ and $M = Vx + c$. The constant c is found from the condition that when $x = 0$, $M = 0$, i.e., $c = 0$. So that in this case for $V = P/2$,

$$M = Vx = \frac{Px}{2} \tag{2.3}$$

At A, B and C, the concentrated loads over an infinitesimally small length, dx, lead to a step change in the SF diagram, and at these points there is a discontinuity in the BM diagram. At B, where V passes through zero, M must go through a maximum, which for $x = l/2$ in Equation (2.3) is $Pl/4$. Although the load is spread over a finite length of beam and the sharp changes are rounded off, calculations can usually be based on a concentrated load.

In a more complicated case, the determination of reactions at the supports of a simply supported beam will require the use of the equations of static equilibrium. These can be stated as

$$\Sigma P + \Sigma R = 0$$

where P are the external loads and R the reactions at the supports, and

$$\Sigma M_P + \Sigma M_R = 0$$

where M_P and M_R are the moments about any point due to the loads and reactions.

Example 2.2

In the case illustrated in Fig. 2.2,
taking moments about B, gives

$$R_A = \frac{P_1 a + P_2 b}{l}$$

taking moments about A, gives

$$R_B = \frac{P_1 (l - a) + P_2 (l - b)}{l}$$

A check on the results can be obtained from

$$\Sigma P + \Sigma R = 0$$

that is, from

$$(P_1 + P_2) + (-R_A - R_B) = 0$$

The reasoning applied in Example 2.1 concerning the relationship between the loading, SF and BM diagrams can be repeated here. A maximum in the BM diagram occurs at C where the SF passes through zero.

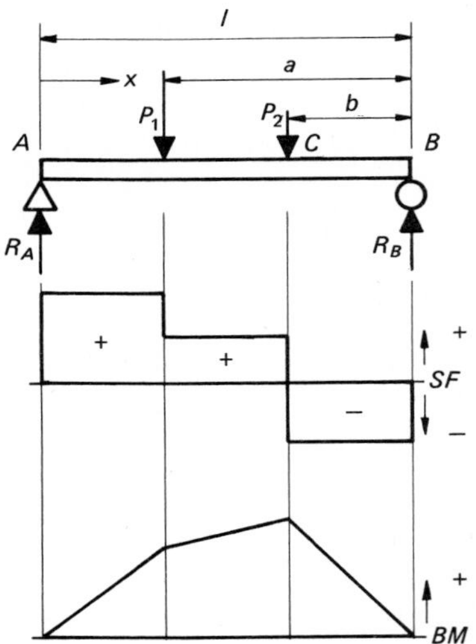

Fig. 2.2. *Simply supported beam.*

In the cases of the statically determinate beams illustrated above, the reactions can be calculated using only the equations of static equilibrium. In the case of statically indeterminate beams these equations provide insufficient information to fix all the reactions and the elastic deformation equation of the beam must be used to reach a solution. A simple illustration of the difference between the two cases is provided by a shaft which is restrained in torsion at the supports.

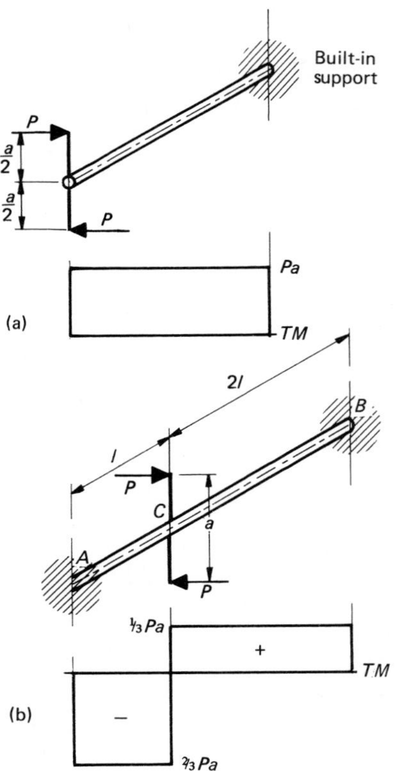

Fig. 2.3. *Torsion on a bar: (a) determinate; (b) indeterminate.*

Example 2.3 (See Fig. 2.3 (a) and (b).)

In case (a), the applied couple results in a torsional moment along the shaft of Pa. In case (b), the torsional moment can be determined only if the stiffness of the shaft along the length is known. If the stiffness is uniform and the lengths are given here as $AC = l$, $CB = 2l$, then the contribution of each length in resisting the couple Pa can be determined. It should be clear that the torque T_{AC} in AC will be twice that in CB. In terms of the shaft stiffness, k, it can be calculated as follows:

$$k = \frac{T}{\theta/l}$$

where θ is the deflection in radians. Since θ is the same for each portion of the shaft at C, and k is the same for AC and CB,

$$T_{AC}\, l = T_{BC}\, 2l$$

$$T_{AC} = 2T_{BC}$$

Example 2.4 (See Fig. 2.4.)

The existence of a couple along a shaft often causes difficulties when the BM diagram is drawn, especially when the diagram is complicated by the presence of other moments. Here a couple is considered on its own; in a more general case such a BM diagram for the couple can be superimposed on the diagram that results from the other forces. Since here the shear force is constant and equal to $-Pa/l$,

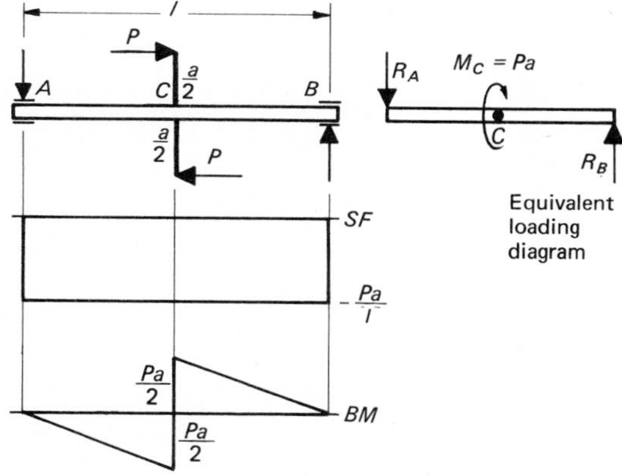

Fig. 2.4. *Shaft with couple.*

$$\frac{dM}{dx} = V = \frac{-Pa}{l}$$

Thus the slope of the BM diagram must be constant throughout, although there will be a step change at the point of application of the couple at C. When $x = l/2$ we have

$$M = -\int_0^{\frac{l}{2}} \frac{Pa}{l}\,dx$$

$$= -\frac{1}{2}Pa$$

Example 2.5

Consider a shaft which is simply supported in the bending mode at the ends A and B, but is restrained from rotation, and is loaded by the force P inclined at 45° as shown in Fig. 2.5.

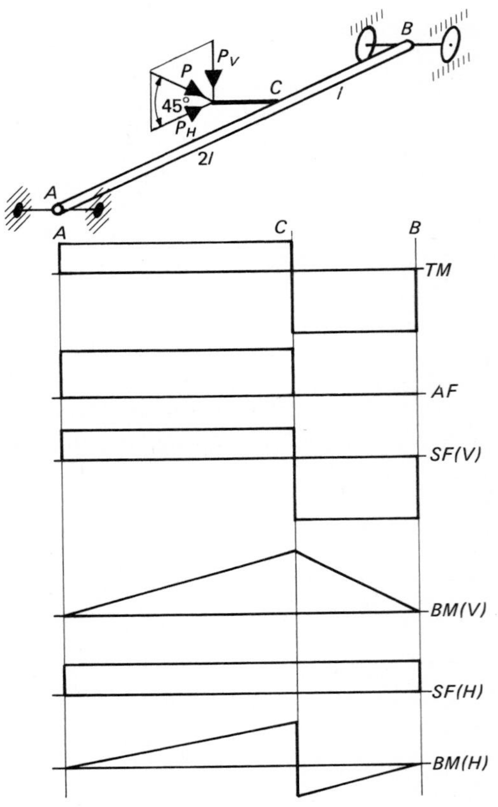

Fig. 2.5. *Shaft with offset load.*

The force P may be resolved into vertical and horizontal components $P_V = P/\sqrt{2}$ and $P_H = P/\sqrt{2}$. If the support at B is free to slide axially, then all the axial load will be carried by AC, and the torsion and axial force diagrams will be as shown.

Both P_V and P_H will cause moments in the shaft, and the moment diagrams for the horizontal and vertical planes are shown separately; these can be resolved at any section between A and B: the resultant will be given by

$$\mathrm{BM(R)} = \sqrt{[\mathrm{BM(V)}^2 + \mathrm{BM(H)}^2]}$$

Clearly the maximum lies at C.

The shear forces in the horizontal and vertical planes are shown in SF(H) and SF(V). Attention should be given to the diagram for BM(H) and the way in which the couple results in a step change at C.

Many of the determinate and indeterminate loadings that are met in practice have been solved and can be found in handbooks.[2, 3, 4] Some useful cases are given in Appendix A2.

LIST OF SYMBOLS

a, b	lengths
F	force
l	length
M	moment
P	force
p	pressure
R	reaction force
T	torque, turning moment
v	velocity
V	shear force
w	load per unit length
x	distance along coordinate axis
θ	angular deflection

REFERENCES

[1] BS 153, *Steel Girder Bridges,* Part 3A: 1972, "Loads" British Standards Institution, London, 1972.

[2] R. J. Roark, *Formulas for Stress and Strain,* McGraw-Hill, New York, 1965.

[3] C. S. Gray, *et al., Steel Designer's Manual,* 4th (metric) ed., Crosby Lockwood, London, 1972.

[4] AISC, *Manual of Steel Construction,* 7th ed., American Institute of Steel Construction Inc., New York, 1970.

PROBLEMS

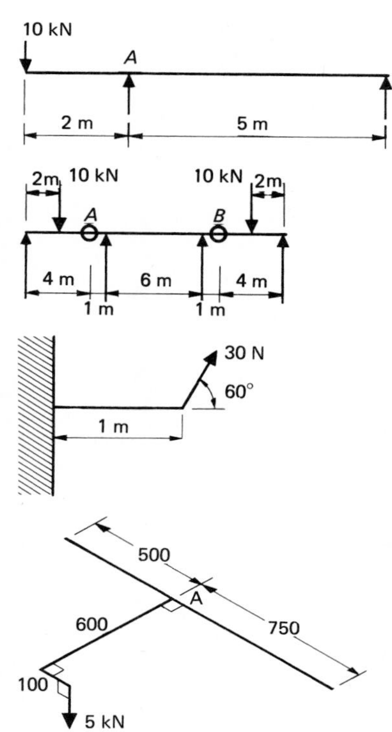

2.1. Sketch the shear force and bending-moment diagrams and the deflected shape for the beam in the adjacent sketch. In particular, what is the shear at A?

2.2. Beams spanning three contiguous gaps are arranged as shown in the margin, with hinges at A and B. Sketch the shear-force and bending-moment diagrams and the deflected shape of the structure. What are the shear force and bending moment at A?

2.3. What are the maximum bending moment and axial force in the cantilever sketched in the margin? What is the deflection of the free end if the cantilever is a 20 mm × 20 mm square steel bar? Take $E = 205$ GPa.

2.4. If the inclined load at the end of the cantilever of Problem 2.3 is replaced by a couple whose moment is 18 N m, compute the slope and deflection at the free end.

2.5. The steel bar shown in the sketch is loaded with a force of 5 kN. Sketch the shear-force, bending- and torsional-moment diagrams. What are the values on each side of the arm at A? The ends of the bar are simply supported in bending and both supports are able to resist a torque.

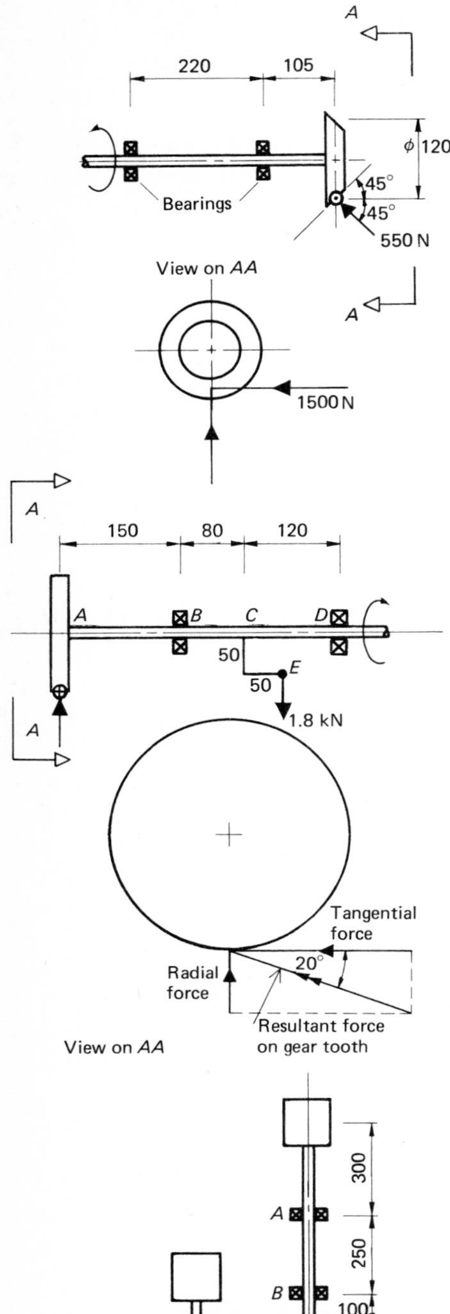

2.6. Draw the shear-force and bending-moment diagrams for a shaft loaded as shown in the sketch. The force is typical of one that would be applied to a bevel gear.

2.7. The shaft *ABCD* rotates in bearings at *B* and *D*. A weight is attached to the shaft at *C* by means of a right-angled bracket and gives rise to an out-of-balance force at *E* of 1.8 kN. The shaft is driven by a 200-mm diameter, 20° involute gear at *A*. The drive torque is 160 N m. Note that the resultant force at the gear pitch circle is as shown in the sketch. Find the maximum bending moment in the shaft.

2.8. A vertical shaft carries a rotor and is located by bearings at *A* and *B* as shown in the sketch, the bearing *A* taking all the axial load. A 0.5-kW motor drives the shaft at 2000 rpm through a V-belt drive. The rotor weighs 0.5 kN and has a radius of gyration of 250 mm. As a preliminary to designing the shaft and bearings, an analysis of the loads and moments is to be made for the following conditions:

(a) at start-up, if the motor torque is four times the rated torque;
(b) at the rated speed, assuming that the rotor centre of gravity is displaced from the centreline by 2% of the radius of gyration.

Assume that there is no tension on the slack side of the belt.

2.9. Select some of the standard loading cases given in Appendix A2 and check the shear-force and bending-moment diagrams by working from first principles.

2.10. Consult several standards or codes (e.g. for cranes, or bridges) and summarise the loading conditions that must be considered. To what extent is the designer left free to select loads and forces?

2.11. You have been retained as a consultant by a circus and fairground proprietor to check the safety of a number of structures and machines. What criteria would you use to determine design load or required strength for the following: roller coaster, merry-go-round, Ferris wheel, lion's cage, elephant's cage, trapeze, stock car barrier.

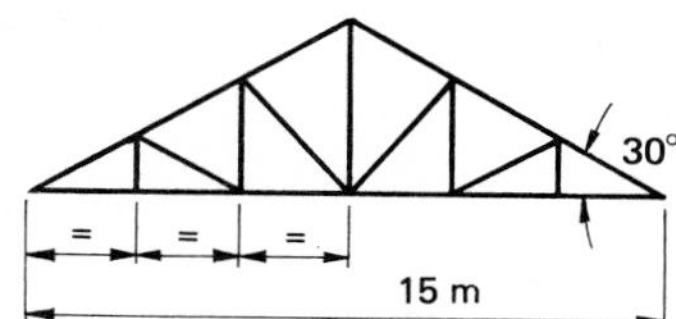

2.12. Calculate the forces in the members of the roof truss sketched in the margin for the following combinations of load:

(a) dead load plus live load;
(b) dead load plus wind load;
(c) dead load plus live load plus wind load.

Use loads specified in your local building code. Note that a reduced live load may be specified for (c) and that the permitted stresses for (b) and (c) may be higher than for (a).

The trusses are spaced 4 m centre-to-centre and the dead load can be assumed to be 400 N/m^2 of plan area.

2.13. When a pin or shaft is loaded as shown in the sketch, the designer must decide on how the load is distributed. Calculate the maximum bending moment on the pin for the three distributions shown; in the third case make suitable simplifying assumptions as to the shape of the distribution curves. How do the proportions of the pin affect the actual distribution?

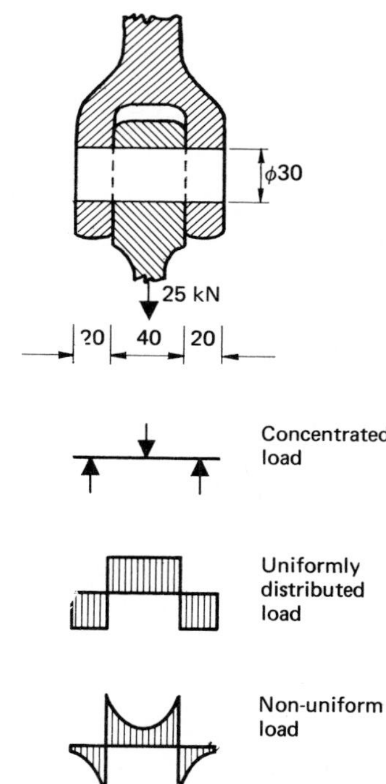

3 Stress Analysis

3.1. INTRODUCTION

When a set of design loads has been established, the designer must determine how the loads are to be distributed in the material and the stress levels they cause at a point. The size and deployment of the structural elements may not yet have been fixed: the equations relating stress to size can be written with some of the dimensions as unknowns, or the optimum arrangement may only become evident after several different trial layouts have been studied.

Given the loading diagrams, the stress analysis is broken down into two phases. First, the stress resulting from each separate load or moment is calculated. Second, all the stresses acting at a point are combined to give a resultant stress which, in conjunction with a failure theory for the material, will indicate whether the margin of safety is adequate.

This chapter is concerned primarily with the second phase, although some of the more commonly used stress equations are summarised in the next section.

3.2. EQUATIONS FOR STRESS

A good deal of design work can be accomplished without the use of any equations more elaborate than the bending and torsion equations for beams and shafts.

In bending, if the load acts along a line of symmetry, the calculation of the moment of inertia of the structural member is straightforward. For more complicated sections, and for those that are unsymmetrical, it is often possible to find tabulated properties (see Appendix A3). The calculation of the maximum bending stress at the extreme fibre follows from the simple bending equation.

The torsional stresses in a circular shaft are also easily calculated; for non-circular sections the problem becomes more complicated, although solutions exist for a number of shapes that are encountered in practice.[1]

The inefficiency of open sections such as channels and angles as torsion members, and the difficulty of calculating torsional stresses in such sections, provides a strong incentive for the use of circular sections.

A summary of the simple bending and torsion equations is given in Table 3.1, along with equations covering other cases that may be encountered. The information supplied is that of greatest interest to the designer – the location and value of the maximum stress. Solutions for other more complicated shapes can be found in the literature.[1, 2]

As the structure becomes more irregular in shape and the loadings become more complicated, the possibility of developing a straightforward closed solution rapidly recedes. In some cases mathematical techniques – for example,

TYPE OF LOADING	STRESS EQUATION	TYPE OF LOADING	STRESS EQUATION	TYPE OF LOADING	STRESS EQUATION
BENDING MOMENT, M, IN PLANE OF SYMMETRY 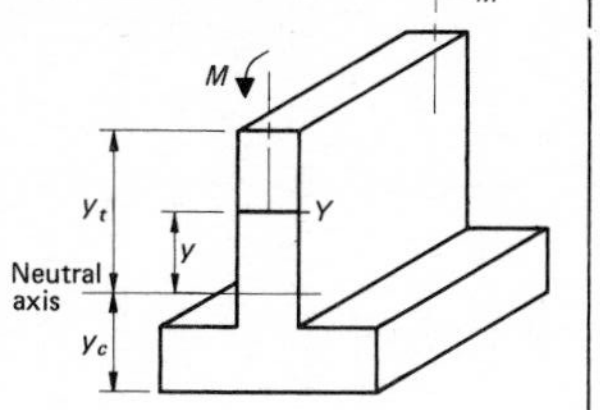	Bending stress at Y: $\sigma_Y = \frac{My}{I}$ Maximum bending stress occurs at extreme fibre where $y = y_t$ (tensile) $y = y_c$ (compressive) Tensile bending stress: $\sigma_{bt} = \frac{My_t}{I} = \frac{M}{Z_t}$, where $Z_t = \frac{I}{y_t}$ Compressive bending stress: $\sigma_{bc} = \frac{My_c}{I} = \frac{M}{Z_c}$, where $Z_c = \frac{I}{y_c}$	SPHERES IN CONTACT, LOAD P 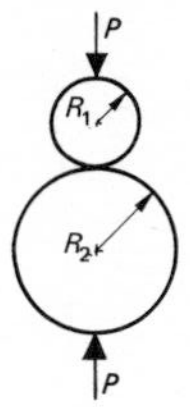	Maximum compressive stress at contact zone $\sigma_{max} = \left[\frac{6P\left(\frac{1}{R_1} + \frac{1}{R_2}\right)^2}{\pi^3 (1-\nu^2)^2 \left(\frac{1}{E_1} + \frac{1}{E_2}\right)^2}\right]^{\frac{1}{3}}$ $= 0.388\left[PE^2 \left(\frac{R_1 + R_2}{R_1{}^2 R_2{}^2}\right)^2\right]^{\frac{1}{3}}$ Assuming $\nu = 0.3$ and E same for both spheres	THIN WALL PRESSURE VESSEL CLOSED ENDS, INTERNAL PRESSURE p 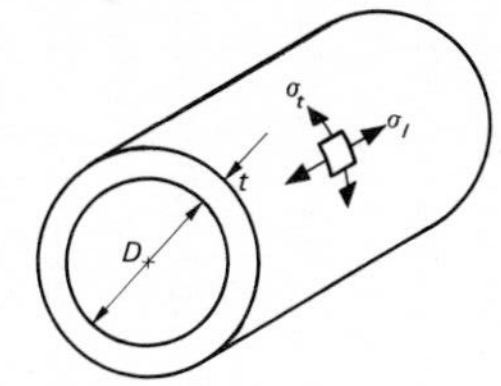	Longitudinal stress $\sigma_l = \frac{pD}{4t}$ Hoop or tangential stress $\sigma_t = \frac{pD}{2t}$
SHEAR FORCE, V 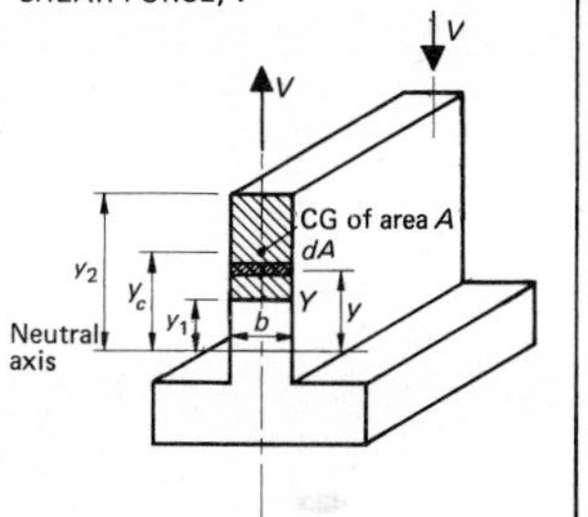	Shear stress at Y: $\tau_Y = \frac{V}{Ib}\int_{y_1}^{y_2} y\,dA$ $= \frac{V}{Ib}\,A y_c$ where y_c is distance from neutral axis to centroid of area A For solid circular section $\tau_{max} = 4/3\,\frac{V}{A}$	CYLINDERS IN CONTACT LOAD P' PER UNIT LENGTH 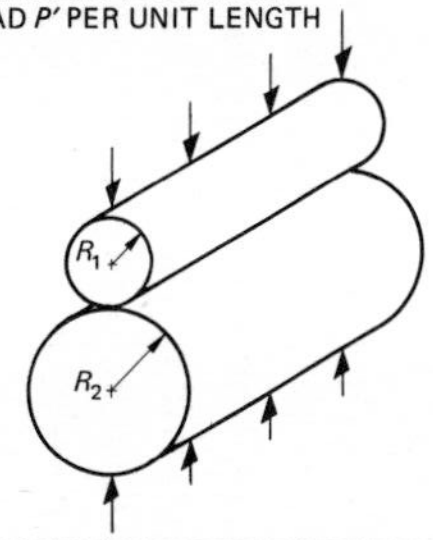	Maximum compressive stress at contact zone $\sigma_{max} = \left[\frac{P'\left(\frac{1}{R_1} + \frac{1}{R_2}\right)}{\pi (1-\nu^2)^2 \left(\frac{1}{E_1} + \frac{1}{E_2}\right)}\right]^{\frac{1}{2}}$ $= 0.418\left[\frac{P'E\,(R_1 + R_2)}{R_1 R_2}\right]^{\frac{1}{2}}$ Assuming $\nu = 0.3$ and E same for both cylinders Maximum shear stress τ_{max} occurs below surface $\tau_{max} = 0.3\sigma_{max}$	THICK WALL PRESSURE VESSEL, CLOSED ENDS, INTERNAL PRESSURE p 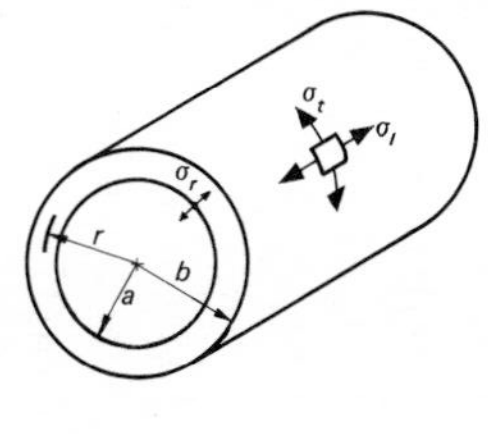	Lamé equations give $\sigma_l = \frac{pa^2}{b^2 - a^2}$ $\sigma_t = \frac{pa^2}{b^2 - a^2}\left(1 + \frac{b^2}{r^2}\right)$ maximum when $r = a$ $\sigma_r = \frac{pa^2}{b^2 - a^2}\left(1 - \frac{b^2}{r^2}\right)$
TORSION, T, ON CIRCULAR SHAFT 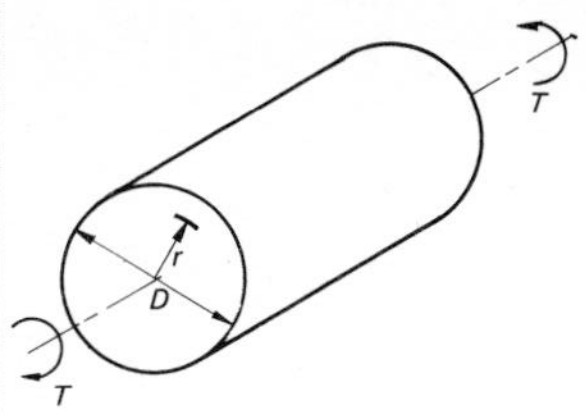	Shear stress at radius r: $\tau = \frac{Tr}{J}$ Maximum stress at extreme fibre, $r = \frac{D}{2}$ $\tau_{max} = \frac{T}{J} \cdot \frac{D}{2}$ $= \frac{16T}{\pi D^3}$ for solid circular shaft with $J = \frac{\pi D^4}{32}$	TORSION, T, ON SQUARE, RECTANGULAR AND TRIANGULAR SECTION SHAFT 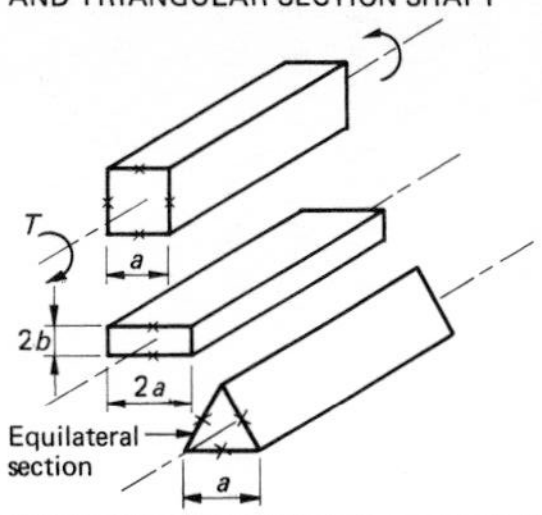 Equilateral section	Maximum shear stresses occur at mid points of sides marked X $\tau_{max} = \frac{4.8T}{a^3}$ $\tau_{max} = \frac{(3a + 1.8b)T}{8a^2 b^2}$ $\tau_{max} = \frac{20T}{a^3}$	**TABLE 3.1** SOME TYPICAL STRESS EQUATIONS USED IN DESIGN. FOR A MORE DETAILED LISTING, SEE R. J. ROARK[1] AND W. GRIFFEL[2]		

the method of finite elements – can be used for a solution, but at some stage the designer must decide how far it is economic to pursue such theoretical treatments, even where computers are available.

When faced with this sort of decision the designer will look at the alternative methods that can be used to achieve a sound structure; these are:

(a) to use simple shapes that are amenable to a theoretical analysis;

(b) to undertake model or full-scale loading tests, using strain gauges to measure stress at locations which are expected to be critical;

(c) to make sweeping assumptions in order to simplify the analysis and use large safety factors to compensate for possible inaccuracies.

The first procedure is commonly adopted and can lead to an efficient and simple structure. Many of the shapes have excellent load-carrying capacity and can be produced economically; the resulting assembly will usually have good visual appeal.

The second approach will often be adopted where the structure defies analysis and where the integrity must be ensured. In such work, the model can be modified and strengthened if weaknesses appear in the early stages of testing. The structure that results from this approach, if care has been taken to start with a well-balanced design, should be efficient and well able to stand up to the specified static loads. However, if fatigue is likely to be a limiting condition, it may be necessary to continue the tests under repeated loading – an expensive and time-consuming operation. The proviso that the design should initially be well-balanced is an important one, for the tests suggested here will do little to indicate those areas which may be grossly overdesigned (unless the strain gauging is taken to great lengths) and the design may still end up being uneconomic with respect to weight.

The third alternative does not at first sight appear to be an attractive one, but unfortunately is one the designer may be forced to use. If limitations on time and resources make this approach unavoidable, he can at least ensure that it is used only where the added material cost and weight do little harm – and there are, of course, some situations where this is so.

The designer must not lose sight of the assumptions that are made when calculating the stress at a point; the material is treated as an isotropic elastic continuum, and the calculated stresses are idealised values.

3.3. STRESS AT A POINT

Stress is defined as force per unit area. The notion of stress at a point has meaning in the sense that any differential quotient has – it is the limit of ΔF divided by ΔA as ΔA tends to zero. Here ΔF is a force distributed in some way over the area ΔA; as ΔA tends to zero, the distribution of ΔF becomes practically uniform and ΔF also tends to zero. No particular problems are encountered as long as the orientations of ΔF and ΔA are known. The element of force can be divided by the magnitude of the area to yield a vector force per unit area and this vector can be resolved into normal and tangential components which are the direct and shear stresses on the known element of area.

However, it is often the case that the direction and magnitude of the internal force ΔF are not known. The designer seeking critical conditions in a body under stress has to locate the relevant area ΔA and to find the direct and shear stresses acting on it. For example, when a round bar is loaded in tension along its axis, the stress σ_{ax} acting on circular cross-sections will be normal to the sections. If an oblique section is taken at an angle θ to the normal section, then σ_{ax} can be resolved, as shown in the sketch, into normal and shear stresses on the face, σ and τ. The way in which these components vary with θ must be determined, since failure in a material is not necessarily related to the maximum normal stress; it may occur when the maximum shear stress, or some combination of the principal stresses, reaches a limiting value. As a preliminary to practical solution of this problem we consider the physical quantity stress at a point in more detail.

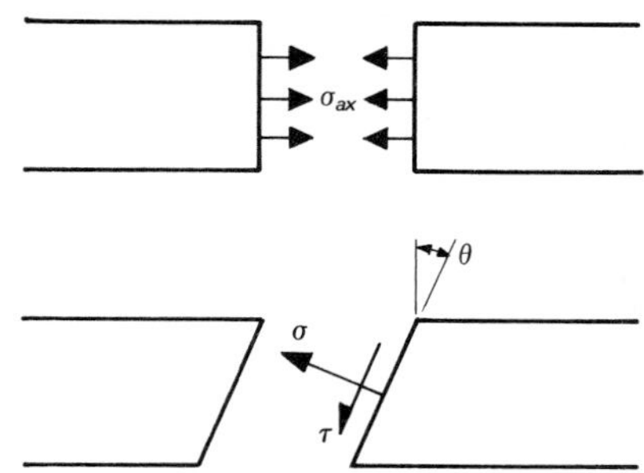

The stress at a point is a second-order tensor – cf. a vector or first-order tensor and a scalar or zero-order tensor. As such, it can be thought of as an entity in itself, as vectors and scalars are, and theoretical analysis of stress fields can be made clearer and more concise by doing so. Alternatively, a second-order tensor can be described completely by nine components, just as a vector can be described by three components, and this fact is used in the computation of direct and shear stresses on arbitrarily chosen areas.

The nine components of a stress tensor are one normal component and two tangential components on each of three elements of area in the coordinate planes. Each component is a force per unit area, either direct or shearing, and each is parallel to one of the axes of coordinates. These nine components and the element of force acting on an arbitrarily oriented element of area can all be shown on a sketch of an infinitesimal tetrahedron. Such a tetrahedron is shown in Fig. 3.1, where the components of stress in the coordinate planes XOY and YOZ have been omitted for clarity.

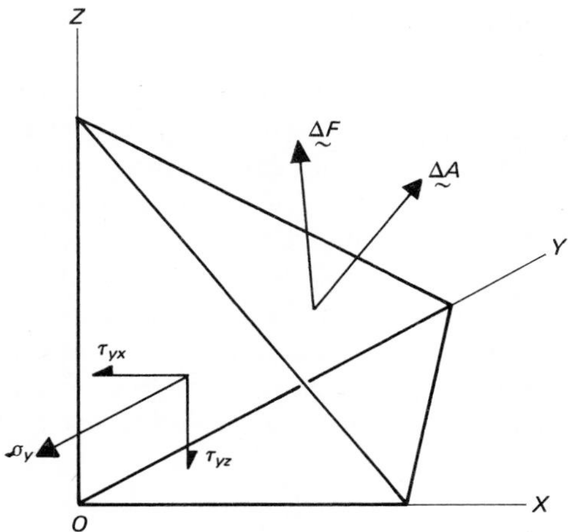

Fig. 3.1. *Stresses on a tetrahedron.*

The notation and sign convention are as follows:

(i) σ indicates a normal component or direct component;

(ii) τ indicates a tangential component or shear component;

(iii) only one suffix is attached to each σ and it indicates the direction of the area vector on which the component acts – e.g., σ_y is the normal component on the face perpendicular to OY;

(iv) there are two suffixes on each τ. The first shows which coordinate plane the component acts in and the second shows the direction of the component in the plane – e.g., τ_{yz} acts parallel to OZ in the face perpendicular to OY;

(v) a component is positive if it acts in the positive coordinate direction on a face whose outer normal is also in its positive coordinate direction, or if it acts in the negative coordinate direction on a face whose outer normal is in the negative direction. For the normal components, this means that tensions are positive.

When the equilibrium of an infinitesimal tetrahedron is considered[3, 4] it can be shown that the set of nine components is equivalent to the force per unit area acting on any inclined face which completes the tetrahedron. Also, it can be shown that only six of the components of the stress tensor are independent and that

$$\begin{aligned} \tau_{yx} &= \tau_{xy} \\ \tau_{xz} &= \tau_{zx} \\ \tau_{zy} &= \tau_{yz} \end{aligned} \tag{3.1}$$

The components of the stress at a point can be written in the form of a matrix as follows:

$$\begin{bmatrix} \sigma_x & \tau_{xy} & \tau_{xz} \\ \tau_{yx} & \sigma_y & \tau_{yz} \\ \tau_{zx} & \tau_{zy} & \sigma_z \end{bmatrix}$$

Each row contains the components acting in one coordinate plane and the above equations show that the matrix is symmetrical.

This discussion of the nature of stress at a point has been undertaken to explain a concept which is sometimes found difficult to grasp. We repeat its essential features:

(a) the stress at a point is a second order tensor which is an entity in itself and can be manipulated as such;
(b) the stress tensor can be described completely by nine components in a Cartesian frame of reference and six of the components are independent;
(c) the problem of finding direct and shear stresses on an arbitrarily chosen element of area is solved in terms of the components of the stress tensor.

Analysis of the stresses in a solid is made by applying the laws of motion to a prismatic element of the solid and, in general, leads to relationships among the gradients of the stress components. We are not concerned with such analyses; reference can be made to the books cited earlier.[3, 4] Our object is to calculate the force per unit area on arbitrarily oriented surfaces, given sufficient information about the components of stress in suitable Cartesian planes of reference.

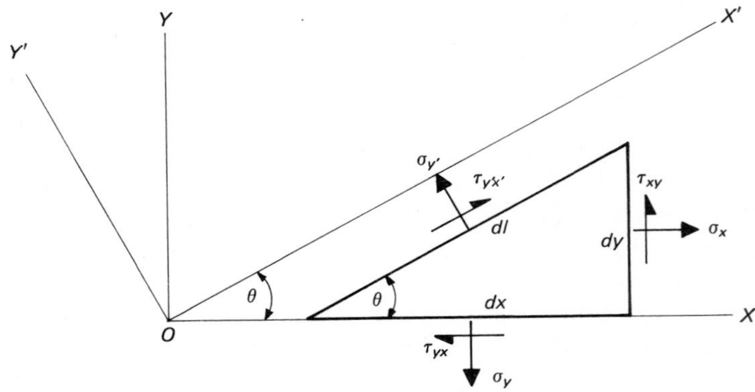

Fig. 3.2. *Two-dimensional stresses on a prism.*

In many cases in design, the axes can be chosen so that the stress components on one of the coordinate planes are small or zero; we will consider the case where $\sigma_z = \tau_{zx} = \tau_{zy} = 0$. This is a two-dimensional state of stress.

Stress components on an elementary triangular prism of unit depth in the Z direction are shown in Fig. 3.2. Two sets of coordinate axes OX, OY and OX', OY' are also shown, the second set being obtained by rotating the first through angle θ counterclockwise. The stresses on the faces dx and dy are known and the stresses on the face dl parallel to OX' are to be found. All the stresses are shown as positive stresses.

The prism is in equilibrium, and an analysis of the forces acting on it yields the following equations:

parallel to OX:

$$\sigma_x\,dy - \tau_{yx}\,dx - \sigma_{y'} \sin\theta\,dl + \tau_{y'x'} \cos\theta\,dl = 0$$

parallel to OY:

$$\tau_{xy}\,dy + \sigma_{y'} \cos\theta\,dl + \tau_{y'x'} \sin\theta\,dl - \sigma_y\,dx = 0$$

Noting from Equation (3.1) that $\tau_{xy} = \tau_{yx}$,

$$\sigma_{y'} = \frac{\sigma_x + \sigma_y}{2} - \frac{\sigma_x - \sigma_y}{2} \cos 2\theta - \tau_{xy} \sin 2\theta \tag{3.2}$$

$$\tau_{y'x'} = -\frac{\sigma_x - \sigma_y}{2} \sin 2\theta + \tau_{xy} \cos 2\theta \tag{3.3}$$

and on a face at 90° to dl,

$$\sigma_{x'} = \frac{\sigma_x + \sigma_y}{2} + \frac{\sigma_x - \sigma_y}{2} \cos 2\theta + \tau_{xy} \sin 2\theta \tag{3.4}$$

Extreme values of σ and τ are found by differentiating Equations (3.2) and (3.3), the results being

$$\sigma_{\text{max, min}} = \frac{\sigma_x + \sigma_y}{2} \pm \left[\left(\frac{\sigma_x - \sigma_y}{2}\right)^2 + \tau_{xy}{}^2\right]^{\frac{1}{2}} \tag{3.5}$$

$$\tau_{\text{max}} = \left[\left(\frac{\sigma_x - \sigma_y}{2}\right)^2 + \tau_{xy}{}^2\right]^{\frac{1}{2}} \tag{3.6}$$

Equations (3.5) and (3.6) relate specifically to extreme values and are of most interest in design. The direction of the maximum or minimum normal stress is found from the condition $d\sigma/d\theta = 0$ to be given by

$$\tan 2\theta = \frac{2\tau_{xy}}{\sigma_x - \sigma_y} \tag{3.7}$$

When values of shear stress are calculated for the planes on which σ has a maximum or minimum value, it is found that they are zero. There are, in fact, always three orthogonal planes through a point on which the shear stresses are zero. The normal stresses on these planes—two are given by Equation (3.5)—are called the principal stresses, and are designated by σ_1, σ_2, σ_3.

The way in which the stresses at a point vary with the angle θ is best visualised by using a graphical method. Such a method—proposed by Mohr in 1882—is very useful in design, since it enables measurement of stresses and

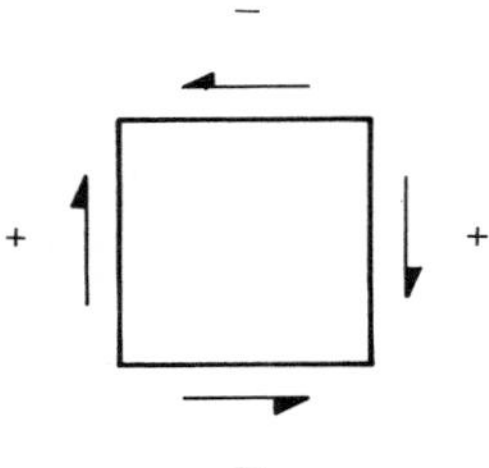

their directions to be made from a scale drawing; also, the critical conditions that initiate failure can be drawn on the same diagram and give a picture of how close the material is to failure. In Mohr's diagram normal stresses are plotted along the horizontal axis and shear stresses along the vertical axis. Normal stresses are plotted to the right of the origin if tensile or positive, to the left if compressive or negative. Several different sign conventions have been proposed for the shear stress, and those most commonly encountered differ from the classical system defined earlier and used in the development of Equations (3.2) to (3.7). We use a convention which defines shear stresses as positive if they tend to rotate the element clockwise, as shown in the sketch. Another statement of the same convention is that positive τ_{xy} stresses are plotted downwards, positive τ_{yx} stresses upwards.

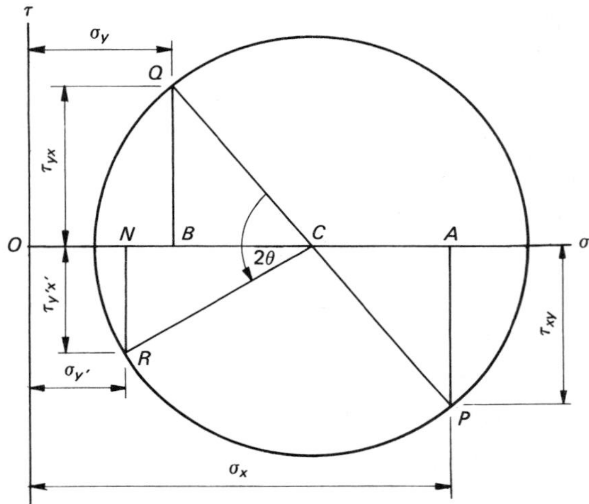

Fig. 3.3. *Mohr's diagram for stresses in Fig. 3.2.*

In Fig. 3.3 the point P represents the stress on face dy in Fig. 3.2 and is located by plotting $OA = \sigma_x$ and $AP = \tau_{xy}$ to some suitable scale; here τ_{xy} is negative and AP is plotted downwards. Similarly, OB represents σ_y and BQ represents τ_{yx} – a positive shear stress by our convention. A circle is drawn through P and Q with its centre at C midway between A and B. Points on the circle represent stresses in the physical plane dl obtained by rotating the axes OX, OY to the new positions OX', OY'. Angles on the Mohr diagram are found to be double the corresponding angles on the XY plane. Thus, the normal and shear stresses on face dl are found by setting out the angle QCR equal to 2θ in the same sense as the rotation of the axes in Fig. 3.2. The line RN is drawn perpendicular to the horizontal axis. It can be shown that ON represents $\sigma_{y'}$ and NR represents $\tau_{y'x'}$, in magnitude and sign, as given by Equations (3.2) and (3.3).

This construction provides a convenient analogue computation method that can be used in design when combined stress problems are encountered. The examples that follow will clarify the construction method used; a justification for the method can be found in Crandall *et al*[5] and a discussion of the sign convention in Wright.[6]

Consider a rod loaded in tension (Fig. 3.4(a)); an element will be stressed as shown. The value of σ_x is determined from P/A, σ_y here is zero. When plotted on the abscissa (since the shear stresses are zero) the points A and B represent the stresses on planes a and b (see Fig. 3.4(b)). A circle centred on the σ axis can be drawn through B and A, and points on this circle will represent stress conditions on various planes that lie between the a and b planes. The correspondence between the planes in the element and the points on Mohr's circle can be determined by noting that in rotating the element 90° between planes a and b, an angle of 180° is swept out between points A and B along path ACB. Thus angles on Mohr's circle correspond to double the angles, measured in the same direction, clockwise or counterclockwise, on the element. For example, the point C 90° from A (or B), where shear stress τ is a maximum, represents the stress on a plane 45° from a (or b) (see Fig. 3.4(c)).

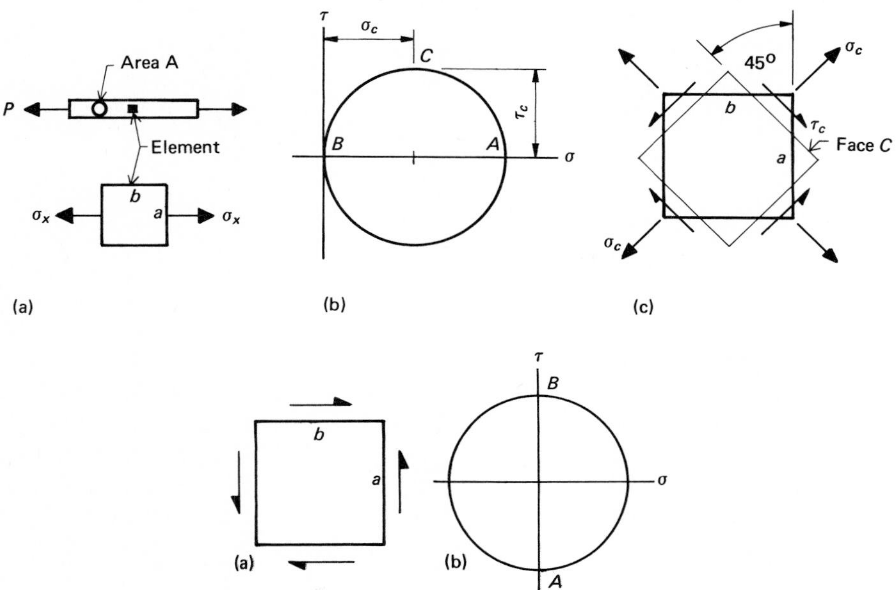

Fig. 3.4. *Mohr's diagram for simple tension:*
(a) stress on element;
(b) Mohr's circle;
(c) plane of maximum shear stress.

Fig. 3.5. *Mohr's diagram for pure shear:*
(a) stress on element;
(b) Mohr's diagram.

The Mohr's diagram for an element in pure shear (Fig. 3.5(a)) can be drawn by locating the positive shear on face b at B on the τ axis and the negative shear on face a at A (Fig. 3.5(b)). The resulting Mohr circle is centred at the origin.

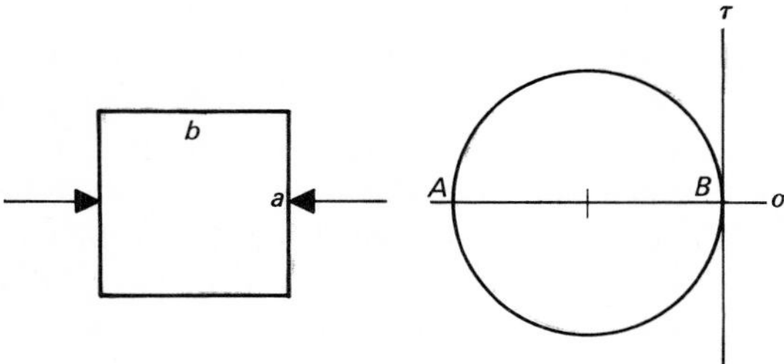

Fig. 3.6. *Mohr's diagram for compression.*

Simple compression is illustrated using similar nomenclature in Fig. 3.6. Here, of course, σ_A is negative.

The stresses for the more general stress condition illustrated in Fig. 3.7(a) are plotted on the Mohr diagram as shown in Fig. 3.7(b), the points A, B representing the stresses on the planes a, b. A point M on the Mohr diagram which represents the stress on some known plane m in the material, can be located as follows. If m is reached from plane a by rotating a through an angle ϕ in a clockwise direction, then M is found by moving around the circle clockwise from A through an angle 2ϕ, subtended at the centre.

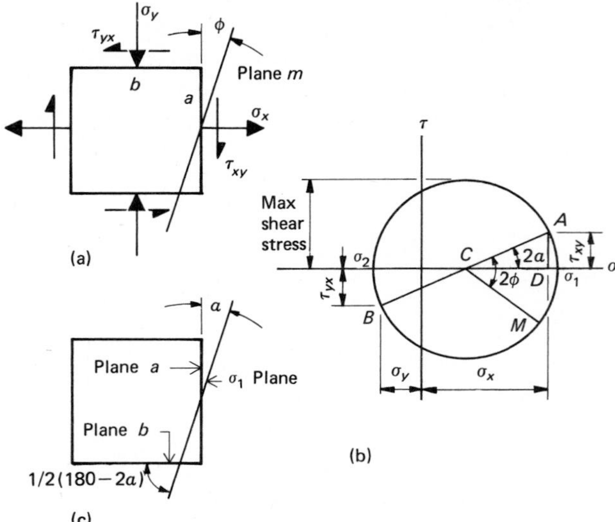

Fig. 3.7. *Mohr's diagram for general two-dimensional stress case: (a) stress on element; (b) Mohr's circle; (c) principal plane.*

In the same way, the plane on which the maximum principal stress σ_1 acts can be located with respect to the known planes a or b. If the angle between A and σ_1 is 2α, the required plane is obtained by rotating plane a clockwise through α. The same plane is reached, of course, by rotating plane b counter-clockwise through $\frac{1}{2}(180 - 2\alpha)$, as shown in Fig. 3.7(c). Note that in this case the maximum shear stress, equal to the radius of the circle, is given by

$$\tau_{\max} = CA = (AD^2 + CD^2)^{\frac{1}{2}} = \left[\tau_{xy}^2 + \left(\frac{\sigma_x - \sigma_y}{2}\right)^2\right]^{\frac{1}{2}}$$

It is important to remember that there are always three principal stresses to consider, even if one of them is zero. In the previous example the third principal stress—σ_3 perpendicular to the XY plane—was zero, and plotting it at the origin on Mohr's diagram would not affect the magnitude of the other stresses, although it would complete the diagram.

Now consider the element stressed as shown in Fig. 3.8(a). Points A and B in Fig. 3.8(b) represent the stresses on faces a and b, and points on the circle through AB represent stresses on planes perpendicular to the Z plane. The principal stresses σ_1 and σ_2 and the shear stress $\tau'_{\max}$ have been identified on Mohr's diagram and the plane on which $\tau'_{\max}$ acts has been located in Fig. 3.8(a). It is the largest shear stress acting in the XY plane. There is, however, a larger shear stress at this point in the material; it is shown when the third principal stress ($\sigma_3 = 0$) is located and the circles through σ_1, σ_3 and σ_2, σ_3 are drawn in. In this case the largest is the σ_1 σ_3 circle, and the maximum shear stress $\tau_{\max}$ is given by $\frac{1}{2}(\sigma_1 - \sigma_3)$. This stress acts on a plane in Fig. 3.8(a) that is oblique to the plane of the paper. To locate this plane, it is first necessary to locate the directions of the principal stresses σ_1 and σ_2; this is done in Fig. 3.8(c). Now consider the σ_1 plane being rotated through 90° to line up with the Z plane;

during this rotation the positions passed through will have corresponding points on the half-circle σ_1 P Q R σ_3. Clearly the plane shown in Fig. 3.8(d) at 45° to the σ_1 and σ_3 planes will be the one, related to point Q, on which the maximum shear stress acts.

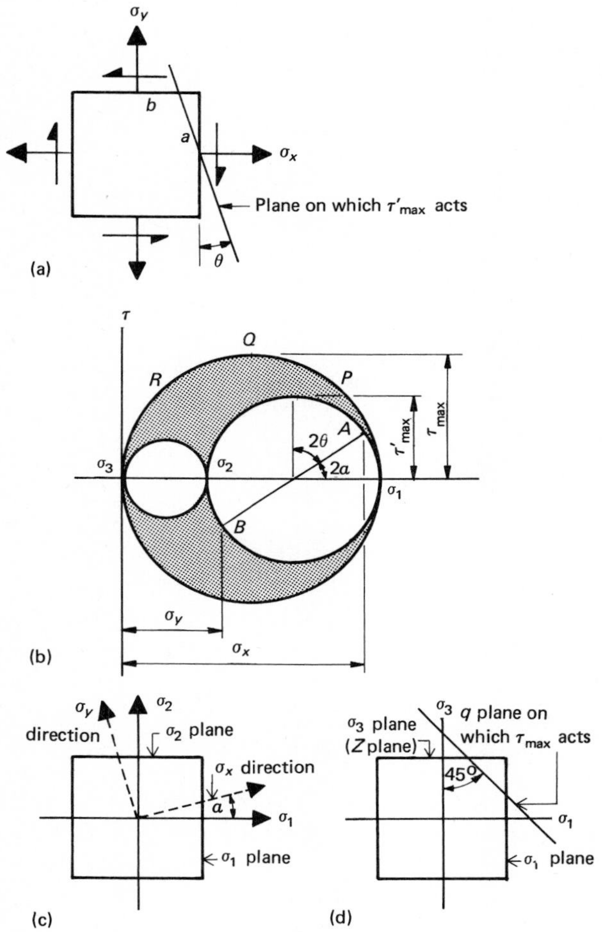

Fig. 3.8. *Mohr's diagram for case where σ_1 and σ_2 have same sign, $\sigma_3 = 0$:*
(a) stress on element;
(b) Mohr's circle;
(c) $\sigma_1\sigma_2$ plane;
(d) plane on which τ_{max} acts.

It should be noted that any point lying in the shaded area on the Mohr diagram in Fig. 3.8(b) represents a state of stress on some plane in the material at the point being considered. We can, fortunately, restrict our attention to the extreme values just discussed. It can be shown that Equations (3.4) and (3.5) still hold when the third principal stress σ_3 is not zero (see Problem 3.5).

Example 3.1

The components of stress at a point are shown in Fig. 3.9(a). Determine the principal stresses from Equation (3.4) and check the answers using Mohr's construction. Calculate the stresses on an element aligned with axes rotated 30° from those shown in the sketch.

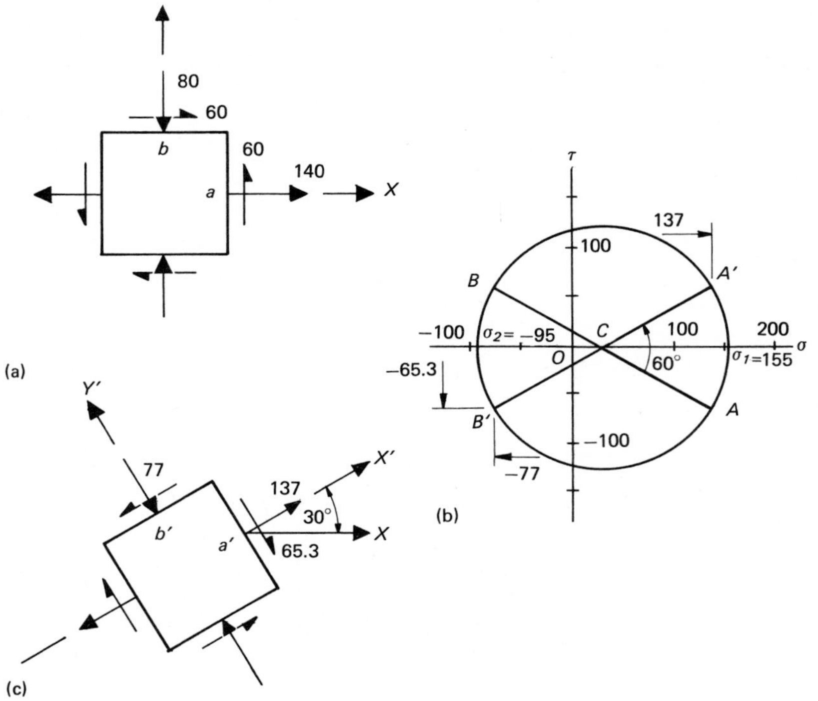

Fig. 3.9. *For Example 3.1:*
(a) stress on element;
(b) Mohr's diagram;
(c) stress on element at 30° to original.

The stresses on faces *a*, *b*, plot as points *A*, *B* on Mohr's diagram (see Fig. 3.9(b)). From Equation (3.5),

$$\sigma_1, \sigma_2 = \frac{\sigma_x + \sigma_y}{2} \pm \left[\left(\frac{\sigma_x - \sigma_y}{2}\right)^2 + {\tau_{xy}}^2\right]^{\frac{1}{2}}$$

$$= \frac{140 - 80}{2} \pm \left[\left(\frac{140 + 80}{2}\right)^2 + (60)^2\right]^{\frac{1}{2}}$$

$$= 30 \pm 125$$

$$\sigma_1 = 155, \sigma_2 = -95$$

These values check with those obtained from the graphical construction.

Calculate stress on element aligned with $X'Y'$ axes, $\theta = +30°$. From Equations (3.2), (3.3), (3.4),

$$\sigma_{x'} = 30 + 110 \cos 60 + 60 \sin 60 = 137$$
$$\sigma_{y'} = 30 - 110 \cos 60 - 60 \sin 60 = -77$$
$$\tau_{x'y'} = -110 \sin 60 + 60 \cos 60 = -65.3$$

The points A', B', represent the stresses on faces a', b' on the element at 30° to the original one (see Fig. 3.9(c)). Note that face a' is located in the physical plane by rotating face a through 30° counterclockwise, and that the point A' on Mohr's circle representing the stress on face a' is obtained by rotating from A through 60° counterclockwise about the centre C to A'.

Example 3.2

Calculate the maximum shear stress in the wall of a 250-mm diameter pipe subjected to pressure of 14 MPa. Wall thickness is 10 mm. Treat as a thin-walled pipe.

$$\text{Tangential stress, } \sigma_t = \frac{pR}{t} = \frac{14 \times 125}{10}$$

$$= 175 \text{ N/mm}^2.$$

$$\text{Longitudinal stress, } \sigma_l = \frac{pR}{2t} = \frac{14 \times 125}{2 \times 10}$$

$$= 87.5 \text{ N/mm}^2.$$

The orientation of these stresses is shown in Fig. 3.10(a).

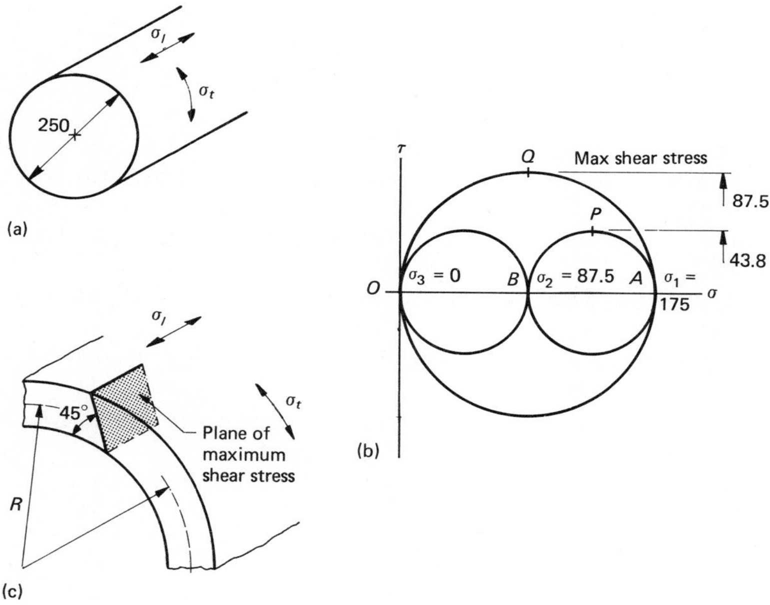

Fig. 3.10. *For Example 3.2:*
(a) stress on shell;
(b) Mohr's diagram;
(c) plane of maximum shear stress.

On $\sigma\tau$ axes (Fig. 3.10(b)) locate point A, 175 MPa,
locate point B, 87.5 MPa.

Draw circle APB, giving maximum shear stress in a plane perpendicular to surface = 43.8 MPa.

Draw circle AQO, giving maximum shear stress = 87.5 MPa.

This acts on a plane inclined at 45° to σ_t direction (see Fig. 3.10(c)).

LIST OF SYMBOLS

A	area
a, b	length; constants in Lamé's equation
D	diameter
E	modulus of elasticity
F	force
I	moment of inertia

J	polar moment of inertia
l	length
M	moment
P	load
p	pressure
R, r	radius
T	torque
t	thickness
x, y, z	distance along coordinate axis
α	angle
ν	Poisson's ratio
θ	angle
σ	normal stress
$\sigma_1, \sigma_2, \sigma_3$	the three principal stresses
τ	shear stress
ϕ	angle

REFERENCES

[1] R. J. Roark, *Formulas for Stress and Strain*, 4th ed., McGraw-Hill, New York, 1965.

[2] W. Griffel, *Handbook of Formulas for Stress and Strain,* Ungar, New York, 1966.

[3] I. S. Sokolnikoff, *Mathematical Theory of Elasticity,* 2nd ed., McGraw-Hill, New York, 1956.

[4] S. Timoshenko and J. N. Goodier, *Theory of Elasticity,* 3rd ed., McGraw-Hill, New York, 1970.

[5] S. H. Crandall, N. C. Dahl and T. J. Lardner, *An Introduction to the Mechanics of Solids,* McGraw-Hill, New York, 1972.

[6] D. K. Wright, *Experimental Mechanics,* **6**, 1966, pp. 19A–25A.

BIBLIOGRAPHY

IME, *Fourth International Conference on Stress Analysis and its Influence on Design,* Institution of Mechanical Engineers, London, 1971.

Lipson, C., and Juvinall, R. C., *Handbook of Stress and Strength,* Macmillan, New York, 1963.

PROBLEMS

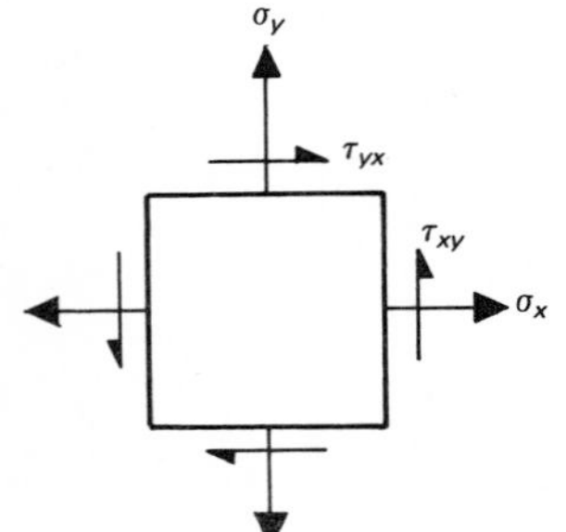

3.1. Draw Mohr's diagram for the following plane-stress cases (see sketch) and determine the maximum shear stress acting at the point. In each case, sketch the element and the planes on which the principal stresses act.

(a) $\sigma_x = 100$, $\sigma_y = 0$, $\tau_{xy} = 50$
(b) $\sigma_x = 100$, $\sigma_y = 50$, $\tau_{xy} = 0$
(c) $\sigma_x = 100$, $\sigma_y = 50$, $\tau_{xy} = 50$
(d) $\sigma_x = 50$, $\sigma_y = 100$, $\tau_{xy} = -50$
(e) $\sigma_x = 100$, $\sigma_y = -50$, $\tau_{xy} = 50$

3.2. Draw Mohr's diagram for the stresses at the inside and outside of a 250-mm diameter cylinder subjected to a pressure of 35 MPa. Wall thickness is 40 mm. Use thick-wall cylinder theory. Find the maximum shear stress acting in the wall.

3.3. A circular shaft of 40-mm diameter is subjected to a bending moment of 1200 N m and a torque of 1500 N m. Determine the maximum principal stress at the surface. Does the maximum shear stress lie in the plane of the surface?

3.4. A cylindrical pressure vessel is subjected to internal pressure which causes a longitudinal stress, $\sigma_x = 35$ MPa, and a circumferential stress, $\sigma_y = 70$ MPa, at a point on the external surface. An external torque applied to the cylinder causes a shearing stress $\tau_{xy} = 12$ MPa at the point.

(a) Find the principal planes and stresses and sketch an element in illustration of the solution.

(b) Find also the maximum shear stress acting in the plane of the external surface.

(c) Does any greater shear stress exist at the outer wall? If so, what are its magnitude and direction?

3.5. The general law for transformation of a second order tensor under rotation of the axes is

$$\sigma_{pq} = l_{ip} l_{jq} \sigma_{ij}$$

where σ_{ij} is the stress component in the original frame of reference; σ_{pq} is the stress component in the new frame; l_{kr} is the cosine of the angle between axis Ok in the old frame and Or in the new frame.

Summation over repeated suffixes is implied. Using this law, derive Equations (3.2) and (3.3).

3.6. The eigenvalues of the matrix of a three-dimensional stress tensor are the principal stresses. Show that, if $\tau_{zx} = \tau_{zy} = 0$ but $\sigma_z \neq 0$, the principal stresses of the three-dimensional tensor are the same as those of the two-dimensional tensor with σ_z also equal to zero.

Explain this physically in terms of the forces acting on an element such as that in Fig. 3.2.

3.7. A rectangular cantilever beam is 80 mm wide and 200 mm deep and is loaded as shown in the sketch. Calculate the stresses on small elements at the points A, B and C, and determine the maximum shear stress in each case.

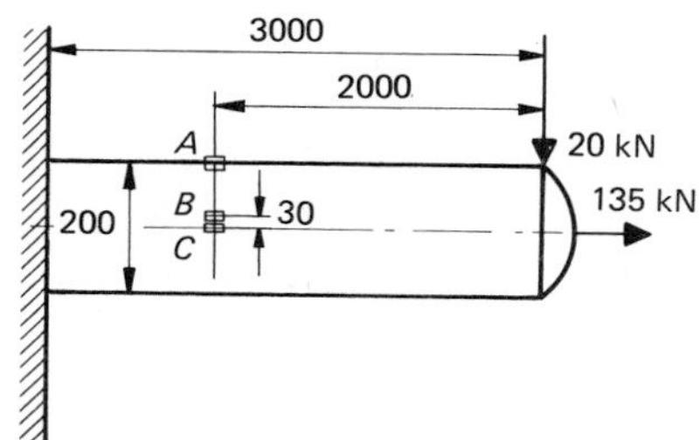

3.8. A rectangular beam is cast in Grade 17 iron (see Table A4.1). If it is subjected to a bending moment of 5.5 kN m and an axial compression of 450 kN, and the depth is to be twice the width, calculate suitable dimensions. Use a safety factor of 3, and assume that the compressive strength is three times the tensile strength. What is the effect of removing the compressive load?

3.9. The grain of a 200-mm square wooden post is inclined at 20° to its length. If a compressive load of 200 kN is applied, calculate the shearing stress parallel to the grain.

3.10. Economies in construction can be achieved by suitable use of prestressing. A pressure vessel, of 750-mm inside diameter and 5 m long, is to operate at a pressure of 70 MPa. Details of two possible construction methods, A and B, are given in the table below. Method A is assumed to give a prestress which ranges from −35 MPa at the inner wall to +35 MPa at the outer wall. Complete the table and comment on the results.

	Method A	Method B
Construction	Laminated 5-mm plates	Solid forging
Material	Carbon steel	3% chrome steel
Yield stress (MPa)	300	400
Safety factor	1.7	1.7
Finished cost ($ per kg)	0.7	1.1
Wall thickness		
Vessel weight		
Vessel price		

4 Failure and Safety Factors

4.1. INTRODUCTION

At various stages in design the loads and stresses in a machine or structure are analysed either to select the size of a component or to assess the likelihood of failure. The problem of setting up limiting conditions which describe failure – and failure may come from many causes other than excessive load – is discussed in the first part of the chapter.

When the limiting conditions have been fixed, the problem arises of reconciling available data with the operating conditions. Many of the data on the behaviour of machines and materials have been determined under standard conditions which may not match the design conditions. For example, the results of a simple tensile test cannot be used, directly, to assess the possibility of failure occurring due to a combined stress arising from tension and torsion. A number of so called "failure theories" are available to extend the usefulness of available test data. The most frequently used theories are discussed here.

Having selected a limiting condition, the designer must further decide what risk of failure can be tolerated; the design factors which are introduced at this stage are measures of the uncertainty in the design parameters and of the extent to which the designer must insure against the occurrence of failure. These factors are considered in Section 4.4.

4.2. LIMITING CONDITIONS

In discussing the mechanical conditions which cause solids in engineering structures to fail, Nadai lists some of the limiting conditions upon which the design of machine parts must be based.[1] These can be summarised as follows:

1. The load or loads under which the first permanent distortion begins to develop if the material has a sharply defined yield stress.
2. The stresses or loads under which the permanent portions of the strains will not exceed certain small more or less arbitrarily chosen limiting strain values.
3. The maximum permissible elastic displacement or deflection in a part of a construction or in the whole construction.
4. A load under which the elastic equilibrium of the stresses in a construction or in parts thereof may become unstable causing elastic buckling or sudden collapse.
5. Similarly instability of the elastic and plastic equilibrium (this may include cases in which a load causing permanent distortion reaches an analytic maximum after which a local deformation develops).

6. The load causing fracture, including failure through fatigue fracture under oscillating loads.
7. Resonance in the vibrations in a construction or parts thereof under oscillating loads.

At elevated temperatures:

8. The maximum permissible permanent deflection or displacement during the expected service time due to long-time creep in metal parts.
9. The maximum permissible load which is sustained continuously in a construction part during the expected life of it without causing fracture.

At very low temperatures:

10. In certain metals which are known to be ductile the danger that a sudden brittle fracture may occur, particularly under impact conditions, when the load increases very rapidly and when sharp notches (concentrators of stress) are present. The design of machines operating at high pressures and speeds, at normal or low temperatures, raises some of these problems.

    (From A. Nadai, *Theory of Flow and Fracture of Solids*, 2nd ed., McGraw-Hill, New York, 1950, by permission of McGraw-Hill Book Company Inc.)

This list of ten limiting conditions provides a good summary of the factors—mainly relating to material strength—which must be taken into

Table 4.1. Limiting conditions to be considered in design of a bearing installation.

Limiting Condition	Typical Method of Specifying Limiting Condition	Result of Reaching Limiting Condition	Possible Cause of Failure*
1. Fatigue failure in bearing material	Design life of 10 000 hours at a specified rotational speed and load	Bearing breakdown: e.g., pitting or spalling of contact surface on race or fracture of ball	End of design life. Design based on incorrect load or speed values.
2. Uncertainty in location	Radial or end play of shaft not to exceed ±0.01 mm.	Reduced working accuracy of machine tools. Noise and vibration. Noisy and worn gearing.	Inadequate location and support of bearing. Failure to allow for preloading. Distortion of housing or shaft under load. Abrasion due to inadequate sealing. Thermal expansion.
3. Excessive noise	Upper limit on noise level in any connected space.	Complaints from occupants of connected space.	Incorrect fits between bearing and shaft or housing. Bearing clearance incorrect. Unexpected loading condition.
4. Excessive temperature	Bearing temperature not to exceed 95°C.	Distortion. Material failure.	Lubricant lack or excess. Friction of seals. Ambient temperature higher than design value. Insufficient heat dissipation from housing.

*It is assumed that the correct bearing type for the purpose is selected in the first place. Incorrect machining, handling and assembly can contribute to failure in all cases: only causes relating to initial design are considered here.

account when considering the different effects of loading. Many failures occur because insufficient attention has been given to one of these conditions at the detail design stage; yet there is no generally recommended approach to this phase of design. It is true that a good deal of design work is concerned with conditions 1, 2 or 6, but the designer must still ensure that the other factors are given adequate attention.

It is useful for students initially to adopt a methodical approach and to list for each problem the relevant limiting conditions. Table 4.1 gives an example of such a list for a bearing installation; in this case, bearing geometry and noise must be considered in the design as well as the material strength property which governs the life of the bearing. The causes and results of failure have also been listed.

The importance of factors other than strength has been mentioned. If an incorrect assumption is made for the limiting condition relating to fatigue loading of a machine part that is produced in quantity, and it fails in service, the failure can often be eliminated by a relatively minor change in process, shape, size or material. The trouble might even be detected and eliminated in the development stage. An error in the limiting condition as regards noise in a machine might have more serious consequences. In the development stage, the tests would probably be made with the same noise limits adopted in design. If consumer reaction to the noise were unfavourable, sales of the product would be jeopardised. The rectification of a fault of this nature would probably require major design changes.

When limiting conditions are listed they should, wherever possible, be given a numerical as well as a phenomenological description; in some cases they may be set out for the designer in a specification. The difficulty of describing precisely all the limiting conditions that an experienced designer will consider before beginning any computations probably explains the absence of such lists in practical design. The use of them, however, should be encouraged and they should be regarded as more than mere aids to memory; the successful introduction of synthesis as a design method must depend to some extent on the adoption of an approach of this type. These limiting conditions will form part of the framework around which a solution can be built.

4.3. FAILURE THEORIES

There are many factors, other than geometry, that determine failure behaviour; some of these are:

temperature of service – classified as low, room, or high;
time dependence of loading – steady, or variable;
nature of material – brittle, or ductile.

In this book, we are mainly concerned with steady and variable loads on ductile materials and with steady loads on brittle materials, all at room temperature. Other combinations of the factors – and of other factors not mentioned, such as high-rate loading – are studied in more advanced courses.

The information available on the strength properties of materials is usually limited and may consist simply of the yield and tensile strengths in tension and the endurance limit for fully reversed bending, all measured at room temperature. The ductility and the elastic constants will also be known. In order to apply this information when the loading conditions are different from those specified—for example, combined bending and torsion—it is necessary to combine the theories of mechanical strength with experimental evidence in order to set up suitable failure theories. "Suitable" implies here that the theory is simple enough to use in practice as well as giving acceptable accuracy.

In the case of fatigue loading, much published information pertains to polished specimens tested in fully reversed bending. The application of these data to design for loads with different ratios of average to alternating values on parts with different surface finish requires the use of a failure theory or failure criterion.

Other such failure theories will be encountered. For example, in the design of rolling bearings some failure criterion is needed so that the one load and life figure that is usually given in catalogues can be applied to other values.

Unfortunately, there is no general theory that will account for all limiting conditions and for all types of materials. In some cases, such as Item 8 in Section 4.2—creep loading—it is necessary to extrapolate the results of short-term tests to longer design lifetimes. If it were not possible to do this, new materials could not be used until they had been under test for at least the life span of the machine, which can be ten years or more. Failure criteria are still being proposed for such cases; their use in design is a specialised one not considered here.

A state of stress in a solid is determined by six quantities—the three principal stresses and the directions of these stresses in the material. If the material is isotropic, the directions are of no consequence. A state of stress which is just necessary to produce failure by plastic yielding or by fracture may be described by the three principal stresses σ_1, σ_2, σ_3. These quantities can be represented by the coordinates of a point P with respect to three mutually perpendicular axes. The totality of points P form a surface which is called the limiting surface of failure—either by yielding or by rupture—of the material.

These three-dimensional failure surfaces may be displayed as solid models, but in describing them on paper it is convenient to project the surface onto a plane containing two of the principal stresses. For design purposes, the limiting lines for the two-dimensional case may be shown on the Mohr's stress plane. Mohr's construction for the design stresses can be drawn on the same diagram, allowing direct comparison of the actual and limiting conditions.

In the work that follows, the reader must distinguish between the plot of σ and τ on Mohr's stress plane and the plot of principal stresses σ_1 and σ_2. This latter display, although not of direct use in design, gives a helpful picture of the relationship between one theory and another.

Materials and failures are classified by engineers as brittle or ductile. The designer must know whether or not he can rely on some redistribution of stress by plastic flow at points where stress concentrations exist; this phenomenon in ductile materials tends to eliminate the effect of such concentrations under steady load conditions and simplifies the design approach. This redistribution of stress cannot occur in brittle materials. Under varying load conditions, of course, the possibility of fatigue failure complicates the problem.

It will be recognised that there are other classifications for fracture failures, and terms such as fibrous or granular, shear or cleavage, are used to describe the detailed visual or crystallographic appearance of the surface. The relation between these various descriptions and the actual failure modes is not always clear-cut, and for design purposes the demarcation between a brittle and ductile material is based on the results of the tensile test.

When the loading corresponds to the simple one-dimensional case of many test data, no failure theory is needed and the test data can be applied directly. This will often be the case in design.

4.3.1. Failure theories for brittle materials

Brittle materials are defined somewhat arbitrarily as those which give less than 5% elongation at fracture on a simple tension test at room temperature. The limiting condition for these materials is taken as fracture: there is usually no other feature in the stress–strain loading diagram which will provide a recognisable and reproducible basis for design.

The maximum stress (Rankine) theory. According to this theory, the principal stress which has the highest numerical value determines failure regardless of the values of the other principal stresses. The limiting surface for this theory consists of a cube centred on the origin of the σ_1, σ_2, σ_3 system of coordinates (see Plate 4.1(a)). Each of the six sides intersects an axis at right angles at a value corresponding to the maximum limiting stress.

The representation of this theory for the two-dimensional case is the projection of the cube onto the plane containing the two of the principal stresses that are not zero (Fig. 4.1(a)). For design purposes, the limiting stresses may be represented on Mohr's stress plane by the two lines aa', bb' (Fig. 4.1(b)). Then, provided that the principal stresses in the design lie within the square in Fig. 4.1(a), or that Mohr's circle for the design stress lies within the limiting lines in Fig. 4.1(b), failure will not, according to this theory, occur. Note that for many brittle materials the compressive strength, f_c, is greater than the tensile strength, f_t, in which case the cube is not centred at the origin.

Rankine's theory provides a simple demonstration of the way in which the failure theories are used in design. Its use is not recommended, since it can be unsafe where both principal stresses are large and of opposite sign (see shaded regions in Fig. 4.1(a)).

Mohr's internal friction theory. The most suitable design theory for brittle materials – for example, cast iron and concrete – is a special case of Mohr's theory. Mohr suggested that materials fail through plastic slip or fracture either when the shearing stress in the plane of failure reaches a limiting value, which will depend generally on the stress normal to the plane, or when the normal tensile stress reaches a limiting value. If the failure is to be a shearing or sliding one, then a picture is called up of some mechanism of internal friction whereby sliding is resisted as increasing (compressive) forces push the two sliding surfaces together.

To establish the precise shape of the envelope of failure requires many difficult experiments for each material. A typical shape is illustrated in the solid model (see Plate 4.1(b)). Designers must establish an envelope which is easy to use and which can be built up from known properties. The envelope illustrated in Fig. 4.2 is often used in design. The circles A and B drawn on Mohr's stress plane correspond to the simple tension and compression failures. These values are easily measured; in the absence of more complete information, the envelope

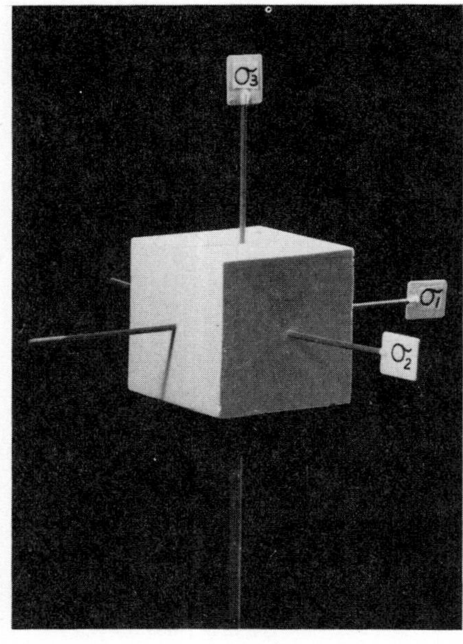

(a)

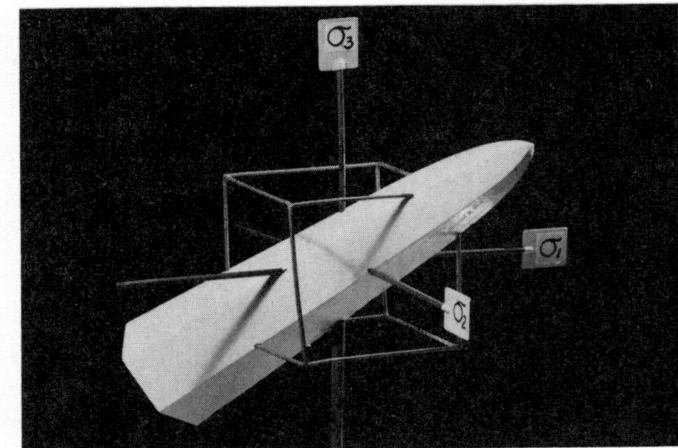

(b)

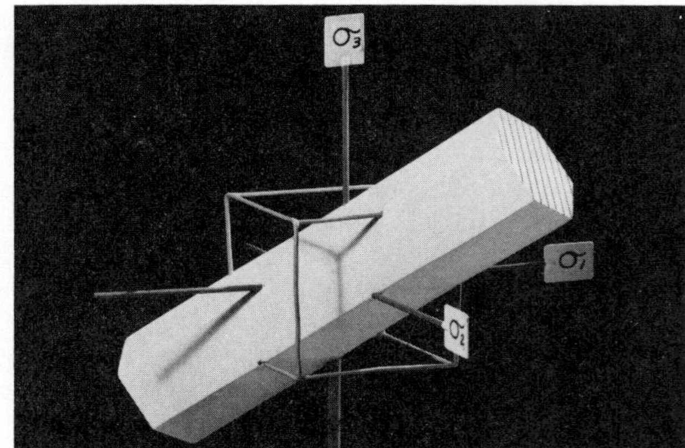

(c)

(d)

Plate 4.1. *Models showing failure envelopes on principal stress axes for yield and fracture theories:*
(a) maximum stress (Rankine);
(b) internal friction (Mohr);
(c) maximum shear stress (Tresca);
(d) maximum distortion energy (von Mises).

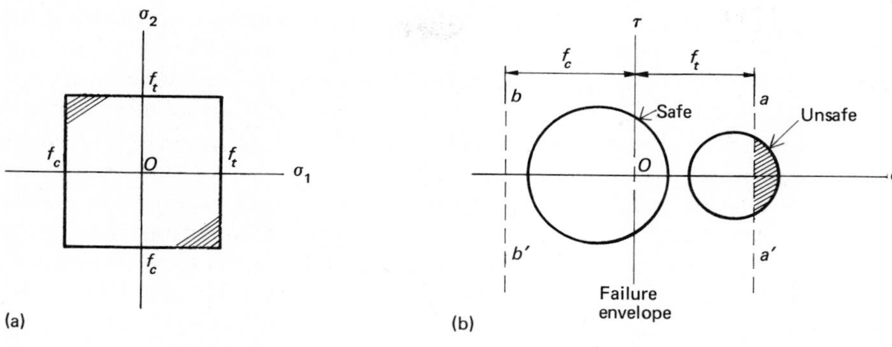

Fig. 4.1. *The Rankine theory displayed:*
(a) on the $\sigma_1\sigma_2$ plane;
(b) on Mohr's diagram.

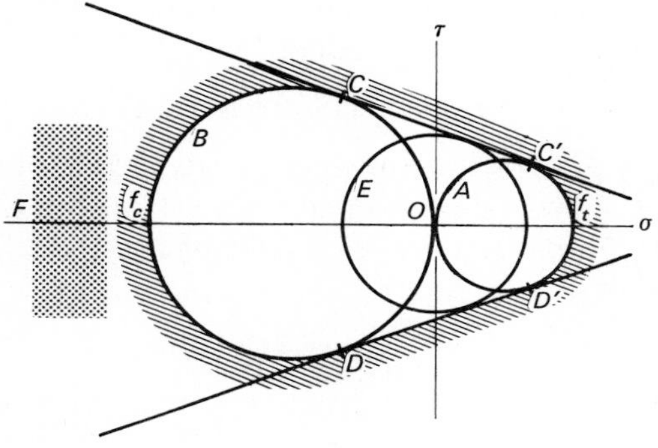

Fig. 4.2. *Mohr's failure theory displayed on Mohr's diagram.*

is completed by drawing the common tangents CC' and DD' to these circles. In Fig. 4.2, the circle E centred on O and tangent to CC' represents failure in pure shear and the radius of E the shear stress at failure according to this theory. This turns out to be in reasonable agreement with experiment in the case of cast iron. The failure envelope usually adopted is shown hatched; note that failure is unlikely to be caused by high three-dimensional or hydrostatic stresses such as exist in the region F although this is outside the envelope.

The safety factor for a given stress state can be expressed in terms of the principal stresses and the material properties f_t and f_c. If the circle F in the sketch is Mohr's circle for the stress in question, the safety factor can be defined as the ratio of the diameter of the largest circle that can be drawn on the same centre within the envelope (circle G), to the diameter of circle F. It can be shown that

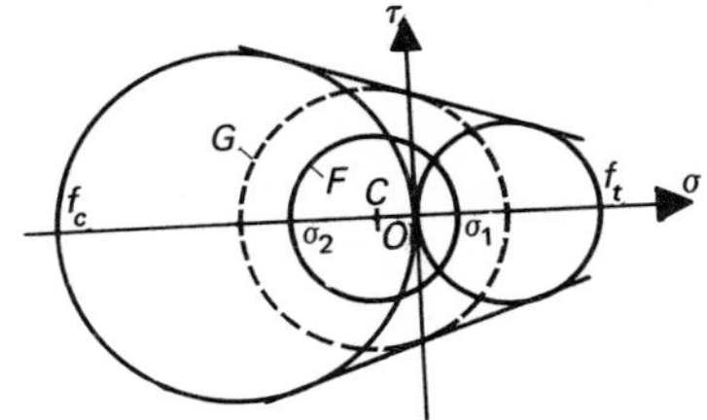

$$\frac{1}{n} = \frac{\sigma_1}{f_t} + \frac{\sigma_2}{f_c} \tag{4.1}$$

A description of the shape of the failure surface for concrete and cast iron is given in Newman and Newman[2] and Paul.[3]

4.3.2. Failure theories for ductile materials

The limiting condition selected as a design criterion for ductile materials is usually the stress at the onset of yield or the proof stress. A failure criterion much used in design is based on Tresca's maximum shear stress theory. Another theory, the maximum distortion energy theory (attributed to von Mises), gives slightly improved accuracy, but is less convenient to use. The failure surface for Tresca's theory is a hexagonal cylinder, and for that of von Mises a circular cylinder (see Plate 4.1). The former could be inscribed in the latter and the difference between the two is never more than 15%. Descriptions of these and other theories are given in Nadai.[1]

Maximum shear stress (Tresca) theory. As a result of experiments on the extrusion of metals Tresca concluded that plastic deformation is initiated when the maximum shear stress q_{max} in the material reaches a critical value. The permanent deformation occurs by slip along planes at 45° to the directions of the principal stresses. If the principal stresses are such that $\sigma_1 > \sigma_2 > \sigma_3$, q_{max} is given by $\frac{1}{2}(\sigma_1 - \sigma_3)$ = constant, irrespective of the absolute values of the principal stresses. The constant in this equation is usually determined from the most readily available test information – the yield strength in simple tension, f_y. Then $\sigma_1 = f_y$ and $\sigma_2 = \sigma_3 = 0$, so that

$$q_{max} = \tfrac{1}{2} f_y = \text{constant} \tag{4.2}$$

In a test under pure torsion, this theory predicts that the yield in shear q_y will be equal to $\frac{1}{2} f_y$. In fact, experimental evidence for steel gives values of q_y of from 0.58 to $0.60 f_y$. Von Mises' theory predicting $q_y = 0.58 f_y$, is in good agreement with this result.

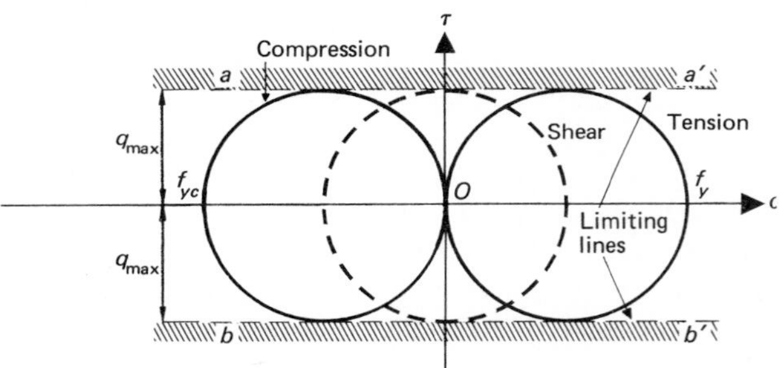

Fig. 4.3. *The limiting lines for the maximum shear stress theory.*

In Fig. 4.3, Mohr's circle for the tensile test is drawn on Mohr's stress plane, with the resulting limiting lines *aa'* and *bb'*. It can be seen that, according to this theory, the compressive and tensile yield stresses must be equal. It is a simple matter in design to determine the maximum shear stress, graphically or by calculation, and to compare it with the limiting value q_{max}. For the two-dimensional stress case, the limiting surface projects on to the σ_1 σ_2 plane as shown in Fig. 4.4. It will be seen that the regions for which the Rankine theory is unsafe (Fig. 4.1(a)) are not included in this figure. The three-dimensional surface is shown on the model in Plate 4.1(c).

It may be noted that this theory is a special case of Mohr's theory, being the case where the internal friction is taken to be zero. Thus the shear stress at failure does not increase as the compressive forces across the shearing plane increase, and the limiting lines run parallel to the σ axis on Mohr's stress plane.

The theory has been much used in practice, in some cases perhaps without being explicitly stated. For example, the well-known ASME Code for steel shafting, which has been in use since 1927, is based on it (see Chapter 8).

Maximum distortion energy (von Mises) theory. It has been shown experimentally that very high hydrostatic pressures ($\sigma_1 = \sigma_2 = \sigma_3 = -p$, where p is

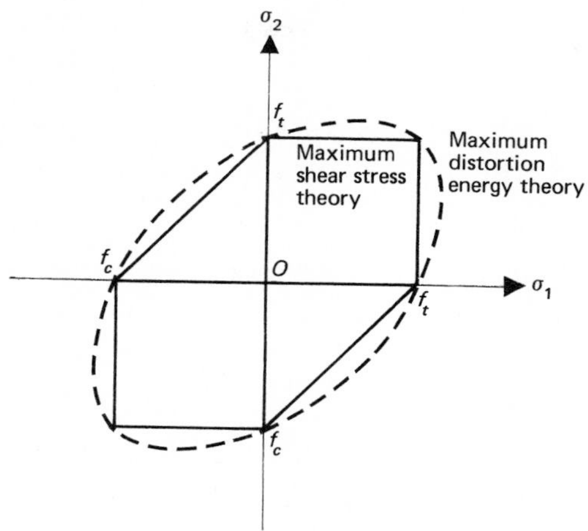

Fig. 4.4. *The maximum shear stress and maximum distortion energy theories.*

the pressure), will not cause a material to yield. This experimental fact has been used to throw doubt on several proposed failure theories; it can be seen from the models of the failure surfaces in Plate 4.1 that the Tresca and von Mises theories are not at variance with it since the points representing this stress state plot along the axis of either of the two solids.

An earlier theory which had postulated that failure would occur when the total strain energy reached a limiting value did not give good agreement with experiment. The theory put forward by von Mises discounts that part of the strain energy causing diminution or dilatation of the element and postulates that failure will occur when the remaining energy of *distortion* reaches a limiting value. In terms of principal stresses this theory states that yield will occur when

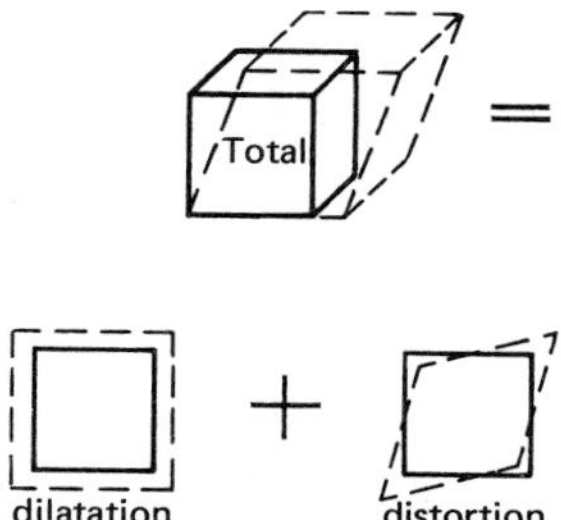

$$(\sigma_1 - \sigma_2)^2 + (\sigma_2 - \sigma_3)^2 + (\sigma_3 - \sigma_1)^2 = 2f_y^2 \tag{4.3}$$

This theory, when displayed as a three-dimensional surface, is represented by a circular cylinder whose axis passes through the origin and is equally inclined to the σ_1, σ_2, σ_3 axes (see Plate 4.1(d)). The section of this cylinder by the $\sigma_1\sigma_2$ plane is an ellipse (see Fig. 4.4) which corresponds to the equation obtained by putting $\sigma_3 = 0$ in Equation (4.3), i.e.,

$$\sigma_1^2 - \sigma_1\sigma_2 + \sigma_2^2 = f_y^2 \tag{4.4}$$

For the pure shear case, $\sigma_2 = -\sigma_1 = q_y$, and substitution in Equation (4.4), gives the value for q_y mentioned earlier

$$q_y = \frac{1}{\sqrt{3}} f_y = 0.58 f_y$$

Another form of Equation (4.4) of more direct use in design can be obtained by substituting the values of σ_1 and σ_2 from Equation (3.5). This gives the yield stress in terms of the combination of stresses σ_x, σ_y and τ_{xy} that just cause failure. Thus

$$f_y^2 = \sigma_x^2 - \sigma_x\sigma_y + \sigma_y^2 + 3\tau_{xy}^2 \tag{4.5}$$

This equation will be encountered in mechanical and structural codes which use the maximum distortion energy (von Mises) theory. Note that the three-dimensional form for this equation is

$$f_y^2 = \tfrac{1}{2}[(\sigma_x - \sigma_y)^2 + (\sigma_y - \sigma_z)^2 + (\sigma_z - \sigma_x)^2 + 6(\tau_{xy}^2 + \tau_{yz}^2 + \tau_{zx}^2)]$$

The use of these failure theories is illustrated in Examples 4.1 and 4.2.

Example 4.1

The hoop and longitudinal stresses in a thin-walled pressure cylinder are given by

$$\sigma_t = \frac{pd}{2t} \quad \text{and} \quad \sigma_l = \frac{pd}{4t}$$

Calculate the pressure at which failure will take place according to
(a) maximum stress theory,
(b) maximum shear stress theory,
(c) maximum distortion energy theory, if the permissible stress is 0.67 of the yield strength.
Cylinder diameter, d = 800 mm.
Wall thickness, t = 20 mm.
Yield strength, f_y = 478 MPa.

The stresses σ_t and σ_l are principal stresses, so that

$$\sigma_1 = \frac{800\,p}{40}, \quad \sigma_2 = \frac{800\,p}{80}, \quad \sigma_3 = 0$$

(a) $\sigma_1 = 0.67 f_y = 320.$

$$p = \frac{40 \times 320}{800} = 16 \text{ MPa}$$

(b) In the first quadrant on the $\sigma_1\sigma_2$ plane (with σ_1 and σ_2 both positive), the maximum stress and maximum shear stress criteria are the same, i.e., allowable pressure = 16 MPa.
(c) Using Equation (4.4),

$$(20\,p)^2 - (20\,p)(10\,p) + (10\,p)^2 = 320^2$$

$$p = \sqrt{341} = 18.5 \text{ MPa}$$

Example 4.2

A cast-iron shaft, diameter 100 mm, is subjected to a torsional moment of 20 kN m. If f_t = 180 and f_c = 650 MPa, calculate the safety factor according to
(a) maximum stress theory,
(b) Mohr's internal friction theory.

The maximum shear stress is given by

$$\tau_{max} = \frac{16T}{\pi d^3} = \frac{16 \times 20 \times 10^6}{\pi \times 100^3} = 102 \text{ MPa}$$

This stress is shown for the pure shear case on Mohr's diagram in the sketch, on which the two failure theories are also drawn.

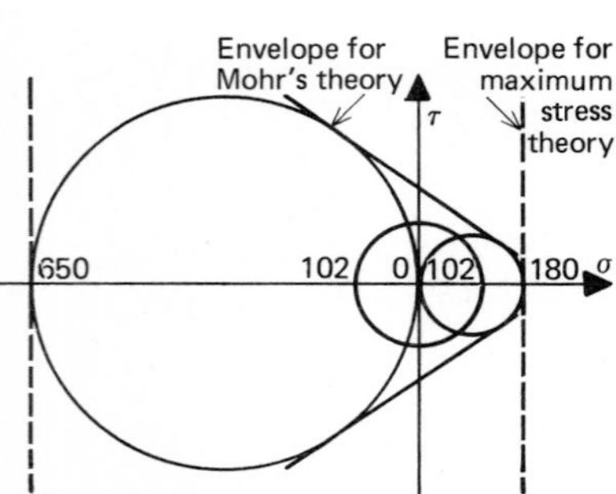

(a) Safety factor $= \dfrac{180}{102} = 1.77.$

(b) From Equation (4.1), $\dfrac{1}{n} = \dfrac{\sigma_1}{f_t} + \dfrac{\sigma_2}{f_c}$

$$= \frac{102}{180} + \frac{102}{650}$$

$$\therefore n = 1.38$$

4.3.3. Fatigue

Much of the information on fatigue strength has been obtained using highly polished specimens tested in fully reversed bending or axial loading. The value of the endurance limit (or of the fatigue strength for the required life) obtained in this way will usually be the starting point for design. The limiting condition used will be that of cracking or fracture – the crack which initiates breakdown can be detected, in some cases, long before final fracture. It is now possible to predict how much energy is required to cause a crack of known size to spread and this energy can be related to the loading; such considerations are not of immediate use in design, but will be more widely used in checking the acceptability of known defects.

The design methods and failure criteria required to apply strength information must take into account differences in the shape and surface finish and in the loading of the machine part from those used in the test. A difference in shape – leading to a stress concentration – is sometimes accounted for by multiplying the alternating stress component of the total stress by a factor which will depend both on the shape of the part and on the sensitivity of the material to stress concentrations. The justification of applying this factor to the alternating component and not to the steady component of stress is not soundly established either by theory or experiment, but the procedure is widely followed in practice.

The failure theories used to extend the test data to more complicated loading conditions encountered in practice are reasonably well established – although the fact that there is more than one in common use indicates that the subject is not completely clear-cut.

The fluctuating stress or load may be broken down into steady and alternating components f_m and f_a respectively. Thus a fully reversed stress has a total amplitude of 2 f_a, f_m being zero (see Fig. 4.5).

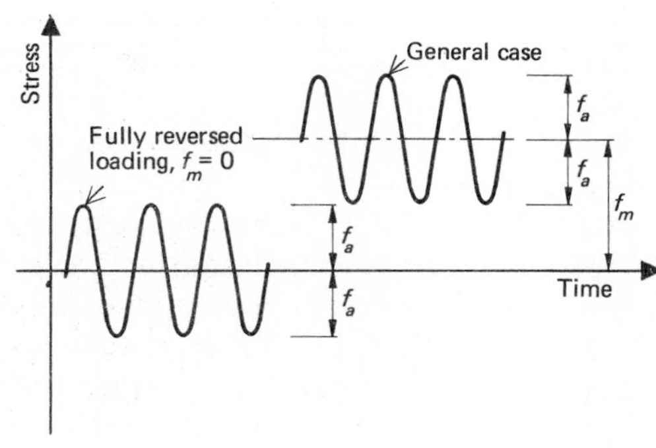

Fig. 4.5. *Fluctuating stress broken down into steady (f_m) and alternating (f_a) components.*

The first fatigue tests were done by Wohler between 1858 and 1870. These covered a wide range of loading conditions, but none of them involved compression. One method of displaying test results for a particular material is illustrated in Fig. 4.6. The values of f_m and f_a are plotted on the horizontal and vertical axes respectively; values of f_m to the left of the origin represent compressive stresses. The endurance limit in fully reversed bending, f_e, will fall on the ordinate as shown. As the value of f_m increases, the safe value of the alternating stress f_a must decrease, and when f_m reaches the ultimate strength of the material, f_a has become zero. Actual test values will fall over a band shown shaded; it is the precise shape of this band that must be established before a failure theory can be set up.

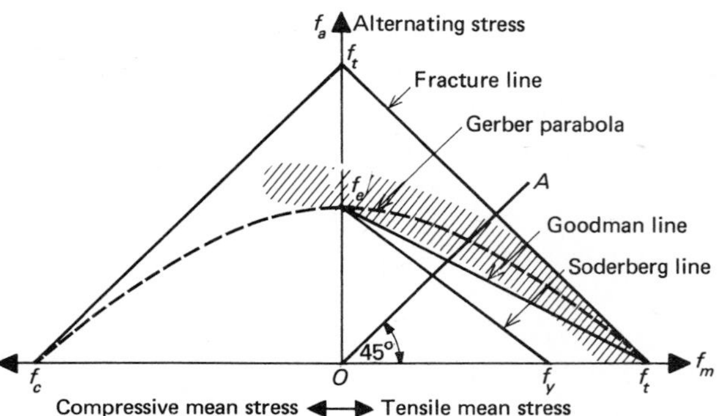

Fig. 4.6. *Relationship between f_m and f_a.*

In order to provide a failure theory for design purposes, Gerber in 1874 drew a parabola through the points f_t, f_c and f_e, justifying the shape in the tension quadrant by Wohler's results. The parabola was continued into the left (compression) half of the diagram even though no test values had been measured there. In 1886, Bauschinger carried out tests in repeated tension giving a point on the line OA (Fig. 4.6) and drew a parabola through f_t and this one fatigue test point. The good agreement of Bauschinger's results to the parabolic shape has caused surprise over the years and it was only recently revealed that they were not experimental values.[4] Such discrepancies, and the fact that many of the earlier test results have been found to be incorrect – the testing machines used did not ensure accurate loading – gives some explanation of the difficulty encountered in selecting a failure theory. The additional scatter introduced by the anisotropy of engineering materials makes it expedient to choose a theory which is safe and simple to use. The failure theories that have been used in design are mostly represented on the f_m–f_a plot by straight lines. The Goodman line is drawn from f_e to f_t, and the Soderberg line extends from f_e to f_y. These lines are shown in Fig. 4.6.

Apart from being subjected to a variety of loading conditions, actual machine parts differ from the specimens tested in laboratories to establish fatigue strengths. These differences relate to size, surface finish and particularly to stress concentrations. The methods used to account for these differences are described in Chapter 7, where the question of failure theories for fatigue is taken up again.

4.4. SAFETY FACTORS

From the discussion of load determination in Chapter 2 it will be clear that uncertainties must exist as to the precise load that will be encountered in many practical problems; the designer must face other uncertainties – for example, in the strength and behaviour of materials. The usual way of dealing with these is to design to a stress which is less than the stress at failure. A safety factor n is applied to the stress at failure to give

$$\text{design stress} = \frac{\text{stress at failure}}{n}$$

In cases where stress is proportional to load, this equation can be written

$$\text{design load} = \frac{\text{load which would cause failure}}{n}$$

There are cases (for example, in the failure of long columns by buckling) where this correspondence does not hold and where the second form should be used. If a plot of stress against load is drawn for a long column, it would take a form such as that shown in the sketch; the relationship between load and stress is not linear. It will be seen that a safety factor of 2.0 on stress corresponds in this case to a factor of only about 1.2 on load.

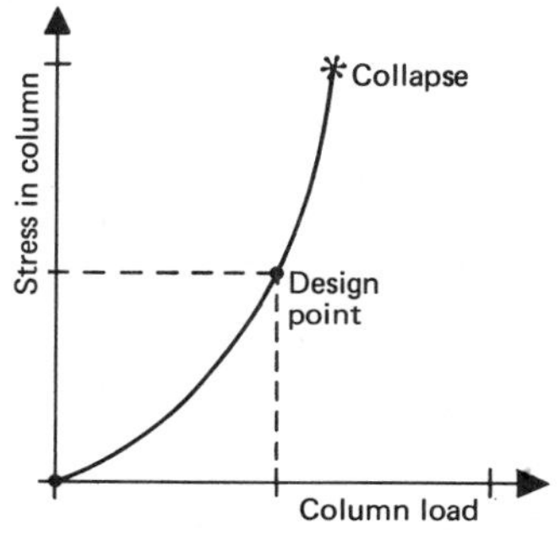

The stress or load at failure must correspond to the particular limiting conditions that have been selected as design criteria. This means, for example, that the stress at failure under steady load for a ductile material will be taken as the yield strength, f_y, whereas the stress at failure for a brittle material in steady tension will be the tensile stress, f_t. In this book permissible design stresses are denoted by p. Exceptions occur where this usage conflicts with a design code which is under discussion.

The value of n – which may range from just greater than unity to 10 or more – can be broken down into two factors which allow for the principal uncertainties encountered in loads and in materials.

$$\text{Safety factor, } n = b\,c$$

where b is a loading or shock factor, and c is a margin for ignorance of material properties.

Values for b and c are suggested by Vallance and Doughtie[5] and typical values are summarised below.

Type of Load	Shock Factor b
Steady or gradually applied	1.0
Light shock	1.25–1.50
Suddenly applied (no impact)	2.0
Heavy shock	2.0 –3.0

Type of Material	Material Factor c
Ductile, uniform microstructure	1.5–1.75
Ductile, non-uniform microstructure	2.0
Brittle	3.0–4.0

Some of the values given above have a theoretical justification; for example, it is shown in textbooks on mechanics that a load suddenly applied to a rod will cause twice the stress caused by a gradually applied load. For this case, $b = 2.0$. In the case of dynamic loading, the kinetic energy of the live load may produce what are described as shock loads. These can be determined theoretically or can be classified as "light" or "heavy" shock and values selected from the table accordingly. When it is possible to determine theoretically or experimentally the precise values of the maximum dynamic loads – and in some cases safety requirements or a code may make this mandatory – then there is no need to apply any further factor (i.e., $b = 1.0$).

The material factor, c, takes into account the variability in the material – for example, a casting whose structure varies from one location to another, depending on cooling rates, should be given a higher factor than a rolled section which is uniform in microstructure. But the main difference here is the higher factor used for a brittle as compared to a ductile material.

The nomenclature described above originated at a time when a good deal of design work was based on the tensile strength. The safety factor was then designated as $n = a\,b\,c$, where $a = \dfrac{\text{tensile strength}}{\text{yield strength}}$ for a ductile material. Now that the yield strength has been accepted as the basis for design there is little need to use the factor a, and the safety factor $n = b\,c$ based on yield strength is entirely analogous to the above definition. Thus

$$p = \frac{f_y}{b\,c} = \frac{f_t}{a\,b\,c}$$

Ductility is a property which plays a most important part – often hidden – in alleviating the effects of a failure. When materials fail in the brittle mode the fracture comes without warning, and is often accompanied by a release of elastic energy that may aggravate the failure. This situation can be contrasted with the behaviour of ductile material. The appearance of a permanent distortion may well render the structure unserviceable, but, provided instability does not arise, a warning has been given and it may be possible to avoid further load increases that will lead to fracture. Thus, the factor c for ductile materials lies in the range 1.5 to 2.0 and for brittle materials in the range 3.0 to 4.0. Particular care must be taken in the selection of ductile materials if they are to be used in conditions which make them susceptible to brittle failure.

It should be noted that no allowance is made at this stage for local stress concentration caused by changes in section or notches. The justification for this is that local yielding will occur in a ductile material at points of high stress concentration, alleviating the effects and even resulting in a strengthening due to the three-dimensional nature of the stress and, in some materials, to work-hardening. In brittle materials the omission of a stress concentration factor cannot be justified, and the stress calculated on the basis of uniform distribution should be multiplied by the stress concentration factor. As will be seen, this practice is conservative.

The effect of a sharp notch on a ductile low carbon steel is shown from the results of tensile tests on the two specimens A and B illustrated in Fig. 4.7(a). Each had the same cross-sectional area, obtained in A by a radius of 25 mm and in B by a sharp rectangular notch. The test results (Fig. 4.7(b)) show that specimen *B* was stronger both at the yield and at rupture; however, the stress–strain diagrams show that the elongation at failure of the sharply notched specimen was less than that of the other specimen. The designer must bear this loss of ductility in mind, even though no specific allowance is made for it in terms of a safety factor. In fatigue, specimen *B* would have a much shorter life than *A* and special account would have to be taken of this, while under impact loading the inability of specimen *B* to absorb energy would greatly impair its usefulness in engineering applications.

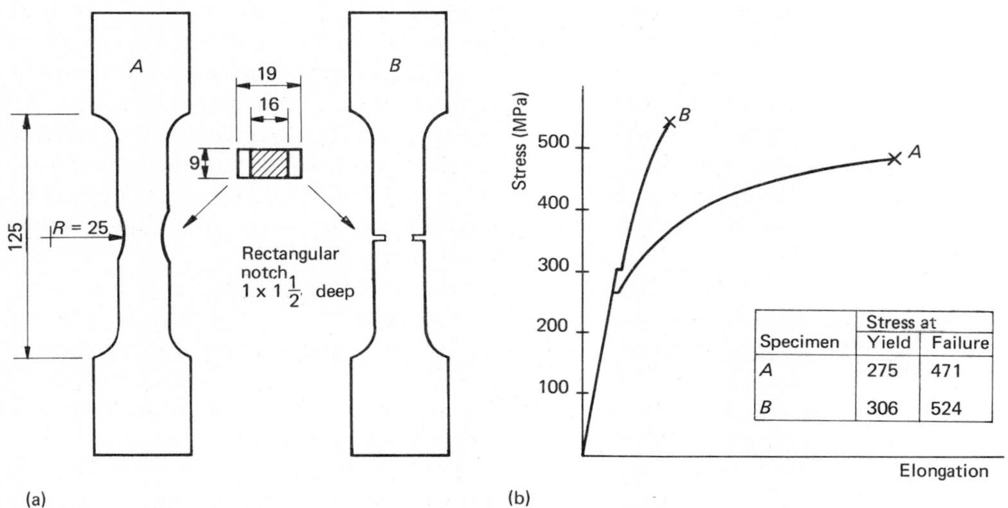

Fig. 4.7. *Test on low carbon steel plate: (a) test specimens; (b) stress–strain curves.*

When done on cast iron specimens, the same type of test produced different results. The three specimens shown in Fig. 4.8 were tested in tension. The

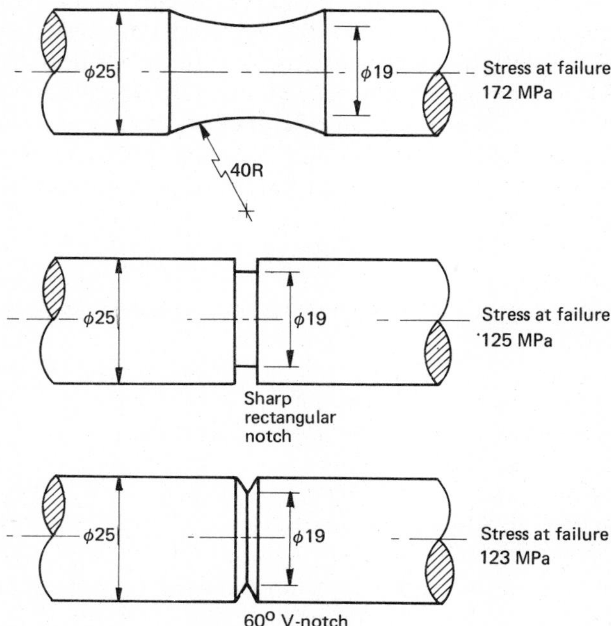

Fig. 4.8. *Cast-iron specimens tested in tension.*

unnotched specimen broke at 172 MPa and the notched specimens at 125 MPa and 123 MPa. Thus there is a decrease in strength due to the notch, although the reduction is considerably less than would be indicated by the theoretical stress concentration factor, which is about 4 for the notches used. In practice, it is common to apply the full stress concentration factor for brittle materials, although in special cases a full investigation may justify a lower value.

A wide variety of safety factors, load factors and design stresses are set out in various codes; some examples are given in Table 4.2. These values are based on a wide experience of the prevailing manufacturing methods and operating conditions and in such cases the designer is relieved of the onus of deciding on a factor. The examples quoted are intended only to show some of the ways in which such values may be presented; a user must be familiar with a code in its entirety before he attempts to use such information from it.

It must be emphasised that any design computation is based on certain loads on the machine or structure or the parts thereof and the strength or deformation of the materials. It must be emphasised as well that a margin of safety is required between conditions in service and those that cause failure. The latter point is the fundamental topic of this chapter.

There are two alternative ways of providing the margin. In one way, service loads are used and the designer is required to ensure that the stresses caused by these loads do not exceed allowable values. The allowable stresses are limiting stresses (yield strength of ductile materials, ultimate strength of brittle materials) divided by a safety factor. The alternative is to multiply the service loads by a load factor and to ensure that stresses (or strains) do not exceed those corresponding to yield of ductile material and failure of brittle materials. In structural design, the former is often called *working strength design* and the latter *ultimate strength design* or *limit design*.

The concept of a safety factor that has been described here is widely used in design, but other approaches may be adopted. For example, when designing for creep or fatigue, a close control of the safety margins can be maintained if the design is based on a life which exceeds the operational life by a suitable margin.

No hard and fast recommendations for values of safety factors can be made to cover all types of loading. The values given in codes usually refer to specific cases, and in other situations the designer must select a factor on the basis of experience and careful analysis of the conditions.

If the designer is to decide on a safety factor, he must consider the following.

(1) The various types of failure that can occur.

(2) The consequences of a failure: these can be described, in descending order of seriousness, as
catastrophic;
resulting in serious danger to life;
endangering life or causing serious economic losses;
causing delays and losses;
causing inconvenience;
rectifiable with little loss or inconvenience.

(3) The extent and accuracy of the loading data and operating conditions, and the expected accuracy of the stress calculations.

Table 4.2. Examples of the way in which design stresses or load factors are specified in codes.

In this code a margin of safety is obtained by the use of load factors. Thus the required strength U provided to resist dead load D and live load L shall be at least equal to $U = 1.4D + 1.7L$.

Other combinations of dead, live and wind loads are specified and must be provided for.

American Concrete Institute
Building Code Requirement for Reinforced Concrete
(ACI 318–71)

Steel to BS 15	230 MPa yield	340 MPa yield
Axial stress, net effective section	140	200
Tension or compression in bending: rolled sections	155	220
Shear—max shear stress	100	145
Compression—short column	140	200
Parts in bearing—pins and axles (no relative movement)	185	265

BS 2573: 1966, *Permissible Stresses in Cranes, Pt.* 1, "Structures".

	Flexure and tension parallel to grain	Compression perpendicular to grain	Compression parallel to grain	Shear parallel to grain
GROUP S1 e.g. fir and pine	13.8	1.72	9.7	1.38
GROUP S2 e.g. spruce, redwood	11.0	1.38	8.3	1.38

Note: The stresses have been rounded off in conversion to SI units.

BS Code of Practice, CP 112: 1971, *The Structural Basic Stresses for Softwoods, MPa.*

Design stress values shall be determined by dividing the appropriate material properties by the factors given below. The least value of design stress obtained by considering each property separately shall apply.

	Factor
Factor for certified or specified minimum yield or 0.2 per cent proof stress at design temperature	1.5
Specified minimum tensile stress at room temperature	2.35
Average stress to produce rupture in 100 000 h at design temperature	1.5
Average stress to produce total creep strain of 1 per cent in 100 000 h at design temperature	1.0

BS 1515: 1965, *Specification for Fusion Welded Pressure Vessels*

(4) The amount of control that can be exerted, by specification and test, over material properties.
(5) The accuracy and uniformity of the manufacturing processes (e.g., welding) and the possibility of deviation from the specified dimensions.
(6) The possibility of the structure deteriorating in service, for example, due to corrosion.
(7) The importance of minimising the size or weight of components.

Some of these factors can be subjected to a statistical treatment, and this approach is considered briefly in the next section.

4.4.1. Statistical approach to safety factor

The approach here is taken from Johnson's *Optimum Design of Mechanical Elements.*[6] In this treatment the statistical distribution curves for the actual load, λ, and for the load capability, L, of a manufactured machine element are considered. Assuming for the moment that the distributions are normal, these variables can be described by a mean and a standard deviation. The actual load and load capability can be plotted against the probability density as shown in Fig. 4.9(a). Here the probable safety factor can be described by $\bar{L}/\bar{\lambda}$; the likelihood of failure occurring is illustrated when the two curves are plotted on the same figure. In the first case (Fig. 4.9(b)), the probability of failure is represented by the overlap area. In the second case (Fig. 4.9(c)), $(L - \lambda)$ is positive in all realistic conditions and failure is avoided. The probability of $(L - \lambda)$ becoming negative can be calculated and a safety factor derived for particular values of standard deviation for L and λ.

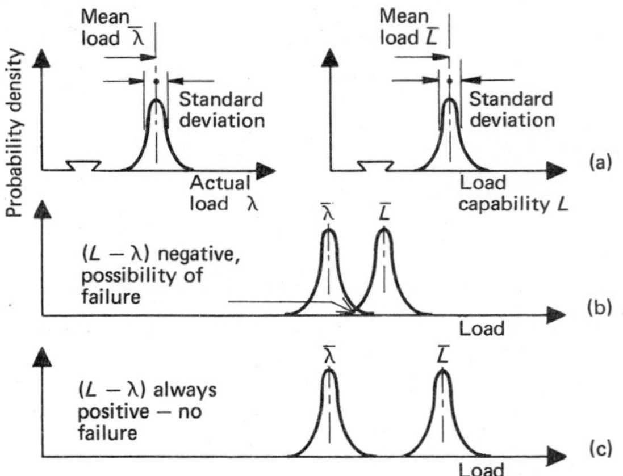

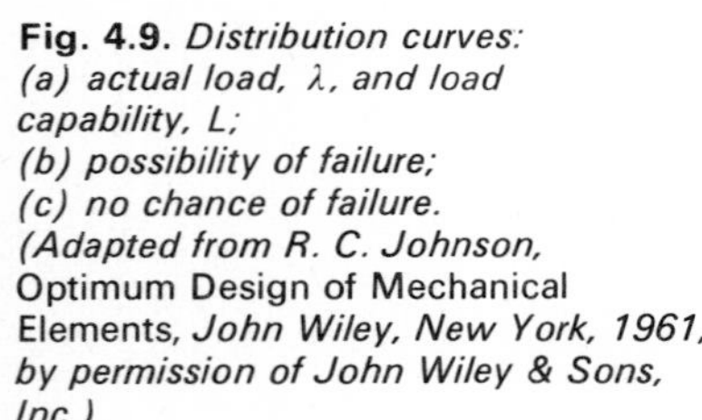
Fig. 4.9. *Distribution curves: (a) actual load, λ, and load capability, L; (b) possibility of failure; (c) no chance of failure. (Adapted from R. C. Johnson,* Optimum Design of Mechanical Elements, *John Wiley, New York, 1961, by permission of John Wiley & Sons, Inc.).*

Johnson shows that the safety factor, n, can be calculated from

$$n = \frac{\bar{L}}{\bar{\lambda}} = 1 + \frac{1.29\sqrt{D_L{}^2 + D_\lambda{}^2}}{A_f{}^{0.128}\,\bar{\lambda}}$$

where D_L and D_λ are the standard deviations for L and λ respectively and A_f is the percentage of failures that will theoretically occur; $\bar{\lambda}$ is the mean value of the actual load. A_f is assumed to be small.

The assumption that the curves of L and λ are normal is often not valid in practice; however, the method described by Johnson does not depend on the values of L and λ being normally distributed.

This approach to design requires statistical loading and strength data that may not be readily available. However, some useful comparisons can be made between values of n obtained using this method and those used in specific problems.

Example 4.3

A design for a part to carry a nominal load of 1000 N has been based on a safety factor of 1.8. A check on some test data shows that the standard deviation for the load capability—based on variations in material and fabrication technique—could be 15% of the design value. For the actual load the corresponding value is 40%. If the consequences of a failure are such that a 1% probability of yielding can be accepted, use the statistical approach to check the suitability of the safety factor.

Here $A_f = 0.01$, $D_L = 150$, $D_\lambda = 400$, $\bar{\lambda} = 1000$ and

$$n = 1 + \frac{1.29\sqrt{D_L^2 + D_\lambda^2}}{A_f^{0.128}\bar{\lambda}}$$

$$= 1 + \frac{1.29\sqrt{150^2 + 400^2}}{0.01^{0.128} \times 1000} = 1.99$$

The safety factor of 1.8 is therefore too low.

LIST OF SYMBOLS

A_f	percentage of failures occurring
D_L, D_λ	standard deviation of L, λ
d	diameter
f	normal stress
f_a	alternating stress
f_c	compressive strength
f_e	endurance limit in fatigue
f_m	mean stress
f_t	tensile strength
f_y	yield strength in tension
L	load capability
n	safety factor
p	permissible or design stress, pressure
q_y	yield strength in shear
t	thickness
λ	actual load
σ	normal stress
τ	shear stress

REFERENCES

[1] A. NADAI, *Theory of Flow and Fracture of Solids*, 2nd ed., McGraw-Hill, New York, 1950, vol. I, pp, 175–6.

[2] K. NEWMAN and J. B. NEWMAN, "Failure Theories and Design Criteria for Plain Concrete", in M. Te'eni(ed.), *Structure, Solid Mechanics and Engineering Design*, Part 2, Wiley Interscience, London, 1971.

[3] B. Paul, "A Modification of the Coulomb-Mohr Theory of Fracture", *Trans. ASME Journal of Applied Mechanics,* **28**, June 1961.
[4] H. C. O'Connor and J. L. M. Morrison, "The Effect of Mean Stress on the Push-Pull Fatigue Properties of an Alloy Steel", *Proceedings of the International Conference on Fatigue of Metals,* Institution of Mechanical Engineers, London, 1956.
[5] A. Vallance and V. L. Doughtie, *Design of Machine Members*, 3rd ed., McGraw-Hill, New York, 1951.
[6] R. C. Johnson, *Optimum Design of Mechanical Elements,* John Wiley, New York, 1961.

PROBLEMS

4.1. Make a list of failure criteria similar to Table 4.1 for the following:
(a) a small footbridge to carry pedestrians over a busy motorway;
(b) an oil-fired steam generator to be used for supplying steam for small-scale food processing;
(c) the electrical control equipment and heating elements for an electric kiln that is to be used for commercial production of ceramics.

4.2. Examine some of the construction codes and summarise the loading conditions and allowable stresses. In each case determine which failure theory is used and estimate the overall safety factor.

4.3. The safety of an important structural component is being investigated. The safety factor is to be 1.5 and, by careful control during manufacture, the standard deviation of the load capability is to be kept below 5% of the design value. How closely must the load be estimated if a 0.1% probability of failure can be accepted?

4.4. Produce a set of design stresses that would be suitable for general construction work where no codes are available.
Materials: mild steel, cast iron, brass, timber (e.g., softwood).
Loading: tension, compression, bearing (without movement).

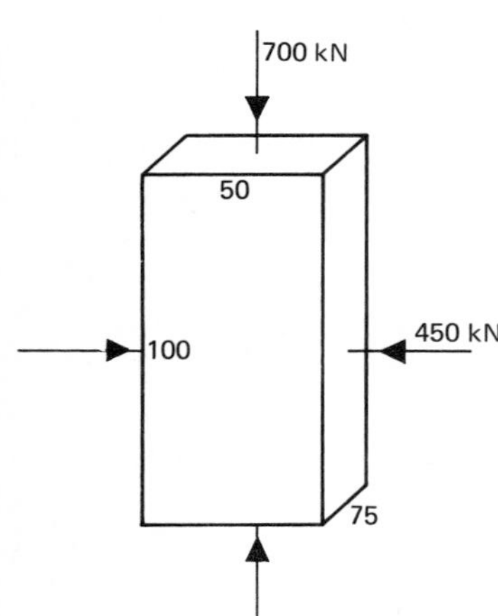

4.5. Table A14.1 in Appendix A14 gives the allowable stress in steel struts of Grade 43 steel. Compare these stresses with the critical stress as given by the Johnson formula and the Euler formula (see Chapter 12) and compute the factor of safety which is implied in this table. Take $f_y = 255$ MPa and $E = 205$ GPa.

4.6. A cast-iron block is loaded in compression as shown in the sketch. A safety factor of 4 is required with material properties $f_t = 140$ MPa and $f_c = 550$ MPa. Check the safety of the design. What single change in dimensions will give the specified safety factor?

4.7. A brittle material has tensile strength, f_t, and compressive strength, f_c. Show that the shear strength, according to Mohr's theory, is given by

$$q = \frac{2f_t}{1 + f_t/f_c}$$

Compare the value of q obtained from this expression with some values for actual materials.

4.8. A 100-mm diameter shaft carries a bending moment of 7 kN m and a torque of 9 kN m. Calculate the safety factor if the yield strength of the material is 300 MPa. Use the maximum shear stress theory, and check with the von Mises theory.

4.9. Use the ASME code (Chapter 8) to design a steel shaft to withstand a bending moment of 190 N m and torque of 130 N m. Assume factors k_b and k_t are both 2.0, and that the material properties are $f_t = 870$ MPa, $f_y = 520$ MPa. Check the actual safety factor in the shaft according to the maximum shear stress theory.

4.10. The Edsel Ford can be described, in economic terms, as a failure. Why did the failure occur, and who was responsible?

4.11. Select a failure (of a household appliance or automotive part, for example) of which you have some direct knowledge and discuss the causes. Comment on the extent to which the designer was responsible and on any remedial action that could have been taken before or after the event.

4.12. Some major engineering failures are described very fully in the literature. Select one such failure and summarise the causes; discuss the part played by the designer in the events. Some suitable topics are: Melbourne West Gate bridge (1970); Hinkley Point power station turbine failure (1969); Ferrybridge natural draught cooling towers (1965); Second Narrows Bridge, Vancouver (1958).

5 Machines and Structures

5.1. INTRODUCTION

In this book our objective is to show how to design the elements of machines and structures. A necessary preliminary is a description of these elements and the way they work together to achieve the objectives of their designer.

Although the number of such elements in engineering is limited, the number of ways in which they can be combined is unlimited, and the variety of engineering works is very great. It is this variety that makes design a creative occupation, for as long as no clear picture of the final assembly exists when the process of synthesis commences, there will be opportunities for creative steps in the design process. This variety also makes it difficult to classify engineering works neatly under the headings "structures" and "machines"; the design skills of several disciplines will often be needed to complete a major enterprise.

A list of a wide range of works carried out by engineers is shown in Table 5.1. In some of the broad classifications of activity that have been adopted are examples that lie wholly within the ambit of the structural or machine designer; in other examples, the design will involve the cooperation of designers from other disciplines. The difficulties of classifying the different structures and machines and describing them in the scope of one chapter will be obvious from the examples selected; in what follows we will concentrate on describing typical applications of the elements of design rather than attempting any complete classification of machines and structures.

Table 5.1. Typical engineering activities and products.

Activity or product	Example
Gaining ores and raw materials	Electric mining shovel (Plate 5.1)
Energy generation, transmission and conversion	Power station (Plate 5.2)
Transport and materials handling	Container crane (Plate 5.3)
Transportation	Train on bridge (Plate 5.4)
Enclosed spaces	Sydney Opera House (Plate 5.5)
Communications	Satellite earth station (Plate 5.6)
Manufacturing plant and machine tools	Production lathe (Plate 5.7)
Process plant and equipment	Fertiliser works (Plate 5.8)
Office equipment and consumer goods	Data processing equipment (Plate 5.9)

Engineers may find themselves involved in the design of machines and structures such as those illustrated at any stage of their careers. In the early stages they will be concerned mainly with the design of component parts, such as are dealt with in this book, under the supervision of other professional engineers. This phase of design is preceded by decisions made at managerial or political levels. These decisions are not considered here, but the engineer working at a lower level must be aware of the impact they have on the cost and efficiency of the complete construction. Some of the factors involved in the decision-making process are discussed in the books quoted in the "Bibliography" and are the subject of problems set at the end of this and other chapters.

Plate 5.1. *This electric mining shovel is working at an open-cast face under the arduous conditions typical of many mining operations. The designer of such equipment encounters structural as well as mechanical and electrical problems. (Reproduced by permission of P. & H. Harnischfeger, Milwaukee, U.S.A.)*

Plate 5.2. *As the demand for power increases, new resources must be tapped and new equipment designed. In this case a geothermal field supplies the power stations on the river with steam. The use of wet steam from the bores presented the pipeline and turbine designers with new problems. (Reproduced by permission of New Zealand Electricity Department.)*

5.2. CLASSIFICATION OF STRUCTURES

From the point of view of the structural engineer, structures can be arranged in four groups—multistorey buildings, single-storey sheds and halls, towers, and bridges. The basic function of a building is to shelter its contents, and that of a tower is to support equipment, while a bridge's function is to carry a transportation route across an obstruction.

The *multistorey building* provides a large area of sheltered floor space on a small area of land. It is therefore built where it is thought desirable to have large numbers of people living or working together and where land values are high. Thus, multistorey buildings are a feature of the centres of cities. They are used as dwellings in apartment buildings, as offices or service buildings (e.g. telephone exchanges, hospitals) and, to a restricted extent, as factories. A modern multistorey building is shown in Plate 5.10.

In many manufacturing processes, materials and components flow more easily horizontally than vertically. A large horizontal floor area is more useful than the same total area in the form of a stack of smaller floors. Also, many such processes require heavy machines and the cost of supporting big loads high above the ground is considerable. For these reasons, multistorey buildings as factories are the exception rather than the rule. Good use can be made of them when very tall items of plant have to be housed, when gravity can be used

Plate 5.3. *Many developments are taking place in road, rail, sea and air transport. The advent of containers has required new ships and handling equipment. This container crane is designed to ensure the quick turn-around of ships. (Reproduced by permission of the Auckland Harbour Board and Cantouris Studios.)*

to move materials from one process to the next (as in a flour mill) or when floor loads are light – for example, in the manufacture of clothing.

There is a wide range of *single-storey*, shed-like buildings. Since the only load to be supported above ground is the roof, relatively long spans are possible leaving unobstructed space for machines, storage or an audience. The low cost of such construction makes it practicable to enclose large areas, provided the cost of land is low enough. It is the kind of construction used in industrial estates for factories and in residential suburbs where it is used for halls, gymnasiums, churches, etc., such as the hall shown in Plate 5.11.

A special subdivision of this class comprises aircraft hangars, exhibition halls and large sports arenas. Roofs of such buildings may be required to span distances of more than 100 metres. They demand extensive skill and experience in structural analysis and provide architect and engineer with opportunities for virtuosity not offered by conventional practice. An interesting survey of these structures is given by Salvadori and Heller[1] and there are frequent papers in the proceedings of professional societies describing this kind of construction.

In this discussion we have referred to the loads carried by the floors and the roof of a building implying that these are vertical forces – the weight of the parts of the building and whatever it supports. Forces that act horizontally, or vertically upwards may also be applied to a building and they often dominate its design. The wind acting on the walls causes a horizontal force to act and, on the roof, it causes pressure to be reduced in some places so that the combined effect of internal and external air pressure is to lift the roof. During earthquakes,

Plate 5.4. *A diesel-hauled train crosses the Waiteti Viaduct on the North Island main trunk railway. Completed in 1889, the 36-m high viaduct still forms a link in the New Zealand transport system. (New Zealand Railways Publicity photograph.)*

the inertial resistance of the building to horizontal motion of its foundations is equivalent to a horizontal load.

In summary then, a building structure may be required to support floor loads in a tall building comprising a stack of floors of relatively modest area, one above the other, or it may be required to support a light roof over an area of many hectares. It must be able to support its own weight and the weight of its contents and such additional horizontal and upwards or downwards loads as may be imposed by wind and earthquake.

Towers are required to support a variety of equipment. In communications, they range from the ubiquitous telephone pole to three-dimensional frameworks supporting antennae of ground stations in satellite communication links. The towers of transmission lines are familiar to all and so are the drilling rigs of oil wells.

Bridges do not call for classification in respect of their function. Every bridge carries a transportation route across an obstruction. At the same time, the bridge must be designed so as to cause minimum interference with whatever it crosses. For example, when a highway has to cross a river, long spans are necessary to keep to a minimum the number of piers which disturb the flow of the river. Similarly, a road bridge over another highway has its supports arranged so that they do not prevent safe use of that highway. In modern motorway or freeway design, emphasis is laid heavily on satisfactory geometric design of the carriageways and this often causes complicated structural problems by restricting choice of pier sites. In Plate 5.12 a highway bridge over a motorway is shown.

Compared with buildings, bridges in general have to carry heavier loads, often moving at high speeds, on longer spans. They have, in common with buildings, the need to resist wind and earthquake forces.

Plate 5.5. *Many problems in design and construction have been met and solved by architects and engineers in the creation of the Sydney Opera House, which is designed to contribute to the cultural life of a large city. The roof structure now adds a new interest to the Sydney Harbour skyline. (Photo by courtesy of John Fairfax and Sons Ltd.)*

5.3. STRUCTURAL FUNDAMENTALS

We can pose the problem faced by the structural designer as that of supporting a load above the ground while leaving the space beneath clear. Loads like this are often described as *suspended loads*, a misnomer to the extent that the load is, in fact, carried by a structure under it. The concept is only partly relevant to towers, since no use, or only limited use, is made of the space beneath the load.

As shown in Fig. 5.1(a), the resultant of the internal stresses in a column directly beneath the load would exert a force vertically upwards on the load, as required to support it. However, the second requirement, for unobstructed space under the load, is not met. Two inclined struts, as in the A-frame in Fig. 5.1(b), enable both requirements to be met and it is apparent that the forces supporting the load now have horizontal components as well as vertical. Further development of this line of thought to improve the space available under the load leads to an arch (Fig. 5.1(c)) or a truss supported on two columns (Figs. 5.1(d) and (e)). Finally, a beam can be used, its internal stresses providing the combination of horizontal and vertical forces necessary to support the load (Fig. 5.1(f)).

This simple illustration leads to the general conclusion that the structure provides horizontal forces or forces with horizontal components by means of which the effects of vertical loads are transferred sideways. Transferred a suitable distance in this way, the effects can be balanced directly by compressions in columns. In all of the structures in Fig. 5.1, a "path of the load" can be traced by following a continuous compression force to the bases of the supports.

Plate 5.6. *The Warkworth Communications Satellite Earth Station is one of many around the world which handle communications via satellites. Although electronic equipment predominates, much structural and mechanical equipment must be specially designed in an installation of this type. (Reproduced by permission of New Zealand Post Office.)*

Plate 5.7. *This capstan lathe is designed for automatic high-speed production of non-ferrous components. Machine movements are programmed on the plugboard, and control circuits actuate valves and air cylinders connected to those machine elements which are normally hand-operated. (Reproduced by permission of Alfred Herbert Ltd.)*

Plate 5.8. *New Zealand relies heavily on artificial fertilisers to maintain the pasture growth required for meat, dairy and wool production. The storage, handling and processing of corrosive products are involved in this chemical plant which produces superphosphate. (Reproduced by permission of Bay of Plenty Co-operative Fertiliser Co. Ltd.)*

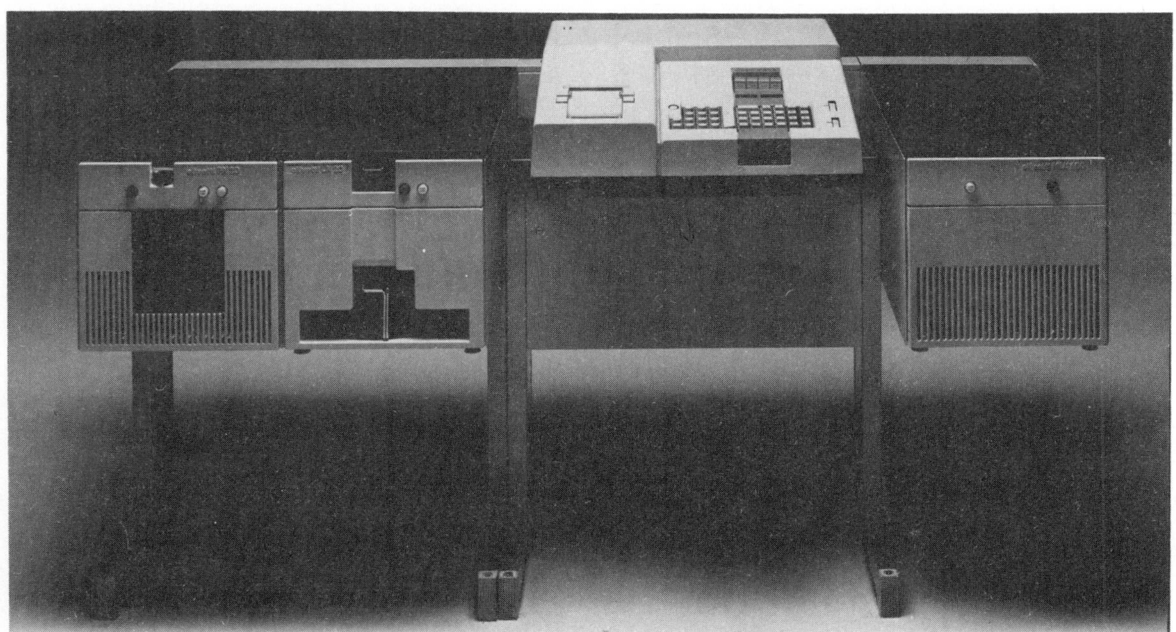

Plate 5.9. *The mini-computer in the illustration is arranged on a modular basis. The principal unit is a computer which, in this case, is supplemented by input and output tape units and by a magnetic memory. The industrial designer is Ettore Sottsass. (Reproduced by permission of Olivetti.)*

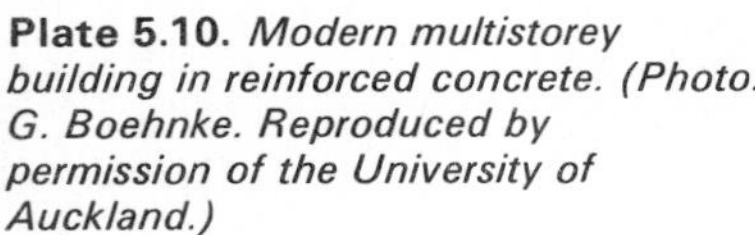

Plate 5.10. *Modern multistorey building in reinforced concrete. (Photo: G. Boehnke. Reproduced by permission of the University of Auckland.)*

Where internal horizontal compressions are used in this path, they must be balanced for internal equilibrium; and in the arch, the truss, and the beam this leads to tensions in parts of the structure.

Some measure of structural efficiency is given by the nature of the internal forces in the members of the structure. Most efficient is the column directly beneath the load. All of the internal stresses in the column act to provide exactly the necessary force in the right place. In all the other structures, additional internal stresses are required by the horizontal forces or components leading to additional material in the structure. This complication goes to the extent of requiring tensile forces in some members of a truss or parts of a beam—a far

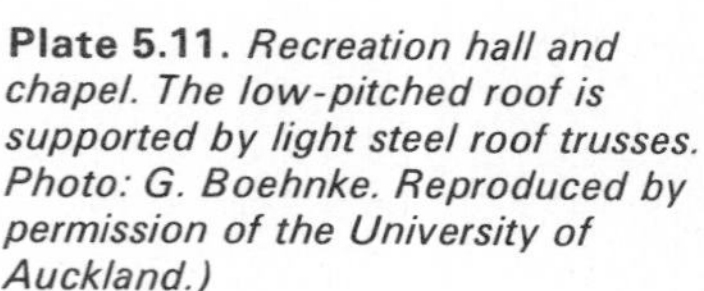

Plate 5.11. *Recreation hall and chapel. The low-pitched roof is supported by light steel roof trusses. Photo: G. Boehnke. Reproduced by permission of the University of Auckland.)*

Plate 5.12. *Twin span bridge over a motorway. The deck slab is supported on steel beams. (Photo: G. Boehnke. Reproduced by permission of the University of Auckland.)*

cry from a member directly under the load and subject to compression only, but necessary to make the horizontal forces balance.

Generally speaking, the most economical structure will be the one which makes direct use of compression to support a load. Next will be a truss in which the members are subject to direct tensions and compressions. However, economy of the finished construction is the real criterion of cost. For example, the increased depth of structure required when trusses are used instead of beams

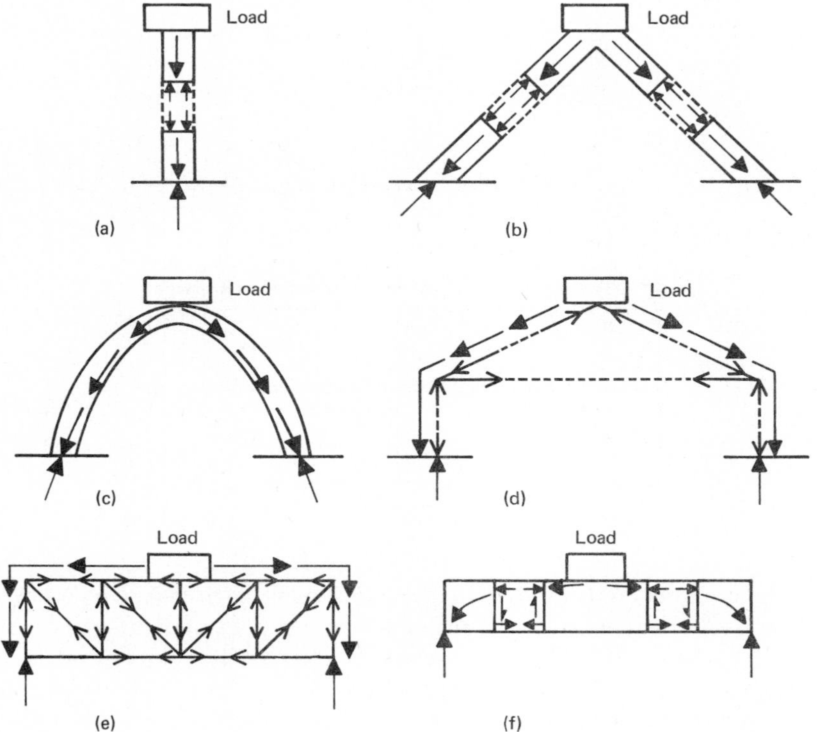

Fig. 5.1. *Various ways of supporting a load:*
(a) column;
(b) raking struts;
(c) arch;
(d) pitched truss;
(e) parallel truss;
(f) beam.

leads to increased costs elsewhere—in walls and partitions. Furthermore, considerations of function and aesthetics often call for beams in preference to trusses.

A comprehensive discussion of structural fundamentals is given by Salvadori and Heller.[1]

5.4. MULTISTOREY BUILDINGS

The frame of a multistorey building comprises a regular array of beams and columns as shown in Fig. 5.2. The *floors* are supported by the *beams* whose ends are, in turn, supported by the *columns*. The bottom of each stack of columns is supported on an enlarged base known as a *footing* to reduce the contact pressure or *bearing pressure* on the soil. Collectively, the footings, or their equivalent, are called the *foundations* of the building. The *roof* may be supported by the same kind of framing as the floors, but frequently it is not. Advantage can be taken of two facts: one, the roof needs a slope or pitch to facilitate drainage and two, it has to support only nominal loads in addition to its own weight, unless snow is likely—this can impose a considerable load on a roof. It may thus be possible for the roof structure to span the whole building. In that event, construction of the top storey can be similar to the lightweight construction of the single-storey shed or hall.

Vertical access through the building is provided by *stairs* and *lifts or elevators*, these being enclosed in a vertical tower inside or adjacent to the building. The tower can be used to support floor beams and to resist horizontal forces as well as to enclose the vertical access.

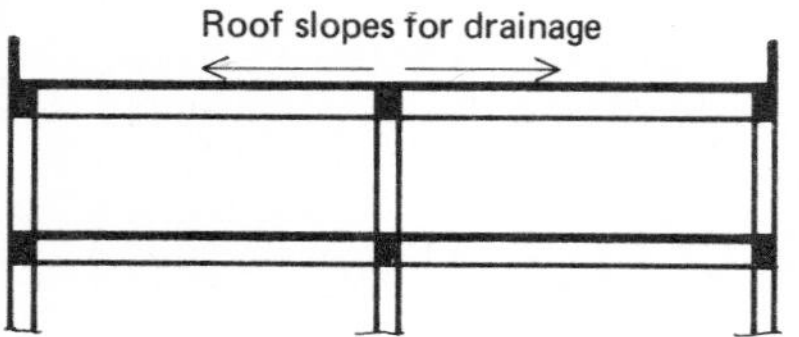

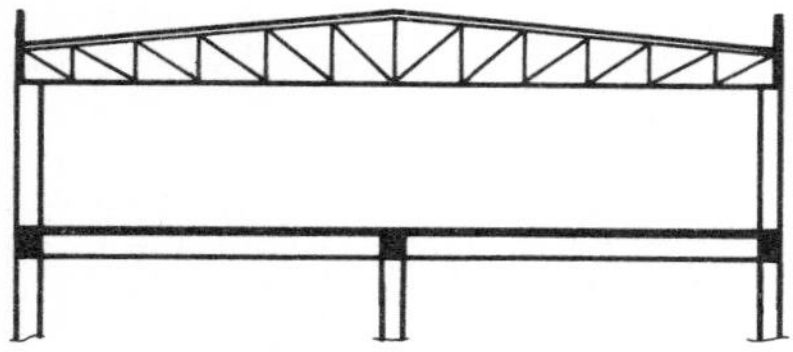

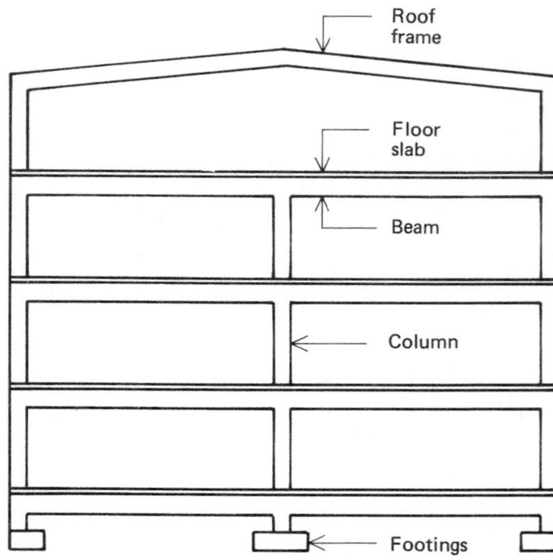

Fig. 5.2. *Cross-section of framed building.*

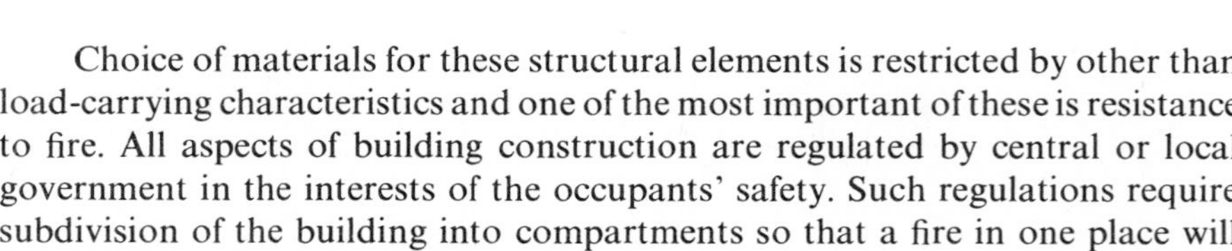

Choice of materials for these structural elements is restricted by other than load-carrying characteristics and one of the most important of these is resistance to fire. All aspects of building construction are regulated by central or local government in the interests of the occupants' safety. Such regulations require subdivision of the building into compartments so that a fire in one place will not spread. It is also necessary that the stairs be isolated to facilitate escape from the building and the structural frame must be able to withstand the effects of heat. These considerations lead to the widespread use of reinforced concrete in multistorey buildings. The thermal conductivity of concrete is low enough to protect the reinforcing steel from the heat of a fire, the strength of concrete is not seriously reduced by heating and concrete does not burn.

Thus, reinforced concrete slabs have excellent properties as floors. They can be designed to carry loads when supported at two or more edges by beams and they prevent vertical spread of a fire. The stair and lift tower is also usually constructed of reinforced concrete. Here also, its excellent load-carrying and fire-resistant characteristics are made use of.

Plate 5.13. *Reinforced-concrete framed building under construction. Concreting of columns and a wall is in progress where the formwork can be seen. Note also reinforcing for the next tier of columns. (Photo: A. Estie. Reproduced by permission of the University of Auckland.)*

The frame may be built of reinforced concrete or steel, and a reinforced concrete frame under construction is shown in Plate 5.13. If reinforced concrete is used, fire resistance is obtained by having sufficient concrete cover over the reinforcing steel. When a steel frame is used, it must be encased in some form of insulating material. Traditionally, this has been done by encasing the steel frame in conventional dense concrete, but there is now increasing use of lightweight materials. These include sprayed asbestos and concrete with lightweight aggregates such as perlite. By these means, adequate fire resistance can be obtained without increasing the self-weight of the structure unduly. We should point out that lightweight floors can also be constructed from steel troughing with light fire-resistant materials applied to it. The saving in weight may be substantial in a tall building but cost and reduced wear resistance may make the alternatives to dense concrete unattractive.

A tall building must be able to resist horizontal forces due to wind or earthquake. This resistance can be provided by the frame, by *bracing* or by *shear walls*. When the beams and columns are joined by rigid joints, so that

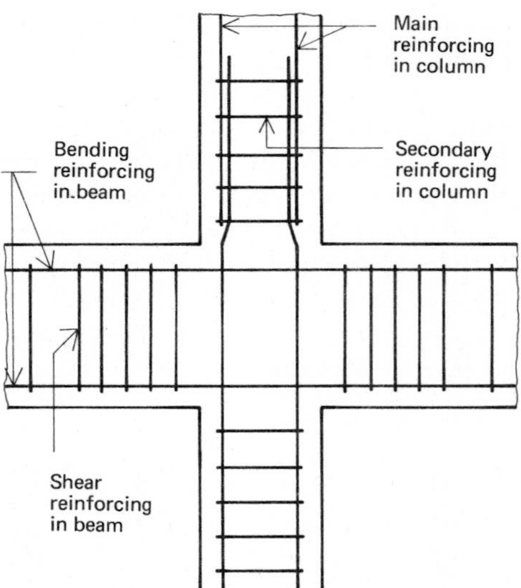

Fig. 5.3. *Elevation of beam–column joint in reinforced concrete.*

there is no relative rotation, the frame is called a *monolithic* or *continuous frame*. By its nature, reinforced concrete lends itself to monolithic construction and reinforced concrete frames are readily made strong in relation to horizontal forces. The reasons for this will be clearer after Chapter 16 on "Reinforced Concrete Beams", 19 on "Columns, Column Bases and Footings" and 23 on "Construction Operations" have been studied. For the present, however, a typical beam to column connection is sketched in Fig. 5.3. The sketch shows how the longitudinal reinforcement of both beam and column passes through the joint. The resultant force on the length of a bar between the column faces is anchored in the concrete common to the beam and the column and the column bars are similarly anchored. An out-of-balance moment in the beams can thus be transferred to the columns.

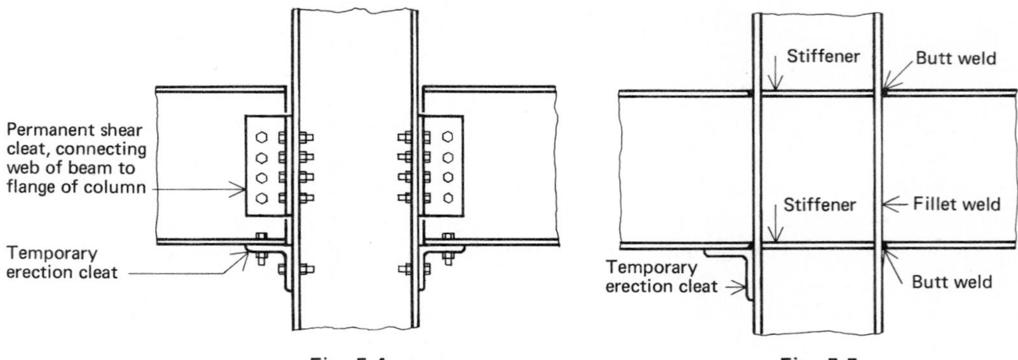

Fig. 5.4. *Bolted beam–column joint in steel, designed for shear only.*

Fig. 5.5. *Welded beam–column joint in steel, designed for shear and moment. The stiffeners are fillet-welded to the web and flanges of the column.*

If a steel frame is to be made monolithic, more work has to be done than is necessary when flexible joints are acceptable. If a joint at the end of a beam has to resist shear only, a simple arrangement of riveted or bolted cleats will suffice (Fig. 5.4). The permanent shear cleats connect the webs of the beams to the flanges of the columns (both being I-sections) but the beam flanges cannot transfer any force to the column. Hence no bending moment can be transferred. When a rigid joint is required, additional flange cleats and reinforcing of the web of the column are necessary. Generally speaking, better details are possible with welded connections (Fig. 5.5). The detail in this drawing makes the beam flanges continuous across the column so that the web of the column does not have to resist the concentrated forces.

Provision of resistance to horizontal forces by bracing multistorey buildings is not common. Although structurally efficient in that the inclined bracing members are loaded in tension and compression, the interference with doors and windows makes it unattractive even though only a few frames in the whole building need be braced.

Shear walls, like braced frames, act as vertical cantilevers when resisting horizontal forces. Their usefulness springs largely from the fact that the subdivision into fire compartments requires walls to be built in certain places. They have to be provided for this purpose and there is an obvious gain if they can be used to brace the building as well. Thus, the stair tower can be designed for this purpose and other walls can also be designed to contribute to lateral strength. Not every frame need have a shear wall in it, just as only a few frames need be braced when bracing is used. The floors then act as horizontal beams to transfer horizontally applied loads to the vertical elements of the structure.

Generally multistorey buildings get their resistance to lateral loads from a combination of monolithic frame and walls, either in the stair tower or as shear walls. The sharing of these loads among the walls and frames is a problem in structural mechanics which we will not cover. It should be noted, however, that the walls and the tower are usually much stiffer than any of the frames. Since the floors, acting as girders in the horizontal plane, make lateral deflections of all walls and frames more or less equal, the stiffer elements carry a bigger share of the load.

Design of a large reinforced concrete building using the stair towers for lateral strength is described by Falconer.[2] A steel-framed building is described by Cooper.[3] The roof of a multistorey building is similar to that of the one storey shed and is not described here (see Section 5.5).

The description we have given of reinforced concrete buildings is in terms of *cast in situ* construction. Increasing use is being made of *precast* beams, often of prestressed concrete. There are advantages of factory manufacture of the beams and simpler erection procedures. However, they are offset to some extent by the difficulty of making connections between precast elements – for example, a beam and a column. The various construction operations are described in more detail in Chapter 23.

5.5. SINGLE-STOREY SHEDS AND HALLS

In its simplest form, the single-storey shed comprises walls and roof supported by columns and roof trusses (Fig. 5.6).

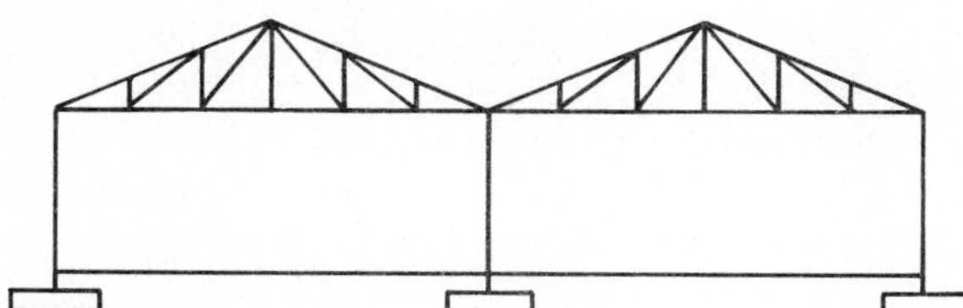

Fig. 5.6. *Cross-section of industrial building with steel frame and roof trusses.*

The enclosed space is sheltered by the *cladding* or *siding* on the walls and the *roofing material* on the roof. Corrugated metal or asbestos sheets, or tiles are used to cover the roof and these materials must be supported at close intervals. For example, when sheets are used they are supported directly by *purlins* which themselves span from truss to truss (Fig. 5.7(a)). When tiles are used, additional support in the form of *rafters* and *tile battens* is necessary (Fig. 5.7(b)). Plate 5.14 shows a steel-framed cargo store with roof trusses.

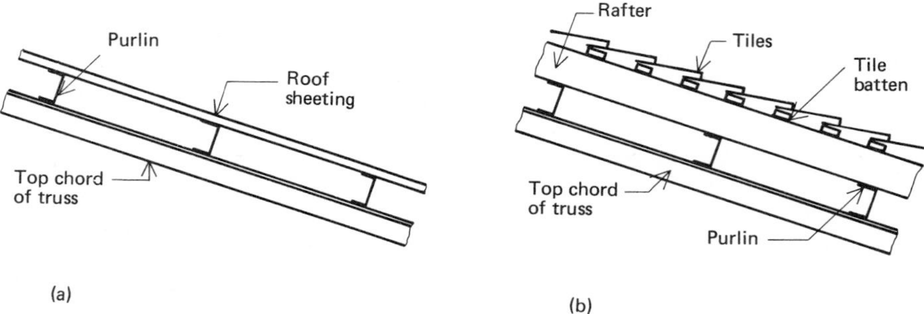

Fig. 5.7. *Details of roofing: (a) sheets used for roof cladding; (b) tiles used for roof cladding.*

When the wall cladding is in the form of sheet material, it is supported on *girts* which span from column to column. Otherwise, the walls may be supported on a light timber frame which is independent of the columns supporting the roof. In such cases the walls may be of sheet material, brick or concrete block. Some walls can be made self-supporting and no framing is required.

Plate 5.14. *Steel-framed store. Roof is supported by steel trusses and purlins. (Reproduced by permission of The Steel and Tube Co. of New Zealand and Wellington Harbour Board.)*

Such a building can usually be made to resist horizontal forces by bracing the end walls. When this is done, bracing in the planes of the roof will also be necessary to transfer wind loads in the middle of the building to the braced end walls. Similarly, the side walls can be braced to resist horizontal forces acting along the building.

It may not be practicable to brace a very long building of this kind by bracing the end walls only. In that case, *knee-braces* can be used to make each pair of columns and the truss supported into a kind of rigid frame (Fig. 5.8). This construction has the disadvantage of obstructing head room in the building.

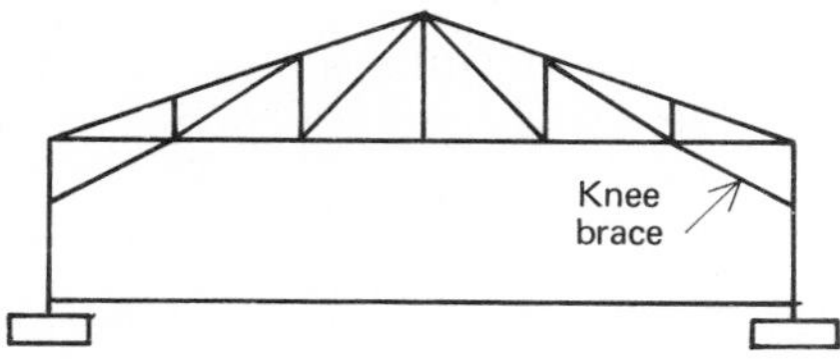

Fig. 5.8.

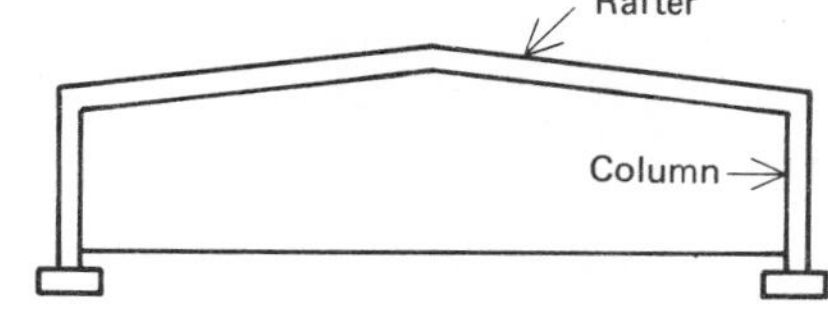

Fig. 5.9.

Fig. 5.8. *Knee braces used to provide resistance to lateral loads.*
Fig. 5.9. *Portal frame, with rigid joint connecting rafter and column.*

When appearance is important, a ceiling may be constructed at the level of the bottom chord of the roof truss to conceal this rather inelegant structure. An alternative often used in halls with low pitched roofs is the *portal frame* (Fig. 5.9). This frame has rigid joints connecting the columns and rafters and may also have the rafters rigidly joined at the apex. A steel-framed warehouse of portal frame construction is shown on Plate 5.15. However, a portal frame can also be built with a hinge at the apex, with the advantage of making the frame three-hinged (assuming rotation is not restrained at the column bases) and so statically determinate. Portal frames are able to resist horizontal forces in the plane of the frame, and bracing is not a problem.

Plate 5.15. *Steel-framed warehouse. The roof is supported by steel purlins on steel portal frames. Girts to support wall cladding can be seen on the right. Diagonal bracing in one bay of the wall and the roof is also visible. (Photo: G. Boehnke. Reproduced by permission of Howden and Wardrop Pty Ltd, Architects and Engineers, Melbourne.)*

Steel is widely used for the structures of single-storey sheds and halls. In most cases light, easily constructed buildings can be obtained at reasonable cost. Fire-proofing of the steel is not usually required. It is easy to escape from a building like this and it will not endanger its surroundings if it collapses in a fire. Both types—that is, column and roof truss, and portal frame—are well adapted to welded construction.

Alternatively, timber can be used. Column and truss construction, possibly with knee-braces, might be used where appearance is not important or if the trusses can be concealed. Using glued laminated timber or plywood box girders, some handsome portal frames have been built for halls and churches. Precast prestressed or reinforced concrete portal frames have been used in buildings like this (see Walley and Adams[4]), but their weight is a disadvantage.

Finally, it should be noted that industrial buildings of this kind are often provided with gantry cranes (see Section 5.8).

5.6. TOWERS

Towers which pose significant problems in structural design are the ones which are three-dimensional frameworks or are very tall enclosed structures.

From a functional point of view, the principal load is the weight of the equipment supported. As far as design is affected, however, other loads are likely to be more significant.

A tall tower has a severe exposure to wind by virtue of its height alone. In addition, many towers are located on windy sites. Consequently, large horizontal forces have to be resisted because of the wind pressure exerted on the tower and what it supports. Other horizontal forces may be brought to bear by the equipment supported—for example, the conductors of a transmission line at a tower where the line changes direction, or the unbalanced tension of a conductor on one side of a tower following a break in the line on the other side. Twisting may be induced by conductor tensions or by wind pressure on equipment. Equipment and tower may undergo vibration and this kind of cyclic loading can cause fatigue failure.

Combinations of vertical and horizontal forces and torques acting on towers make the three-dimensional framework desirable for stability and economy, if not for beauty. Each member of the frame is a strut or a tie, and joints are usually regarded as imposing little restraint on the rotation of the ends of the members.

5.7. BRIDGES

The elements of a bridge are its *deck*, *girders*, *piers* and *foundations*.

A continuous surface is required for the deck of a highway bridge, since the deck forms the pavement of the highway. Reinforced concrete slabs are obviously well suited. Supporting the slab there is a system of longitudinal and transverse beams which transfer the loads to the girders and these span from pier to pier. When a continuous deck is not required, as may be the case with a railway bridge, the slab may be omitted and the track supported on the deck beams. Figure 5.10 shows a typical highway bridge in cross-section, using a steel structure and Fig. 5.11 shows a similar cross-section using prestressed concrete for the girders.

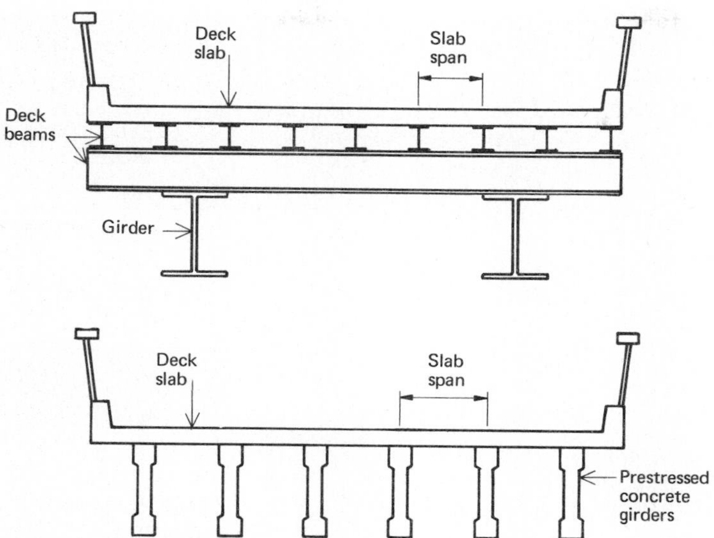

Fig. 5.10. *Cross-section of bridge with steel beams and girders.*

Fig. 5.11. *Cross-section of bridge with prestressed concrete girders.*

The piers function as large columns supporting the ends of the girders and they, in turn, are supported by the foundations. Bridge foundations may be footings similar to the footings of a building when the subsoil is suitable and there is no risk of scour. This may be the case at other than river crossings. However, when a bridge is built across a river with an alluvial bed, scour round the piers is certain to occur during floods. The foundations must be built sufficiently below the deepest scour to avoid being undermined and these deep foundations often pose difficult problems of construction. *Piles* may be used in the form of precast concrete or steel piles driven into the ground. Alternatively, holes may be bored and concrete piles cast *in situ*. In another form of construction, larger excavations within cofferdams are undertaken.

In the typical concrete bridge shown in Fig. 5.11, the deck slab is supported directly by the girders. A larger number of girders close together is shown than in the steel bridge of Fig. 5.10. In both constructions, the span of the slab is kept small; the sketches make the point that short slab spans are possible with or without deck beams. Slab spans are kept down to one metre or less because of the relatively high wheel loads which have to be supported.

It should be noted that the arrangement of Fig. 5.11 is well adapted to the use of precast prestressed concrete girders. These can be manufactured in a factory and lifted into place on the piers, where they will function as simply supported beams. The slab can then perform a second function as a flange to the girders. The junction between the slab, which is cast *in situ*, and the top of the precast beam is designed to transmit a shear stress to ensure this composite action. A highway bridge of this kind is shown under construction in Plate 5.16.

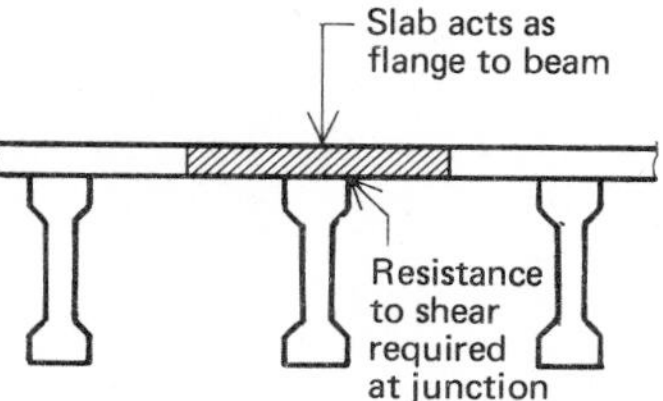

A long-span steel bridge may have trusses for its girders. In this way gaps up to 300 metres can be crossed. These principal structural elements may be made continuous over several supports, with a gain in strength and stiffness. (See, for example, Roberts and Kerensky[5] and Smith and Pain.[6])

When shorter spans are bridged in steel, plate girders are often used as girders. These I-shaped beams have web and flanges made of steel plates assembled by riveting or welding. Plate girders have been used extensively for

Plate 5.16. *Motorway viaduct under construction. In the background, pier caps for the support of deck beams can be seen. On the left, precast prestressed concrete beams have been placed for one of two carriageways. Note the spiral reinforcing half embedded in the beams. This ensures effective transfer of shear between the deck slab and the precast beam. (Reproduced by permission of the Ministry of Works, New Zealand.)*

single-track railway bridges. With one girder under each rail, wheel loads are carried directly by the girders, without intermediate structure. The girders must, however, be fastened to each other by transverse members.

5.8. MACHINES

It is not easy to classify machines in the way that has been done for structures; there are several reasons for this. One is that they serve a far wider range of uses than do structures; another is the great variety of loading conditions associated with their dynamic characteristics – some divisions of machines may be further subdivided into "low-speed" and "high-speed" classes, each with special design problems. This variation in loading is often accentuated by variations in the environment in which the machine operates – in temperature and pressure, for example – which can be far wider than those encountered in structural applications. Nevertheless it is useful to describe some typical machines and some of the problems encountered in building them up from the basic elements described later in the book.

A machine tool – a lathe or a milling machine – conforms well to the idea of a machine that many people hold. The elements that make up such a machine are basically simple, consisting of a power input, a transmission system, and an output drive in which the prescribed motions are given to the workpiece and tool. The various components are mounted on a frame or bedplate.

The elaboration in controls required in advancing from a simple workshop machine to a high-speed automatic or numerical control (NC) machine is very great, and the control system will account for a large proportion of the design effort. At the outset we will look for principles that are common to all types, and will not concern ourselves with any special control equipment. The increase in sophistication of control equipment is just one result of the demand for increased production rates.

One of the most important principles in such machines is the need to build great stiffness into the structural framework. The accuracy of machining operations will depend in the first place on the designer's ability to minimise movement under static and dynamic loads; for this reason the frame sizes will be selected on the basis of stiffness rather than strength. The high stiffness should put the natural frequency of the frame above the range of the larger exciting forces, and the massive structure will make the whole machine stable and immune from random disturbing forces.

As production rates increase there is a need for higher cutting forces, and in some cases the deflections may exceed the tolerance expected on the finished part. Such deflections can be allowed for, but there will be a great incentive to design a structure which will minimise them. Since strength is not an important criterion, cast iron has been a frequent choice for the frames of machine tools. Apart from its low cost, it is easy to cast and machine, and exhibits low wear under sliding friction; it also has higher internal damping to vibration than does steel. Its main deficiency in this application is its low value of Young's modulus, E (130 GPa as compared with 205 GPa for steel). Since the deflection of a given beam is inversely proportional to E, designers are seeking to make use of this favourable property in welded steel frames. Thus the changes in production rates that follow the new control systems call for further changes in the design approach.

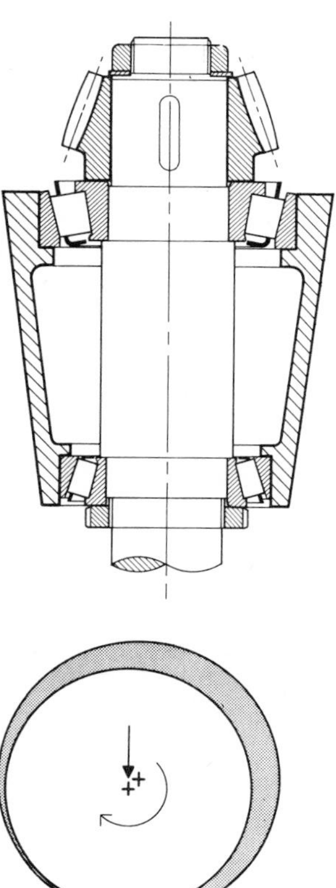

The stiffness of the supporting structure must be matched by the stiffness of the rotating and translating elements. The spindle on which the tool or work-piece rotates will be larger than is dictated by the stresses set up in bending and torsion; the bearings – rotating or translating – must have all clearances and backlash reduced to a minimum. A common method of eliminating clearance in a rolling bearing is to pre-load it against another – this can be done using two taper roller bearings as shown in the sketch. Similar but more complicated assemblies are used to support the rotating tool or workpiece in machine tools; in some designs the bearings are supported in such a way that the radial and axial loads can be carried on separate bearings. Hydrodynamic film bearings cannot be used where the demands for accuracy are very high; they depend for their load-carrying capacity on a small deflection of the shaft forming a wedge-shaped annulus in the clearance space between the journal and bearing.

The rotational motion in machine tools is often provided by one or more alternating-current electric motors. Such motors are essentially constant-speed devices, while a range of cutting speeds and feed rates will be required as an input to the machine. Belt and chain drives and gears can be used to transmit power with some change in speed.

The first two provide a simple drive which allows speed ratios up to about 6 to 1 in one step. At medium speeds (1500 rpm on the faster shaft) powers of about 10 kW can be transmitted by a single chain or belt, but since such devices are torque limited, their power capability will be directly proportional to the shaft speed. Actual design methods require wear and fatigue on the belt or chain to be taken into account. Chain drives give a positive speed ratio between the two shafts, and can be used where a timing function is needed; however, some backlash is unavoidable and they do not lend themselves readily to changes in ratio. V-belts can be used with adjustable pulleys which provide a simple means of changing speed but, because of slip, no positive timing can be relied on between drive and driven shafts; synchronous belts can be used to provide this service, although again there are limits to the precision that can be achieved. Although belts can be crossed to connect two shafts at right angles, gears are usually used for driving between shafts whose axes are not parallel.

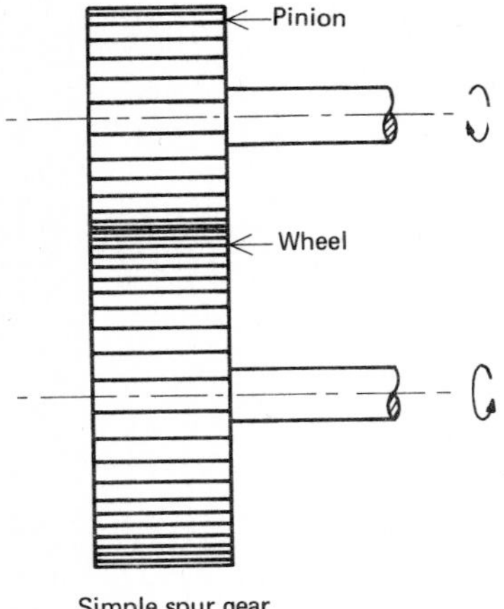

Simple spur gear

Gears provide great flexibility in terms of speed ratio and power, and can be built with the small backlash that is essential in some machine tool applications. The reduction between a pair of gears is limited mainly by the size of the wheel that can be accepted; where ratios greater than about 6 to 1 are called for, the layout may be simplified by going to a train of several gears.

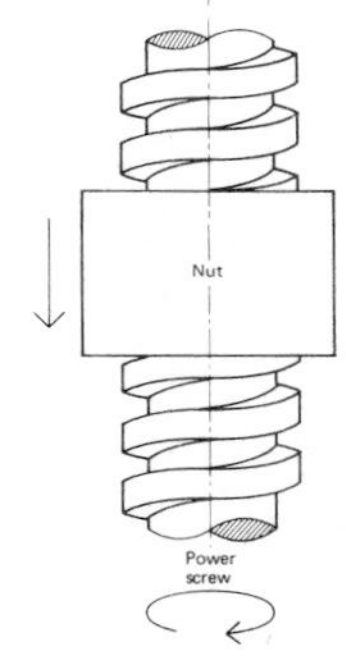

Gearboxes can be designed to provide a range of output speeds; the arrangement may call for gears which slide or pivot in order to change engagement. Alternatively, several gears may be permanently in mesh, with the output speed being altered by the operation of brakes or clutches. Once again changes in production techniques and control methods are influencing the design of other components; there is a demand for a wider choice of speed ratios that can be readily met using conventional gearboxes, and new drive systems are being introduced. Ward-Leonard drives and electronic frequency changing give a wide speed range, as does the hydraulic drive. In the latter system a hydraulic pump is driven at constant speed and the volume output to a hydraulic motor is altered by varying the pump stroke or by bypassing some of the total flow. The concepts behind these various methods are not new; they have been applied in a new way in response to a changing need, and have brought the advantage of stepless speed changes that can be achieved with simple controls.

If linear movements are required – for example, of the slide carrying the tool – these can be obtained from the rotational output of an electric motor by means of a power screw. This arrangement gives a positive drive with high accuracy and small backlash. The rate of the translational motion is controlled by the screw; the direction is controlled by machined surfaces which slide over each other.

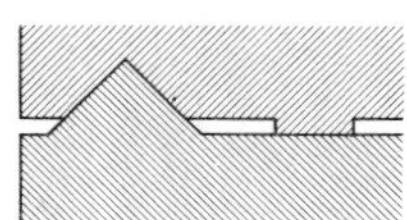

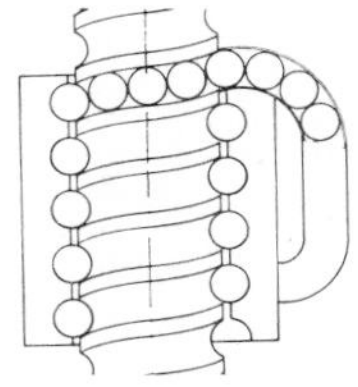

In both the power-screw and the guiding slide surfaces frictional forces are high; the efficiency of the drive is low and stick-slip effects limit the accuracy with which the motion can be controlled. To better match the accuracy of the new control systems, some machines are being equipped with recirculating ball screws; when the balls in this device reach the end of the nut, they are recirculated to the other end through external tubes. The predominantly rolling action achieved gives a high efficiency as compared with the ordinary power screw.

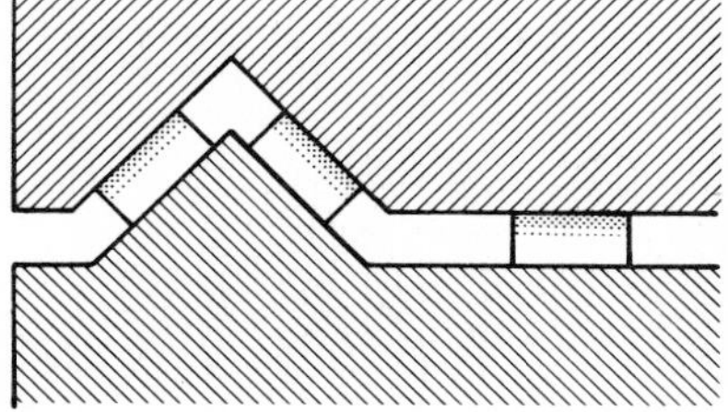

In the same way the slideways are being redesigned to incorporate rolling elements (see sketch) or else hydrostatic slides are being used. The sliding surfaces in this case are separated by a high-pressure oil film; the thickness of the film is very small and variations under load are negligible. In such installations the good frictional properties of cast iron are no longer a direct advantage and the designer is left with a wider choice of materials.

The power screw just described is not suitable for high-speed action, since the inertia of the system is high. When a rapid control action is required, the direct translational output of hydraulic or pneumatic actuators can be used. These mechanical devices will often be found in control systems which are otherwise predominantly electrical; the electrical solenoid, which provides a similar function, is less versatile as far as control of speed, damping and available length of stroke are concerned.

The preliminary description of the machine tool given – power input, transmission and output drive – can be generalised in terms of input, transformation and output. This gives us a very broad definition of a dynamic machine or system which will cover anything from an amplifier to a power station. This broadening of the description is not much help to us here, but will serve as a reminder that on a machine tool there are a number of inputs; the power inputs that we have described must be controlled, and this control function may be exerted by a human being or by a tape input. In the first case, improvements in the design of the input system may be based on a study of ergonomics – in the second, on a study of automatic control systems.

Let us turn now to a different type of machine – one in which deflections under load must be minimised, but in which this cannot be achieved by massive construction. In high-speed machines the dynamic loads become very large, and increase in the size of components is a self-defeating procedure. Consider a gas turbine as an example. In order to increase the work output per unit size of rotor, the rotational speed must be as high as possible, and will approach the upper limit set by the strength of material at the rotor tip under centrifugal loading. If the limiting tip speed is taken as 500 m/s, the rotational speeds of rotors with diameters of 0.15 m and 0.75 m will be 64 000 and 13 000 rpm. These speeds are so high that, even for the larger machine, a reduction will be necessary for most applications, and gearing will be required to transform the speed to an acceptable value. There will clearly be a considerable incentive to hold the masses of all rotating parts to a minimum, and this will extend to components in which maximum possible stiffness is also required. Here some compromise between strength and stiffness will have to be made, or else some way be sought to minimise the effect of deflections under load.

A good example of this approach is found in the design of epicyclic gears, which are often used in gas turbines. The main components are the sun and annulus wheels, which have external and internal teeth respectively, and the three (or more) equally spaced planet wheels which are carried in the space between the other two on the planet carrier. Any one of the sun gear, annulus gear or planet carrier may be fixed, and each arrangement will give different speed ratios between the remaining two, which become the input and output.

The relative rotation of the gears is shown in Fig. 5.12(a) which also shows the forces that act on one of the planet wheels. The force on a gear tooth has a radial component, since the surfaces are curved and the forces between teeth

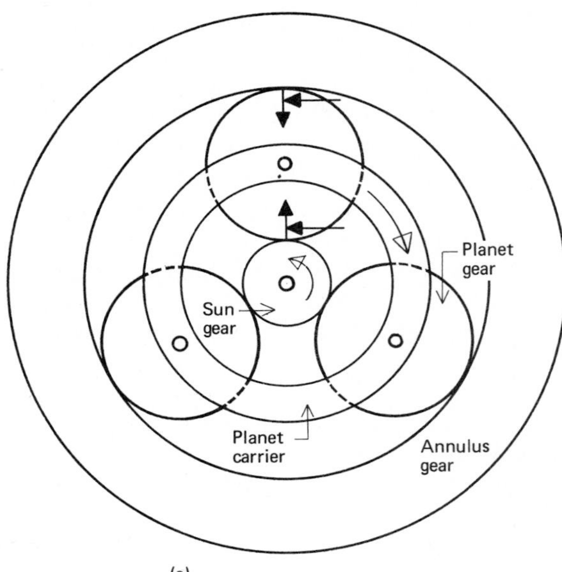

Fig. 5.12. *Epicyclic gear:*
(a) arrangement of sun, planet and annulus gears;
(b) force on a gear tooth;
(c) arrangement of planet gear.

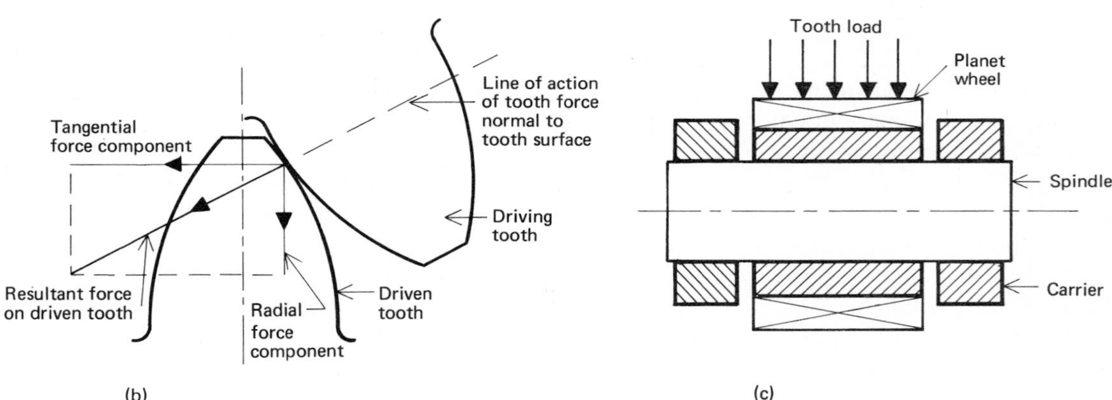

are normal to the surface (Fig. 5.12(b)). The planet wheel turns on a bearing spindle (Fig. 5.12(c)). The resultant of the planet-wheel tooth forces acts on the spindle as shown in Fig. 5.13. In order to avoid excessive deflection of the spindle, it has been common practice to build the carrier up from two plates and support the spindle at each end. To make the carrier plates rigid, they are connected in the spaces between the gears by webs, and this restricts the number of planet gears that can share the load and thus the capacity of a given size of drive. Since the loading on the central sun gear is symmetrical, it is advantageous to mount this gear on a slender cantilevered shaft; the flexibility of the mounting allows the gear to deflect and even out any maldistribution of load from the three planet gears.

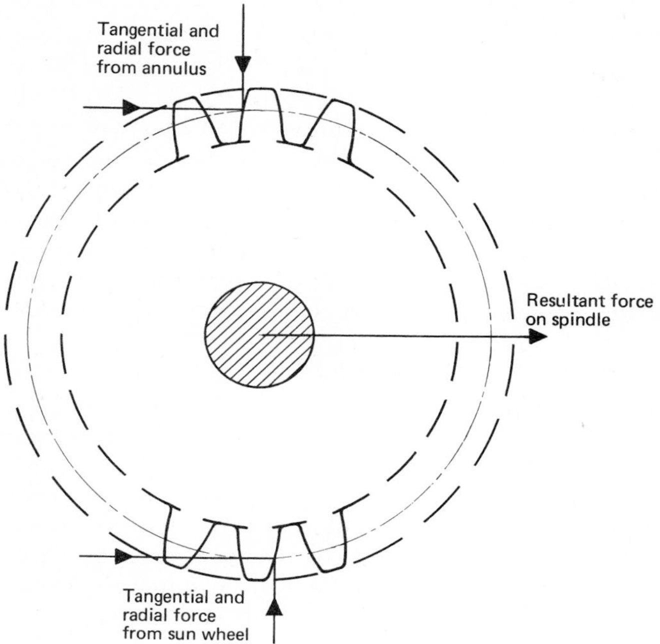

Fig. 5.13. *Resultant force on planet spindle.*

If the planet shaft is cantilevered from a single plate, more space is available but the deflection of the shaft leads to uneven load distribution on the teeth. Hicks in *Gearing in 1970*, describes the use of a flexible supporting pin which maintains correct tooth alignment.[7] The pin has an interference fit in the plate at one end and in the spindle on which the gear runs at the other end (Fig. 5.14); these fits are sufficiently tight so that the pin can be regarded as built in at each end. Under load the pin deflects but the gear tooth alignment does not change (see sketch). It is useful to be able to calculate the pin deflection in the situation shown; this is left as an exercise to Chapter 8, (Problem 8.5), where an additional problem is to find what deflection occurs when the gear is loaded at one end as shown in the sketch. It will be seen that in the latter case the tendency is for the deflection to relieve the effect of the concentrated load.

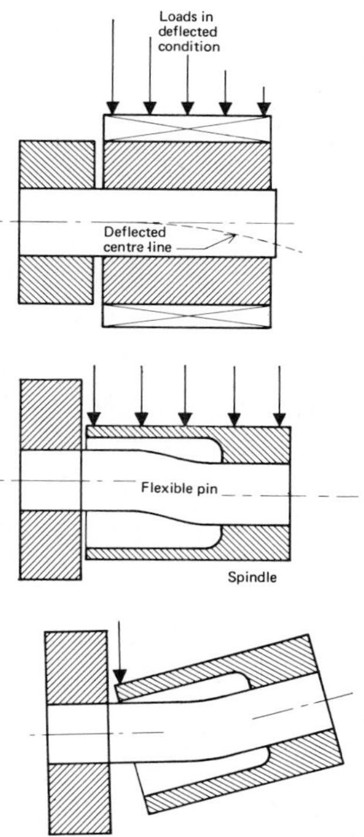

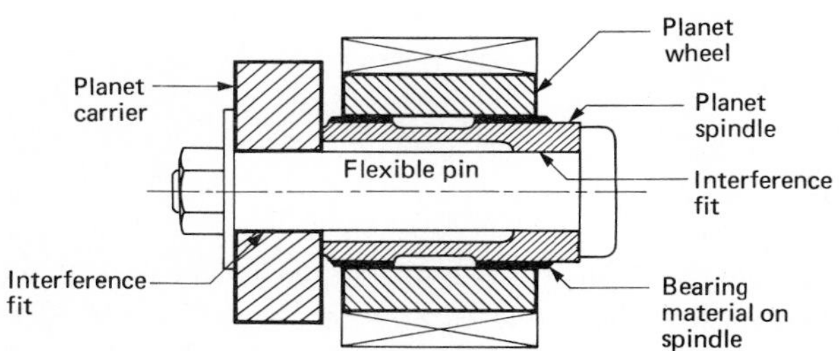

Fig. 5.14. *Arrangement of flexible pin for planet gear.*

In the last example we have been concerned mainly with deflection, but the pin will be highly stressed, and will have to be proportioned to withstand high static loads and also any fatigue loads that arise due to, say, imperfect machining of the gears.

The question of design for strength has not been overemphasised in this discussion, since it is so often the first choice for a limiting condition. For example, in a winch or crane it is clear that the first requirement for the lifting rope is that it shall have an adequate reserve of strength under all likely operating conditions; the safety factor selected must ensure an adequate margin as the rope wears. We will conclude by considering some of the mechanical components that go to make up the lifting machinery for a crane. A type that can be seen in many factories and engineering shops is an overhead travelling crane; it provides an example of the way in which structural, mechanical and electrical components are combined in one piece of equipment. Some design requirements for these components are set out in the relevant British Standards, BS 466 and BS 2573.[8, 9]

A typical layout for an overhead crane is shown in Fig. 5.15. The main structural girder or bridge spans the space to be served and the hoisting gear is carried on this. In order to allow the lifting hook to be positioned over any point on the factory floor, the hoisting gear is mounted on a carriage or crab which can traverse along the girder, while the girder can travel along rails which run the length of the building. In small installations, the position of the lifting hook can be controlled by an operator on the floor; larger installations require a driver who is housed in a cabin from which he has an unrestricted view of the lifting operation.

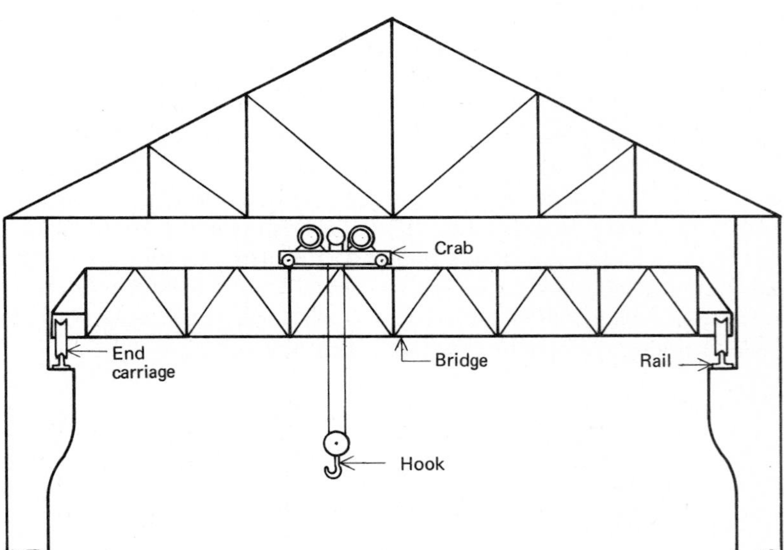

Fig. 5.15. *Layout of travelling overhead crane.*

The electric motor that provides the input power to the hoisting motion can be reversed electrically; it is connected through gearing to a drum on which the hoisting rope is wound. This drum is grooved, and must be of sufficient length to carry the full lift of rope in one layer. A brake must be provided to prevent rotation of the drum when the motor is not energised. It is clear that there are many components between the motor input and the output – the lifting hook and sling – whose strength is vital to the safe operation of the equipment. The failure of a gear, key, shaft or brake could allow the load to run down out of control. At first sight, the best location of the brake would appear to be as close to the rope as possible – acting perhaps directly on the rope drum. However, the torque on this shaft is the highest that exists through the transmission system, and to control this, the brake will have to be very massive and the actuating gear correspondingly large and unwieldy. The alternative is to mount the brake on the same shaft as the motor, where it has a large mechanical advantage. In this case the brake is small and can readily be operated by light electromechanical actuators. This solution is often adopted in positioning the brake, but the actual arrangement must depend on the type of operation and the demands of the user; the problem involved in designing a brake to operate against the high torque of the low-speed output shaft must be clearly understood before this is demanded in a specification.

One solution to the layout of the brake and gearing is shown in Plate 5.17, which illustrates the arrangement for a small hoist. The wire rope is fed to the grooved drum by a guide. The electric motor is installed to the right of the drum, and drives it through a flexible coupling and a gearbox. The electric rotor and stator are arranged to give an axial force as well as a torque when the motor is running. The cone brake at the right-hand end of the assembly engages in a drum in the fixed housing, under the action of a spring. The small axial movement that occurs when the motor is energised releases the brake and allows rotation. A power failure will, of course, allow the brake to return to its "on" position under the action of the spring.

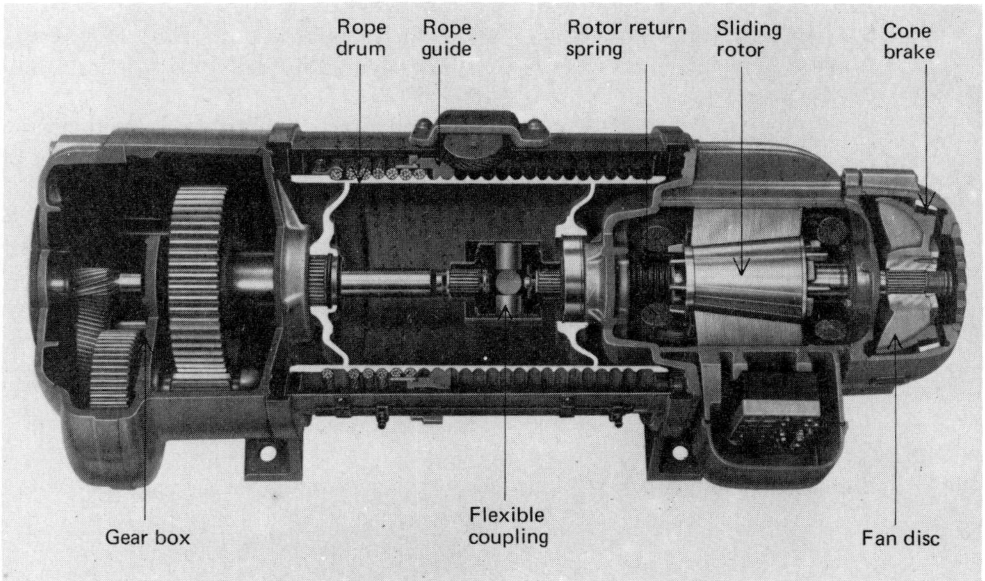

Plate 5.17. *Cutaway view of electric rope hoist. (Reproduced by permission of Demag Fördertechnik.)*

The sectional arrangement shown in Plate 5.17 illustrates the operating principles clearly. The complete design requires attention to many details such as bearings, seals and switches, and to the correct shape of the castings that make up the gearbox and motor housings.

5.9. CONCLUSION

Some typical structures and machines have been described with the object of illustrating the way in which they are built up from the elements available to the designer. It must be recognised that the examples given are selective—especially so in the case of machines—and that they illustrate some of the solutions that have been adopted at different times. As social, economic and technical ideas and pressures alter, new solutions—or old ones in new forms—will be developed.

REFERENCES

[1] M. SALVADORI and R. HELLER, *Structure in Architecture*, Prentice-Hall, Englewood Cliffs, 1963.

[2] B. H. FALCONER, "Departmental Building, Bowen Street, Wellington: Structural Aspects", *New Zealand Engineering,* **13**, November 1958, pp. 390–9.

[3] G. COOPER, "A 19-storey All-welded Structure Utilising Rolled Sections to BS 968:1962", *New Zealand Engineering,* **22**, June 1967, pp. 246–53.

[4] F. WALLEY and H. C. ADAMS, "Prestressed Concrete as Applied to Building Frames", *Proceedings of the Institution of Civil Engineers, Part III,* **5**, April 1956, Institution of Civil Engineers, London, pp. 70–91.

[5] G. ROBERTS and O. A. KERENSKY, "Auckland Harbour Bridge Design", *Proceedings of the Institution of Civil Engineers,* **18**, April 1961, Institution of Civil Engineers, London, pp. 423–58.

[6] H. S. SMITH and J. F. PAIN, "Auckland Harbour Bridge Construction", *Proceedings of the Institution of Civil Engineers,* **18**, April 1961, Institution of Civil Engineers, London, pp. 459–78.

[7] R. J. HICKS, "Experience with Compact Orbital Gears in Service", *Gearing in 1970: Proceedings of the Institution of Mechanical Engineers 1969–70*, **184,** Part 30.

[8] BS 466:1960, *Electric Overhead Travelling Cranes,* British Standards Institution, London, 1960.

[9] BS 2573:1960, *Permissible Stresses in Cranes*, British Standards Institution, London, 1960.

BIBLIOGRAPHY

Books

APPLEYARD, D., K. LYNCH and J. R. MYER, *The View from the Road,* The Technology Press & Harvard University Press, Cambridge, Mass., 1960.

Deals with the aesthetics of highways (U.S.A.) and the way they look to the driver and his passengers.

BEAKLEY, G. C., and E. G. CHILTON, *Introduction to Engineering Design and Graphics*, Macmillan, New York, 1973.

An introductory text that describes in some detail the factors that influence a designer, and the way he sets about his work.

BOHEN, P. and A. KREESE (eds.), *The Economics of Environment*, Macmillan, London, 1971.

Economic analysis of environmental pollution.

BURNS, T. and G. M. STALKER, *The Management of Innovation*, 2nd ed., Tavistock Publications, London, 1966.

Based on studies in industry, this book examines the way in which different kinds of management systems affect the ability of the organisation to cope with change.

COMMONER, B., *The Closing Circle*, Jonathan Cape, London, 1972.

Professor Commoner's book spells out the urgency of the environmental crisis and places some measure of blame on the technologists for the narrowness of their vision in solving engineering problems. There is much in what he has written for engineering designers to think about.

DAVENPORT, W. H. and D. ROSENTHAL (eds.), *Engineering: Its Role and Function in Human Society*, Pergamon Press, New York, 1967.

A collection of articles that illustrate the attitudes of humanists and engineers to science and engineering.

FALK, R. A., *This Endangered Planet*, Random House, New York, 1971.

Discusses war, overpopulation, depletion of resources and deterioration of the environment.

FORRESTER, J. W., *World Dynamics*, Wright-Allen Press, Cambridge, Mass., 1971.

Describes the model used in *The Limits of Growth* (*see* MEADOWS).

GABOR, D., *Innovations: Scientific, Technological and Social*, Oxford University Press, 1970.

Describes inventions and innovations existing and expected, and the way in which they will affect society—for better or for worse.

GIBBERD, F., *Town Design*, The Architectural Press, London, 1959.

Historic and contemporary design aspects of the various residential, commercial and industrial components of urban areas.

KRICK, E. V., *Technology and Society*, Feffer and Simons, New York, 1974.

Aims at exposing the student to the design concerns and problem-solving techniques of engineering.

MEADOWS, DONELLA H., *et al.*, *The Limits to Growth*, Potomac Associates, Earth Island Ltd., London, 1972.

A computer model is used to simulate the effect of growth on world resources and population. The conclusion of the study is that uncontrolled growth cannot continue without disastrous consequences for civilisation as it is now known, and that a fundamental revision of human behaviour and of the fabric of present-day society must occur.

See also a critique of this work, *Thinking about the Future: A Critique of the "Limits to Growth"*. Edited for the Science Policy Research Unit of Sussex University by H. S. D. COLE (and others). Chatto and Windus for Sussex University Press, London, 1973.

MIDDLETON, M., *Group Practice in Design*, The Architectural Press, London, 1967.

A wide-ranging discussion of the methods of design, and of the results of architects, industrial designers and engineers practising in groups.

RUDD, D. F. and C. C. WATSON, *Strategy of Process Engineering*, John Wiley, New York, 1968.

Although it refers mainly to chemical engineering topics, this book gives many examples of optimisation methods and of engineering design in the face of uncertainty.

THOMPSON, D'ARCY W., *On Growth and Form*, vols I and II, 2nd ed., Cambridge University Press, 1952.

This book was first published in 1917. Thompson investigates geometry and structure in nature and their relation to engineering structures.

WILSON, I. G. and M. E. WILSON, *From Idea to Working Model*, John Wiley, New York, 1970.

Discusses the steps taken in system design by the innovator or designer, mainly in terms of "black boxes".

Journals

Development and Change, The Institute of Social Studies, The Hague.

A journal, published twice-yearly, dealing with social, economic, and technical problems faced by non-industrial countries.

Harvard Business Review, Cambridge, Mass.

This journal frequently contains articles on decision-making under conditions of uncertainty and also on the human relations problems that are involved. *See*, for example,

SCHOEN, D. R., "Managing Technological Innovations", *Harvard Business Review,* May–June 1969.

Journal of Socio-Economic Planning Sciences, Pergamon Press, Oxford.

A bi-monthly journal concerned with the systems approach to planning.

PROBLEMS

5.1. Select one of the activities or products described in Table 5.1 and discuss in relation to it any of the following that are relevant:

(a) the primary design objectives;

(b) the skills and qualities needed in the person or team responsible for the design;

(c) conflicts that may occur with the environment or other interests – for example, the established labour force;

(d) the criteria that would be used in assessing the merit and success of the design.

5.2. Discuss the following topics, which relate to the activities and products described in Table 5.1.

(i) What forms of drives and controls are suitable for working and moving shovels of the type shown in Plate 5.1?

(ii) What are the environmental problems of open-cast mining?

(iii) What special demands are made on the structural engineer by the need to shield large equipment in nuclear power stations?

(iv) The bridge illustrated in Plate 5.4 is over 80 years old. How can an appropriate design life be determined for such a structure?

(v) What are some of the difficulties involved in predicting the costs of a structure such as the Sydney Opera House?

(vi) What problems are encountered in steering the disk of a large aerial such as that shown in Plate 5.6?

(vii) What materials are used for handling corrosive fluids in plant such as that illustrated in Plate 5.8?

5.3. In what ways will the increasing shortages of fuel and other raw materials affect the approach of the engineering designer?

6 Selection of Materials

6.1. INTRODUCTION

In this chapter, the factors that a designer must consider when selecting a material are discussed and typical properties of a variety of materials are described. We are concerned at this stage with properties that typify some generic class of material and will not detail the values that would be given in a full specification. More complete data are given for specific materials in Chapter 20 and Appendix A4, which can be regarded as complementary to this chapter. The reason for presenting data in this manner is that the actual selection process takes a similar course: first, a generic class is selected whose properties best match those demanded by the application; second, constituent materials in the class are studied and a specific selection is made.

Some properties of a variety of engineering materials are given in Table 6.1; these values are intended partly for comparative purposes in this chapter, but they are sufficiently accurate for preliminary design calculations that have to be made to check the suitability of a class of material.

A great deal of design work can be done using static room temperature strength properties – yield strength, for example. In the discussion that follows, there is some preoccupation with the effects of unusual loading conditions on such properties; our objective is to ensure that any limitations to their use are made clear. Once these are understood, the designer can use them in normal loading conditions with confidence.

In some cases, there will exist no doubt about the best material for a job. If previous experience, availability and proven economy all favour one material, it will be chosen on the basis of its known fitness for the purpose. However, progress in materials technology is rapid, and previous experience will not always be a reliable guide; changes from established practice will occur, although they will usually be preceded by detailed investigations.

In Chapter 2 it was suggested that the limiting conditions that govern a design should be listed to ensure that all likely failure modes are accounted for. Such a list will also be useful when selecting a material, and the relevant limiting conditions can head the list that will dictate the properties. They are not necessarily the most important: the limiting conditions will be supplemented by other requirements set by performance in manufacture and service. For example, ease of casting will influence the selection of a material for an intricately shaped component. In most cases material cost is important, and the choice of material will significantly affect machining and service costs. The judgement of the designer plays an important part in assessing the relative importance of these various items, particularly when no precise numerical value can be given to them. The factors that influence the designer in the decision-making process are set out in Fig. 6.1. Basically, the design factors in the first block determine

Table 6.1. Typical properties for some commonly used engineering materials. (All strength values are given in MPa.)

	MATERIAL—GENERIC TYPE										
	Cast iron	**Cast iron**	**Steel**	**Steel**	**Steel**	**Aluminium alloy**	**Aluminium alloy**	**Copper alloy**	**Concrete**	**Timber**	**Plastic**
Description	Grey iron	Nodular or SG iron (ferritic)	Mild steel 0.23% carbon	Low alloy weldable steel	18–8 stainless steel	Heat-treatable alloy	Weldable alloy	Gunmetal or red brass (leaded)	Normal strength dense concrete	Typical softwood	ABS thermoplastic
Condition or form	Cast	Cast; annealed	Wrought	Wrought; hardened and tempered	Wrought; cold-worked	Solution heat treated and aged	Cold worked; ½ hard	Cast	Cured test block at age 28 days	Air dried	Moulded or extruded
Typical alloying elements (%)	3.0 C	3.5 C + trace Mg or Ce	0.25 C max	0.3 C, 1.5 Mn + Si, Ni	18 Cr, 8 Ni	4 Cu + Si, Mn, Mg	2 Mg	5 Sn, 5 Zn, 5 Pb	—	—	—
Applications	General casting purposes where high strength is not demanded and brittleness is not a disadvantage	Castings with better toughness than grey iron	General construction work where medium strength and good ductility are required	Construction work where weldable steel with good strength and toughness is required	Where high strength combined with corrosion resistance is essential	High strength wrought products where light weight is required	For structures where light weight and corrosion resistance are called for	Machine parts where good corrosion resistance is essential. Other alloys available in wrought form	General construction work	Light structural work	Used for many moulded and extruded domestic and light engineering components
Yield or proof stress, f_y	—	250	240	460	650	410	200	100	—	—	—
Tensile strength, f_t	250	430	450	620	1200	470	250	240	—	30	60
Elongation (%)	0.8	12	20	20	22	10	7	30	0.3†	—	20
Compressive strength, f_c	900								20		
Endurance limit, f_e	$0.35f_t$	$0.45f_t$	$0.45f_t$	$0.45f_t$	490	130	125	76	—	—	—
Modulus of elasticity (GPa)	115	170	205	205	195	70	70	90	15–20	9	2.5
Modulus of rigidity (GPa)	40	70	77	77	77	28	26	23	—	—	—
Poisson's ratio	0.25 at low loads	0.17	0.28	0.28	0.25	0.32	0.32	0.35	0.08–0.16	—	—
Specific gravity	7.3	7.2	7.8	7.8	7.9	2.8	2.7	8.8	2.5	0.5–0.7	1.1
Specific heat (kJ/kg°C)	0.5	0.5	0.5	0.5	0.54	0.92	0.92	0.5	0.84	1.5	0.4
Electrical resistivity (μohm cm)	25–100	65	15	20	75	4.5	5.0	12	—	—	Good insulator
Thermal conductivity (W/m°C)	46	34	46	46	17	150	140	72	1–1.7	0.15	0.3
Coefficient of expansion (1/°C)	13×10^{-6}	12×10^{-6}	11×10^{-6}	11×10^{-6}	17×10^{-6}	24×10^{-6}	24×10^{-6}	18×10^{-6}	12×10^{-6}	3.5×10^{-6} along $30\text{–}60 \times 10^{-6}$ across grain	7×10^{-5}
Weldability	—	—	Good	Good	Some grades with precautions	No	Good	Solder or braze	—	Glued joints excellent	Can be welded or glued
Machinability	Good	Good	Good	Good	Fair	Very good	Very good	Very good	—	Good	Very high
Maximum temperature for general use (°C)	350–400	500	300	400	650	150	250	250	—	—	80
*Cost (cents/kg)	8.5	—	22	—	—	110	—	38	1	25	35

*Varies with time and place.

† Ultimate compressive strain for design.

the material to be used, which in turn determines the construction or manufacturing process. The way in which the total cost to provide a service is related to these decisions is shown. In practice, the process is complicated by feedback from the later stages, and by cross-linking between the various blocks.

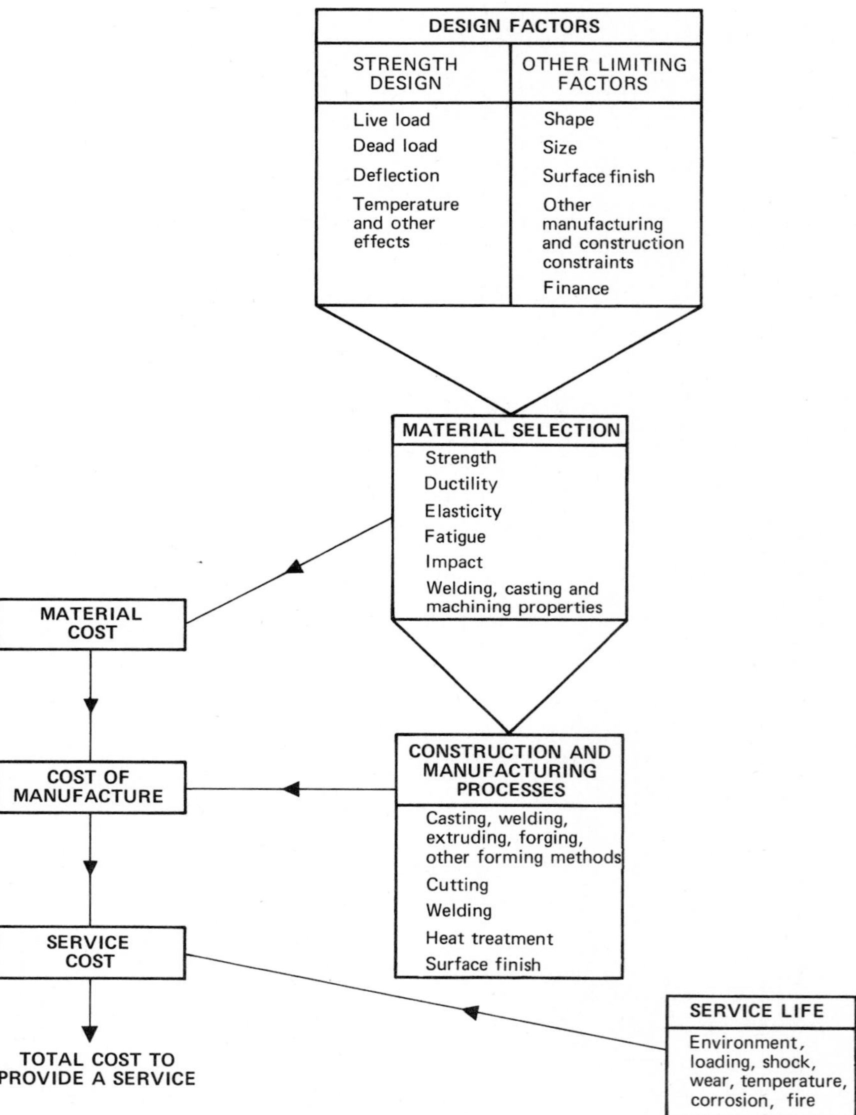

Fig. 6.1. *Factors to be considered in selection of materials.*

6.2. MATERIAL PROPERTIES

It is convenient for the purpose of this chapter to identify three types of property and to discuss them in separate classes.

(1) Physical and mechanical properties which can be described by numerical values used directly in design. Examples are yield strength, endurance limit, elastic modulus, density and thermal conductivity.

(2) Properties which have a numerical value but which are not of immediate use in design, though they may give a guide to performance. Examples are impact and hardness test values and chemical analyses.
(3) Properties which are of importance in the selection process but which are difficult to describe and account for directly in terms of a numerical value. Examples are resistance to wear, corrosion and fire, machinability and weldability.

The second block in Fig. 6.1 can now be enlarged to show the properties subdivided into these classes; this has been done in Table 6.2. It will be appreciated that the terms in the third class do not readily lend themselves to analytical treatment, and that the engineer may have to collaborate with a metallurgist or an architect in assessing what value should be assigned to these terms.

Table 6.2. Classification of material properties.

CLASS 1	CLASS 2	CLASS 3
Yield strength	Elongation	Cost
Tensile strength	Reduction in area	Machinability
Endurance limit	Impact strength	Weldability
Creep strength	Hardness	Formability
Density	Chemical analysis	Visual properties
Young's modulus		Acoustic properties
Shear modulus		Corrosion resistance
Thermal conductivity		Fire resistance
		Resistance to chemicals
		Resistance to wear

We will first discuss these classes in isolation. However, the designer must usually be concerned with the way the various properties interact, and a brief review of combined properties such as cost per unit of strength or mass is given at the end of this chapter.

6.2.1. Class 1 properties

The first class can be subdivided into properties which are used in strength design and those (like the thermal and electrical properties) which are used in other aspects of design; the latter are not of direct importance in this book. The strength properties are identified and used frequently in subsequent chapters dealing with analysis; their significance is usually self-evident, since their application is so direct.

We will deal first with strength properties on their own. A wide range of strength values can be obtained for each generic type of material, and new processes, alloys and heat treatments are extending this range. Improvement in strength is often achieved at the expense of other properties such as weldability, machinability, ductility and cost, and the principal difficulty in the selection process lies in attaining a satisfactory balance between these.

The yield strength, f_y, of a material is defined as the stress at which elongation of a test piece first takes place without increase in load. For ductile materials which do not exhibit the yield phenomenon – as may be the case with aluminium alloys and heat treated steels – the proof strength may be specified and used in place of f_y in design; this is the stress at which a non-proportional elongation, equal to a specified percentage of the gauge length, occurs. This value is often taken as 0.2%. The yield strength in compression f_{yc} is usually taken to be equal to that in tension; this is inherent in the failure theories used for ductile materials and is confirmed by test for many materials. The yield strength in shear, q_y, is usually about 0.6 f_y; the maximum shear stress theory predicts it as 0.5 f_y, and the maximum distortion energy (von Mises) theory as 0.58 f_y.

The tensile strength, f_t, is used for control purposes in manufacture and heat treatment. It is of minor importance as a failure criterion for ductile materials, since a structure has to undergo such large deflections before fracture that other limiting conditions will prevail. However, for brittle materials it is the only property that can be readily identified and measured in a tensile test, and it is used directly in design calculations. For brittle materials, the ultimate strength in compression, f_c, may be greater than f_t. For cast iron, the ratio f_c/f_t lies between 3 and 4, and for concrete is about 10.

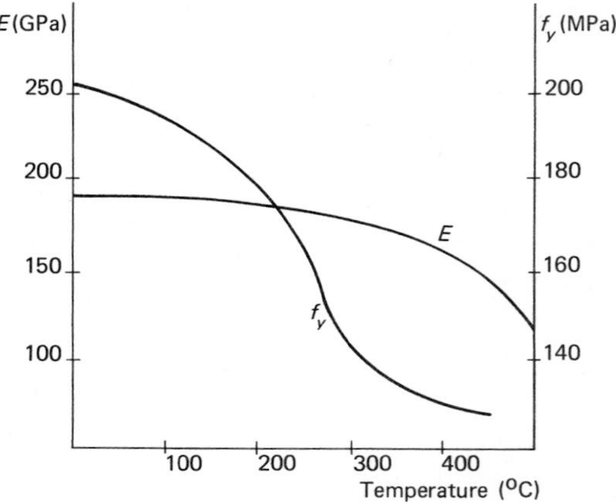

Fig. 6.2. *Variations of yield stress, f_y, and Young's modulus, E, with temperature. Material: 0.2% carbon steel. (Extract from BS 1501,* Steel for Fired and Unfired Pressure Vessels, *is reproduced by permission of the British Standards Institution, 2 Park Street, London W1A 2BS.)*

Physical properties change with temperature and upper limits must be set at which materials can be used without loss of strength or stiffness. The effect of temperature on yield stress and Young's modulus for a low-carbon steel[1] is shown in Fig. 6.2. Although there is no sharp fall off, it is clear that the designer must be concerned about the loss of strength and stiffness at temperatures above about 300°C. In any case the material would be unsuitable for use at higher temperatures because of surface oxidation.

A similar reduction in strength occurs for materials intended for high temperature application, as is shown in Fig. 6.3 for a Type 403 stainless steel.[2] In this case there is good oxidation resistance up to 650°C, but strength design at this temperature must be based on creep properties rather than yield strength. For this particular steel the time to rupture at a stress of 45 MPa would vary from about 1000 hours at 650°C to 50 hours at 700°C. A comparison of these figures with the yield strength information in Fig. 6.3 will show that it is essential to consider creep data when designing for high temperatures. This topic is not taken further in this book, and the reader is referred to works of Greenfield[3] and Lubahn and Felgar[4] for information on use of steels at high temperatures.

Most materials can be produced in a range of strengths, and this provides an alternative to increasing material thickness. An example quoted by Haaijer[5]

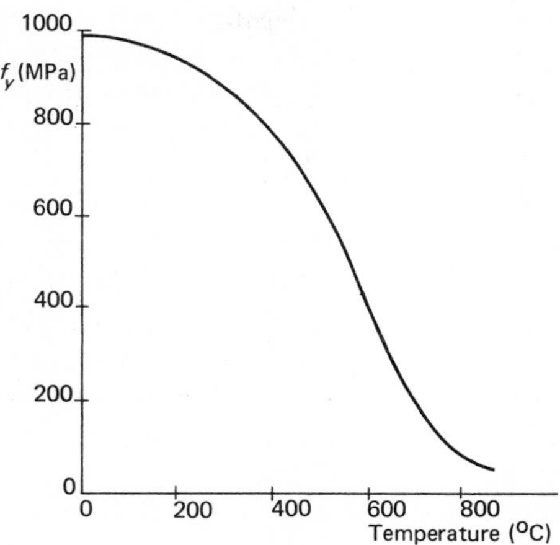

Fig. 6.3. *Variation of yield stress of a martensitic (Type 403) stainless steel with temperature. (From V. Weiss (ed.),* Ferrous Alloys, *Syracuse University Press, Syracuse, 1965, by permission of the Syracuse University Press.)*

is in the construction of storage tanks, where the higher pressure loads towards the base of the tank can be carried on plates of higher strength instead of increased thickness. Apart from making use of the improved properties of thinner section plates, this design results in a cleaner appearance and simpler weld details.

Most of the materials described in this book are assumed to be isotropic, although in some cases the difficulty of obtaining uniform properties in thicker sections is acknowledged. Strength values—for example, yield and tensile strengths and notch toughness—are often specified for a range of sizes. It is easier to obtain uniformity of temperature and grain size when forming thin sections than when thicker slabs are being handled, and this is reflected in higher strengths. The variation to be expected is illustrated in Fig. 6.4 by values of f_y for two grades of wrought steel taken from a structural steel specification.[6]

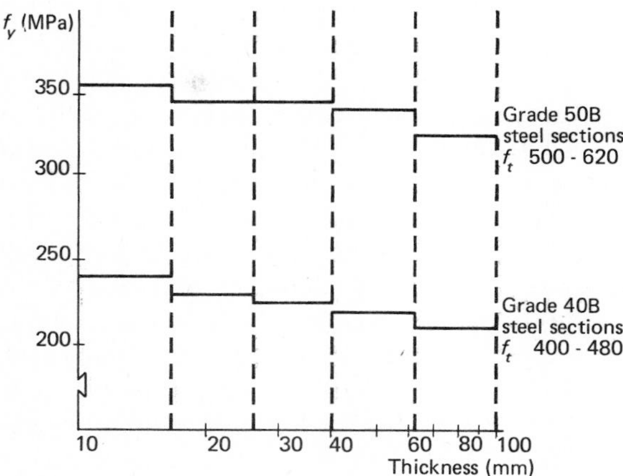

Fig. 6.4. *Variation of strength of steel with thickness.*

Although there are differences in the metallurgical structure of wrought (e.g., rolled or forged) and cast materials, these are not generally taken into account in preliminary design. However, when a specific material has been chosen that is suitable for a particular production process, it will be found that differences in the specification do exist, depending on what production process and heat treatment are used. Forged products display anisotropy and advantage should be taken of this in the geometrical design of a component, even if it is not allowed for in the design stress. Timber is highly anisotropic and allowable stresses are defined depending on their orientation to the grain.

6.2.2. Class 2 properties

Properties in the second class have limitations which usually arise from their being determined by simple tests which – in default of better methods – are used to give the manufacturer and user of the material a guide to some complex behaviour characteristics. Values given in some specifications may be of little direct service to the designer and may be intended mainly for use as a production control by the manufacturer. The tensile strength of a ductile material is an example of such a value.

An important property in this class is the elongation, measured in a low strain rate tension test over a specified test length. When using a ductile material the designer is able to ignore stress concentrations at notches on the assumption that some plastic flow will occur at these locations and distribute the load to the less highly stressed regions. Notches (or triaxial stresses) and shock loading tend to make a material behave in a brittle fashion, and a large elongation before fracture is regarded as one important safeguard against such failure. The reduction in area at fracture should be considered in conjunction with the elongation, since it is possible for a very ductile material that does not work-harden to have low elongation, all the plastic flow occuring at the neck where failure finally takes place; in such a case there would be a very large reduction in area. When material that work-hardens is tested, plastic flow will occur more uniformly along the test length until necking starts at the location of final failure. Although these two measurements that are made in the tensile test are often referred to as properties, McClintock and Argon point out that "reduction of area cannot be considered to be a property of the material, but only its characterisation in a certain mode of testing".[7] The reader will find other cases where values are spoken of as "properties" when some other term would be more appropriate.

There are several factors which tend to make a normally ductile material behave in a brittle fashion and result in cleavage rather than shear failures, and the designer must be sensitive to the possibility of their occurrence. Brittle fractures can occur at very low nominal stresses, of the order of a quarter of the yield strength, and a large elongation in the tensile test is not an adequate safeguard. The state of stress exerts a controlling influence on the mode of fracture, and this state is modified by changes in section and by cracks or notches. Temperature also plays an important part in determining behaviour. Since, after the formation of a crack, the three-dimensional stress pattern in a thick plate is different from that in a geometrically similar thin plate, size effects are involved that make the thick plate more susceptible to cleavage failures.

Impact tests on notched specimens measure the energy absorbed (usually in bending) for some particular notch configuration and temperature; they give an indication of behaviour under triaxial stresses – not necessarily impact ones – and are a measure of "toughness". When determined over a range of temperatures the test values fix the ductile–brittle transition, if it exists, and this provides a useful guide to the designer in material selection. Since behaviour is influenced by the type and rate of loading and by notch shape (Boyd[8]) the specification of minimum test values for a material must be based on a wide experience of actual service performance. This approach is often adopted in design codes, a common requirement being a minimum energy absorption of 27 J in the Charpy V-notch test at the anticipated service temperature, with specimens cut in the direction of rolling, or 20 J in the transverse direction.[8] However, these values must be increased for higher strength materials and for thick sections.

In some cases, the dangers of brittle fracture can be lessened during manufacture – for example, by careful control and inspection of welding and by carrying out a stress-relieving heat treatment after welding. In the case of high strength materials, the methods of fracture mechanics make it possible to predict the maximum permissible crack or flaw size, or, for a given size crack, to determine the permissible operating stress. This approach, which requires a knowledge of the material fracture toughness K_{1c}, is discussed in Ruiz and Koenigsberger[9] and the I.S.I. publication on "Fracture Toughness".[10]

6.2.3. Class 3 properties

One of the principal difficulties in design is that so few properties can be placed in the first class, while so many belong in the third. Cost has already been mentioned, and it has a major influence on selection; it has been put in this class since it varies with time and place and the physical condition of the material. For example, the cost of round steel bar will differ depending on whether it is supplied in the hot-rolled or cold-worked condition; the physical properties and the surface finish will also have changed; a grinding process – which transforms it into "bright bar" – will improve the surface finish and tolerance on diameter and may obviate the need for further machining. It is not always easy to evaluate the effects of such cost variations on manufacturing costs, but they must be considered if optimum economy is sought, and the material specification must include details of any special finishes or surface treatment.

Machinability, weldability and corrosion resistance are other properties in this third category. The first two properties are important in determining manufacturing costs and the third affects service life and cost.

The factors that contribute to good machinability are high speed of chip removal, low tool wear and good surface finish. It may be necessary to distinguish between machinability and the finish that is attainable since the two are not directly related. The addition of about 0.25% sulphur to a steel will improve its machinability and formability, but there will be some adverse effect on its strength properties. Free-cutting steels are also manufactured with up to 0.35% lead, which does not affect the strength properties. Other metals, such as copper and brass, are also supplied in free-cutting grades. The extra cost of these materials will only be offset where machining costs form a high proportion of the finished costs.[11]

Metals can be classified in various ways to indicate their suitability for use in welded structures; apart from the base metal properties, weldability is affected by the filler metal properties and the process used. In the tables in this book we simply indicate a metal as being suitable, suitable subject to special processes or precautions, or unsuitable for welding in routine construction or production.

As carbon and alloy content increase, it becomes more difficult to produce a crack-free weld, and alloy steels and associated welding and heat-treatment must be carefully selected and specified (see Section 13.4.3). When specifying steels for welding, an upper limit is set on the carbon equivalent (CE). For example, for structural steels with tensile strengths in the range 400 MPa to 430 MPa the requirement given in BS 4360[6] is that CE should be not greater than 0.4 where CE is specified in terms of the percentage of alloying elements as follows:

$$CE = C + \frac{Mn}{6} + \frac{Cr + Mo + V}{5} + \frac{Ni + Cu}{15}$$

6.2.4. Combined properties

If a structure or machine is to be proportioned for strength, while other criteria relating to weight, cost or deflection are met, the designer must consider parameters such as specific cost (cost per unit mass) or cost per unit of strength. Some interesting comparisons between materials can be made using these parameters, and some examples are studied here. Only material costs will be taken into account and, unless otherwise stated, secondary modes of failure such as buckling are ignored. The self-weight of the structure will not be considered in the analysis.

We will examine several aspects of the problem of supporting a load of 10 kN over a distance of 3 m. A first look at a general problem of this type was taken in Chapter 5, where it was made clear that it is structurally more efficient to carry a load directly in tension or compression than in bending.

In tension, the cross-sectional area required for steel with a yield stress, f_y, of 250 MPa and a design stress, p_t, of 125 MPa is 80.0 mm^2, that is a circular bar 10.1 mm in diameter. If other steels are used in tension, their economic merit can be found by comparing the cost : strength ratio. Some typical values are given by Lay[12] and these are shown in Table 6.3. It will be seen that the strength used in the above calculation corresponds to that of an A 149 steel. If an A 151 steel were used the material costs would be reduced by a factor of 0.88.

Table 6.3. Economics of high yield stress steels.

Steel	Approximate cost ratio	Yield stress f_y (MPa)	Relative cost per yield stress
A 149	1.00	250	1.00
A 135B	1.15	250	1.15
A 151	1.25	355	0.88
AusTen 50	1.45	345	1.04

Source: M. G. Lay, "Steel—the Designer's Fourth Dimension"[12], by permission of The Broken Hill Proprietary Co. Ltd.

Although the strength of steel is the same in compression and tension, a rod of these proportions would fail by buckling long before the design load in compression was reached. The parameter used to describe the slenderness of a compression member is the ratio of length to radius of gyration, l/r. In this case the value of l/r is 840, and the value needed to ensure that buckling does not reduce the safety margin is of the order of 20. This value would be obtained if a 300 mm diameter tube was used, but the wall thickness to maintain the same cross-sectional area would be only 0.08 mm, and local buckling would certainly take place in such a thin wall. Some additional material is needed to produce a sound structure of adequate stiffness in compression and material strength is of less consequence than cost.

If different materials are to be compared when loaded in tension, the differences in density must be taken into account. A measure of the merit of different materials is the ratio (specific cost) : (strength/relative density). Some typical values of this ratio are given in Table 6.4, along with other combined properties which are of interest in design.

Table 6.4. Some combined properties used in the selection of materials. Typical values have been taken (see Table 6.1) and the relative density (RD) has been used.

Ratio	Units	MODE OF FAILURE				
		Steel	Aluminium	Timber	Cast iron	Concrete
		Tension (yield)	Tension (yield)	Tension (rupture)	Compression (rupture)	Compression (rupture)
$\frac{\text{COST}}{\text{MASS}}$	cents/kg	22	110	25	8.5	1
$\frac{\text{STRENGTH}}{\text{RD}}$	MPa	45	110	50	120	8
$\frac{\text{STIFFNESS}}{\text{RD}}$	GPa	26	26	15	16	7
$\frac{\text{COST/MASS}}{\text{STRENGTH/RD}}$	$\frac{\text{cents/kg}}{\text{MPa}}$	0.5	1.0	0.5	0.07	0.13

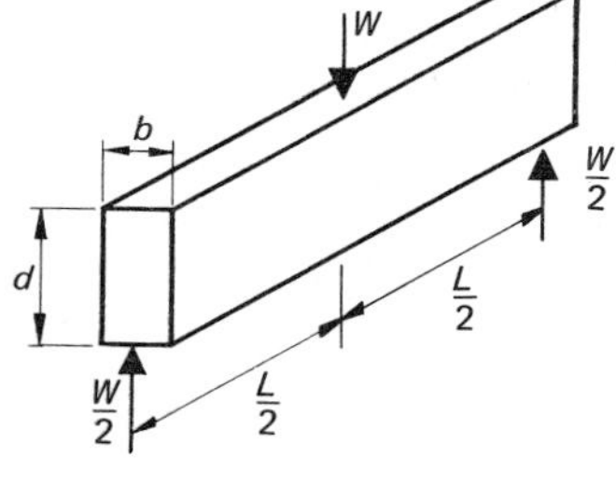

Now consider a uniform rectangular steel beam carrying the 10 kN load at its centre as shown in the sketch. We will determine the relative costs for two of the structural steels (A 149 and A 151) whose properties are given in Table 6.3. The steels are identified by subscripts 1 and 2, where steel 2 has the higher strength. It is assumed that the compression side of the beam is restrained from buckling.

The maximum bending stress, f_b, is given by

$$f_b = \frac{M}{Z} = \frac{WL}{4}\frac{6}{bd^2}$$

and for some fixed ratio $\frac{b}{d}$

$$d \propto \left(\frac{1}{f_b}\right)^{\frac{1}{3}} \tag{6.1}$$

The ratio of the volumes for materials 1 and 2 is

$$\frac{V_1}{V_2} = \frac{Ld_1b_1}{Ld_2b_2} = \left(\frac{f_{b2}}{f_{b1}}\right)^{\frac{2}{3}}$$

If the specific cost is denoted by c, the ratio of material costs for two beams of density ρ_1 and ρ_2 would be

$$\frac{C_1}{C_2} = \frac{\rho_1 V_1 c_1}{\rho_2 V_2 c_2} = \left(\frac{f_{b2}}{f_{b1}}\right)^{\frac{2}{3}} \frac{\rho_1 c_1}{\rho_2 c_2}$$

Substituting values from Table 6.3

$$\frac{C_1}{C_2} = \frac{1}{1.25}\left(\frac{355}{250}\right)^{\frac{2}{3}} = 1.01$$

According to this analysis, the higher-strength material is marginally more economical. Note that no account was taken of the dead weight of the structure; this will be lower for the stronger material and will confer an added advantage in practice.

In the last example no stiffness requirements were specified. In some structural standards an upper limit to the deflection is imposed – a value of $\frac{1}{360}$ of the span would be typical. For the centrally loaded beam of the last example the deflection δ is given by

$$\delta = \frac{WL^3}{48EI} \tag{6.2}$$

where I is the second moment of area of the beam. The stress is given by

$$f_b = \frac{M}{I}\frac{d}{2}$$

or $$f_b \propto \frac{M}{I}\frac{1}{f_b^{\frac{1}{3}}}$$ from Equation (6.1)

Thus $$I \propto \frac{1}{(f_b)^{\frac{4}{3}}}$$

The ratio of the deflections for the two beams is given by Equation (6.2)

$$\frac{\delta_1}{\delta_2} = \frac{I_2}{I_1} = \left(\frac{f_{b1}}{f_{b2}}\right)^{\frac{4}{3}} = \left(\frac{250}{355}\right)^{\frac{4}{3}} = 0.6$$

This shows that, when taking advantage of higher allowable stresses, provision must be made for larger deflections. Steel has the highest value of modulus of elasticity among common construction materials (Table 6.5) and should not be replaced by other materials without considering the changes in stiffness involved.

If stiffness is sufficiently important to warrant a detailed study, the designer must seek materials with a high specific stiffness – Young's modulus : density ratio. Unfortunately, the variation among common constructional materials is not very great – it can be seen from Table 6.5 that the values for metals used for structural work fall within quite a narrow band. Large increases in specific stiffness occur for some of the ceramics and cermets[13], but use of these can be justified only in special applications.

Table 6.5. Modulus of elasticity, density and specific stiffness for a range of construction materials.

Material	Modulus of elasticity (GPa)	Density ρ (kg/m^3)	$\frac{E}{\rho}$ (MN m/kg)
Beryllium	290	1700	170
Steel	205	7800	26
SG iron	170	7200	24
Titanium	110	4500	25
Brass	100	8800	11
Cast iron	95	7300	13
Aluminium	70	2700	26
Magnesium	45	1800	25
Concrete	18	2500	7
Glass reinforced plastic	15	1500	10
Timber	9	600	15
Plastic	3	1050	3

A more general analysis for a beam in bending will now be given, which takes account of the proportions of the cross-section of the rectangular beam in addition to the strength and deflection. We consider the loading used in the previous example.

To satisfy strength requirements, we can write

$$\frac{WL}{4}\frac{6}{bd^2} < f_b$$

Expressing the material volume as $V = Lbd$, this gives

$$V > 1.5\,\frac{W}{f_b}\left(\frac{L^2}{d}\right) \tag{6.3}$$

This requirement can be shown on a plot of V against L^2/d; if the volume corresponding to any value of L^2/d lies above the line, the stress is acceptable (see Fig. 6.5).

The maximum deflection in this case is given by

$$\delta = \frac{WL^3}{48EI} = \frac{WL^3}{4Ebd^3}$$

If the maximum allowable deflection δ_m is specified, we can write

$$\frac{WL^3}{4Ebd^3} < \delta_m$$

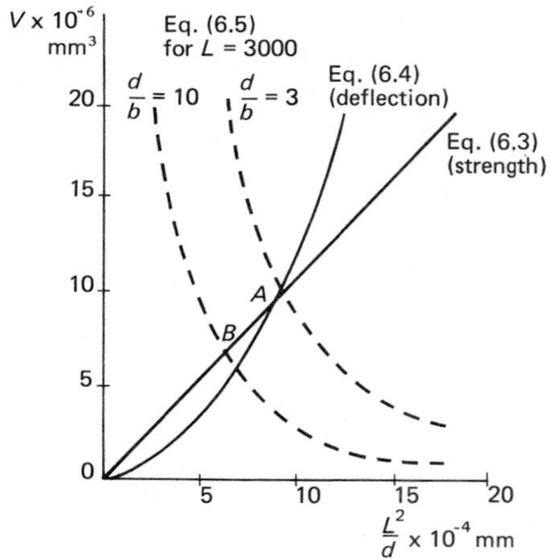

Fig. 6.5. *Volume of a rectangular steel beam as governed by strength and deflection.*

and

$$V > \frac{W}{4\,E\delta_m}\left(\frac{L^3}{d}\right)^2 \tag{6.4}$$

This inequality can also be shown on the $V \sim (L^2/d)$ plot, and gives a region above the parabola where the deflection is acceptable.

If Inequalities (6.3) and (6.4) must both be satisfied, the combinations of L, V and d that give acceptable values of stress and deflection can be found from Fig. 6.5. The point A corresponds to the only point at which neither excess stiffness nor strength is built into the beam, but does not necessarily correspond to the best design. For example, typical design values for a steel beam may give a d/b ratio that is unsuitable for practical applications.[9]

We can increase the usefulness of Fig. 6.5 by adding to it curves which represent limiting values of d/b. Putting $d/b = k$, and selecting lower and upper limits k_1 and k_2 respectively, we can write, by eliminating b from the equation $V = Lbd$,

$$\frac{L^5}{k_1}\,\frac{1}{(L^2/d)^2} > V > \frac{L^5}{k_2}\,\frac{1}{(L^2/d)^2} \tag{6.5}$$

The inequalities given in (6.3), (6.4) and (6.5) have been plotted in Fig. 6.5 for the following values

$W = 10$ kN;
$L = 3000$ mm;
$k_1 = 3,\ k_2 = 10$;
$\delta = \dfrac{\text{span}}{300}$;
$f_b = 140$ MPa.

The point A here corresponds to a d/b ratio of just over 3, and in this case could be regarded as a possible solution; however, all other points lying in the region between the two hyperbolas and above the straight line and parabola also satisfy the requirements, so that it is possible to select other beam proportions which will have a lower volume than that represented by point A. At point B, the beam will have a d/b ratio of 10 and the minimum volume.

6.3. CONCLUSION

Not all the properties listed in Table 6.2 have been discussed here, and indeed the acquisition of some information needed by the designer – for example, acoustic properties – may require considerable research. Other properties, not mentioned specifically in this chapter, are dealt with elsewhere; frictional properties would be a case in point.

Many journals and handbooks are available which summarise information for use in design, and some of these are listed in the references and bibliography in this chapter and Chapter 20. When a clear idea of the important properties has been established, it is possible to make intelligent use of such sources of information; to attempt to use them without some preliminary investigation can be confusing and discouraging.

LIST OF SYMBOLS

b	beam width
C	cost
c	cost per unit mass, specific cost
d	beam depth
f_b	bending stress
f_c	compressive strength
f_t	tensile strength
f_y	yield strength
k	ratio d/b
K_{1c}	fracture toughness
l	length of strut
L	length of beam
M	moment
p_t	design stress
r	radius of gyration
V	volume
W	load
Z	modulus of beam
δ	deflection
ρ	density

REFERENCES

[1] BS 1501:Pt 1:1964, *Steels for Fired and Unfired Pressure Vessels,* British Standards Institution, London, 1964.

[2] V. Weiss (ed.), *Ferrous Alloys* (Aerospace Structural Metals Handbook), vol. 1, Syracuse University Press, Syracuse, 1965.

[3] P. Greenfield, *Creep of Metals at High Temperatures,* Monograph ME/9, Mills and Boon, London, 1972.

[4] J. J. Lubahn and R. P. Felgar, *Plasticity and Creep of Metals,* John Wiley, New York, 1961.

[5] G. Haaijer, "Selection and Application of Constructional Steels", *Metallurgical Society Conferences,* vol. 40, Gordon & Breach, New York, 1968.

[6] BS 4360:Pt 2:1969, *Weldable Structural Steels,* British Standards Institution, London, 1969.

[7] F. A. McClintock and A. S. Argon, (eds), *Mechanical Behavior of Materials*, Addison-Wesley, Reading, 1966.

[8] G. M. Boyd (ed.), *Brittle Fracture in Steel Structures*, Butterworths, London, 1970.

[9] C. Ruiz and F. Koenigsberger, *Design for Strength and Production*, Macmillan, London, 1970.

[10] I.S.I., *Fracture Toughness*, I.S.I. Publication 121, The Iron and Steel Institute, London, 1968.

[11] A.S.M., "Machining", *Metals Handbook*, vol. 3, 8th ed., American Society for Metals, Ohio, 1967.

[12] M. G. Lay, "Steel—the Designer's Fourth Dimension", *BHP Technical Bulletin*, **13**, 3, 1969.

[13] J. F. Lynch, C. G. Ruderer and W. H. Duckworth (eds), *Engineering Properties of Selected Ceramic Materials*, The American Ceramic Society, Ohio, 1966.

BIBLIOGRAPHY

A.S.M., *Metal Progress Databook*, American Society for Metals, Ohio, 1968. (Data sheets published regularly.)

A.S.M., "Forming", *Metals Handbook*, vol. 4, 8th ed., American Society for Metals, Ohio, 1969.

A.S.M., "Welding and Brazing", *Metals Handbook*, vol. 6, 8th ed., American Society for Metals, Ohio, 1971.

Probst, E. H. and Comrie, J., *Civil Engineering Reference Book*, Butterworths, London, 1951.

Rabald, E., *Corrosion Guide*, 2nd ed., Elsevier, Amsterdam, 1968.

The Society of the Plastics Industry, *Plastics Engineering Handbook*, 3rd ed., Reinhold Publishing Corporation, New York, 1960.

PROBLEMS

6.1. Select one of the generic types of materials in Table 6.1 and find the range of engineering and physical properties recorded in the literature. Ascertain the strength range that is readily available in your locality and the variation in price.

6.2. Produce a plot of the type shown in Fig. 6.5 for a uniform aluminium beam to support a central load of 10 kN on a span of 3 m. Assume a suitable value for the allowable bending stress.

6.3. Alternative forms of construction for the roof structure of a factory are light steel roof trusses supported on concrete-block walls and solid glulam portal frames (see Chapter 5) with concrete-block walls between the columns.

Assess the performance of these alternatives from the point of view of fire resistance.

6.4. Consider the advantages and disadvantages of using lightweight insulating materials instead of dense concrete for the fire protection of the beams and columns of a steel-framed multistorey building.

6.5. Compare the merits of steel and aluminium for the construction of a long-span footbridge. The principal load-carrying members will be trusses fabricated from angles.

6.6. What is meant by the term *toughness*? How is it measured, and what is its engineering significance?

6.7. Compare the merits of steel and aluminium for the construction of the superstructures of ships.

6.8. List, in order of importance, the criteria used in selecting materials for the following applications:
(a) automotive panels, load-bearing;
(b) automotive panels, non-load-bearing;
(c) automotive fenders;
(d) heat-exchanger tubes for salt water;
(e) low-pitch roofing material;
(f) thermal insulation for cold store;
(g) non-load-bearing partitions in a block of
(i) offices,
(ii) apartments.

6.9. Bring the cost information in Table 6.4 up to date and add data for other materials, such as stainless steel. Discuss the following:
(a) the limitations to using the cost per unit strength ratio as a criterion for selecting construction materials;
(b) other criteria of the type tabulated—for example, those relevant in the selection of materials for long columns which fail by buckling.

PART II

7 Fatigue Design

7.1. INTRODUCTION

Estimates given in the literature on fatigue indicate that from 80% to 90% of service failures in machine parts arise from fatigue. Therefore, it is important that, where machine elements and structures are subjected to varying stresses, the strength and form be determined on the basis of the fatigue performance of the material. The strength under static load will give little if any guidance as to the fatigue strength of a part; indeed the notched steel bar described in Section 4.4, which displayed an increase in strength over the plain specimen in static loading, would be many times weaker if the comparison were made with alternating loads.

Although a good deal of the large and expanding literature on fatigue covers a wide range of shapes, sizes and loading conditions, the bulk of the information used directly in design comes from fatigue tests on small polished specimens subjected to fully reversed loads. The central problem in design is to relate the data from such tests to a part which has different shape, size and surface finish and is subjected to a different loading pattern. The relationship between these various factors is extremely complicated and no general failure theory is yet in sight.

In these circumstances, it is important for the designer to be circumspect in the choice of a design method and to be well informed as to its basis. The first part of this chapter is taken up with descriptions of the factors that influence fatigue behaviour and which must be accounted for in design. On the basis of this work two design methods are described, the first suitable for preliminary design purposes and the second for cases where a more elaborate treatment is warranted. The chapter concludes with a description of some of the measures that can be taken in design and production to improve fatigue life.

Fatigue strength is greatly affected by local loading conditions, and the designer must take pains to shape the part and distribute the load so as to minimise stress concentrations. Time is as well spent on such refinements as on strength calculations, but both approaches are essential for a successful result.

7.2. DESIGN FACTORS

7.2.1. Basic material properties

Fatigue strength test data are presented in the first place as a plot of the alternating stress, f_a, that will cause failure at various numbers of repetitions, N, (the S–N diagram). This information is usually obtained from tests in fully reversed tension or bending (the mean stress level being zero) on small specimens with

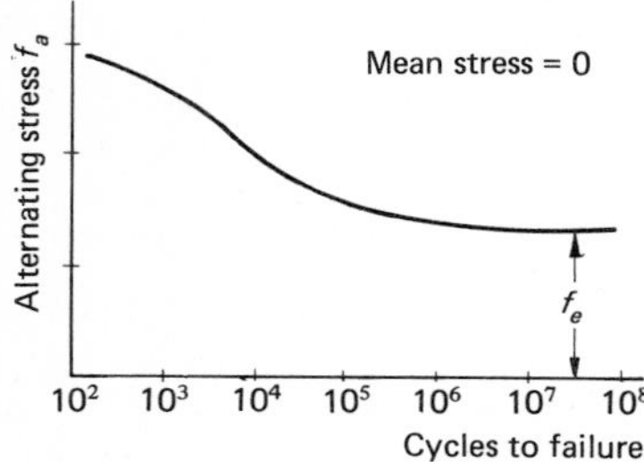

a high quality surface finish. Some materials – notably low-carbon steels – show a fatigue or "endurance" limit, f_e, as illustrated in the sketch; this is the stress level at which a further increase in cycles will cause no decrease in strength. Others – aluminium alloys, for example – show no such limit and the strength continues to fall off, albeit at a reduced rate, as the number of cycles increases. The fatigue strength of such materials can be typified by the value at some extended life – for example, 10^7 cycles. To avoid circumlocution, this value will also be included in the scope of the symbol f_e.

In the past, much design has been based on this endurance limit. However, where a part is subjected to infrequent load variations or is intended only for a limited life this procedure is uneconomic, and it is preferable to design to some specific life or number of repetitions. This applies particularly to aluminium alloys which have low fatigue strength compared to their tensile strength. When a full S–N curve is available for the material, it is a simple matter to read off the alternating stress that will cause failure at the chosen design life N_D. In the absence of this full information, the designer must seek some generalised relationship for the material being used; two such relations, one for steel and one for aluminium, are given below. They relate the fatigue strength to the tensile strength; the correlation between these two properties is close, but by no means perfect.

For steel, Heywood[1] suggests that A_0, the ratio of the fatigue strength at zero mean stress in axial loading to the tensile strength, is given for any value of N by

$$A_0 = \frac{f_a}{f_t} = \frac{1 + 0.0038\,N_L^{\,4}}{1 + 0.008\,N_L^{\,4}} \tag{7.1}$$

where $N_L = \log_{10} N$. Note that $f_a = f_e$ when $N = \infty$.

For aluminium alloy the corresponding equation is

$$A_0 = \frac{1 + \{0.0031N_L^{\,4}/(1 + 0.0065f_t)\}}{1 + 0.0031N_L^{\,4}} \tag{7.2}$$

These equations describe quite accurately most of the materials they represent but there are limitations to their use; for example, the fatigue strengths of very high tensile steels are not accurately predicted by Equation (7.1). The designer should seek wherever possible to check their validity for any specific material, for example, by substituting a known set of values of f_e, f_t and N. Note that stresses are expressed in MPa.

Although it is simpler to test specimens in fully reversed bending than in axial loading, the latter test information is more useful since a size effect shows up in the bend tests. This size effect appears to be associated with the stress gradient in bending, and results in small specimens – less than 5 mm diameter – having an endurance limit up to 20% higher than that given from the axial tests. The difference between the two types of test is negligible for specimens over 20 mm diameter. The fatigue strength from axial tests is referred to as the *intrinsic fatigue strength.* Equations (7.1) and (7.2) give values for this intrinsic stress at zero mean stress. Material specifications giving fatigue values do not always state which test has been used but since a common size for specimens in the bend test is 7.5 mm, there is some justification for reducing the fatigue strength by 15% when designing for axial loads.[2] This reduction is not necessary when using intrinsic strength data.

7.2.2. Failure theories

When the mean stress is other than zero, the designer must, in the absence of tests at the actual loading levels, fall back on a failure theory. Mention was made of several such theories in Chapter 4, where the method of plotting the failure envelope on mean and alternating normal stress axes (f_m–f_a axes) was introduced.

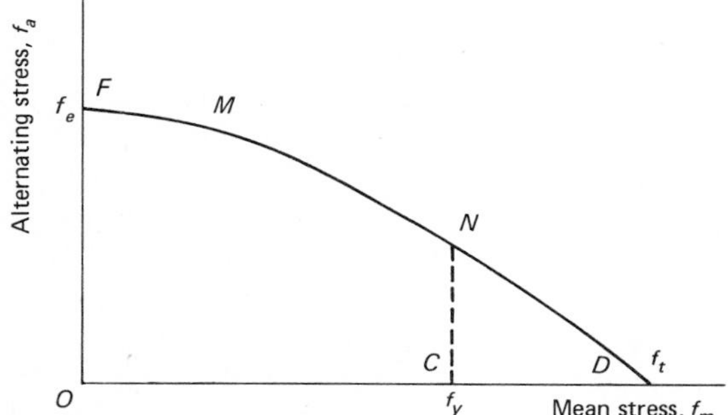

Fig. 7.1. *The failure envelope, f_m–f_a plot.*

For zero mean stress, $f_m = 0$, the endurance limit can be used to establish a point on the failure envelope; it will lie on the vertical axis (point F on Fig. 7.1). As the mean stress is increased, the value of the allowable alternating stress will decrease. Thus the failure envelope will fall off along some such line as $FMND$ which must be established experimentally. When the mean tensile stress reaches the yield strength (point C), the material will still withstand a varying stress CN – it being noted that the failure criterion in fatigue is the formation of a crack, or fracture, and not yield. Thus, right up to a mean stress equal to the tensile strength of the material (point D) some varying stress can be accommodated without fracture. At high mean stresses, work-hardening occurring in the first few load cycles may enable test specimens to withstand an alternating stress even when the mean stress is equal to f_t.

Although cases of combined stresses are studied in a more general way later in this chapter, we will now consider this special case of normal stress in some detail since it is the best documented and gives a good understanding of the design approach used. More complex loading situations are handled in the same way, but design techniques are not well established.

How best to represent the failure envelope $FMND$ (Fig. 7.1) for design purposes presents a problem which has not been satisfactorily solved. For preliminary design work, a simplified form of the failure envelope is assumed; some that have been proposed for this purpose are shown in Fig. 7.2. They are usually referred to by the names of their originators.

The Goodman line, DF, is a straight line joining f_t on the mean stress axis to f_e on the varying stress axis. The Soderberg line, CF, joins f_y to f_e. Another approach is to draw a line CK from f_y at 45° to the f_m axis to intersect the Goodman line at K. Then at any point along CK the sum $(f_m + f_a)$ will be equal to the yield strength; the line CKF allows a portion of the Goodman line to be used without at any stage exceeding the yield strength.

Although the Goodman line represents the behaviour of steel and light alloys with reasonable accuracy, it has the disadvantage from our viewpoint of taking fracture as the failure criterion for a steady load, i.e. at point D in Fig. 7.2. Since we have defined the safety factor for a steady load in terms of the yield strength, the use of this criterion would lead to an inconsistency. The Soderberg

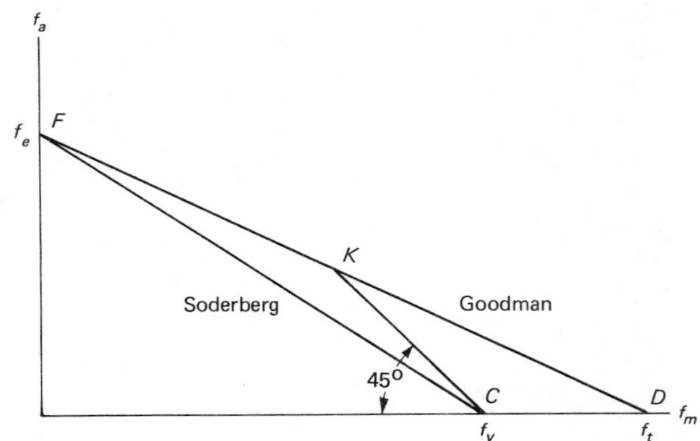

Fig. 7.2. *Failure envelopes for design.*

criterion does not suffer from this disadvantage, although, as will be seen when typical material behaviour is discussed, it is rather conservative at high mean stresses. The third alternative described above, which sets the line *FKC* as a failure envelope, is based on the Goodman line but has the added requirement that

$$f_{\max} = f_m + f_a < f_y \tag{7.3}$$

This envelope forms a useful compromise between the other two.

The equation for the Goodman line can be written as

$$\frac{f_a}{f_e} + \frac{f_m}{f_t} = 1 \tag{7.4}$$

and that for the Soderberg line as

$$\frac{f_a}{f_e} + \frac{f_m}{f_y} = 1 \tag{7.5}$$

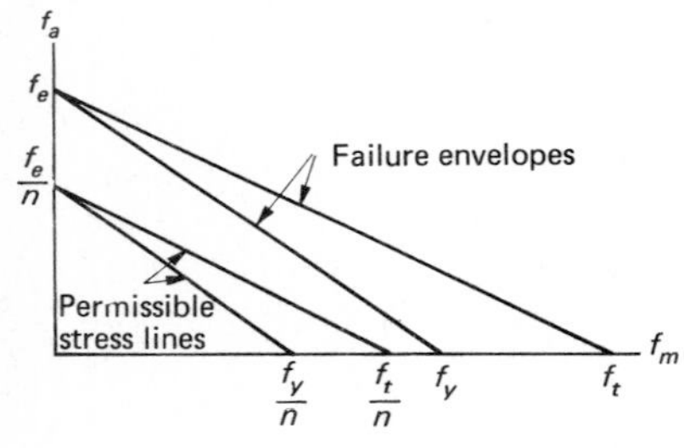

In either of these equations, f_a and f_m will be the combination of alternating and mean stress required to give failure. If we introduce a safety factor, *n*, and divide both material properties that occur in each equation by it, the permissible stress lines will be those shown in the sketch, and their equations will be

Goodman line $$\frac{f_a}{f_e} + \frac{f_m}{f_t} = \frac{1}{n} \tag{7.6}$$

Soderberg line $$\frac{f_a}{f_e} + \frac{f_m}{f_y} = \frac{1}{n} \tag{7.7}$$

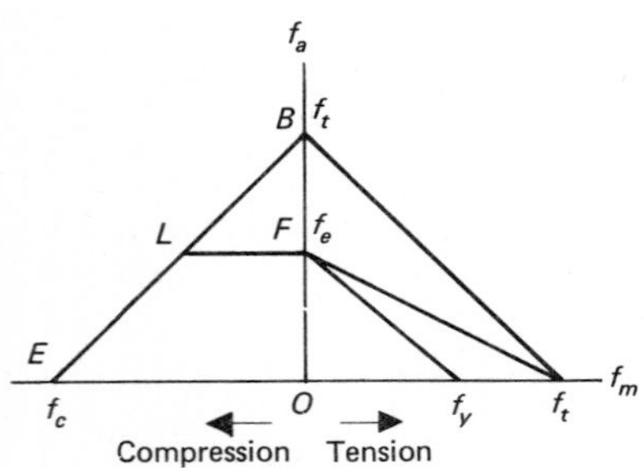

If, as is recommended, stresses in fatigue are selected from a graphical construction of the f_m–f_a diagram, the use of any of the design criteria is a simple matter. Other design envelopes are described in Marin's *Engineering Materials*[3] and in *Design Against Fatigue*.[4]

On the compression side of the diagram, the design envelope may be assumed to be a horizontal line *FL*, extending from *F* to the line *BE* which limits the maximum stress to the (compressive) strength of the material.

It is often convenient to make the f_m–f_a diagram non-dimensional by dividing all the stresses by f_t. In this way, a range of steels having different tensile strengths can be compared; it is found that, at the fatigue limit, their behaviour can be represented with good accuracy by a single line. The usefulness of the

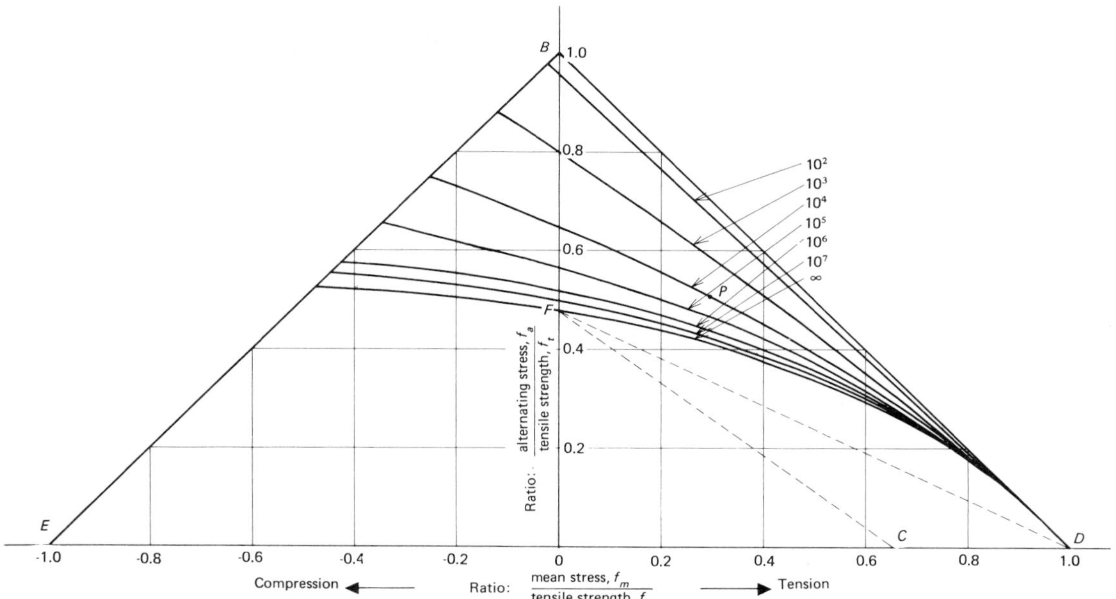

Fig. 7.3. *Generalised f_m–f_a diagram for steel. (From R. B. Heywood,* Designing Against Fatigue, *Chapman and Hall, London, 1962, by permission of the Associated Book Publishers.)*

diagram can be extended by plotting also the lines for values of N lower than that corresponding to the fatigue limit.

Figure 7.3 illustrates such a diagram for unnotched steel specimens subjected to mean and alternating stresses. The curves on this diagram are plotted from the formula put forward by Heywood[1]

$$\frac{f_a}{f_t} = (1 - f_m/f_t)\{A_0 + \gamma(1 - A_0)\} \tag{7.8}$$

where A_0 is defined in Equation (7.1) and

$$\gamma = f_m(2 + f_m/f_t)/3f_t$$

Under static loading conditions, where failure takes place in effect in quarter of a cycle, the tensile strength is represented by point B on the f_a axis or equally well by point D on the f_m axis, since no differentiation can be made between a mean and varying stress for this case. The point C, representing the yield strength of a typical steel having $f_y \approx \frac{2}{3}f_t$, can also be located on the f_m axis. Under these conditions of static loading, any point on the line DB represents a static tensile failure. In the same way any point on the line EB represents a compressive failure; this compressive side of the diagram does not have immediate significance for the designer since a compressive mean stress will increase the resistance of a material to a varying stress – an effect that is made use of in surface treatment (see Section 7.4.2).

When the curves for increasing values of N are plotted, it will be seen that for the first 100 repetitions the reduction in strength is small. However, there is a rapid decrease between 10^2 and 10^4, especially where the ratio f_a/f_t is high. Choosing some specific point, such as P on Fig. 7.3, we see that for a steady tensile stress equal to one-third of the tensile strength, a varying stress of one-half the tensile strength will cause failure after 10^4 cycles. The peak stress for this loading is $\frac{5}{6}f_t$. After 10^5 cycles there is not much further decrease before the line ∞, representing the endurance limit, is reached.

Since the curves on this diagram are drawn correctly to scale, it is possible to judge the extent to which the Soderberg (line CF) and Goodman (line DF) failure theories underestimate the strength of a typical steel.

The f_m–f_a diagram for aluminium alloys has the same general shape as that described above, although in this case no one diagram represents all alloys. The expression equivalent to Equation (7.8) is given by Equation (7.9) and, since the equation for the factor A_0 given earlier (Equation (7.2)) contains f_t, there will be a different family of curves for each value of f_t

$$f_a/f_t = \{1 - f_m/f_t\}\{A_0 + \gamma(1 - A_0)\} \tag{7.9}$$

where $$\gamma = \frac{f_m}{f_t\{1 + (f_t N_L/2200)^4\}}$$

Heywood illustrates several families of such curves for a range of tensile strengths.[1]

The design of components subjected to combined stresses, one or more of which are alternating, presents a difficult problem, for data are not plentiful and are sometimes conflicting. The failure criteria described in Chapter 4 for static loads on ductile material generally hold for fatigue loading, the von Mises criterion giving the most accurate description of behaviour with the Tresca criterion being conservative but sufficiently accurate for most design purposes.

Equations are developed here for two dimensional stresses in plain (i.e. unnotched) specimens, and when the method of dealing with stress raisers has been considered these equations can be modified to allow for the reduction in strength caused by them. As will be seen, these modifications are not always well supported by experimental work.

First, the failure theories can be used to decide what should be the endurance limit in pure shear. According to the Tresca criterion the value of q_e should be

$$q_e = 0.5 f_e \tag{7.10}$$

while the von Mises criterion predicts a value of

$$q_e = \frac{1}{\sqrt{3}} f_e = 0.58\, f_e \tag{7.11}$$

The latter value for q_e agrees well with experimental results for ductile materials.

For two-dimensional fluctuating stresses it is helpful to think in terms of an equivalent static or mean stress, f_{eq}, derived from whichever failure theory is in use. We consider the relationship between the point P on Fig. 7.4 representing alternating and mean stress components f_a and f_m, and the Goodman line FD. If a line KR is drawn through P parallel to the Goodman line FD, the stress on the f_m axis at R will be the static tensile stress having the same safety factor as the stress represented by P.

If the safety factor is n, this will require that $f_t/f_{eq} = n$; this can be shown by extending the line OP to cut the Goodman line at M. All points on the line OM have the same ratio f_m/f_a; since M represents the failure condition the safety factor n at P is defined by OL/f_a or ON/f_m where L and N are the points of intersection of the f_m–f_a axes with the perpendiculars from M to these axes.

Now $f_t/f_{eq} = MN/PS$ since triangles OMD, OPR are similar, and hence it follows that

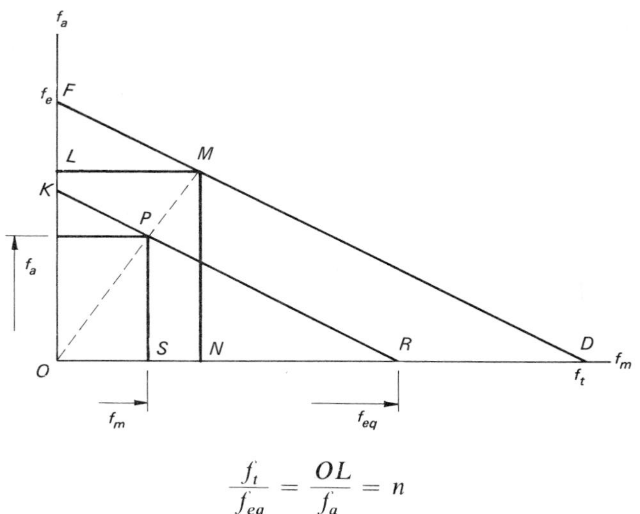

Fig. 7.4. Calculation of equivalent stress, f_{eq}.

$$\frac{f_t}{f_{eq}} = \frac{OL}{f_a} = n$$

We can determine f_{eq} by noting in Fig. 7.4 that $f_{eq} = f_m + SR$, and since triangles FOD and PSR are similar, $SR/OD = PS/FO$

$$\therefore\ f_{eq} = f_m + \frac{OD \times PS}{FO} = f_m + \frac{f_t}{f_e} f_a \tag{7.12}$$

For the Soderberg criterion the equivalent stress is given by

$$f_{eq} = f_m + \frac{f_y}{f_e} f_a \tag{7.13}$$

The equivalent shear stress can be expressed in the same way as

$$q_{eq} = q_m + \frac{q_y}{q_e} q_a = q_m + \frac{f_y}{f_e} q_a \tag{7.14}$$

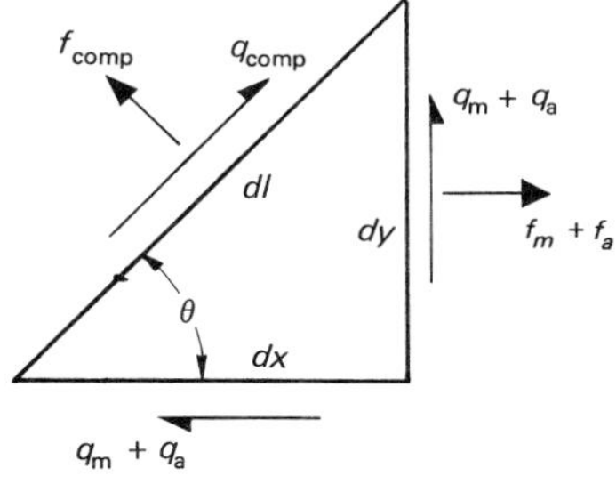

We now consider the case of two-dimensional alternating stresses shown in the sketch. Selecting the maximum shear stress (Tresca) criterion, the shear stress and hence the safety factor on an arbitrary plane, dl, can be determined. By differentiating with respect to θ, the plane on which the safety factor is a minimum can be found and n expressed in terms of the component stresses.

From Equation (3.3) we write for the value of the complementary shear stress on dl

$$q_{comp} = -\frac{f_m + f_a}{2} \sin 2\theta + (q_m + q_a) \cos 2\theta$$

Separating into steady and alternating parts,

$$q_{comp} = \left(-\frac{f_m}{2} \sin 2\theta + q_m \cos 2\theta\right) + \left(-\frac{f_a}{2} \sin 2\theta + q_a \cos 2\theta\right)$$

Using the failure relationship given in Equation (7.13) and the Tresca criterion, Equation (7.10),

$$\frac{1}{n} = \frac{\left(-\frac{f_m}{2} \sin 2\theta + q_m \cos 2\theta\right)}{f_y/2} + \frac{\left(-\frac{f_a}{2} \sin 2\theta + q_a \cos 2\theta\right)}{f_e/2} \tag{7.15}$$

If this expression is differentiated with respect to θ and the angle for maximum $1/n$ (minimum n) found, the value of θ can be substituted in Equation (7.15) to give

$$\frac{1}{n} = \left[\left(\frac{f_m}{f_y} + \frac{f_a}{f_e}\right)^2 + 4\left(\frac{q_m}{f_y} + \frac{q_a}{f_e}\right)^2\right]^{\frac{1}{2}}$$

This is the design equation using the Soderberg and Tresca criteria. Note that it can be written in terms of the equivalent stress from Equation (7.13) as

$$\frac{1}{n} = \left[\left(\frac{f_{eq}}{f_y}\right)^2 + \left(\frac{q_{eq}}{f_y/2}\right)^2\right]^{\frac{1}{2}} \tag{7.16}$$

The equivalent expression using the von Mises criterion is

$$\frac{1}{n} = \left[\left(\frac{f_{eq}}{f_y}\right)^2 + \left(\frac{q_{eq}}{f_y/\sqrt{3}}\right)^2\right]^{\frac{1}{2}} \tag{7.17}$$

The use of this approach to calculate the safety factor is illustrated later (Example 7.2).

We have set up equations describing the fatigue performance of a material in terms of the failure theories used for static stresses. There is no *a priori* reason why a common failure theory should hold, but until theoretical and experimental studies provide a better alternative, these criteria will continue in use. Sines has suggested a useful rule which enables the designer to establish the effect of combining an alternating and a static stress.[5] He examined the static normal stresses that exist on the planes of greatest alternation of shear stress in order to establish their effect on the permissible alternating stresses for the following cases:

(1) static tension and axial alternating stress;
(2) static compression and axial alternating stress;
(3) static torsion and alternating torsion;
(4) static torsion and alternating bending;
(5) static tension and alternating torsion.

Each of these cases is identified with the corresponding illustration in Fig. 7.5; the loading combination is shown and an enlarged view drawn of the material element which carries the maximum alternating shear stress. The effect of the mean stress on the permissible amplitude of the alternating stress is also illustrated in the right hand column on an f_m–f_a diagram – a horizontal line indicating that the mean stress does not affect the permissible alternating stress levels. Sines[5] gives experimental data which confirm the hypothesis that, if the sum of the normal stresses, $N_1 + N_2$, on the planes of maximum alternation of shear stress is positive, an increasing static stress reduces the permissible alternating stress (Cases (1) and (5)). When the sum $N_1 + N_2$ is negative (Case (2)), the permissible alternating stress is increased, and when $N_1 + N_2$ is zero (Cases (3) and (4)) the static stress has no effect. We have, of course, already studied the first two cases in some detail; they are included here for completeness. The third and fourth are important since they are frequently met in practice.

In Case (3), it is clear that the steady and alternating stresses on the element in question are shear stresses and the normal stresses are zero. In Case (4), the alternating bending stress (acting in the axial direction) causes an alternating shear stress on a plane at 45° to the axis. On the 45° planes, the steady torsional

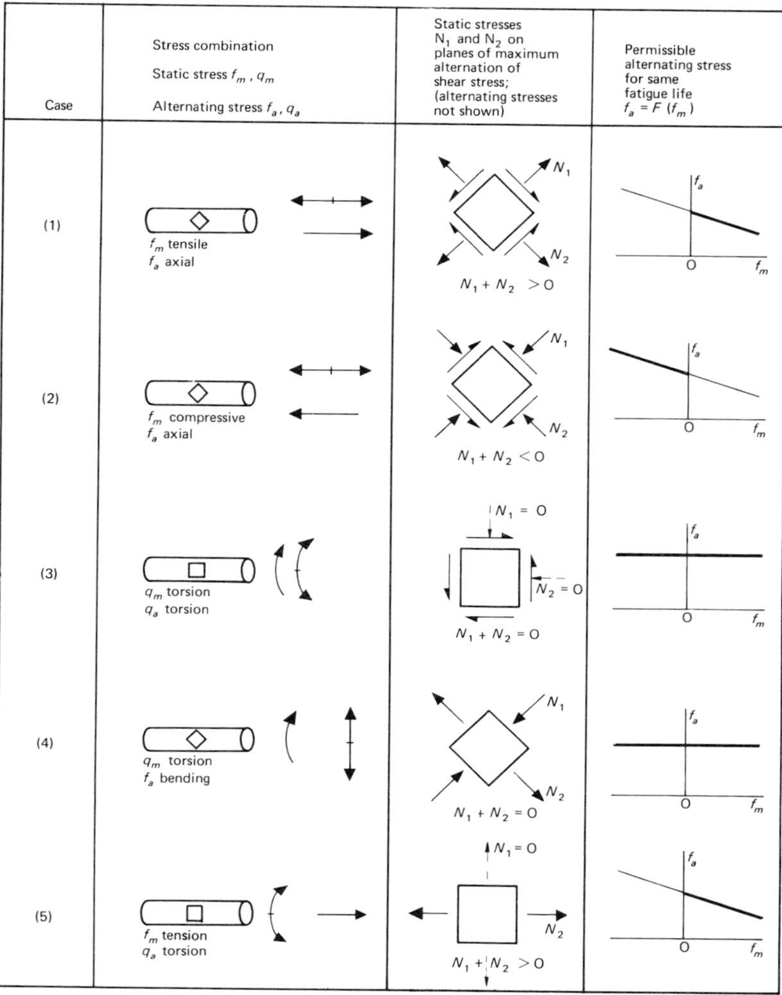

Fig. 7.5. *Effect of different combinations of static and alternating stresses on fatigue life. (From Sines and Waisman (eds.)* Metal Fatigue, *McGraw-Hill, New York, 1959, by permission of McGraw-Hill Book Company.)*

moment causes normal tensile and compressive stresses N_1 and N_2; and as shown in Fig. 7.5, $N_1 + N_2 = 0$. Thus the allowable alternating stress is not affected by the steady torsion. This result (which is well established experimentally) holds, provided the maximum shear stress does not exceed the torsional yield strength.

This criterion, which is based on test results, may be used in design to decide whether or not the average stress need be taken into account. It must be noted that some stress raisers will sufficiently alter the stress pattern in a material so that the *local* stresses must be examined in detail before it can be accepted that the above analysis holds. Sines quotes the case of a circular bar with a transverse hole, loaded with steady and alternating torque. A detailed analysis of the stress distribution around the hole shows that the simple stress pattern given in Case (3) of Fig. 7.5 is altered so that the mean torsion does now reduce the allowable alternating torsion load.[5]

It may be noted that this criterion is not in agreement with Equations (7.16) and (7.17) for all cases. For example, Case (3) indicates that the shear stress term in these equations should be zero.

7.2.3. Stress concentration

While the fatigue strength of a plain specimen will be of the order of 30% less than the static (yield) strength, the addition of a stress raiser can further reduce the strength by up to 70%. The presence of stress raisers (or "notches") must therefore be taken into account; as might be expected, this does not turn out to be a simple matter.

Accounting for a stress raiser starts with the determination of a theoretical stress concentration factor, K_t. This factor, which can be determined by a theoretical treatment in simple cases or by photoelastic or some other experimental methods, is defined as

$$K_t = \frac{\text{maximum stress at stress raiser}}{\text{nominal or engineering stress}}$$

Since the maximum stress in this definition is based on the behaviour of an ideal elastic material, K_t does not accurately describe what occurs in a real material; nevertheless it forms the usual starting point in design since it has been very fully investigated and extensive information exists for a wide range of shapes[6, 7]; some of this is summarised in Appendix A11.

The main difficulty in quantifying what occurs in a real material in fatigue must still be faced. The alternating strength reduction factor, K_a, relates the strength of a notched specimen to that of a plain specimen. Thus

$$K_a = \frac{\text{alternating stress in plain specimen, } f_a}{\text{nominal alternating stress in specimen with notch, } f_{an}}$$

both these stresses being determined for failure at *the same number of cycles*. When evaluated at the fatigue limit, this factor is given the symbol K_A. Values of K_a may range from K_t to 1, depending on the material and on the number of cycles.

Several theories have been advanced to explain the difference between K_t and K_a. It appears that two factors influence the probability of a fatigue crack being initiated – the presence of a stress gradient at the location of maximum stress and the total volume of material (and therefore the number of inherent flaws) at this location. The latter gives a possible explanation of the existence of a marked size effect; tests on notched specimens in axial loading and on plain specimens in bending have shown that for small specimens, K_a approaches a value of 1, and for large specimens, K_a tends towards K_t. In each of these cases a stress gradient occurs. This size effect is present in plain specimens in bending but not in axial loading where no stress gradient exists.

Various relationships between K_t and K_A have been put forward; one that has sound theoretical justification suggests that all materials have inherent flaws, typified by a length a_f, and that the larger these flaws are in relation to the notch, considered to have radius R, the lower will be the ratio of K_A/K_t. Thus a material which has a large number of built-in flaws – for example, cast iron – will have a low value for this ratio. At the other extreme, titanium has a value close to unity; such a material is described as "fully notch-sensitive".

An expression proposed by Heywood[1] is

$$K_A = \frac{K_t}{1 + 2\left(\frac{K_t - 1}{K_t}\right)\left(\frac{a_f}{R}\right)^{\frac{1}{2}}} \qquad (7.18)$$

where R is the notch radius (mm), and a_f is the notch alleviation factor, corresponding to the length of equivalent inherent flaws.

Table 7.1. Typical values of a_f.

Material	Type of Notch	a_f mm
Steel		$(175/f_t)^2$
		$(140/f_t)^2$
		$(105/f_t)^2$
Aluminium alloys		$(280/f_t)^6$
Cast iron (flake graphite)		0.36
Cast iron (spheroidal)		$(175/f_t)^2$
Cast steel		0.046
Magnesium alloys		0.005
Titanium alloys		Very low
Reinforced plastics		Very high
		f_t expressed in MPa

Source: Adapted from "Engineering Sciences Data Item 71010"[4], by permission of the Engineering Sciences Data Unit.

Equation (7.18) accounts for the size effect and satisfies the requirement that, for a given material, K_A should approach K_t for large specimens (when a_f/R is small).

Typical values for a_f are given in Table 7.1. It will be seen that the factor is dependent on strength, especially for aluminium alloys; it can be remarked here that this dependence means that an increase in tensile strength does not result in a proportional increase in useful fatigue strength. Where the notch is severe, there may be no improvement at all. This important point will be taken up again. Note that the lower limit for K_A is taken as 1.0. Values of R below the size of the equivalent flaw are not meaningful.

A relation between K_A and K_t can also be expressed in terms of the notch sensitivity factor q, defined by

$$q = \frac{K_A - 1}{K_t - 1} \tag{7.19}$$

Values of q are given in the literature for various materials and notch shapes[5] and these may be used to calculate K_A from K_t. Figure 7.6 illustrates

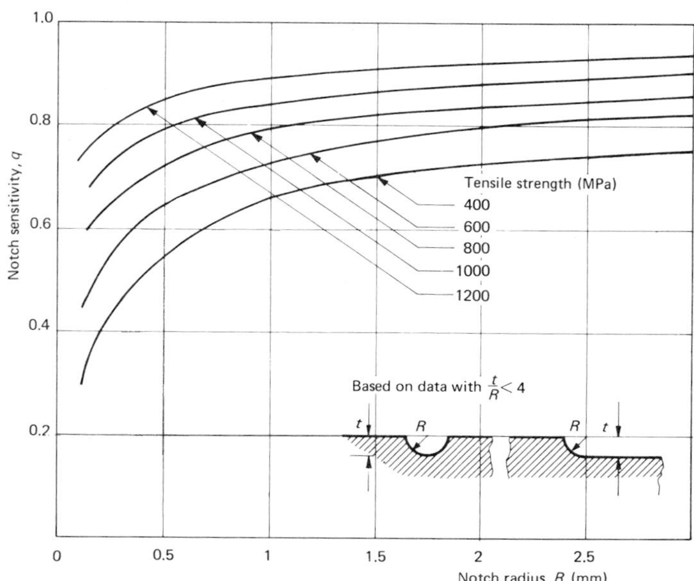

Fig. 7.6. *Values of notch-sensitivity, q. (From Sines and Waisman (eds.),* Metal Fatigue, *McGraw-Hill, New York, 1959, by permission of McGraw-Hill Book Company.)*

the way in which q is usually presented. Note that the ratio on the right-hand side of Equation (7.19) is used in preference to the simple form K_A/K_t so that the variation of q is contained in the range 0 to 1. When $q = 0$, $K_A = 1$ and the material is insensitive to notches. When $q = 1$, $K_A = K_t$ and the material is fully notch-sensitive. A disadvantage of this method is that it does not account for size effect.

Because of the large scatter in available data and the high notch-sensitivity of the material, the design of aluminium alloy parts is often based on the assumption of full notch-sensitivity. For severe notches this is conservative and the full theoretical factor need not be used; the value of a_f from Table 7.1 can be applied in Equation (7.18).

The calculation of K_A using the approaches described above is illustrated in Example 7.1. In this case, the two methods do not give good agreement and it is clear, from the differences in the basic data, that such differences are to be expected. Both methods have been described because it is difficult to find information to cover all materials and shapes. It should be noted that there is some information[2] to indicate that, however large K_t may be, the value of K_A will not exceed 3 or 4. As will be seen, this does not appear to hold for corroded parts.

When the strength reduction factor is used in design, the nominal or engineering stress, f_{an}, for the part is calculated, and the value of f_a used – for example, in Equations (7.12), (7.13) – is $K_A\, f_{an}$.

It is often assumed that the mean stress can be treated like a static stress, to which, as we have seen in Chapter 4, a stress concentration factor need not be applied. If f_{mn} is the nominal mean stress, Equations (7.12) and (7.13) then become

$$f_{eq} = f_{mn} + K_A\, f_{an}\frac{f_t}{f_e} \qquad \text{(Goodman)} \qquad (7.20)$$

$$f_{eq} = f_{mn} + K_A\, f_{an}\frac{f_y}{f_e} \qquad \text{(Soderberg)} \qquad (7.21)$$

The use of these equations is illustrated in Example 7.2.

The information presented so far is sufficient for preliminary design purposes; a suitable method is summarised in Section 7.3.1 and is illustrated by several worked examples.

The practice of neglecting the effect of the stress concentration on the mean stress is widely adopted. Several authors have pointed out that the basis for this practice is not well-established; in fact, it is at variance with published information, which indicates that a strength reduction factor – but not necessarily the same value – should be applied to both f_m and f_a. We will at this stage distinguish also between the value of the factor at some finite life and at the fatigue limit taken as 10^7 cycles. The nomenclature for these various strength reduction factors is set out in Table 7.2. (Note that K_s is the strength reduction factor in static tensile loading.) In the following paragraphs, the procedure used to determine these factors is taken from Heywood's *Designing Against Fatigue.*[1]

When designing for finite life it is clear that the value of K_a can be expected to change progressively from a value of K_s at a few reversals to K_A at a very large number of reversals. We can write

$$K_a = K_s + q_A(K_A - K_s) \qquad (7.22)$$

where q_A is a notch-sensitivity coefficient with values $q_A = 0$ at static failure and $q_A = 1.0$ at 10^7 cycles. Fair agreement is obtained with published

Example 7.1 Sheet 1 of 1

Calculate the strength reduction factor in bending for the shaft fillet shown in the sketch.

Quenched and tempered steel shaft, $f_t = 700$ MPa.

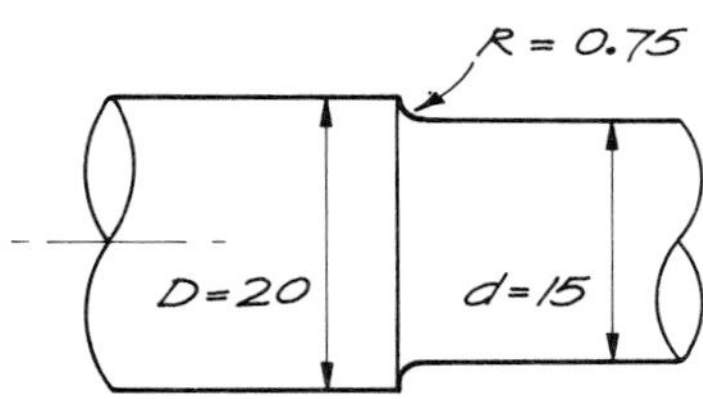

$\frac{R}{d} = 0.05, \quad \frac{D}{d} = 1.33$

From Fig A 11.7 $K_t = 2.01$

From Table 7.1 $\sqrt{a_f} = \frac{140}{700} = 0.2$

$$K_A = \frac{2.01}{1 + \frac{2 \times 1.01}{2.01} \times \frac{0.2}{\sqrt{0.75}}} = 1.63$$

Note that if the size is doubled and the proportions remain the same, $K_A = 1.73$

Using the information from Fig 7.6, $q = 0.73$

$$K_A = 0.73 \times 1.01 + 1 = 1.74$$

Example 7.2 Sheet 1 of 1

A shaft with a keyway is acted on by loads which give the following nominal stresses:

$$f_{mn} = 70 \qquad q_{mn} = 45$$

$$f_{an} = 150 \qquad q_{an} = 60$$

The strength reduction factors at the keyway are:
bending, $K_A = 1.6$ torsion, $K_{Aq} = 2.0$

Determine the safety factor if the yield strength of the material is 800 MPa. Use Soderberg criterion and the von Mises failure theory.
Take $f_y = 800$, $f_e = 500$.

For normal stress, Equation (7.20) gives

$$f_{eq}' = 70 + 1.6 \times 150.\ \frac{800}{500} = 454$$

For shear stress, Equation (7.14) with $q_a = 2.0 q_{an}$ gives

$$q_{eq} = 45 + 2.0 \times 60.\ \frac{800}{500} = 237$$

From Equation (7.17), $\frac{1}{n} = \left[\left(\frac{f_{eq}}{f_y}\right)^2 + 3\left(\frac{q_{eq}}{f_y}\right)^2\right]^{\frac{1}{2}}$

$$= \left[\left(\frac{454}{800}\right)^2 + 3\left(\frac{237}{800}\right)^2\right]^{\frac{1}{2}} = 0.77$$

$$n = 1.3$$

Table 7.2. Nomenclature for strength reduction factors (srf)

Condition	Alternating srf	Mean srf
Static failure ($N = 1$)	K_s	K_s
General condition, N specified	K_a	K_m
Limiting condition, $N = 10^7$ cycles	K_A	K_M

experimental results if q_A is expressed as

$$q_A = N_L{}^4/(b + N_L{}^4) \tag{7.23}$$

where, as before, $N_L = \log_{10}N$.

This relationship is suitable for general application in design, since b is a parameter whose value is dependent only on the material and, for simple geometric notches, is independent of the type of notch. Heywood suggests the values for b given in Table 7.3.

Table 7.3. Values for factor b in Equation (7.23)

Material	b
Steel	$\left(\frac{12\,000}{f_t}\right)^2$
Aluminium alloy, Al–Zn–Mg	25
Aluminium alloy, Al–Cu	60
Magnesium alloy	80

Source: Adapted from R. B. Heywood, *Designing Against Fatigue,* Chapman and Hall, London, 1962, by permission of the Associated Book Publishers.

Note that the value of K_a progressively approaches K_s, which is usually close to unity, as the number of cycles decreases. Some of the economy in designing to a specific life, less than the fatigue limit, comes from this reduction.

Values for K_s are given in Table 7.4; the notched strength differs significantly from the plain strength only for the second case.

Table 7.4. Values of static strength reduction factor, K_s

Type of notch	Steels	Aluminium alloys
Transverse hole, unloaded	0.95	1.05
Transverse hole, loaded through hole	1–1.7	1–1.8
Shoulder in plate	1.0	1.05
Shoulder in bar	1.0	1.0
Groove in plate	0.95	1.05
Groove in bar	0.75	0.9

Source: From R. B. Heywood, *Designing Against Fatigue*, Chapman and Hall, London, 1962, by permission of the Associated Book Publishers.

We now seek a method for determining K_m, which can be written in the same form as Equation (7.22) for K_a

$$K_m = K_s + q_M(K_M - K_s) \tag{7.24}$$

where q_M again will have the same limits as q_A at $N = 1$ and $N = 10^7$. It is assumed that at the fatigue limit, $K_m = K_A$, and for other lives, Equation (7.25) can be used to determine K_m

$$K_m = \frac{f_m}{f_{mn}} = K_s + (K_A - K_s)\left(1 - \frac{f_{mn} + f_{an}}{f_{tn}}\right)^2 \tag{7.25}$$

where f_{tn} is the static tensile stress at failure for the notched part and f_{mn} and f_{an} are the nominal mean and alternating stresses.

Note that when the peak stress in a cycle $(f_{mn} + f_{an})$ reaches the static strength of the part f_{tn}, Equation (7.25) satisfies the requirement that $K_m = K_s$.

The use of these factors to determine the safe life of a component is not straightforward. The way in which K_m is calculated is illustrated in Example 7.3. Reference should be made to Heywood, *Designing Against Fatigue*[1], for further details.

7.2.4. Surface finish

The reduction in the endurance limit for fully reversed bending that can be expected for various surfaces on a plain specimen is illustrated in Fig. 7.7. The surface finish factor K_F, defined by Equation (7.26), is plotted for various grades of surface finish against tensile strength

$$K_F = \frac{\text{endurance limit, polished surface}}{\text{endurance limit, actual finish}} \tag{7.26}$$

As with other strength reduction factors, K_F increases markedly as the material strength increases; it is defined only at the endurance limit. Information given in the literature is often conflicting, and the curves in Fig. 7.7 must be regarded as giving typical results only. Further information can be found in Lipson and Juvinall, *Handbook of Stress and Strength*, and Kravchenko, *Fatigue Resistance.*[2, 8]

In all cases where fatigue is a problem, every attempt must be made to start with and to maintain a good finish. Where this can be done, and where pronounced stress concentrations exist, special attention need not be directed to the effect of the microfinish in the stress calculation, and it would be unrealistic to reduce the material strength by both K_A and K_F. However, where stress raisers have been largely eliminated, K_F must be taken into account, and the strength reduction factor taken as the product of these two terms.

It is interesting to note that if oxidation scale exists, K_F increases by a factor of 2 as the material strength is increased by a factor of 3, resulting in very little overall gain. For corroded surfaces, the higher-strength material gives a lowered fatigue strength—assuming that the same corrosion damage occurs to the higher-strength material.

Kravchenko[8] suggests that surface finish is less important in torsion and that a factor K_{Fq} be used where

$$K_{Fq} = \frac{K_F}{0.4K_F + 0.6} \tag{7.27}$$

Example 7.3 Sheet 1 of 1

The part shown in the sketch has been designed for a life of 10^4 cycles under the following factored loads:

Mean load, 36 kN; alternating load, 55 kN.

Check the life, if the steel has a tensile strength of 750 MPa.

Take $K_t = 2.35$, $K_A = 2.10$

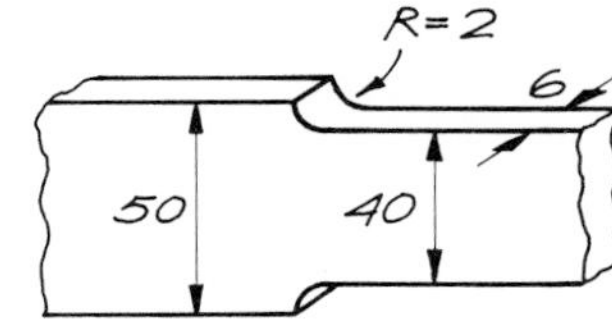

Mean stress $f_{mn} = \frac{36000}{6 \times 40} = 150$ MPa

Alternating stress $f_{an} = \frac{55000}{240} = 229$ MPa

$K_a = K_s + \frac{N_L{}^4}{b + N_L{}^4}(K_A - K_s)$. Take $K_s = 1.0$

$b = \left(\frac{12000}{f_t}\right)^2 = 256$, $K_a = 1 + \frac{4^4}{256 + 4^4}(2.10 - 1) = 1.55$

$K_m = K_s + (K_A - K_s)\left(1 - \frac{f_{mn} + f_{an}}{f_{tn}}\right)^2$, and for

$K_s = 1.0$, $f_{tn} = f_t = 750$

$K_m = 1.0 + (2.10 - 1.0)\left(1 - \frac{379}{750}\right)^2 = 1.275$

$f_a = K_a f_{an} = 1.55 \times 229 = 357$, $\frac{f_a}{f_t} = 0.476$

$f_m = K_m f_{mn} = 1.275 \times 150 = 191$, $\frac{f_m}{f_t} = 0.255$

Plotting these points on Fig 7.3 gives a life of $N = 10^5$. The part is therefore safe for a life of 10^5 cycles.

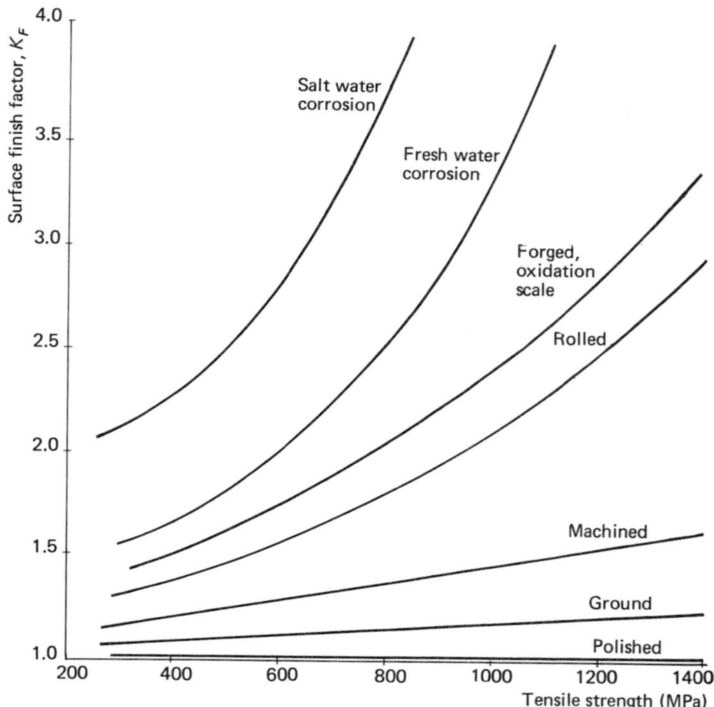

Fig. 7.7. *The effect of surface finish and corrosion on endurance limit in bending.*

7.2.5. Cumulative damage

It has been assumed so far that the whole of the fatigue life of a structure is spent at fixed values of mean and alternating stress; it is often the case that the stress levels will vary and, if the pattern of this variation is known, the life expectancy can be predicted.

Consider an element which is subjected to N_1 cycles at a stress level at which its life would be N_{1_t}; it is postulated that this loading has used up a fraction N_1/N_{1t} of the total life. Further cycling at different stress levels can be continued until the summation of the fractions is equal to unity when, according to the cumulative damage rule, the structure will fail. This rule is simple to apply and is widely used in practice[9]; other, more complicated hypotheses have not given a greatly improved accuracy. It will be assumed therefore that, provided Equation (7.28) is satisfied, the structure is safe under a range of different loadings. This approach was first suggested by Palmgren; it is usually referred to as the *Palmgren-Miner rule*

$$\frac{N_1}{N_{1_t}} + \frac{N_2}{N_{2_t}} + \frac{N_3}{N_{3_t}} + \ldots \leqslant 1 \qquad (7.28)$$

7.3. SUMMARY OF DESIGN METHODS

7.3.1. Simplified method

It is considered that even the simplified method must bring out the fact that higher strength materials tend to suffer from increased notch-sensitivity and therefore do not always bring improved performance. The suggested method

Example 7.4 Sheet 1 of 1

A flat steel plate with a cross-sectional area of 250 mm² is subjected to a load that varies from 30 kN to 70 kN. Calculate the safety factor if the material properties are

$$f_t = 420, \quad f_y = 280, \quad f_e = 220 \text{ MPa}$$

Mean load = 50 kN, f_m = 200 MPa

Varying load = 20 kN, f_a = 80 MPa

Plot these on the $f_m - f_a$ diagram at P

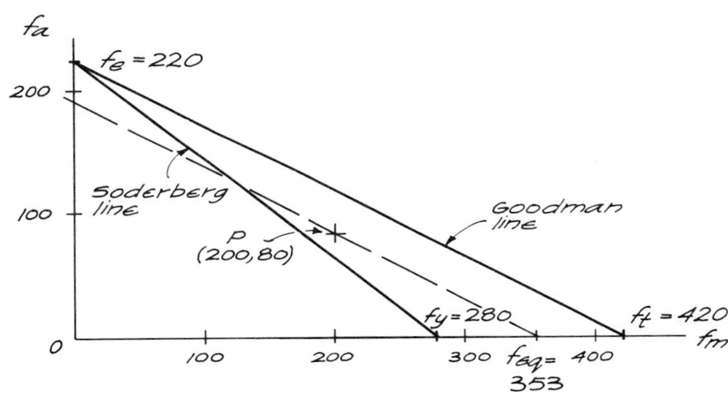

According to Soderberg criterion, loading is unsafe

According to Goodman criterion

$$f_{eq} = f_m + \frac{f_t}{f_e} \cdot f_a = 200 + \frac{420}{220} \cdot 80 = 353$$

$$n = \frac{420}{353} = 1.19$$

Note that the maximum stress of (200+80) MPa is equal to the yield strength of the steel in this case. The safety factor of 1.19 would be too low for most purposes.

Example 7.5 Sheet 1 of 1

A hole is drilled in the plate in Example 7.4 and the net cross-sectional area reduced to 240 mm^2 The resulting srf is 2.0. Determine the material properties to maintain the same safety factor.

$f_{mn} = 50000/240 = 208$

$f_{an} = 20000/240 = 83; \quad f_a = f_{an} \cdot K_A = 166$

Plot the point P(208, 166) on the $f_m - f_a$ diagram

Assume ratio (f_t/f_e) remains unchanged at $\left(\frac{420}{220}\right)$

$f_{eq} = 208 + \left(\frac{420}{220}\right)166 = 525$

For same safety factor 1.19
Using Goodman criterion
required $f_t = 1.19 \times 525 = 625$

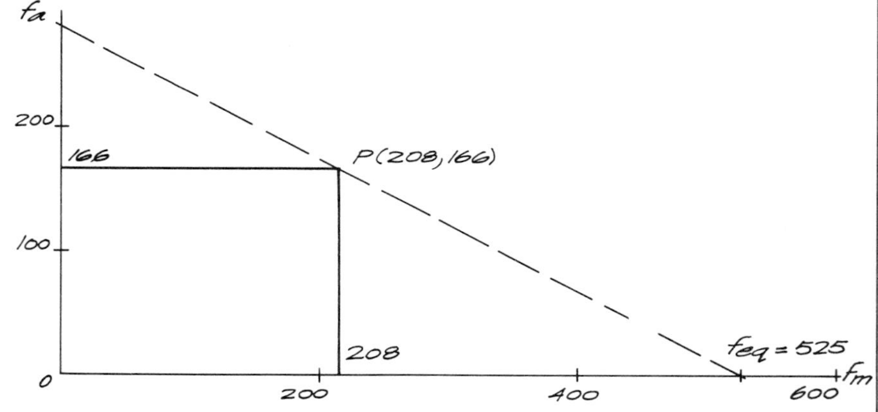

Example 7.6 Sheet 1 of 1

Calculate the diameter of a circular steel tie rod subjected to a repeated load (i.e., $f_{mn} = f_{an}$) of 15 kN. There is a shoulder on the shaft of height 5 mm, with a fillet radius of 1 mm.
Given $f_t = 400$, $f_e = 200$ MPa, $n = 1.5$
Use Goodman criterion.

Since K_t is a function of d a trial-and-error solution will be used.

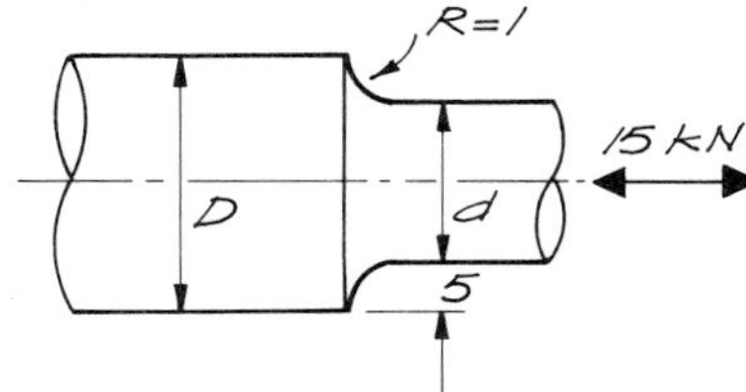

Try $d = 15$ mm $D = 25$ mm
$D/d = 1.67$, $R/d = 0.067$

From Appendix A11 $K_t = 2.18$
From Table 7.1, $a_f = (140/400)^2 = 0.35^2$

Equation (7.18) $$K_A = \frac{2.18}{1 + \frac{2 \times 1.18 \times .35}{2.18 \times 1}} = 1.58$$

Equation (7.6) with $f_{an} = f_{mn}$ for repeated loading
$$\frac{1}{1.5} = \frac{1.58}{200} f_{an} + \frac{f_{an}}{400} = \frac{4.16}{400} f_{an}$$

Allowable $f_{an} = 64$, $f_{max} = 128$

Area required at critical section, diameter d,
$$= \frac{15000}{128} = 117 \text{ mm}^2, \quad d = 12.2$$

Now check value of K_t for $d = 12.2$ $D/d = 1.82$

$K_t = 2.09$ $R/d = 0.082$

This altered value of K_t for new shaft diameter makes very little difference to d. A recalculation gives

d = 12.1 mm diameter.
D = 22.1.

will give an accuracy that in most cases will be in keeping with the precision with which the loads can be estimated. An alternative and more detailed approach is described in the next section. The following steps are involved, assuming that a preliminary estimate of sizes has been made on the basis of the static loads:

(1) determine the material properties, f_e, f_y, f_t;
(2) decide on a suitable safety factor, n;
(3) draw the Goodman diagram (or other selected failure criterion);
(4) establish the theoretical stress concentration factor, K_t, for any notch in the critically stressed region;
(5) calculate the actual strength reduction factor, K_A, using Equation (7.18);
(6) from the known dimensions calculate the nominal stresses, f_{an} and f_{mn};
(7) plot on the diagram $K_A f_{an}$ vertically, f_{mn} horizontally, and check safety;
(8) repeat if necessary with altered dimensions or material.

While the graphical method is recommended, particularly since it simplifies checking a number of alternative arrangements, the safety factor can be determined by using Equations (7.6) or (7.7).

Examples 7.4, 7.5 and 7.6 illustrate this method, which can also be used to design to a specific life, N, by replacing f_e with the fatigue strength at N cycles. However, the more elaborate treatment described below is probably warranted in this case.

7.3.2. Detailed design approach

In most cases, this method will be used to check the safety of a part which has been sized on the basis of static loading or by using the simplified method just described. Some of the steps involved in the design can be carried out quite quickly if an f_m–f_a chart (see Fig. 7.1) is available for the selected materials for a range of N values; calculation of individual values of f_m and f_a at a given value of N (for example, from Equation (7.3)) can be very time-consuming. It is assumed here that such a chart is available and that the determination of safety of the structure under N_1 cycles at loading 1 and N_2 cycles at loading 2 is to be confirmed. The detailed approach involves the following steps:

(1) establish the theoretical stress concentration factor, K_t;
(2) calculate the nominal stresses, f_{an} and f_{mn};
(3) calculate the strength reduction factor, K_A. Use Equation (7.18);
(4) calculate the strength reduction factor, K_a, for each value, N_1, N_2 (Equation (7.22));
(5) calculate the strength reduction factor, K_m, for each value, N_1, N_2 (Equation (7.25));
(6) locate for each loading the point representing $K_a f_{an}$ and $K_m f_{mn}$. Read the life expectancy at each point, N_{1_t} and N_{2_t};
(7) calculate $(N_1/N_{1t}) + (N_2/N_{2t})$. Failure should not occur if this is less than unity.

The designer must exercise his judgement in deciding whether the calculated value of

$$\frac{N_1}{N_{1_t}} + \frac{N_2}{N_{2_t}}$$

is sufficiently below unity for the design to be accepted as safe. The experience of other designers in this field may be called on to back up this judgement.[10, 11] Example 7.7 illustrates the use of this method for a simple loading case, using data from Table 7.5.

If the design has reached a stage where major size changes are hard to make, a design that is regarded as only marginally safe may be improved by some of the methods discussed in the next section.

7.4. IMPROVING FATIGUE PERFORMANCE

Just as important as design calculations are the practical steps that can be taken to minimise the possibility of a fatigue failure. These can be classified under two main headings – the correct shaping of parts and their surface treatment.

7.4.1. Correct shaping

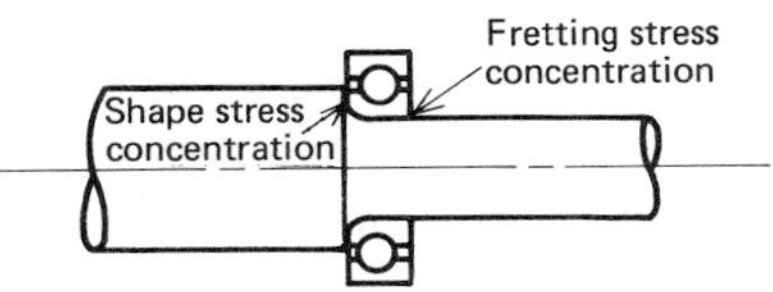

When a change of shape is required it should be made as gradually as possible, the radius of the transition curve being as large as possible. It should be noted that the omission of a change in shape is often not a complete solution; for example, when a bearing is to be fitted on a shaft, the omission of the shoulder will not remove the problem of a stress concentration completely: the inner ring of the bearing causes fretting at the shaft surface which will act as a stress raiser. The radius at the shoulder is governed by the corner radius on the bearing ring but should be as large as possible, should have a smooth finish free from machining marks and, for maximum strength, can be surface-treated by rolling as suggested in the next section. The fillet radius can sometimes be increased by machining into the shoulder as shown in the sketch, or even into the shaft diameter.

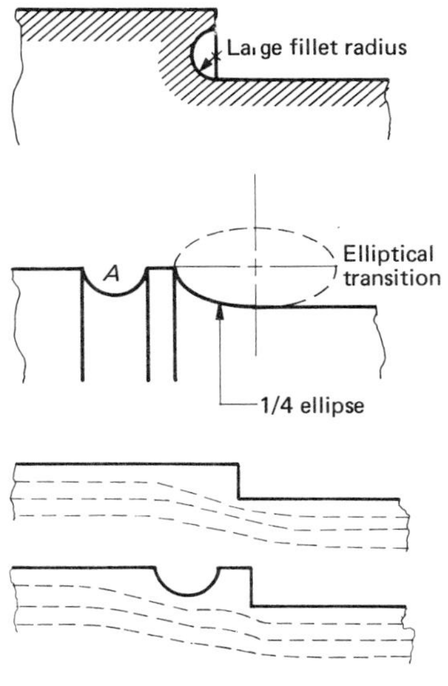

An improvement on the plain fillet is the elliptical transition and this may be further improved by machining a radius into the major diameter just behind the shoulder, at *A*. The radius should be placed so as to shepherd the stress lines away from critical region; the marginal sketch illustrates this, and it is suggested that in many cases a sketch of this type will enable the designer to improve the shape of a part by visualising how the stress lines can be induced to flow more smoothly. It must be remembered, however, that stress flow occurs in three dimensions, and also that torsional stresses are difficult to represent in this way.

Many examples are given in the literature on fatigue of the way in which various parts can be strengthened, and the reader should study these and develop a feel for the approach used.[1, 7, 12, 13] Typical of the approach is that taken to improve the fatigue performance of a bolt, the standard pattern of which is likely to fail at points *a*, *b* or *c* in the sketch. The stress concentration at *a* can be reduced by increasing the radius at the fillet. The weakness at *b* where the threads run out can be eliminated by blending the threaded part into a reduced diameter using a large radius. An improved bolt is shown in Fig. 7.8. The fillet radius at the head can be increased because of the smaller shank diameter, while the undercut nut gives more uniform transfer of load along the length of the threads. The static strength of the bolt may be slightly reduced, but its fatigue strength is enhanced. Note that the stiffness of the bolt has been

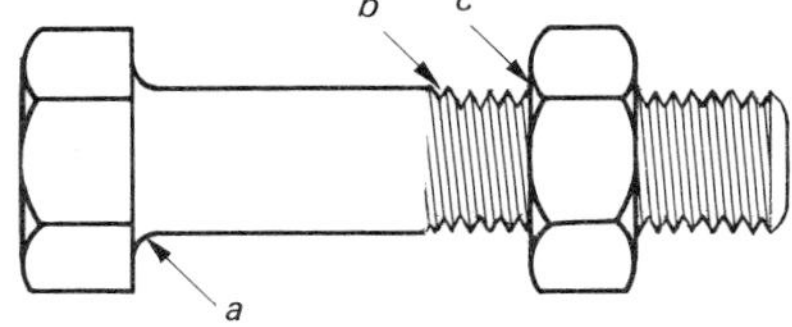

Example 7.7 Sheet 1 of 1

A beam in a welded steel structure has the following stress cycles superimposed on a steady stress of 100 MPa:

1. Varying stress ± 100 MPa, 5×10^5 cycles
2. Repeated stress $^{+}_{-}$ 150 MPa, 5×10^5 cycles
0

Use the Palmgren - Miner rule to check the safety of the structure; assume welding is to Class A, BS 153, take the data from Table 7.5

Loading 1. $f_{max} = 200$, $f_{min} = 0$, $\dfrac{f_{min}}{f_{max}} = 0$

From Table 7.5 by interpolation

$N_{1_t} = 1.1 \times 10^6$ cycles

Loading 2. $f_{max} = 250$, $f_{min} = 100$, $\dfrac{f_{min}}{f_{max}} = 0.4$

$N_{2_t} = 1.7 \times 10^6$ cycles

$$\frac{N_1}{N_{1_t}} + \frac{N_2}{N_{2_t}} = \frac{5}{11} + \frac{5}{17} = 0.74$$

Since this is less than 1.0 the beam would be considered safe.

Table 7.5. Values of design stress for Class A constructional welds.

f_{min}/f_{max} \ N	p or f_{max} tensile (MPa)					p or f_{max} compressive (MPa)				
	100 000 cycles	600 000 cycles	2 000 000 cycles	10 000 000 cycles	100 000 000 cycles	100 000 cycles	600 000 cycles	2 000 000 cycles	10 000 000 cycles	100 000 000 cycles
1.0										
0.8				335	318					
0.6		306	288	275	254					
0.4	294	265	248	232	210					
0.2	255	232	216	201	179					
0	227	207	193	177	156		−345	−322	−293	−256
−0.2	204	184	173	159	139	−286	−258	−241	−221	−193
−0.4	184	167	156	144	125	−227	−207	−193	−176	−155
−0.6	167	153	141	130	114	−190	−173	−161	−148	−130
−0.8	155	141	131	120	105	−162	−148	−139	−126	−110
−1.0	142	130	120	110	97	−142	−130	−120	−110	−97

Note: The ratio f_{min}/f_{max} is positive or negative respectively if the maximum and minimum stresses are of like or unlike sign.

Source: Extract from BS 153: 1954, *Steel Girder Bridges*, is reproduced by permission of the British Standards Institution, 2 Park Street, London, W1A 2BS.

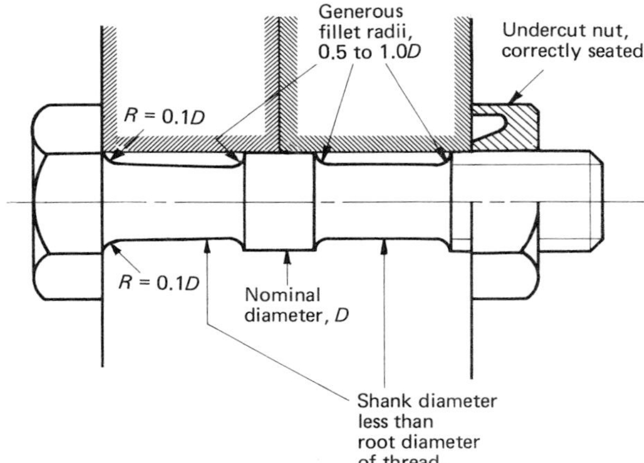

Fig. 7.8. *Design features of special bolt to improve fatigue life.*

reduced; this will mean that a smaller fraction of the alternating load on the joint will be transferred to the bolt (see Section 13.3.2).

7.4.2. Surface treatment

Two types of treatment will be considered here – the various types of surface protection used to prevent damage and the surface treatment used to induce a compressive stress at the surface.

The first type can be covered briefly: any treatment which reduces damage and corrosion will extend the life of the part under conditions where damage is likely. Cadmium plating is widely used for this purpose, but it should be noted that in itself it reduces the fatigue strength of steels – especially the higher strength steels. This effect is also evident in chromium and nickel plating and in galvanising. However, the strength reduction is small compared with that caused by the severe corrosion it is intended to prevent (see Fig. 7.7). Plating may also be used to reduce fretting effects.

The second type includes the various treatments used to set up a surface compressive stress; these are useful in that they provide an improvement without changes in the size of a part. The f_m–f_a diagram (Fig. 7.1) shows that improvements in strength or life can be expected if the mean stress is reduced or is shifted into the compression half of the diagram. This compressive stress may be induced at the surface (where most fatigue cracks are initiated) by

(1) shot peening,
(2) rolling,
(3) case hardening,
(4) nitriding.

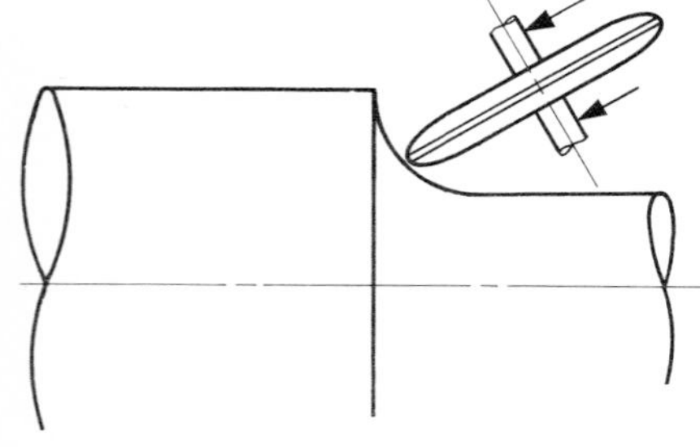

Shot peening. This process of bombarding the surface with chilled cast-iron shot will induce surface hardening and residual compressive stresses in a steel part. Notched sections show a gain of 20% or more in the fatigue limit and rough surfaces which are not to be machined will also benefit from this treatment. Any improvements will be lost after heat treatment.

Rolling. A deep layer of residual compressive stresses can be set up by pressing a roller onto a circular section as it is rotated. High contact forces are needed, such that the stresses, when calculated from the Hertzian stress equation (Table 3.1), are of the order of 1500 MPa or more. Strength improvements of from 30% to 60% can be obtained, but this is at a large number of cycles approaching the endurance limit. As with the previous method, little gain is made at high stresses and lower lives, probably because the yield strength is exceeded at notches and the favourable residual stress wiped out.

Case hardening. The quenching operation in this process induces compressive stresses at the surface which give large strength gains. The quenching process can cause distortion and is usually followed by machining (grinding). If excessive material is removed, the beneficial effects are lost.

Nitriding. In this process no quenching is required; lapping or honing will leave the surface with a greatly improved fatigue strength.

Once a designer has decided that an improvement may be gained by applying one of these methods, he should consult the literature for details of the most effective treatment.[1, 13]

7.5. FASTENINGS

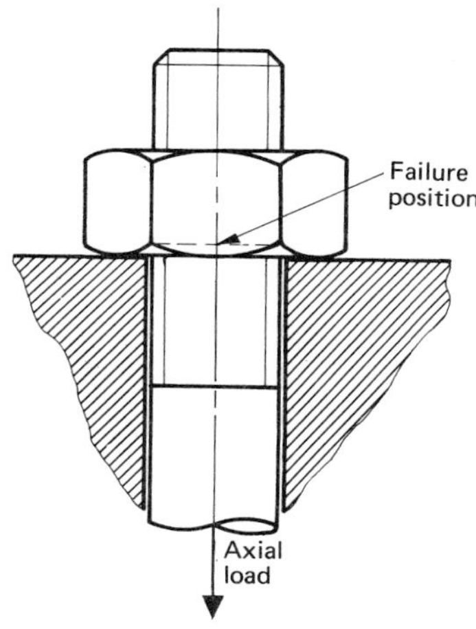

Because of the nature of the duty they perform and their location at points of high stress concentration, fastenings of various sorts are particularly susceptible to fatigue. We deal here briefly with bolted and welded fastenings.

It is difficult to assign meaningful values of K_t and K_A to a bolt, particularly for the most common failure location across the threads just inside the nut. The value of K_t for various types of threads differs, but a maximum value of 3.85 is quoted by Hetényi on the basis of photoelastic tests on a standard bolt-nut assembly.[14] The strength reduction occurring in an actual bolt will depend on the type of thread and the method of manufacture, and the stress will be affected by practical factors such as misalignment at the nut seating.

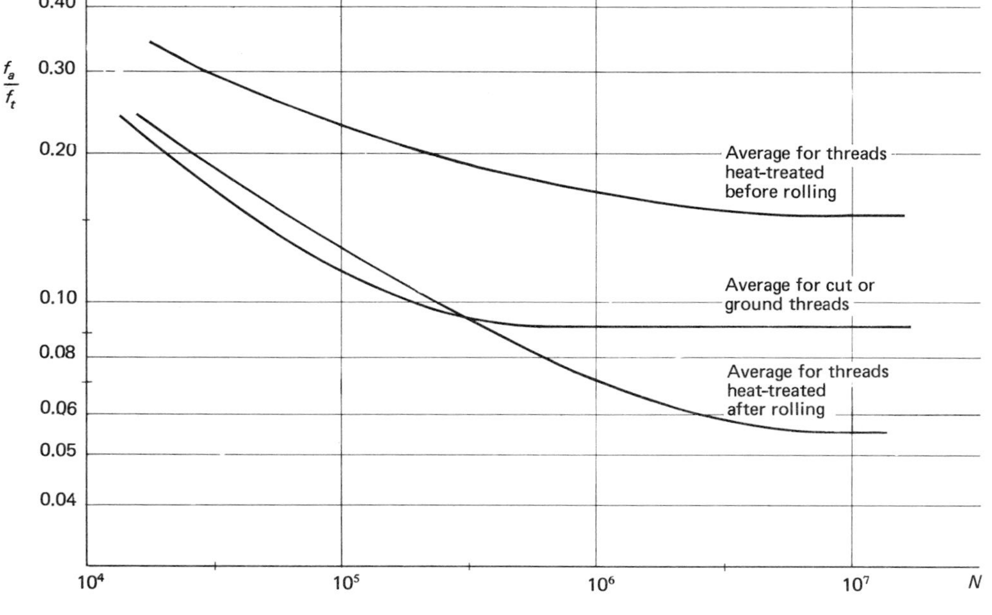

Fig. 7.9. *Fatigue strength of threads. (Adapted from "Engineering Sciences Data", Item 69001, by permission of the Engineering Sciences Data Unit.)*

Results from a large number of tests on bolts below 25 mm diameter are given in Engineering Sciences Data[15] and the mean curves drawn through the relatively large scatter bands are shown in Fig. 7.9. The ratio f_a/f_t for the standard production methods of rolling and cutting or grinding is 0.15 for one and 0.09 for the other at 10^7 cycles.* Repeated loading ($f_m = f_a$) was used for these tests, which indicate that the strength reduction factor ranges from 2.2 to 3.7 for the abovementioned processes. It appears that the beneficial residual stresses induced during the rolling operation are lost in heat treatment, and these threads showed a particularly low fatigue resistance.

It is pointed out in Chapter 13 that the alternating load on a bolt in a joint can be minimised by making the ratio (joint stiffness/bolt stiffness) as high as possible. There is clearly a good deal to be gained by attention to this and other detailed aspects of joint design. Elaborate strength calculations are not generally warranted, and preliminary design can be based on Fig. 7.9.

The design of welded joints for static loads is described in Chapter 13. Under fatigue conditions there are many stress raisers in a welded structure at which fatigue will be initiated. For example, the reinforcement at a butt weld in a flat plate, which may strengthen the plate in static loading, will weaken it in fatigue. The extent of this weakening has been shown by Gurney who plotted strength against reinforcement angle (Fig. 7.10).[12] Other tests have shown that the fatigue strength of the plate is not diminished if the reinforcement is machined flush with the plate surface. It is evident that the reduction of fatigue strength at a weld is due very largely to shape effects.

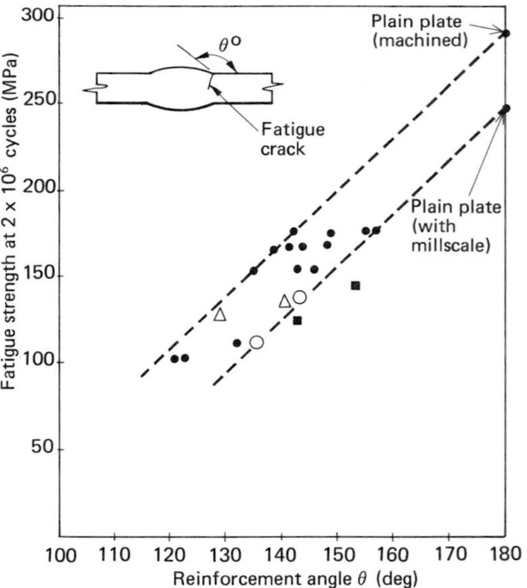

Fig. 7.10. *The relationship between reinforcement angle and fatigue strength of transverse butt welds. (Adapted from T. R. Gurney,* Fatigue of Welded Structures, *Cambridge University Press, 1968, by permission of Cambridge University Press.)*

It must be understood that the mere presence of a weld on a load-bearing plate will initiate fatigue; it is not necessary for the weld itself to carry the main load or any load. Thus temporary brackets and cleats must be located with care if the structure is to carry fluctuating loads.

*Assuming $K_M = K_A$.

We have noted that increased tensile strength does not always give an improvement in fatigue performance. Figure 7.11, taken from Gurney's *Fatigue of Welded Structures*[12], illustrates this point: there is no correlation between tensile strength and fatigue strength for transverse butt welds.

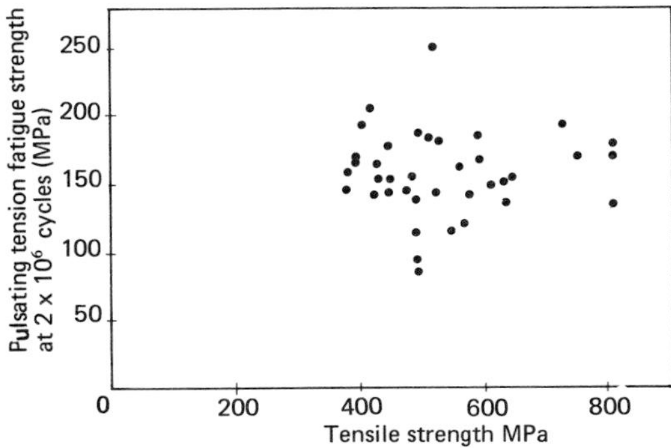

Fig. 7.11. *Relationship between fatigue strength of transverse butt welds and ultimate strength of parent material. (Adapted from T. R. Gurney,* Fatigue of Welded Structures, *Cambridge University Press, 1968, by permission of Cambridge University Press.)*

It is not possible in structural design to consider each weld individually as a possible stress raiser, and the procedure adopted in practice is to set stress levels that are in keeping with the methods used in manufacture. In BS 153, *Steel Girder Bridges*, seven classes of constructional details are described.[16] Class A, for which the highest design stresses are allowed, is for plain steel in the as-rolled condition with no gas cut edges, or for fabricated members in which the full penetration welds are machined or ground flush with the plate surface. At the other end of the scale, Classes F and G describe members with T-type butt welds, fillet welds that are load-carrying, or other intermittent welds which are not load-carrying. The allowable stresses for each class are tabulated in terms of the ratio f_{min}/f_{max}. Table 7.5 gives an extract from BS 153, showing the design stresses for Class A constructional details.

7.6. CONCLUSION

In this chapter some of the factors that are important in designing for fatigue have been discussed. No standardised design approach exists, and several different ones are described in the literature.[1, 2, 4, 6, 16] These should be examined so that the importance of the various factors that determine strength can be assessed.

Many other design aspects have not been mentioned. For example, the designer may accept that fatigue cracks will occur during the lifetime of a component, and design the structure so that sufficient static strength remains after the crack appears to ensure safety until the next planned inspection. This can be done by providing multiple paths for the loads through independent structural members. This "fail-safe" approach is adopted in aircraft design.[9, 11]

The accuracy of fatigue design calculations is not always good and, in some cases, approximations – such as taking the strength reduction factor as being equal to the stress concentration factor – are justified. Even in such cases, the more complete design calculations are important for the improved appreciation they give of the notch-sensitivity of different materials.

LIST OF SYMBOLS

A_0	ratio f_a/f_t
a_f	notch alleviation factor, Equation (7.18)
b	factor in Equation (7.23)
D	shaft diameter
f	normal stress
f_a	alternating stress
f_{an}	nominal alternating stress of notched specimen
f_{comp}	complementary stress
f_e	endurance limit in fatigue
f_{eq}	equivalent static stress
f_{max}	maximum stress $= f_m + f_a$
f_{min}	minimum stress $= f_m - f_a$
f_m	mean stress
f_{mn}	nominal mean stress of a notched specimen
f_t	tensile strength
f_{tn}	nominal static tensile strength of a notched specimen
f_y	yield strength
K_A	strength reduction factor at 10^7 cycles, alternating stress
K_a	strength reduction factor for finite life, alternating stress
K_F	surface finish factor, bending
K_{Fq}	surface finish factor, torsion
K_M	strength reduction factor, at 10^7 cycles, mean stress
K_m	strength reduction factor for finite life, mean stress
K_s	strength reduction factor, static loading
K_t	theoretical stress concentration factor
l	length
N_1, N_2	normal stress (Fig. 7.5)
$N_1, N_2,$	number of repetitions of stress cycle
N_{1_t}	number of repetitions allowed at given stress level
N_D	design value for N
N_L	$\log_{10} N$
n	safety factor
p	permissible design stress
q	notch sensitivity factor
q_e	endurance limit in shear
q_{eq}	equivalent static shear stress
q_A	factor defined by Equation (7.23)
q_a	alternating shear stress
q_{comp}	complementary shear stress
q_m	mean shear stress
q_M	factor defined in Equation (7.24)
R	radius at root of notch
γ	term in Equations (7.8), (7.9)
θ	angle

REFERENCES

[1] R. B. Heywood, *Designing Against Fatigue*, Chapman & Hall, London, 1962.

[2] C. Lipson and R. C. Juvinall, *Handbook of Stress and Strength*, Macmillan, New York, 1963.

[3] J. MARIN, *Engineering Materials*, Prentice-Hall, New York, 1952.
[4] ENGINEERING SCIENCES DATA, 71009, 71010, "Design Principles", *Design Against Fatigue*, Engineering Sciences Data Unit, London, 1971.
[5] G. SINES and J. L. WAISMAN (eds), *Metal Fatigue*, McGraw-Hill, New York, 1959.
[6] R. E. PETERSON, *Stress Concentration Design Factors*, John Wiley, New York, 1953.
[7] ENGINEERING SCIENCES DATA, 65004 *et seq.*, *Stress Concentration Data*, Engineering Sciences Data Unit, London, 1971.
[8] P. Y. KRAVCHENKO, *Fatigue Resistance*, Pergamon, Oxford, 1964.
[9] C. C. OSGOOD, *Fatigue Design*, Wiley Interscience, New York, 1970.
[10] J. A. POPE (ed.), *Metal Fatigue*, Chapman & Hall, London, 1959.
[11] W. BARROIS and E. L. RIPLEY (eds), *Fatigue of Aircraft Structures*, Pergamon, Oxford, 1963.
[12] T. R. GURNEY, *Fatigue of Welded Structures*, Cambridge University Press, London, 1968.
[13] R. CAZAUD, *Fatigue of Metals*, Chapman & Hall, London, 1953.
[14] M. HETÉNYI, "The Distribution of Stress in Threaded Connections", *Proceedings of the Society for Experimental Stress Analysis*, **1**, 1, 1943.
[15] ENGINEERING SCIENCES DATA, 69001, "Fatigue Strength of Steel Screws Under Axial Loading", *Design Against Fatigue*, Engineering Sciences Data Unit, London, 1971.
[16] BS 153:1954, "Stresses", Amendment No. 4, 1962, Part 3B, *Steel Girder Bridges*, British Standards Institution, London, 1954.
[17] J. H. OSCAR (ed.), *Metals Engineering Design* (ASME Handbook), 2nd ed., McGraw-Hill, New York, 1965.

BIBLIOGRAPHY

ENGINEERING SCIENCES DATA, "Fatigue", vol. 1; "Stress Concentrations", vol. 3, *In Stress and Strength*, Engineering Sciences Data Unit, London, various dates.
FROST, N. E. and MARSH, K. J., "Designing to Prevent Fatigue Failures in Service", Paper 7, *Proceedings of IME*, **184**, Part 3B, The Institution of Mechanical Engineers, London, 1969–70.

PROBLEMS

Unless otherwise stated, solutions to these problems should be based on the Goodman failure criterion.

7.1. Collect strength data for a range of hardened and tempered steels, and plot f_e and f_e/f_t against f_t. Differentiate between different types of steel (carbon, low-alloy and high-alloy). Check the accuracy of Equation (7.1) for the endurance limit for the various steels.

7.2. The design stresses for fatigue in Table 7.5 are given in terms of f_{max} and f_{min}/f_{max}. Plot the data for 6×10^5 and 10^8 cycles on an f_m–f_a diagram. If the information is for a Grade 43 steel (Table A17.2), draw the failure line on the same diagram. What value of safety factor will result from the use of the design stresses? What values of strength reduction factor can be tolerated?

7.3. Determine the maximum allowable fully-reversed bending moment at the shoulder on a shaft where the diameter steps from 40 mm to 30 mm, if a safety factor of 1.5 is required. The fillet radius at the shoulder is 2.0 mm and the steel properties are f_t = 550 MPa, f_y = 415 MPa, and f_e = 280 MPa.

7.4. A 30-mm bolt is subjected to a tensile mean load of 130 kN and alternating load of 50 kN. What strength of steel is required to give a safety factor of 1.5? Assume a strength reduction factor of 3.2 for the threads.

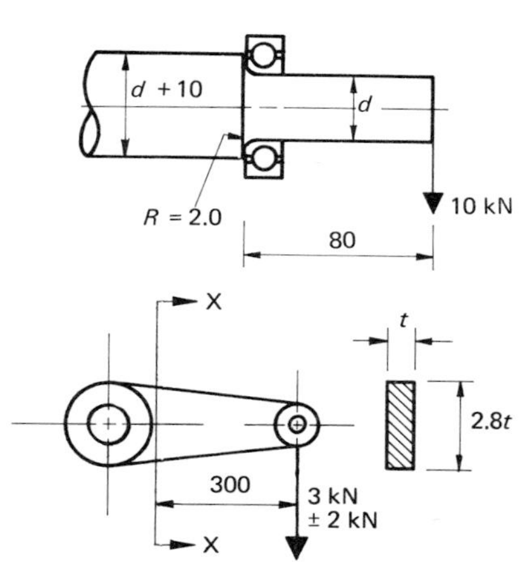

7.5. The overhung rotating shaft shown in the sketch carries a fixed load of 10 kN. Using a safety factor of 2.0 and properties for mild steel, determine a suitable shaft diameter.

7.6. The lever shown in the sketch carries a steady load of 3 kN and a superimposed alternating load of ±2 kN. Determine suitable dimensions for the rectangular section at XX if the lever is made of steel with tensile strength of 550 MPa. Use a safety factor of 1.5.

7.7. Repeat Problem 7.6 using an aluminium alloy having a tensile strength of 280 MPa, and assuming a life of 10^7 cycles. Is there any saving if a life of only 10^5 cycles is required?

7.8. A plain shaft is subjected to the following mean and alternating components of torque (T) and bending moment (M):

T_m = 4.2 kN m T_a = 1.4 kN m M_m = 6.0 kN m M_a = 3.0 kN m.

Determine the safety factor if the material is steel with yield strength of 360 MPa and endurance limit of 310 MPa. Use
(a) the maximum shear stress criterion, Equation (7.16), and
(b) the von Mises criterion, Equation (7.17).
Shaft diameter is 80 mm.

7.9. It is stated in BS 3580: 1964, *The Strength of Screw Threads*, that improved fatigue strength can be obtained in a bolt by using a nut which has a lower modulus of elasticity than the bolt. Explain the reason for this, and describe other ways in which the fatigue strength of bolted fastenings can be improved.

7.10. A plain rod is to be subjected to a fully reversed tension load of ±35 kN. It is designed to have a machined surface with a safety factor of 6 in a steel with a tensile strength, f_t, of 450 MPa. What value of safety factor might be expected if the rod became heavily corroded in service? Repeat the calculations for a high strength steel with f_t = 900 MPa.

7.11. A circular mild steel beam, 600 mm long, is simply supported at its ends and is subjected to a central pulsating load varying from 0 to 5 kN. What diameter bar would be required if it has a hot-rolled finish and an ultimate strength of 420 MPa? Would anything be gained by specifying a turned finish?

7.12. The double-acting piston of a large marine diesel engine is subjected to inertia and pressure forces that amount to ± 2.5 MN. The piston rod has a diameter of 250 mm and fails in fatigue at the threaded connection at the cross-head. It is known that the yield strength of the rod material is 300 MPa. Would the failure be expected?

7.13. Check the safety of a welded beam in fatigue for the following loads:
bending moment varying from 0 to 68 kN m for 3×10^5 cycles;
bending moment varying from 0 to 60 kN m for 2×10^6 cycles.
The modulus Z for the beam is 33×10^4 mm^3. Use stresses from Table 7.5, and apply the Palmgren–Miner rule.

8 Shaft Design

8.1. INTRODUCTION

The shaft is perhaps the most frequently used machine element that has to be designed in detail – the bearings in which it rotates can often be selected directly from a catalogue. A shaft usually rotates and transmits power, although in some cases the rotation is limited and it will simply transmit a torque or axial force. Bending loads are often superimposed on the torsional loads.

It might be expected that well-established design codes would exist for an item so commonly encountered: in fact, although these do exist, none of them is suitable for general application and the designer will frequently have to start from first principles. The lack of an all-embracing standard is explained by the great diversity in shape, size and loading conditions met in practice.

Shaft design will be based on one or more of three limiting conditions: static loading, deflection and fatigue loading. The first two are considered in detail in this chapter; the third has been discussed in Chapter 7, but a code for fatigue design will be examined here. Since the shaft size may be predicated by the bearing size rather than strength considerations, the design of the two components must go hand in hand. Bearing design is taken up in Chapter 9. The shape will also be affected by other parts that are fixed to the shaft – keys and couplings, for example – and these will be dealt with briefly at the end of this chapter.

8.2. STATIC LOADING

The methods used to support shafts and to transmit torque to them often result in bending as well as torsional loads. A general design method will also allow for axial loads. At the outset the designer must decide if he is going to design for fatigue or not. If not, the loading is treated as a static loading case. We consider fatigue loading in the next section.

8.2.1. Design from first principles

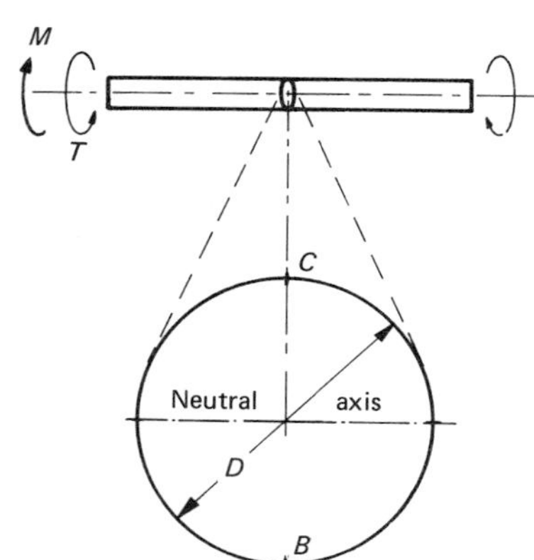

Shafts are usually made from a ductile material and the limiting condition will be yielding under static load. If the bending moment, torsional moment and axial force are designated M, T and F respectively, the stresses for a circular shaft of diameter D shown in the sketch are:

$$f_{bt}, f_{bc} = \frac{M\,D/2}{I} = \frac{32M}{\pi D^3}$$ tension or compression, maximum at locations B and C furthest from neutral axis; (8.1)

$$q = \frac{TD/2}{J} = \frac{16T}{\pi D^3}$$ maximum at extreme fibre – the outer circumference; (8.2)

$$f_a = \frac{4F}{\pi D^2}$$ tension or compression, uniformly distributed across cross-section.

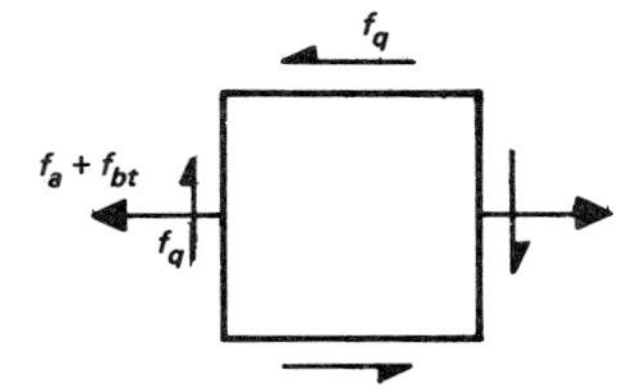

Note that direct shear stresses are not taken into account here; these stresses reach a maximum at the neutral axis of the shaft and are zero where f_b is a maximum; they can be neglected when considering the conditions at B and C but should be added to q if conditions at the neutral axis are being checked.

The stresses are shown in the sketch as they act on an element of material for the case where f_b and f_a are both positive; the resulting Mohr's circle construction is also illustrated. Since, for a ductile material, we are interested in the maximum shear stress that exists, this can be measured directly from the diagram or calculated from the geometry. Thus, the radius of Mohr's circle $r = q_{max}$ and

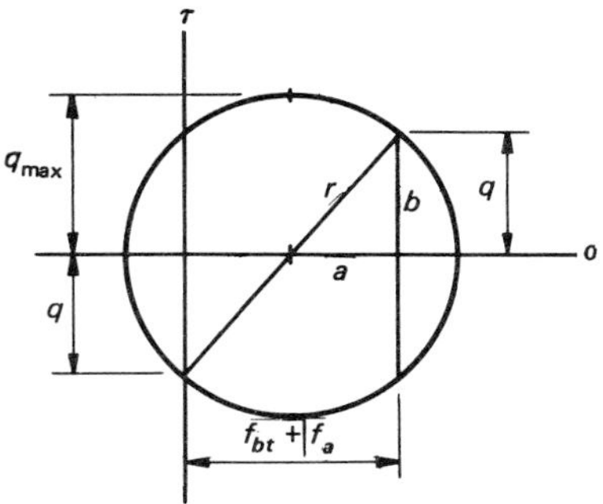

$$q_{max} = (a^2 + b^2)^{\frac{1}{2}}$$

or

$$q_{max} = \sqrt{\left(\frac{f_{bt} + f_a}{2}\right)^2 + q^2}$$

$$= \frac{16}{\pi D^3}\sqrt{\left(M + \frac{FD}{8}\right)^2 + T^2}$$

More generally, for a hollow shaft having outer and inner diameters D_o and D_i respectively, putting $D_i/D_o = k$,

$$q_{max} = \frac{16}{\pi D_o{}^3}\sqrt{\left\{\left(M + \frac{FD_o(1 + k^2)}{8}\right)^2 + T^2\right\}\left(\frac{1}{1 - k^4}\right)} \tag{8.3}$$

This equation provides the basis for design, and can be used with a safety factor, n, and some maximum expected values of M and T to find the diameter; putting $q_{max} = f_y/2n$, i.e. using the maximum shear stress theory and a factor of safety, n, and considering a solid shaft, we have

$$D = \left(\frac{32n}{\pi f_y}\sqrt{\left(M + \frac{FD}{8}\right)^2 + T^2}\right)^{\frac{1}{3}} \tag{8.4}$$

In many applications the term $FD/8$ is small or zero. If it is not, the equation is readily solved for D by a trial-and-error or graphical solution. If the term is large and the shaft long, the possibility of a buckling failure must be investigated (see Chapter 12). In general, if the ratio of the length of the shaft between supports to the radius of gyration is less than 20, buckling need not be considered in this type of loading.

8.2.2. ASME Code

Although this code was first put forward in 1927 and withdrawn in 1955, it will no doubt be encountered for some time to come. Equation (8.1) is modified by the inclusion of shock and fatigue factors and a permissible shear stress p_q specified to give

$$D_o = \left(\frac{16}{\pi p_q}\sqrt{\left(K_b M + \frac{FD}{8}\right)^2 + (K_t T)^2}\right)^{\frac{1}{3}}$$

When commercial steel shafting is used, the recommended value of p_q is 55 MPa; when a steel under specification is used, the value is the lower of 0.18 f_t and 0.3 f_y. In all cases the allowable stress is reduced by 25% when there is a keyway at the point where the shaft is being designed.

To the extent that a lower stress is used in the presence of a keyway, the equation could be said to account for fatigue; however, no other stress raisers are allowed for and it remains basically a static design equation. It is included here for the sake of completeness and because it is convenient for preliminary design work. Recommended values for K_b and K_t are given below. They should be selected with care, since some safety margin has been applied already in determining the allowable stress p_q.

LOADING	K_b	K_t
Stationary shafts:		
Load gradually applied	1.0	1.0
Load suddenly applied	1.5–2.0	1.5–2.0
Rotating shafts:		
Load gradually applied	1.5	1.0
Load suddenly applied, minor shock	1.5–2.0	1.0–1.5
Load suddenly applied, heavy shock	2.0–3.0	1.5–3.0

The full code also gives factors which take into account the end fixity of the shaft when the axial loads are not negligible. We will not discuss them here.

The use of the ASME Code is illustrated in Example 8.1.

8.2.3. Use of von Mises criterion

If the maximum-distortion energy theory is used, the basis of a design equation is

$$p^2 = \left(\frac{f_y}{n}\right)^2 = f_x{}^2 - f_x f_y + f_y{}^2 + 3q_{xy}{}^2$$

For a combination of bending and torsion this becomes

$$\left(\frac{f_y}{n}\right)^2 = f_{bt}{}^2 + 3q^2$$

Substituting Equations (8.1) and (8.2) gives

$$\frac{f_y}{n} = \frac{32}{\pi D^3}\left(M^2 + \frac{3}{4}T^2\right)^{\frac{1}{2}} \tag{8.5}$$

or

$$D = \left[\frac{32n}{\pi f_y}\left(M^2 + \frac{3}{4}T^2\right)^{\frac{1}{2}}\right]^{\frac{1}{3}} \tag{8.6}$$

This is the equivalent to Equation (8.4) with the axial load omitted. It is used in some codes in preference to Equation (8.4). For example, the design equation given in the Australian Standard, AS B249, *Design of Shafts for Cranes*

Example 8.1 Sheet 1 of 1

Use the ASME code to design a shaft to withstand bending moment 190 Nm, torque 130 Nm

Material: steel with $f_t = 700$; $f_y = 520$ MPa
Rotating shaft; minor shock: use $K_b = 2$, $K_t = 1.5$

p_q = lesser of 0.18×700, 0.3×520

$p_q = 156$

$$D = \left\{\frac{16 \times 1000}{\pi \times 156}\left[(2 \times 190)^2 + (1.5 \times 130)^2\right]^{1/2}\right\}^{1/3} = 24.0 \text{ mm}$$

Check the safety factor obtained by using Equation 8.4

$$n = \frac{f_y \cdot \pi \cdot D^3}{32}\left[M^2 + T^2\right]^{-1/2}$$

$$= \frac{520 \times \pi \times 24.0^3}{32 \times 1000}\left[190^2 + 130^2\right]^{-1/2} = 3.07$$

$n = 3.07$

This high safety factor results from the low design stress in shear – 156 – compared with the shear strength of the material – $\frac{520}{2}$ – according to the max. shear stress theory, and the use of the factors K_b and K_t

Note: in presence of a keyway design stress would have been $156 \times 0.75 = 117$ MPa, giving an overall safety factor of $3.07/0.75 = 4.09$

and Hoists[1], specifies the following equation for cases where fatigue is not a limiting condition

$$D^3 = \frac{20}{f_y}\left[\left(M_p + \frac{F_p D}{8}\right)^2 + \frac{3}{4}T_p^2\right]^{\frac{1}{2}} \tag{8.7}$$

Comparing this with Equation (8.6), it will be seen that a value of n just under 2 has been used. This relatively low safety factor is justified in the standard by the use of peak values of loads when calculating M_p, T_p and F_p.

8.3. FATIGUE LOADING

The designer must choose between working from the methods and equations developed in Chapter 7 and using a code. The Australian Standard[1] quoted above is one of the few examples of such a code. Because of the complexity of the problem, it is difficult to produce a standard which is suitable for general design use, and this one relates specifically to shafts for cranes and hoists. The code results can be compared with the design methods used in Chapter 7.

The two relevant design equations in AS B249 are

$$D^3 = \frac{12C_sK}{f_e}\left(M_p^2 + \frac{3}{4}T_p^2\right)^{\frac{1}{2}} \tag{8.8}$$

$$D^3 = \frac{12}{f_e}\left((C_sKM_p)^2 + \frac{3}{16}[(1 + C_sK)T_p]^2\right)^{\frac{1}{2}} \tag{8.9}$$

We have omitted the axial load term and used our notation where applicable. Equation (8.8) is used for shafts of power-operated drives generally, while Equation (8.9) is intended for drives in which the torque may be considered non-reversing. In a discussion of the code, Borchardt makes it clear that while both equations are based on fully reversed bending, the first is suitable for fully or partially reversed torque, and the second is for cases where the torque fluctuates between zero and some value up to the maximum.[2] Comparing these equations with Equation (8.4), it will be found that n here has a value of 1.2. The low safety factor is justified, again, by requiring the detailed calculation of peak loads (for example, those due to inertia effects in speeding up and braking). The other factors in the equations are C_s (see Fig. 8.1), a size factor which takes care of the size effect discussed in Chapter 7, and K, which is a "stress-raising factor" equivalent to the strength reduction factor, K_a. The values of these factors are obtained from charts (see Figs 8.2 to 8.5). Figure 8.3 provides a correction factor, Δ, which, when added to the ratio R/D, is used in conjunction with the tensile strength to find K for a shaft with a shoulder from Fig. 8.2. Figures 8.4 and 8.5 give K values for a variety of stress concentrations; those in Fig. 8.4, for example, relate to the stress raiser caused by the fretting action of a bearing ring on the shaft.

The code is used in Example 8.2 for the design of a shaft subjected to fatigue loading.

8.4. SLOPE AND DEFLECTION

Slope or deflection may be a limiting condition in the design of shafts and beams, and although the discussion of their calculation is centred on shaft design, the methods developed here are for general application. The limits on slope and

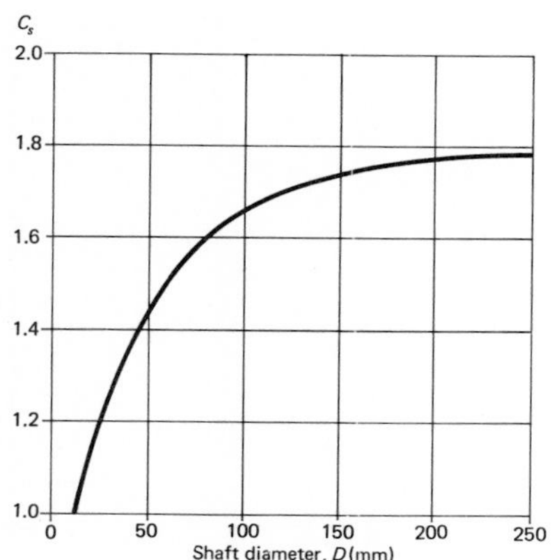

Fig. 8.1.

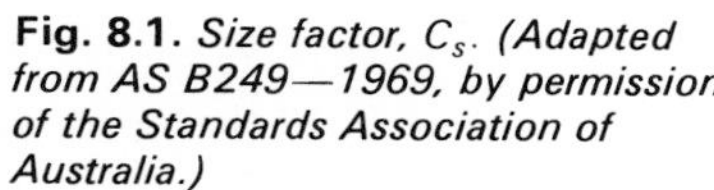

Fig. 8.1. *Size factor, C_s. (Adapted from AS B249—1969, by permission of the Standards Association of Australia.)*

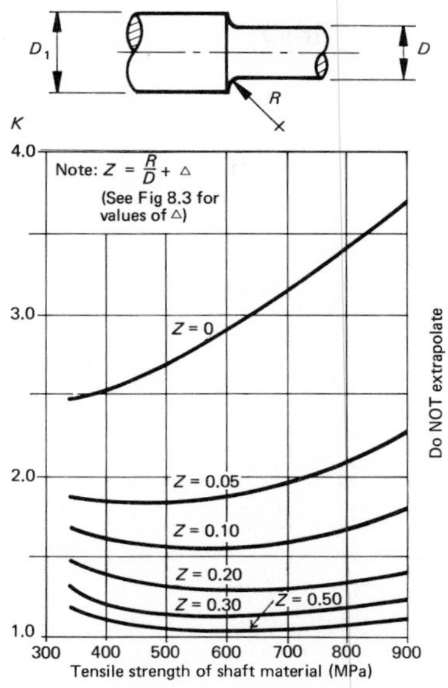

Note: For annular grooves select value of K from the $Z = 0$ curve. The diameter thus calculated is the root diameter of the groove.

Fig. 8.2.

Fig. 8.2. *Stress-raising factor, K, for stepped shafts and annular grooves. (Adapted from AS B249—1969 by permission of the Standards Association of Australia.)*

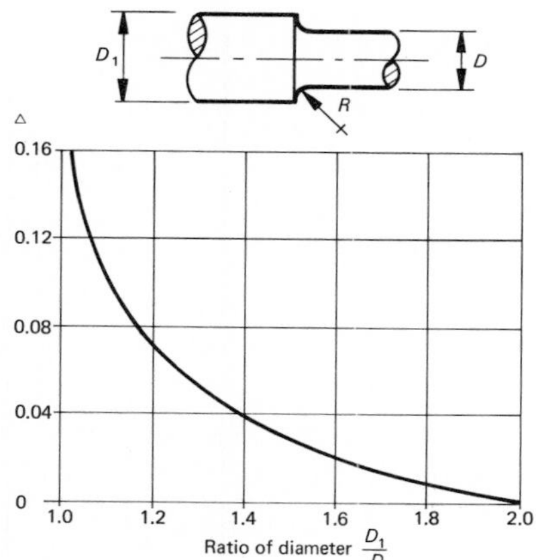

Fig. 8.3.

Fig. 8.3. *Correction factor, Δ. (From AS B249—1969, by permission of the Standards Association of Australia.)*

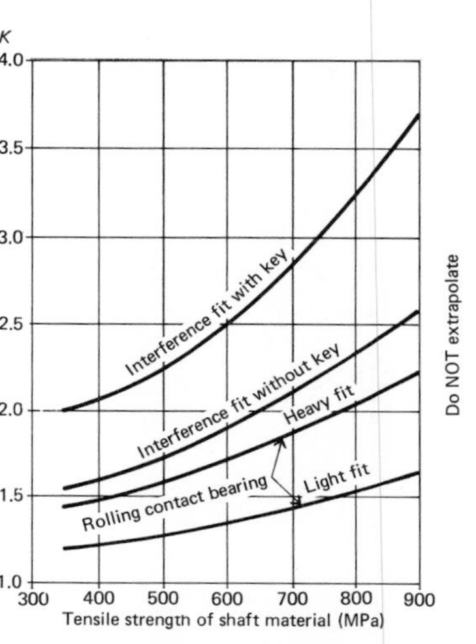

Note: Radii at the bottoms of keyways to be not less than those specified in BS 46: Part 1, "Keys and Keyways".

Fig. 8.4.

Fig. 8.4. *Stress-raising factor, K, for shafts with fitted components. (Adapted from AS B249—1969, by permission of the Standards Association of Australia.)*

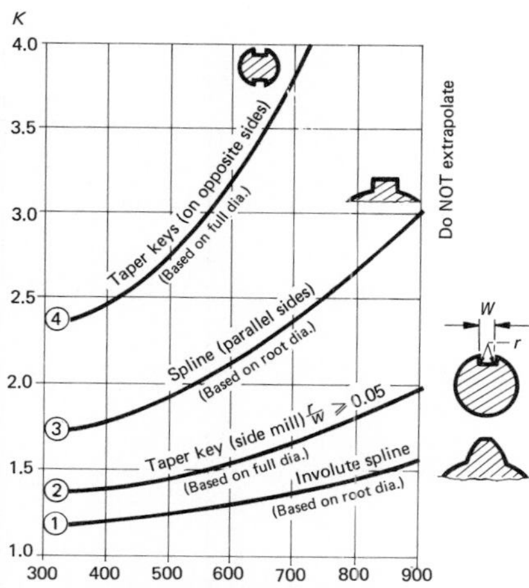

Fig. 8.5.

Fig. 8.5. *Stress-raising factor, K, for shafts with splines or keyways. (Adapted from AS B249—1969, by permission of the Standards Association of Australia.)*

Example 8.2 Sheet 1 of 1

Use AS B249 to determine the required endurance limit for steel for a shaft, given the following peak loads :

M_p = 210 Nm (fully reversed bending)
T_p = 160 Nm (fully reversed torque)

Shaft dimensions as shown in sketch

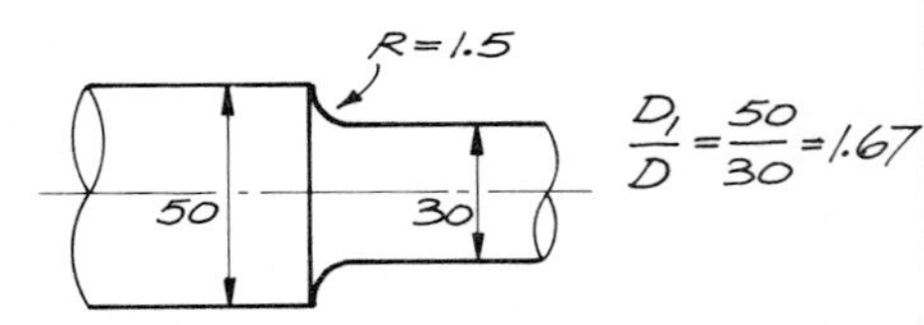

$\frac{D_1}{D} = \frac{50}{30} = 1.67$

Take $f_e = 0.45 f_t$
Try $f_t = 480$ MPa
$f_e = 216$

From Figure 8.1, size factor $C_s = 1.22$
From Figure 8.3, correction factor $\Delta = 0.02$ for $\frac{D_1}{D} = 1.67$
From Figure 8.2, with $z = \frac{R}{D} + \Delta = 0.05 + 0.02 = 0.07$

$$K = 1.70$$

From Equation (8.8),

$$f_e = \frac{12 \times 1.22 \times 1.7}{3.0^3} \sqrt{\left(210^2 + \frac{3}{4}(160^2)\right)} = 232$$

i.e. Steel required $f_e = 232$
$f_t = 515$

Check initial assumption : $f_t = 480$. The change to $f_t = 515$ will not change K significantly
Specify steel $f_t = 515$ MPa.

deflection in machine design usually come from a need to preserve clearances or uniformity of loading between moving parts. Gear teeth are sensitive to changes in slope and clearance, while the load capacity of a journal bearing will be affected if the clearance is reduced by a change in the slope of the shaft. Limits are sometimes laid down in codes and catalogues, and in other cases the designer will have to work out reasonable values. For example, in the design of a pump the impeller and seal clearances may fix the allowable deflections on the basis that metal-to-metal contacts must not occur. Roller bearings are particularly sensitive to changes in slope, and the limit may be as low as 2 minutes of arc. This value would also be typical of the allowable relative slope for shafts carrying spur gears.

The equation for flexure of a beam or shaft is

$$\frac{d^2y}{dx^2} = -\frac{M}{EI} \tag{8.10}$$

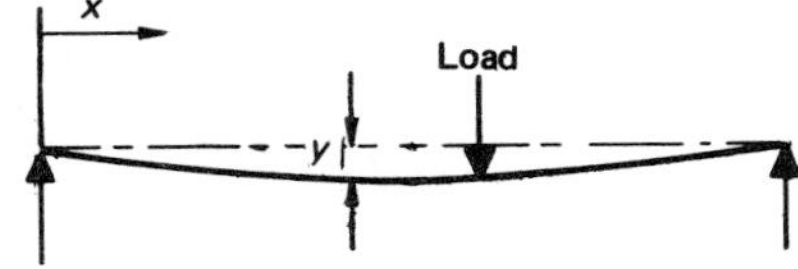

where the deflection y is positive downwards. The first integration of this equation gives dy/dx or θ, the slope, and the second gives y. When the value of EI is constant along the shaft, there are several suitable methods for determining the deflection, and many standard loading cases are given in structural handbooks. A selection is given in Table A2.1 (see Appendix A2).

It is often difficult to integrate Equation (8.10) from first principles, especially if I changes, and there are several alternative approaches. One is the moment-area method (see Chapter 15). We will describe here a graphical integration method which gives slope and deflection in successive integrations. We will show how the graphical integration is done, and then give examples of the method, which is quite simple to use in practice[3].

The curve $y = f(x)$ on axes OX, OY, in Fig. 8.6(a) is to be integrated, and the resultant integral curve $\int_0^{x'} y\,dx$ to be plotted on the axes $O'X'$, $O'Y'$ in Fig. 8.6(b). The curve is drawn to scales S_h, S_v on the horizontal and vertical axes, and is divided into segments de, ef at intervals on the abscissa that do not need to be equal. We consider the areas A, B, on Fig. 8.6(a) which represent the value of the integral between the selected ordinates. These ordinates da, eb, fc are produced downwards to cut the $O'X'$ axis in a', b', c'.

From the centre point of each segment de, ef a line is drawn horizontally to meet the OY axis in g, h. Selecting any convenient pole, P, on XO produced, draw Pg, Ph. If $O'e'$ is drawn parallel to Pg, and $e'f'$ parallel to Ph, the line $a'e'f'$ is an approximation to the required integral curve, to an as yet unknown scale. This is shown as follows. Area A can be approximated by $ag\,ab$. Since the triangles $a'e'b'$ and Pga are similar,

$$\frac{ag}{p} = \frac{b'e'}{a'b'}$$

where p is the pole distance OP.
Thus

$$b'e' = \frac{ag\,ab}{p} \approx \frac{\text{area } A}{p} \tag{8.11}$$

and, in the same way

$$c'f' - b'e' = \frac{\text{area } B}{p} \quad \text{and} \quad c'f' = \frac{\text{area } (A + B)}{p}$$

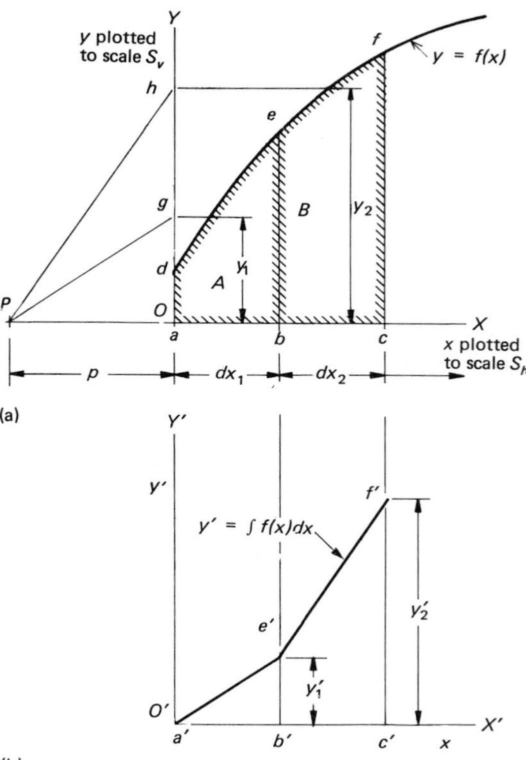

Fig. 8.6. *Graphical integration.*

In deriving the scale factor we must distinguish between the lengths *ag*, *ab*, etc., measured on the diagram (in mm) and the actual values of the variables y and x that they represent. The scale factors are used to convert, for example, ag (mm) to $ag\ S_v = y_1$ (units).

The scale factors are determined below. The error involved in putting $A = ag\ ab$, or $\int f(x)dx = ydx$, is neglected.

$$A = ag\ ab = \frac{y_1}{S_v}\frac{dx_1}{S_h}$$

$$y_1 dx_1 = A\ S_v\ S_h$$

In Fig. 8.6(b), $b'e' = y'_1/S'_v$ so that the required scale factor, $S'_v = y_1'/b'e'$

$$y_1' = \int_{x=a}^{x=b} f(x)dx = y_1 dx_1$$

$$\therefore\ S'_v = \frac{y_1 dx_1}{b'e'} = \frac{AS_vS_h}{b'e'}$$

and from Equation (8.11), $S'_v = S_vS_h p$.

The calculation of the scale factors is clarified when an actual example is worked out.

In Example 8.3, the scale chosen for x, the distance along the beam, is 1 mm = 40 mm, i.e., S_h = 40 mm/mm. This scale factor remains constant in the load, bending moment, slope and deflection diagrams. In the BM diagram, the variable on the ordinate has been taken as M/EI instead of M. This is not strictly necessary in the problem, since EI is constant; however, the merit of the

Example 8.3 Sheet 1 of 2

Find the slope at the right-hand end and the deflection at the centre of the circular steel shaft shown in the sketch. Compare the deflection with the value calculated from the formula in Table A2.

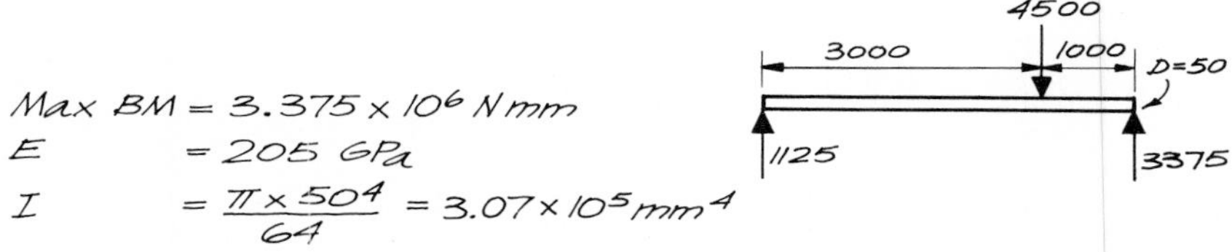

Max BM $= 3.375 \times 10^6$ N mm

$E = 205$ GPa

$I = \dfrac{\pi \times 50^4}{64} = 3.07 \times 10^5 \text{ mm}^4$

For maximum BM, $\dfrac{M}{EI} = \dfrac{3.375 \times 10^6}{205 \times 3.07 \times 10^8} = 5.37 \times 10^{-5}$ 1/mm

Select scales : S_h, 1mm = 40 mm

For $\dfrac{M}{EI}$ S_v, 1mm = 1×10^{-6} 1/mm

Draw slope and deflection diagrams

$S_\theta = 2 \times 10^{-3}$ rad/mm

Maximum slope $= 32.5 \text{ mm} \times 2 \times 10^{-3} \text{ rad/mm} = 0.065$ radians

$S_y = 4.0$ mm/mm

Deflection at centre = 16.5 mm × 4.0 mm = 66 mm

From Table A2

Deflection at $x = \dfrac{Pbx}{6EI\ell}(\ell^2 - b^2 - x^2)$

$$= \frac{4500 \times 1 \times 2\,(16 - 1 - 4) \times 10^{12}}{6 \times 205 \times 10^3 \times 3.07 \times 10^5 \times 4000}$$

$= 65.4$ mm

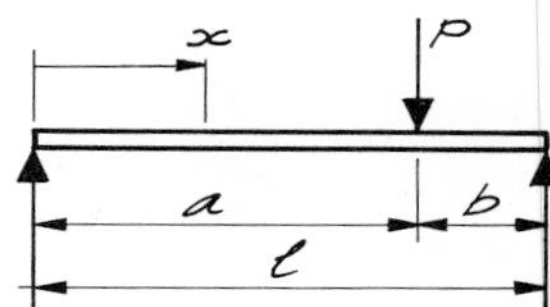

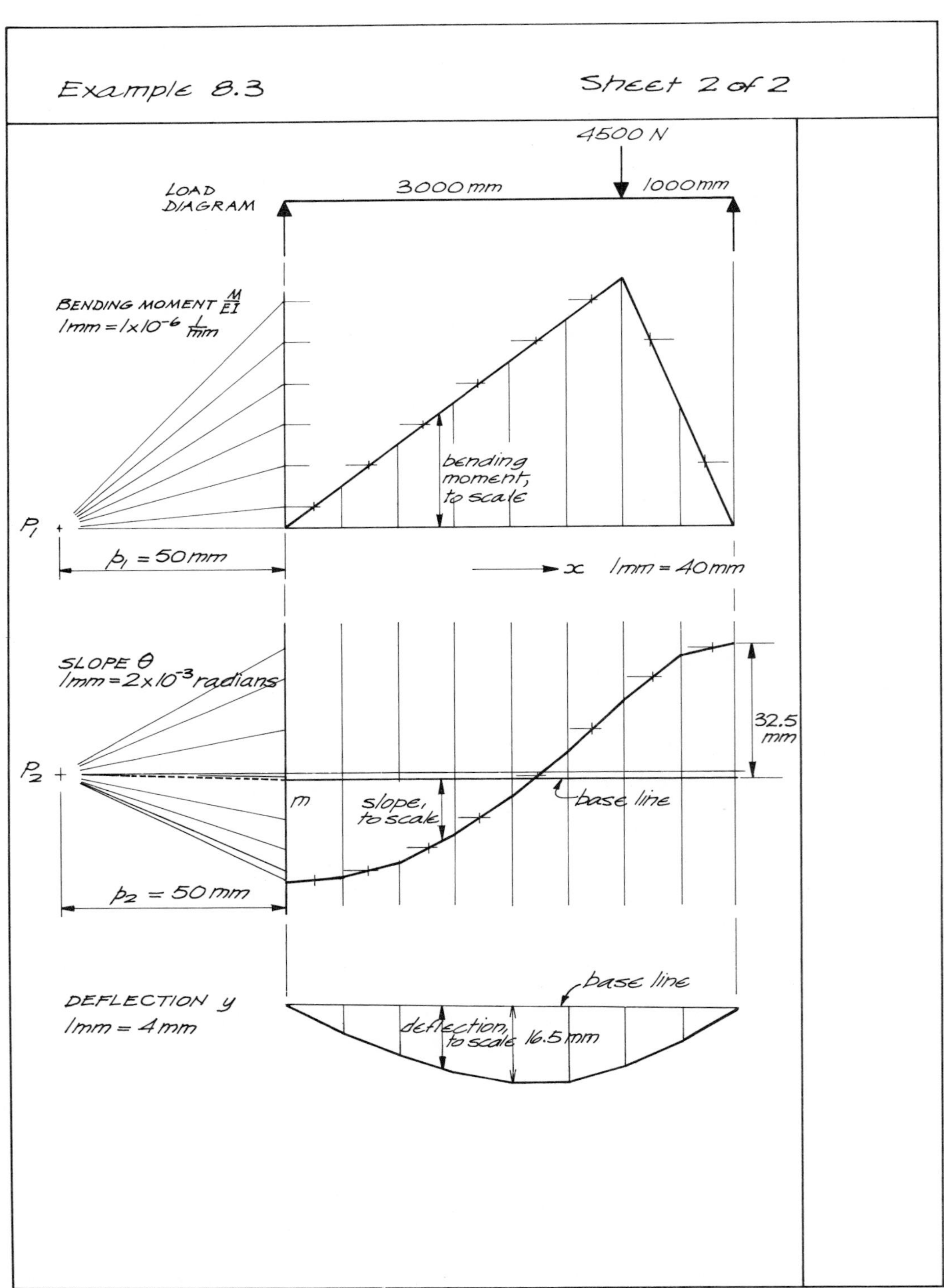
Example 8.3
Sheet 2 of 2
4500 N
LOAD DIAGRAM
3000 mm
1000 mm
BENDING MOMENT M/EI
1mm = 1×10⁻⁶ 1/mm
bending moment, to scale
P₁
p₁ = 50 mm
x 1mm = 40 mm
SLOPE θ
1mm = 2×10⁻³ radians
32.5 mm
P₂
m
slope, to scale
base line
p₂ = 50 mm
DEFLECTION y
1mm = 4 mm
base line
deflection, to scale
16.5 mm

graphical method is that it can be used when I varies, and we take M/EI as the variable here to illustrate the approach. The scale chosen is 1 mm = 1×10^{-6} 1/mm, i.e., $S_v = 1 \times 10^{-6}$ 1/mm^2. The pole distance p_1 in the M/EI diagram is 50 mm, so that for the slope diagram,

$$\begin{aligned} S_\theta &= S_v\, S_h\, p \\ &= 40 \times 1 \times 10^{-6} \times 50 = 2 \times 10^{-3} \text{ rad/mm} \end{aligned}$$

In the same way the scale for deflection

$$\begin{aligned} S_y &= S_\theta\, S_h\, p \\ &= 2 \times 10^{-3} \times 40 \times 50 = 2.00 \text{ mm/mm} \end{aligned}$$

Note that in the θ and y diagrams, the base line is not initially known; it can be drawn on the y diagram from the knowledge that $y = 0$ at each end of the beam. If a line P_2m in the graphical construction in Example 8.3 is drawn parallel to this base line, the base line for θ is the horizontal line through m. Clearly, the graphical integration does not require that the pole lies on the datum line.

A rather wide spacing of ordinates has been adopted in the worked examples for clarity. Even with this spacing, sufficient accuracy for many design problems is obtained.

In Example 8.4 the deflection for a stepped shaft is calculated using this method.

8.5. KEYS, SPLINES AND SHRINK-FIT COLLARS

In this section we describe methods used to transmit torque between a shaft and some fitting on the shaft – such as a pulley or gear – which we will refer to as the *hub* or *collar*.* It is difficult to transmit torque in this manner without setting up a stress concentration and, as the loads become heavier, more attention must be paid to the form of the drive. It is necessary that care be taken to ensure that both the shaft and collar are strong enough for the required duty – we are concerned at this stage mainly with the interface between them.

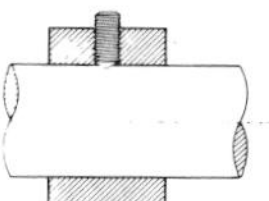

Very light torques may be transmitted by means of a set screw with a hardened point which is tightened so as to grip the shaft. A pin driven through both shaft and collar can also be used, and in this case can be designed so that it will protect the drive unit by shearing in the event of excessive loads.

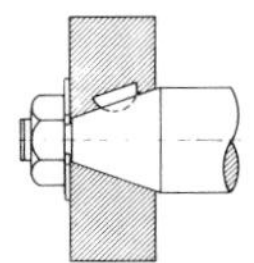

The Woodruff key is used for light loads, and lends itself to the location of a hub on a tapered shaft with some particular angular orientation.[4] When the hub is tightened on the taper, friction between it and the shaft will transmit most of the torque. This is a good arrangement, even though the key is usually designed like a rectangular key (see below) to transmit the full torque.

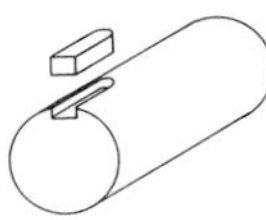

The square or rectangular key, which is most commonly used for medium to high torques, is shown in the sketch. It consists of a steel bar which fits into a keyway cut in the shaft and hub.[5] The parallel key has a top clearance in the hub but fits closely at the sides. The shaft fit is usually tighter than the hub fit, and it is possible to arrange for axial freedom of the hub; this type is best suited to unidirectional drives. Heavy unidirectional or reversing drives can be transmitted by a key which has a top taper and is driven into the keyway to give a

*The terms "hub" and "collar" are used here interchangeably.

Example 8.4 Sheet 1 of 2

Draw the deflection diagram for the shaft shown in the sketch and find the maximum deflection.

30 mm dia shaft, $I = 7.85 \times 10^3$
40 mm dia shaft, $I = 39.8 \times 10^3$

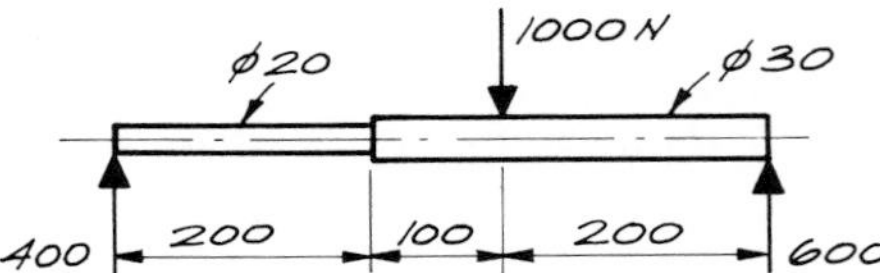

Select S_h, 1mm = 5 mm

Draw BM diagram, scale 1mm = 4000 N mm

Draw $\frac{M}{EI}$ diagram, scale 1mm = 1×10^{-6} 1/mm

Draw slope θ diagram, scale = $5 \times 30 \times 10^{-6}$
= 1.5×10^{-4} rad/mm

Draw deflection y diagram, scale = $5 \times 30 \times 1.5 \times 10^{-4}$
= 2.25×10^{-3} mm/mm

Maximum deflection occurs when slope is zero: 25 mm scaled off diagram

Maximum deflection = $25 \times 2.25 \times 10^{-3}$
= 0.56 mm

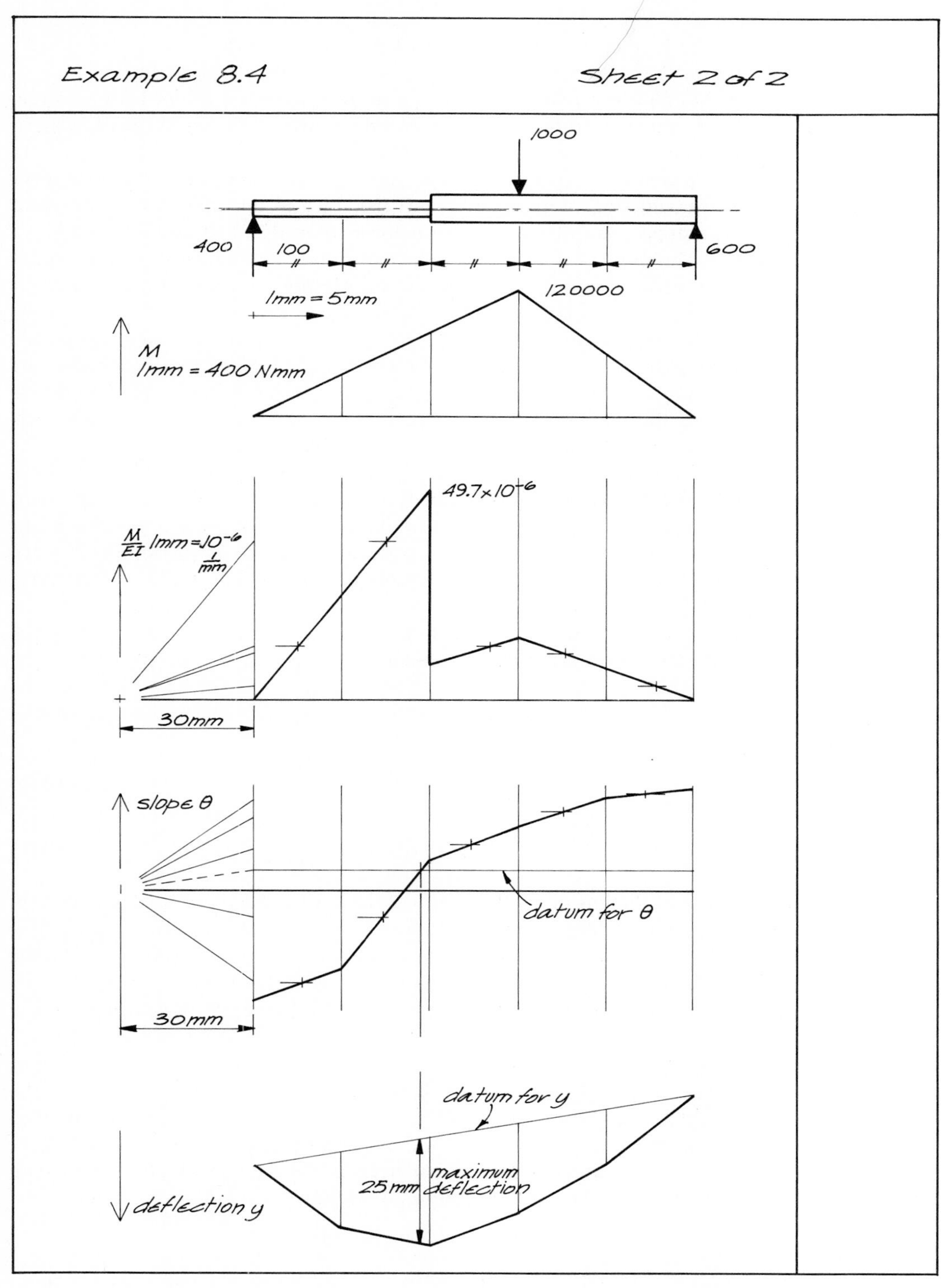
Example 8.4
Sheet 2 of 2
1000
400
100
600
1mm = 5mm
120000
M
1mm = 400 Nmm
49.7×10⁻⁶
M/EI 1mm = 10⁻⁶ 1/mm
30mm
slope θ
datum for θ
30mm
datum for y
maximum deflection
25mm
deflection y

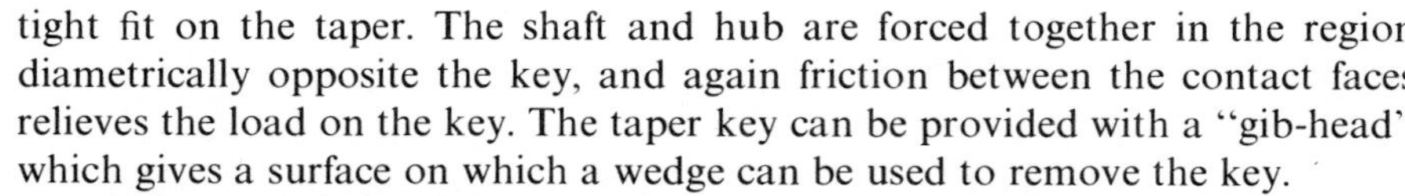

tight fit on the taper. The shaft and hub are forced together in the region diametrically opposite the key, and again friction between the contact faces relieves the load on the key. The taper key can be provided with a "gib-head" which gives a surface on which a wedge can be used to remove the key.

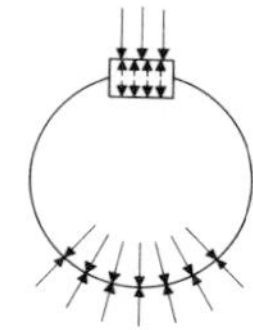

Although it is possible to fit two or more keys to a shaft for heavy torque, it is better practice to resort to splines, which are in effect multiple keyways. The sizes for these are defined in BS 3550.[6]

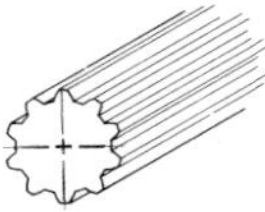

The increased load-bearing areas and the more uniform load distribution permit transmission of high torques. If the sides of the spline are of involute form, the stress concentration factor is lower than that of a single keyway (see Fig. 8.5). The spline shapes are complicated to manufacture, although in large-scale production the hub can be produced rapidly by a broaching operation, while the shaft splines can be rolled.

We describe the design of a key here; the same approach is used for splines. Under load, the sides of the key and keyway are subjected to a bearing stress, while the key is in shear. Failure in bearing and shear are taken as the limiting conditions, the usual approach being to select a size suitable for the shaft diameter and to design the key length to withstand the torque.

Keys are cut from key bar which is supplied in sizes that are standardised in BS 4235.[5] A selection of seven of the twenty-six available sizes is given in Table A13. This is an example of a local design standard being drawn up, using a national standard as a basis, to give a restricted but for most purposes adequate selection. The result can be a considerable saving in stock-holding of material and tools.[7]

For a design stress, $p = f_y/n$, the following equations can be written, taking design shear and bearing stresses of $p/2$ and $1.5\,p$ respectively:

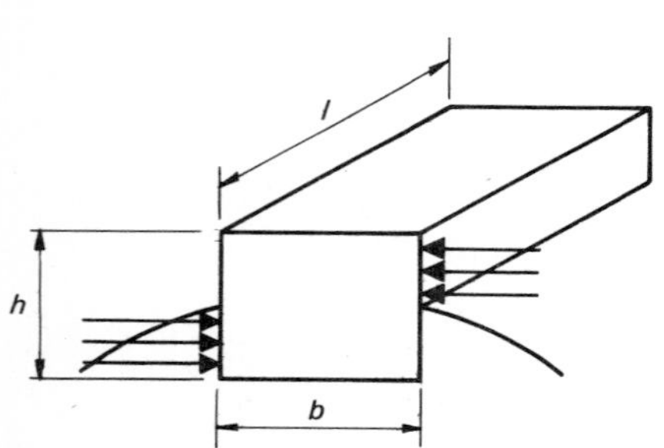

$$\text{Tangential force, } F = \frac{\text{Torque}}{D/2} \tag{8.12a}$$

$$\text{Bearing area, } l\frac{h}{2} = \frac{F}{1.5\,p} \tag{8.12b}$$

$$\text{Shear area, } lb = \frac{F}{p/2} \tag{8.12c}$$

Uniform loading along the length of the key has been assumed; this cannot be the case for loading in the elastic range, and it is likely that some plastic flow takes place. The bearing stress should pertain to whichever is the weaker of the shaft, hub and key materials. Note that because of the three-dimensional compressive nature of the bearing stress, a value 50% higher than the design tensile stress is used.

It will be seen from Table A13.1 that for small shaft sizes – up to about 22 mm – a square key is used. The Table gives tolerances that ensure that the key is a tight fit in the shaft keyway. If it is necessary to give the hub freedom to slide on the shaft, the key fitting in the hub can be chosen as a sliding ($D10$) fit.

A typical specification for key-bar steel[8] is 070M20, for which the following strength values can be used in preliminary design:

$$f_t = 615; \qquad f_y = 355 \text{ MPa}$$

Keys and splines can be eliminated by shrinking the collar onto the shaft, and relying on the friction forces at the interface to transmit torque. Although stress concentrations are not eliminated by this method they can be minimised by using the geometry shown in the sketch. A close estimate of the allowable torque can be made, provided that accurate machining and a good surface finish on the contacting surfaces are ensured.

A design method is given in *Engineering Sciences Data* 68003, "Shafts with Interference Fit Collars"[9], where fatigue performance is also discussed. We will consider here a steel shaft and collar assembly – full details for hollow shafts of different materials can be found in *Engineering Sciences Data.*[9] This method is finding increasing use where major components must be fitted to a shaft.

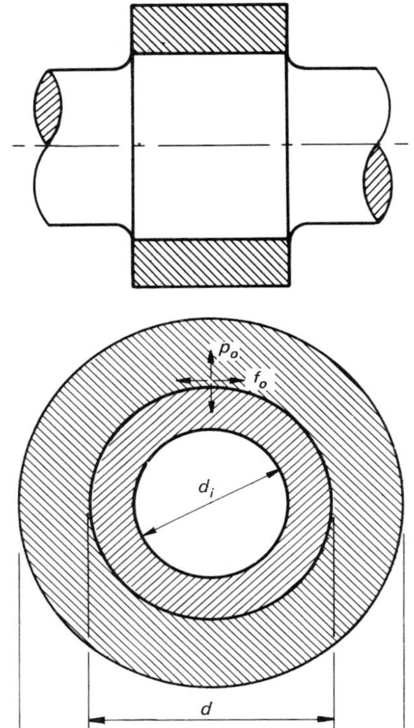

Thick cylinder theory[10] can be used to derive the following equations for the radial and tangential stresses at the shaft-collar interface for an interference fit. The diametral interference is defined as ε = interference/diameter. The interface pressure, p_o, is given in non-dimensional form by

$$\frac{p_o}{E_c \varepsilon} = \left[\frac{E_c}{E_s}\left(\frac{1 + (d_i/d)^2}{1 - (d_i/d)^2} - \nu_s\right) + \frac{1 + (d/D)^2}{1 - (d/D)^2} + \nu_c\right]^{-1} \tag{8.13}$$

and the circumferential stress, f_o, at the bore of the collar by

$$\frac{f_o}{E_c \varepsilon} = \frac{p_o}{E_c \varepsilon}\left(\frac{1 + (d/D)^2}{1 - (d/D)^2}\right) \tag{8.14}$$

E_s and E_c are the moduli of elasticity for the shaft and collar respectively, and ν_s and ν_c are Poisson's ratio.

When the shaft and collar materials have the same value of E, $E_c/E_s = 1.0$ and the non-dimensional interface pressure $p_o/E\varepsilon$ and circumferential stress $f_o/E\varepsilon$ can be plotted in terms of the ratio d/D for a range of values of d_i/d (see Fig. 8.7(a) and (b)). The value of ν has been taken as 0.3.

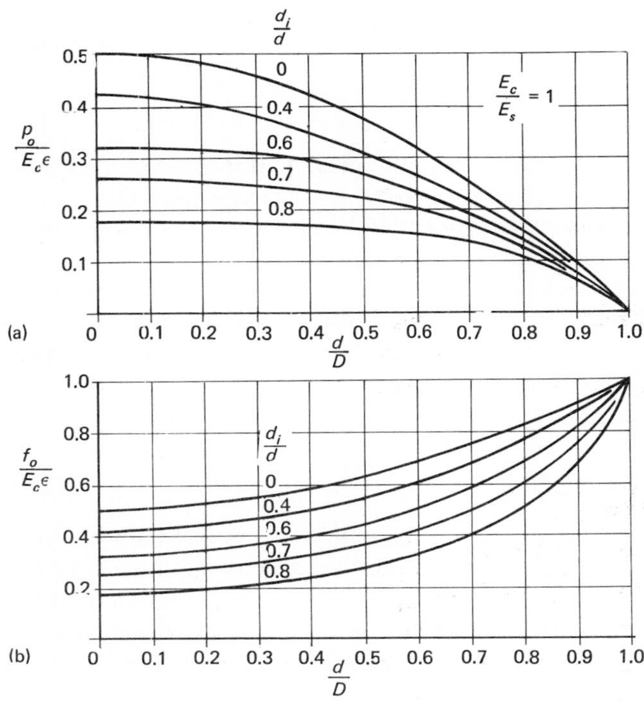

Fig. 8.7. *Components of stress for interference fit shaft and collar:*
(a) radial stress;
(b) tangential stress.
(From "Engineering Sciences Data", Item 68003, by permission of the Engineering Sciences Data Unit.)

At high rotational speeds, the interface pressure is reduced and the circumferential stress increased; a method is given in *Engineering Sciences Data*[9] for calculating these changes, which are small for normal rotational speeds.

The slipping torque for an interference fit may be estimated from

$$T_s = \tfrac{1}{2}\pi\mu p_o d^2 l \tag{8.15}$$

Suggested values of μ given in *Engineering Sciences Data*[9] for cleaned and dried surfaces range from 0.12 for press fitting to 0.31 for cold shrinking in liquid nitrogen.

It is necessary to check that the yield strength in the collar and shaft has not been exceeded. The actual stress distribution in the collar or shaft close to the interface will be a complicated one and will vary along the length. We will assume that the stresses to be taken into account are those tabulated below. The nomenclature shown in the sketch is used.

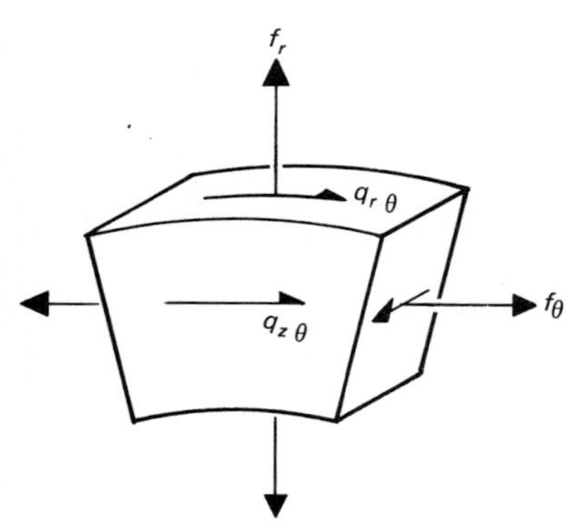

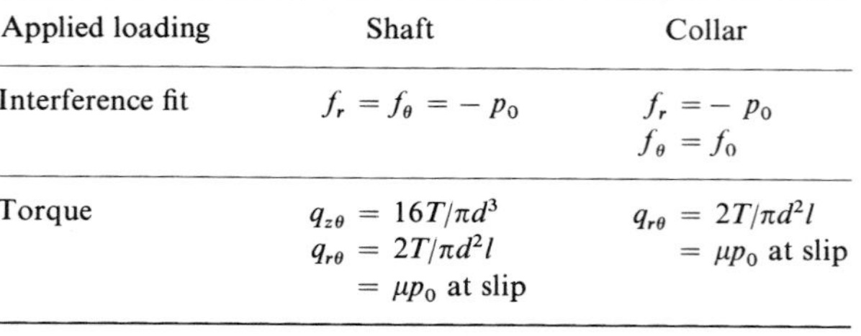

Applied loading	Shaft	Collar
Interference fit	$f_r = f_\theta = -p_0$	$f_r = -p_0$ $f_\theta = f_0$
Torque	$q_{z\theta} = 16T/\pi d^3$ $q_{r\theta} = 2T/\pi d^2 l$ $= \mu p_0$ at slip	$q_{r\theta} = 2T/\pi d^2 l$ $= \mu p_0$ at slip

The stress $q_{z\theta}$ is the shear stress due to torsion in the shaft, and is assumed to be negligible in the collar. The stress $q_{r\theta}$ comes from the frictional forces at the interface and is limited to μp_o. The assumption made here is that this shear stress $q_{r\theta}$ is uniformly distributed over the whole interface.

The equivalent stress in the collar for the von Mises criterion of failure can be written

$$f_{eq} = (f_r^2 - f_r f_\theta + f_\theta^2 + 3\,q_{r\theta}^2)^{\frac{1}{2}} \tag{8.16}$$

To ensure that yield does not occur, f_{eq} should be smaller than f_y. A factor of 1.5 can be used to allow for stress concentration effects. The shaft is not likely to yield under the influence of the interference-fit stresses before the collar does.

Example 8.5 compares the torque of a key with that of an interference-fit collar. The lower limit for the fit used in this problem does not give very high stresses and, clearly, if a high interference is used with the upper limit for μ, the value of T_s will be higher than the safe value for a keyway of typical proportions.

8.6. COUPLINGS

Many different types of flexible couplings can be purchased from stock; before making a selection, the designer must be aware of the characteristics of the drive — in terms of torque, shock and vibration loading — and of the degree and type of misalignment that may occur. As the drive speed increases, the coupling must be able to accommodate some misalignment without introducing out-of-balance forces or variations in angular velocity.

Example 8.5 Sheet 1 of 2

A 75mm diameter shaft carries a collar 160mm outside diameter, 100mm long. If they are assembled with a force fit (H6–t5), calculate the slipping torque corresponding to the maximum and minimum interference. Compare this with the torque transmitted by a key of standard proportions (20×12×100). Check the maximum stress at the shaft–collar interface, and determine the safety factor for the critical component.

Take μ at interface = 0.31; f_y for key material and collar = 355 MPa.

For the proposed fit, H6: ES = +19, EI = 0

t5: es = +88, ei = +75

Maximum interference = 88 μm, minimum = 56 μm

From Figure 8.7(a) and (b), with $d_i/d = 0$, $d/D = 0.47$

$$\frac{p_o}{E_c\epsilon} = 0.38, \qquad \frac{f_o}{E_c\epsilon} = 0.61$$

Calculate the maximum interface pressure:

$$\epsilon = \frac{88 \times 10^{-3}}{75}$$

$$= 1.173 \times 10^{-3} \text{ mm/mm}$$

$p_o = 0.38 E_c\epsilon = 0.38 \times 205 \times 10^3 \times 1.175 \times 10^{-3} = 91.4$ N/mm²

Maximum torque $= \frac{1}{2}\pi\mu p_o d^2 \ell$

$= \frac{1}{2}\pi \times 0.31 \times 91.5 \times 75^2 \times 100 \times 10^{-3}$

$= 25000$ Nm

Minimum torque $= 25000 \times \frac{56}{88} = 15900$ Nm

Torque with key : take $p_q = \frac{0.58 f_y}{n} = \frac{0.58 \times 355}{1.5}$

For shear of key, torque $= 20 \times 100 \times \frac{0.58 \times 355}{1.5} \times \frac{75}{2} \times 10^{-3}$

$= 10300$ Nm for safety factor of 1.5

For crushing of key, take $p_c = \frac{1.5 f_y}{n}$

Torque $= 6 \times 100 \times \frac{1.5 \times 355}{1.5} \times \frac{75}{2} \times 10^{-3} = 8000$ Nm

Check stresses at collar-shaft interface, for maximum interference

$f_o = 0.61 E_c\epsilon = 147$ N/mm²

$q_{r\theta} = 0.31 \times 91.4 = 28.3$

Example 8.5 Sheet 2 of 2

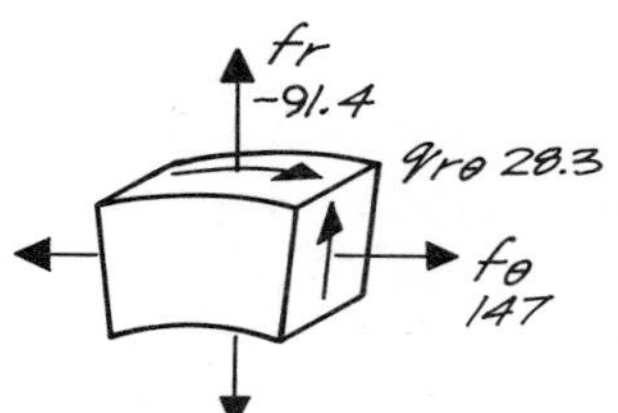

Stresses on element at collar interface are shown on sketch

Using von Mises criterion

$f_{eq} = [91.4^2 + 91.4 \times 147 + 147^2 + 3 \times 28.3^2] = 214$

Safety factor $= 355/214 = 1.66$ This is greater than the value of 1.5 suggested in the Data Sheets

[Using maximum shear stress criterion

$q_{max} = \left[\left(\frac{147 + 91.5}{2}\right)^2 + 28.4^2\right]^{\frac{1}{2}} = 123$

Safety factor $= 355/2 \times 123 = 1.44$]

At shaft interface, $f_r = f_\theta = -91.5$, and equivalent stress is lower.

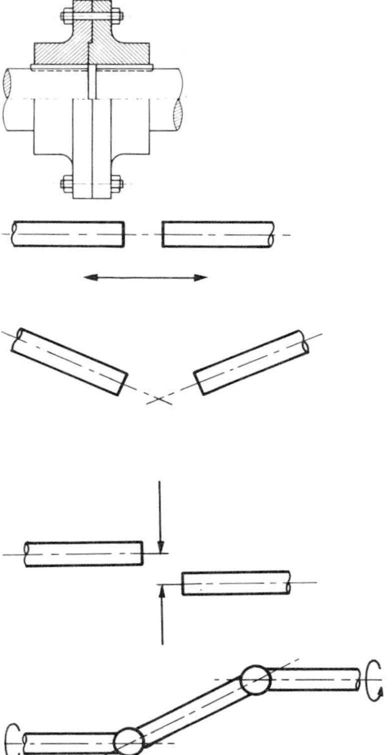

If it is simply a matter of providing a connection between two shafts that can be correctly aligned and will stay in alignment, a flanged coupling[11] keyed onto the shafts will be satisfactory. The spigot on the face of the coupling ensures that the shafts are concentric. If any misalignment occurs after this coupling has been assembled, high loads may be put on the bearings of the machines it connects.

The types of movement and misalignment that are of concern to the designer are illustrated in the sketches. Axial movement can be accommodated by drives which use keys or, preferably, splines to transmit the torque, and presents no great difficulty in design.

Angular misalignment, with the shaft axes intersecting, can be accommodated by one universal joint, such as a Hooke's coupling. However, this joint introduces a variation in angular velocity which, at large misalignments or high speeds, is not acceptable. If these joints are correctly oriented in pairs, the velocity variation is eliminated.

Constant-velocity universal joints are available. They are designed so that the points of contact between the members that transmit the torque always lie in the plane bisecting the angle between the shafts. The Bendix-Weiss and Rzeppa constant-velocity joints are examples of these.[12]

If the shafts are laterally misaligned, an Oldham type coupling may be used. Two discs with rectangular slots engage with a central floating block which has keys oriented at 90° protruding from each surface. This type is not suitable for high speeds or powers, and two constant-velocity joints may be used when this misalignment is encountered.

Many types of couplings use rubber bushings or some type of shaped rubber insert to provide the interface between two metal flanges. These couplings have the capacity to take up small misalignments that may occur during service due to thermal expansion or small movement in machine mountings. A design method for couplings with elastomeric flexible elements is given in J. R. Gensheimer, "Flexible Couplings".[13]

LIST OF SYMBOLS

b	width of key
C_s	size factor
D	diameter
D_i, D_o	inner and outer diameter
d	diameter
d_i	inner diameter
E_s, E_c	Young's modulus, shaft and hub
F, F_p	force, peak force
f	normal stress
f_a	axial stress
f_{bc}, f_{bt}	bending stress, compression and tension
f_o	tangential stress
f_r	radial stress
f_θ	tangential stress
f_t	tensile stress
f_y	yield strength
f_{eq}	equivalent stress

LIST OF SYMBOLS (Continued)

h	height of key
I	moment of inertia
J	polar moment of inertia
K	stress raising factor
K_b, K_t	load factors in bending and torsion
k	ratio D_i/D_o
L	length
l	length of key
M, M_p	moment, peak moment
n	safety factor
p, p_q	permissible normal and shear stress
p_o	interface pressure
p	pole distance
q	shear stress
q_{max}	maximum shear stress at a point
$q_{r\theta}$	shear stress
R	fillet radius
S_h, S_v	scale factors
T, T_p	torque, peak torque
T_s	slipping torque
x	distance along shaft
y	shaft deflection
Δ	correction factor, Fig. 8.3
ε	diametral interference, mm/mm
θ	deflection angle of shaft
μ	coefficient of friction
ν_s, ν_c	Poisson's ratio, shaft and hub

REFERENCES

[1] AS B249–1969, *Design of Shafts for Cranes and Hoists*, Standards Association of Australia, Sydney, 1969.

[2] H. A. Borchardt, "Design of Shafts for Cranes and Hoists Using AS B249–1969", *Transactions of The Institution of Engineers, Australia*, **MC5**, May, 1969.

[3] W. Abbott, *Practical Geometry and Engineering Graphics*, 3rd ed., Blackie, London, 1944.

[4] BS 46:Part 1:1958, "Keys and Keyways", British Standards Institution, London, 1958.

[5] BS 4235:Part 1:1972, "Parallel and Taper Keys", British Standards Institution, London, 1967.

[6] BS 3550:1963, *Involute Splines*, British Standards Institution, London, 1963.

[7] J. A. Bachrich, *Standardisation in Machine Tool Production*, Birniehill Institute, Glasgow, 1970.

[8] BS 970:Part 1:1972, *Wrought Steels*, British Standards Institution, London, 1972.

[9] Engineering Sciences Data, 68003, "Shafts with Interference Fit Collars", Engineering Sciences Data Unit, London, 1968.

[10] C. T. Wang, *Applied Elasticity*, McGraw-Hill, New York, 1953.

[11] BS 2715:1956, *Shaft Coupling Flanges,* British Standards Institution, London, 1956.
[12] R. M. Phelan, *Fundamentals of Mechanical Design*, 3rd. ed., McGraw-Hill, New York, 1970.
[13] J. R. Gensheimer, "Flexible Couplings", *Machine Design*, **33**, 19, 1961.

BIBLIOGRAPHY

Broersma, G., *Couplings and Bearings,* Stam International, Culemborg, 1968.
Hopkins, R. B., *Design and Analysis of Shafts and Beams,* McGraw-Hill, New York, 1970.
The Pennsylvania State University, "High Speed Flexible Couplings", *Engineering Proceedings P-41,* The Pennsylvania State University Press, Pennsylvania, 1963.

PROBLEMS

8.1. Use the information given in Problem 2.6 to design the shaft according to the ASME code, taking the factors k_b and k_t to be equal to 2.0 and 1.5 respectively.

8.2. What shaft diameter would be required for the data in Problem 2.7 if the shaft were designed for static loading conditions using Equation (8.5)? Use a safety factor of 1.2 and a value of f_y that would be expected for a steel with a tensile strength of 500 MPa.

8.3. The rear axle of a diesel-engined tractor is illustrated in the sketch. It is estimated that it carries a maximum torque of 10 kN m; the wheel load is 9 kN. If the tensile and yield strengths of the axle material are 800 MPa and 630 MPa respectively, check the safety factor of the shaft. Make any assumptions necessary to determine the loads and stresses.

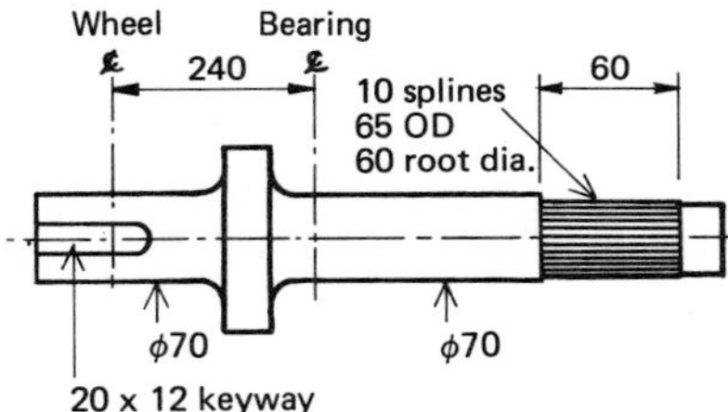

8.4. A couple M is applied at the end of a cantilever beam, length L. Show that the slope θ and deflection y at the end of the beam are given by

$$\theta = \frac{ML}{EI}, \qquad y = \frac{ML^2}{2EI}.$$

8.5. The 30-mm diameter pin shown in the sketch carries gear-tooth loads whose components are F_R and F_T; the arrangement was described in Chapter 5 (Fig. 5.14). Calculate the slope and deflection at the end of the pin if the loads are applied:
(a) in the plane A, and
(b) in the plane B.
The value of F_T is 3.5 kN.

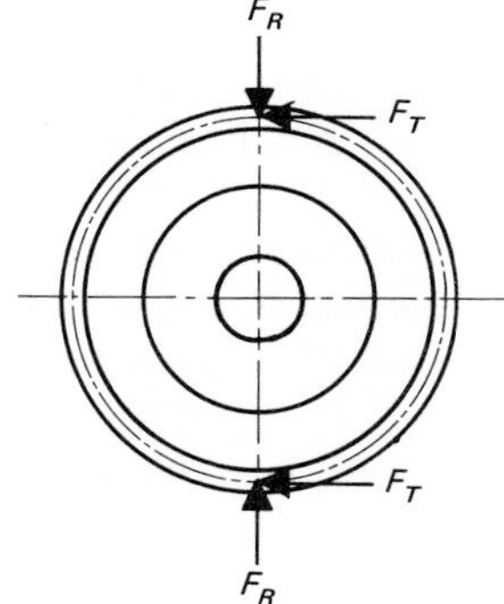

View on AA

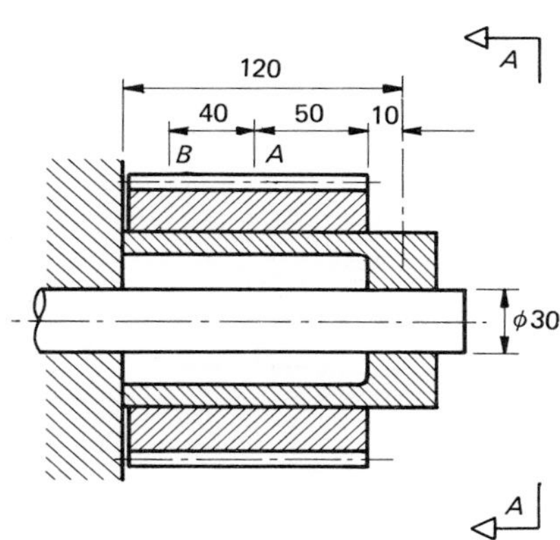

8.6. An overhead crane runs on 500-mm diameter wheels which fit on a shaft that overhangs its bearings by 200 mm. Design the shaft diameter if the yield strength of the material is 310 MPa and a safety factor of 2 is required. Assume that when the crane is running, a side load of 10% of the vertical load can occur at the wheel flange. The wheel load is 250 kN. Check the effect on the safety factor of starting and braking loads. What diameter would be given by the ASME code?

8.7. Check the design of the shaft in Problem 8.6 for fatigue. Assume a suitable value of fatigue strength for the steel. A fillet at the plane of maximum bending moment gives a strength reduction factor of 2.2.

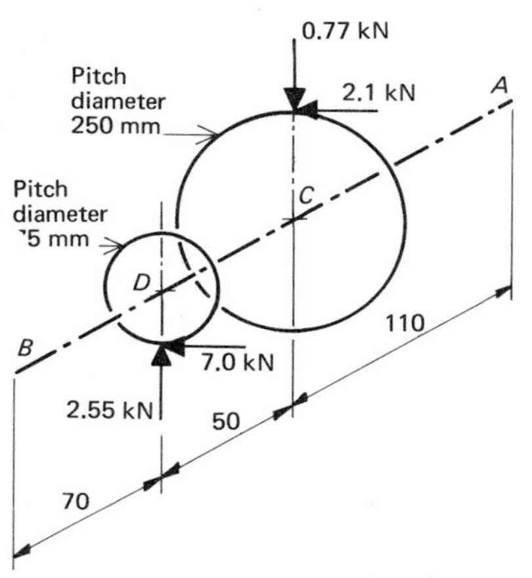

8.8. A shaft is supported in bearings at A and B and carries gears at C and D. The tooth forces on the gears for peak loads are shown in the sketch. Draw the bending-moment diagram for the shaft and design the shaft diameter at the location of maximum bending moment. Use Australian Standard B249. Assume that the torque is fully reversed, and that the gears are keyed to the shaft with an interference fit. The shaft material has a tensile strength of 500 MPa.

8.9. Repeat Problem 8.8 with the contact point on gear D shifted through 180°.

8.10. (a) When calculating the stress in the hub of an assembly in which torque is transmitted by an interference fit, the shear stress on the plane perpendicular to the axis is assumed to be zero. What is the justification for this assumption?

(b) A 90-mm diameter steel shaft carries a collar of the same material which has an outside diameter of 110 mm. Consider the extremes of fit that can be obtained with an **H7–s6** fit (see Appendix A5) and calculate the length of collar required to transmit safely the torque that the shaft can carry with a safety factor of 2. The yield strength of the material is 520 MPa. Would a rectangular key of this length be satisfactory?

(c) Repeat part (b) using a more extreme fit, e.g. **H5–t5**. Values for this fit will be found in BS 4500: 1969, *ISO Limits and Fits*.

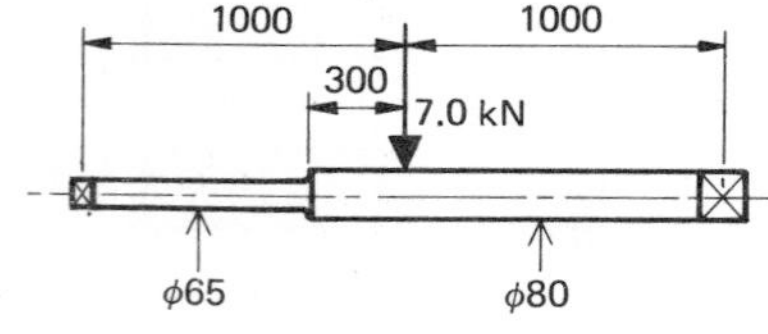

8.11. Calculate the maximum slope and deflection for the steel shaft loaded as shown in the sketch. Would the slope at the left-hand end of the shaft rule out the use of a roller bearing there?

8.12. Consult Australian Standard B249 and calculate the shaft size required for one of the several examples for which loads are given.

9 Rolling Bearings

9.1. INTRODUCTION

Bearings are used to reduce friction and to ensure accurate alignment between moving parts. The wide range of bearing types can be broadly classified as "rolling" or "sliding": in the former, an element having rotational symmetry is interposed between two moving surfaces and rolls with them; in the latter, the two surfaces, usually separated by a fluid film, slide over each other.

The first step in design will be to establish which of the various categories is best suited to the problem in hand. In so far as load and speed will affect the choice, a chart such as Fig. 9.1 will help in making a preliminary selection[1]; however, many other factors such as the type and direction of load and the allowable eccentricity must be considered. The figure shows how different regions of the load-speed regime are suited to different types of bearing. In the upper right-hand region of the diagram, the limit set by the bursting stress in a rotating steel shaft is indicated. Moving to the left from this line, hydrodynamic oil-film bearings are recommended for high speed. At lower speeds there is difficulty in maintaining the hydrodynamic film, and rolling bearings are an appropriate choice. Oil-impregnated bearings are suitable for low loads in the medium speed range. With a combination of low load and speed, rubbing bearings are acceptable; in these bearings the supply of oil or grease is, at best, intermittent, and breakdown of the lubricating film will occur.

In the case of hydrodynamic bearings, a rough preliminary size may be estimated from information such as that shown in Fig. 9.1, where the curves are labelled with the shaft diameter, based on a bearing with a length equal to the diameter.

When the hydrodynamic film breaks down, and the metallic surfaces rub together in the presence of some lubricant, boundary friction conditions are said to exist and significant wear will take place. In such cases it is common to specify the surface rating as a pV term where p is the pressure (defined as load/projected area) in MPa and V, the velocity in m/s. Typical design values of pV range from 0.2 to 2.0 depending on the materials used and also, since there is finite wear, on the tolerable increase in clearance during the design life. Manufacturers' catalogues should be consulted for more detailed information of this type.

The value of the coefficient of friction for bearings with boundary lubrication will be in the range from 0.05 to 0.15, while in oil-lubricated hydrodynamic bearings a value of 0.01 can readily be obtained; the minimum value can be as low as 0.001. The lowest values of friction factor, f, that can be expected for rolling bearings will be in the range 0.001 to 0.003. In this case, the friction torque is evaluated from $T = fP\,d/2$, where P is the radial load and d, the

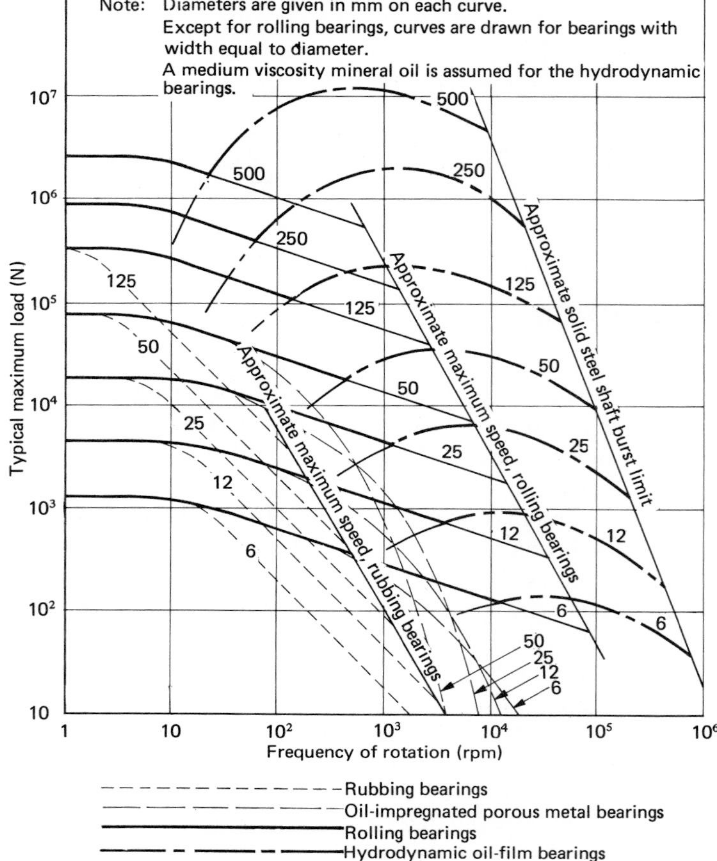

Fig. 9.1. *Selection chart for bearing types. (Adapted from "Engineering Sciences Data", Item 65007, by permission of the Engineering Sciences Data Unit.)*

bearing bore. It should be noted that seal friction, which may be greater than that of the bearing, must also be allowed for.

The advantage of low frictional torque on rolling bearings is maintained down to the lowest speeds, whereas the oil film in the hydrodynamic type will break down, with an increase in friction and wear; this may only occur when starting and stopping.

Not every available bearing type is shown on Fig. 9.1; for example, hydrodynamic (or hydrostatic) bearings can use air, and the air slide is familiar as a piece of laboratory equipment in which low frictional forces are maintained down to very low speeds.

The regimes drawn on Fig. 9.1 are by no means hard and fast; since rolling bearing installations are relatively simple to design, there is a tendency for them to be used in conditions more suited to sliding bearings.

The design of sliding bearings is based on the equations for laminar flow in a wedge; by integrating the expressions for pressure distribution, the supporting force on a rotating shaft can be calculated. Details of several design methods have been published in which solutions to the equations are displayed on charts. The procedures are not difficult, but cut and try methods must be adopted. The load that can be supported and the friction in the oil film are functions of viscosity, which changes markedly with temperature; since the mean temperature of the oil film depends on the friction, the viscosity can only be estimated approximately at the outset. These methods are not described here, and the interested reader is referred to F. T. Barwell, *Lubrication of Bearings*[2], "Engineering Sciences Data" Item 66023[3] and H. C. Rippel, *Cast Bronze Bearing Design Manual*.[4]

In the first part of this chapter, the various types of rolling bearings are described; then, following a discussion of selection procedures, some practical aspects of installation are discussed. These aspects are particularly important, since they can have a profound effect on the layout of the other components.

9.2. BEARING CONFIGURATION

Bearings can be classified in terms of the shape of the rolling element – ball, roller, taper roller and so on – or in terms of the duty they perform – whether the loading is predominantly radial or axial, or mixed. The former classification is used here. The types described are those most commonly encountered in practice; most bearing manufacturers will include them in their range. They will also supply more specialised types, for information on which the reader should consult manufacturers' catalogues and brochures.

Since preliminary selection of a bearing will be based largely on the type of load it must carry, the descriptions that follow will make a point of this factor. The principal components of a typical bearing are illustrated in the sketch. The rolling elements – rollers in this case – are prevented from making sliding contact with each other by a separating cage; steel pressings are commonly used since they are light and require little lubrication. The rollers are restrained from sideways movement by lips on the outer race. In this particular configuration the inner race has no lips and can be removed.

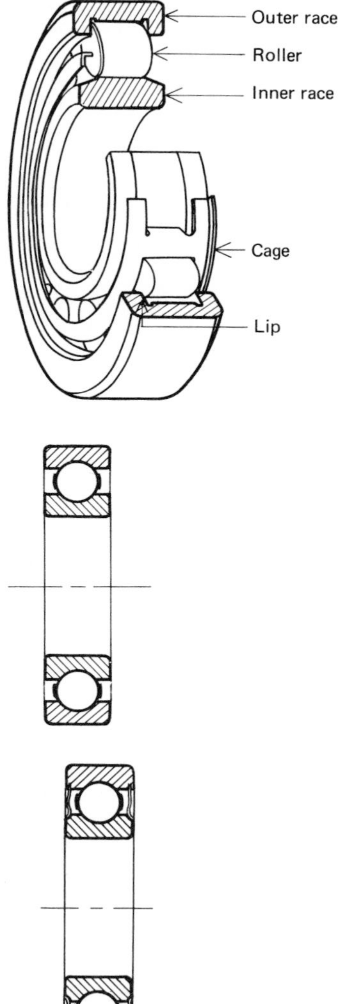

Deep groove ball bearing. This is a general-purpose bearing which is suitable for light and medium load applications. Its capacity to carry both radial and axial loads at high or low speeds and its comparatively low price result in its being widely used. The balls roll in grooves in the inner and outer races; a high degree of conformity between the ball and groove radius results in contact over an arc rather than at a point, with a correspondingly high load capacity; under combined axial and radial loading, the races are displaced slightly and the point of maximum pressure moves on to the shoulder of the groove. Some radial clearance between the races must be allowed to permit this movement to take place. This bearing is usually specified where silent running is important.

Since the races of a ball bearing cannot be separated, it is assembled and must be removed as a unit; the latter operation should be borne in mind at the planning stage.

This type of bearing can usually be supplied with side seals on one or both sides; when two seals are fitted, sufficient grease is sealed-in to last for the life of the bearing. The seals will exclude dust and moisture but under unfavourable conditions supplementary protection is required; also, the seals make a significant contribution to the friction of the assembly.

Bearing with two seals

Self-aligning ball bearings. The track of the outer ring is a spherical surface centred on the shaft axis, and some misalignment – from 1 to 3 degrees,

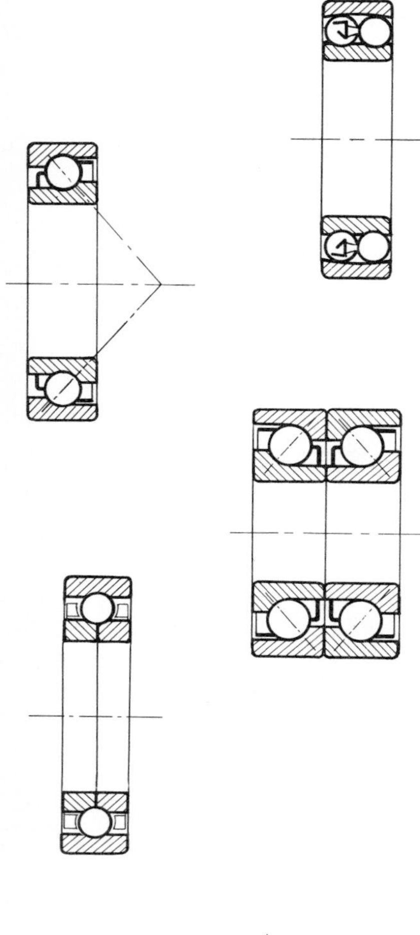

depending on the type – between shaft and housing can be accommodated. The inner race carries two grooves for two rows of balls; for a given shaft diameter, this bearing is wider, more costly and has a smaller load capacity than the equivalent deep groove bearing, and should only be used where shaft deflections or possible mounting errors justify it. Both radial and axial loads can be supported. It is often used in plummer block installations where accurate alignment cannot be guaranteed. For this purpose it can be supplied with a tapered bore which allows it to be secured on a shaft with the tapered adapter sleeve described in Section 9.4.2.

Angular contact bearing. In this ball bearing, which is not separable, the contact region is moved from the radial position so that the load through the balls acts obliquely; this makes the bearing suitable for heavier axial loads than can be supported on a deep groove bearing.

A radial load imposed on this type induces an axial load which must be counteracted, and the bearings are often installed as pairs which can be adjusted against each other. This adjustment may be done with the aid of suitable packing or shims, but the bearings can be supplied in pairs so the correct pre-loading is obtained without special adjustment. The sketch shows such a pair mounted "back-to-back" so that an axial force can be taken in either direction. The double row angular contact bearing is a variation of this arrangement in which the two rows of balls are carried in one-piece inner and outer rings, giving a bearing which will carry radial loads and heavy axial loads in either direction.

Another variation is the "duplex" bearing (see sketch), which has a split inner or outer ring with one row of balls, and can carry an axial load in either direction. This type is suitable for use where the axial load predominates. It has a high load capacity, since the split ring allows a larger number of balls to be fitted than in the deep groove bearing. Although these two types may be similar in general outline, they differ in the manner in which they support a load, and the angular contact bearing should not be run unloaded so as to avoid having the balls track along the dividing line between the rings.

Single thrust ball bearings. This bearing is designed to deal with a thrust load along the shaft in one direction only. The balls rotate in tracks in two washers; the shaft washer has a flat seat which butts up against a shoulder while the housing washer can be flat or, if some degree of self-alignment is needed, spherical. Because the centrifugal load on the rolling elements is carried on the sides of the groove, the maximum allowable speeds for this and similar configurations is limited, as compared with radial bearings, and they must not run unloaded. To carry bidirectional axial thrusts, similar bearings with two rows of balls are available.

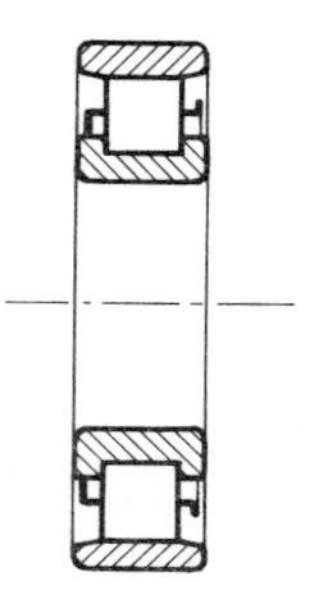

Cylindrical roller bearing. This is a widely used type which, because of the line contact between roller and race, can take heavier radial loads than comparably sized ball bearings. The permissible angular misalignment is small, and may be as low as 2 minutes of arc.

Any axial load on this bearing would have to be transmitted across the sliding contact face between the roller and the lip of the race; this is a duty that the bearing is not primarily intended for, but modern roller bearings are able to sustain considerable axial loads under certain conditions. We will not consider this aspect.

The lips may also be used to prevent axial movement of the races, and various arrangements of the lips are available to suit different installations. Where a lip is omitted, the race may be separated from the roller and cage; this facility sometimes simplifies assembly operations.

When there are space limitations, the race may be omitted and the rollers run directly on the shaft; this requires a suitable shaft material and surface finish and is a matter for consultation with the bearing supplier.

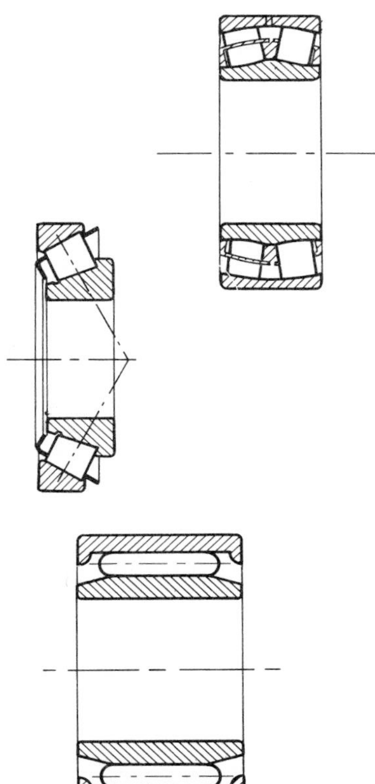

Spherical roller bearings. With two rows of rollers in a common spherical outer track, this bearing has self-aligning properties and will carry very heavy radial loads; axial loads can also be carried in either direction. Since the barrel-shaped rollers make contact with the track at different radii, there is some sliding and they will run at higher temperatures than cylindrical rollers.

This roller bearing is manufactured mainly in larger sizes; it is widely used where heavy loads must be carried.

Taper roller bearing. The rolling element in this type is a frustrum of a cone, whose apex lies on the axis of the shaft and bearing assembly. It is well-suited to carry radial and axial forces acting simultaneously. When subjected to a radial load component, the races tend to separate axially; it is therefore invariably mounted in opposition to a similar bearing, against which it can be adjusted and which will prevent this movement. The friction between the rollers and the inner ring guide flange makes this type run at higher temperatures than cylindrical roller bearings.

Needle roller bearing. Most rolling bearings require more space in the radial direction than do sliding bearings; an exception is the needle roller type which uses long rollers of comparatively small diameter. A further saving in space may be achieved by omitting the inner ring and running the rollers directly on the shaft.

9.3. BEARING SELECTION

Bearings are produced in very large quantities, and the selection must be made from the range described in catalogues. The designer loses very little by having to follow this procedure, since the available size range is very wide, and he has the advantage of using products that have undergone a great deal of development and testing. The basic selection procedures are generally well standardised between manufacturers, but it is essential that the designer is familiar with the handbooks specific to the particular bearing he intends to use. Information on specialised topics – such as the design of high reliability installations described in Snare's "How Reliable are Bearings?"[5] – can usually be obtained from the manufacturers.

For any given shaft size there are produced, in the more widely used types, several sizes of bearing; as the outside diameter and width increase, the load capacity increases. The designer will take advantage of this and attempt to balance the bearing and shaft strength; it may well happen, however, that one of these components must be larger than strength requirements call for in order to match the other. It should be noted that only a selection of the available size range is given in the tables in this book.

The stresses set up when two curved surfaces are pressed together were studied by Hertz, and he developed expressions for the elastic stresses and deflections in terms of the curvature of the surfaces and the moduli of elasticity of the materials. The Hertz equations for balls and rollers on a curved surface are given below (see also Table 3.1). Hertz assumed that the materials are isotropic and homogeneous, that the contact area is flat, and that the proportional limit is not exceeded.

For contact between spheres, the maximum pressure occurring at the centre of the contact area is

$$f_{max} = 0.388 \left[\frac{PE^2 (R_1 + R_2)^2}{R_1^2 R_2^2}\right]^{\frac{1}{3}} \tag{9.1}$$

where P is the force pressing the spheres together; it is assumed that both materials have the same value of E and that $\nu = 0.3$.

For contact between rollers,

$$f_{max} = 0.418 \left[\frac{P'E (R_1 + R_2)}{R_1 R_2}\right]^{\frac{1}{2}} \tag{9.2}$$

Here P' is the force per unit length of roller. When the curvature of both surfaces is in the same sense, R_1 and R_2 have opposite signs.

For both these cases, the maximum shear stress occurs below the surface and has a value of about $0.3 f_{max}$. This stress is frequently the cause of fatigue failures in bearing races, and the result of the sub-surface failure is a "spalling" or breaking-away of a surface layer in the failure region.

If the load distribution between the load-carrying rolling elements in a bearing can be calculated, the Hertz equations can be used to find the stresses set up at each contact area. It is found that extremely high stresses can be tolerated by bearing elements loaded in this way – stresses of 1500 MPa would not be uncommon. The ability of the materials to withstand these stresses is accounted for by the three-dimensional nature of the stresses. In practice, capacities of bearings that are given in catalogues are determined experimentally, since the Hertz stresses only give general guidance to the loads they can carry. However, it is interesting to calculate these stresses for a typical case.

Most failures in bearings result from fatigue; this comes from the repeated loading and unloading as a rolling element passes over the race at the point of maximum load. Any minor defect or uneveness in the track at this location will accelerate the failure. This fact points up the need to avoid even the smallest damage when a bearing is being handled. It also provides an indication that the life of members of a batch of nominally identical bearings will be a random variable, and that selection will have to be based on some probable life.

The life will be directly related to the number of stress reversals, which, in turn, can be expressed in terms of the number of revolutions. If a sample of 100 bearings is tested under fixed loading conditions, the plot of number of bearings failed against number of revolutions to failure will be similar to that shown in Fig. 9.2. This type of information is used by the manufacturer when producing design data, and some of the definitions used in tabulating these data in catalogues will now be considered.

The *nominal life* is defined as the number of revolutions that will be endured by 90% of the sample without any sign of failure. It will be noted from Fig. 9.2 that, typically, 50% of the sample will survive for about five times the nominal life. The nominal life is taken as the basis for design, and maintenance practices

Example 9.1 Sheet 1 of 1

The sketch shows the dimensions of a typical bearing which has 17 rollers 14 mm in length. Calculate the surface stress in the roller when the bearing is loaded to half its static carrying capacity of 50 kN. Assume that the load P_{max} on the most heavily loaded roller is given approximately by $P_{max} = \frac{5F_r}{Z}$, where Z is the number of the rollers.

The static carrying capacity is defined as the load required to give a specified small deformation. The Hertz equation is derived on the basis of elastic behaviour, and will not be valid for loads as high as the static capacity. We will assume it does hold for a load of 25 kN

d=11

17 rollers

radius = 30

Load on one roller $= \frac{25 \times 5}{17}$ kN

Load per mm of roller $= \frac{25 \times 5 \times 10^3}{17 \times 14}$ N

$$f_c = 0.418 \left[\frac{P'E(R_1 + R_2)}{R_1 R_2} \right]^{\frac{1}{2}}$$

$$= 0.418 \left[\frac{25 \times 5}{17 \times 14} \times 10^3 \times 205 \times 10^3 \times \frac{35.5}{165} \right]^{\frac{1}{2}}$$

$= 2020$ N/mm² or MPa

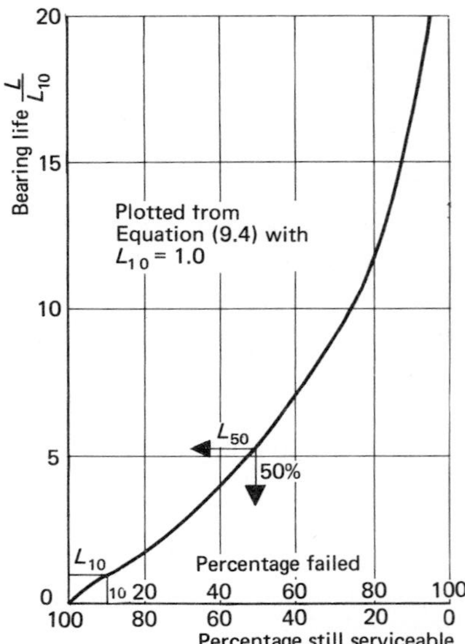

Fig. 9.2. *Life curve for bearings.*

must be considered when fixing this life. Suggested values are given in Table 9.1, which has been taken from a manufacturer's catalogue.

When using this approach, it is clear that some probability of failure is accepted; in many installations the failure will be made evident by an increase in noise and vibration some time before any major breakdown of the running surfaces takes place, and preventive measures can usually be taken in time to avoid this. If this is not acceptable and a much lower probability of failure is required, other design methods must be used (see Section 9.3.1).

The *basic dynamic load rating*, C, is the constant radial load that will result in a nominal life of 10^6 revolutions of the inner ring. In the case of bearings which are subjected to a combined axial and radial load, it is necessary to cater for all possible combinations of these loads; this is done by defining an *equivalent load*, P, which will give the same life as the combined load. The *static carrying capacity*, C_0, is defined as the static radial load which causes, in the contact between rolling element and raceway, a permanent deformation of 0.0001 of the diameter of the rolling element. Experience has shown that such a load can be tolerated in most bearing applications without impairment of operation. This static load capacity is tabulated, along with the basic dynamic capacity, to define the carrying capacity. Where a bearing is to oscillate in rotation about some fixed mean position, the static capacity provides a good guide to the load that can be carried.

Values of C and C_0 for a range of bearing types are given in Appendix A10, along with the principal bearing dimensions and other design data. This information has been taken from the *SKF General Catalogue* by permission of the SKF Bearing Company. Only the most commonly used types are described here; the following have been selected:

deep groove ball bearing, Series 62 and 63;
self-aligning ball bearing, Series 12 and 13;
single row angular contact bearing, Series 72 and 73;
cylindrical roller bearing, Series 2 and 3.

Table 9.1. Guide to values of nominal life, L_h, for different classes of machines

Class of Machines	Hours of Service (L_h)
Machines used for short periods or intermittently and whose breakdown would not have serious consequences.	
Hand tools, domestic machines, agricultural machines, lifting tackle in workshops, foundry cranes	4 000 to 8 000
Machines working intermittently whose breakdown would have serious consequences.	
Electric motors for agricultural equipment and domestic heating and refrigerating appliances, auxiliary machinery in power stations, conveyor belts, lifts, workshop cranes or machine tools infrequently used . . .	8 000 to 12 000
Machines for use 8 hours per day and not always fully utilised.	
General purpose gear units	12 000 to 20 000
Stationary electric motors	16 000 to 24 000
Machines for use 8 hours per day and fully utilised.	
Machine tools, wood processing machinery, machines for the engineering industry, cranes for bulk materials, ventilating fans	20 000 to 30 000
Machines for continuous use 24 hours per day.	
Compressors, pumps, stationary electric machines and mine hoists . .	50 000 to 60 000
Ships' propeller shaft thrust bearings, stationary electrical machines, tunnel shaft bearings	60 000 to 100 000
Machines to work with high reliability 24 hours per day.	
Pulp and papermaking machinery, power station plant, water works machinery, mine pumps	> 100 000

Source: SKF General Catalogue,[6] by permission of the SKF Ball Bearing Company.

The reader should consult manufacturers' catalogues for more complete information on these and other types.

Since handbooks tabulate only the basic dynamic load rating (i.e., the load for a life of 1 million revolutions), some failure theory is required to design for other loads and lives. It has been found that the life, L, expressed in millions of revolutions, is related to the basic dynamic capacity, C, and the equivalent load, P, by an equation of the form

$$L = \left(\frac{C}{P}\right)^p$$

where $p = 3$ for all ball bearings and 10/3 for roller bearings. We do not differentiate here between rotation of the inner and the outer ring.

When only a radial component, F_R, is taken on a bearing, as in a roller bearing, P is equal to F_R. When an axial component is also carried (for example, by a ball bearing) the equivalent load, P, is found from

$$P = XF_R + YF_A$$

The values of X and Y relevant to each bearing type are given in Tables A10. It will be seen that X and Y are functions of the ratio of F_A/F_R and sometimes also of F_A/C_0. Since the latter term contains the bearing parameter, C_0, the selection in this case must involve a trial-and-error approach. When F_A is small, it does not need to be considered; the criterion is that F_A/C_0 should be smaller than a limiting value which is tabulated as e.

The design method just described is the basis for the selection methods recommended by most manufacturers and is the one used in this book. In practice, it is also necessary to study catalogue and handbook information to determine any minor variations from this method and to establish the procedures recommended for dealing, for example, with shock loads and varying loads. For example, in the design of bearings that support gears, one manufacturer suggests the use of load factors ranging from 1.1 for precision gears up to 1.3 for commercial gears; these factors account for the shock loading caused by inaccuracies in the gears.

Examples 9.2 and 9.3 illustrate the way in which bearings are selected to carry specified loads.

A study of the way in which the stresses are set up at the contacting surfaces of rolling bearings will explain one of the variations in design method that may be encountered. Since the contact stresses between two curved surfaces are greater in the case where the curvatures are opposed than they are when they are in the same direction, it is clear that the highest stresses will act where the rolling element touches the inner ring. If the load acts in such a way that the point of highest stress is stationary relative to the inner ring, there is clearly a high probability that any fatigue failure occurring will start at this point. In some design methods the increased severity of loading in this condition is allowed for by multiplying the radial load by a factor of, for example, 1.2.

9.3.1. Design for high reliability

The problems involved in establishing the life curve for bearings are discussed by Snare in "How Reliable are Ball Bearings?" from which this discussion is taken.[6] The life curve shown in Fig. 9.2 is representative of the way in which parts fail in fatigue loading; an analytical expression which reproduces this behaviour is the Weibull distribution equation, which has the form:

$$\frac{\ln S}{\ln 0.9} = \left(\frac{L - L_0}{L_{10} - L_0}\right)^{\varepsilon} \tag{9.3}$$

where S is the fraction of still serviceable bearings, and the distribution index, ε, has been found to have a value of about 10/9. L_0 is the minimum life – that is to say, the life below which no failures take place – and L_{10} is the life at which 10% of the sample has failed. To determine L_0 with any certainty requires tests on very large batches of bearings. Tests made to measure the rating life of a bearing usually involve a batch size of about 30; such tests will not give precise values of L for $S > 0.9$ or for $S < 0.1$, since very few failures can be expected in these ranges. The practice has been to assume $L_0 = 0$, and this does not significantly affect the shape of the life curve in the central portion which is of most general interest. Putting $L_0 = 0$ in Equation (9.3) gives

$$\ln S = \left(\frac{L}{L_{10}}\right)^{\frac{10}{9}} \ln 0.9 \tag{9.4}$$

and it is from this equation that the curve in Fig. 9.2 has been drawn, with $L_{10} = 1.0$.

Example 9.2 Sheet 1 of 1

Select bearings for the following duties

(a) Roller bearing (series 3) for radial load 5.5 kN
(b) Ball bearing (series 62) for radial load 4.0 kN
axial load 1.1 kN
Given life requirement is 8000 hours at 1800 rpm

Number of revolutions in 8000 h $= 8 \times 10^3 \times 1.8 \times 10^3 \times 60$
$= 865 \times 10^6$

ie $L = 865$

(a) Roller bearing $L = \left(\frac{C}{P}\right)^{\frac{10}{3}}$ $\frac{C}{P} = (865)^{0.3} = 7.6$
Here $P = F_r = 5.50$, $C = 5.5 \times 7.6 = 41.8$
Select NU 307 which has $C = 42.9$ $d = 35$
$D = 80$

(b) Ball bearing $L = \left(\frac{C}{P}\right)^3$ $\frac{C}{P} = (865)^{0.33} = 9.52$
As a first approximation assume $P = F_r$
Then $C \approx 9.5 \times 4 = 38$
Nearest bearing (6213) has $C_o = 34$
Now find X and Y $\frac{F_a}{C_o} = \frac{1.1}{34} = 0.032$
From Table A10.1, $X = 0.56$, $Y = 1.9$, $e = 0.23$
Now compare $\frac{F_a}{F_r}$ with e: $\frac{F_a}{F_r} = \frac{1.1}{4} = 0.275$
Since $\frac{F_a}{F_r} > e$, $X = 0.56$ $Y = 1.9$ and
$P = 0.56 \times 4 + 1.9 \times 1.1 = 4.33$
$C = 4.33 \times 9.52 = 41.2$
From bearing tables confirm that bearing 6214 has $C = 47.1$ which will give required life
Select bearing 6214, $d = 70$
$D = 125$

Example 9.3 Sheet 1 of 2

Bearings are required to support the shaft shown in the sketch at A and B. Select suitable bearings so that the shaft sizes are sufficient to support the specified loads. Use ASME code for shaft design, with $p_q = 55$ MPa and $K_b = 1.5$, $K_t = 1.0$
Torque on shaft = 0.1 kN m

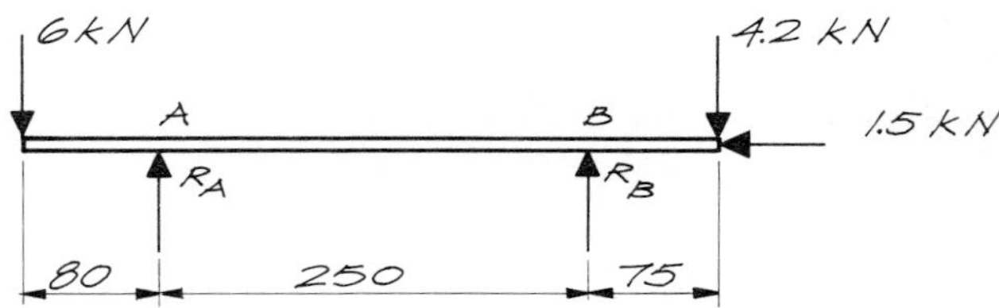

Calculate reactions at A and B $R_A = 6.66$
$R_B = 3.54$

Check minimum shaft sizes at A and B

$$D_A = \left\{ \frac{16 \times 10^6}{\pi \times 55} \left[(1.5 \times 0.48)^2 + 0.1^2 \right]^{1/2} \right\}^{1/3} = 40.6 \text{ mm}$$

$$D_B = \left\{ \frac{16 \times 10^6}{\pi \times 55} \left[(1.5 \times 0.315)^2 + 0.1^2 \right]^{1/2} \right\}^{1/3} = 35.4 \text{ mm}$$

Select bearing type at A and B. Use ball bearing to take axial load and locate it at B where the radial load is smaller.
Use roller bearing at A where the radial load is heavy

$$\text{Life } L = \frac{2 \times 10^4 \times 1.6 \times 10^3 \times 60}{10^6} = 1920 \text{ million revolutions}$$

Example 9.3 Sheet 2 of 2

Roller bearing at A

$\frac{C}{P} = L^{3/10} = 9.6$

In this case $P = F_r = 6.66$ $\therefore C = 6.66 \times 9.6 = 64$ kN

Select NU309, $d = 45$ mm → this matches the shaft size for strength

$C = 69.4$

Note that lighter bearing NU209 with same value of d has $C = 40$ which would be inadequate

Ball bearing at B

Preliminary check shows that bearing 6312 will give the required life. This has shaft diameter 60 mm and strength requirement at B is for 35 mm i.e. bearing size is governing factor

For $L = 1920$, $\frac{C}{P} = (L)^{0.333} = 12.4$

For 6312 bearing, $C_o = 48.0$, $F_a/C_o = 1.5/48 = 0.0313$

From bearing tables $e = 0.22$ and since

$F_a/F_r = 0.42 > e$

$X = 0.56$, $Y = 1.95$, and $P = XF_r + YF_a = 4.91$

$\therefore$ Required $C = 12.4 \times 4.91 = 62.2$

Bearing 6312 gives $C = 62.2$ – Select 6312,

$d = 60$ mm

In practice, with a wider choice of bearings, a heavier duty bearing at B would result in a smaller shaft diameter and a better match with shaft strength requirement

In most bearing-selection problems, it is reasonable to work to the L_{10} life, and accept the small probability of a failure occurring. If this is not acceptable, L_0 must be measured before a logical design basis can be set up.

To establish L_0 with any precision, a very large test batch must be used. Recent work (for example, that of Snare, using a batch of 500 bearings[5]) has given detailed information on the values for $S > 0.9$, and the existence of a minimum life has been confirmed; it appears that, during this critical running-in period, the strengthening effect of cold working offsets fatigue effects. A value of $L_0 = 0.05\,L_{10}$ is suggested for the minimum life; putting this value in Equation (9.1) gives

$$\frac{L}{L_{10}} = 0.95\left(\frac{\ln S}{\ln 0.9}\right)^{0.9} + 0.05 \qquad (9.5)$$

When extremely high reliability is required, Equation (9.5) can be used in place of Equation (9.4). The work on which this equation is based involved heavy loads and relatively short lives, but it is often under such circumstances that high reliability is required; the general applicability has yet to be confirmed.

9.4. BEARING INSTALLATION

The correct selection of type and size is only the first step in obtaining satisfactory operation; correct detailing of the installation is equally important. We will now consider some of these details, selecting those that have the greatest influence on the overall layout and that are likely to affect shaft strength, assembly methods and space requirements—all important factors even at the preliminary design stage. The following points will be taken up:

(i) shaft and housing tolerances;
(ii) support or location of inner and outer rings;
(iii) lubrication and sealing.

9.4.1. Shaft and housing tolerances

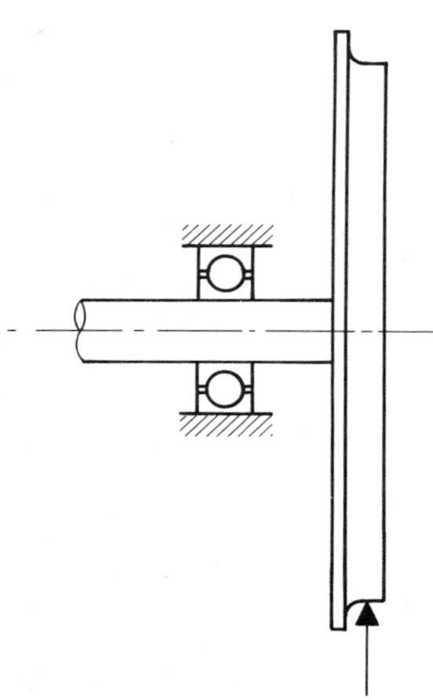

The diametral fit between the bearing rings and the shaft or housing is set by the requirement that the rings must be properly supported and that they must not rotate on the shaft or in the housing; any relative movement can cause damage to both the shaft and the bearing. When the load is fixed in direction relative to the mating parts there is no tendency for movement, or "creep" between them, and a tight fit is not required; the outer ring of a bearing carrying a wheel axle subjected to steady load fits this case (see sketch). When the load rotates with respect to a fixed point on a ring there is a tendency for that ring to rotate; the inner-ring/axle combination just illustrated is subjected to this type of load, and accordingly requires a tight fit.

Where the load direction is indeterminate, a tight fit is also used. In general, the more severe the loading the tighter the fit. For example, one manufacturer recommends a toleranced dimension of $50.000\,{}^{+0.039}_{+0.020}$ for a 50 mm nominal size shaft subjected to a "rotating inner ring load", with heavy loads. Note that the interference fit arising from the oversize shaft specified here will lead to a significant increase in the inner ring diameter on assembly; a larger than normal internal radial clearance on the bearing may have to be specified to ensure that this does not cause excessive tightness. In such a case, the shaft would be cooled, and the bearing heated in an oil bath, before assembly.

There is a tendency to assume that the rotating load condition occurs on the rotating element. However, in the example shown in the sketch, a heavy centrifugal load on the rotating shaft causes a rotating load on the stationary outer ring, and it is this ring which, in this case, must be restrained from rotating. The load is fixed in direction relative to the shaft and inner ring, which do not need to have a tight fit.

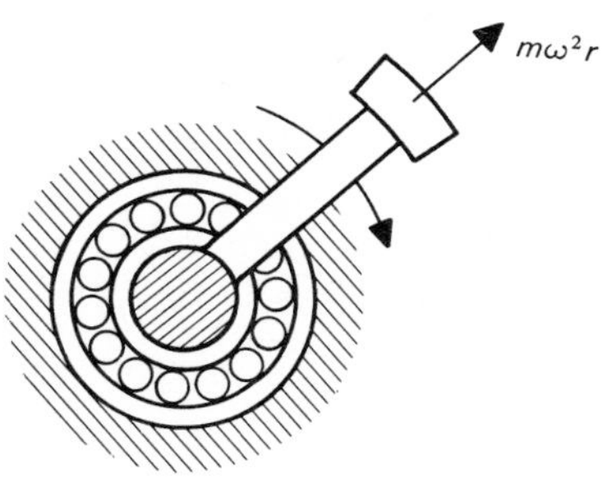

If axial loads on a bearing are to be avoided, one of the rings must be given freedom to move axially; a sliding fit is required at this location, and a clearance must be maintained between the bearing and any parts intended to provide axial retention. In the case of a cylindrical roller bearing, the omission of a lip allows some axial freedom for the roller.

Some information is given in Table 9.2 on the type of fit to be specified for different running conditions. The fits are described in terms of the nomenclature of Chapter 11, "Limits and Fits".

Table 9.2. Typical fits for shaft and housing under various load conditions. (For detailed recommendations, e.g., variation of fit with shaft size, consult manufacturer's catalogue.)

Fits	Example	Tolerance
For Steel Shafts		
Stationary Inner Ring Load		
Axial displacement of ring on shaft desirable	Wheel on non-rotating axle	g6
Axial displacement not necessary		h6
Rotating Inner Ring Load or Direction Indeterminate		
Light and variable loads	conveyors	j6
Normal to heavy loads	electric motors, pumps	k5, m5
Heavy shock loads	traction motors, rolling mill	n6, p6
For Steel Housings		
Stationary Outer Ring Load		
All types of loads, ring can be displaced axially	general application	H7
Direction Indeterminate		
Light loads, ring can be displaced axially	electric motors, pumps	J7
Normal loads, ring cannot be displaced		K7
Heavy shock loads	electric traction motors	M7
Rotating Outer Ring Load		
Light loads		M7
Normal and heavy loads		N7
Heavy shock loads		P7

Bearing rings are comparatively light and easily distorted. The shafts and housings used with them must be accurately machined and, most important, truly circular. If a split housing is used, care must be taken to ensure that it will always keep its correct form.

9.4.2. Axial location

The methods used to support and locate bearing rings are important to the successful operation of an installation, and the need to provide shoulders, lips and grooves for this purpose influences the selection of a layout for the parts associated with the bearing.

When two bearings are used to support a shaft, the designer must decide which will be the "locating" bearing if significant axial loads exist. This one must be fully supported to take the axial loads, and the other "non-locating" bearing must be installed so that it can take no axial load. This is done either by having one of the rings free enough to move axially, or by using a roller bearing which has internal axial freedom.

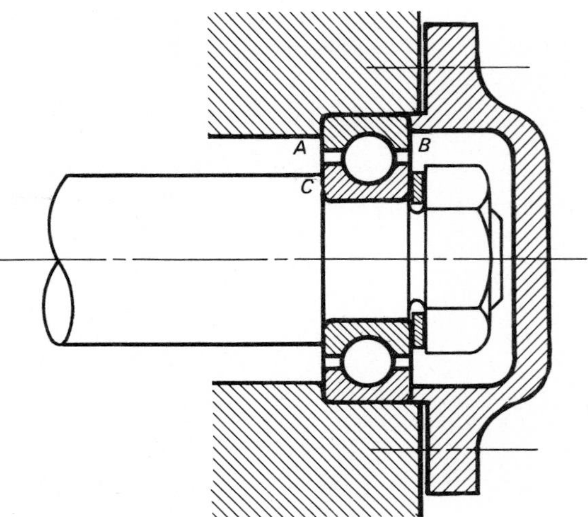

Fig. 9.3. *Bearing location using shoulders and nut.*

Varying levels of axial loading can be given different treatment. Heavy loads are best carried by shoulders that provide a rigid and uniform support. In Fig. 9.3, the outer ring is supported at *A* by a step in the housing and at *B* by a spigot on the cover fastened to the housing. A shoulder on the shaft at *C* locates the inner ring in one direction, and a locknut with washer locates it in the other direction. Such an installation will support axial loads in either direction. The shoulder heights must be correctly proportioned and are determined by the dimensions d_a and D_a (Tables A10) under the heading "abutment and fillet dimensions". The fillet radius r_a at the shaft shoulder is important in determining the fatigue strength of the shaft at this point. It should be as large as possible, but not greater than the radius at the corner of the ring. Values of r_a (max) are also given in tables of Appendix 10.

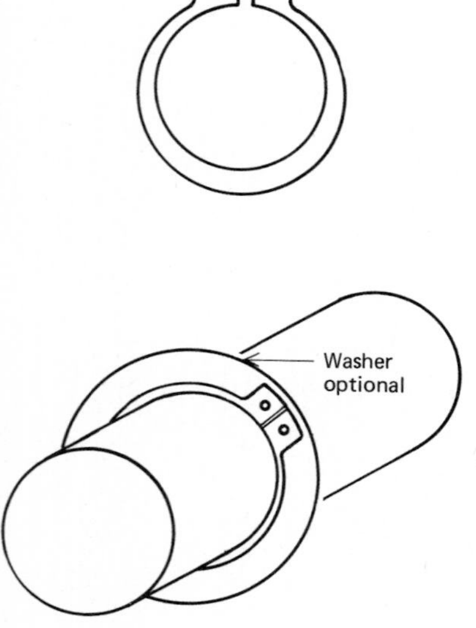

If the axial loads are not heavy, bearing location can be simplified by the use of circlips. These spring rings fit into grooves in the housing or shaft; a washer may be interposed between the bearing and circlip. It will be noted that the circlip groove forms a substantial notch that should be avoided where fatigue loading occurs. Another method of locating the inner ring is shown in Fig. 9.4, where the shoulder on the shaft is displaced axially by using a spacing collar that butts up against a component further along the shaft.

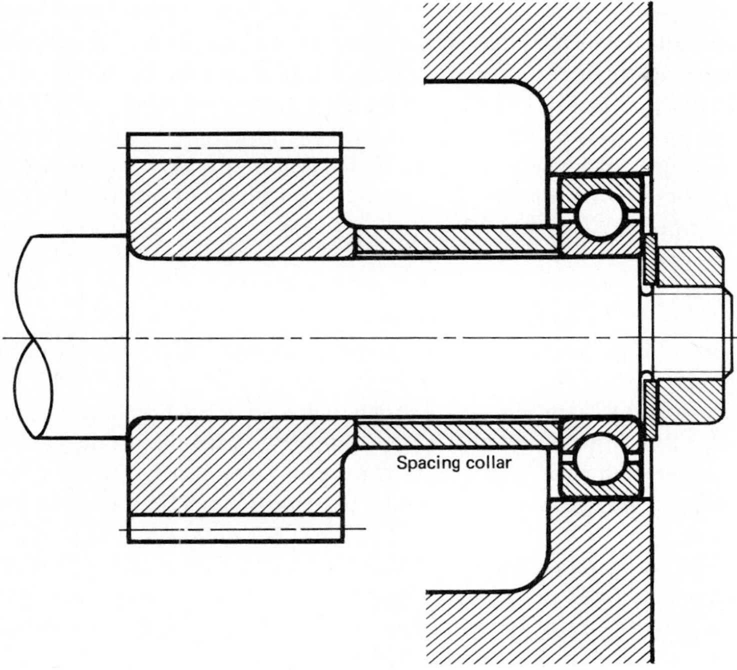

Fig. 9.4. *Bearing location illustrating the use of a spacing collar.*

It should be noted that in all cases both rings of a bearing must be retained, in the sense that they must be prevented from moving out of position. For example, in Fig. 9.5(a) the outer ring of the roller bearing would be free to move to the right in the housing; a circlip at A would be sufficient to retain it. Where a ring is given an interference fit to prevent rotation (Section 9.4.1), it can be regarded in most circumstances as adequately retained.

In the case of a non-separable bearing such as the ball bearing shown in Fig. 9.5(b), the inner ring needs no retention since it cannot move axially.

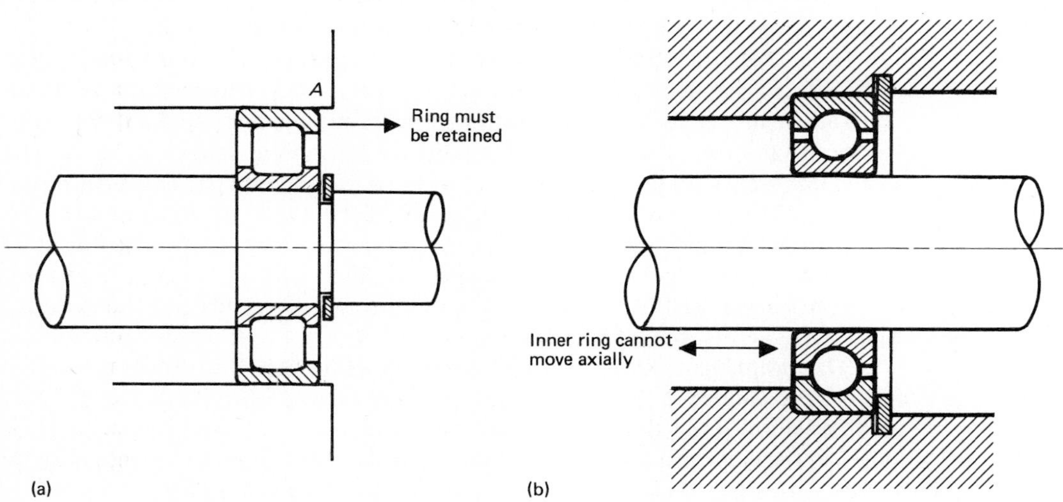

Fig. 9.5. *Bearing location: (a) required; (b) not necessary.*

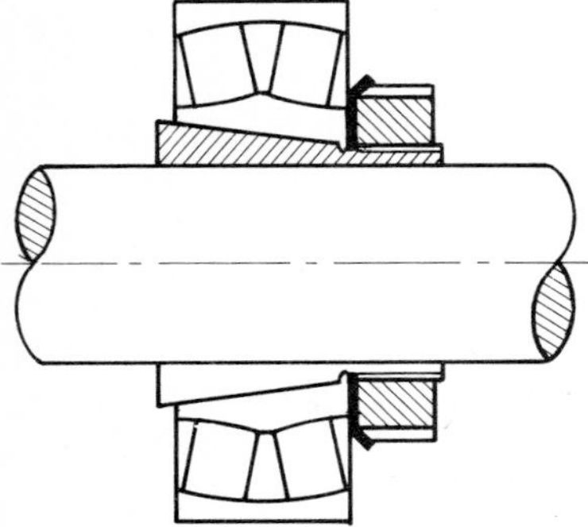

Some types of bearing, notably the self-aligning types, can be supplied with a taper bore which is used in conjunction with a tapered adapter sleeve. This sleeve, shown in the sketch, is pulled into the bore by means of a nut threaded onto one end of the sleeve; the sleeve has a slit machined in it which gives a tight grip on the shaft. This is a useful component, since it provides the means of fitting a bearing anywhere along the length of a shaft without requiring steps or grooves to be machined. It is used in conjunction with the plummer blocks described in Section 9.4.4.

9.4.3. Lubrication and sealing

The amount of lubricant required for correct and smooth running of a rolling bearing is very small. However, it must be provided at all times, and in many cases this is done by maintaining a volume of grease in the housing surrounding the bearing; this can be replenished periodically by feeding grease through a nipple. This internal supply of grease can serve a secondary purpose in filling up the passages in a groove or labyrinth seal and obstructing the entry of dirt and moisture.

For this type of lubrication, reasonably generous space around the bearing must be available to act as a grease reservoir, and the greasing nipple must be located so that the fresh grease will tend to flush the old grease out of the bearing.

When a bearing is subjected to particularly onerous duties – in terms of rotational speed or operating temperature – it is necessary to lubricate it with oil. This oil may be circulated around the bearing space by means of a flinger ring which is mounted on the shaft and which dips into an oil reservoir; in the case of high-speed bearings a flow of oil must be maintained. Oil circulation requires a pressure feed pump. In this latter arrangement, the oil will also serve to remove heat whether this is generated by friction in the bearing or conducted along the shaft from some high-temperature region; the oil can be cooled in a heat exchanger. Whatever means is adopted to circulate the oil, it is necessary to ensure that the bearing does not rotate continuously at high speed in an oil bath, since the churning action would add to the heat generation and the oil would tend to froth. It will be seen from the tables of Appendix A12 that a higher limiting speed is specified for oil than for grease lubrication.

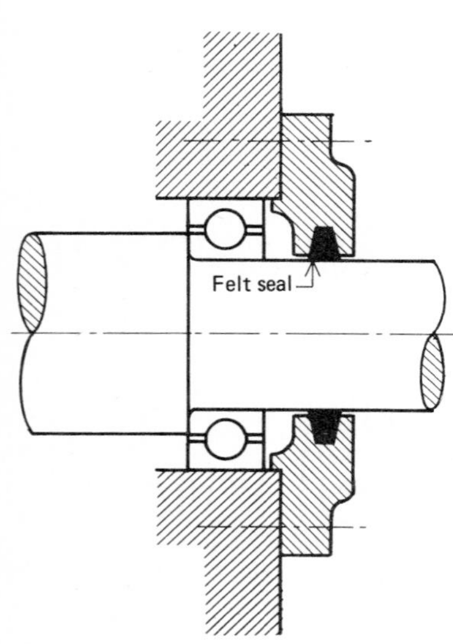

Seals are required to prevent bearings being contaminated with dust, dirt or moisture; they also prevent loss of lubricant. A simple form is the rubbing seal shown in the sketch; it consists of a felt ring which is soaked in grease before installation. This type is unsuitable for peripheral speeds above about 4 m/s because frictional heat generated at the surface could cause overheating.

In the labyrinth seal, there is no direct contact between the moving surfaces, the tortuous passage serving as a barrier to foreign matter. The action of a labyrinth seal is enhanced if the passages are filled with grease, and passages can be arranged for this purpose. Some designs of labyrinth seal are shown in Fig. 9.6.

The simple gap type in Fig. 9.6(a) is suitable for dry, dust-free conditions. A groove machined in the cover-plate (b) improves the effectiveness; this groove may be helical if the shaft always rotates in the same direction. Increasing the length of the labyrinth gives a further improvement; the tongues can be arranged axially with solid housings (c) and radially with split housings (d).

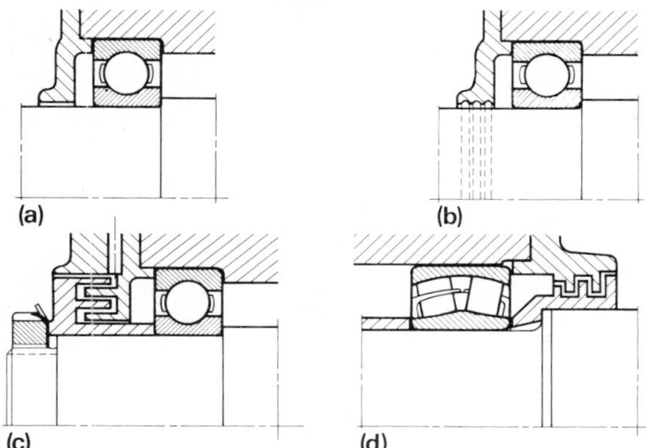

Fig. 9.6. *Labyrinth seals:*
(a) plain;
(b) grooved;
(c) for solid housing;
(d) for split housing.
(From SKF General Catalogue, *by permission of the SKF Ball Bearing Company.)*

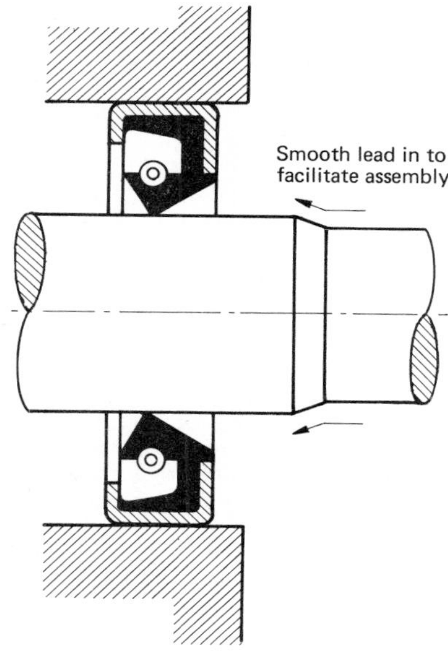

Contact seals are manufactured in which the sealing action is performed by a rubber lip held onto the shaft by means of a light spring. A typical configuration is shown in the sketch. This type of seal is pressed into a recess in the housing and does not usually require any retention. It must be carefully handled, since the lip is easily damaged; the surface on which it runs must have a smooth ground finish, and it must not be forced over rough edges when being assembled. These seals must not run dry, and a supply of oil or grease should be available on one side. A wide choice of sizes is available; for each shaft size there are usually several different outside diameters, so that a range of housing diameters can be fitted.

9.4.4. Built-up assemblies

Most manufacturers produce a variety of assemblies of bearings complete with housings, ready for use. If these are to carry a shaft, some means of securing the inner ring must be provided. Some examples of these assemblies are given here.

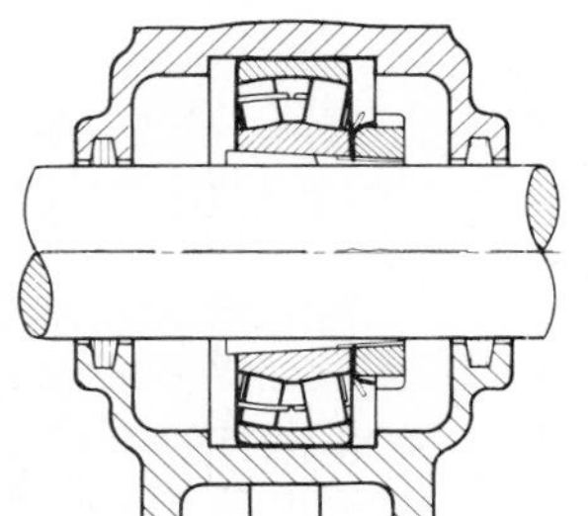

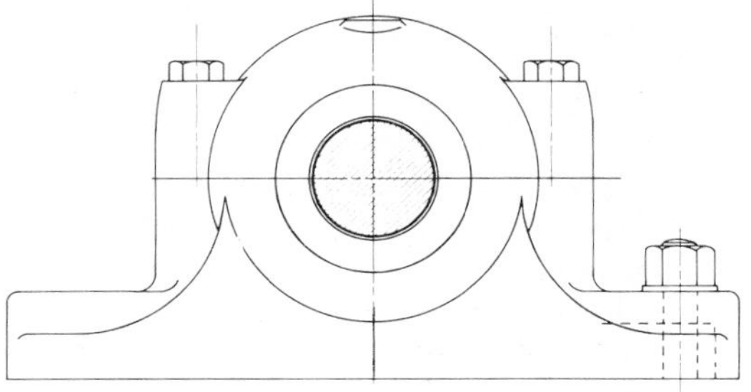

Fig. 9.7. *Plummer block with spherical roller bearing. (From* SKF General Catalogue, *by permission of the SKF Ball Bearing Company.)*

In Fig. 9.7, a spherical roller bearing installed in a plummer block is held onto the shaft by an adapter sleeve; this assembly can be fitted anywhere along the length of a parallel shaft.

A sealed deep groove ball bearing mounted in a cast iron flange housing[7] is shown in Fig. 9.8; in this case the inner ring is locked onto the shaft by means of an eccentric locking collar. The cartridge can be bolted directly onto a

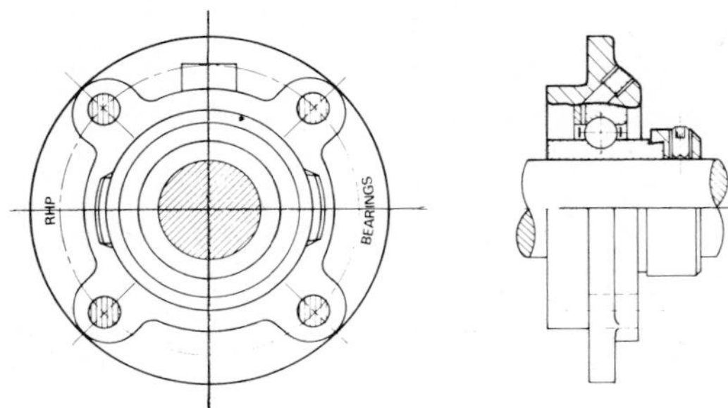

Fig. 9.8. *Flange-mounted bearing. (From* RHP Self-Lube Bearing Unit Catalogue, *by permission of the Bearing Service Company Limited (Ransome Hoffman Pollard Ltd).)*

support plate or bulkhead, and the spherical outside diameter of the outer race will accommodate some misalignment.

In aircraft control systems, there is a need for high-strength, low-friction control rod bearings; these rod ends are supplied by manufacturers in a variety of shapes and sizes. A typical one, taken from an *RHP Catalogue*[8], is illustrated in Fig. 9.9. The sealed self-aligning bearing will allow up to 10° misalignment in either direction. The saving in space and weight that results from using this specialised fitting is made clear if an attempt is made to design one using standard bearings, seals and locating methods.

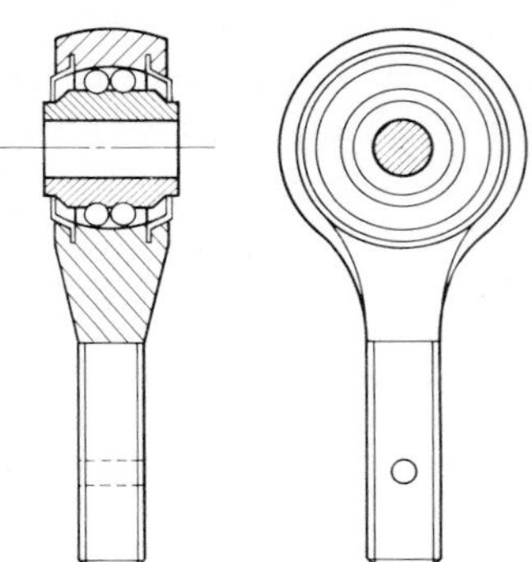

Fig. 9.9. *Aircraft-type control rod end. (From* RHP Bearing for Aircraft Catalogue, *by permission of the Bearing Service Company Limited (Ransome Hoffman Pollard Ltd).)*

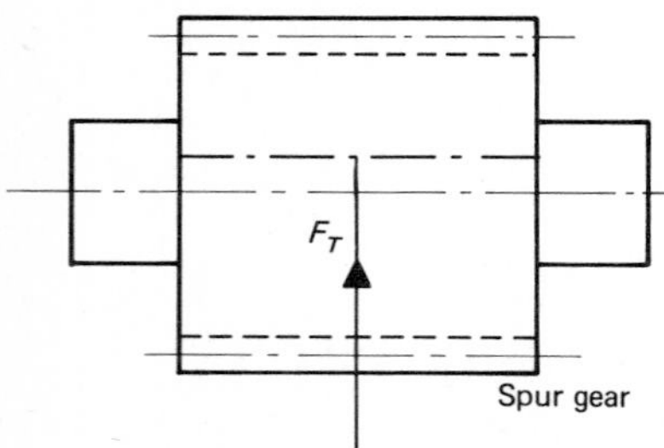

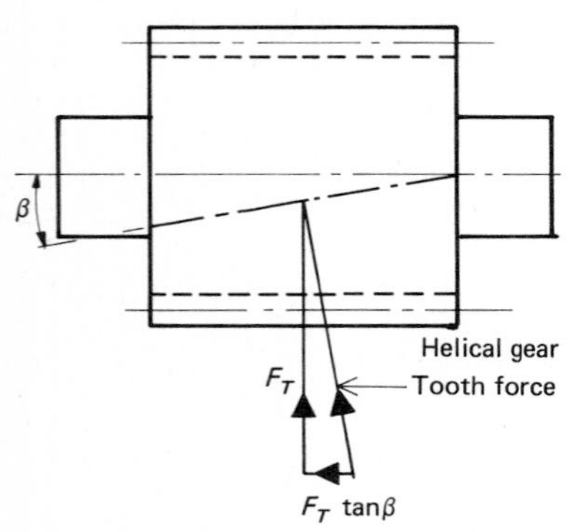

9.5. TYPICAL BEARING LAYOUT

Some examples are given here of the layout of bearings and associated components.[9] These are discussed in relation to spur and helical gear shafts, which can operate under a wide range of loads and speeds.

It can be seen from the sketches that while there is no axial load on a spur gear, the value of the axial load for a helical gear is given by $F_T \tan \beta$, where β is the helix angle of the tooth and F_T is the tangential force on the gear. This axial force will act in one direction under normal running conditions, but may reverse when the drive is decelerating. The axial force on the helical gear can be balanced out by the use of double helical teeth.

The arrangement shown in Fig. 9.10 is suitable for light or average loads and for speeds in the range of ordinary industrial practice. The two centrelines on the gear face indicate that it can be either a spur or helical gear. Oil splash from the gear lubricates the bearings, a lip-type seal retaining the oil. A small clearance on the end covers allows axial location without any possibility of preloading the bearings.

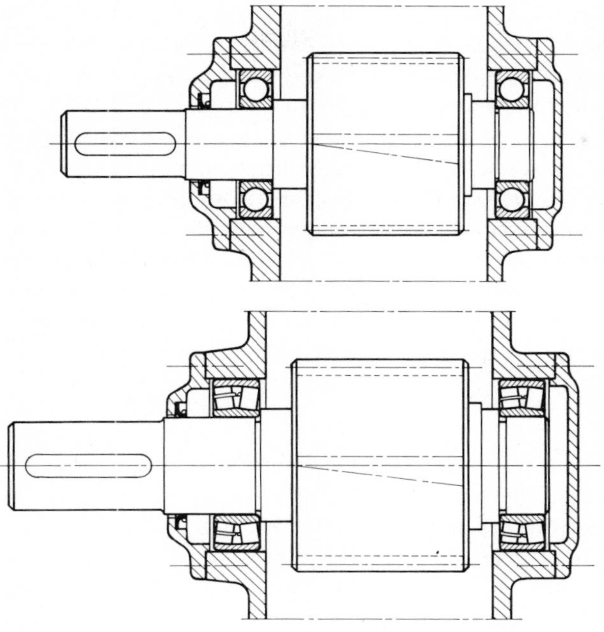

Fig. 9.10. *Spur or helical gear mounted in ball bearings. (From* SKF Publication 2462E, *by permission of the SKF Ball Bearing Company.)*

Fig. 9.11. *Spur or helical gear mounted in spherical roller bearings. (From* SKF Publication 2462E, *by permission of the SKF Ball Bearing Company.)*

A similar arrangement using spherical roller bearings (Fig. 9.11) is suitable for medium or low speeds and heavy loads.

For high speeds and loads, a more elaborate layout is shown in Fig. 9.12. The upper half of the drawing illustrates a single helical gear, the axial force

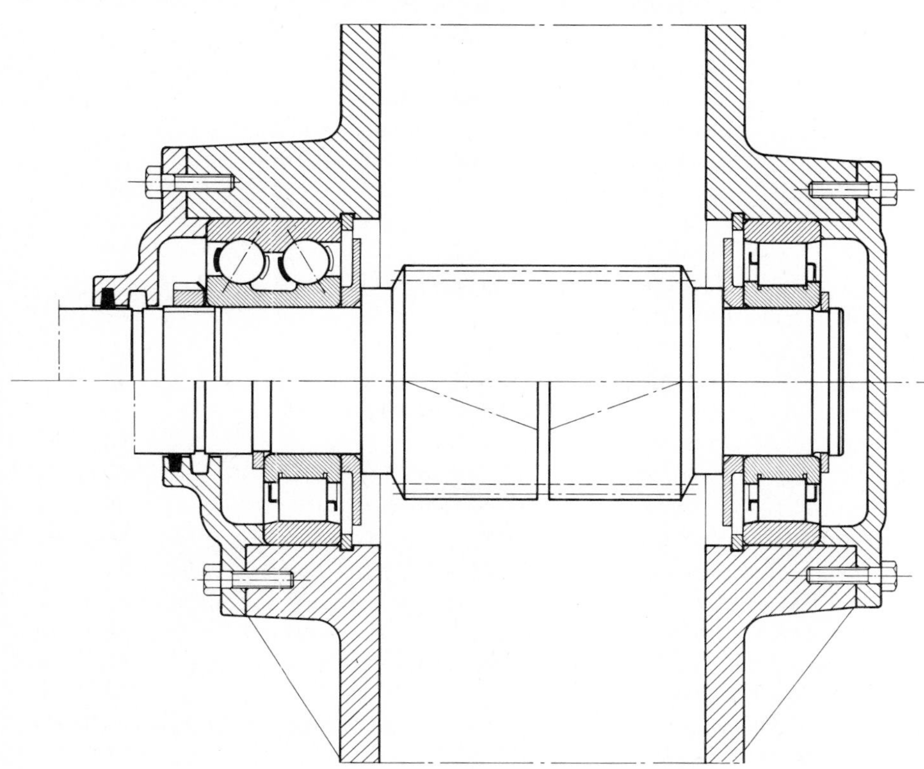

Fig. 9.12. *Alternative arrangement for spur gear and double helical gear. (From* SKF Publication 2462E, *by permission of the SKF Ball Bearing Company.)*

being taken on a double row angular contact bearing. In the lower half of the drawing, the double helical gear shown can be carried on two cylindrical roller bearings; there is no axial load, and axial freedom is provided for the gear to take up the correct meshing position with its mating gear. Some of the details of this installation are of interest. The angular contact bearing is retained by a locknut which presses it onto a washer interposed between it and the shaft shoulder. The washer protects the bearing from excessive oil splash at high speeds, and also permits the use of a larger fillet radius than would be possible if the bearing butted directly against the shoulder. Oil is prevented from reaching the seal in excessive quantities by means of an oil-slinger groove which throws the oil off the shaft into an annulus in the cover plate connected with the housing by means of a drain hole. The roller bearings are located throughout with circlips.

LIST OF SYMBOLS

C	basic dynamic load rating
C_0	static carrying capacity
d	bore diameter
D	outer ring diameter
e	factor in tables A10
E	Young's modulus
f	friction factor
F_A	axial bearing load
F_R	radial bearing load
F_T	tangential force at gear tooth
L	life (millions of revolutions)
L_h	life (hours)
L_0	life below which no failures occur
L_{10}	life for which 10% of sample fail
p	bearing pressure, exponent in life equation
P	equivalent dynamic radial load
P'	force per unit length
R, r	radius
S	fraction of sample still serviceable
T	torque
X, Y	factors in life equation
β	helix angle of gear tooth
ε	exponent in Weibull distribution equation
ν	Poisson's ratio

REFERENCES

[1] ENGINEERING SCIENCES DATA, 65007, *General Guide to the Choice of Journal-Bearing Type*, Engineering Sciences Data Unit, London, 1965.

[2] F. T. BARWELL, *Lubrication of Bearings*, Butterworths, London, 1956.

[3] ENGINEERING SCIENCES DATA, 66023, *Calculation Methods for Steadily Loaded Pressure Fed Hydrodynamic Journal Bearings*, Engineering Sciences Data Unit, London, 1966.

[4] H. C. RIPPEL, *Cast Bronze Bearing Design Manual*, Cast Bronze Bearing Institute, Illinois, 1960.

[5] B. SNARE, "How Reliable are Bearings?", *The Ball Bearing Journal*, 162, 1970 (SKF Ball Bearing Company).
[6] SKF BALL BEARING COMPANY, *SKF General Catalogue*, 2800E/GB600 Edition, Aktiebolaget Svenska Kullagerfabriken, Gothenburg, Sweden, April 1970.
[7] RANSOME HOFFMAN POLLARD LTD, *RHP Self-Lube Bearing Units Catalogue*, RHP Transmission Bearing Division, Knottingley, 1972.
[8] RANSOME HOFFMAN POLLARD LTD, *RHP Bearings for Aircraft*, RHP Aerospace Bearings Division, Stonehouse, 1972.
[9] SKF BALL BEARING COMPANY, "Bearings for Gear Shafts", *SKF Publication 2462 E*, Aktiebolaget Svenska Kullagerfabriken, Gothenburg, Sweden, 1964.

BIBLIOGRAPHY

ALLAN, R. K., *Rolling Bearings*, 2nd ed., Pitman, London, 1956.
HARRIS, T. A., *Rolling Bearing Analysis*, John Wiley, New York, 1966.
TRUMPLER, P. R., *Design of Film Bearings*, Macmillan, New York, 1966.

PROBLEMS

9.1. Complete the table at the locations marked*. Assume a life of 10 000 hours where necessary.

Case	Load (kN) Radial	Load (kN) Axial	Speed (rpm)	Life (10^6 revs)	Bearing size
(a)	*	0	800	—	Series 6308
(b)	*	0	800	—	Series NU 308
(c)	4.0	2.5	800	*	Series 6308
(d)	5.0	3.0	—	10	Series 63*
(e)	6.0	1.0	—	10	Series 63*
(f)	4.0	1.1	1200	—	Series 62*

9.2. The forces acting on a shaft are shown in the sketch. Select suitable bearings for the locations A and B; the life requirement is 50×10^6 revolutions. Check for strength, using the ASME code, the shaft sizes that are determined by the bearing selection, assuming factors for light shock loading. Shaft torque = 2.5 kN m.

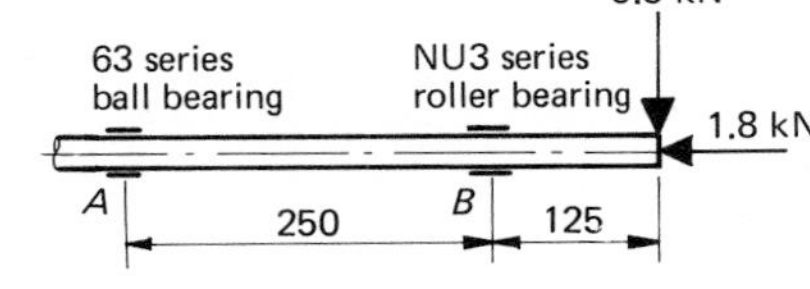

9.3. Check the effect on the design of interchanging the bearing types at A and B in Problem 9.2.

9.4. Select suitable bearings to support a shaft at C and D. A life of 8000 hours at 1500 rpm is required. Check the shaft strength using the ASME code, with p_q = 55 MPa, and K_b = 1.5, K_t = 1.0. Torque on shaft = 0.1 kN m.

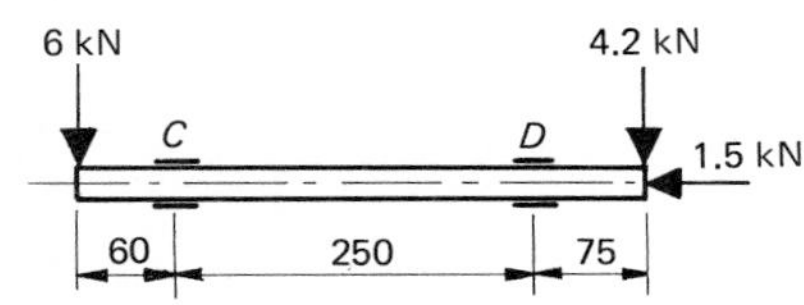

9.5. Use Equation (9.5) to select a Series NU 3 bearing to carry a radial load of 30 kN. Life required is 100 hours at 3000 rpm, with a probability of failure of 0.005.

9.6. The bearings on electric motors are designed to carry an overhung load on the output shaft as well as the weight of the motor. Manufacturers' catalogues give guidance on the estimation of allowable loads. Select a motor from a catalogue which gives shaft size and bearing type, and estimate the life of the bearing, assuming some worst loading condition due to a belt or gear drive on the output shaft. Check also the strength of the shaft in fatigue.

9.7. If a bearing operates at different loads and speeds during its life, the Palmgren-Miner rule (Equation (7.28)) can be used to estimate the expected life. A Series NU 308 bearing has a dynamic load rating of 50.7 kN (Table A10.4). Estimate the life for the following load cycle:

radial load 5 kN for 60% of running time;
radial load 10 kN for 30% of running time;
radial load 20 kN for 10% of running time.

9.8. Make a scale sketch of the installation shown in Fig. 9.10, using a Series NU 306 bearing on the left-hand side. The bearing on the right-hand side must be a fully located ball bearing to carry the axial load from the helical gear. It must be large enough to allow the gear with an outside diameter of 90 mm to pass through the housing for assembly.

9.9. Select suitable bearings for the shaft described in Problem 2.6. Design for a life of 10 000 hours at 1200 rpm, using one ball bearing and one roller bearing. Apply a load factor of 1.3 to the specified loads.

9.10. Select suitable bearings for the installation shown in the upper half of Fig. 9.12, given the following information:

Power transmitted	12 kW
PCD of gear	65 mm
Pressure angle	20°
Helix angle	18°
Face width	100 mm
Rotational speed	1200 rpm

Life required is 10 000 hours. Check the strength of the shaft that matches the bearing sizes.

10 Power Transmission

10.1. INTRODUCTION

In this chapter we discuss the design and selection of some of the components used in the transmission and control of power. We will first consider the bearing that the type of driver and driven machine have on the choice of transmission components. When a prime mover or power unit is selected, the first consideration must be the way in which its output characteristics – for example, the power and torque over a range of speeds – match the needs of the machine it is to drive. It will often be the case that the two are not well matched, and in such cases it is necessary to alter the speed ratio so as to balance the available and demanded torque. The picture is usually complicated by the fact that accelerating as well as steady-state conditions are involved, and by the need to keep the power unit operating in the range of its maximum economy. The type of load imposed by the driven machine on the driver will also have implications in the choice of components, some being better suited than others to take shock loads.

After looking at some of the components used in power transmission and the types of drive available, the design and selection of several of these will be considered in more detail.

10.2. A TYPICAL PROBLEM

We illustrate the design of a transmission system by considering briefly a familiar problem – that of matching the engine output to the demand at the wheels of a road vehicle.

The characteristics of the power unit determine the need here. In the first place, since the internal combustion engine has a minimum running speed (the idling speed) there must be some way of disconnecting it from the wheels when the vehicle is stationary. A mechanical clutch or fluid flywheel (see Sections 10.7 and 10.8) can be used to do this. If a certain class of electric motor or a steam engine is used, the high torque available at zero speed may allow the clutch to be dispensed with. Since the direction of rotation of the latter two types of power unit can be changed, the provision of reverse drive can be arranged without the reverse gear required by an internal combustion engine.

When the vehicle is running at a steady road speed, the traction force can be measured or estimated. This force – or what is its equivalent, the torque at the wheels – can be plotted for different road speeds and several gradients, as shown in Fig. 10.1. If the engine output that reaches the wheels is now plotted on this diagram for some specific type of transmission, the equilibrium running speeds can be found. We will consider only the full throttle output which will

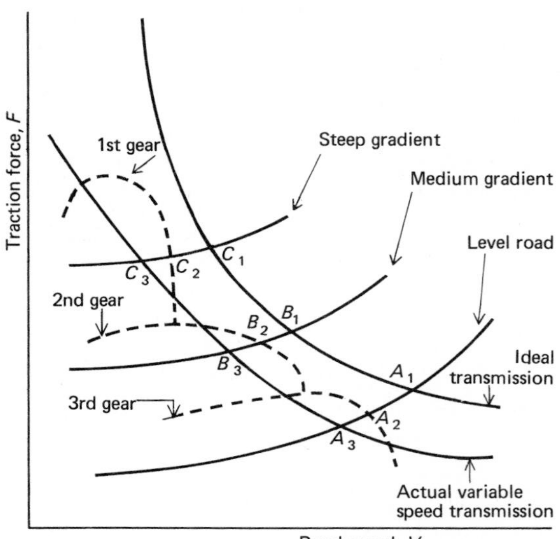

Fig. 10.1. *Traction force required and available for a road vehicle.*

give the maximum speeds, but curves for some specified fraction of the engine output can also be displayed.

The plotting process involves selecting an engine speed, N_1, and finding the corresponding torque, T_1, from the engine performance characteristic (Fig. 10.2). Knowing for a given drive the speed and torque ratios, this and other similar points can be transferred to Fig. 10.1. In making this transfer, the drive efficiency must be taken into account. In the case of a gear drive, the speed ratio is fixed but the output torque will be less than the ideal value by some predictable amount. For a fluid coupling, the input and output torque are the same but the output and input rpm will differ by an amount (the slip) that can be determined from the coupling characteristics (see Section 10.8).

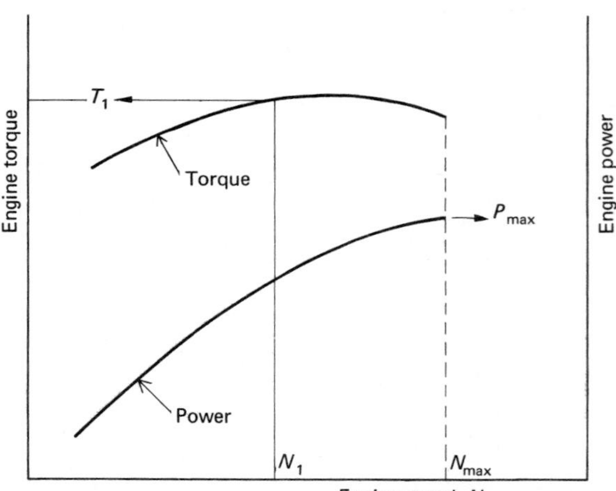

Fig. 10.2. *Full-throttle engine performance.*

Consider first an infinitely variable speed drive with 100% efficiency. This will enable the engine to run at its maximum speed for all road speeds, so that the power produced by the engine—and used at the wheels—would be the maximum value, P_{max}. Since the power used in driving the vehicle is given by

FV, where F is the traction force and V is the speed, the equilibrium running points will be located at $A_1B_1C_1$ where the hyperbola $FV = P_{max} =$ (constant) intersects the curves for the various gradients.

The traction force available for various speeds for a variable speed transmission with losses is also shown on Fig. 10.1, and the running points are located at A_3, B_3, C_3. If a three-speed gear drive is used, each gear will give a separate curve, and again the intersection of these curves with the power required lines (points A_2, B_2, C_2) will determine the maximum attainable speeds. For the case represented in Fig. 10.1, the efficiency of the variable speed drive is lower than that of the gear drive, but each fixed gear can only supply maximum available power at one ground speed, while it can be utilised at all ground speeds with the variable speed drive.

These curves can also be used to determine the accelerating force available. Consider the third-gear and level-ground curves which are reproduced in the sketch. At each of the speeds, V_1, V_2 and V_3, the excess traction force to accelerate the vehicle is given by the intercepts CD. From these values, the acceleration at full throttle in third gear can be calculated up to the limiting speed, V_{max}.

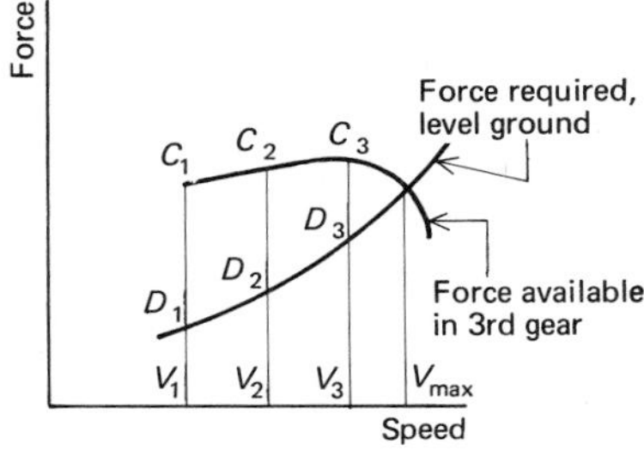

A great variety of transmission devices have been developed for road vehicles, and some of these are described in papers given at a conference on "Drive Line Engineering".[1] The drive line – the components from the engine flywheel to the road-tyre contact – includes clutches, gears, fluid couplings and chain and belt drives. Several of these – for example, fluid coupling, clutches and gears – may be used in building a transmission system for a vehicle. We will consider some aspects of the design and selection of each of these components, although, since their use in road vehicles is rather specialised, this particular application will not be stressed.

The selection procedures are relatively simple, and are set out in a range of handbooks and standards. Correct detailing is important for economy and performance, and some of the design features of the various installations will be discussed. Perhaps the most important decision to be made concerns the initial selection of the power unit and control method, and we will first consider this aspect briefly.

10.3. THE POWER UNIT

One of the most common power units in use is the squirrel-cage electric induction motor. It is robust, has a low first cost and its output speed remains fairly constant over a range of loads. Low-inertia machines can be directly coupled to it, if one of the standard speeds coincides with the required value. V-belt drives (Section 10.4) provide a flexible and economic method of obtaining other speeds, and in the low power range (up to about 5 kW) the variable-speed V-belt drive gives a speed range of from 5 to 15 to 1.

Before the design of a drive for a motor can be undertaken, its operating characteristics must be known. These are shown for a squirrel-cage induction motor in Fig. 10.3. The rated values of torque and current are given at a speed typically about 5 % below the synchronous speed, and although the motor will run at other speeds, the highest efficiency will be achieved close to the rated value. The speeds available are those corresponding to synchronous speeds; for example, with a 50-Hz supply, the speeds are 3000, 1500, 1000 and 750 for 2-, 4-, 6- and 8-pole machines respectively.

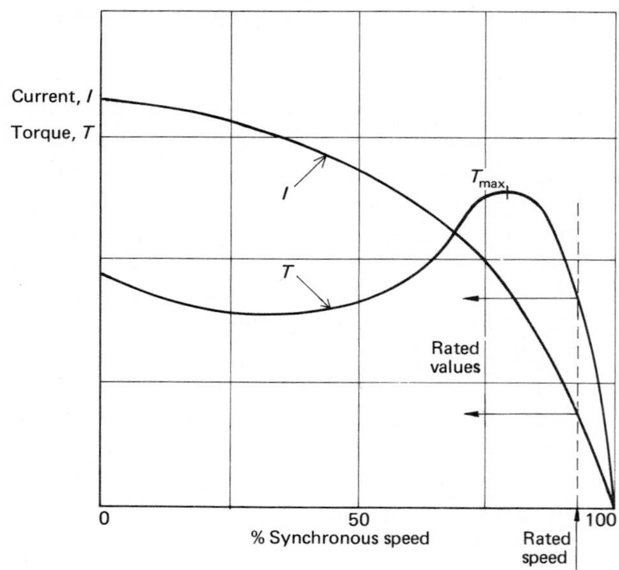

Fig. 10.3. *Typical torque and current characteristics for an induction motor.*

The shape of the torque curve can be varied to suit different applications by appropriate design of the windings, but a typical value of the ratio $\frac{T_{start}}{T_{rated}}$ is from 1.5 to 2, while the value of $\frac{I_{start}}{I_{rated}}$ may range from 5 to 7 with "direct-on-line" starting. The designer must take these conditions into account since they may provide the limiting conditions in the design of associated equipment.

When a motor of this type drives a high-inertia load, the large starting current will be drawn for a long period and special starting arrangements – for example, star-delta or auto-transformer starter connections – may be needed. Recourse can be made to the wound-rotor type of induction motor which is suitable for starting high-inertia loads, drawing current within acceptable limits while maintaining high starting torque. It is generally more expensive than the squirrel-cage type both in starting equipment and in the motor itself. An alternative is to use a fluid coupling (Section 10.8) which will allow the motor to come to speed quickly and will accelerate the load over a longer period, permitting direct-on-line starting for the motor. This coupling will also insulate the motor and drive from machines which have heavy shock characteristics.

A number of speed change units based on belt and other friction devices are produced and described in catalogues. Although not usually well suited to precise or automatic control, they will provide a service at a lower cost than the electrical or hydraulic methods that must be used where precise control over a wide speed range is needed. If variable speed is an essential feature, the designer must establish the speed range and accuracy and must specify other requirements such as braking and reversing. These may then be compared with the characteristics that typify various systems, such as

speed range;	overall efficiency;	cost of motor;
speed regulation;	braking;	cost of control system;
response rate;	reversing;	total cost.

The Ward-Leonard control system mentioned in Chapter 5, "Machines and Structures", uses the characteristics of the direct current motor to provide accurate speed control over a range of up to 15 :1. A block diagram for such a system is shown in Fig. 10.4. Because of the high electrical and mechanical inertia of the machines, the rate of response is not high and it must be expected that the cost will be at least four times the cost of a constant-speed drive of the same output.

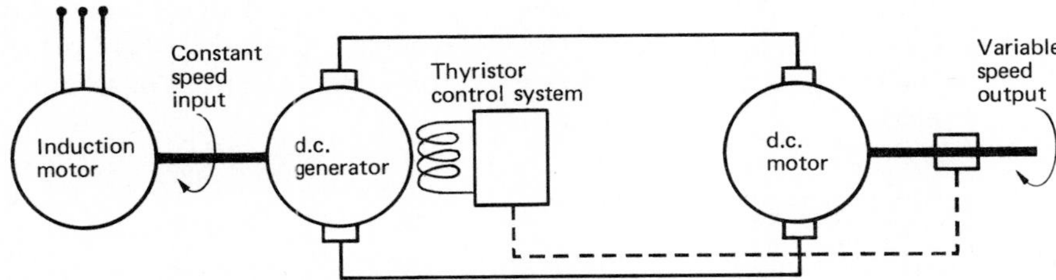

Fig. 10.4. *Block diagram for Ward-Leonard drive.*

Thyristors were first used in this type of equipment to provide the field supply. These solid-state devices switch a rectified alternating current power supply in such a way that only the fraction of the waveform needed to give the required power is used (see Fig. 10.5). High-power thyristors are now used to supply the armature and field circuits of large machines. Thyristors lend themselves to fast response control and the overall efficiency of the system is high. The power shovel illustrated in Plate 5.1 uses motors with this type of control.

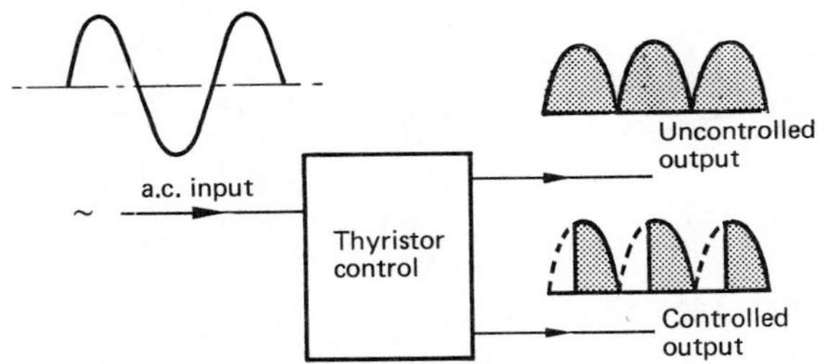

Fig. 10.5. *Thyristor-controlled power supply.*

The hydraulic variable-speed system, also mentioned in Chapter 5, provides speed variation and control over a wide range. A typical arrangement, in which a variable-stroke pump drives a hydraulic motor, is shown in Fig. 10.6. Since leakage of the working fluid across the units is small, the speed of the motor is closely proportional to the pump output or stroke length. An electric motor or power take-off from an engine can be used to drive the pump. These systems operate at pressures of 20 MPa or more and form a compact drive with efficiencies in the range from 70 % to 80 %.

These three examples illustrate the variety and complexity of variable-speed drives. When they are used, their cost must be justified by the improved performance of a drive that is well matched to the demands of the driven machine.

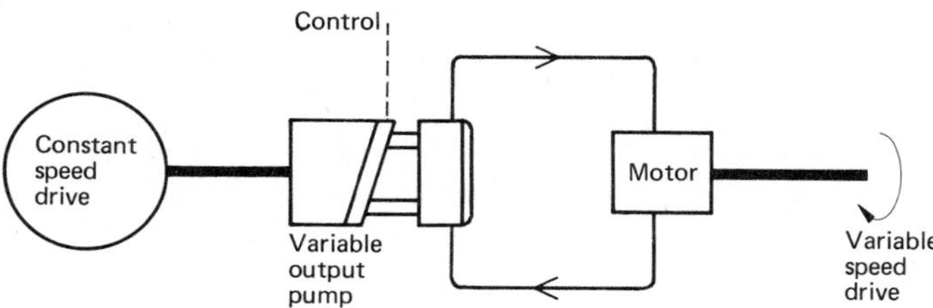

Fig. 10.6. *Block diagram for variable-speed hydraulic drive.*

For further information on variable speed drives the reader is referred to works of Cotton[2], Fitzgerald *et al*[3] and "Mechanical Drives"[4] where details of electric motors—such as the Schrage motor—and of various mechanical drives are given.

10.4. V-BELT DRIVES

Belts are used to transmit power between shafts with parallel axes where an exact speed ratio is not necessary. When a driver and driven pulley are connected by a belt as shown in Fig. 10.7(a), the torque is transmitted by virtue of the difference in tension on the tight and slack sides. The variation of stress along the length of the belt is shown in Fig. 10.7(b), the value on slack side T_L being determined, for a given torque, by the pre-tension when the shafts are stationary and by the centrifugal loads set up as the belt follows a curved path.

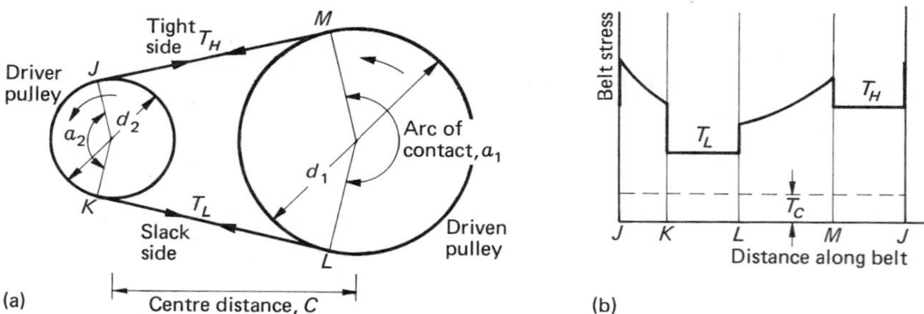

Fig. 10.7. *Belt drive: (a) layout; (b) variation of belt stress.*

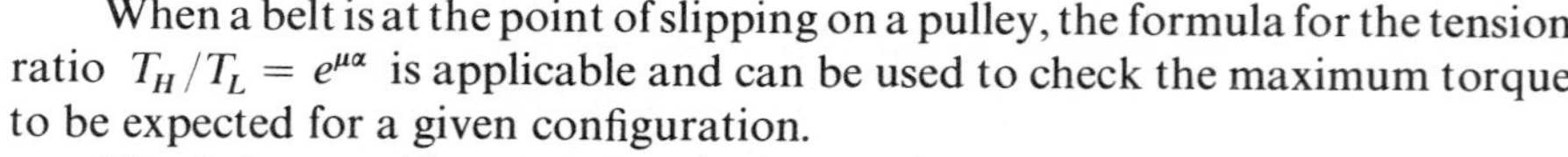
When a belt is at the point of slipping on a pulley, the formula for the tension ratio $T_H/T_L = e^{\mu\alpha}$ is applicable and can be used to check the maximum torque to be expected for a given configuration.

Flat belts provide an economical method of transmitting power between two shafts whose axes are relatively far apart, but the V-belt is far more widely used since it provides a compact installation with smaller loads on the shaft. Individual belt manufacturers supply tables or calculators for selecting flat belt sizes.

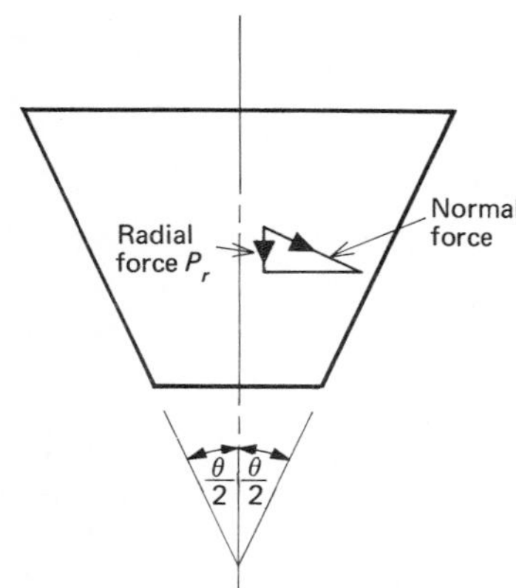

In a V-belt drive, the belt runs in a straight-sided groove in the pulley. The normal pressure between the belt and pulley surface is increased, for a given radial force or belt tension, as the angle Θ of the V of the belt is decreased (see sketch). A lower limit for this angle is set by the tendency of the belt to jam in the groove and, in practice, an included angle of 40° is accepted as standard. If P_r is the radial force, the normal force will be $P_r/\sin 20$ or $2.9\,P_r$. The increased surface pressure that results allows greater belt forces to be applied on the tight side before slip and increases the capacity of a V-belt above that of a flat belt.

The size of belts has been standardised and Fig. 10.8(a) shows the cross-section of the most commonly used sizes to scale; these are designated A, B, C and D in BS 1440, *Endless V-Belt Drives*[5]; full descriptions and design information are given for these and for two lighter grades, designated Y and Z. The latter, which are intended for light domestic applications, are not considered here.

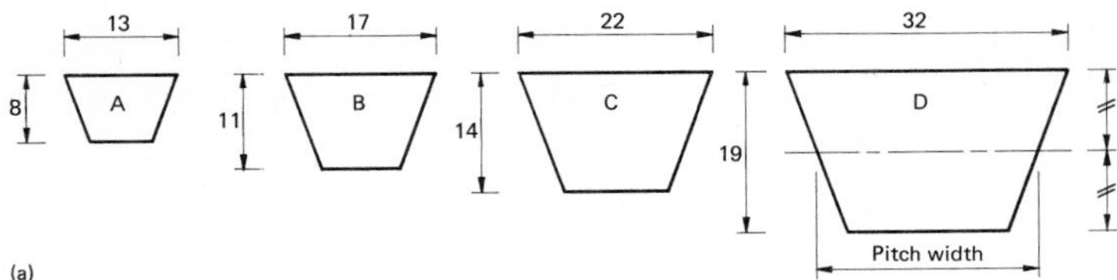

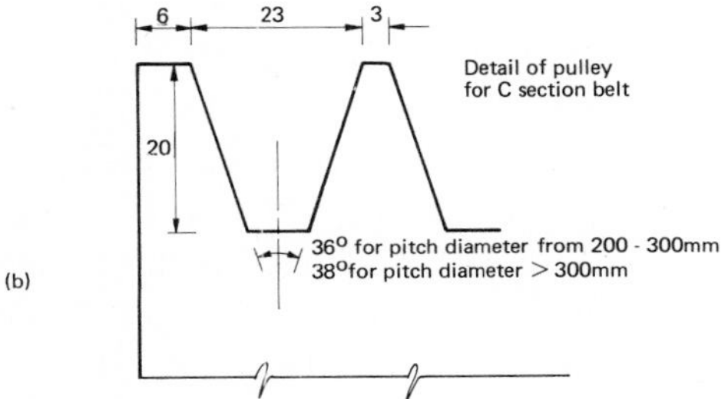

Fig. 10.8. *V-belt proportions drawn to scale:*
(a) belt cross-sections;
(b) pulley detail.

Some typical pulley dimensions are also shown in Fig. 10.8(b); it will be seen that the groove angle is less than 40° so as to accommodate the change in belt cross-section that occurs owing to the Poisson effect as the belt bends.

Standard sizes of pulley which can be purchased from stock follow very closely the preferred number series, and these are given in Table 10.1 in terms of the pitch diameter of the pulley. Cast iron is specified as a suitable material for belt speeds below 30 m/s. To keep belt wear to a low value, the surface roughness on the sides of the grooves should not be greater than 0.1 μm.

Table 10.1. Recommended standard pulley pitch diameters.

Belt size	Pulley pitch diameter (mm)
A section	75–200, follow R40 series*
	224–800, follow R20 series*
B section	125–224, follow R40 series
	250–1120, follow R20 series
C section	200–315, follow R40 series
	355–1600, follow R20 series
D section	355–630, follow R40 series
	710–2000, follow R20 series

*See Table 11.2 for R20 and R40 preferred number series.

The pitch length, L_p, of a V-belt is measured around the circumference at half the height of the cross-section, and standard lengths are given in Table 10.2. It is quite usual for a drive to be made up of several V-belts running on the same pulley, and to ensure equal sharing of load in such multi-belt drives it is necessary to specify a matched set. Normal tolerances on pitch length are about $+2\%$ to -1%, but a system of coding is used so that matched sets with much closer tolerances can be selected. Construction is typically of rubber that is bonded to reinforcing fibres placed at the most highly stressed regions.

The following equations can be derived for the centre distance, C (see Fig. 10.7(a)) and the pitch length, L_p

$$L_p = 2C + \frac{\pi}{2}(d_1 + d_2) + \frac{(d_1 - d_2)^2}{4C} \quad (10.1)$$

$$C = A + (A^2 - B)^{\frac{1}{2}} \quad (10.2)$$

where

$$A = \frac{L_p}{4} - \frac{\pi}{8}(d_1 + d_2)$$

$$B = \frac{(d_1 - d_2)^2}{8}$$

and d_1, d_2 are the pitch diameters of the large and small pulleys respectively.

A simple method of obtaining a variable speed ratio with a V-belt is to use pulleys whose sides can be displaced to vary the effective pitch diameter. The two pulleys making up a drive are arranged so that as one increases the other decreases in effective diameter. A description of this type of drive can be found in "Mechanical Drives".[4]

10.4.1. Design of V-belts

The power transmitted by a belt will be limited at high torque by slip, but the limiting condition for reasonable life is fatigue. The changes in stress in the belt as it travels one pitch length are shown in Fig. 10.7(b), and the significant feature of the loading is the repeated bending that occurs at each encounter with a pulley. The belt life (in hours) will be lowered as the pulley diameters are decreased and as the number of loadings is increased at higher rpm. Lengthening the belt will have a favourable effect since, at a given speed, the number of repetitions is decreased. Any design method should take these factors, as well as the intrinsic strength of the belt, into account. Although a V-belt drive helps to insulate the driver from shock loads in the driven unit, the effect of these loads on the life of the belt must be accounted for.

Power ratings given in BS 1440[5] are derived using an allowable stress determined by deducting a bending and a centrifugal stress from the fatigue strength of the material. The fatigue strength is taken as proportional to $S^{-0.09}$, where S is the belt speed; and the allowable stress, p_t, is given by an equation of the form

$$p_t = c_1 S^{-0.09} - \frac{c_2}{d} - c_3 S^2$$

Here c_2/d is the bending and $c_3 S^2$ is the centrifugal stress in the belt. For example,

Table 10.2. Standard belt pitch lengths and power correction factors for length.

Belt cross-section symbol							
A		**B**		**C**		**D**	
Pitch length (mm)	Factor	Pitch length (mm)	Factor	Pitch length (mm)	Factor	Pitch length (mm)	Factor
630	0.80	930	0.81	1 560	0.82	2 740	0.82
700	0.82	1000	0.83	1 760	0.84	3 130	0.86
790	0.84	1100	0.85	1 950	0.87	3 330	0.87
890	0.86	1210	0.87	2 090	0.88	3 730	0.90
990	0.88	1370	0.90	2 190	0.90	4 080	0.92
1100	0.90	1440	0.90	2 340	0.91	4 620	0.94
1250	0.93	1560	0.92	2 490	0.92	5 400	0.97
1430	0.96	1690	0.94	2 720	0.94	6 100	1.00
1550	0.98	1760	0.95	2 800	0.95	6 840	1.03
1640	0.99	1950	0.97	3 080	0.96	7 620	1.05
1750	1.00	2070	0.98	3 310	0.98	8 410	1.07
1940	1.02	2180	0.99	3 520	0.99	9 140	1.09
2050	1.04	2300	1.00	3 710	1.00	10 700	1.12
2200	1.05	2500	1.02	4 060	1.02	12 200	1.16
2300	1.06	2700	1.04	4 450	1.04	13 700	1.18
2480	1.08	2870	1.05	4 600	1.05	15 200	1.20
2570	1.09	3090	1.07	5 010	1.07		
2700	1.10	3200	1.08	5 380	1.08		
2910	1.12	3500	1.10	6 100	1.11		
3080	1.13	3700	1.11	6 860	1.14		
3290	1.14	4060	1.13	7 600	1.16		
3540	1.16	4430	1.15	9 100	1.21		
		4610	1.16	10 700	1.24		
		5000	1.18				
		5370	1.19				
		6070	1.20				

Note: Manufacturing tolerances on the nominal lengths amount to about +2 to −1% of the length. In a multi-belt drive, matched belts which have tolerances of about one-fifth of these values should be used.

Source: Extract from BS 1440, *Endless Belt Drives*, is reproduced by permission of the British Standards Institution, 2 Park Street, London W1A 2BS.

the equation for power for the B-section belt is

$$\underset{(\text{kW})}{\text{Power}} = Nd\left[0.0585(Nd)^{-0.09} - \frac{2.626}{d} - 19.12 \times 10^{-9}(Nd)^2 + 2.626N\left(1 - \frac{1}{F}\right)\right] \tag{10.3}$$

where N is the rpm/1000 of the fastest shaft; d is the pitch diameter of smaller pulley (mm); and F is the small diameter factor (dependent on speed ratio). The last term in the equation allows for an increase in belt stress. This arises from the fact that, as the speed ratio increases, the bending becomes less severe on the larger of the two pulleys.

Table 10.3. (a). Power ratings (kW) for A-section V-belts.

Speed of faster shaft (rpm)	Smaller pulley pitch diameter (mm)								
	75	80	85	90*	100*	106*	112*	118*	125*
200	0.19	0.22	0.24	0.26	0.31	0.33	0.36	0.39	0.42
400	0.33	0.37	0.42	0.46	0.55	0.60	0.66	0.71	0.77
600	0.46	0.52	0.59	0.65	0.78	0.85	0.93	1.00	1.08
800	0.57	0.66	0.74	0.82	0.98	1.08	1.18	1.27	1.38
1000	0.68	0.78	0.88	0.98	1.18	1.29	1.42	1.54	1.66
1200	0.78	0.90	1.02	1.13	1.37	1.50	1.64	1.78	1.93
1400	0.88	1.01	1.15	1.28	1.54	1.69	1.86	2.02	2.19
1600	0.97	1.12	1.27	1.42	1.72	1.87	2.06	2.24	2.43
1800	1.05	1.22	1.39	1.55	1.88	2.07	2.26	2.46	2.66
2000	1.13	1.31	1.50	1.68	2.03	2.23	2.44	2.65	2.87
2200	1.21	1.40	1.60	1.79	2.17	2.38	2.61	2.84	3.07
2400	1.28	1.48	1.70	1.90	2.31	2.54	2.78	3.02	3.26
2600	1.34	1.57	1.80	2.01	2.43	2.67	2.92	3.17	3.42
2800	1.40	1.64	1.88	2.10	2.55	2.80	3.06	3.31	3.57
3000	1.46	1.71	1.95	2.19	2.66	2.92	3.18	3.45	3.72
3500	1.57	1.85	2.12	2.38	2.87	3.15	3.43	3.70	3.97
4000	1.65	1.95	2.24	2.51	3.03	3.30	3.58	3.84	4.10
4500	1.70	2.01	2.31	2.59	3.10	3.35	3.63	3.85	4.07
5000	1.71	2.05	2.32	2.60	3.09	3.32	3.56	3.77	—

*Preferred pulley diameters.

Source: Extract from BS 1440, *Endless V-belt Drives,* is reproduced by permission of the British Standards Institution, 2 Park Street, London W1A 2BS.

Table 10.3. (b). Power ratings (kW) for B-section V-belts.

Speed of faster shaft (rpm)	Smaller pulley pitch diameter (mm)								
	125	132	140*	150*	160*	170*	180*	190	200*
200	0.57	0.63	0.69	0.77	0.85	0.92	1.00	1.08	1.15
400	1.01	1.11	1.23	1.38	1.52	1.67	1.81	1.96	2.10
600	1.40	1.55	1.72	1.93	2.14	2.35	2.55	2.76	2.96
800	1.75	1.94	2.16	2.44	2.70	2.97	3.24	3.50	3.78
1000	2.08	2.31	2.58	2.91	3.23	3.55	3.87	4.18	4.49
1200	2.38	2.66	2.96	3.35	3.72	4.09	4.46	4.81	5.17
1400	2.66	2.97	3.32	3.75	4.17	4.59	4.98	5.39	5.78
1600	2.92	3.26	3.65	4.12	4.58	5.04	5.48	5.91	6.33
1800	3.15	3.52	3.94	4.45	4.95	5.44	5.90	6.36	6.80
2000	3.36	3.76	4.21	4.75	5.28	5.78	6.27	6.74	7.19
2200	3.54	4.00	4.44	5.01	5.55	6.07	6.58	7.05	7.50
2400	3.70	4.14	4.63	5.22	5.78	6.31	6.81	7.28	7.72
2600	3.82	4.28	4.79	5.39	5.95	6.48	6.97	7.42	7.83
2800	3.93	4.40	4.91	5.52	6.08	6.60	7.06	7.48	7.85
3000	4.00	4.48	4.99	5.59	6.14	6.63	7.07	7.44	—
3500	4.04	4.50	5.00	5.55	6.01	—	—	—	—
4000	3.85	4.28	4.70	—	—	—	—	—	—

*Preferred pulley diameters.

Source: Extract from BS 1440, *Endless V-belt Drives,* is reproduced by permission of the British Standards Institution, 2 Park Street, London W1A 2BS.

Table 10.3 (c). Power ratings (kW) for C-section V-belts.

Speed of faster shaft (rpm)	Smaller pulley pitch diameter (mm)									
	200*	212*	224*	236*	250*	265*	280*	315*	355*	400*
200	1.66	1.83	2.00	2.16	2.36	2.57	2.77	3.25	3.79	4.39
400	2.93	3.24	3.56	3.87	4.23	4.62	5.00	5.88	6.87	7.96
600	4.04	4.49	4.94	5.38	5.90	6.44	6.98	8.21	9.59	11.09
800	5.04	5.61	6.18	6.74	7.39	8.07	8.75	10.28	11.96	13.75
1000	5.93	6.61	7.29	7.95	8.71	9.51	10.29	12.05	13.94	15.90
1200	6.71	7.49	8.25	9.00	9.85	10.74	11.60	13.50	15.48	17.43
1400	7.37	8.23	9.06	9.87	10.79	11.74	12.64	14.59	16.52	18.27
1600	7.91	8.83	9.71	10.56	11.52	12.48	13.39	15.28	17.00	—
1800	8.32	9.28	10.19	11.05	12.00	12.95	13.82	15.52	—	—
2000	8.59	9.56	10.47	11.32	12.24	13.12	13.89	—	—	—
2200	8.71	9.67	10.55	11.35	12.19	—	—	—	—	—
2400	8.67	9.58	10.40	11.12	—	—	—	—	—	—
2600	8.45	9.38	—	—	—	—	—	—	—	—
2800	8.05	—	—	—	—	—	—	—	—	—

*Preferred pulley diameters.

Source: Extract from BS 1440, *Endless V-belt Drives,* is reproduced by permission of the British Standards Institution, 2 Park Street, London W1A 2BS.

Table 10.3 (d). Power ratings (kW) for D-section V-belts.

Speed of faster shaft (rpm)	Smaller pulley pitch diameter (mm)									
	355*	375*	400*	425*	450*	475*	500*	530	560*	600*
200	6.04	6.61	7.32	8.02	8.72	9.42	10.11	10.93	11.75	12.83
400	10.57	11.61	12.19	14.19	15.45	16.70	17.94	19.40	20.85	22.74
600	14.34	15.79	17.56	19.30	21.01	22.68	24.32	26.23	28.09	30.49
800	17.39	19.14	21.26	23.32	25.31	27.22	29.06	31.17	33.16	35.62
1000	19.64	21.57	23.88	26.07	28.14	30.07	31.86	33.82	35.57	—
1200	20.98	22.96	25.26	27.36	29.25	30.92	—	—	—	—
1400	21.29	23.15	25.21	—	—	—	—	—	—	—
1600	20.46	—	—	—	—	—	—	—	—	—

*Preferred pulley diameters.

Source: Extract from BS 1440, *Endless V-belt Drives,* is reproduced by permission of the British Standards Institution, 2 Park Street, London W1A 2BS.

The constants used are empirical and are derived from extensive laboratory and field-performance information. The equations for power are clearly not convenient for design use and, in the Standard, the power ratings are tabulated.

The ratings are given in Tables 10.3(a), (b), (c) and (d) and are for smooth-running conditions; where shock loading or long running times exist, the drive power should be multiplied by a service factor selected from Table 10.4 to determine the design power. A more detailed description of these factors is given in BS 1440 which also gives more comprehensive power ratings than those shown in Table 10.3.*

*Design information for V-belts in this chapter has been taken from BS 1440: 1971, *Endless V-belt Drives*, and is reproduced by permission of the British Standards Institution, 2 Park Street, London, W1A 2BS.

Table 10.4. Service factors for V-belt drives.

Types of driven machines	Type of drive: Smooth-drive normal torque electric motors, multicylinder internal combustion engines — Operational hours per day			Type of drive: High torque motors, brakes or clutches in drive; single-cylinder internal combustion engines — Operational hours per day		
	10 and under	10–16	Over 16	10 and under	10–16	Over 16
Light duty: e.g. fans and blowers up to 7.5 KW	1.0	1.1	1.2	1.1	1.2	1.3
Medium duty: machine tools and conveyors	1.1	1.2	1.3	1.2	1.3	1.4
Heavy duty: piston compressors, pulverisers	1.2	1.3	1.4	1.4	1.5	1.6
Extra-heavy duty: crushers, mills	1.3	1.4	1.5	1.5	1.6	1.8

Source: Adapted from BS 1440, *Endless V-belt Drives,* by permission of the British Standards Institution, 2 Park Street, London W1A 2BS.

The tabulated power ratings are based on a standard belt length and 180° contact angle on both pulleys, that is a 1:1 speed ratio. This means that the last term in Equation (10.3) has been left out and must be considered separately. Hence the corrections to be applied for other conditions are

(a) the speed ratio correction which allows additional power for speed ratios other than 1.0. This correction is plotted in Fig. 10.9 (a), (b), (c) and (d);
(b) a belt length factor, which is greater than 1.0 for belts longer than the standard, and less than 1.0 for belts shorter than the standard length. This is given in Table 10.2;
(c) an arc of contact factor which decreases as the arc of contact on the smaller pulley falls below 180°. For convenience this factor is plotted as a function of $(d_1 - d_2)/C$ (see Fig. 10.10).

A preliminary choice of belt section can be made from Fig. 10.11 which illustrates speed and power regimes for each section. The selection procedure is as follows:

(i) multiply drive power by service factor from Table 10.4. Make preliminary choice of belt size from Fig. 10.11;
(ii) select small and large pulley pitch diameters (Table 10.1) to give required speed ratio, bearing in mind that the smaller the pulley size the smaller the allowable power for a given life;
(iii) select centre distance, C, and calculate belt length, L_p, from Equation (10.1). Choose nearest available length from Table 10.2, and recalculate C from Equation (10.2);

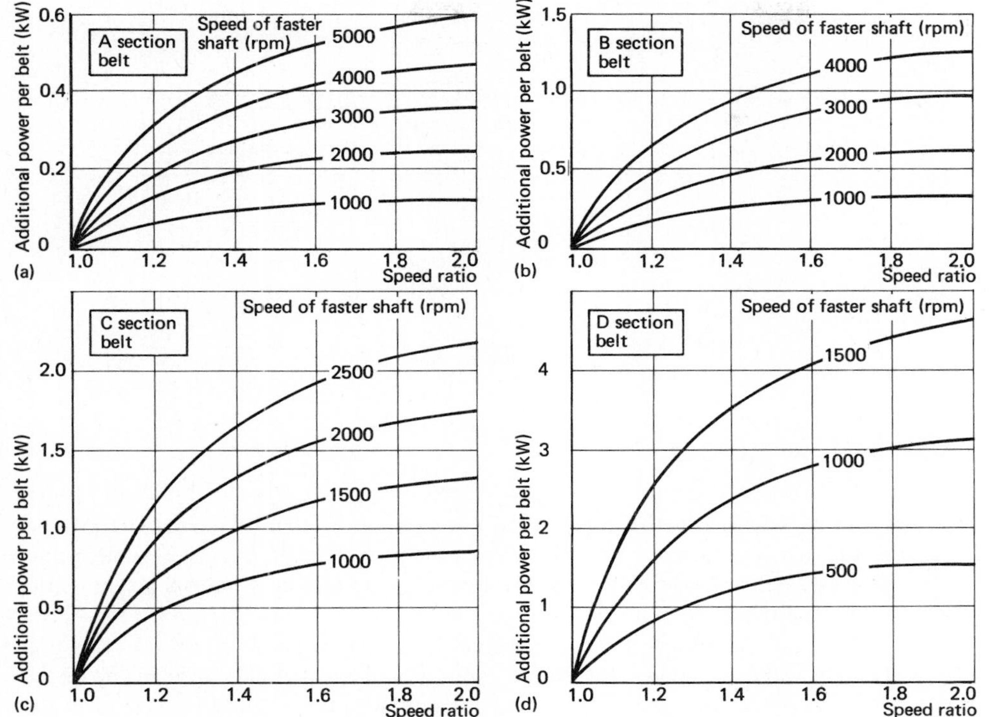

Fig. 10.9. *Speed-ratio correction factor:*
(a) A-section belt;
(b) B-section belt;
(c) C-section belt;
(d) D-section belt.
(Adapted from BS 1440, Endless V-belt Drives, *reproduced by permission of the British Standards Institution, 2 Park Street, London, W1A 2BS.)*

(iv) from Table 10.3, read the basic power per belt, and add the additional power for speed ratio from Fig. 10.9;

(v) multiply the value from (iv) by the arc of contact (Fig. 10.10) and belt length (Table 10.2) factors;

(vi) compare the value from (v) with the design power to determine the number of belts.

The selection of a V-belt drive is illustrated in Example 10.1.

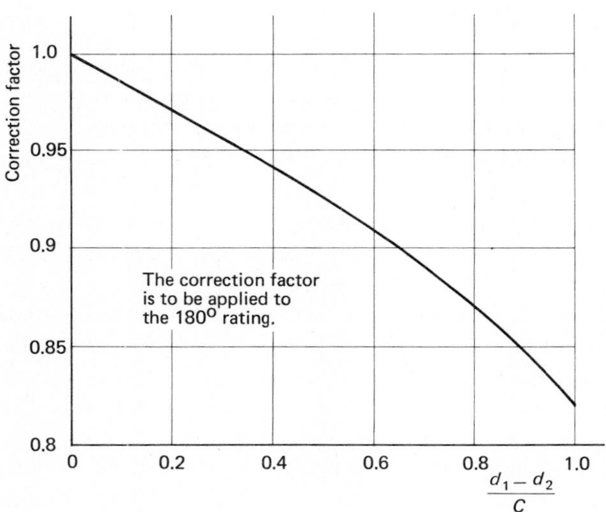

Fig. 10.10. *Arc of contact correction factor. (Adapted from BS 1440,* Endless V-belt Drives, *by permission of the British Standards Association, 2 Park Street, London W1A 2BS.)*

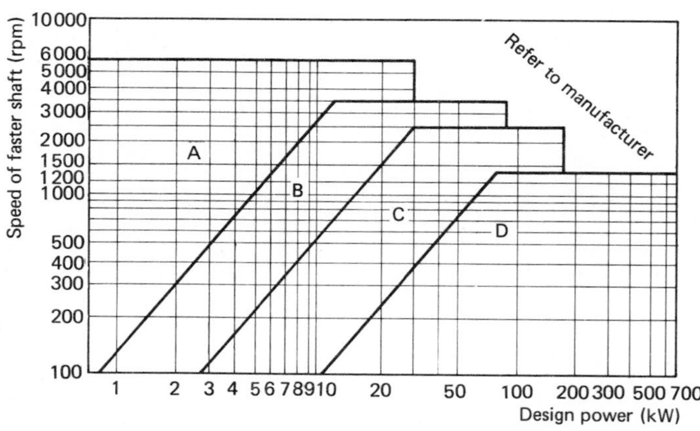

Fig. 10.11. *Preliminary selection chart for V-belts. (Extract from BS 1440,* Endless V-belt Drives, *is reproduced by permission of the British Standards Institution, 2 Park Street, London W1A 2BS.)*

10.4.2. Installation

The belt drive is specified by the size, length and number of belts, the pulley pitch diameters, and the adjustment required on centre distance. The latter is usually taken as minus 1.5%, plus 3% of the pitch length, L_p, the first allowance being to facilitate installation of the belt and the second to allow stretch to be taken up during the belt life.

The belt tension imposes a bending moment on the shaft and, as this may be of the same order as the torsion, it must be taken into account in the design of the shaft. It will also be required for calculations of bearing life, although where the bearing is preselected, as in an electric motor, the manufacturer's catalogue should be consulted for guidance as to what overhung load can be put on the motor shaft. An approximate estimate of the belt forces can be made by assuming that the tension on the slack side is one-quarter of the tension on the tight side. The sketch shows how the belt forces can be replaced by a torque on the shaft and a force through the shaft centreline. Thus if F is the force necessary to give the torque T, i.e. $F = 2T/d$, the belt tensions will be as shown in the sketch.

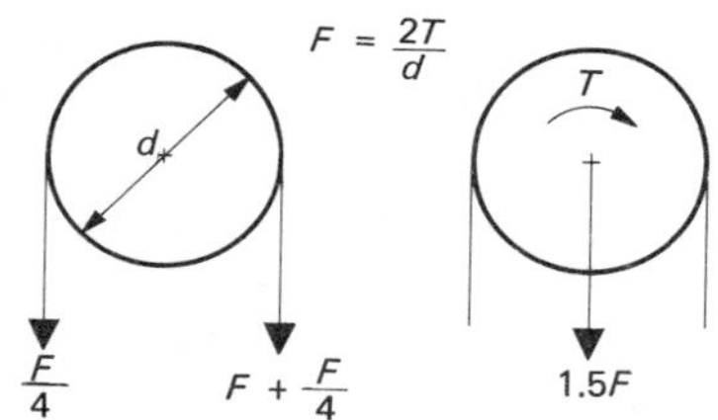

10.5. GEAR DRIVES

A gear transmission has advantages when high reliability, efficiency and compactness are sought, and in particular where a fixed and constant velocity ratio is required. Although we shall not consider them here, gears can be designed to transmit power between shafts whose axes intersect (bevel gears) and shafts whose axes do not intersect and are not parallel (e.g., crossed helical gears). We shall confine our attention to the design of spur gears in which the shafts are parallel and the tooth surfaces are parallel to the shafts. The nomenclature used for spur gears is shown in the sketch.

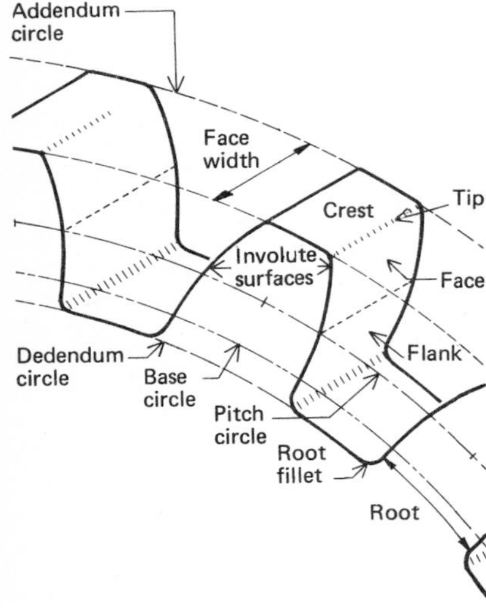

Although in theory many tooth-forms can be used, the involute form has so many advantages as to make it practically universally adopted. An important advantage is that gears with involute tooth-forms maintain a constant velocity ratio even when they are operated at centre distances other than the design value.

Several design methods for gears are well established[6, 7]; the one described here is set out in BS 436, "Spur and Helical Gears".[8] In this method, the teeth

Example 10.1 Sheet 1 of 1

A squirrel-cage induction motor rated at 10 kW at 1400 rpm is to drive a centrifugal fan at 2700 rpm. Design a V-belt drive for a 12-hour/day service, keeping the centre distance to a low value.

Service factor = 1.2 ; Design power = 12 kW.
In Fig 10.11 a B-section belt is indicated

Small pulley : select minimum size 125 mm
Large pulley : nearest to $125 \times \frac{27}{14} = 241$;
Select 250 mm

Centre distance : must be greater than (sum of pulley outside diameters) ÷ 2

Choose C = 200,
L = 1009 from Equation (10.1)
Select pitch length 1100, recalculate C
C = 248 mm

Basic power per belt (Table 10.3)	3.88 kW
Correction for speed ratio of 1.93	+0.76
	4.64

Arc of contact factor for $\frac{D-d}{C} = 0.504$ is 0.93

Belt length factor = 0.85
Power per belt = 4.64 × 0.93 × 0.85 = 3.67 kW
Number of belts 12/3.67 = 3.3, use 4 belts

Use 4 - B1100 belts, centre distance $248^{+7.5}_{-4}$ mm

Pulley pitch diameters 250, 125 mm
Outside diameters 258, 133 mm
(from BS 1440)
Note that using longer belt – 5000 mm
larger pulleys – 200 and 400 mm
Basic power per belt = 7.86 + 0.76 = 8.62
Arc of contact factor = 0.99
Belt length factor = 1.18
Number of belts $= \frac{12}{8.62 \times 1.18 \times 0.99} = 1.2$

Use 2 belts

in both meshing gears are designed to meet both strength and wear criteria; this is typical of most methods, although details vary considerably. A full description is given in works of Tuplin[7], Merritt[9] and Houghton[10]. The same basic data are used in the design of helical gears, but these are not considered here.

In order to analyse the forces on gear teeth, it is necessary to have an understanding of the geometry of the involute and the way it is used to form the working surface of the teeth. A brief description of this is given in the next section, and the picture can be clarified further by building a model in which a rack can be used to generate involute tooth shapes (see Problem 10.6). In order to illustrate the geometry clearly on a diagram it is usually necessary to distort the tooth proportions, and the suggested exercise will give a better idea of actual tooth shapes.

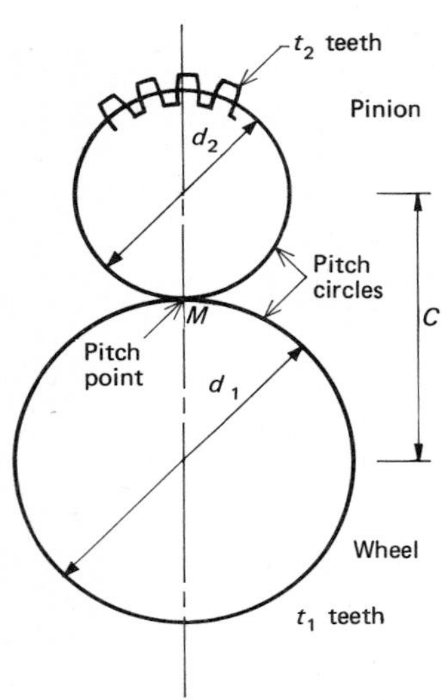

10.5.1. Geometry of involute gears

The action of two gears in mesh may be considered in terms of two cylinders which roll on each other and whose diameters are proportional to the number of teeth. We consider two gears, 1 and 2, having t_1 and t_2 teeth; by convention the smaller of the two is called the *pinion* and the other the *wheel.* When one gear drives the other, the cylinders can be thought of as rolling together without sliding. These cylinders are called the *pitch cylinders*, and the transverse section of the cylinder normal to the axis, the *pitch circle.* The *pitch point* is defined as the point of tangency of these circles (see sketch). For correct engagement, the teeth must be uniformly disposed around these circles, whose diameters in terms of the centre distance, C, will be

$$d_1 = 2C\frac{t_1}{t_1 + t_2}$$

$$d_2 = 2C\frac{t_2}{t_1 + t_2}$$

The spacing of the teeth is defined by the *circular pitch*, *p*, which is the length of arc of the pitch circle between similar faces of successive teeth. Other quantities used to define the spacing are the *module*, *m*—the pitch diameter divided by the number of teeth—or the *diametral pitch*, *P*, which is the reciprocal of the module. We have used both terms, since, although BS 436 uses the diametral pitch, the module will be preferred with SI units. Preferred values of *m* are given in the standard.

While over a number of turns the velocity ratio, G_r, of this drive will be t_1/t_2, for small rotations the tooth profiles must have special properties to ensure that the ratio remains constant at all phases of the engagement. This is the most important kinematic condition to be satisfied by the profile selected for the working surface of the tooth. If the wheels turn with angular velocities ω_1 and ω_2, where $\omega_1/\omega_2 = t_2/t_1$ and ω_1 and ω_2 are in opposite directions, the relative velocity of gear 1 with respect to gear 2 is $(\omega_1 + \omega_2)$; the relative motion is a rotation of gear 2 about the instantaneous centre, which is the pitch point, M. Now consider the relative motion between the two surfaces at a point of contact. It is seen from Fig. 10.12 that the velocity of profile 2 in relation to profile 1 is $PM(\omega_1 + \omega_2)$, and that the direction is perpendicular to PM. The sliding motion of the two surfaces at P will be possible only if the profiles have

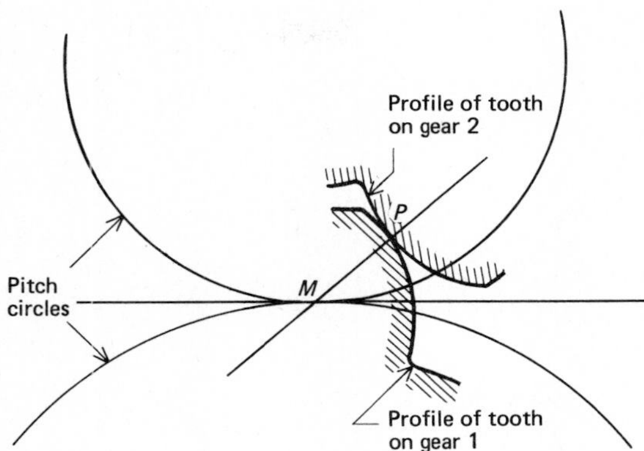

Fig. 10.12. *Relative motion between conjugate surfaces.*

a common normal MP. Surfaces which meet this requirement are said to be *conjugate.* Note that the sliding velocity is proportional to the distance of P from M, and that there is pure rolling only when the contact point is at M.

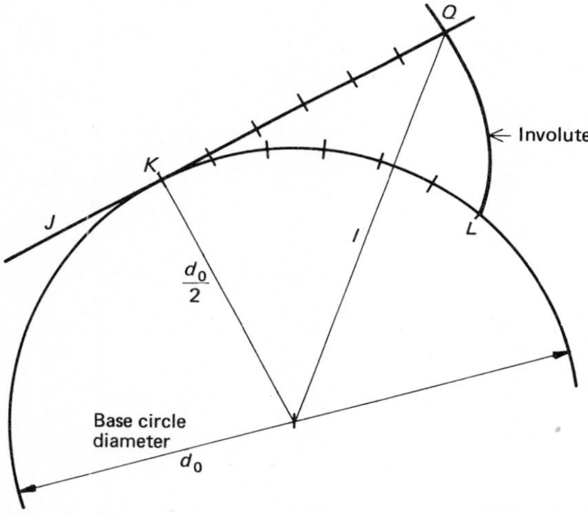

Fig. 10.13. *Generation of involute.*

An involute curve is traced out by a point on a generator which rolls without slipping on a circle diameter, d_0, called the base circle. In Fig. 10.13, the point Q on the generator JKQ is tracing the involute LQ. The radius of curvature at Q is KQ and this line is normal to the involute at Q. If two involutes forming tooth surfaces on gears 1 and 2 are in contact at Q, the common normal will be a continuation of KQ and will have to be tangent to the base circle of the second gear. The involute surfaces will fulfil the requirement for constant angular velocity. A special case of the two surfaces in contact—at the pitch point M—is shown in Fig. 10.14. If the gears are rotated in either direction from this position, the point of contact will move along the curve of contact LMN—this is a straight line for involute teeth.

The angle ψ between the line of contact and the common tangent AB to the pitch circles is called the *pressure angle.* A standard value for ψ is selected for any specific gear system—the British Standard uses 20°, while another widely used value is $14\frac{1}{2}°$. The angle affects mainly the tooth proportions, but

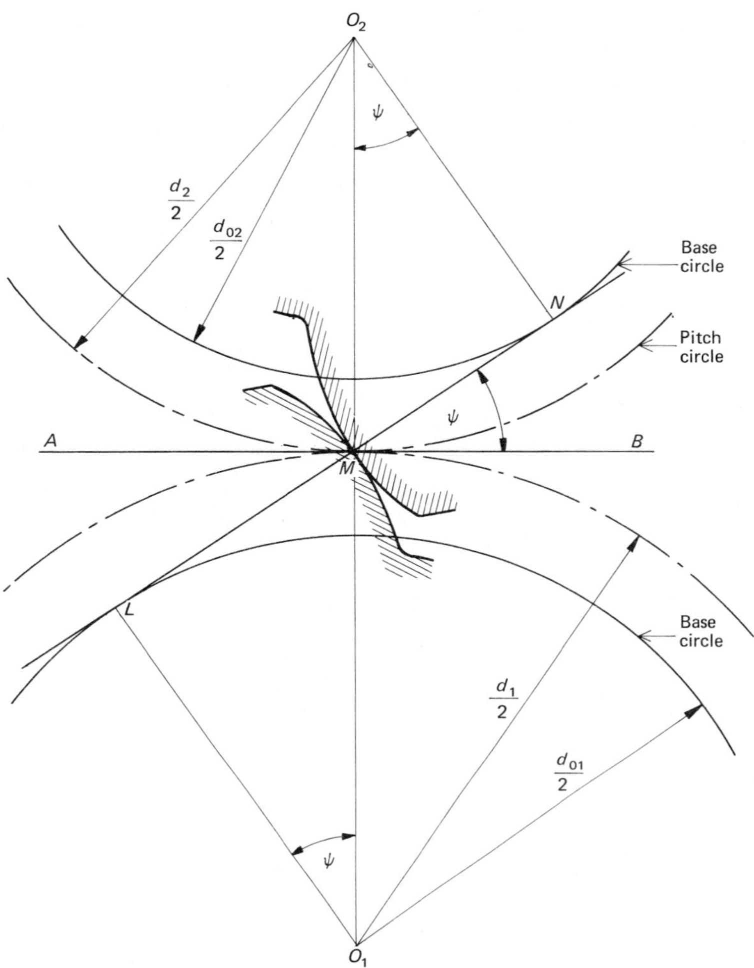

Fig. 10.14. *Involute surfaces in contact.*

the normal reaction force between geared shafts increases with increase of the pressure angle.

If the centres O_1, O_2 of the base circles in Fig. 10.14 are moved apart radially while maintaining the lengths O_1M, O_2M in the same ratio, the involutes generated from the two base circles can still meet at M – at a point now further along the involutes – and conjugate action is still obtained. This is an important property of involute teeth. If this separation does occur, the pressure angle will have increased. Therefore, the pitch circles and pressure angle as defined depend on the centre distance between two meshing gears. Tuplin points out the difficulties inherent in this approach and proposes a more rational terminology.[7]

A rack is a gear in which the pitch line is straight, that is to say, the pitch radius is infinite. The basic rack is important in the specification of tooth proportions and, since the sides of the teeth are straight lines, it is easy to reproduce. The proportions of a 20° basic rack specified in BS 436 are given in Fig. 10.15. In gear manufacture, involute surfaces can be generated using a cutting tool based on this shape, and actual racks when engaged with a pinion give straight line motion.

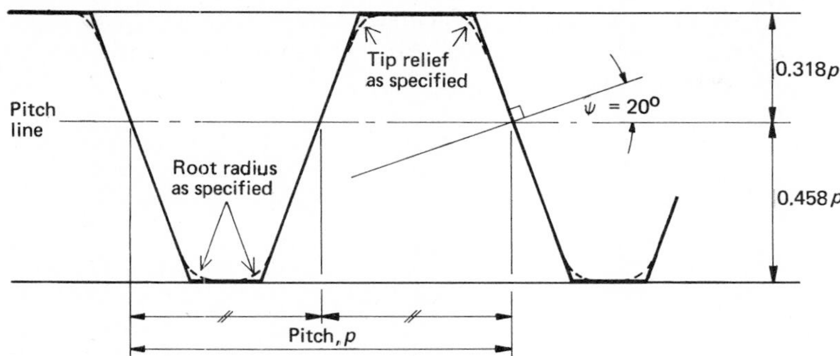

Fig. 10.15. *The British Standard basic rack for precision ground gears.*

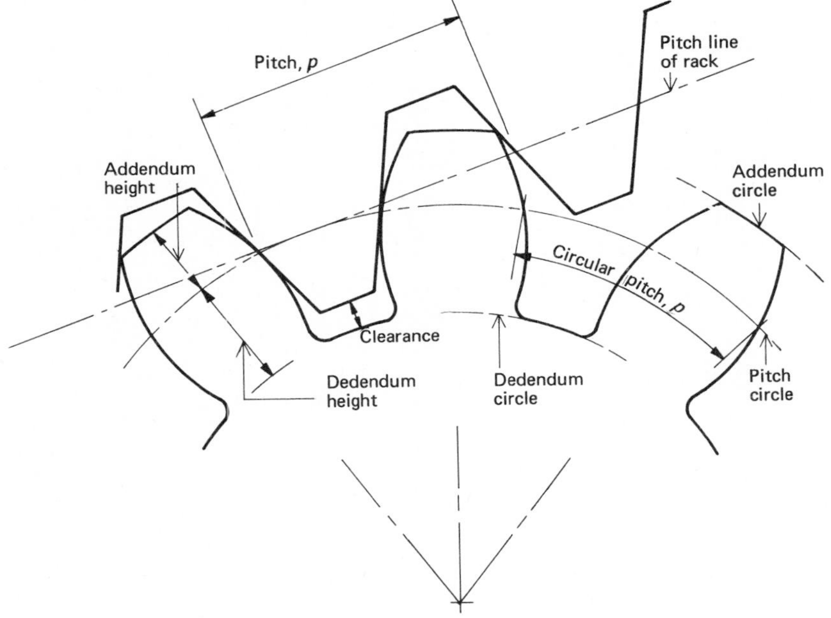

Fig. 10.16. *Gear nomenclature.*

Some other terms used to describe the geometry of spur gear teeth are shown in Fig. 10.16, in which a rack is shown in engagement with a pinion. If the pitch line of this rack rolls on the pitch circle of the pinion, the flat faces of the rack generate the correct involute surfaces for the other gear.

The radius of curvature of the involute at Q in Fig. 10.13 is KQ and has a value $\sqrt{l^2 - (d_0/2)^2}$. Since the stress at the point of contact of a curved surface increases as the radius of curvature decreases, it is clear that the tip of a gear should not make contact with the flank of a mating gear at the base circle. Correct proportioning of the gear teeth will ensure that this does not happen. The geometry of gears must also be examined to ensure that at least one pair of teeth is always in contact. Consider the situation in Fig. 10.17 where two pairs of teeth are shown in contact at C and D on the line LMN which is the common tangent to the base circles and on which contact points must lie for correct tooth action. For the lower gear rotating in a clockwise direction, the tooth $1A$ will lose contact with the tooth $2A$ at Q where the addendum circle 1 cuts the line of contact LMN. We must check that, when this separation occurs, the tooth $1B$ has already taken up the drive on tooth $2B$. This will occur where the addendum circle 2 cuts LMN at R.

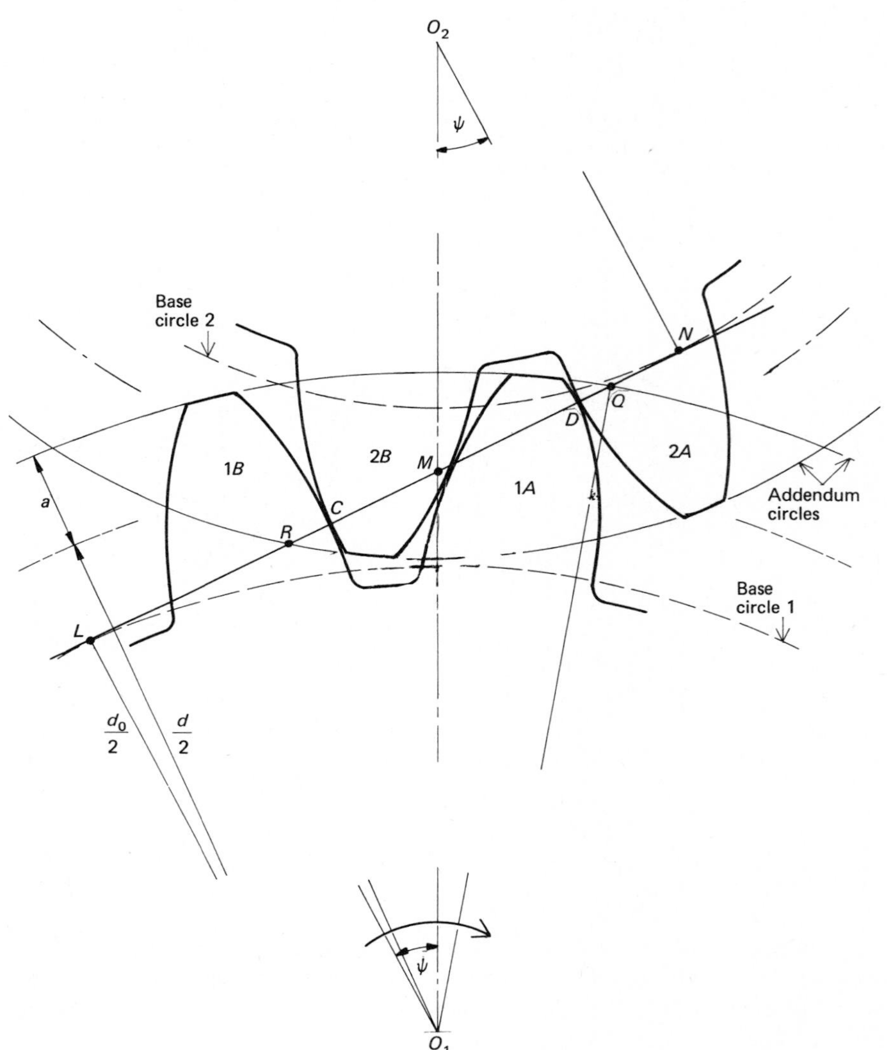

Fig. 10.17. *Line of contact for involute teeth.*

The contact path from M to Q for tooth $1A$ can be calculated as follows

$$MQ = LQ - LM = (O_1Q^2 - O_1L^2)^{\frac{1}{2}} - LM$$

$$= \left[\left(\frac{d}{2} + a\right)^2 - \frac{d^2}{4}\cos^2\psi\right]^{\frac{1}{2}} - \frac{d}{2}\sin\psi$$

where a is the addendum height for gear 1. On simplification this gives

$$MQ = \frac{a}{2}\left\{\left[\left(\frac{d}{a}\sin\psi\right)^2 + 4\left(\frac{d}{a} + 1\right)\right]^{\frac{1}{2}} - \frac{d}{a}\sin\psi\right\}$$

The contact path, RM, for tooth $2B$ can be found in the same way. The requirement for continuous contact can be written as a ratio

$$\frac{RM + MQ}{\text{circular pitch}} > 1$$

noting that in Fig. 10.17 the circular pitch is equal to the length CD. This ratio usually has a value between 1.4 and 2 for teeth with a pressure angle of 20°, and no check need be made of its value for gears with standard proportions. If it has a value greater than 2 in special gears the sharing of load may be taken into account in strength design. It is usual to apply "tip relief" to a gear; this entails shaving away part of the tip of the gear (as indicated on the rack in Fig. 10.15) so that the taking up of load is more even. This easing of the tip helps to reduce the dynamic load on the teeth, and means that the actual contact ratio will be less than that calculated above.

As the number of teeth in a pinion is reduced, there will come a time when the tip of the mating gear extends beyond the point of tangency of the base circle and the line of action. In order to avoid the resulting interference it may be necessary to cut away part of the involute on the flank of the pinion. Such undercutting weakens the pinion tooth, and other remedies – such as using non-standard proportions for the teeth – may be called for. With a 20° pressure angle, interference only occurs when the pinion has less than 17 teeth, and if this lower limit is respected, the designer need not be immediately concerned about the problem. The proportions of the teeth may be altered (even though no interference is involved) by extending the addendum of the pinion and reducing that of the wheel if $(t_1 + t_2)$ is less than 60. This gives better zone and strength factors, and is reflected by a discontinuity in the design curves that are given in the next section. Gears modified in this way are said to be "corrected" and details are given in BS 436[8] and in Merritt's *Gears*.[9]

10.5.2. Design for strength and wear

The designer's task is usually to determine suitable gear proportions, given the torque and speed of the pinion and wheel. The main variables are gear materials and geometry – pitch diameter, number of teeth and face width. At high speeds the number of stress reversals may exceed 10^6 per hour and, for a design life which is commonly taken as 26 000 hours for industrial gears, it is clear that fatigue will be a limiting factor. At the same time the teeth are subjected to high contact stresses under a sliding and rolling action, and the possibility of abrasion, fretting and wear must be considered. Since the properties that control strength and wear are not closely related, it is necessary to check them both. We will first establish the design equations and then describe their use.

The gear tooth is designed for strength by treating it as a cantilever beam, and the first step is to fix dimensions, such as those shown in the sketch, so that the bending moment at the worst loaded section can be found. The load on unit width of tooth will be considered.

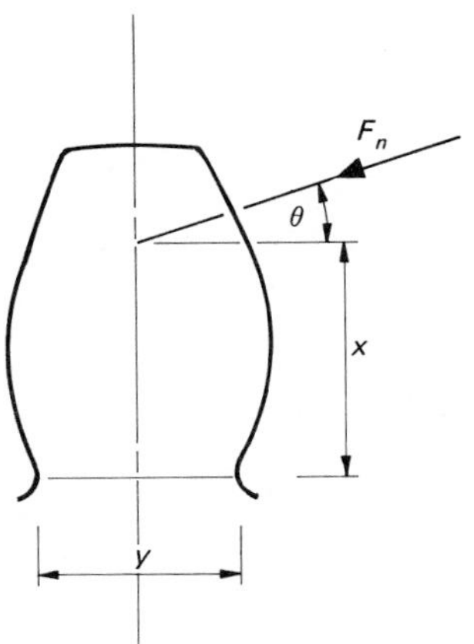

In Fig. 10.18, two teeth are in contact at Q and the load F_n acts on tooth $1A$ normal to the surface. At this special position, the tooth $1B$ is just coming into contact at S; as the contact point on $1A$ moves upwards the load will be shared between two teeth. Thus, although the lever arm for bending at section GH may become larger, the load decreases and the configuration shown is taken as the critical one.

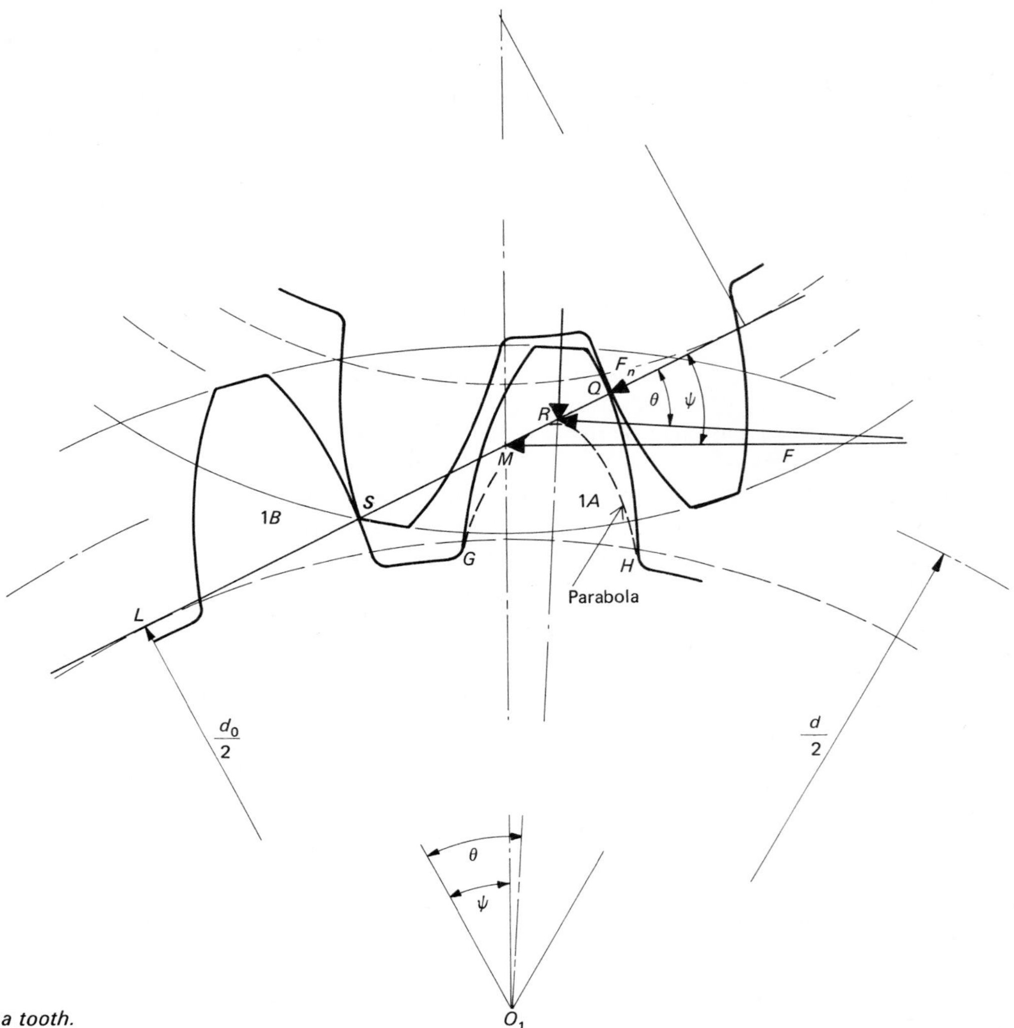

Fig. 10.18. *Forces on a tooth.*

The line of action of F_n at Q is tangential to the base circle at L and intersects the tooth centreline O_1R at R. Note that the angle θ between F_n and the normal to O_1R is larger than the pressure angle ψ. The critical bending section at the root of the tooth can be estimated, or can be found by drawing a parabola with vertex at R so that it is tangent to the tooth form at the root. Since the parabola represents a cantilever of uniform bending strength, the tangent points G and H determine the critical section and hence the length of the lever arm from R to the root section which we call x. The stresses at G and H for unit width of tooth are

$$f_{bt} = \text{tensile stress at } H \quad = \frac{F_n \cos\theta\, x}{y^2/6} - \frac{F_n \sin\theta}{y} \tag{10.4}$$

$$f_{bc} = \text{compressive stress at } G = \frac{F_n \cos\theta\, x}{y^2/6} + \frac{F_n \sin\theta}{y} \tag{10.5}$$

Our knowledge of fatigue would lead us to expect that the tensile loading would be critical, and this is confirmed in practice. However, when the behaviour of gears was first investigated, the maximum numerical value of stress was regarded as critical and strength design was based on Equation (10.5). Following this approach

$$f_{bc} = F_n \bigg/ \left(\frac{y^2}{6x \cos \theta + y \sin \theta} \right)$$

This formula for the bending stress is more useful if it is expressed in terms of the tangential force, F, at the pitch circle. The torque on the gear is

$$F\frac{d}{2} = F_n \cos \theta \times O_1 R$$

and since

$$O_1 R = \frac{d_0}{2 \cos \theta}, \qquad Fd = F_n \times d_0$$

or

$$F_n = F\frac{d}{d_0} = \frac{F}{\cos \psi}$$

This gives

$$f_{bc} = F \bigg/ \frac{y^2 \cos \psi}{6x \cos \theta + y \sin \theta} = F/Y \tag{10.6}$$

where

$$Y = \frac{y^2 \cos \psi}{6x \cos \theta + y \sin \theta} \tag{10.7}$$

Y is called the *strength factor* and for given tooth proportions is a function of the number of teeth in the gear and pinion. It can be determined by measuring x, y and θ from accurate scale drawings, but for the British Standard 20° tooth form it has been plotted for a module of unity (see Fig. 10.19).

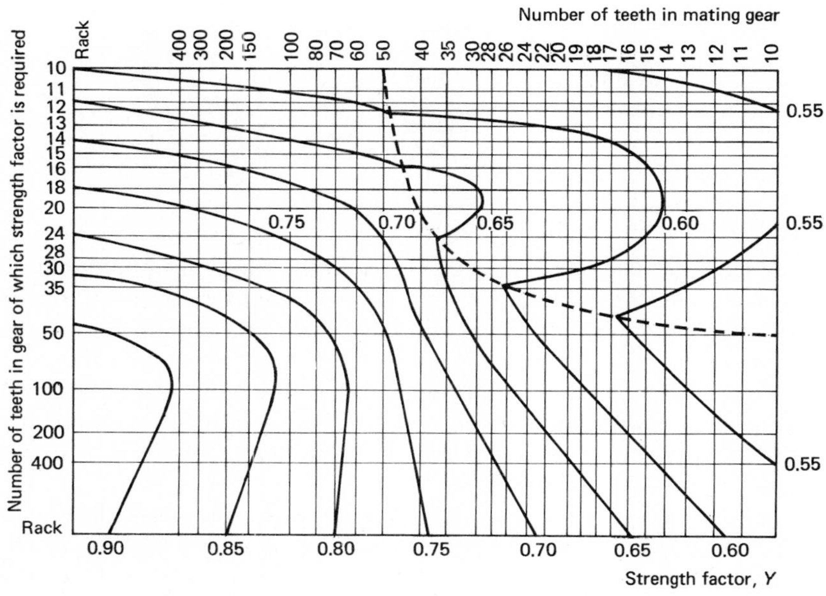

Fig. 10.19. *Strength factor, Y. (Extract from BS 436,* Machine Cut Gears, *is reproduced by permission of the British Standards Institution, 2 Park Street, London W1A 2BS.)*

Over the period that suitable design values for f_{bc} have been standardised by evaluating test results, this Y factor, rather than one derived from Equation (10.4), has been in use. No doubt the allowable stresses take this into account. If the tensile stress is in the future accepted as the basis of design, the higher Y values derived from Equation (10.5) will have to be compensated by a decrease in the allowable stress. It is important to note, therefore, that there is some empiricism in the design stresses, particularly since no explicit stress concentration factor is applied for the fillet radius at the root. Recommended values of this stress for a selection of materials have been taken from BS 436, and are given in Table 10.5. Attention is called to the empirical nature of the stress by calling it a *bending stress factor*; we replace the symbol f_{bc} by the one used in the British Standard, S_b.

For values of m (or P) other than unity, the allowable tooth load is determined by increasing Y in proportion to the linear dimensions, that is

$$F = \frac{S_b Y}{P} = S_b Y m \tag{10.8}$$

To account for fatigue, this equation is modified by introducing a factor X_b which is called a *speed factor for strength* (Fig. 10.20). The fatigue life of a gear is reduced with increasing speed because of the increased number of stress reversals in a given time and also because of the increased dynamic loads. The first of these factors is related to rotational speed, the second to linear speed. It will be seen from Fig. 10.20 that the X_b values are plotted as a function of running time and rpm, and therefore only the first is accounted for explicitly. The picture is a complicated one, for the value of X_b falls off more quickly with rpm than would be required for fatigue only. The standard value of X_b is based on a running time of 12 hours per day for 6 years, giving a life of 26 000 hours. Other curves are also plotted, giving equivalent running times from 0.01 to 24 hours per day. Taking values of X_b for the same number of revolutions but for different speeds, we have, for example, the values tabulated below:

Hours per day	rpm	X_b
0.1	10 000	0.25
1	1 000	0.38
10	100	0.42

It can be seen that the value of X_b decreases as the rotational speed increases, although in each case the number of reversals is the same. Some recognition is therefore given to dynamic effects, although there may be some doubt about the validity of the method used.

It is made clear in BS 436 that the accuracy of manufacture of gears must be commensurate with surface speed if excessive dynamic loads are to be avoided. The reader is referred to Tuplin[7] for a design method which accounts for dynamic loads on gear teeth.

Excessive surface stresses will show up in various ways. Pitting and scuffing – the latter is a local welding and smearing of the surface in the direction of sliding – may be the first signs of a situation which is dealt with in BS 436

Table 10.5. Selected values for S_c and S_b for gear design.

Material	BS Number	Condition	Minimum tensile strength (MPa)	Basic surface stress factor S_c	Basic bending stress factor S_b
Forged '30' carbon steel	970 En 5	N	495	9.65	117
Forged '30' carbon steel	970 En 5	H & T	540	11.0	145
Forged '40' carbon steel	970 En 8	N	540	9.65	131
Forged '40' carbon steel	970 En 8	H & T	620	13.8	169
Forged '55' carbon steel	970 En 9	N	695	15.9	165
Forged '55' carbon steel	970 En 9	H & T	770	18.3	210
Forged '60' carbon steel	970 En 11	H & T	850	20.7	231
Forged carbon-manganese steel	970 En 14B	H & T	620	13.8	169
Forged 3% nickel steel	970 En 21	H & T	695	15.9	186
Forged 4¼% nickel-chromium steel	970 En 30A	H & T	1545	41.4	342
Forged 3% nickel-chromium molybdenum steel	970 En 27	H & T	850	20.7	231
Forged surface hardened '40' carbon steel	970 En 8	Hardened in water	540	19.3	117
Forged surface hardened '55' steel	970 En 9	Hardened in air	695	27.6	148
Case hardened 3% nickel steel	970 En 33	Case hardened	695	72.5	325
Cast iron medium grade	821	As cast	247	9.32	52.5
Phosphor bronze casting	1400 PB2–C	Chill cast	232	6.07	62.8

N: normalised.
H & T: hardened and tempered.
Note: A more comprehensive list is given in BS 436: 1940, *Machine Cut Gears*, Appendix A, 1960.[8]

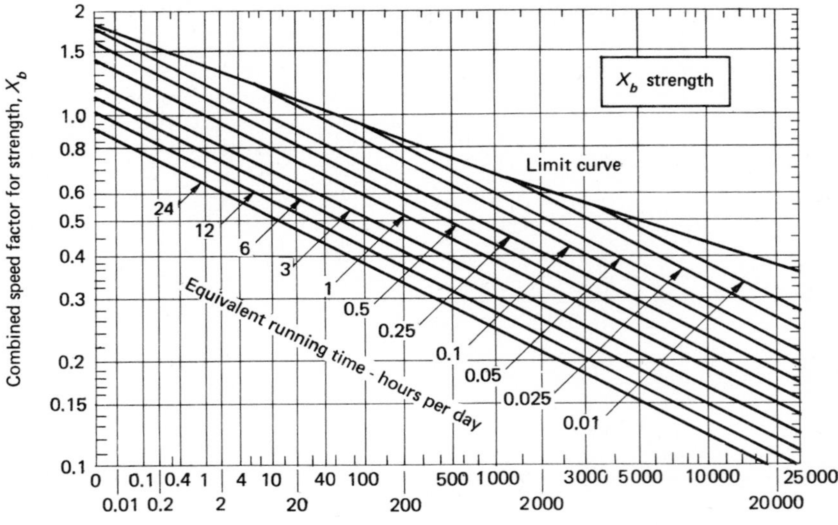

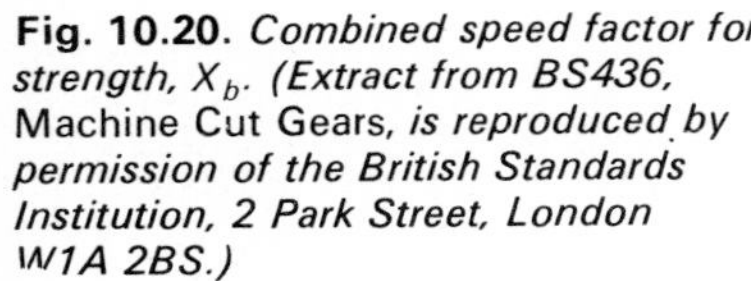

Fig. 10.20. *Combined speed factor for strength, X_b. (Extract from BS436,* Machine Cut Gears, *is reproduced by permission of the British Standards Institution, 2 Park Street, London W1A 2BS.)*

under the broad heading of "wear". The design method used in this case is based on the Hertz stress equation which, on the basis of assumptions described in Chapter 9, shows that the maximum compressive stress at the contact area of two cylindrical surfaces is

$$f_{\max} = \left[\frac{F_n \left(\dfrac{1}{R_1} + \dfrac{1}{R_2} \right)}{\pi (1 - \nu^2) \left(\dfrac{1}{E_1} + \dfrac{1}{E_2} \right)} \right]^{\frac{1}{2}} \tag{10.9}$$

where R_1 and R_2 are the radii of curvature of the surfaces and F_n is the normal force per unit contact length. The properties ν and E are constant for some combination of materials in contact, and

$$f_{\max} \propto \left[F_n \left(\frac{1}{R_1} + \frac{1}{R_2} \right) \right]^{\frac{1}{2}}$$

or

$$f_{\max} \propto \left(\frac{F_n}{R_r} \right)^{\frac{1}{2}}$$

where R_r is the equivalent radius, $\dfrac{R_1 R_2}{R_1 + R_2}$.

The correlation between the performance of different materials and their physical properties has not been found good enough to warrant the use of this equation directly. The discrepancies have been partly explained by the work-hardening that occurs under the sliding and rolling contact, and by the presence of a lubricant. The ratio F_n/R_r is retained as a measure of the maximum contact stress, but has been modified so as to reduce the rate at which the tooth loads increase with size. The British Standard thus uses the criterion

$$S_c = \frac{F_n}{R_r^{0 \cdot 8}} \tag{10.10}$$

where we again use the symbol for stress employed in the British Standard. The modification is empirical, although it has been explained on the basis that the load-carrying ability of larger gears suffers due to disproportionately large inaccuracies in manufacture.

Design values of S_c for a high grade steel are about 35 MPa, and since the use of Equation (10.9) gives a value for the Hertzian stress of about 1100 MPa, it is clear that it does not represent an actual stress. S_c is usually termed a *surface stress factor*.

The contact stress would be expected to be highest at the lowest point of single-tooth contact on the pinion, for here the radius of curvature has its lowest value for single-tooth loading. In practice, it is found that pitting frequently occurs close to the pitch line, and it is here that the stress calculation will be made. At this location the values of R_1 and R_2 are $(d_1/2)\sin\psi$ and $(d_2/2)\sin\psi$ (see Fig. 10.14). Since the module $m = d/t$, we have

$$R_1 = \frac{t_1 m \sin\psi}{2}, \qquad R_2 = \frac{t_2 m \sin\psi}{2}$$

that is

$$R_r = \frac{R_1 R_2}{R_1 + R_2} = \frac{m \sin\psi\, t_1 t_2}{2(t_1 + t_2)}$$

Using this value of R_r in Equation (10.10), and putting $F_n = F \sec\psi$ as before,

$$F = S_c\, m^{0 \cdot 8} \left(\frac{t_1 t_2 \sin\psi}{2(t_1 + t_2)} \right)^{0 \cdot 8} \cos\psi$$

or

$$F = S_c\, m^{0.8}\, Z,$$

where

$$Z = \left(\frac{t_1 t_2 \sin\psi}{2(t_1 + t_2)} \right)^{0 \cdot 8} \cos\psi \tag{10.11}$$

For a given ψ, Z can be plotted as a function of the number of teeth in each gear, and values for $\psi = 20°$ are shown in Fig. 10.21.

Introducing a factor for wear, X_c, the final design equation is

$$F = X_c\, Z S_c\, m^{0 \cdot 8} \tag{10.12}$$

X_c (see Fig. 10.22) has approximately the same numerical values as X_b for a running time of 12 hours, and it fills much the same purpose. For low values of running time, when wear is unlikely to be critical, the upper limit for X_c approaches 3, while the upper limit for X_b is about 2. The values of X_b and X_c are called *combined speed factors* since they account both for rotational speed and for running time per day. Note that where one gear makes multiple contacts – for example, in an epicyclic gear – the running time is raised by a factor equal to the number of separate contacts (see Example 10.2).

10.5.3. Application to design

The power equations for wear and strength are obtained by multiplying Equations (10.8) and (10.11) by the pitch line velocity. For face width W, this

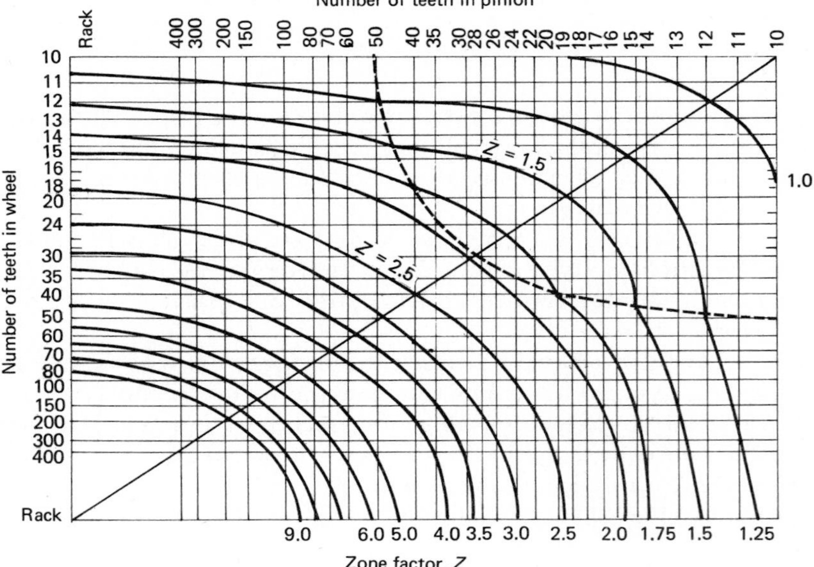

Fig. 10.21. *Zone factor, Z. (Extract from BS 436,* Machine Cut Gears, *is reproduced by permission of the British Standards Institution, 2 Park Street, London W1A 2BS.)*

gives

$$KW_c = \frac{X_c S_c Z W N t m^{1 \cdot 8}}{10^7} \text{ kW (wear)} \tag{10.13}$$

$$KW_b = \frac{X_b S_b Y W N t m^2}{1.9 \times 10^7} \text{ kW (strength)} \tag{10.14}$$

For given values of S_b and S_c, the pinion will be the critical gear. In practice, a material of lower strength may be used for the wheel, so that all values must be checked. It is difficult to find a starting point in design, and it must be expected that some trial solutions will be required before the final layout is adopted. The labour involved can be reduced if it is noted that, for given proportions and stress factors

$$KW_c \propto L^{2 \cdot 8} \quad \text{and} \quad KW_b \propto L^3 \tag{10.15}$$

where L is a linear dimension. This is one consequence of X_b and X_c being determined by the rotational speed and not the linear speed, which makes it possible to predict the effect of scaling the size of a gear up or down.

Sketching out some tooth shapes to scale for typical values of m or P will help the designer get a feel for the proportions of a gear. Since tooth strength increases with increasing size – i.e. with increasing m – there is a tendency to put the smallest possible number of teeth in a pinion. The number selected may range from 15 to 20, but values below 17 will require action to avoid interference. If 17 is taken for a preliminary trial, and the first estimate of pinion pitch diameter is taken as being twice the shaft diameter, the material strength can be calculated from Equations (10.13) and (10.14). Further adjustments to the size can be made using Equation (10.15) if necessary. The preferred values for the module run from 1 to 6 in $\frac{1}{2}$ mm steps and from 6 to 16 in 1 mm steps. The use of the British Standard design method is illustrated in Example 10.2.

Example 10.2 Sheet 1 of 2

The reduction gear for a turbine auxiliary power plant is to transmit 140 kW and is to give a reduction from 20000 to about 2000 rpm in a two stage reduction.
In a proposed layout for the first stage, the high speed pinion has 17 teeth and a module of 2.0 mm. It engages with three planet gears (72 teeth) on a fixed carrier. The design of the second stage is considered in problem 10.7. Determine suitable proportions for the first stage gearing. Design for 2 hours per day.

Take the face width $W \approx 4p = 4\pi m = 25$ mm
For pinion and wheel select case-hardened 3% nickel steel (En33)
$S_c = 72.5$, $S_b = 325$

Each driving tooth on pinion 1 must transmit 140/3 kW, but makes 3 contacts per rev, i.e. design for 6 hours/day

Wheel 2: design for 140/3 kW, 2 hours/day

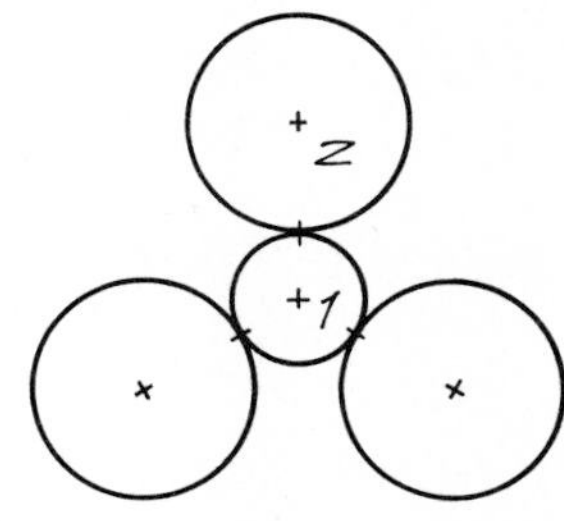

Example 10.2 Sheet 2 of 2

Design for wear	1 (6h/day)	2 (2h/day)	Design for strength	1 (6h/day)	2 (2h/day)
X_c	0.132	0.305	X_b	0.118	0.215
S_c	72.5	72.5	S_b	325	325
Z	1.82	1.82	Y	0.682	0.577
W	25	25	W	25	25
N	20000	4720	N	20000	4720
t	17	72	t	17	72
m	2	2	m	2	2
$m^{1.8}$	3.48	3.48	m^2	4	4

$$KW_{c1} = \frac{0.132 \times 72.5 \times 1.82 \times 25 \times 20000 \times 17 \times 3{\cdot}48}{10^7} = 51.5$$

$$KW_{c2} = \frac{0.305 \times 72.5 \times 1.82 \times 25 \times 4720 \times 72 \times 3{\cdot}48}{10^7} = 119.0$$

$$KW_{b1} = \frac{0.118 \times 325 \times 0.682 \times 25 \times 20000 \times 17 \times 4}{1.9 \times 10^7} = 46.8$$

$$KW_{b2} = \frac{0.215 \times 325 \times 0.577 \times 25 \times 4720 \times 72 \times 4}{1.9 \times 10^7} = 72.1$$

The strength of the pinion limits the power to 3×46.8 or 140 kW – which meets the requirements

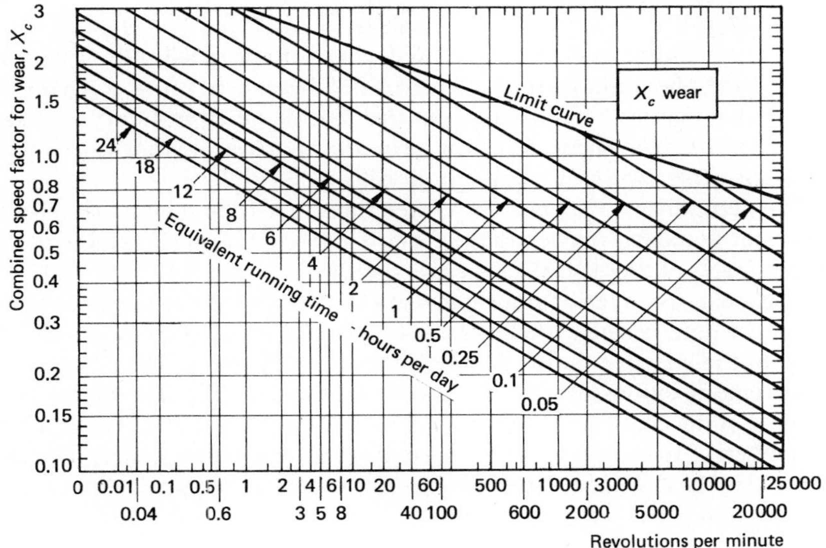

Fig. 10.22. *Combined speed factor for wear, X_c. (Extract from BS436,* Machine Cut Gears, *is reproduced by permission of the British Standards Institution, 2 Park Street, London W1A 2BS.)*

BS 436 provides a sound basis for the design of industrial gears. It has been used for the design of automotive gears, but is now generally accepted as being somewhat conservative for this purpose.[1] Although it is simple to use, it is difficult to resolve clearly the nature of the empirical stress factors. A knowledge of the way in which the strength and wear equations are derived will be helpful in understanding any revision of the standard that is produced.

10.5.4. Installation

To operate correctly, gears must be correctly aligned, and they must not deflect excessively under the heavy loads to which they are subjected. Close attention must be paid to the design of shafts and bearings with this in mind, especially where the face width of the gears is high. For normal service, the ratio W/p is in the range 3 to 4, although higher values may sometimes be used.

When the tangential force at the pitch point has been calculated, the resultant force is given by $F \sec \psi$ (see sketch). This resultant force is used in calculating the bending moment on the shaft and the reaction at the bearings. Deflections are kept to a minimum if the gear is mounted as close as possible to the supporting bearing.

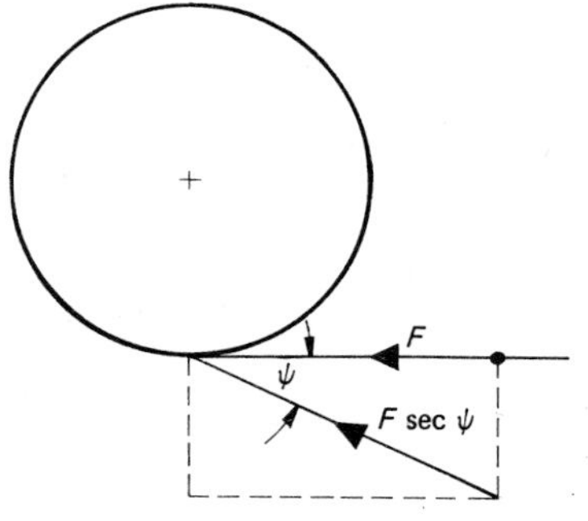

With spur gears there are no significant axial forces on the shaft; however, with helical and bevel gears the bearing system must be arranged to carry the axial forces that occur.

The designer must ensure that adequate lubrication of the right type is provided at the gear tooth. The requirements become more demanding as speed increases; for preliminary design purposes the following will serve as a guide:

low speed, lightly loaded gears up to 12 m/s	grease, dip or splash oil lubrication
12 m/s to 50 m/s	forced feed spray

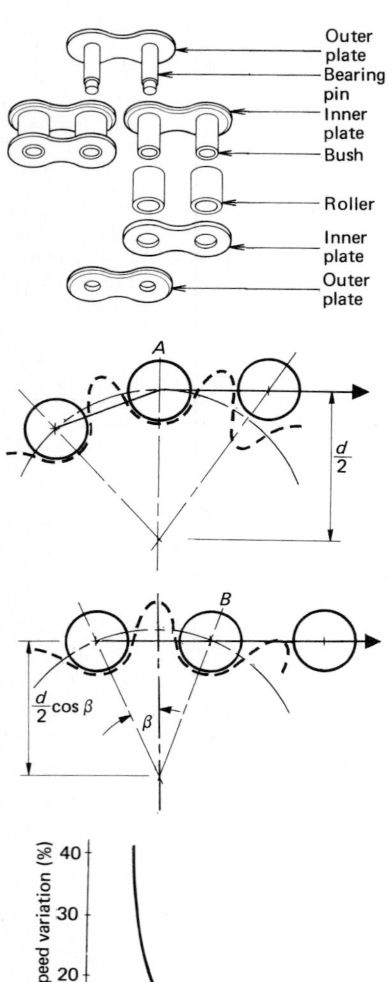

10.6. CHAINS AND SYNCHRONOUS BELTS

Chains or synchronous belts can be used when a fixed speed ratio is required. The mode of operation of the two drives is different, but it is convenient to consider them under one heading, and this affords an opportunity to make a comparison between them. Selection procedures are straightforward, and we will describe them without going into details.

The components that make up three links in a roller chain are shown in the sketch. Standard pitches range from 8 mm to 115 mm; the larger sizes transmit forces of up to 3000 kN at low speeds.

The mean speed ratio for a chain drive is fixed, but the instantaneous ratio varies with time because of the so-called *chordal effect*. This is caused by a change in the effective radius at which the roller operates. The sketch shows two extreme positions of a roller, and it will be seen that the radius to the centreline of the chain changes from $d/2$ to $d/2 \cos \beta$ where $\beta = 360/2t$, t being the number of teeth. If the wheel is rotating at constant speed ω, the chain speed at A in the tangential direction is $\omega d/2$ and at B is $\omega d \cos \beta/2$. This variation in chain speed is plotted in the sketch: it is less than $1\frac{1}{2}\%$ for 19 or more teeth. For compact drives at low speeds fewer teeth can be used, provided the power rating of the chain is decreased. The dynamic effects associated with chordal action result in noise and wear, and the standard roller chain is not suitable for high speeds. The "silent" chain which has an inverted involute tooth form can be used in this application.[1]

Table 10.6. Data for British Standard roller chains.

ISO chain number	Pitch (mm)	Recommended centre distance (mm)	Minimum breaking load (kN)		
			Simple	Duplex	Triplex
05B	8.00	400	4.51	7.9	11.2
06B	9.53	450	8.92	17.0	24.9
08B	12.70	600	17.9	31.2	44.5
10B	15.88	750	22.3	44.5	66.8
12B	19.05	900	28.9	57.9	86.8
16B	25.40	1050	42.3	84.5	126.8
20B	31.75	1200	64.5	129.1	193.6

Source: Extract from BS 228, *Transmission Precision Roller Chains and Chainwheels*, is reproduced by permission of the British Standards Institution, 2 Park Street, London W1A 2BS.

Chains are made in the simple, duplex (2-row) and triplex (3-row) types. Leading dimensions for these chains are given in Table 10.6.[11] The sprockets they run on are standardised and stock sizes are usually available with a keyway machined in the bore. The British Standard and American Standard chain forms differ slightly. We deal only with the former.

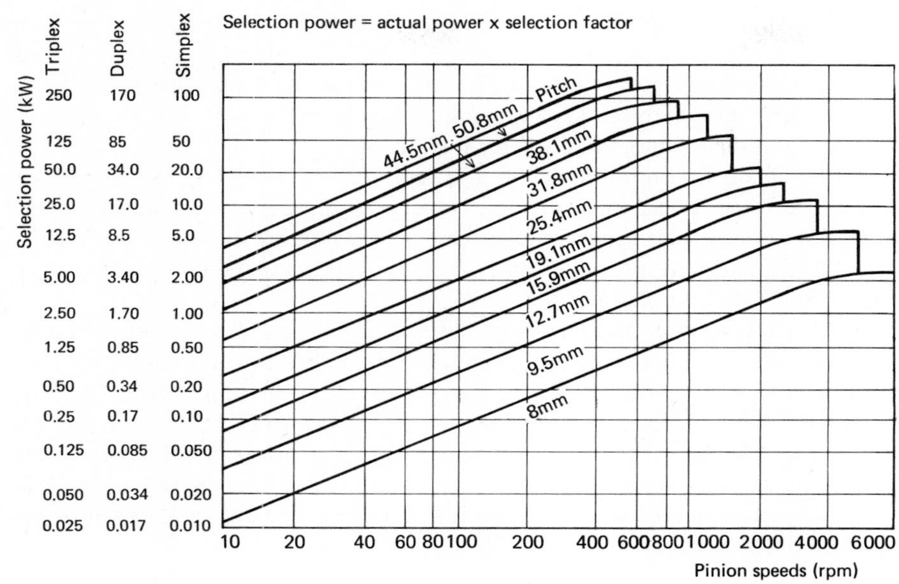

Fig. 10.23. *Power rating curves for chain drives. (Adapted from "Renold Chain Transmission", Ref. CT681 in* Renold Ltd Catalogue, *1968.)*

The power rating curves[12] illustrated in Fig. 10.23 are based on the following conditions

19-tooth pinion,
steady loading,
centre distance 30 to 80 pitches,
approximately 15 000 hours life with correct lubrication.

The selection power is obtained by multiplying the drive power by the selection factor given in Table 10.7 and the pitch and type of chain best suited to the application are found from Fig. 10.23. The selection process is completed by specifying, from a manufacturer's catalogue, the pinion and wheel teeth numbers and the chain length. The chain is made up from any selected number of links, preferably an even number since an odd number requires a cranked link.

Table 10.7. Power selection factor for chains.

Machinery characteristics	Number of teeth in pinion 17	19	21	23	25
Steady	1.1	1.0	0.9	0.85	0.75
Medium impulsive	1.7	1.5	1.4	1.3	1.2
Highly impulsive	2.2	2.0	1.8	1.7	1.6

In the installation some tension adjustment is needed to take up wear on the chain—either by varying the centre distance of one of the wheels or by running an idler or jockey pulley on the slack side of the chain.

Lubrication is essential. At low speeds, a drip feed or oil bath arrangement will suffice; for high speeds and powers over 35 kW a pumped supply should be used. In any case a protective housing, which must be dust- and oil-tight, will be required.

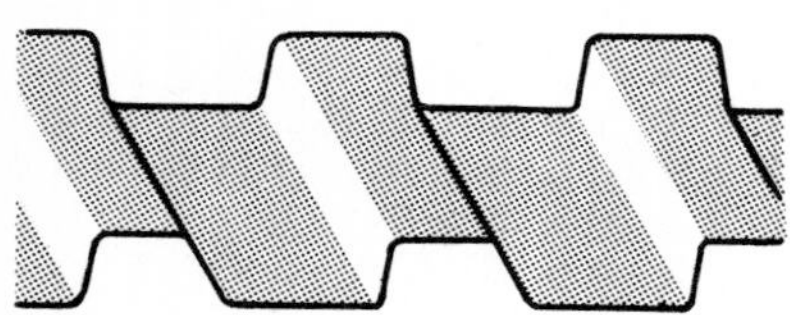

The synchronous belt, also called a *timing belt*, gives a high-speed drive with full synchronisation between shafts. The belt, which runs on grooved pulleys, consists of a nylon or flexible steel backing to which neoprene teeth are integrally moulded (see sketch). Stretch has been virtually eliminated and no take-up adjustment is needed after installation.

Apart from its timing function and lack of stretch, it has the advantage over a V-belt that it can operate in wet or dry conditions and, unlike the chain drive, does not require lubrication. It can be used on small pulleys because it is so flexible, and requires only small initial tension. Its installed cost is usually higher than V-belt or chain drives.

The nomenclature used to describe the belt and pulley is given in Fig. 10.24, in which standard values of pitch and pitch line differential are also given. The sides of the groove may be straight or of involute form. For low powers, moulded plastic pulleys can be used but aluminium or cast iron pulleys are necessary for higher powers, and machining the grooves (a hobbing process is used for the involute form) contributes to the higher cost. Since the pulleys are usually designed for some specific ratio, they are selected and specified on the basis of the number of grooves.

Pitch code		Pitch (mm)	Pitch line differential PLD(mm)
XL	(extra light)	5.080	0.254
L	(light)	9.525	0.381
H	(heavy)	12.700	0.686
XH	(extra heavy)	22.225	1.397
XXH	(double extra heavy)	31.750	1.524

Note: PCD = OD + 2 PLD

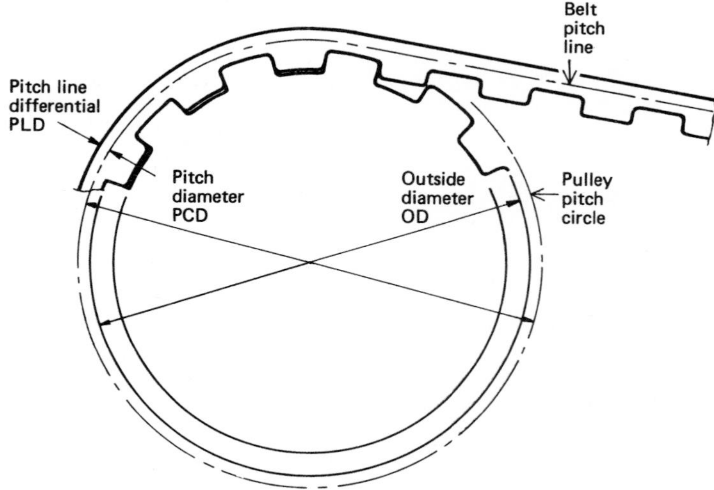

Fig. 10.24. *Nomenclature for synchronous belt drive.*

There are as yet insufficient data to allow fatigue to be taken into account in design so that power ratings are based on an equation of the form

$$KW \propto (T_p - T_c)DN \quad \text{(kW)}$$

where D is the pitch diameter, N is the rpm, and T_p and T_c are the allowable belt load and centrifugal load respectively. The latter term is proportional to N^2D.

Detailed power rating tables are given in BS 4548 for a 25 mm wide belt.[13] Preliminary selection is based on a plot of power as a function of rpm for the various sections (see Fig. 10.25) and the tables are then used to find the actual belt-width required. The full strength of the belt is developed when six or more teeth on the smaller pulley are in mesh with the belt. The use of smaller numbers of teeth in engagement is permissible, with a reduction in power.

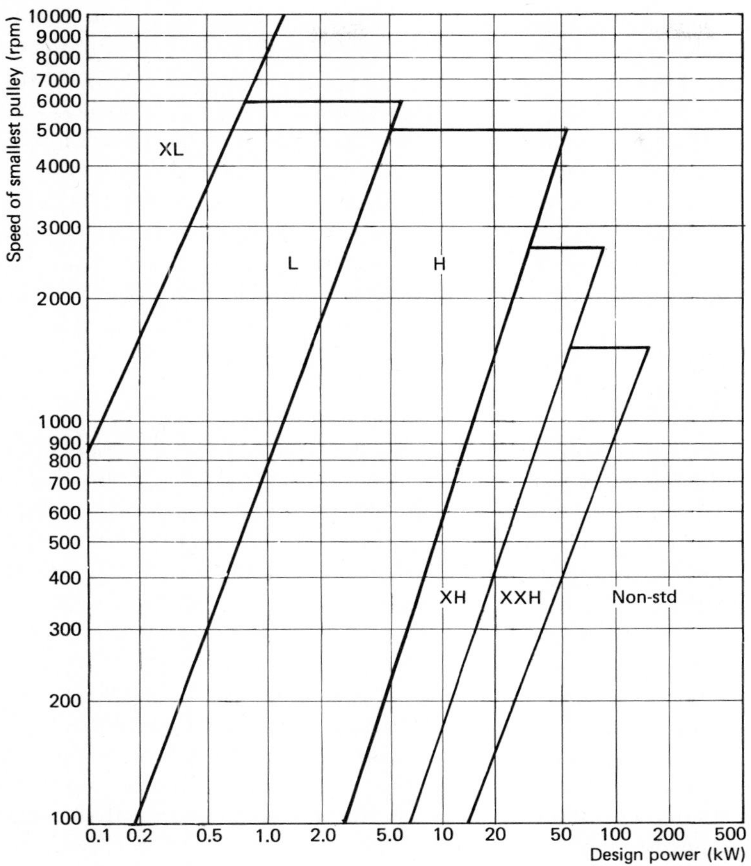

Fig. 10.25. *Preliminary selection chart for synchronous belts. (Extract from BS 4548,* Synchronous Belt Drives, *is reproduced by permission of the British Standards Institution, 2 Park Street, London W1A 2BS.)*

Installation presents few problems since belt tensions are low and no take-up is needed. Belts must be retained by fitting side plates on the pulleys, the minimum requirement being two flanges on one pulley, or one flange on each pulley on opposite sides.

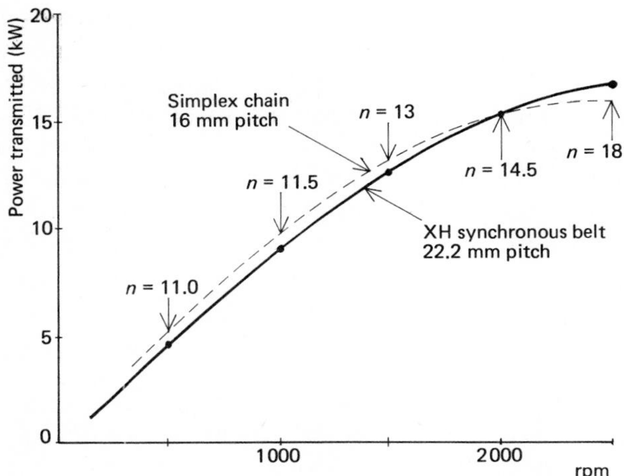

Fig. 10.26. *Power curves for synchronous belt and chain.*

A comparison between the drive characteristics of chain and synchronous belt drives is given in Fig. 10.26. For the sizes selected, the numerical values and shape of the two curves are very similar, the allowable power for the chain falling off more quickly at high rpm. It is interesting to calculate the safety factor at which the chain operates. The breaking load specified for a 15.9 mm pitch chain is 22 260 N, and the safety factor, n, can be defined as the ratio of this value to the force needed at the pitch radius to transmit the specified power, no allowance being made for centrifugal loading. It is plotted at several points on the power curve on Fig. 10.26, from which it will be seen that the chain operates at well below its tensile strength. The large safety factor is necessary to achieve a long life under the rolling wear conditions that prevail.

10.7. CLUTCHES

Where semi-permanent connections are made between shafts, bolted flanges or some of the flexible couplings described in Chapter 8 can be used. If the two shafts must be connected and disconnected when one is rotating, some type of clutch is used. Equalising the speed between the two shafts involves the transmission and the dissipation of energy—and the clutch must be designed so that it will perform the operation in a specified time and will not overheat or be subject to excessive wear.

We restrict our attention to solid friction surfaces which, when pressed together, transmit a frictional torque which will bring the speed of the driven shaft up to the speed of the driver. The design approach for clutches and brakes is similar, but the details of the mechanisms differ. Brakes, in particular, often have some degree of servo-action; this is avoided in clutches since it would make them uncontrollably fierce in operation. We consider only flat-plate and cone clutches in detail; multi-plate clutches are a simple development of the former, although the increase of the number of friction surfaces does not necessarily increase the capacity in proportion. It is more difficult to dissipate heat from multi-plate clutches, and if they are operated while the plates have a torque applied to them, the axial actuating force will fall off in successive plates owing to friction at the splines.[14]

In a simple application, a constant-speed motor would be connected by means of a clutch to a stationary rotor of known inertia, which is brought up to the speed of the motor by the torque provided by the clutch plates as they slip. It is important to note that, if friction losses in the rotor are negligible, the heat generated in this case is not a function of the type of friction surface or of time. This point is illustrated in Problem 10.10. However, if the clutch must supply a constant torque to the driven shaft over and above that needed to overcome the inertia, then the energy dissipated is a function of the clutch capacity and can be controlled by the designer by altering the torque capacity of the clutch. A treatment of this problem is given by Jania in "Friction Clutch Transmission".[15]

10.7.1. Basis of design

In the simple flat plate clutch shown in Fig. 10.27 the friction surfaces are normal to the shaft axis and are brought together with a force F_n. The torque transmitted at any radius R by an annulus dR is $2\pi R dR\, \mu\, p\, R$ where p is the surface pressure. If the pressure is uniformly distributed over the bearing surface, $p = F_n/\pi(R_2^2 - R_1^2)$ = constant and T, the torque, is given by

$$T = \frac{2\mu F_n}{R_2{}^2 - R_1{}^2} \int_{R_1}^{R_2} R^2 dR = \frac{2\mu F_n (R_2{}^3 - R_1{}^3)}{3(R_2{}^2 - R_1{}^2)} \qquad (10.16)$$

where μ is assumed to be constant.

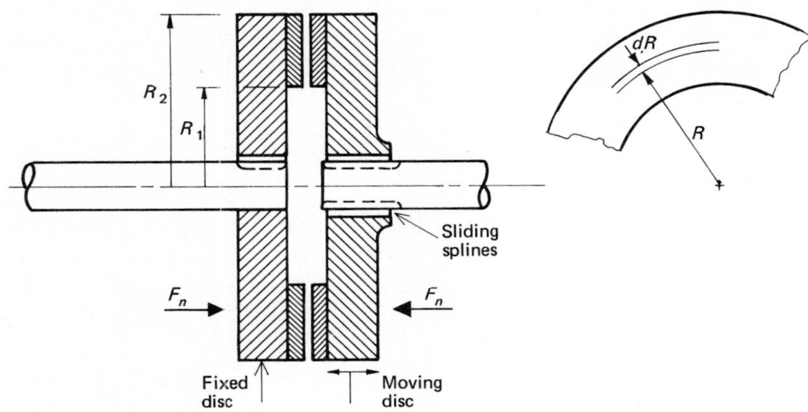

Fig. 10.27. *Simple flat-plate clutch.*

The assumption of uniform pressure may be reasonable in certain circumstances, but it does not account for the wear characteristics of sliding surfaces. Once the surfaces have settled down to a steady wear rate with time, the rate must be the same at any radius. Also it is accepted that for the materials and surface speeds used in clutches the wear rate is proportional to the surface speed, V, as well as to the pressure. Thus for uniform wear across the clutch surface,

$$\text{wear} \propto pV \propto p\omega R$$

or $p = k/R$ where k is a constant.

If the maximum allowable pressure on the clutch surface is $p_{\max}$, then

$$k = p_{\max} R_1$$

$p_{\max}$ occurring at the smallest radius. The torque is given by

$$T = \int_{R_1}^{R_2} 2\pi R^2 \mu p dR$$

$$= 2\pi\mu \int_{R_1}^{R_2} kRdR$$

or

$$T = \pi\mu k (R_2{}^2 - R_1{}^2) = \pi\mu p_{\max} R_1 (R_2{}^2 - R_1{}^2) \qquad (10.17)$$

Since

$$F_n = \int_{R_1}^{R_2} 2\pi R \frac{p_{\max} R_1}{R} dR$$

$$= 2\pi (R_2 - R_1) R_1 p_{\max}$$

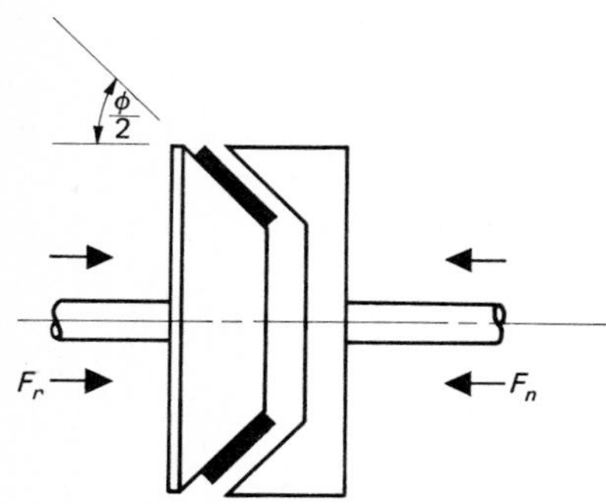

we have

$$T = \frac{1}{2}(R_1 + R_2)\, \mu\, F_n \tag{10.18}$$

In this case the effective force, μF_n, can be regarded as acting at the mean radius.

The uniform-wear assumption is generally the more appropriate for use in design, and will be used here; the difference between it and the uniform pressure case is small – less than 2% for a typical configuration.

In the cone clutch (see sketch) use is made of the higher surface pressures that result from a given axial force to increase the capacity. The wedging action obtained from a cone angle of ϕ is proportional to $1/\sin(\phi/2)$, so that Equation (10.18) becomes

$$T = \frac{(R_1 + R_2)\mu F_n}{2 \sin(\phi/2)} \tag{10.19}$$

The cone angle must be large enough so that the surfaces will not lock together. Simple theory shows that this requires that $\tan^{-1}(\phi/2) < \mu$; in practice, ϕ is usually taken between 24° and 36° for values of μ ranging from 0.3 to 0.5.[16] A cone clutch has greater capacity to dissipate heat than a plate clutch, and therefore may be more heavily rated.

In the design of a clutch to transmit a given torque using Equation (10.17), values of p_{max} and μ must be known. Other criteria are also involved. Since μ for most friction materials falls off with increasing temperature while wear increases markedly, the clutch must be designed so that the friction material does not overheat or wear too rapidly. This can be done by calculating the heat generated and balancing it against the energy absorbed or dissipated by conduction and convection.[15, 16] An alternative approach often used is to classify clutches in terms of the duty they perform and their capacity to absorb or dissipate heat, and to limit the power dissipated per unit area to a suitable value. Recommended design values range from 0.6 W/mm^2 for light duty to 0.1 W/mm^2 for heavy duty applications. Some manufacturers specify suitable $p_{max}V$ values for their products. For a given value of μ this is the same as specifying a rating in terms of power per unit area (see Equation (10.21)).

A variety of friction materials are in use and they must be selected to fit the application.[16] The main groups are woven cotton, woven asbestos, moulded asbestos and sintered metal. The second and third groups can be manufactured with a metallic inclusion that improves heat transfer and wear characteristics. The wear of a friction material is quantified in terms of a life value, L, having units J/mm^3, and a typical curve for L for a woven friction material is given in Fig. 10.28. Values of the coefficient of friction range from 0.5 at low temperatures to 0.3 at high temperatures.

Oil-immersed clutches have an application where smooth engagement and low wear are required. Special materials are produced for this use, in which the cooling obtained from the oil allows higher power ratings which offset the lower values of μ on the lubricated surfaces.

10.7.2. Design method

In simple cases, the clutch proportions can be determined for some specified torque from Equation (10.17), $p_{max}V$ being selected to suit the material and the

Example 10.3 Sheet 1 of 1

A clutch is required to accelerate a rotor ($I = 0.9$ kg m^2) from 0 to 1000 rpm in 0.25 seconds. If the maximum allowable outside radius is 150 mm determine the axial force required for a flat plate clutch. Take $\mu = 0.35$

Suitable proportion for clutch is

$$\frac{R_2}{R_1} = 1.5 \qquad \text{i.e. } R_1 = 100 \text{ mm}$$

$$\text{Required torque} = I\ddot{\theta} = \frac{0.9 \times 1000 \times 2\pi}{60 \times 0.25}$$

$$= 377 \text{ Nm}$$

From Equation 10.17

$$p_{max} = \frac{377 \times 1000}{\pi \times 0.35 \times 100 \times 12500} = 0.274 \text{ MPa}$$

This is within acceptable limits

From Equation 10.18

$$F_n = \frac{377 \times 1000 \times 2}{250 \times 0.35} = 8620 \text{ N}$$

This force could be reduced by using a cone clutch; for a cone angle of 36° the force would be 8620 sin 18 = 2660 N

Example 10.4 Sheet 1 of 1

Check the rating for the friction material for the clutch in Example 10.3 and calculate the wear expected in a life of 10^5 applications.

R_p is based on a mean speed i.e. V_m is calculated at 500 rpm at 125 mm radius

$$V_m = \omega R = \frac{2\pi \times 500 \times 125}{60 \times 1000} \text{ m/s}$$

$$R_p = \mu \, p_{max} \, V_m = \frac{0.35 \times 0.274 \times 2\pi \times 500 \times 125}{60 \times 1000}$$

$$= 0.63 \text{ W/mm}^2$$

This is slightly above the recommended range of 0.6 to 0.1

Check wear

Energy dissipated per application (See problem 10.4) is given by $1/2 \, I\omega^2$

i.e. Energy per application $= \frac{0.9}{2}\left(\frac{2\pi \times 1000}{60}\right)^2$

$= 4930$ Nm

$$\text{Wear} = \frac{n_a R_p t_s}{1000 L} = \frac{10^5 \times 0.63 \times 0.25}{4000}$$, using value of L from Fig. 10.29 of 4 kJ/mm^3

i.e. Wear = 4 mm. This would not be excessive. A thickness of about 8 mm would be specified.

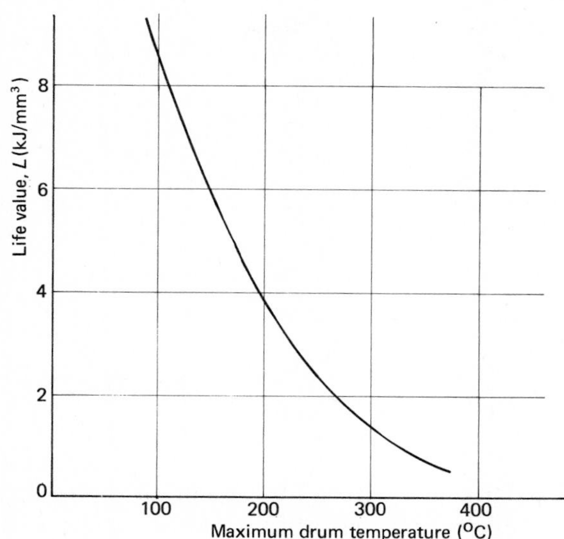

Fig. 10.28. *Typical life–temperature curve for woven friction material (From "Friction Materials for Engineers",* Ferodo Publication 1051, *1968, by permission of Ferodo Ltd.)*

surface speed. Note that V is the mean velocity during the engagement. Recommendations given by the manufacturers of friction materials should be followed in this case; typical values for p_{max} range from 0.1 MPa to 0.4 MPa.

A method that takes a more direct account of some of the important wear and energy criteria will now be described. Let R_p be the rating of the material in W/mm^2. The following equations can be written

$$V = \frac{\pi D N}{60\,000} \text{ m/s} \tag{10.20}$$

$$R_p = \mu p_{\max} V \text{ W/mm}^2 \tag{10.21}$$

Here V is the mean relative speed between the surfaces during the engagement, i.e., D is the mean diameter and N the mean relative rpm between the sliding surfaces. Combining these equations gives an expression for the maximum contact pressure

$$p_{\max} = \frac{60\,000\, R_p}{\mu\, \pi D\, N} \text{ MPa} \tag{10.22}$$

For wear, if n_a is the number of applications expected during the life of the friction surface and t_s the slipping time during each engagement, the energy, E_d, dissipated during the lifetime will be

$$E_d = \frac{n_a\, R_p\, A\, t_s}{1000} \text{ kJ} \tag{10.23}$$

where A is the area of one wearing surface. The expected wear volume is given by E_d/L and the wear thickness, t_w, by

$$t_w = \frac{E_d}{LA} = \frac{n_a\, R_p\, t_s}{1000L} \text{ mm} \tag{10.24}$$

A typical plot of permissible pressure at varying speeds is shown in Fig. 10.29. This has been derived from Equation (10.22) using values of $R_p = 0.6$ W/mm^2 and $\mu = 0.3$.

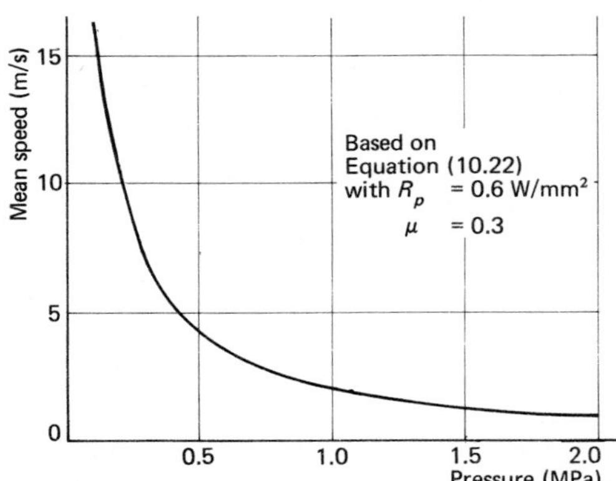

Fig. 10.29. *Permissible pressure at varying slipping speeds.*

In many cases, the geometry of the clutch will be predetermined and a knowledge of suitable proportions helps reduce the number of variables—for example, it is common practice to use a ratio R_2/R_1 of about 1.5. The design approach based on Equations (10.22) and (10.24) is illustrated in Example 10.4, in which values from Example 10.3 are checked. Note that in this case a specific value of lining thickness is calculated for the specified life. This approach does not differ fundamentally from the one previously illustrated, but the data are used in a more methodical manner.

10.8. FLUID COUPLING

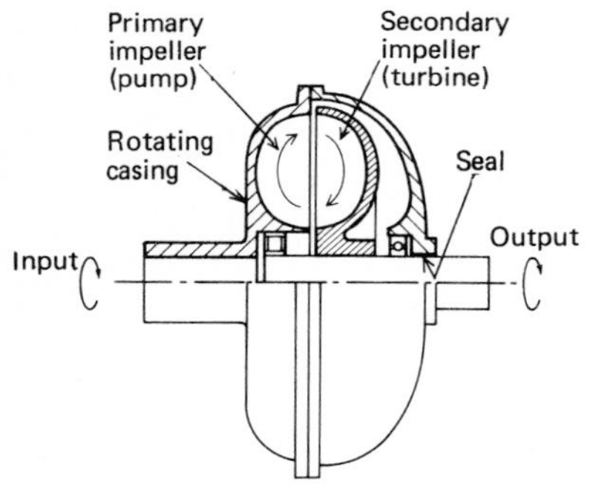

The fluid coupling provides a link between a driver and driven machine which under steady load will transmit power smoothly and with low losses, but which will automatically slip if the unit is overloaded and when it is being started up.

The coupling consists of a casing in which vanes connected to the input and output shafts rotate in a fluid. The vanes on the input or primary shaft (see sketch) act as a pump, causing a flow to the vanes on the secondary element which acts as a turbine. The torques on the primary and secondary shafts are equal, since no other external torque is applied. A torque converter operates in a similar manner, but stationary guide vanes provide an external torque reaction which allows an increase or decrease in torque and speed between the primary and secondary elements. We deal here only with the fluid coupling.

The moment of momentum of the fluid increases as it flows outward in the primary vanes, which are radial and resemble the segments of an orange cut in half. This moment of momentum is transmitted to the output shaft during inward flow in the secondary vanes. The generation and absorption of moment of momentum are equal, although there is a loss of energy due to shock losses and friction. If the efficiency is defined as the ratio of secondary to primary power, the fact that $T_p = T_s$ gives

$$\eta = \frac{\omega_s}{\omega_p}$$

The slip, S, is defined as

$$S = \frac{\omega_p - \omega_s}{\omega_p} = 1 - \eta$$

The operating characteristics of a coupling are usually given in terms of the slip, which is about 3% to 4% under normal operating conditions.

The way in which the characteristics of a coupling of this type can be combined with the torque curve of an induction motor is shown in Fig. 10.30. The points of intersection of the motor characteristic with the converter slip curves give the running point of the combination—when it has accelerated to steady speed. For example, the rated torque is transmitted with 4% slip—an efficiency of 96%. If the output shaft is stalled, the 100% slip curve shows that 200% of the rated torque will be transmitted. Under prolonged operation at these conditions the energy dissipated would cause overheating unless the converter were cooled in some way.

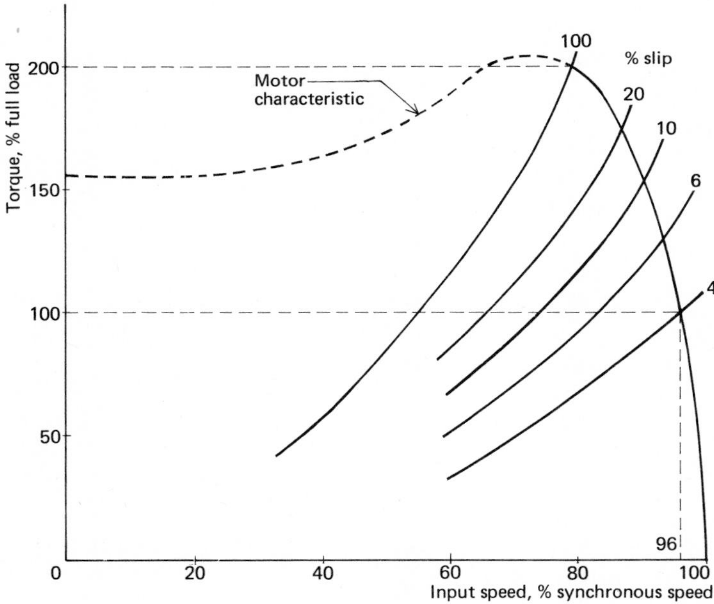

Fig. 10.30. *Motor-coupling characteristics.*

A method used to give greater flexibility in speed control is to vary the quantity of fluid in the coupling. In this way, the coupling can be used to control the speed of a machine, such as a fan or pump, or to regulate the torque transmitted.

A number of standard fluid couplings are available for use, one type being built into the pulleys of a V-belt drive. Some of these units are described in "Drive Line Engineering"[1] and Elderton's *Principles, Characteristics and Applications of Fluid Couplings.*[17] Some simple design calculations for couplings and torque converters are given by W. Annand in *The Mechanics of Machines.*[18]

LIST OF SYMBOLS

a	addendum height
A, B	constants
A	surface area of clutch
C	centre distance of shafts
C_1, C_2, C_3	constants in V-belt power equation
d, d_1, d_2	pitch diameter
d_0	base circle diameter
E	modulus of elasticity
E_d	energy dissipated during life of clutch surface (kJ)
F, F_n	tangential and normal forces on gear tooth, per unit width
F_n	clutch engagement force
F	traction force, small-pulley diameter factor
f_{bt}, f_{bc}	tensile and compressive bending stresses
$f_{\max}$	Hertzian contact stress
G_r	velocity ratio
I	current, amps
KW_b, KW_c	gear power for strength, wear
L	life value for clutch material (kJ/mm^3)
L_p	pitch length of V-belt
l	length
m	module
N	rotational speed
n	safety factor
n_a	number of applications in life of clutch surface
P	diametral pitch
$P_{\max}$	maximum engine power
P_r	radial force on belt
p	circular pitch (mm)
$p, p_{\max}$	pressure on clutch surface, maximum pressure
p_t	tensile stress in belt
R	radius
R_p	rating of friction material (W/mm^2)
R_r	equivalent radius
S	slip in fluid coupling
S_b	bending stress factor
S_c	surface stress factor
S	belt speed (m/s)
T_L, T_H	belt tensions
$T, T_{\max}$	torque, maximum torque (N m)
T_p, T_s	torque on primary and secondary impellers of fluid coupling
t	number of teeth
t_s	slipping time for clutch engagement
V	velocity
$V_{\max}$	limiting velocity
W	face width of gear
X_b, X_c	combined speed factors for strength and wear
x, y	tooth measurements
Y	strength factor
Z	zone factor

α	angle of wrap
β	angle on chain sprocket
η	efficiency
Θ	V-angle of belt
θ	tooth force angle
μ	coefficient of friction
ν	Poisson's ratio
ϕ	cone angle of clutch
ψ	pressure angle for involute gear tooth
ω	angular velocity

REFERENCES

[1] IME, "Drive Line Engineering", Book 1, *Proceedings,* vol. 184, Part 31, The Institution of Mechanical Engineers, London, 1969–70.
[2] H. Cotton, *Electrical Technology*, 7th ed., Pitman, London, 1957.
[3] A. E. Fitzgerald, C. Kingsley and A. Kusko, *Electrical Machinery*, 3rd ed., McGraw-Hill, New York, 1971.
[4] "Mechanical Drives", *Machine Design Reference Issue,* **41**, 29, 18 December, 1969.
[5] BS 1440:1971, *Endless V-Belt Drives,* British Standards Institution, London, 1971.
[6] AGMA 210.02, *Surface Durability of Spur Gear Teeth*: AGMA 220.02, *Rating the Strength of Spur Gear Teeth,* AGMA Standards, Washington, 1966.
[7] W. A. Tuplin, *Gear Load Capacity*, Pitman, London, 1962.
[8] BS 436:1940, *Machine Cut Gears,* British Standards Institution, London, 1940.
[9] H. E. Merritt, *Gears,* 2nd ed., Pitman, London, 1946.
[10] P. S. Houghton, *Gears,* 3rd ed., Technical Press, London, 1970.
[11] BS 228:1970, *Transmission Precision Roller Chains and Chainwheels,* British Standards Institution, London, 1970.
[12] Renold Ltd, "Renold Chain Transmission", Ref. CT681, *Renold Ltd Catalogue*, Manchester, 1968.
[13] BS 4548:1970, *Synchronous Belt Drives*, British Standards Institution, London, 1970.
[14] T. P. Newcomb and H. E. Merritt, "Effect of Spline Friction on the Torque Capacity and Interface Temperature Reached During a Multi-disc Clutch Engagement", *Journal of Mechanical Engineering Science,* **4**, 4, 1962.
[15] Z. J. Jania, "Friction Clutch Transmission", *Machine Design*, **30**, 23, 24, 25, 26, 1958.
[16] Ferodo Ltd, "Friction Materials for Engineers", *Ferodo Publication 1051,* 2nd ed., Chapel-en-le-Frith, 1968.
[17] J. Elderton, *Principles, Characteristics, and Applications of Fluid Couplings,* Fluidrive Engineering, Middlesex, 1963.
[18] W. J. D. Annand, *The Mechanics of Machines,* Heinemann, London, 1966.

BIBLIOGRAPHY

Broersma, G., *Couplings and Bearings,* Stam International, Culemborg, 1968.
Cauling, S. A., *Industrial and Marine Gearing*, Chapman & Hall, London, 1962.

IME, "Gearing in 1970", *Proceedings,* vol. 184, Part 30, The Institution of Mechanical Engineers, London, 1969–70.
NEWCOMB, T. P., "Temperatures Reached in Friction Clutch Transmissions' *Journal of Mechanical Engineering Science,* **2**, 4, 1960.

PROBLEMS

10.1. Apart from the main drive line on a motor vehicle, there are many auxiliary drives which use components such as belts, gears, flexible drives, etc. List as many of these as you can for a vehicle with which you are familiar.

10.2. The methods used to transmit movements from the control column of an aircraft to the control surface, e.g. the elevator or aileron, depend on the size of the aircraft and the forces involved. Make a list of the various methods that could be used and describe the main advantages and disadvantages of each.

10.3. A 5-kW electric motor drives a rock crusher through a V-belt installation which is required to reduce the speed from 750 rpm to about 180 rpm. Select a suitable V-belt drive using:
(a) the smallest possible pulleys and centre distance, and
(b) a 200-mm pulley on the motor shaft, with a centre distance double that used in (a). Assume 10 hours per day operation.

10.4. A two-cylinder air compressor is driven by a 2-kW electric motor through a V-belt drive that reduces the speed from 1500 rpm to 800 rpm. Specify suitable sizes to give the smallest possible centre distance. Assume operation is less than 8 hours per day.

10.5. A drive in a food-packaging machine must transmit 0.6 kW and reduce the speed from 750 rpm by a factor of 2 exactly. There are no restrictions on space. Investigate the suitability of a chain drive and synchronous belt drive for this application.

10.6. The geometry and shape of involute gear teeth and the occurrence of interference can be demonstrated by generating the teeth using a basic rack. The rack (only one full tooth is needed) is cut out of thick cardboard according to the proportions of Fig. 10.15. A straight piece of board is fastened to the back of the rack along the pitch line, with a spacer setting a gap between the two. A segment of a circle with an outside diameter corresponding to the addendum circle of the gear fits in this gap, and has a segment fastened to the back of it representing the pitch circle. When the pitch line rolls on the pitch circle, pencil lines drawn on the circular segment along the face of the rack tooth will generate a gear tooth. The use of different pitch diameters, corresponding to, say, 12 and 25 teeth, will illustrate the variation of tooth shape with numbers of teeth.

10.7. Design the second stage of the reduction gear described in Example 10.2. The output wheel is to be concentric with the input pinion 1 of the worked example, and is to be driven by three pinions mounted on the same shafts as the 72-tooth gears.

10.8. A single-reduction spur gear is required to transmit 30 kW with a reduction of 5 to 1 from a speed of 950 rpm. If the pinion and wheel are to have 18 teeth and 90 teeth respectively, determine a suitable module, face width and centreline distance for the gears.

Equivalent running time: 12 hours

Selected face width: $W = 10 \times$ module

Materials: pinion $S_c = 28$; $S_b = 148$;

wheel $S_c = 9.7$; $S_b = 131$.

10.9. Use the solution to Problem 10.8 to find suitable gear proportions for a 60-kW drive.

10.10. A high-inertia drive motor rotates at constant speed ω and is connected to a rotor of inertia, I, by a clutch. When the clutch is engaged, the speed of the rotor is brought up to the speed of the motor in a short time interval. Show that, if frictional losses in the rotor shaft are ignored, the energy dissipated in the clutch is $\frac{1}{2}I\omega^2$.

10.11. A drying machine has a rotor of 50-kg mass and 0.3-m radius of gyration. It is accelerated to a speed of 1200 rpm in 1.5 seconds by a constant speed motor which is connected to it by a clutch.

(a) Using a value of μ of 0.3 and a ratio R_2/R_1 of 1.5, calculate the clutch dimensions and the axial load required, if the maximum surface pressure is 0.2 MPa.

(b) What value of power dissipation per unit area results from the values selected in (a)?

(c) What will be the wear per 1000 applications of the clutch, if the life value is 4 kJ/mm^3?

11 Limits and Fits

11.1. INTRODUCTION

It is not generally possible to manufacture a number of parts precisely to a single specified size; this makes it necessary for the designer to set limits of size within which the manufacturer can work. The outer bounds of such limits are set by functional requirements – for example, the fit of mating parts – and the inner bounds by the economics of the manufacturing processes available. The closer the limits are set (perhaps because of demands for improved interchangeability) the more demanding and costly will be the process. The way in which such limits are selected and specified is discussed in this chapter. It should be noted that alternative methods of obtaining correct fits are available (for example, by selective assembly or by hand fitting) and a decision must be made as to which system is to be adopted.

The principles that are developed here are relevant to shop production where accurate mating of parts is a basic aspect of assembly. Most civil engineering construction does not call for such mating. When two parts of a structure are built in different places, probably by different people, and have to be joined together on site, practice is to allow for a substantial gap between them and to fill the gap after assembly with packing and to cover it or, if it is unimportant, to leave it. An example of this way round the problem is described in Section 23.3. There, the bolting of a steel column to its foundations is considered; it is a typical example of the interface between shop and site operations. Similar techniques are used when rails or machines are bolted to concrete foundations.

Dimensioning of drawings is discussed in Section 11.3 and the principles described there in relation to shopwork apply also to civil engineering construction. Chapter 24, "Communications and Organisation", should also be consulted. There, the use of surveyor's pegs, string lines, stretched wires and so on in setting-out the work is discussed in relation to dimensioning.

11.2. THE SELECTION OF FITS

The system to be described provides a guide to the selection of fits for machine parts, but the ideas and methods can be applied wherever it is necessary to ensure the correct assembly of mating parts or where the size of a production item – for example, bar stock – must be controlled.

The term *fit* is defined in BS 2517 as the relationship existing between two mating parts with respect to the amount of clearance or interference which is present when they are assembled.[1]

If a 30-mm shaft is to run smoothly and without excessive play in a bush, the fit could be achieved by labelling the drawing "machine to give smooth running fit". This presupposes that the same workman will make both parts; he could finish the shaft to a nominal 30-mm diameter and then produce the bush on a lathe, opening out the bore of the bush until what he judges to be the required fit on the shaft is achieved. Or he might ream out the bush with a 30-mm reamer and then turn down the shaft to give the fit. The second would be the better method since it is easier to control size on a lathe. This approach has several disadvantages. The responsibility of deciding what constitutes the correct fit has been delegated to the workman, who may have little knowledge of how the parts function. The exact size of the parts produced will not be recorded, and provision of standard spare parts would be impossible.

Another approach would be for the designer to specify, from his experience that a diametral clearance of 1/1000 of the diameter gives a good running fit, that the minimum clearance should be 0.03 mm. To simplify the workman's job he would translate this into actual dimensions. If the bush is to be reamed, the workman will have little control on the size of the hole, which will usually turn out to be slightly oversize. The hole can accordingly be dimensioned as $30\,^{+0}_{+0.04}$ meaning that the hole must lie somewhere between the design size of 30.00 mm and the upper limit of size of 30.04 mm, the 0.04 margin having been selected on the basis of a knowledge of the accuracy attainable with a reamer. The shaft must be turned down to a size below the basic 30-mm dimension – the design size here will be 29.97 mm, allowance being made for the 0.03 mm clearance. Adding some margin for error, the specified size could be $29.97\,^{+0}_{-0.02}$ giving the lathe operator a variation or tolerance of 0.02 mm below the desired size. The final dimensions are shown in the sketch; it will be seen that, if the parts are made within the specified limits, the smallest possible clearance will be 0.03 and the largest 0.09.

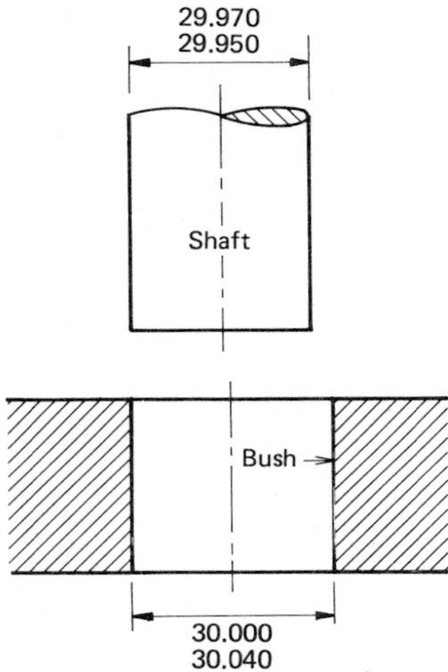

The system that is described below follows a similar procedure in fixing the limits of size, but the methods have been adopted as an International Standard. The British Standard that will be referred to here, BS 4500: 1969, conforms to this.[2]

The terminology is rather specialised and will be summarised before the use of the standard is described; it will be seen that some of the terms have already been used in this section. The system will be more readily understood if it is remembered that the desired type of fit is obtained by applying an allowance to the size of one or other of two mating parts; each part must also have a *tolerance* to give some margin for error. We will discuss fits in terms of a shaft and a hole, but any parts which must fit together can be dimensioned using this approach.

11.2.1. Definitions

The *basic size* is the theoretical size of a part, and is the basic dimension to which the tolerances and allowances that determine the fit are applied. It is the same for both members of a fit. Since it is usually easier to vary the size of the shaft (for example, in turning) than the size of the hole (in drilling or reaming), the allowance is usually applied to the shaft, and in this case we talk about a *hole-basis* system of fits. However, in some cases where the shaft size is predetermined, the allowance must be applied to the hole dimension (*shaft-basis* system); such a case arises in specifying the hole size for the outer ring of a rolling bearing (which is in this case the "shaft"). The bearing is supplied by the manufacturer within specified size limits, and the fit of the bearing in its housing can only be adjusted by varying the allowance for the hole in the housing.

The *limits of size* are the maximum and minimum sizes permitted for a dimension. The *tolerance* is the amount of variation permitted for the size of a dimension. It is the difference between the maximum and minimum limits of size and is an absolute value without sign. *Bilateral tolerances* (e.g. 90 $\pm$ 0.2) are normally used for dimensioning non-mating features. Where mating parts are involved, *unilateral tolerances* are normally used; these tolerances lie wholly above the design size for the hole and below the design size for the shaft.

The *deviation* is the algebraic amount by which a size is greater (+) or less (−) than the basic size. The *upper deviation* is the algebraic difference between the maximum limit of size and the corresponding basic size; this is designated **ES** for a hole and **es** for a shaft. The *lower deviation* is the algebraic difference between the minimum limit of size and the basic size, and this is designated **EI** for a hole and **ei** for a shaft. These letters, which are used in the tables of tolerances, stand for the French *écart supérieur* and *écart inférieur* respectively. Some of these terms are displayed in diagrammatic form in Fig. 11.1. It will be noted that the line drawn to represent the basic size is called the line of zero deviation and that the deviations are all referred to it. The *fundamental deviation* is that one of the two deviations, being the one nearest to the zero line, which is conventionally chosen to define the position of the tolerance zone in relation to the line of zero deviation.

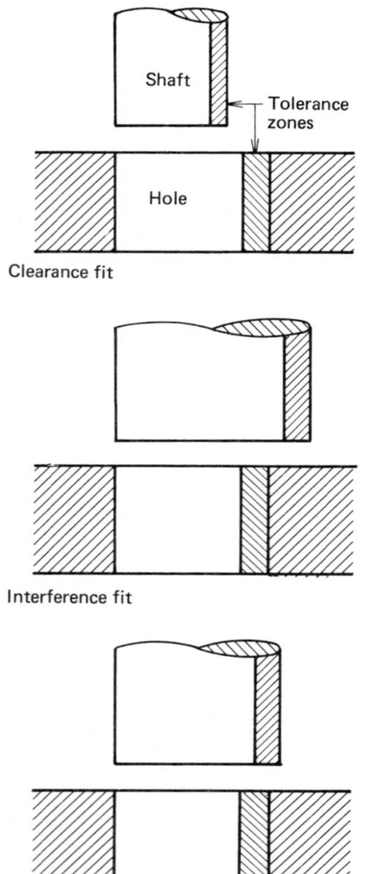

Although we have talked loosely about using an allowance to determine the type of fit, it is the fundamental deviation that is actually defined in the standard for each fit by an empirical equation. Full definitions of the other terms can be found in BS 2517: 1954[1] and BS 4500: 1969.[2]

When showing a tolerance graphically, it can be represented by the zone between the limits of tolerance. If the tolerance zones on the shaft and hole do not overlap, the fit is either a clearance fit (where the fundamental deviation is positive) or an interference fit (where the fundamental deviation is negative). If the tolerance zones overlap, then either a clearance or an interference can result, depending on what size the parts are made within the allowed tolerance; this is called a transition fit. The three types of fit are shown in the sketch.

Values of upper and lower deviations are given in the standard and are added algebraically directly to the basic size and no attempt is made to distinguish between the fundamental deviation and the tolerance.

11.2.2. Description of fits

In the standard, a letter and a number are used to describe the combination of some particular deviation and tolerance. For the shaft, lower case letters **a** through **h** to **z** describe fits from very large clearances (undersize shaft) to very large interferences (oversize shaft); the letter **h** signifies zero deviation, the larger limit of size corresponding to the basic size. In normal engineering practice only the letters **d** through to **s** are commonly encountered. Capital letters are used in the same way to describe the disposition of the deviation for the hole, **A** giving a large clearance (oversize hole) and **Z** a large interference.

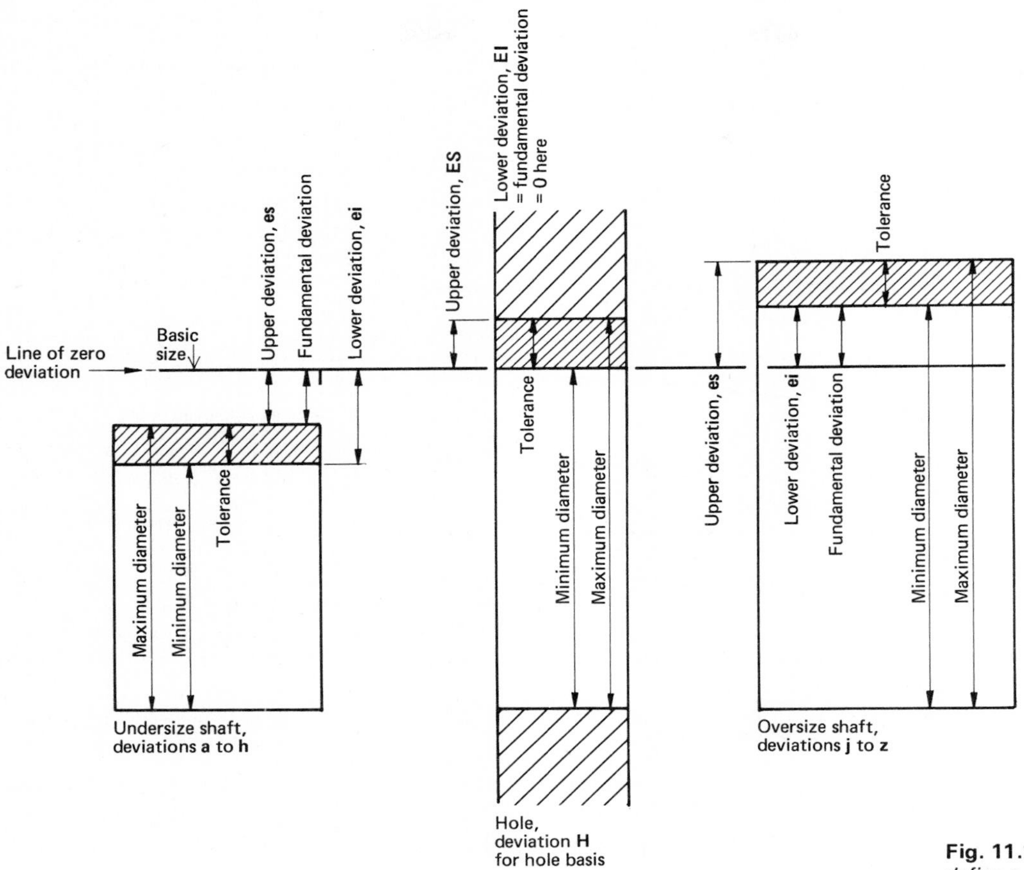

Fig. 11.1. *Illustration of terms used to define a fit.*

Using a hole basis system, only the letter **H** is used, different fits being obtained by varying the deviation on the shaft. Some special types of fit have been designated by double letters – e.g. **za zb zc** for high interference fits. These are not considered here.

The suffix numbers are used to describe the tolerance, and run from 01, 0, 1 through to 16. The numbers below 5 correspond to high precision work not usually encountered outside the tool room or metrology laboratory. The tolerance quality 16 would be achieved in sand casting or flame cutting. Figure 11.2 illustrates the relationship between these numbers and the accuracy to be expected from various manufacturing processes under average conditions; also plotted are actual values of the tolerance for a range of diameters.

Numerical values for the fundamental deviation have been fixed by practical considerations, its variation with size of a component being determined from an empirical equation.[2] The values for tolerances are based on a geometric series of preferred numbers (see Section 11.5). A tolerance unit, I, is defined by

$$I = 0.45\ \sqrt[3]{D} + 0.001D \qquad (11.1)$$

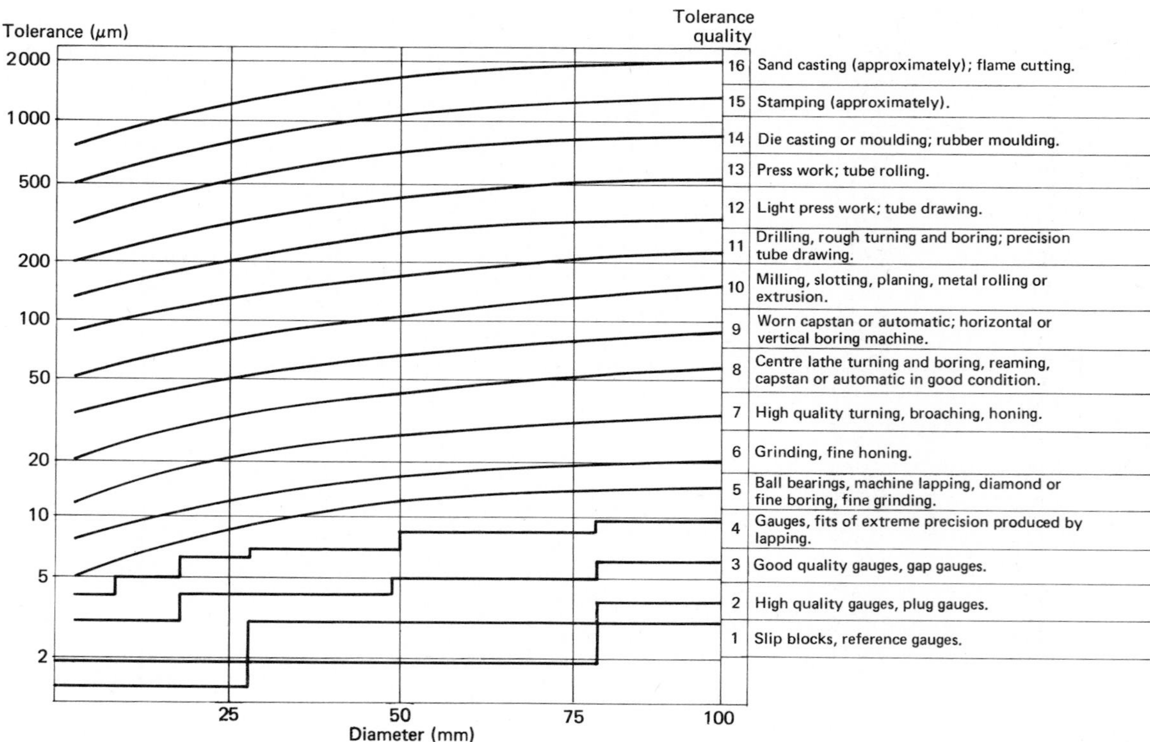

Fig. 11.2. *The accuracy of various manufacturing processes. (Extract from BS 1916,* Guide to the Selection of Fits, *reproduced by permission of the British Standards Institution, 2 Park Street, London W1A 2BS.)*

where D is the shaft diameter in mm and I is in μm. For tolerance grades from 6 to 16 (they are labelled **IT**6, **IT**7, etc., where **IT** stands for "ISO tolerance") the values of the tolerance are as follows:

Grade	**IT**6	**IT**7	**IT**8	**IT**9	**IT**10	**IT**11	**IT**12	**IT**13	**IT**14	**IT**15	**IT**16
	$10I$	$16I$	$25I$	$40I$	$64I$	$100I$	$160I$	$250I$	$400I$	$640I$	$1000I$

The tolerances given by Equation (11.1) are matched to the size of the part; the first term accounts for the difficulty of manufacturing parts to a given accuracy (this is found in practice to be proportional to the cube root of the size) and the second term accounts for the difficulty of making accurate measurements, which increases with the size of the part.

11.2.3. Selected fits

Out of the many possible combinations in the ISO system, a selection has been made in BS 4500 which will provide for most normal engineering products. The following hole and shaft tolerances are used.[2]

Hole tolerances: **H7**; **H8**; **H9**; **H11**.
Shaft tolerances: **c11**; **d10**; **e9**; **f7**; **g6**; **h6**; **k6**; **n6**; **p6**; **s6**.

The way in which these tolerances may be combined to give a small but flexible range of fits is shown in Figs 11.3 and 11.4. Numerical values for these selected holes and shafts are given in Appendix 5 (Tables A5.1, A5.2). The tolerances given relate to a hole-basis system, since only **H** or zero allowance fits are specified for the hole. As has been explained, some cases are encountered where a shaft-basis system must be used; for this purpose, additional values for holes are given in Table A5.3. While these will be useful mainly when selecting tolerances for housings of rolling bearings, it should be noted that the same fit can be obtained (provided the tolerance grades are the same) by interchanging letters in any given combination, thus **H8–f8** in the hole-based system can be replaced by **F8–h8** in the shaft-based system to give the same fit. This approach would be useful when a number of different components have to be fitted to the same shaft.

Type of fit	Shaft tolerance	Hole tolerances			
		H7	**H8**	**H9**	**H11**
Clearance	**c11**				▨
	d10			▨	
	e9			▨	
	f7		▨		
	g6	▨			
	h6	▨			
Transition	**k6**	▨			
	n6	▨			
Interference	**p6**	▨			
	s6	▨			

Fig. 11.3. *Selected fits, hole basis. (Extract from BS 4500,* ISO Limits and Fits, *reproduced by permission of the British Standards Institution, 2 Park Street, London W1A 2BS.)*

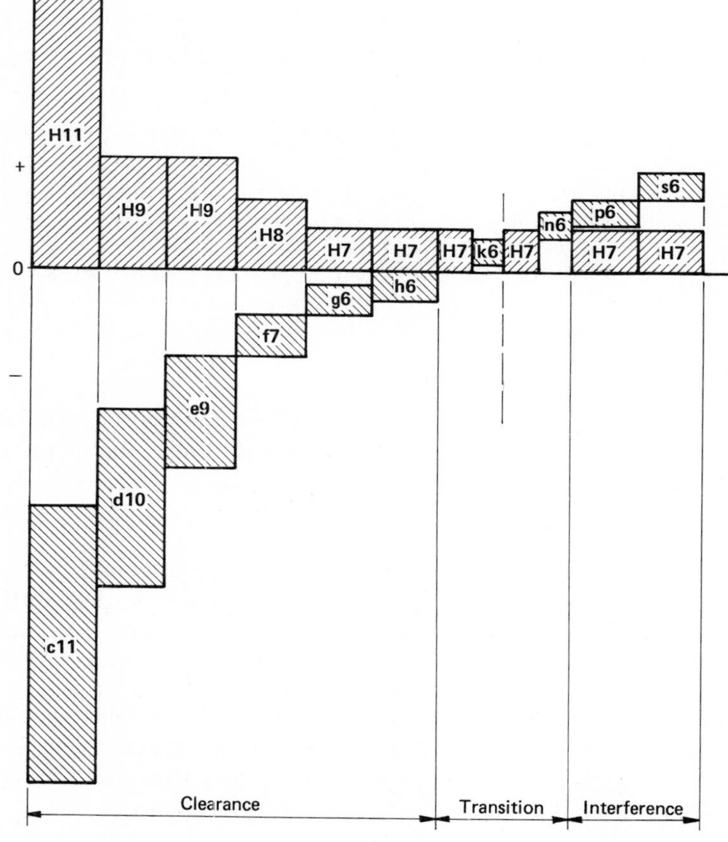

Fig. 11.4. *Relative sizes of tolerance bands for selected fits. (Extract from BS 4500,* ISO Limits and Fits, *reproduced with permission of the British Standards Institution, 2 Park Street, London W1A 2BS.)*

It will be noted that a given hole tolerance is usually associated with a shaft tolerance that is one grade finer, e.g. a "7" hole with a "6" shaft. This reflects the fact that, for most production methods, higher accuracy can be achieved on the shaft than on the hole.

An indication of typical applications for the selected tolerances has been summarised from BS 1916[3] and is given below. Note that the numbers in brackets following each description give the maximum and minimum clearance or interference in μm that would apply to a 50 mm nominal size shaft-hole combination.

H11–c11 This combination would only be used to give a very large clearance where ease of assembly must be assured. (Clearance 130 to 450.)

H9–d10 This fit gives a large clearance for such applications as gland seals and could be used for slack running fits. (Clearance 80 to 242.)

H9–e9 This combination gives an easy running fit where an appreciable clearance is permissible, for example, for widely separated bearings, or several bearings in line. (Clearance 50 to 174.)

H8–f7 This is the most commonly used normal running fit and is suitable for the majority of applications requiring a good quality fit which is relatively easy to produce. Typical uses would be for shaft bearings in gear boxes and for pump bearings and medium and light rotating and sliding mechanisms. (Clearance 25 to 89.)

H7–g6 The clearance provided by this fit is small and is suitable for lightly loaded precision bearings and for pins and spigots where a precision clearance fit is required. (Clearance 9 to 50.)

H7–h6 Although the minimum clearance will be zero, a small clearance will usually be present. This is the closest available clearance fit and is mainly used for non-running assemblies such as spigots, pins and other location fits. (Clearance 0 to 41).

H7–k6 This transition fit averages only a small clearance, and can be used to locate wheels, couplings and gears that are keyed to shafts and for spigots requiring accurate location. (Interference 18 to clearance 23.)

H7–n6 A clearance still results from the lower deviations of shaft and hole, but the fit generally gives tight assembly fits suitable for bushes, fitted bolts and shaft–hub combinations. (Interference 33 to clearance 8.)

H7–p6 This combination provides the first true interference fit, and with ferrous parts, a press fit results which can be dismantled and assembled when required. It can be used for bushes, gears and hubs which are to be pressed into place. (Interference 42 to 1.)

H7–s6 A heavy drive fit results from this combination, which can be used for the permanent or semi-permanent assembly of ferrous parts and for bearing bushes in alloy housings. Assembly by shrinkage may be required to avoid damage. (Interference 59 to 18.)

Example 11.1 | Sheet 1 of 1

(a) Determine suitable tolerances for 45 mm diameter steel pump shaft, running in a bronze bearing.
Select H8 – f7 fit

For shaft, f7, es = −25, ei = −50 (Table A5.1)
For hole, H8, ES = +39, EI = 0 (Table A5.2)

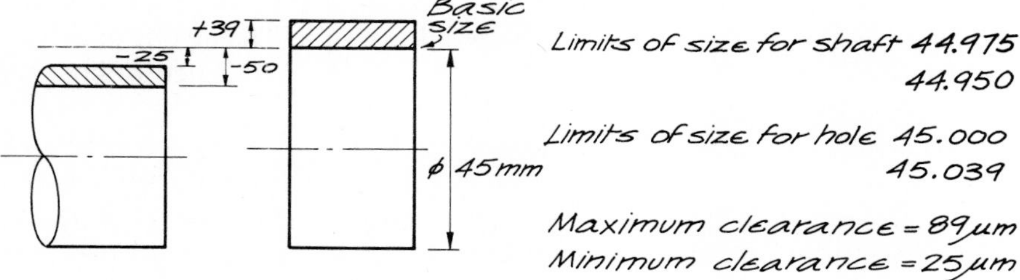

Limits of size for shaft 44.975
44.950

Limits of size for hole 45.000
45.039

Maximum clearance = 89 μm
Minimum clearance = 25 μm

(b) Determine suitable tolerances for the steel shaft and pump rotor, which are assembled with an interference fit, but must be separated for maintenance. Diameter, 60 mm.
Select H7 — p6 fit

For shaft, p6, es = +51, ei = +32
For hole, H7, ES = +30, EI = 0

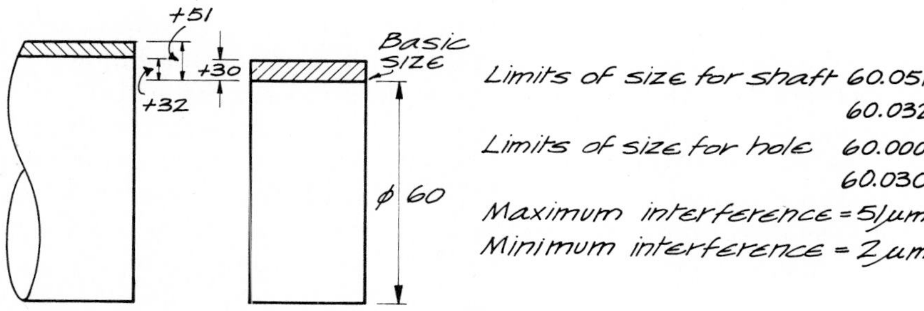

Limits of size for shaft 60.051
60.032
Limits of size for hole 60.000
60.030
Maximum interference = 51 μm
Minimum interference = 2 μm

It should be noted that, although the designer may show a shaft in a sketch as having an **H8–f7** fit, the final drawings will always be labelled with the actual limits of size. The problems involved in setting these dimensions out on drawings will now be considered.

11.3. DIMENSIONING

While it is important to select correctly the basic size and tolerances of a component, it is equally important to display the sizes as dimensions on a drawing in such a way that the task of the workman is simplified and the accumulation of errors during construction or manufacture is avoided. To do this the designer must be aware of the way in which jobs are set out; the most important factor here is the selection of a suitable datum for the dimensions on the drawing – preferably the same datum as will be used when the job is set out.

Consider a simple problem first: that of locating a pattern of bolts that are set in a concrete foundation so that they will line up with the holes in the bedplate of a machine that is to be bolted down to it. The layout of the holes, which we will assume are drilled precisely to size and are located without error, is shown in Fig. 11.5(a).

It is difficult to position bolts in a concrete slab with any precision and we must decide what tolerance can be assigned to their location while ensuring that the 10 mm diameter bolts will fit the 12 mm diameter oversize holes in the plate. If only one bolt in the concrete is off-centre by two units, the plate will fit but all the bolts will be hard up against the sides of their holes; if several bolts are incorrectly located, there is a probability that the errors will be cumulative and, in order to ensure that the plate can be assembled, all the bolts must be positioned within ± 1 unit from their correct location.

In Fig. 11.5(b), the bolt centres are located by dimensions running from the sides of the foundation; given this drawing, the workmen will be encouraged to set the bolts up with reference to the sides, which will be the boxing for the concrete, or a concrete face if the bolts are to be centred in oversize holes after the concrete has been poured. It is unlikely that the accuracy and straightness of the sides will be much better than ± 5 units, in which case the location of the bolts can be in error by at least this amount. In Fig. 11.5(c), the dimensions have been based on the centrelines xx', yy'. Although these centrelines are not located by any specific physical feature – it happens here that some of the bolts are centred on them – they can be drawn in on the foundation or can be represented by straight edges or wires. If the datum is accurately located, each dimension that positions a bolt on the left half of Fig. 11.5(c) can be given a tolerance of ± 1. On the right-hand half of the figure, the dimensions for holes A and B in the x direction are such that the errors are cumulative and, with a tolerance of ± 1, hole B could be out of position, in the worst case, by ± 2.

Another illustration of the importance of correctly selecting the datum for dimensions is given in Fig. 11.6. The lengths AB, BC, CD and DE are all 100 units and must fall within the tolerances shown in Fig. 11.6(a). If face A is used as a datum, a small tolerance must be put on AB because any inaccuracy in the location of face B when machining will affect the length BC; the tolerances shown in Fig. 11.6(b) are the largest that can be used while still meeting the tolerances. If face C is used as the datum, the tolerances are less demanding –

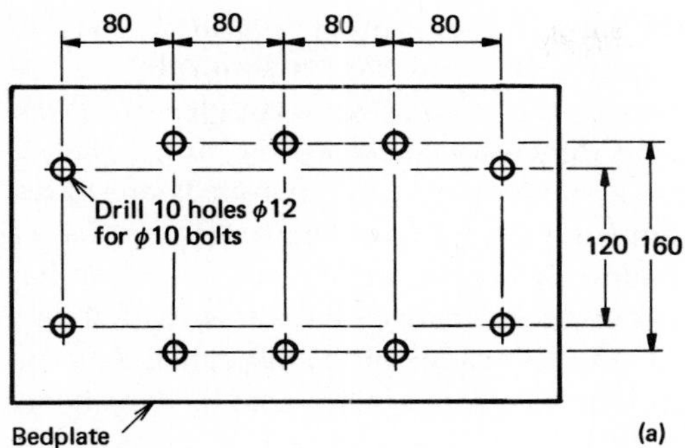

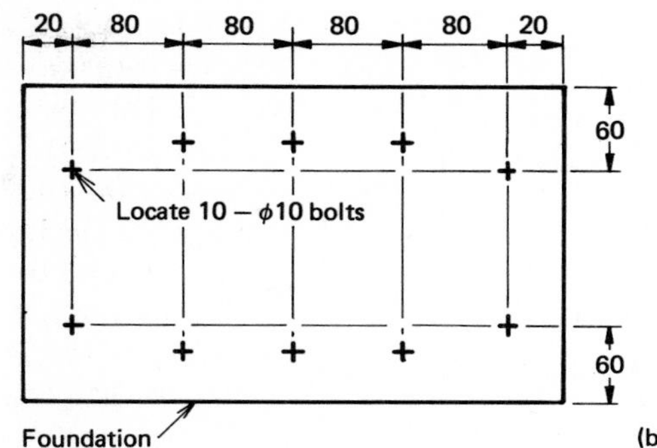

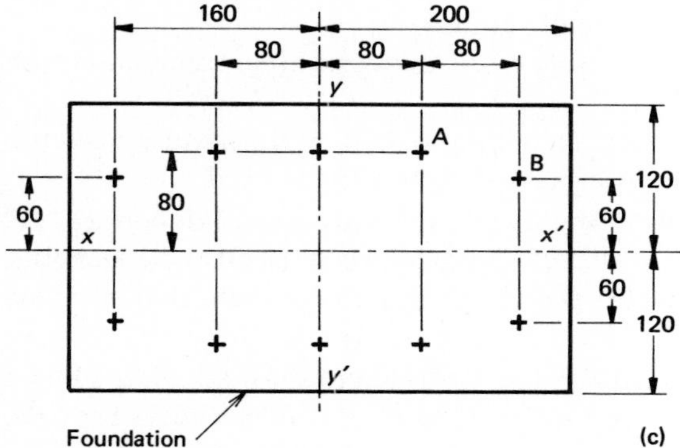

Fig. 11.5. *Location of bed-plate on a concrete foundation:*
(a) holes in bed-plate;
(b) bolts dimensioned from edge of foundation;
(c) bolts dimensioned from centrelines.

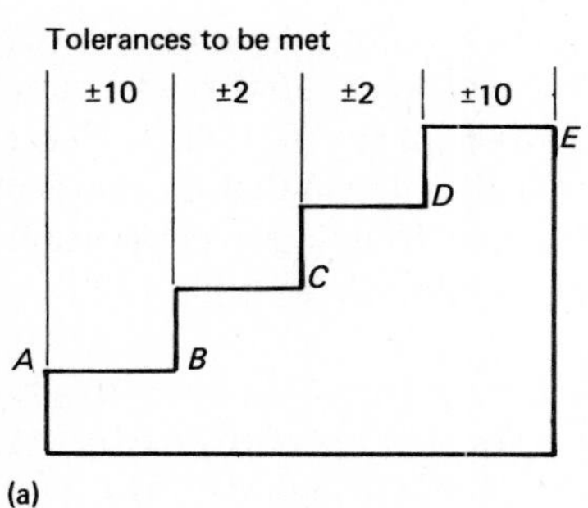

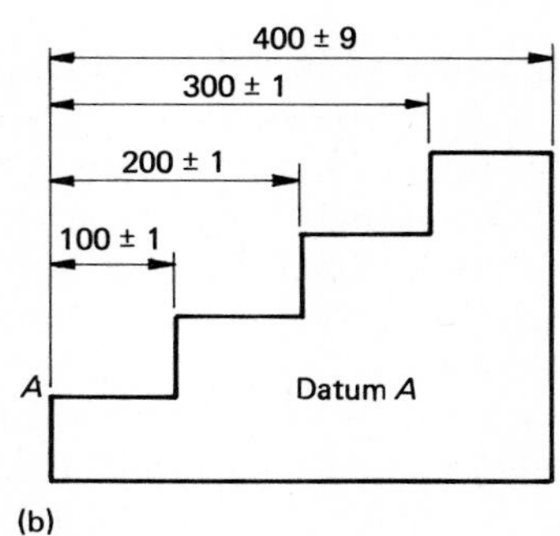

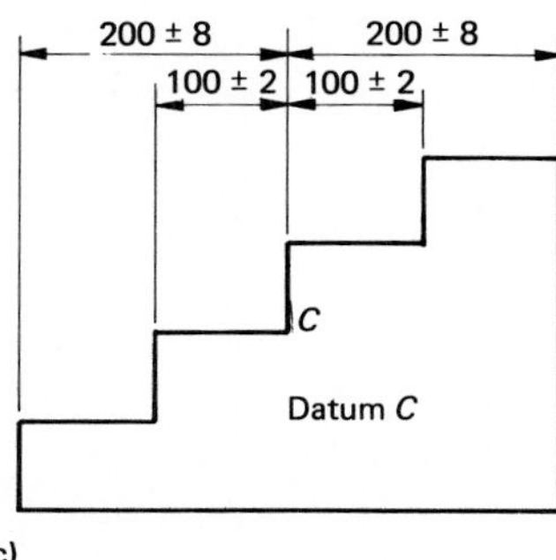

Fig. 11.6. *Selection of a datum:*
(a) tolerances to be met;
(b) dimensions using face A as datum;
(c) dimensions using face C as datum.

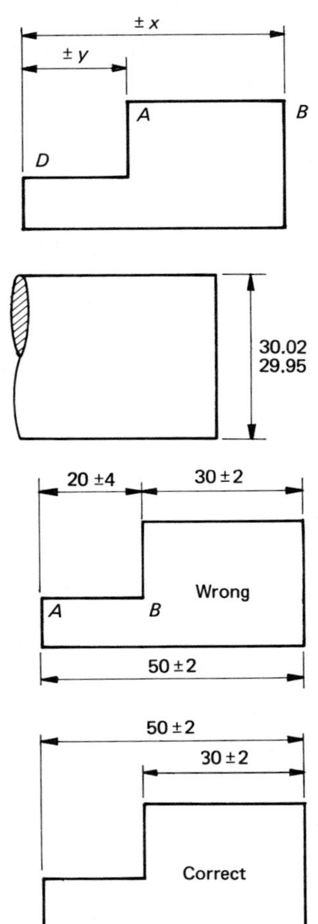

those shown in Fig. 11.6(c) are the largest that can be accepted. Smaller tolerances can, of course, meet the requirements, but less economically.

The two problems just discussed are simplified examples of those encountered when drawings are being dimensioned. It will be noted that in discussing these problems it has been assumed that, when two ends of a given face AB are dimensioned from a datum D, the error in the length AB is $\pm(x + y)$. The worst possible case has been assumed to occur and no consideration has been given to the probability of its occurrence. These cases can be treated on a statistical basis (see, for example, E. T. Fortini[4]). One of the difficulties encountered when this approach is taken is that the errors are not usually distributed in a normal manner. When faced with the tolerances shown in the sketch, a careful worker would produce all the parts close to the 30.02 limit. Another worker might produce a more random distribution, while the variations from an automatic machine would depend on the machine setter and on the degree of sharpness of the tool. These variations, and other difficulties, must be faced if a statistical approach is to show useful results.

Other points to be given attention when dimensioning are summarised below:

(a) a convenient datum for dimensions along a shaft is the feature—often a shoulder—against which the axially locating bearing is fixed;

(b) avoid redundant dimensions: in the sketch, the least critical dimension AB should be allowed to fall out as the difference between the 50 and 30 lengths. Redundant dimensions can be shown in brackets when they provide additional information;

(c) when inserting the tolerances it helps the workman to show maximum metal size first; he will usually work to the first dimension, and will then have the tolerance in hand as a margin for error both for holes and for shafts;

(d) make sure that finished drawings are checked; this should be done systematically, with each dimension being marked when it is accepted as correct.

11.4. SURFACE FINISH

It is to be expected that there will be a relation between the tolerance and the surface finish demanded for a part; a rough surface will wear rapidly, and there would be little point in machining it to a fine tolerance in the first place. There is also a connection between the accuracy and the surface finish that is obtained in various machining operations. If the degree of surface roughness is important to the correct functioning of a part, the designer should specify it in addition to the tolerance.

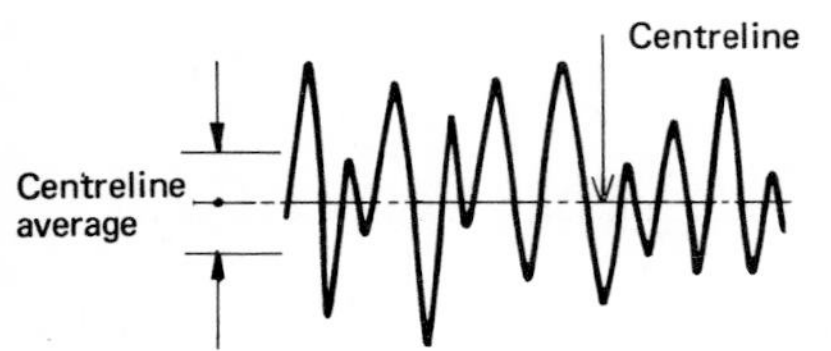

Surface roughness is expressed in terms of the centreline-average height (CLA height or **Ra** value). This is the arithmetic average value of the departure of the surface profile above and below the centreline of the roughness. This and other terms that are used to assess surface texture are defined in BS 1134.[5]

The way in which the recurrent irregularity of roughness is distributed is called the *lay* of the surface. Different machining operations show characteristic differences, although they may have the same value of **Ra**. It is only when the lay is important to the functioning of a part that the machining operation must be specified in detail.

Plastic and stainless steel samples showing different values of roughness and types of lay are manufactured, and can be used by the designer and machine operator to get a feel for various finishes. Accurate measurements can be made on electronic meters.

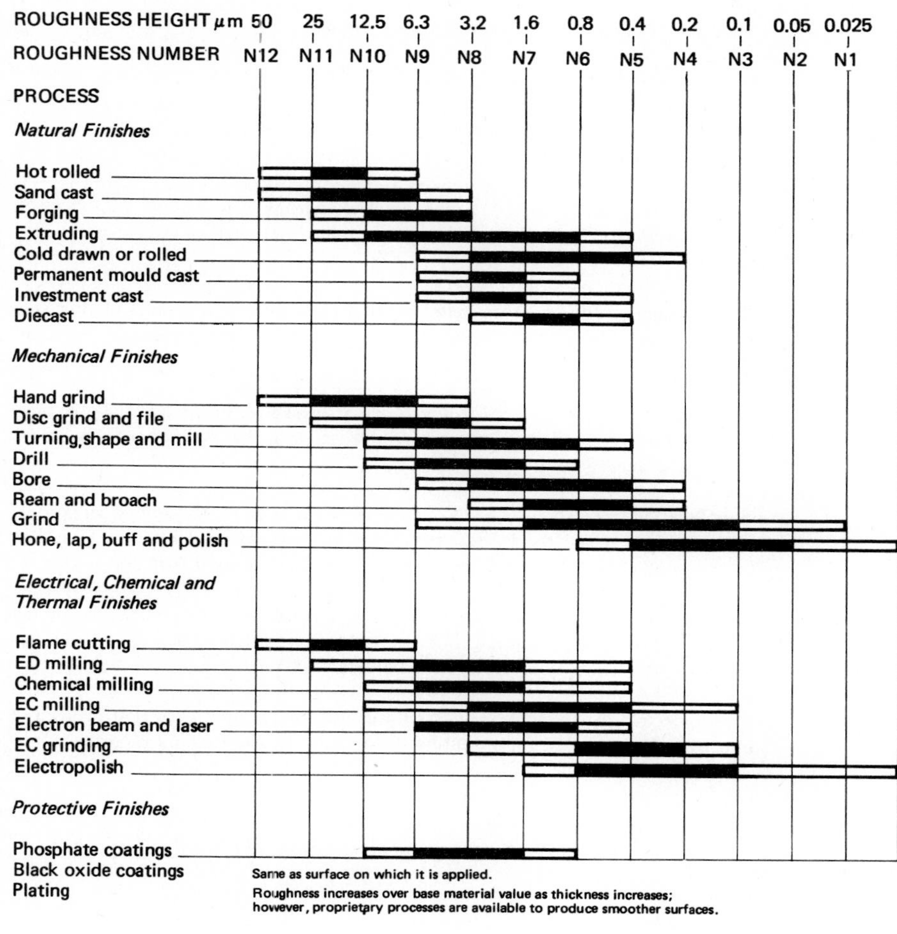

Fig. 11.7. *Surface roughness produced by common production methods. (From* Machining Data Handbook[6] *by permission of Metcut Research Associates.)*

Figure 11.7 gives details of the surface finish that can be expected from various machining operations. The designer now has the option of specifying the finish he requires by calling for a surface that is "rough turned" or "fine ground", or by specifying the surface roughness. The latter method is precise and measurable, and is to be preferred. The use of roughness (**N**) numbers on drawings is recommended in BS 308[7]; the relationship between these numbers and **Ra** values is shown on Fig. 11.7.

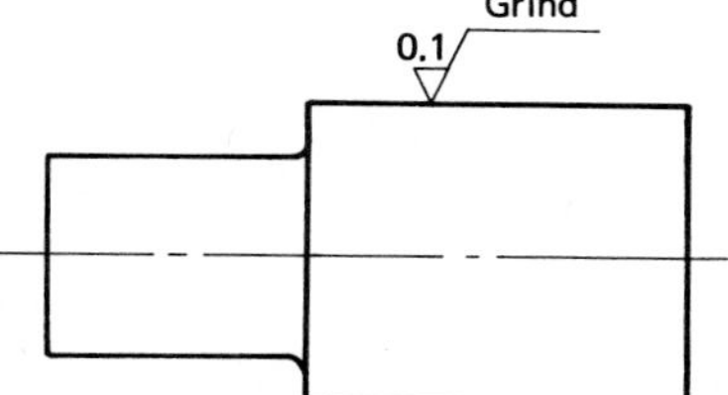

Where surface finish is important, it is usual for handbooks and standards to specify a suitable value for the roughness. Examples of this are found in the standards on V-belts (for pulleys) and gears. A general guide to suitable values is given in Table 11.1, taken from Fortini's *Dimensioning for Interchangeable Manufacture.*[4]

Table 11.1.
Typical average roughness values, **Ra**, for surfaces on mechanical parts.

Ra (μm)	Applications
0.025	Micrometer anvils, mirrors, master-gauge faces
0.05	Shop-gauge faces, comparator anvils
0.1	Vernier caliper faces, wrist pins, hydraulic pistons, lapped ball and roller bearings, carbon-seal mating surfaces
0.2	Cam and crankshaft journals, connecting rod journals, valve stems, cam faces on engine camshafts, hydraulic cylinder bores, honed ball and roller bearings, rolled threads
0.4	Bronze bearings for precision spindles, engine cylinder bores, faces of precision gear teeth, ground power screws, lead screws, shafts running against oil seals, O-ring dynamic seals
0.8	Brake drums, ground ball and roller bearings, gasket seals for hydraulic fittings, bronze bearings for electric motors, cylindrical push fits, sliding surfaces, worm gears, cam faces
1.6	Gear locating faces, screw threads of less than 2.5 mm pitch, splines and serrations, teeth of ratchets and pawls, gear bores, engine cylinder-head surfaces, gear box faces, piston crowns, rotating labyrinth seals, accurate datum surfaces
3.2	Datum surfaces, general mounting surfaces, surfaces for soft gaskets, die or tap cut screw threads, screw threads of more than 2.5 mm pitch, spotfaces and counter-bores for bolt heads and nuts 20 mm diameter or less
6.3	Clearance surfaces, wrench slots, spotfaces and counterbores for bolt heads and nuts over 20 mm diameter, mounting surfaces on rough mechanical parts
12.5	Clearance surfaces on heavy machinery

Source: Adapted from E. T. Fortini, *Dimensioning for Interchangeable Manufacture*,[4] by permission of the Industrial Press Inc.

11.5. PREFERRED NUMBERS

The designer is not always free to specify the exact sizes that result from his design calculations: he must have regard to the sizes of materials and tools that are produced and stocked. In practice, this is not a severe restriction; the availability of a limited selection of stock sizes allows the designer to concentrate on the important factors that govern the integrity of the design, and frees him from the need to investigate a wide combination of possible sizes. In exceptional cases – for example, where weight is at a premium – a special run of material must be produced. This may be the precursor of an extension of the range; it would be preferable to have such an addition planned to fit into an accepted range of sizes.

The fact that stock items have usually undergone intensive appraisal and use can give some assurance that they will perform satisfactorily. To take a simple example, when the designer specifies a bolt for a tension load on the

Table 11.2. Basic series of preferred numbers.

Serial number	Basic series R5	R10	R20	R40
0	**1.00**	**1.00**	1.00	1.00
1				1.06
2			1.12	1.12
3				1.18
4		**1.25**	1.25	1.25
5				1.32
6			1.40	1.40
7				1.50
8	**1.60**	**1.60**	1.60	1.60
9				1.70
10			1.80	1.80
11				1.90
12		**2.00**	2.00	2.00
13				2.12
14			2.24	2.24
15				2.36
16	**2.50**	**2.50**	2.50	2.50
17				2.65
18			2.80	2.80
19				3.00
20		**3.15**	3.15	3.15
21				3.35
22			3.55	3.55
23				3.75
24	**4.00**	**4.00**	4.00	4.00
25				4.25
26			4.50	4.50
27				4.75
28		**5.00**	5.00	5.00
29				5.30
30			5.60	5.60
31				6.00
32	**6.30**	**6.30**	6.30	6.30
33				6.70
34			7.10	7.10
35				7.50
36		**8.00**	8.00	8.00
37				8.50
38			9.00	9.00
39				9.50
40	**10.00**	**10.00**	10.00	10.00

Source: Extract from BS 2045, *Preferred Numbers*, reproduced by permission of the British Standards Institution, 2 Park Street, London W1A 2BS.

basis of the stress area in the plane of the threads, he does not normally need to check the bearing and shear stresses in the head and nut; he knows from experience that they have proportions that give balanced strength under normal loading conditions.

If the designer must select elements – beams, bars, tubes, bearings, drills and construction machines – from a limited range of sizes, care must be taken to ensure that he is provided with as wide a choice as possible, subject to the need to achieve economy in holding stock. It is found in practice that a range of sizes best satisfies consumer preference when it follows a geometric progression.

Details of the way in which several ranges of preferred numbers can be built up are given in BS 2045.[8] The four principal ones are termed the **R5**, **R10**, **R20** and **R40** series, and are sometimes referred to as *Renard numbers* after the French engineer who proposed their use.

The numbers are derived from geometric series having one of the following four common ratios:

$$\sqrt[5]{10} \qquad \sqrt[10]{10} \qquad \sqrt[20]{10} \qquad \sqrt[40]{10}$$

or 1.58 1.26 1.12 1.06

The basic series are given in Table 11.2. Although some rounding-off has been done, this amounts nowhere to more than 1.26%.

A wide variety of products is produced to conform to these series, and several examples are to be found in tables in this book.

Designers are being forced to select from preferred numbers. When setting up new size ranges, they should attempt to conform to such numbers. Even if only one product is initially planned, it should, if possible, match one value of a series so that later production may fit into the series.

REFERENCES

[1] BS 2517:1954, *Definitions for Use in Mechanical Engineering,* British Standards Institution, London, 1954.

[2] BS 4500:1969, *ISO Limits and Fits,* British Standards Institution, London, 1969.

[3] BS 1916:Part 2:1953, *Guide to the Selection of Fits,* British Standards Institution, London, 1953.

[4] E. T. Fortini, *Dimensioning for Interchangeable Manufacture,* Industrial Press, New York, 1967.

[5] BS 1134:1961, *Centreline–Average Height Method for the Assessment of Surface Texture*, British Standards Institution, London, 1961.

[6] Metcut Research, *Machining Data Handbook*, Metcut Research Associates, Ohio, 1966.

[7] BS 308:Part 2:1972, "Dimensioning and Tolerancing of Size", *Engineering Drawing Practice*, British Standards Institution, London, 1972.

[8] BS 2045:1965, *Preferred Numbers,* British Standards Institution, London, 1965.

BIBLIOGRAPHY

BS 308:Part 3:1972, "Geometrical Tolerancing", *Engineering Drawing Practice*, British Standards Institution, London, 1972.

Gladman, C. A., *Manual for Geometric Analysis of Engineering Design,* Australian Trade Publications, Sydney, 1966.

PROBLEMS

11.1. List a number of items that are manufactured in a range of sizes, and compare the available sizes with the series of preferred numbers. Do they conform to the series? What advantages and disadvantages are there in following the preferred-number system?

11.2. Machinists sometimes follow a rule of thumb that allows a diametral clearance of one five-hundredth of the shaft diameter for a running fit in a journal. Does this practice conform to any particular ISO fit?

11.3. Consider an interference fit of a 70-mm diameter aluminium shaft and a collar of 120 mm outside diameter. If an **H6** hole is used, which Grade 6 shaft fit will give a stress in the bore that will reach the yield stress of 200 MPa? (Assume the maximum possible interference for any given fit).

11.4. Make a drawing of the components shown in the bearing installation of Fig. 9.3 and where possible provide fully toleranced dimensions. A type 6310 deep groove ball bearing is to be used, and the component is to be produced in quantity. The shaft is machined to an **n6** fit for the bearing, and the housing to an **M7** fit.

11.5. Procure a drawing of

(a) a component that is to be produced in large quantities (e.g., an automotive piston), and

(b) a machine that is to be built on a "one-off" or small-batch basis (e.g., a laboratory testing machine).

Examine the tolerances critically. Comment on whether they are well-matched to the function of the machine, and on the control that has been achieved on quantities such as concentricity, weight, surface finish, etc. What machines do you think the designer had in mind for the various manufacturing operations required?

12 Ties and Struts

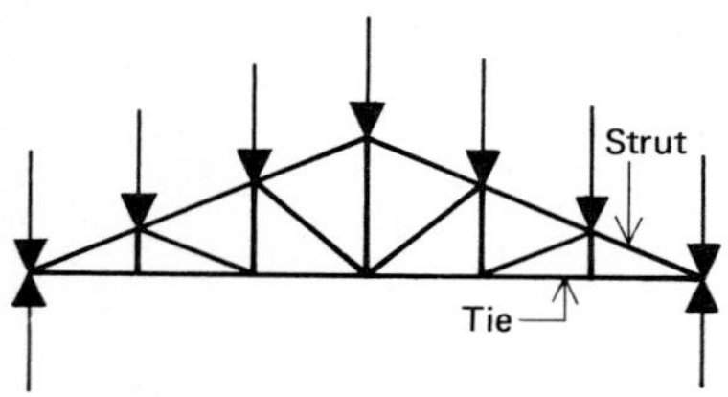

12.1. INTRODUCTION

A *tie* is a structural element carrying a tensile load and a *strut* is one which supports a compressive load. They have considerable significance in structural engineering because they may be assembled into frames which can support loads on long spans. Plane frames, or *trusses*, are of great importance in everyday practice. Ties and struts also appear in various guises in machine parts—for example, in a hydraulic cylinder, the piston rod is designed as a strut.

The tensile or compressive load in each member of a frame can be calculated by using the principles of statics and it is normal practice to assume that all joints are pinned—i.e. that the members meeting at a joint may rotate relative to each other without restraint (see Chapter 13, "Connections"). Although welded and even bolted joints do restrain relative rotation of the ends of members meeting at a joint, this restraint is usually small. It has the effect of superimposing bending stresses on the tensile or compressive stress in the adjacent members and, because of their smallness, these bending stresses are known as secondary stresses. We will assume that secondary stresses are negligible.

In this chapter, we examine methods of selecting suitable members for ties and struts in machine parts and in trusses of steel and timber. The discussion covers axially loaded struts and struts which have to resist bending moments as well as direct loads. Such bending moments may be caused by transverse loads or eccentricity of the longitudinal load.

12.2. BASIS OF DESIGN

Current practice in structural engineering, both for ties and for struts, is to satisfy the requirement that stress in service shall not exceed a specified stress which is less than the yield stress or ultimate stress by an adequate margin. Thus, the designer uses loads to be carried in service and failure stress divided by a factor of safety. In fact, derivation of the design criteria is related more closely to the way these elements fail and it would be more logical to base design on service loads multiplied by a load factor and failure of the member.

Superficially, the only difference between a strut and a tie is the sign of the stress imposed on the material. Although this is the only significant difference at small loads, at loads large enough to cause failure, the behaviour of a tie is quite different from that of a strut. This difference can be shown with simple desk top experiments.

Failure of a tie can be demonstrated by pulling a strip of paper which has a hole through it. The tie fails by parting and the characteristics of the failure are:

(a) it cannot be reversed by unloading the tie;
(b) it occurs locally;
(c) it occurs where the effective area of cross-section is least.

In a similar experiment, a thin wooden batten can be loaded in compression. As far as its ability to serve in a structure is concerned, failure occurs when the batten buckles. Owing to its curvature in this state, its ends are brought closer together by a relatively large distance; large deflections are rarely acceptable and a strut in this state has, for practical purposes, failed. The failure has the following characteristics:

(a) there is no parting of the material, and it may be possible for the structure to recover on being unloaded (elastic buckling);
(b) the strut as a whole fails – failure does not occur locally;
(c) the presence of a hole at some location does not dominate the mode of failure.

These notions are incorporated into the procedure for design of ties and struts.

12.3. DESIGN OF TIES

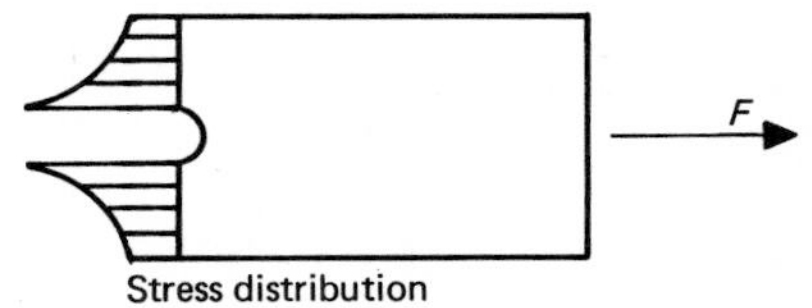

If the smallest effective area of any cross-section is A_E, the allowable stress is p_t and the load in service is a tension of F, then the tie must be chosen to satisfy the rule

$$p_t A_E \nless F$$

Attention must first focus on the uniform distribution of stress which is implied. Consider a tie of rectangular cross-section, with holes through it as would be required to bolt it to another part of the structure. Assuming one line of holes, the smallest cross-section will be one through the centre of a hole. It is well known from mechanics of materials that, for stresses within the elastic range, the hole acts as a stress-raiser and the distribution of stress is not uniform. Distributions of stress and strain would be approximately as sketched in the margin, provided

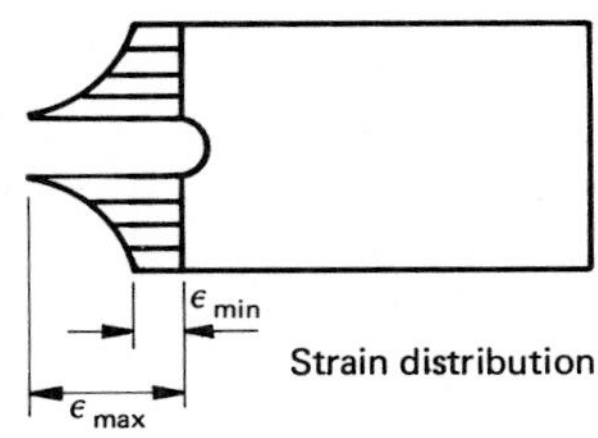

$$\varepsilon_{max} < \varepsilon_y = \text{strain at yield.}$$

If the load is increased, all the strains will increase, not necessarily with a similar distribution but certainly with greatest strain near the hole and least strain at the edge of the tie. For a ductile material like steel, this has the important effect of making the stress distribution more uniform. Using an idealised stress–strain relationship, it is apparent that

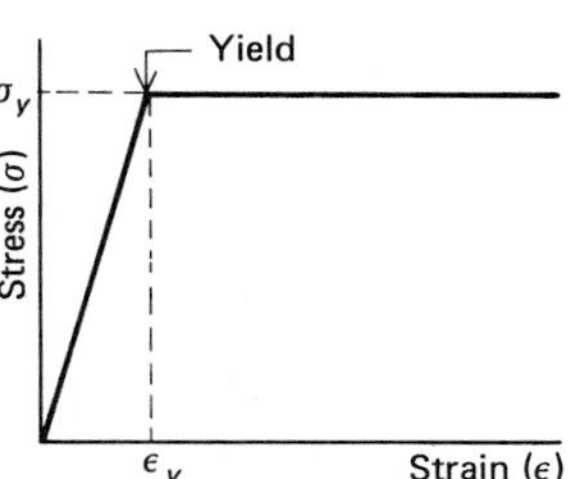

$$\sigma = \sigma_y = \text{constant}$$

for

$$\varepsilon > \varepsilon_y$$

Thus, when the load is such that

$$\varepsilon_{min} > \varepsilon_y$$

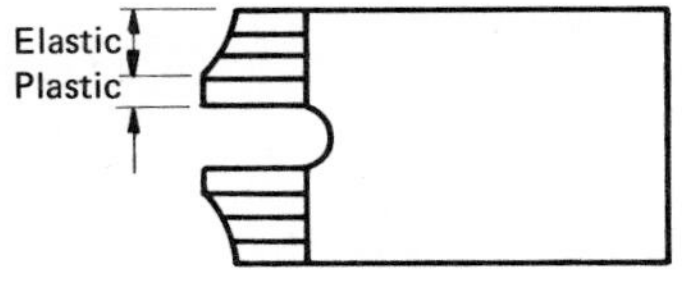

the stress is uniformly distributed and equals σ_y. For an intermediate load, part of the cross-section will have yielded and will be subject to a uniform tensile stress equal to σ_y, while the balance will be subject to a varying elastic stress.

In summary then, the design basis is seen to imply local yielding of a ductile material as a mode of failure. It is made applicable to service conditions by requiring the computed average stress to be less than yield stress, but it is important to note that service loads may in fact cause local yielding of the material.

The same procedure is used to design timber ties, although it is not so readily justified for this material.

The nett area, A_E, for an axially loaded tie is the area of the smallest cross-section of the tie. It is computed by deducting the projected area of holes for bolts or recesses for other fasteners, such as timber connectors, from the gross area. Steel sections are rolled to standard dimensions and the properties of the sections are published in various places (e.g., BS 4: 1972, *Specification for Structural Steel Sections*[1]). Gross areas are available from such sources, and properties of selected shapes from BS 4 are given in Appendix A3.

In the case of timber ties, cross-sections are rarely other than rectangular and gross section properties are easily calculated. Care is necessary, however, to use actual dimensions, because they are different from nominal dimensions. Sawmillers set their saws to nominal sizes and timber lost in saw cuts and planing may reduce dimensions by as much as 10 mm.

Calculations for the design of a steel tie of rectangular cross-section are shown in Example 12.1.

The distribution of stress in a tie is also affected by the position of the line of action of the load relative to the centre of area of the cross-section. An axially loaded tie has the line of action of the load passing through this centre of area. Except as modified by holes or other local irregularities, the tensile stress is uniformly distributed. If the tie is not axially loaded, bending stresses are superimposed on the uniformly distributed tension and the maximum tensile stress exceeds the average. Bending stresses can be minimised by placing fasteners as close as possible to the gravity axis – the locus of the centres of area.

Such eccentricity of loading occurs frequently in light trusses, and can be explained as follows. During erection, or under the action of wind loads, a member which is in tension under normal loading may have to resist a compression force. Rods and flats which are well able to perform as ties may fail by buckling under reversed load and, for this reason, they do not always make good ties. A sounder choice is an angle. It has the advantage of a good distribution of area from the point of view of stiffness while providing a flat surface for easy fastening. The main disadvantage in using angles is the eccentricity of loading. The load is applied to one leg of the angle where it is fastened to another part of the truss.

One course of action open to the designer is to apply the laws of mechanics and select an angle such that the maximum tensile stress due to bending and tension does not exceed the allowable stress. This requires too much computation to be a practicable procedure in routine design and most codes of practice allow a simplified approach. Some permit the eccentricity of loading to be ignored. The British code, BS 449, *The Use of Structural Steel in Building*[2], has special rules for the design of angles as ties. In these, the load is assumed to be applied axially and the eccentricity is allowed for by reducing the effective area of the cross-section. For example, in the case of a single angle, the effective

Example 12.1 Sheet 1 of 1

Select a suitable rectangular cross-section for a steel tie to carry a load of 108 kN. Allow for one hole, 12 mm diameter. Permitted stress for axially loaded tie = 215 MPa.

$$p_t A_E \not< F$$

$$A_E \not< \frac{1.08 \times 10^5}{2.15 \times 10^2} = 503 \text{ mm}^2$$

Try flat 10 mm thick

$$\text{Nett width} = \frac{503}{10} = 50.3 \text{ mm}$$

Gross width = 50.3 + 12 = 62.3 mm

Say, 65 mm × 10 mm flat.

area A_E is given by

$$A_E = a_1 + \frac{3a_1}{3a_1 + a_2} a_2 = a_1 + a_2/(1 + \frac{1}{3}\frac{a_2}{a_1})$$

where a_1 is the nett area of the connected leg, allowing for holes, and a_2 is the area of the unconnected leg.

Other rules are given for pairs of angles in different arrangements. It is clear that such a rule leads to reduced effective areas as the ratio $a_2 : a_1$ increases.

Example 12.2 shows the design of a single angle as a tie using the rule from BS 449. For comparison, the stress in the tie in Example 12.2 may be calculated allowing for eccentricity of the applied load with respect to both axes of inertia. It will be found that the computed maximum tensile stress exceeds the yield stress. This is another case where the designer relies on the ductility of the steel to reduce the significance of errors arising from incomplete knowledge of structural behaviour and the use of simplified procedures.

12.4. DESIGN OF STRUTS

The concept of a critical load on a strut is based on the strut's stability. An ideal strut is one which is initially perfectly straight and uniform in cross-section. Such a strut will not buckle if loaded axially and buckling arises from imperfections in the strut and in the symmetry of its load. In a simple analysis, these imperfections can be idealised in the form of an infinitesimal transverse load W. A critical load, P_{cr}, can be found, using the criterion that the transverse deflection of the strut should tend to infinity as the axial load tends towards this critical value.

In Fig. 12.1, the bending moment at the point X is given by

$$M_x = \frac{W}{2}x + Py \qquad (12.1)$$

For elastic deformation under these loads

$$M_x = \frac{1}{R}EI = -\frac{d^2y}{dx^2}EI \qquad (12.2)$$

where R is the radius of curvature of the deflected strut; E is the modulus of elasticity of the material of the strut; and I is the second moment of area of the strut's cross-section.

Substitution of Equation (12.2) into Equation (12.1) leads to a differential equation for y as a function of x and its solution for a uniform strut is

$$y = \tfrac{1}{2}\frac{W}{P}x\left(\frac{\sin(\alpha x)}{(\alpha x)\cos(\alpha \frac{L}{2})} - 1\right)$$

$$\alpha = \sqrt{\frac{P}{EI}}$$

The constants of integration have been determined by the conditions that y and M_x are both zero at $x = 0$ and $x = L$. (For details, see Bresler, Lin and Scalzi.[3])

Example 12.2 Sheet 1 of 1

Select a suitable angle for a tie to resist a tensile load of 200 kN. Use the rule from BS 449 and $p_t = 215$ MPa. Allow for one 12 mm diameter hole in the connected leg.

Try 89 mm × 64 mm × 8 mm angle
(3½ in × 2½ in × 5/16 in)

Long leg to be connected
Take dimensions from Appendix A3:
88.9 mm × 63.5 mm × 7.8 mm

$$a_1 = \left(88.9 - \frac{7.8}{2}\right) \times 7.8 - 12 \times 7.8$$
$$= 85.0 \times 7.8 - 12 \times 7.8$$
$$= 570 \text{ mm}^2$$

$$a_2 = \left(63.5 - \frac{7.8}{2}\right) \times 7.8$$
$$= 59.6 \times 7.8$$
$$= 465 \text{ mm}^2$$

$$1 + \frac{1}{3}\frac{a_2}{a_1} = 1 - \frac{465}{3 \times 570}$$
$$= 1.272$$

$$A_E = 570 + \frac{465}{1.272}$$
$$= 935 \text{ mm}^2$$

$$p_t A_E = 215 \times 935 \text{ N}$$
$$= 201 \text{ kN}$$
$$> 200 \text{ kN} \quad \text{OK}$$

Use 89 × 64 × 8 angle, long leg connected.

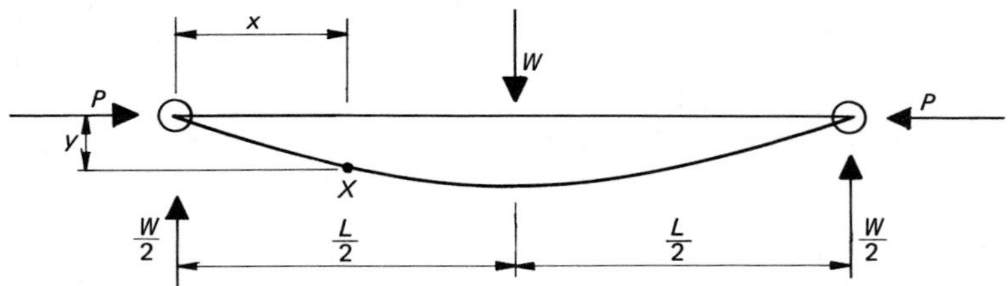

Fig. 12.1. *Definition sketch for an axially loaded strut with a transverse load applied at mid-length.*

It follows that $y \to \infty$ as $\cos(\alpha L/2) \to 0$ even though W is infinitesimal. The smallest value of P which makes

$$\cos\left(\alpha \frac{L}{2}\right) = 0$$

is clearly the critical load P_{cr}. Hence

$$\alpha \frac{L}{2} = \frac{\pi}{2}$$

and

$$P_{cr} = \frac{\pi^2 EI}{L^2} \tag{12.3}$$

The critical load on a strut is determined by a property of the material and two properties of the strut's shape and dimensions. This load is known as the *Euler critical load.*

In fact, no transverse load is applied in a buckling experiment. The critical load then is the smallest load for which the strut is in equilibrium while slightly curved. A rather more difficult theoretical argument based on this definition leads to the same expression for the critical load. For a load smaller than critical, the strut would return to its original alignment if disturbed and would collapse if so treated while carrying a load greater than the critical load.

It is emphasised that what is critical is the total load in relation to stability – not the average intensity of load in relation to the strength of the material. This emphasis is necessary, because Equation (12.3) can be made more useful by dividing through by the *gross* area of the cross-section. Then the critical stress σ_{cr} is given by

$$\sigma_{cr} = \frac{P_{cr}}{A_g} = \pi^2 E \frac{I}{A_g L^2} = \pi^2 E \frac{1}{(L/r)^2} \tag{12.4}$$

where A_g is the gross area of cross-section, and $r = \sqrt{I/A_g}$ is the radius of gyration. The advantage of this adjustment is that critical stress becomes a function of only one geometrical property (L/r) and hence, for struts of a given material, a function of a single variable.

Equation (12.4) is sketched on Fig. 12.2 and it is clear that small values of L/r, the slenderness ratio, give very large values of σ_{cr}, the critical stress. However, a perfectly ductile material will fail by crushing if the stress exceeds the yield stress. Also, the analysis based on elastic behaviour is not relevant in that event. We would expect the critical stress of a ductile material to be given by a curve like $AB'C$, which takes into account failure by yielding and by elastic buckling.

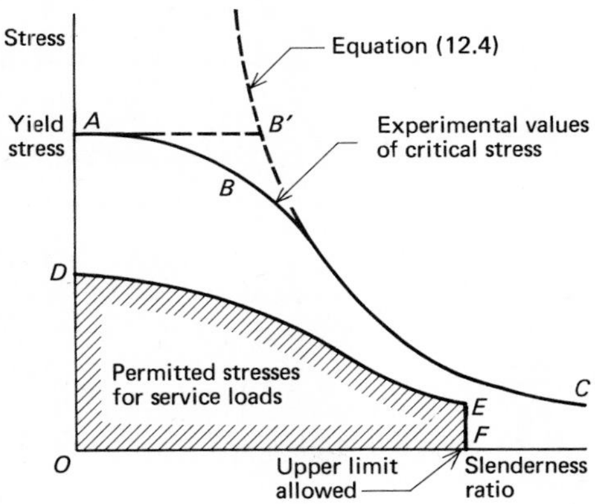

Fig. 12.2. *Stress versus slenderness ratio for an axially loaded strut.*

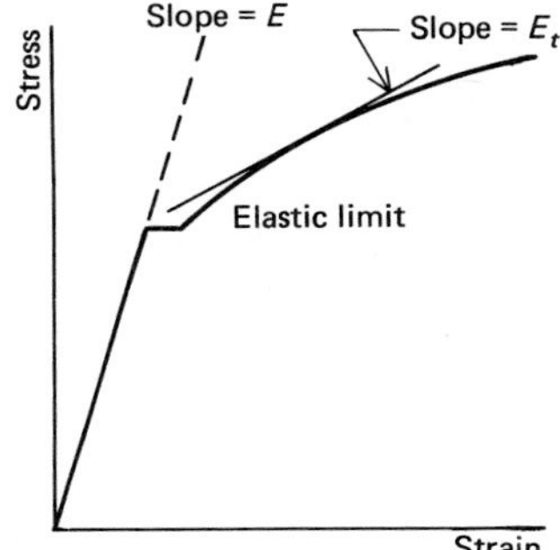

Tests on struts support these ideas, and experimental results follow a curve like ABC. The difference between ABC and $AB'C$ is due to the occurrence of plastic buckling over a range of slenderness ratios and may be explained by the fact that the modulus of elasticity is not constant for strains beyond the elastic limit. For any small range of strains beyond the elastic limit, the true stress–strain curve can be replaced by its tangent and the slope of this tangent, known as the *tangent modulus*, E_t, can be used in place of E in Equation (12.4). The resulting curve for critical stress will agree with curve ABC in Fig. 12.2.

Before the tangent modulus theory was advanced, several empirical equations had been suggested for this range of slenderness ratio. The parabolic formula of J. B. Johnson is extensively used in practice in machine design for which there are no codes. It is the equation of a parabola which has its vertex at

$$\sigma_{cr} = f_y$$

$$\frac{L}{r} = 0$$

and is tangent to the Euler curve at

$$\sigma_{cr} = \frac{1}{2} f_y$$

The equation of the parabola is

$$\sigma_{cr} = f_y \left[1 - \frac{f_y}{4\pi^2 E} \left(\frac{L}{r}\right)^2\right]$$

and the slenderness ratio at the point where it touches the Euler curve is

$$\frac{L}{r} = \sqrt{\frac{2\pi^2 E}{f_y}}$$

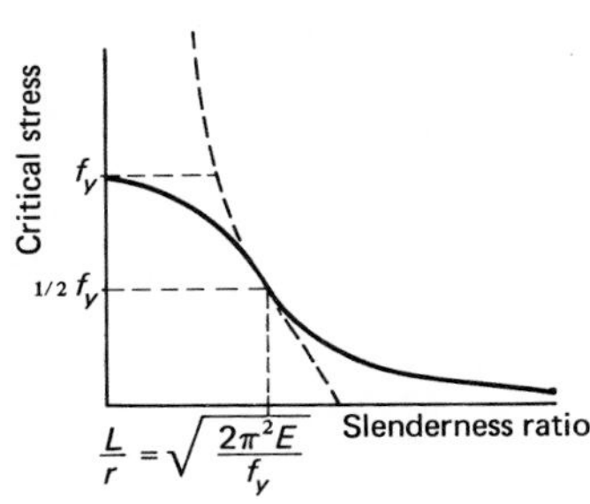

Note that the Johnson parabola and the Euler curve give the critical stress—critical load divided by gross area of cross-section. A factor of safety or a load factor must be used when they are applied in design. Design of a strut on this basis is illustrated in Example 12.3.

Example 12.3 Sheet 1 of 3

Select the diameter for a solid circular strut to carry an axial load of (a) 20 kN (b) 40 kN using the Johnson formula or the Euler formula,

$f_y = 275$ MPa $\quad E = 207$ GPa

Factor of safety = 1.8

Effective length : (a) 1.0 m (b) 0.7 m

Johnson $\quad \sigma_{cr} = 275\left[1 - \frac{275}{4\pi^2 \times 207 \times 10^3}\left(\frac{L}{r}\right)^2\right]$

$$= 275\left[1 - 3.36 \times 10^{-5}\left(\frac{L}{r}\right)^2\right]$$

Euler $\quad \sigma_{cr} = \frac{\pi^2 \times 207 \times 10^3}{\left(\frac{L}{r}\right)^2}$

$$= \frac{2.04 \times 10^6}{\left(\frac{L}{r}\right)^2}$$

Tangent point $\sigma_{cr} = 137.5$ MPa

$$\frac{L}{r} = \sqrt{\frac{2 \times \pi^2 \times 207 \times 10^3}{275}}$$

$$= 122$$

Allowable stresses :

$$p_c = \frac{\sigma_{cr}}{1.8}$$

Johnson $\quad p_c = 153\left[1 - 3.36 \times 10^5\left(\frac{L}{r}\right)^2\right]$ MPa

Euler $\quad p_c = \frac{1.13 \times 10^6}{\left(\frac{L}{r}\right)^2}$ MPa

Tangent point $p_c = 76.5$ MPa

(a) Try 25 mm diameter round

$$r = \frac{D}{4} \qquad = 6.25\text{ mm}$$

$$\frac{L}{r} = \frac{1000}{6.25} = 160 \quad \text{— use Euler}$$

$$p_c = \frac{1.13 \times 10^6}{1.6 \times 1.6 \times 10^4} = 43.1\text{ MPa}$$

$$A = \frac{\pi}{4} \times 25^2 \qquad = 490\text{ mm}^2$$

$$p_c A = 41.3 \times 4.90 \times 10^3\text{ N} = 21.2\text{ kN} \qquad \text{OK}$$

Example 12.3 Sheet 2 of 3

(b) Try 25 mm diameter round

$$r = \frac{D}{4} = 6.25\ mm$$

$$\frac{L}{r} = \frac{700}{6.25} = 112 \quad \text{—— use Johnson}$$

$$p_c = 153\,(1 - 3.36\times10^{-5}\times1.12\times1.12\times10^4)$$

$$= 153\times0.578$$

$$= 88.4\ MPa$$

$$A = \frac{\pi}{4}\times25^2 = 490\ mm^2$$

$$p_c A = 8.84\times4.90\times10^3\ N$$

$$= 43.3\ kN \qquad \text{ok}$$

Use 25 mm diameter round for both

Alternatively, these struts can be designed graphically on a plot of the Johnson and Euler formulas. The sketch adjacent shows p_c vs L/r as derived from the Johnson and Euler formulas. We also have

$$P \nless p_c \frac{\pi}{4} D^2 = p_c 4\pi r^2$$

r = radius of gyration = $\frac{D}{4}$

Hence $$p_c \ngtr \frac{P}{4\pi L^2}\left(\frac{L}{r}\right)^2$$

For Case (a) $\frac{P}{4\pi L^2} = 1.59\times10^{-3}$ MPa

Case (b) $\frac{P}{4\pi L^2} = 6.50\times10^{-3}$ MPa

These parabolas are superimposed on the Johnson - Euler diagram and yield

for (a): $p_c \ngtr 42$ MPa $\quad D \nless \sqrt{\frac{4P}{\pi p_c}} = \sqrt{\frac{4\times2\times10^4}{\pi\times42}}$

$$= 24.6\ mm$$

for (b): $p_c \ngtr 86$ MPa $\quad D \nless \sqrt{\frac{4\times4\times10^4}{\pi\times86}}$

$$= 24.4\ mm.$$

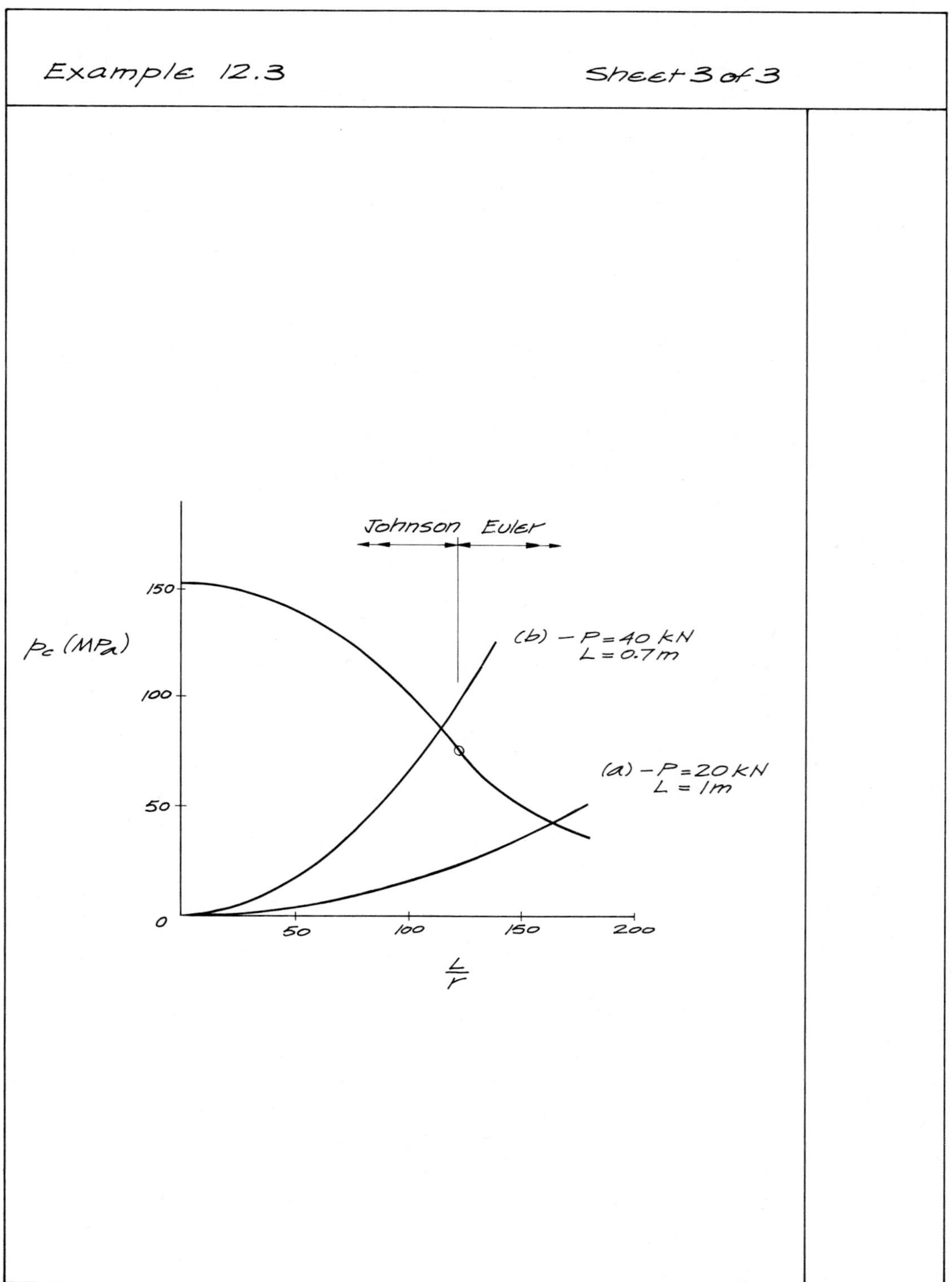
Example 12.3
Sheet 3 of 3
Johnson
Euler
150
100
50
0
50
100
150
200
p_c (MPa)
(b) – P = 40 kN
L = 0.7m
(a) – P = 20 kN
L = 1m
$\frac{L}{r}$

The permitted stress for structural design can now be defined by dividing the ordinates of the curve *ABC* on Fig. 12.2 by a factor of safety. This is the stress which must not be exceeded by loads in service. Codes often also specify an upper limit to the slenderness ratio to give some protection against the effects of unforeseen loads in service or during erection. These concepts all lead to definition of an area *ODEF* to control the selection of a strut's cross-section. Ordinates of the curve *DE* are usually given in a table in the design code for a range of values of the slenderness ratio. Table A14.1 in Appendix A14 is typical.

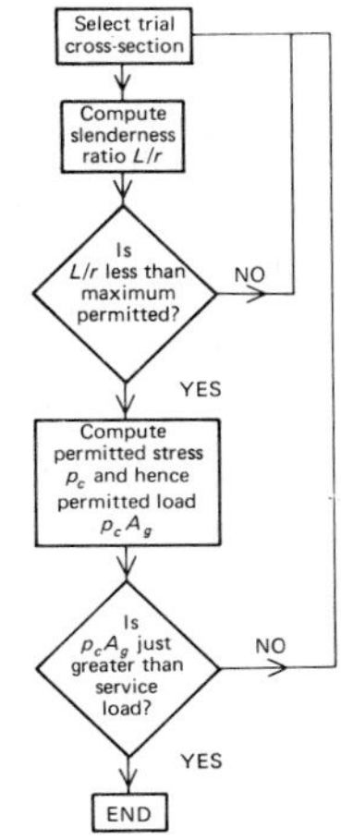

Design of a strut is complicated by the fact that one of the quantities which determines the permitted stress (the radius of gyration) is not known in advance. It is necessary to proceed by trial. The steps necessary are illustrated in the marginal sketch. As indicated in this sketch, a satisfactory selection is one which enables the design rules to be obeyed without being uneconomically over-designed.

Evaluation of the slenderness ratio requires the idea of an equivalent length or effective length to be introduced. Equation (12.4) and the corresponding table in a design code are based on certain conditions being satisfied at the ends of the strut. These are that the ends are restrained against lateral translation, but are free to rotate. The transverse deflection and the bending moment at the ends are zero. With different kinds of end fixing, the restraint applied to the ends will be different. For example, rotation as well as translation may be restrained. When such a strut buckles, its deflected shape will be as sketched in the margin, and it is clear that the curvature of the strut reverses at two points. These points, the *inflection points*, are places where the bending moment is zero. Since a hinge or a pin can transmit no bending moment, the inflection points are structurally like pins and we arrive at the notion of an equivalent pin-ended strut. The effective length of the strut under consideration is the distance between points of inflection on its deflected axis, i.e. the length of the equivalent pin-ended strut.

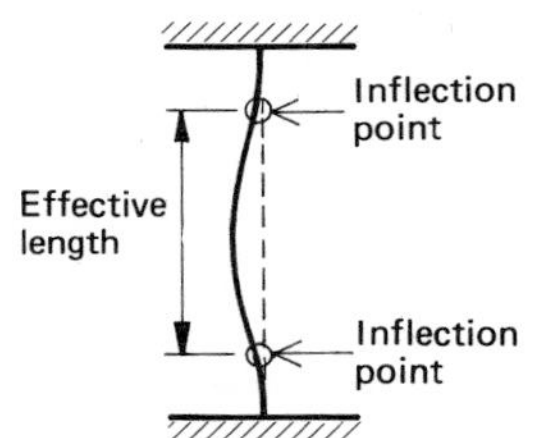

The effective length is found by multiplying the geometrical length by a factor which depends on the restraint at the ends of the strut. This factor is usually given in design codes for a few straightforward examples. These do not cover all possible kinds of end restraint and it is sometimes difficult to decide what the effective length is. When a designer has to design a strut with end conditions which do not fit a standard description, the best he can do is to sketch the curve that he thinks the axis of the strut would follow if the strut were to buckle. A conservatively long estimate of the effective length can usually be made from such a sketch and the standard examples. As an illustration of this process, buckling of the piston rod of a hydraulic lifting cylinder is sketched in the margin. The theoretical derivation for the effective length in this case is given in Spotts.[4]

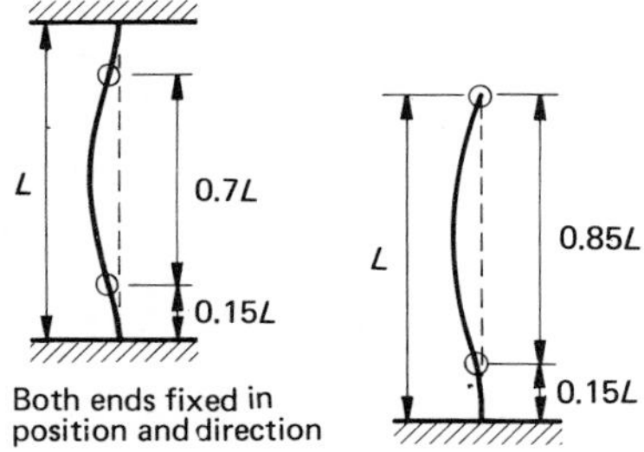

Both ends fixed in position and direction

One end fixed in position and direction, one end fixed in position but free to rotate

The radius of gyration to be used is obviously the smallest one – it will give the lowest critical load for the cross-section. If the section is unsymmetrical (e.g., a steel angle), the smallest radius of gyration will be about a principal axis of inertia – not necessarily one of the more obvious axes. Still using the steel angle as an example, the obvious and convenient axes are those parallel to the legs of the angle. The principal axes are inclined to them.

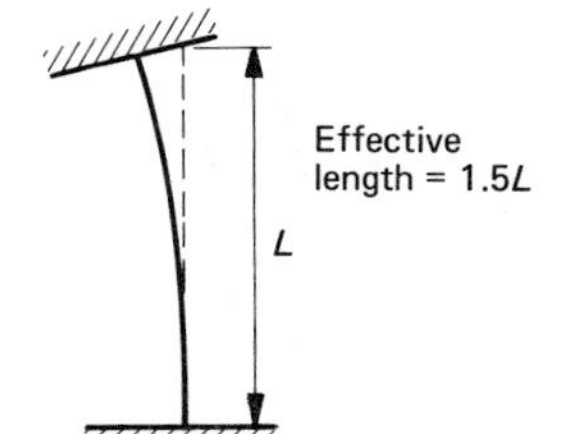

One end fixed in position and direction, one end partially restrained in position and direction

The statements above must be qualified in two respects. First, the mode of buckling may be restrained in some way. In the case of two angles back to back, each restrains the other against buckling about a principal axis and the

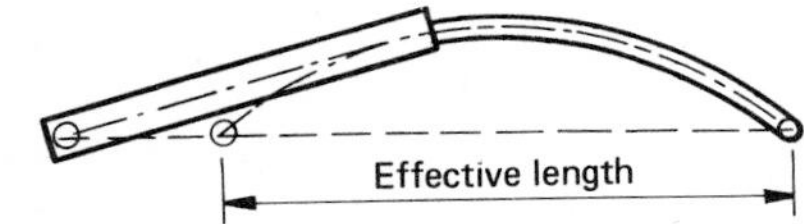

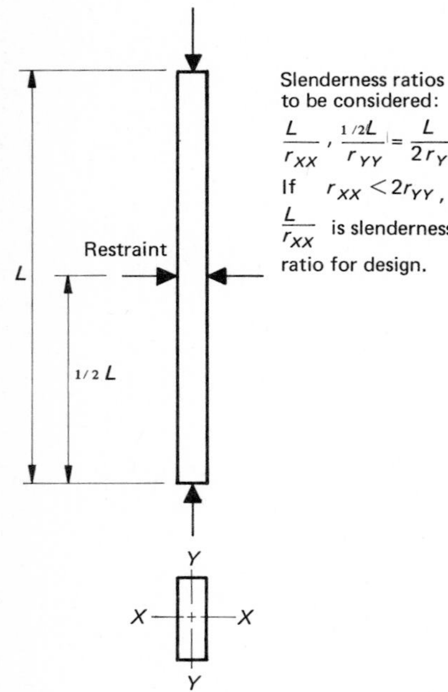

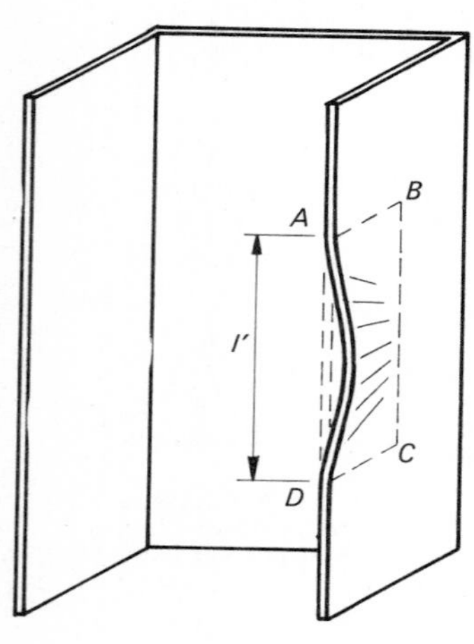

radius of gyration of the pair about one or other of the axes parallel to the legs is required. Secondly, deflection of a strut may be restrained at points along its length. Thus, the ratio of effective length to lesser radius of gyration may be less than the slenderness ratio for buckling about an axis with greater radius of gyration (see marginal sketch).

As is the case for ties, some codes (e.g., BS 449) have special rules for the design of angles as struts. These are relevant mainly to trusses, and specify the effective length to be used.

The foregoing discussion of buckling has been in terms of buckling of the strut as a whole, or its components if it is a built-up strut. Local buckling under the action of compressive stress can also occur. For example, the compression flange of a beam carrying a transverse load may buckle because of the compressive stress induced by bending. The web of a beam may buckle locally under a concentrated load or as a result of inclined compressive stresses due to the shear in the web. These aspects of beam design are treated in Chapter 15, "Steel Beams". In a similar way, the wall of a tubular strut, or an outstanding part of one with an open cross-section may buckle locally. In a flange, for example, this is equivalent to the buckling of a shorter column $ABCD$ which has length l' and some constraint along the edge BC. Such a problem does not yield to simple analysis and, in practical design, limits on the ratios of significant dimensions are used. Standard hot-rolled steel sections can be assumed to be stable in this respect. In thin-walled steel and aluminium tubing, local buckling is likely when thickness to diameter ratio is less than 0.03; and for cold formed sections, the ratio of flat width to thickness for an outstanding part continuous on one edge with another part of the strut should not exceed 60.[5]

Examples of the design of a single angle strut, using BS 449, and a timber strut, using BS CP 112, *The Structural Use of Timber*[6], are given in Examples 12.4 and 12.5 respectively.

12.5. BUILT UP TIES AND STRUTS

It is often convenient to use two components alongside each other to build up a tie or a strut. Neater joints in both steel and timber trusses can be made and asymmetry of loading is less. Good design requires the individual members to be connected together at intervals along their length by rivets, bolts or welds in the case of steel, by bolts or connectors in the case of timber. Also, if they are spaced apart at the ends, packing is required at these intermediate connections. Arbitrary rules are used to design the connections.

The need to join the components together in this fashion is related to buckling. The greatest slenderness ratio for the whole strut can be made smaller than the slenderness ratio its components would have if of equal effective length. Consequently, the permitted stress is higher. To ensure that the two components act together and that the critical load will be determined by buckling of the whole strut, two conditions must be satisfied:

(a) the shear strength of the intermediate connections must be such as to prevent slip of the components past each other if buckling occurs;
(b) the spacing of connections along the strut must be small enough to keep the slenderness ratio of the individual components down.

These criteria are covered by rules in design codes. For example, BS 449, *The Structural Use of Steel in Building*[2], in relation to struts comprising pairs of angles back to back, requires at least two intermediate connections along the strut and sets out how they are to be made. It also requires the maximum slenderness ratio for each angle to be not more than 40, or 0.6 times the greatest slenderness ratio of the strut as a whole.

Similar rules are given in BSCP 112, *The Structural Use of Timber*.[6] First, the clear space between the shafts of a built-up strut should not be greater than three times the thickness of one shaft. In practice, the individual shafts are usually of equal thickness and the thickness of the packing is also the same. The packings at the ends are required to be fastened to the shafts by glue, bolts, screws or connectors of sufficient strength to resist a shear equal to 1.5 times the average compressive load. An exception to this rule is made for spaced struts in trusses. In that case, other members of the truss are interlaced with the shafts of the strut. These members serve as packing and are, in any case, connected to the shafts of the strut.

At least two intermediate packings are required, unless the length of the strut is less than 30 times the thickness of one shaft, when one packing is sufficient. The greatest slenderness ratio for part of any shaft between packings must not exceed 0.7 times the greater slenderness ratio for the column. The effective length for parts of shafts between packings is the distance between centres of the connections at the packings.

The permissible load on the strut is then taken to be the smallest of the following (see Fig. 12.3):

(a) the allowable load for a solid strut whose area is that of the whole cross-section, buckling being about XX;

(b) the allowable load for a solid strut whose cross-section is that of one shaft, multiplied by the number of shafts. For this purpose, the effective length is the mean centre-to-centre spacing of the packings;

(c) the allowable load for a strut buckling about YY, its geometrical properties being those of the built-up section. For this calculation, the effective length is multiplied by a modification factor which, for joints made with bolts or connections is as follows:

Ratio of space to thickness of thinner member	0	1	2	3
Bolts	1.7	2.4	2.8	3.1
Connectors	1.4	1.8	2.2	2.4

These rules are relevant to struts in timber trusses, with some additional guidance in relation to effective length. With continuous compression members, such as the top chords of trusses, the effective length may be taken as from 0.85 to 1 times the distance between node points for buckling in the plane of the truss and from 0.85 to 1 times the distance between purlins or similar lateral restraints for buckling perpendicular to that plane. The purlins must be fastened to the top chord and effectively anchored by bracing. In the case of non-continuous struts, such as web members, the general rules are applied – e.g., if there is a single bolt or connector at each end, the class of end fixity is "restrained at both ends in position, but not in direction".

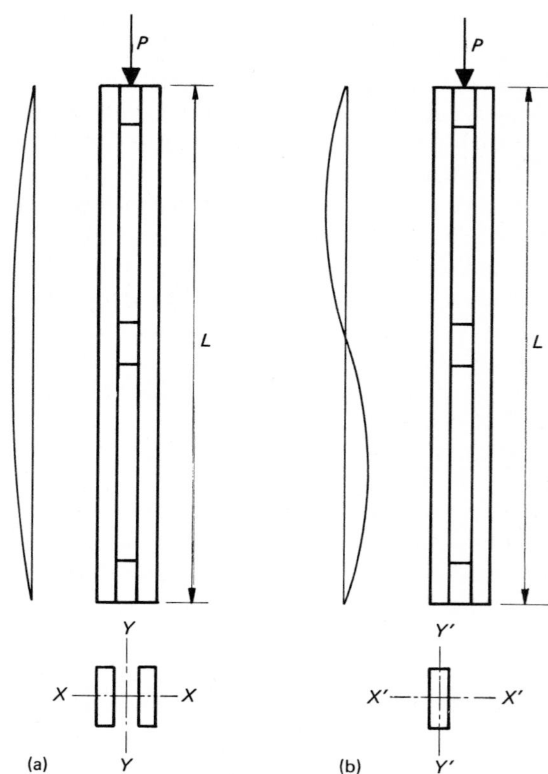

Fig. 12.3. *Buckling of spaced timber strut:*
(a) for buckling of strut as a whole slenderness ratio $= \frac{L}{r_{xx}}$ *or* $\frac{L}{r_{yy}}$. *(Cases (a) and (c) in text.*
(b) for buckling of individual components slenderness ratio $= \frac{\frac{1}{2}L}{r_{y'y'}}$. *(Case (b) in text.)*

Detailed rules for the design of spaced struts are also given in a design code published in the United States by the National Lumber Manufacturers' Association (*National Design Specification for Stress-Grade Lumber and its Fastenings*[7]).

Design of a spaced strut is illustrated in Example 12.6.

12.6. COMBINED BENDING AND COMPRESSION

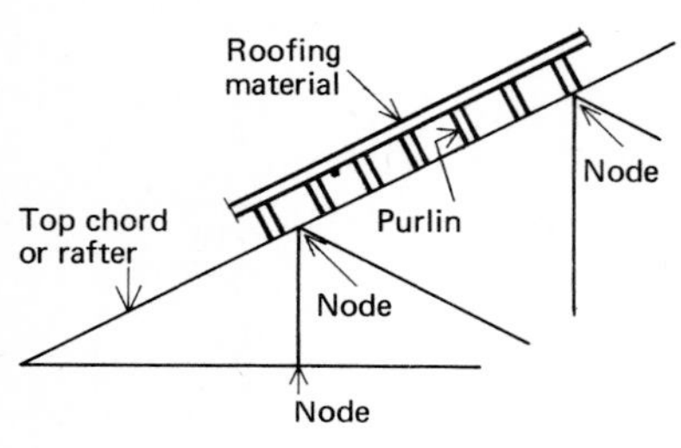

In a truss supporting a pitched roof, the top chord is in compression as a result of its function in the truss. Since roofing materials are generally not able to span the full distance between nodes, the roof load is applied to the top chord by means of small beams called purlins. The roof spans from purlin to purlin and the purlins span the gaps between the trusses. The top chord is thus loaded in bending as well as in direct compression. The compressive load is found by analysing the truss assuming that all loads are concentrated at the nodes. The true distributed nature of the load is used to calculate the bending moment in the top chord, or rafter.

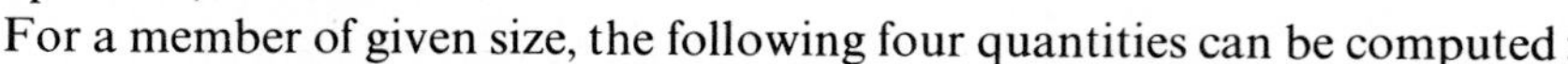

For a member of given size, the following four quantities can be computed:

f_c = average axial compressive stress;
p_c = permitted compressive stress for an axially loaded strut;
f_{bc} = total compressive stress due to bending (about both axes, if relevant);
p_{bc} = permitted compressive stress in bending (see Chapter 14, "Timber Beams," and Chapter 15, "Steel Beams").

Example 12.4 Sheet 1 of 1

Select a single angle for a compressive load of 32 kN, the length between centres of end connections being 1.5 m. Design to BS 449, using Grade 43 steel.

For single angle struts with each end connected by at least two bolts or rivets or equivalent welding, BS 449 allows the strut to be designed as if axially loaded with effective length = 0.85 × length of strut. If the end connections are single bolted or riveted, the full length centre to centre of end connections is used and the allowable stress reduced by 20%

Try double bolted connections

$\ell = 0.85 \times 1500 = 1275$ mm

Try 51 × 51 × 6 angle

$r_{min} = 9.9$ mm $\quad A = 608$ mm^2

$\frac{\ell}{r} = \frac{1275}{9.9} = 129 \longrightarrow p_c = 53$ MPa

$p_c A = 32.2$ kN OK

Try single bolted connections

$\ell = 1.5$ m $= 1500$ mm

Try 64 × 64 × 6 angle

$r_{min} = 12.5$ mm $\quad A = 759$ mm^2

$\frac{\ell}{r} = \frac{1500}{12.5} = 120 \longrightarrow p_c = 60$ MPa

$p_c A = 45.5$ kN OK

Example 12.5 Sheet 1 of 1

Find the cross-section of a single plank for a long term compressive load of 32 kN, the length being 1.5 m with both ends restrained in position and direction. Use Grade 50 S1 timber (see BSCP 112)

From BSCP 112 : Grade stress for this species group and grade = 1.93 MPa

Effective length = 0.7 × 1.5 = 1.05 m

Try 75 mm × 300 mm

$r = \frac{t}{\sqrt{12}} = 0.289 \times 75 = 21.6$ mm

$\frac{\ell}{r} = \frac{1050}{21.6} = 48.7$

From BSCP 112 : Modification factor for long term loading and slenderness ratio of 48.7 = 0.87

$p_c = 0.87 \times 1.93 = 1.68$ MPa

$A = 75 \times 300 = 2.25 \times 10^4$ mm²

$p_c A = 37.8$ kN OK

N.B. If timber is sold locally by nominal size, order plank such that actual dimensions are not less than standard basic size calculated.

Example 12.6 Sheet 1 of 2

Determine the safe long term load for a timber strut comprising 2/75 mm × 150 mm shafts with 75 mm spacers. The strut is 5 m long and its ends are restrained in position and direction. The species and grade of timber are the same as in Example 12.5 and packings will be fastened with connectors.

Length of strut : shaft thickness $= \frac{5000}{75} = 66.7$
At least two intermediate packings required

Slenderness ratios :
(1) Individual shafts :
Effective length $\approx \frac{1}{3} \times 5 = 1.67\,m$

$r = \frac{75}{\sqrt{12}} = 21.7\,mm$

Slenderness ratio $= \frac{1670}{21.7} = 77.0$

(2) Complete strut about XX :
Effective length $= 0.7 \times 5 = 3.5\,m$

$r = \frac{150}{\sqrt{12}} = 43.4\,mm$

Slenderness ratio $= \frac{3500}{43.4} = 80.6$

(3) Complete strut about YY :

75 | Y | 150 | X | X | 75 | Y

$I = 2 \times \frac{1}{12} \times 150 \times 75^3 = 1.06 \times 10^7$
$2 \times 150 \times 75 \times 75^2 = \underline{12.68 \times 10^7}$
$13.74 \times 10^7\,mm^2$

$A = 2 \times 150 \times 75 = 2.25 \times 10^4\,mm^2$

$r = \sqrt{\frac{1.37 \times 10^8}{2.25 \times 10^4}} = 78.0\,mm$

Slenderness ratio $= \frac{3500}{78} = 44.9$

Slenderness of individual shafts too great
Reduce to $0.7 \times 80.6 = 56.4$
Spacing of packings $= 56.4 \times 21.7 \times 10^{-3}$
$= 1.22\,m$
Say, 3 packings, at midpoint and quarterpoints

Example 12.6 Sheet 2 of 2

Safe loads :

(1) Complete section, buckling about XX :

$$\frac{\ell}{r} = 80.6$$

Stress modification factor $= 0.69$

$$p_c = 0.69 \times 1.93$$
$$= 1.33 \text{ MPa}$$
$$A = 75 \times 150 \times 2$$
$$= 2.25 \times 10^4 \text{ mm}^2$$
$$p_c A = 30.0 \text{ kN}$$

(2) Individual shafts :

$$\frac{\ell}{r} = 56$$

Stress modification factor $= 0.84$

$$p_c = 0.84 \times 1.93$$
$$= 1.62 \text{ MPa}$$
$$A = 75 \times 150$$
$$= 1.125 \times 10^4 \text{ mm}^2$$
$$p_c A = 18.2 \text{ kN}$$

Total safe load $= 36.4$ kN

(3) Complete section, buckling about YY :

$$\frac{\ell}{r} = 1.8 \times 44.9 = 80.7$$

Safe load as for case 1 $= 30.0$ kN

Required safe load $= 30.0$ kN

In general,

$$p_{bc} \neq p_c$$

and direct comparison of stress occurring and stress allowed cannot be made. A simple criterion for suitability of a chosen cross-section is

$$\frac{f_c}{p_c} + \frac{f_{bc}}{p_{bc}} \ngtr 1$$

Clearly it reduces to the appropriate rule at each extreme ($f_c = 0$ and $f_{bc} = 0$) and, for combined stresses, is simple to use.

A variation of this interaction formula is given for timber struts in BSCP 112, *The Structural Use of Timber*[6], as follows:

$$\frac{f_c}{p_c} + \frac{f_{bc}}{p_{bc}} \ngtr 1 \qquad \text{if } \frac{l}{r} \ngtr 20$$

$$\frac{f_c}{p_c} + \frac{f_{bc}}{p_{bc}} \ngtr 0.9 \qquad \text{if } \frac{l}{r} > 20$$

For additional discussion of formulas of this type, see Section 19.2.2.

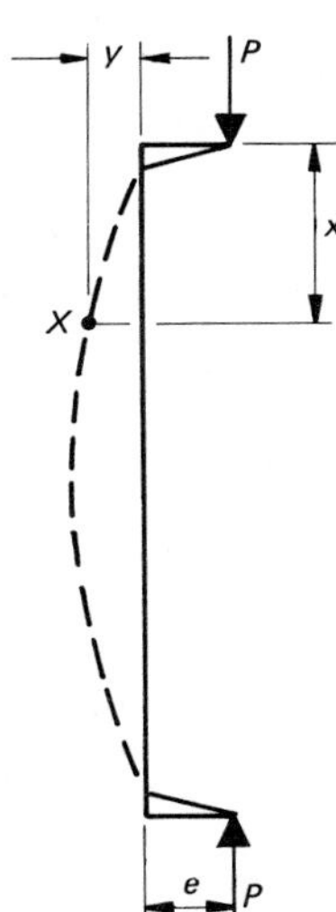

Combined bending and compressive stresses also occur in an eccentrically loaded strut (see marginal sketch). The bending moment at any point X is given by

$$M_x = P(e + y) \tag{12.5}$$

and, using Equation (12.2), a differential equation for y is obtained

$$\frac{d^2y}{dx^2} = -\frac{P(e + y)}{EI}$$

The solution is

$$y = e\left(\tan \alpha \frac{L}{2} \sin \alpha x + \cos \alpha x - 1\right) \tag{12.6}$$

where

$$\alpha = \sqrt{\frac{P}{EI}}$$

It should be noted in passing that the Euler critical load (see Section 12.4) makes y tend to infinity even for very small values of e. In other words, a stability analysis could be based on a nominal eccentricity instead of a nominal transverse load.

Applied to analysis of an eccentrically loaded strut, Equation (12.6) is used in Equation (12.5) to yield

$$M_x = Pe\left(\tan \alpha \frac{L}{2} \sin \alpha x + \cos \alpha x\right).$$

The maximum bending moment is given by

$$M_{\max} = Pe\left(\tan \alpha \frac{L}{2} \sin \alpha \frac{L}{2} + \cos \alpha \frac{L}{2}\right)$$

$$= Pe \sec\left(\alpha \frac{L}{2}\right)$$

and hence the maximum combined stress is

$$\sigma = \frac{P}{A} + \frac{Pe \sec\left(\alpha \frac{L}{2}\right) c}{I}$$

where A is the area of cross-section of strut; I is the second moment of area of cross-section of strut; and c is the distance of extreme fibre from centroidal axis. After some manipulation, this becomes

$$\sigma = \frac{P}{A}\left(1 + \frac{ec}{r^2}\sec\frac{L}{r}\sqrt{\frac{P}{4AE}}\right)$$

This equation is known as the secant formula. With $P = P_{cr}$, a critical load, and σ equal to a limiting stress, it can be used to calculate the critical load for a given strut eccentrically loaded. Clearly, a trial and error method is necessary.

Although widely used in practice and mechanical engineering, the secant formula sometimes underestimates the critical load. More accurate design methods can be found in Bleich[8], for example.

Design of an eccentrically loaded strut using the secant formula is illustrated in Example 12.7.

LIST OF SYMBOLS

A	area of cross-section
A_E	effective area of cross-section of a tie
A_g	gross area of cross-section of a strut
a_1, a_2	areas of legs of an angle
c	distance of extreme fibre from centroidal axis
E	modulus of elasticity
E_t	tangent modulus
e	eccentricity of applied load
F	tensile load in a tie
f_{bc}	compressive bending stress
f_c	compressive stress due to axial load
I	second moment of area of cross-section
L	length of strut
M_x	bending moment at distance x from one end of strut
P	load applied to strut
P_{cr}	critical load on strut
p_{bc}	allowable compressive stress, bending
p_c	allowable compressive stress, axial load
p_t	allowable tensile stress
R	radius of curvature
r	radius of gyration
W	transverse load on strut
x	distance along strut
y	transverse deflection of strut
α	a parameter, $\alpha = \sqrt{\frac{P}{EI}}$
ε	strain
$\varepsilon_{min}, \varepsilon_{max}$	minimum and maximum strain
ε_y	yield strain
σ	stress
σ_y	yield stress
σ_{cr}	critical stress

Example 12.7 Sheet 1 of 1

Select the diameter of a circular steel strut to carry a load of 20kN at an eccentricity off 100 mm.

f_y = 275 MPa E = 207 GPa
Factor of safety = 1.8
Effective length of strut = 1.0 m

Allowable stress $= \frac{275}{1.8} = 153$ MPa

By trial and error :

d	c	r	A	$\frac{P}{A}$	$\frac{ec}{r^2}$	$\frac{L}{r}\sqrt{\frac{P}{4AE}}$	$\sec\frac{L}{r}\sqrt{\frac{P}{4AE}}$	σ
50	25	12.5	1960	10.2	16.0	0.278	1.04	180
55	27.5	13.75	2380	8.4	14.5	0.232	1.03	134

Use 55 mm diameter solid shaft.

REFERENCES

[1] BS 4:Part 1:1972, "Hot Rolled Steel", *Specification for Structural Steel Sections,* British Standards Institution, London, 1972.

[2] BS 449:Part 2:1969, "Metric Units", *The Use of Structural Steel in Building,* British Standards Institution, London, 1969.

[3] B. BRESLER, T. Y. LIN and J. B. SCALZI, *Design of Steel Structures,* 2nd ed., John Wiley, New York, 1968.

[4] M. F. SPOTTS, *Design of Machine Elements,* 3rd ed., Prentice-Hall, London, 1961.

[5] BS 449:Addendum 1:1961 (PD 4064), "The Use of Cold Formed Steel Sections in Building", British Standards Institution, London, 1961.

[6] BSCP 112:Part 2:1971, "Metric Units", *The Structural Use of Timber,* British Standards Institution, London, 1971.

[7] NLMA, *National Design Specification for Stress-Graded Lumber and its Fastenings,* 2nd ed., National Lumber Manufacturers' Association, Washington, 1957.

[8] F. BLEICH, *Buckling Strength of Metal Structures,* McGraw-Hill, New York, 1952.

BIBLIOGRAPHY

BOWEN, L. P., *Structural Design in Aluminium*, Hutchinson, London, 1966.
Describes British practice.

GERSTLE, K. H., *Basic Structural Design*, McGraw-Hill, New York, 1967.
A general textbook on structural design. Includes a good introduction to the effects of yielding and buckling.

GRAY, C. S., KENT, L. E., MITCHELL, W. A., and GODFREY, G. B., *Steel Designers' Manual*, 4th ed., Crosby Lockwood, London, 1972.
A comprehensive text covering all aspects of design of steel structures to satisfy the provisions of BS 449.

HANSEN, H. J., *Modern Timber Design*, John Wiley, New York, 1948.
Covers all aspects of timber design.

PROBLEMS

12.1. A threaded steel tie rod is required for a load of 60 kN. Assume a working stress of 215 MPa and calculate the required diameter. The effective area is at the root of the thread, which is to be a ISO metric screw thread. (See Appendix A9).

12.2. Calculate the allowable tensile load for a tubular steel tie with outside diameter of 50 mm and wall thickness of 3 mm. Allow for a 5-mm bolt through the tie on a diameter. Allowable stress in tension is 215 MPa.

12.3. Calculate the allowable axial compressive load for the tube of Problem 12.2, assuming an effective length of 1 m. Use the allowable stresses in your local building code or those in Table A14.1. Repeat the calculation using the Johnson formula.

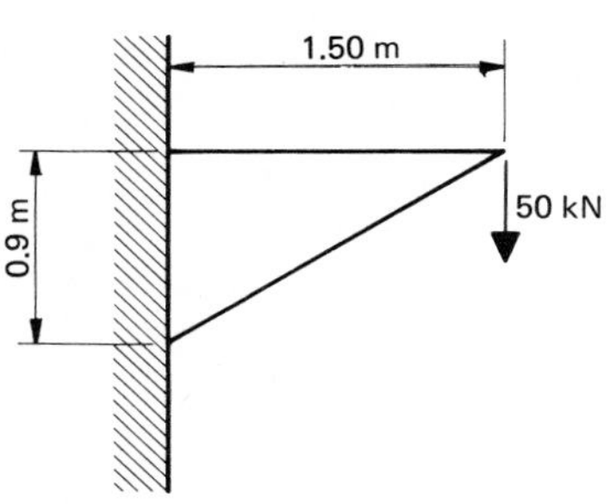

12.4. Select suitable members for the frame sketched in the margin, using
(a) single angles,
(b) pairs of angles back to back.
Use an allowable tensile stress of 215 MPa and take the allowable compressive stress from Table A14.1. Alternatively, use your local building code. The strut can be assumed fixed in position, but not in direction, at both ends.

12.5. Compute the safe load for the strut of Problem 12.3, assuming an eccentricity of 10 mm. Use the secant formula with an allowable total stress of 215 MPa.

13 Connections

13.1. INTRODUCTION

Fabrication of a structure or a machine is the process of assembling it from its components. These components may be castings or they may be manufactured from pieces of plate, tubes, rolled steel sections or timber selected from a range of standard shapes and sizes. This chapter is concerned with design of the joints necessary to make sub-assemblies from the components and to complete the construction from the sub-assemblies.

Connections can be classified in several ways. In the first place, they may be shop joints or field joints, depending on the place where the work is done. Shop joints made in a workshop are the better of the two. The workshop provides superior working conditions, closer supervision, better equipment for handling and holding components and a wider range of tools so that workmen can produce high quality products at lower cost. Field joints are made on construction sites where working conditions are often arduous and equipment and tools often cruder and less versatile.

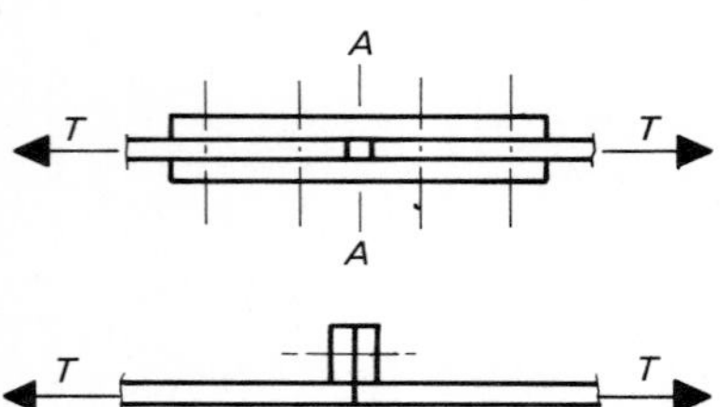

Another classification uses the direction of the load transferred in relation to the axes of the fasteners. For example, tension in a plate might be transferred in either of two ways. In one, a plate is fastened to each side of the plates to be joined. Consideration of a free body to the left of AA, say, shows that in a joint like this, the fasteners are subject to shearing. Alternatively, the plates might be fitted with flanges, and the joint made by fastening the flanges together. In this joint, the fasteners are subject to tension by the load T.

In a third classification, connections are grouped according to the kind of fastener used and the operations performed in making the joint. The commonly used fasteners are pins, rivets, bolts and welds for metals and timber connectors for timber. Synthetic resins are also used as adhesives to make joints in timber and metals, but they will not be considered here.

The third classification is used in this chapter to subdivide the topic. It will be recognised, of course, that any of these fasteners may be used for shop joints and for field joints, and most of them can transfer a tensile load or a shearing load or a combination.

13.2. DESCRIPTIONS

In this section, we describe pins, rivets, bolts, welds and timber connectors, and how joints are made.

Pins, rivets and bolts have one feature in common. They all require matching holes to be made in the parts to be joined and the joint is made by passing the fasteners through the holes and securing them in place. There are, however, considerable differences in the details of the procedure for making a joint and in the basic components used.

A *pin* is used in a shear type joint to avoid restraining the relative rotation of the parts joined. Only one pin can be used and friction in the joint must be reduced to a minimum. To illustrate this, consider a beam supported on two columns and carrying a load at midspan. Deflection of the beam makes its ends turn through a small angle. If the beam is rigidly connected to the columns, this rotation of the ends of the beam is restrained by the columns but is not, in general, eliminated. As a result, the columns are made to bend and they have to resist bending moments as well as direct compressive loads. It may be desirable to prevent this from happening and a single frictionless pin at each end of the beam will support the ends of the beam without requiring the tops of the columns to rotate.

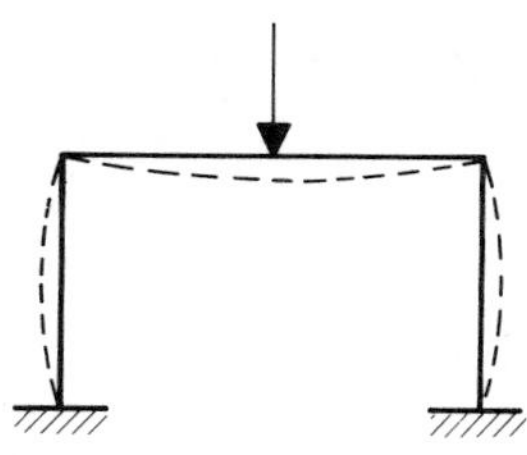

There are several disadvantages in the use of pins, and these are often such that it is better to allow for rigid joints than to try to avoid them. Inadequate maintenance may result in rusting, in which case the joint will be far from frictionless. Numerous cycles of loading and unloading will cause wear unless special materials are used. The single pin will generally be a substantial member. Furthermore, the structure as a whole may gain in stiffness by having rigid joints and, in some cases, the secondary stresses arising from joint rotation may be negligible. Joints in light roof trusses are relevant examples, as discussed in Chapter 12.

The pin itself is a steel cylinder, turned to the required diameter. It may be prevented from working its way out of the holes by a bolt passed through the pin to hold caps on its ends, or by a head turned on one end with a locking pin at the other, or by two locking pins.

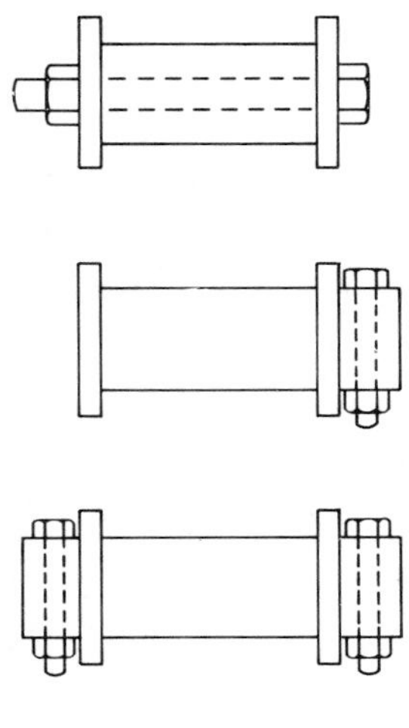

A *rivet* is also a metal cylinder, but its diameter is much smaller than that of a pin. In steel fabrication for machines and structures, steel rivets 10 mm to 25 mm in diameter are commonly used. Aluminium alloy rivets are used in the fabrication of light alloy structures for aircraft and superstructures of ships and, in the case of aircraft, diameters of 2 mm upwards are used.

The load-carrying capacity of a single rivet is relatively small and a riveted joint usually requires several rivets. No attempt is made in construction to avoid rigidity of the joint. If the parts connected are flexible, joint rotation will cause only small bending moments which can be ignored, as is the case in a typical light roof truss. When stiff members are connected, as in a portal frame comprising a beam and two columns, rigidity of the joint is taken into account and the resulting bending moments are calculated.

As supplied by the manufacturers, the rivet has a head formed on one end. To make a joint, a steel rivet is heated, inserted in the matching holes and a head is forged on the other end. A pneumatic riveting hammer is used, the rivet being held by a heavy tool or anvil pushed against the manufactured head. After driving, the hot rivet fills the hole – being made of a ductile metal, it deforms plastically while being driven. As it cools, it shrinks longitudinally and laterally. The longitudinal shrinkage is restrained by the parts being joined, so that the rivet is in tension and the parts are clamped together in the completed joint. Owing to lateral thermal contraction, and a secondary strain due to the longitudinal tension, the diameter of the rivet in service is less than that of the hole by a small margin. Aluminium rivets are similar to steel rivets, but they are driven cold.

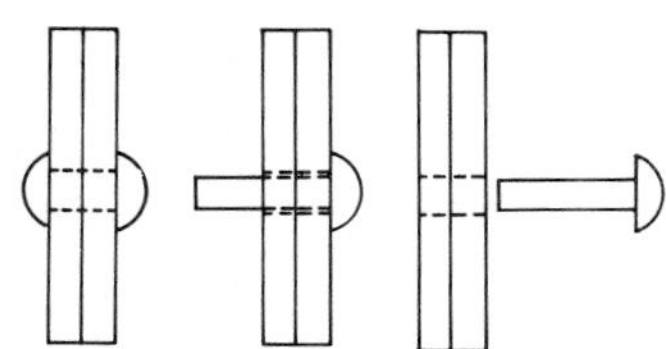

Holes for rivets are made by punching or drilling. Sometimes, holes are punched to a size smaller than that finally required, which is called sub-punching. After the parts have been set up for riveting, the sub-punched holes are opened out to full size by reaming. This process is costly, but has the advantages of removal of damaged metal near the punched hole and accurate alignment of the full-sized holes. In any case, the holes as prepared for larger rivets are usually 2 mm greater in diameter than the rivet as supplied – these are *clearance holes*. This clearance is provided to facilitate insertion of the hot rivet.

A *bolt*, like a rivet, has a cylindrical barrel, with a head on one end. The other end, however, is threaded to take a nut and both head and nut are either square or hexagonal to permit holding with a spanner. A joint is made by passing bolts through matching holes and tightening nuts on the threaded ends. The tools used are less cumbersome than those required for riveting and the bolt does not have to be heated. It is also an advantage that a bolted joint can be readily dismantled, making bolts suitable for temporary connections.

Since the industrial revolution, independent development in Britain, Europe and U.S.A. has led to an extremely wide and complex range of types of threaded fasteners. They vary in thread form, pitch, diameters, head and nut form, and production tolerances, far beyond the technical requirements for various applications.

Several attempts have been made to achieve some unification and standardisation at least between Britain and the U.S.A. and also within individual countries. These moves have met only with partial success. Some more gradual effect may be achieved by encouraging young designers and draughtsmen to concentrate on a narrower range of accepted forms and there has been a tendency to prefer the unified American system at the expense of the British Whitworth and BSF range. However, the change to the metric system will once more aggravate the situation but will in the long term lead to the complete disappearance of the older British system and leave, one hopes, two main ranges of American and metric forms. These in themselves are sufficiently diverse to perpetuate considerable chaos in the future.

Students in all branches of engineering must therefore have a working knowledge of the ranges in U.S., metric and British forms. A comprehensive reference is *Machinery's Screw Thread Book*.[1] It gives a brief summary of the evolution of the screw thread, the attempts at standardisation and a comprehensive specification for a large range of screw threads.

In general, thread forms are described by the following measurements (see also Fig. 13.1):

Nominal diameter. The diameter used to describe the bolt and usually equal to the diameter of the unthreaded part of the barrel.

Major diameter and minor diameter. The greatest and least diameters of the threaded part.

Effective diameter. The diameter of a cylinder which gives equal intercepts between successive tooth flanks.

Pitch. The distance gained along the bolt by one complete turn of the thread. It may be given as the number of threads per unit of length or as the distance gained.

As noted above, increasing use of the metric system of measurements is expected to lead to displacement of most British threads by metric threads. As

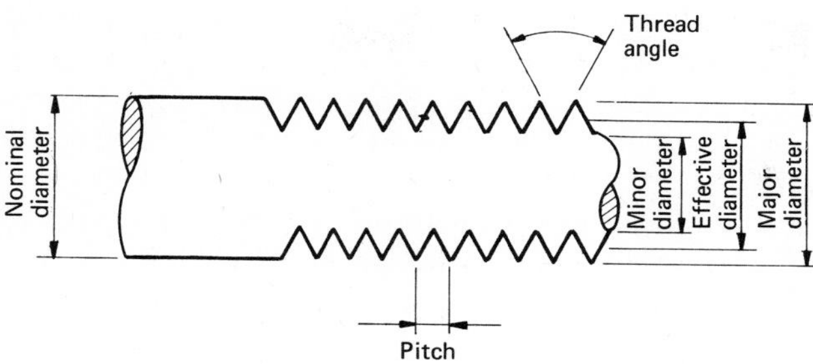

Fig. 13.1. *Significant dimensions of thread forms.*

yet the unification of all metric systems is by no means complete. The International Organization for Standardization (ISO) is engaged in this task and has put forward recommendations some of which have already been adopted. The operation is by no means finalised and even in Europe there are several ranges of metric threads which have not been unified. The more important of these are French, German and Swiss, all of which have a similar basic thread form and range of nominal sizes. It is expected that adoption of ISO standards will ultimately remove these differences.

The basic metric thread form has a 60° thread angle similar to the American thread and all dimensions are expressed in millimetres.

ISO recommendation R262 provides for coarse and fine series. Various classes of fit are obtained by manufacture within certain standard tolerance zones, three for bolts and two for nuts. These zones are defined relative to the basic size and classes of fit can be obtained by combinations of the tolerance zones in nut and bolt; three class-combinations are suggested. Quite large clearances are required where, for example, the working parts are to be plated.

The basic range of nominal diameters runs from 1 mm to 300 mm listing preferred sizes. The coarse series with graded pitches is the only one so far recognised but provision is also made for a fine series and a constant-pitch series. A list of some of these sizes is given in Appendix A9.

The recommended method of designation for nuts and bolts with ISO metric threads (from BS 3643 : Part 2 : 1966, "ISO Metric Threads") is as follows:

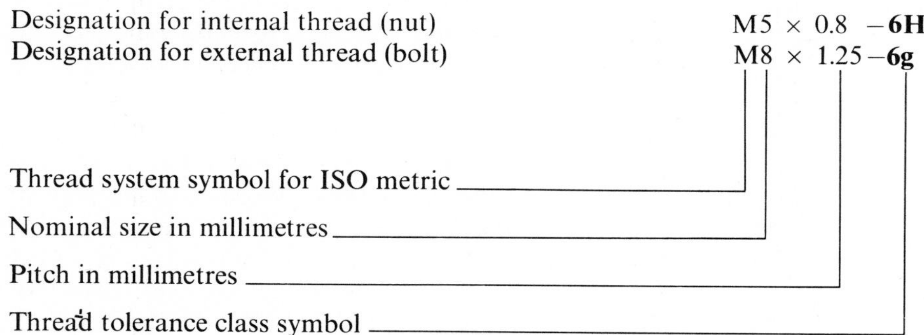

Black bolts are forged from round bar stock and may be supplied without any machining at all (black all over) or partially machined (faced under the head or faced under the head and turned on the shank). The name "black bolt" has nothing to do with the colour of the bolt – it signifies that relatively large tolerances are allowed in manufacture. For holes, a clearance of up to 2 mm is common for bolts 24 mm in diameter and smaller and 3 mm for larger bolts.

Bolts known as *close-tolerance bolts* are partially machined black bolts that are faced under the head and turned on the shank. Holes for close-tolerance bolts are drilled after the parts are set up or they may be sub-punched and, after setting-up, reamed. A typical clearance allowed on the basic size of the bolt is 0 to +0.15 mm.

The following dimensions from BS 4190: 1967 are typical tolerances on diameter for black bolts and close-tolerance bolts.[2]

Basic size (mm)	Tolerance on diameter	
	Black all over (mm)	Close tolerance (mm)
6	±0.48	0 to −0.18
10	±0.58	0 to −0.22
20	±0.84	0 to −0.33
30	±0.84	0 to −0.33
48	±1.00	0 to −0.39

Note that the tolerances for close-tolerance bolts are approximately **h13**.

Residual tension in black bolts, including close-tolerance bolts, may be small with little reliable clamping action. When clamping is required in the completed joint, care must be taken to strain the bolts to some suitable extent, which can be calculated.

High-strength friction-grip bolts (HSFG) rely entirely on friction to transfer shearing loads. They are made of high tensile steel and hardened washers are fitted under the head of the bolt and the nut. When the nut is tightened, it is turned far enough to induce a large tension in the bolt. There is a correspondingly large strain and, even though deferred strain (or creep) acts to release some of the tension as time passes, a considerable permanent tension remains. As a result, there is a reliable clamping force to give the necessary friction.

Clearly, control of the initial strain is of vital importance in the use of high-strength friction-grip bolts. In British practice, the manufacturing standard specification for the bolts requires each bolt to have a specified ultimate load in tension and to be able to carry a proof load.[3] The specified proof load is approximately 70% of the ultimate load and there must be no permanent elongation on unloading the specimen. In fabrication, each nut is tightened sufficiently to induce a tension not less than the proof load. This can be done by turning the nut with a pneumatic impact wrench which can be adjusted to stall at a selected torque. Once or twice in each shift, during fabrication, the wrench is calibrated by tightening a bolt in a load cell and the stalling torque is adjusted so that the wrench stalls for the required bolt load. Otherwise, the "turn of the nut" method is used. All nuts are first tightened so that the parts to be joined are pulled up snug against each other. Each nut is then rotated a further specified fraction of a turn, care being taken to prevent the head of the bolt from turning. Plate 13.1 shows a pneumatic wrench being used to tension a joint made with HSFG bolts.

High-strength friction-grip bolts have many advantages when compared with other kinds of bolts, or rivets. Clearance holes can be used with the advantage of straightforward preparation of parts, yet there is no slip. Also,

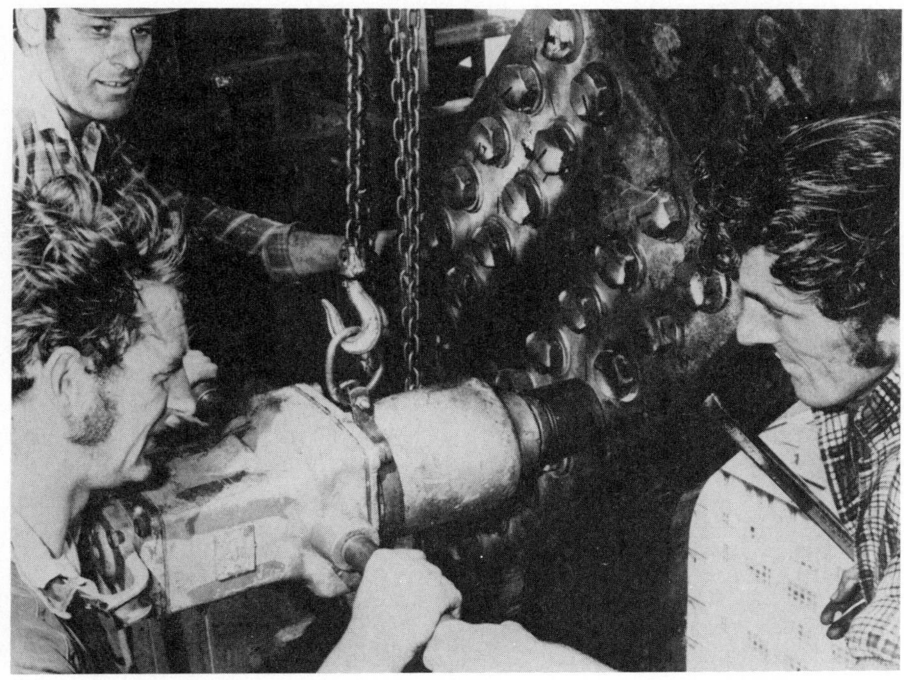

Plate 13.1. *Tightening HSFG bolts with an air wrench. (Reproduced by permission of The Broken Hill Proprietary Coy Ltd.)*

since the load is distributed by the hardened washer concentrations of stress are reduced. There is the further advantage that the effective section of the parts joined is greater (see Chapter 12). In practice, the washers, by distributing the load round the holes, tend to prevent a tension force from breaking a tie through the holes. Failure is more likely to take place away from the holes. Although this is not taken into account in design, it leads to a stronger joint. Finally, being free from slip, such joints serve better under reversing or fluctuating loads – the load on the bolt remains constant and the joint is stiffer.

The disadvantages of HSFG bolts, compared with black bolts, are the higher cost of the material and the more elaborate tightening procedure. Although not warranted when small loads have to be transferred, as in light trusses, the advantages outweigh the disadvantages when the loads are large.

Welding is a process whereby metal parts can be joined together by the application of heat. There are several kinds of welding, of which only electric arc-welding is considered here. It is applied to steel, principally low-carbon steel, and the necessary heat is generated by means of an electric arc struck between an electrode and the workpiece. As a result, the workpiece melts near the arc, a pool of molten steel forming there. The electrode also melts and magnetohydrodynamic forces transfer the molten steel from the electrode to the pool on the workpiece. The arc is made to travel along the joint under construction and the molten steel cools and solidifies behind the arc. In this way, two parts are joined together by the melting, mixing and solidification of steel from the parts and the electrode.

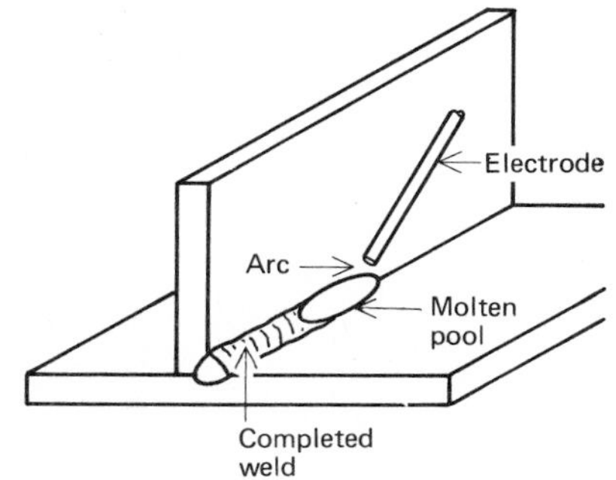

Electrodes are usually coated with a material which vaporises and shields the arc and also forms a slag on the completed weld. Thus, it protects the arc and the molten metal from the atmosphere and prevents atmospheric gases from entering the liquid steel. The slag also reduces the rate of cooling of the hot solid steel.

Arc-welded joints may be made manually or by machine. Welding positions may be downhand, vertical or overhead, as illustrated in Fig. 13.2. Downhand welding is easiest for manual welding, and the work should be planned so as to have as much welding as possible done in this position.

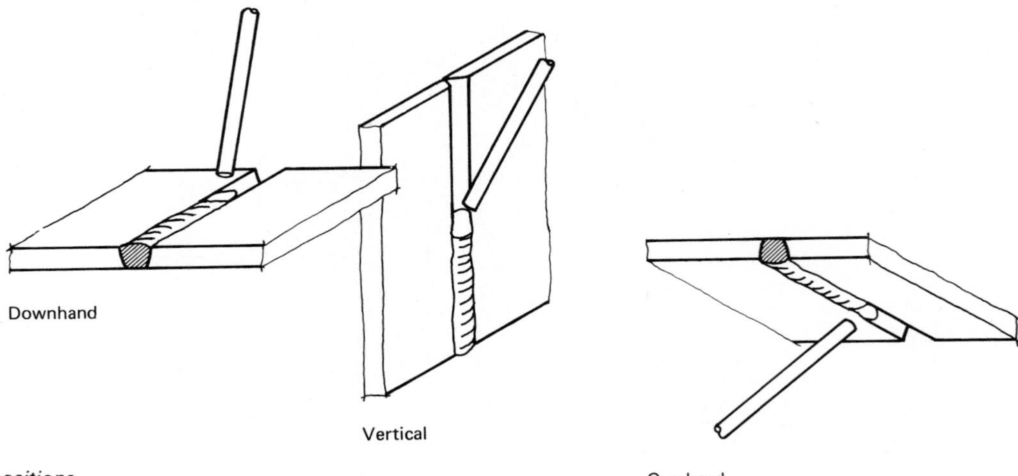

Fig. 13.2. *Welding positions.*

Welds can also be classified as fillet welds or butt welds. A fillet weld is placed in a corner formed by the parts to be joined, whereas a butt weld is a continuation of one or both of the parts to be joined. (See Fig. 13.3).

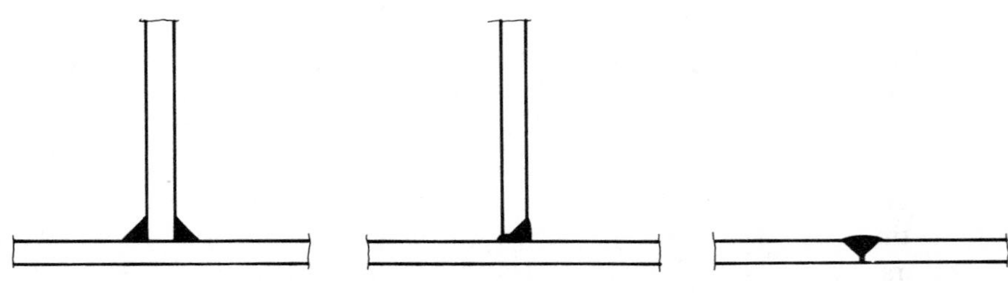

Fig. 13.3. *Fillet welds and butt welds.*

No edge preparation is required for fillet welds or for butt welds in light plate. For butt welds in heavy plate, the edges must be prepared by cutting back in some form or other—for example, single V-edge or double V-edge preparation as illustrated in Fig. 13.4. The edge preparation is necessary to ensure complete fusion through the whole thickness of the plate and it must be wide enough for the welder to get his electrode into the joint.

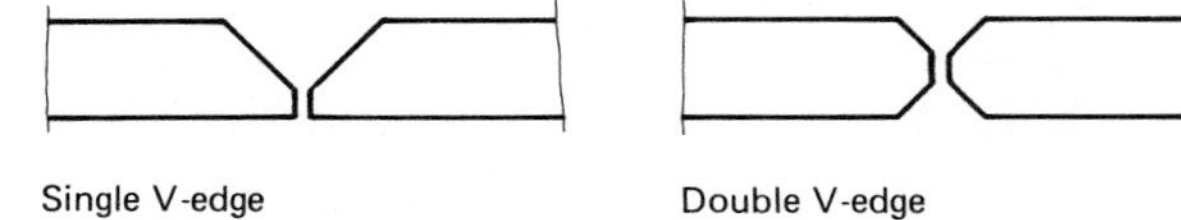

Fig. 13.4. *Typical edge preparation for butt welds.*

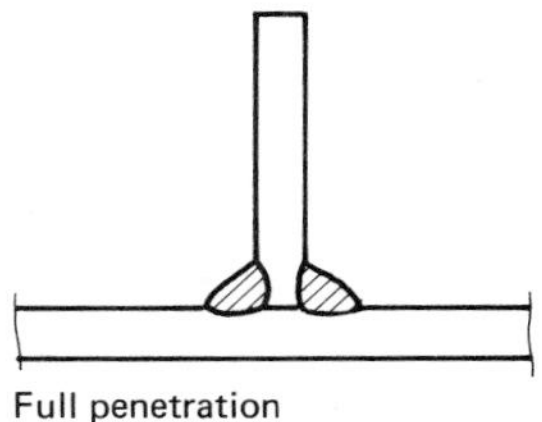

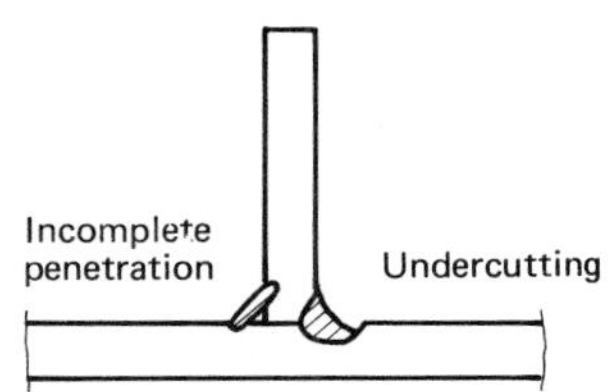

Sound welding requires full penetration of molten metal from the parts or the electrode into the joint. Poor workmanship can cause defects such as incomplete penetration, undercutting and inclusions of gas or slag. Some of these defects are particularly dangerous because they cannot be detected by visual inspection of the finished weld. Much reliance must be placed on the training and skill of the welder.

When the strength and quality of welding are particularly important, visual inspection of the process and the finished job is insufficient. Radiographic inspection is then used; examination of X-ray photographs reveals defects not normally visible.

Timber connectors are fastenings used to join timber parts together, mostly in trusses, or to join steel and timber parts. They are designed to transfer relatively large loads by spreading the transfer over a considerable area of timber. There are four kinds of connectors – the split-ring connector, the shear-plate connector, the toothed-ring connector and the Gang-Nail.

The split ring is a strip of steel bent into a circular ring. A shear plate comprises a circular plate with a central hole and a one-sided flange on its circumference. The toothed ring is made from a circular disc by pressing triangular teeth normal to the disc, on one side of the disc or alternately one side and the other, at its perimeter; it has a hole in the centre of the disc. Each of these is used in conjunction with a bolt. The Gang-Nail is a rectangular plate of galvanised steel from 1 mm to 2 mm thick with teeth formed on one side by punching.

Split-ring connectors are used with hard timbers. Matching holes are drilled in the parts to be joined and, in each face which is to be in contact with another part, a concentric groove is cut with a special tool. The depth of the groove is half the height of the connector. To make the joint, the connector is placed in one groove, the groove in the matching part is fitted over the connector and the parts are pulled together with the bolt. Square steel washers are required under the head of the bolt and the nut.

Shear plates are used in a similar way when one of the members at the joint is a steel part. The flat side of the connector fits against the steel and a groove in the timber takes the flange of the connector. A circular recess in the timber is cut so that the flat side of the connector will be flush with the face of the timber. Shear plates can be used in pairs back-to-back to make joints between timber members, although with an obvious cost penalty. Toothed-ring connectors are used with soft timbers. There is no need to prepare grooves for the connector, which can be forced into the timber by tightening the bolt.

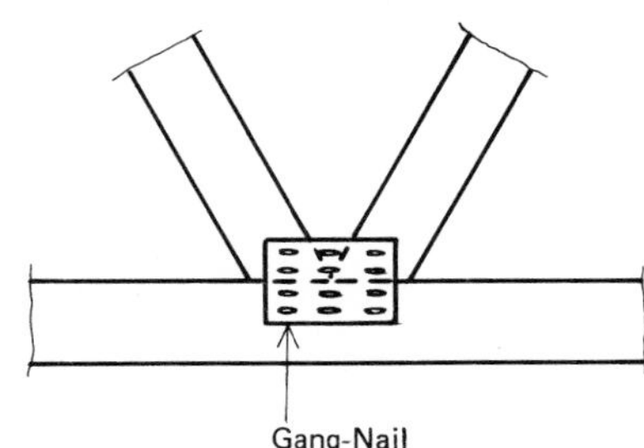

Gang-Nail joints are made by pressing Gang-Nails into the timber on both sides of the joint. Construction is a factory operation. Members to be joined are assembled on a jig, with the connectors located as required. A press passes over the jig and drives the Gang-Nails into the timber. Loads are transferred by the teeth embedded in the timber from the timber to the Gang-Nail. Thus, several members are joined to the Gang-Nail and the strength of the connector keeps them fastened together. Generally, timber of 50 mm nominal thickness is used. Where one truss so built is not strong enough to carry the loads, two can be used alongside each other.

These connectors can lead to improved appearance and reduced timber costs in trusses. There is no need for the members to overlap at a joint, as there may be when split-ring or toothed-plate connectors are used, and member sizes are not controlled by the need to maintain edge distances and connector spacing. Such considerations often override the load-carrying capacity of the members when split rings or toothed plates are used.

Detailed design procedures for Gang-Nails will not be given here. Fabrication is carried out only by operators accredited by the patent holders and these operators employ or have access to specialist design staff. Design manuals are available; for example, the one published by Automated Building Components (N.Z.) Limited.[4]

Plate 13.2 shows some of the more common fastenings.

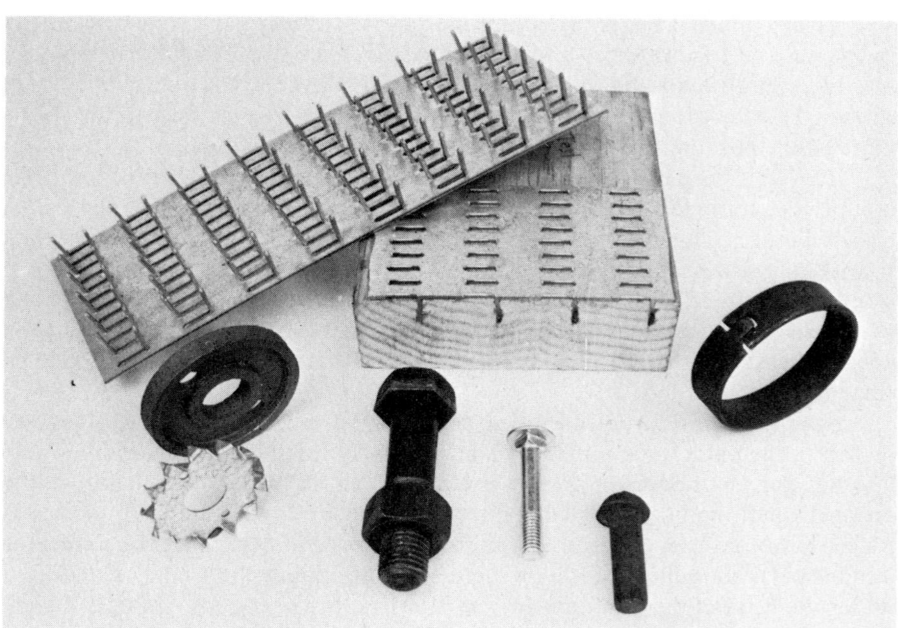

Plate 13.2. *A selection of fasteners. In the background, two Gang-Nails, one cut through after being pressed into a piece of timber. In the foreground, left to right: a toothed-ring connector in front of a shear-plate connector, an engineer's bolt and nut, a coach bolt, a rivet and a split-ring connector.(Photo: G. Boehnke. Reproduced by permission of the University of Auckland.)*

13.3. STRESSES IN FASTENERS

Stresses are induced in fasteners during fabrication and by the applied loads. The transfer of load also causes stresses in the parts joined and these may dominate the design, rather than the stresses in the fastener. For example, there is a wide disparity between the strengths of steel and timber, so that the strength of a joint using timber connectors depends on the parts joined and not on the connector. Again, the load applied to a joint sealed by a gasket will depend on the stress required in the gasket to ensure that it will not leak.

In this section, the stresses in pins, rivets, bolts and welds are examined. The first three have a good deal in common and we begin by looking at the stresses in a pin.

13.3.1. Stresses in pins

In Fig. 13.5(a) and (b), a pin supporting the end of a steel beam is shown. To ensure free rotation, a clearance, c, is provided between the parts of the bearing fixed to the beam and those fixed to the foundation. Figure 13.5(c) shows idealised loading, shear force and bending moment diagrams.

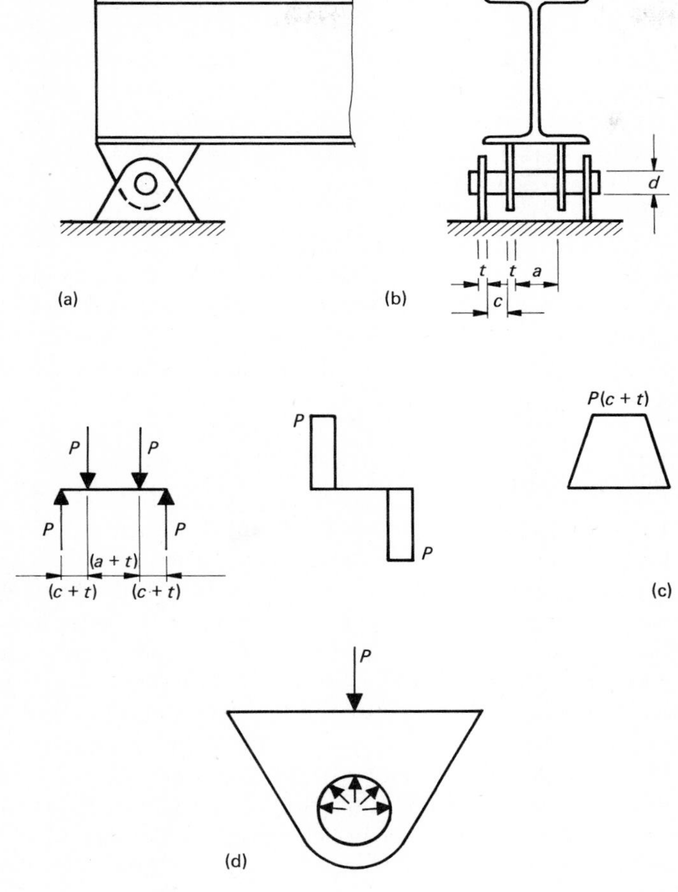

Fig. 13.5. *Design of a pin to support a beam:*
(a), (b) details of support;
(c) loads, shears and bending moments;
(d) bearing of pin on plates.

The stresses in the pin are as follows:

(i) Longitudinal tension and compression due to the bending moment $P(c + t)$. These can be estimated from the simple theory of bending:

$$f_{bt} = f_{bc} = \frac{M}{Z} = \frac{P(c + t)}{\pi d^3/32}$$

(ii) Shear stress due to the external shear P. This can also be estimated using the simple theory of bending, taking into account the variation of shear stress across the section:

$$f_q = \frac{4}{3}\frac{P}{\pi d^2/4}$$

However, the assumptions on which this formula depends are of doubtful validity in a beam whose depth-to-span ratio is large. Normal practice is to use the mean stress in design of a pin,

$$f_q = \frac{P}{\pi d^2/4}$$

(iii) Bearing stress, where the plates of the bearing are in contact with the pin (Fig. 13.5(d)). The distribution of this stress is not known and an average bearing stress is used as a measure. This average is calculated by dividing the load transferred by the area of contact projected on a plane normal to the line of action of the load:

$$f_b = \frac{P}{td}$$

Note that this stress also occurs in the plates.

There will be combined bearing and bending stresses where the plates attached to the beam are in contact with the pin. Shear stress is also a maximum at this section, but can be examined separately. It does not reach its maximum value at the point in the cross-section where bearing and bending stresses are the largest.

13.3.2. Rivets and bolts

Rivets and bolts are subject to axial tension during fabrication and stresses due to the loads are superimposed on this axial tensile stress. The ability of a rivet or bolt to transfer a shearing load is affected by the axial tension, since the tension in the fastener leads to a friction force between the parts joined which may contribute to the resistance to shearing loads. Stresses in a single fastener and the friction available between two parts are investigated here.

Consider first a single fastener joining two parts together and loaded in tension. There is a soft packing, or gasket, between the parts to seal the joint. The initial tension after fabrication and before loading is P_I. The load applied to the fastener is P_A and the resultant tension in the fastener, P_F, is sought.

The load P_A is considered to be applied directly to the ends of the fastener, but this is only for convenience in the analysis. In practice, it is more likely to be applied to the parts joined – the flanges, for example – in which case it is transferred to the fastener indirectly.

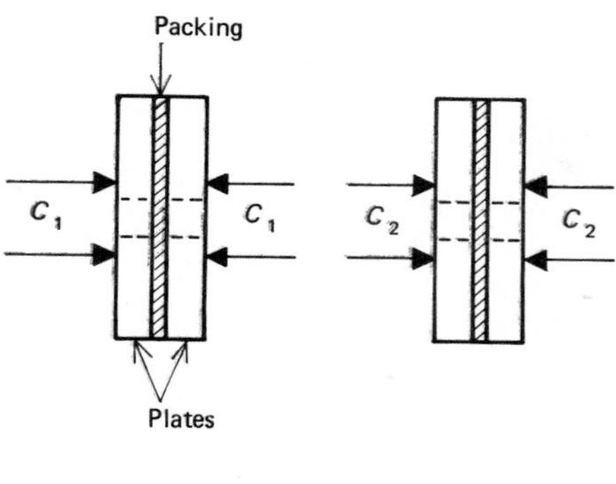

Fig. 13.6. *Forces on fastener and plates: (a) before external load applied; (b) after loading.*

The problem is solved by considering the changes in deflections that take place when the load P_A is added. In Fig. 13.6, the fastener and the plates are shown separated from each other, part (a) showing the forces acting after fabrication, while part (b) shows those acting after the load P_A has been applied. Clearly

$$C_1 = P_I \tag{13.1}$$

and

$$C_2 + P_A = P_F \tag{13.2}$$

The change in tension in the fastener is $P_F - P_I$ and causes an increase in length equal to

$$\frac{P_F - P_I}{k_F}$$

where k_F = stiffness of bolt = load which causes unit deflection = AE/l. Here A is the gross area of cross-section assuming no thread on the part of the bolt in the hole; E is the Young's modulus; and l is the length of fastener.

This change in deflection is equal to the change in deflection of the plates and the packing due to a change in clamping force from C_1 to C_2, provided

$$C_2 \nless 0 \tag{13.3}$$

that is, provided the joint remains in compression. This change in deflection is

$$\frac{C_1 - C_2}{k_P}$$

where k_P is the stiffness of packing and plates.

However, from Equations (13.1) and (13.2),

$$C_1 - C_2 = P_I + P_A - P_F$$

so that

$$\frac{P_I + P_A - P_F}{k_P} = \frac{P_F - P_I}{k_F}$$

Solving for P_F yields

$$P_F = P_I + \frac{k_F}{k_F + k_P} P_A$$

or

$$\frac{P_F}{P_I} = 1 + \frac{1}{1 + (k_P/k_F)} \frac{P_A}{P_I} \tag{13.4}$$

The Condition (13.3) is the same as

$$P_F - P_A \nless 0$$

or

$$\frac{P_F}{P_I} \nless \frac{P_A}{P_I} \tag{13.5}$$

Equation (13.4) and the Inequality (13.5) are plotted on Fig. 13.7. Up to the limit given by the Inequality (13.5), the straight lines on Fig. 13.7 show how P_F varies with P_A. Clearly, the increase in P_F caused by an increment in P_A is quite small, except for joints with soft packing. Beyond that limit, the clamping force C_2 is zero, and an increase ΔP in P_A causes an equal increase in P_F. Thus,

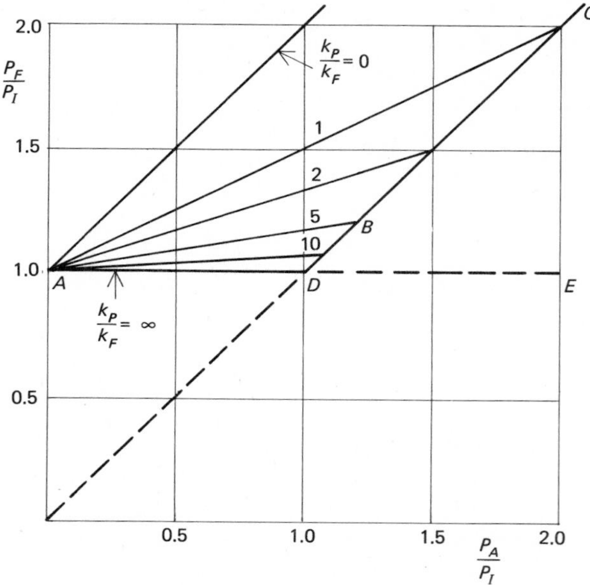

Fig. 13.7. *Relationship between fastener tensions and relative stiffness.*

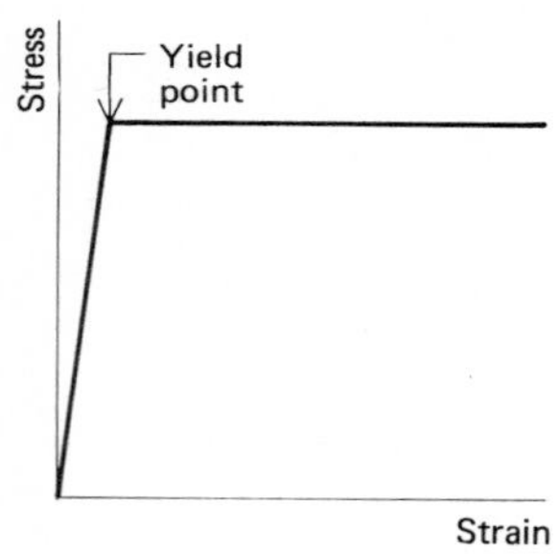

for any value of k_P/k_F (say, 5) P_F/P_I is given by the line ABC as P_A/P_I changes. Along BC, C_2 is zero and the joint is open.

The consequences of bolt strains greater than yield strain can also be seen on Fig. 13.7. With an idealised stress–strain relationship, $k_F = 0$ for such strains and $k_p/k_F = \infty$. The ratio P_F/P_I is given by ADE on Fig. 13.7. Thus, as P_A increases, there is no change in P_F and $P_F = P_I$. This is a reflection of the fact that increased strain in the fastener causes no increase in stress. The point D corresponds to vanishing of the clamping force, C_2, and for the segment DE, the joint is open.

In practice, it is difficult to assign values to k_P/k_F. In structural joints without packing,

$$\frac{k_P}{k_F} \sim 10$$

and the limiting value of P_F/P_I is given by solving Equations (13.2) and (13.4) when C_2 vanishes. Hence

$$\frac{P_F}{P_I} = 1 + \frac{1}{11}\frac{P_F}{P_I}$$

and

$$\frac{P_F}{P_I} = \frac{P_A}{P_I} = 1.10.$$

Thus, provided the applied tension does not exceed the initial tension by more than 10%, the plates will remain in contact and the final bolt tension, P_F, will not exceed the initial tension by more than 10%.

A point of general significance to be emphasised is that the final tension, P_F, is not equal to the sum of the initial and applied tensions. As the applied tension is increased, only part of it acts to increase the load in the fastening. The rest of it acts to reduce the clamping force.

It will be seen that, when varying loads are applied to a joint, the fluctuations in the load in the fastener can be reduced by making k_P/k_F large. Thus some control over the fatigue of the fastener can be exerted.

As an aid to estimation of bolt tensions, some approximate values of k_P/k_F are tabulated here.

Type of joint	k_P/k_F
Soft packing with studs	0.00
Soft packing with through bolts	0.25
Asbestos gasket	0.67
Copper-asbestos gasket	0.67
Soft copper gasket	1.12
Hard copper gasket	3.00
Lead gasket	9.00
Metal to metal joint	10.00 or more

For important joints which are not described in this table, experimental determination of the ratio will be justified. For less important joints, k_P and k_F may be estimated from the properties of the materials and the dimensions of the plates, packings and fasteners. Thus,

$$k_F = \frac{AE}{l}$$

as defined above. Note that A here will be shank area, not the stress area. To find k_p, dispersion of the bolt tension through the plates and packing at an angle of 45° may be assumed.[5] The equivalent stiffness, being the force required to cause unit deflection, can be calculated by using average areas for the plates and packing as follows:

$$\text{Plate 1:} \qquad A_1 = \pi(r_1{}^2 - r_h{}^2)$$

$$\text{Gasket:} \qquad A_g = \pi\left\{\left(\frac{r_1 + r_2}{2}\right)^2 - r_h{}^2\right\}$$

$$\text{Plate 2:} \qquad A_2 = \pi(r_2{}^2 - r_h{}^2)$$

where r_1, r_2 and r_h are defined on Fig. 13.8. Then, if E_1, E_g and E_2 are the

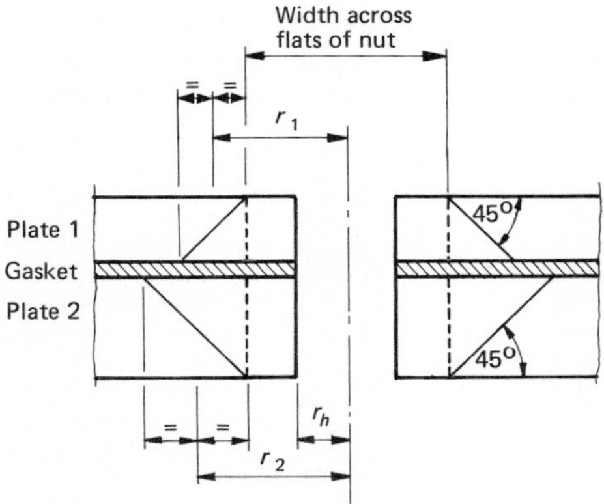

Fig. 13.8. *Estimation of k_p.*

respective elastic moduli and t_1, t_g and t_2 are the thicknesses, k_P is given by

$$\frac{1}{k_P} = \frac{t_1}{A_1E_1} + \frac{t_g}{A_gE_g} + \frac{t_2}{A_2E_2}.$$

An alternative to using this fundamental approach to calculation of bolt tension is given in BS 1515: Part 1: 1965, "Fusion Welded Pressure Vessels".[6] It is an empirical method and requires that the bolts in a flanged circumferential joint be able to sustain the larger of two loads P_1 and P_2 given by

$$P_1 = \pi D_m by \tag{13.6}$$

$$P_2 = \frac{\pi}{4} D_m{}^2 p + \pi D_m b\, mp \tag{13.7}$$

where D_m is the mean gasket diameter; b is the effective width of gasket; y is the compressive stress required to seat the gasket; p is the internal pressure in the boiler or pressure vessel; and m is the overpressure factor.

Thus, P_1 is the load required to seat the gasket under zero pressure. In Equation (13.7), the first term on the right is the pressure force tending to open the joint, so that $(P_2 - \pi/4\ D_m{}^2p)$ is the nett clamping force. Divided by πD_mb, it yields the nett pressure on the gasket and this is clearly mp. This leads to the description of m as an *overpressure factor*. Note that in Equation (13.7) the full pressure term is used, i.e. it is assumed that $k_P/k_F = 0$.

This method is useful in that it provides a means of deciding the bolt tension required initially to ensure yielding of the gasket—a requirement for effective sealing of the joint. A table showing typical values of m and y, adapted from BS 1515[6], is given below.

Gasket material	m	y (MPa)
Hard rubber sheet	1.00	1.24
Asbestos composition, 1.5 mm thick	2.75	25
Soft copper, corrugated	2.75	25
Copper, asbestos filled	3.50	45
Soft copper, solid	4.75	90

When a shearing load is applied to a riveted or bolted joint, additional stresses will be imposed on the fastener only if the parts slip relative to each other; and slip will occur only if the shearing load exceeds the friction.

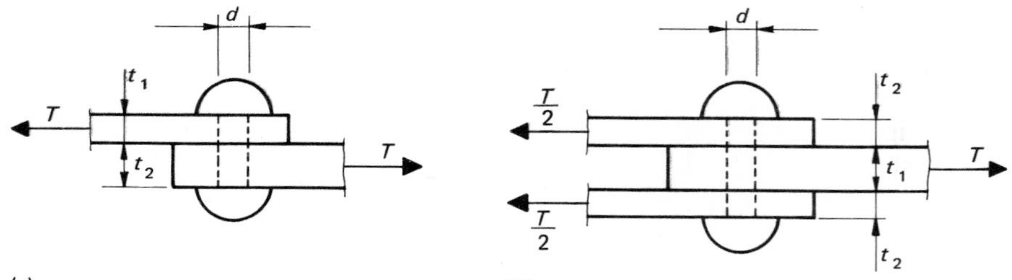

Fig. 13.9. *Effective interfaces at a joint: (a) one effective interface; (b) two effective interfaces.*

Effective interfaces are between those parts of the joint which tend to slip past each other under the action of the loads. For example, the joint in Fig. 13.9(a) has one effective interface, while that in Fig. 13.9(b) has two. If the tension in the fastener is P and the coefficient of friction for the parts in contact is μ, the total friction to be overcome is

$$j \mu P$$

where j is the number of effective interfaces.

When a large and reliable residual tension can be created in the fastener, as is the case with high-strength friction-grip bolts, the friction force is used to determine the safe shearing load per fastener. Otherwise, it is ignored. Thus, high-strength friction-grip bolts are designed for no slip, but friction is omitted from strength calculations for rivets, black bolts and close-tolerance bolts.

The amount of slip that takes place when a shearing load is applied to a joint depends on the kind of fastener. A well-driven rivet can be assumed to fill its hole—it fails to do so only because of thermal contraction and secondary strain. Consequently, a very small slip is sufficient to bring the parts joined into bearing contact with the fastener. In the case of bolts, the amount of slip depends on the clearance and may be as much as 2 mm for black bolts. When the parts have slipped under the action of a shearing load, the fasteners will be subject to additional stresses by the load.

When the parts joined have slipped, a rivet or a bolt resembles a pin, except that bending stresses can be ignored. There is no clearance between the parts joined and it would be impossible to apply any straightforward evaluation of bending of the fastener. The fasteners are designed as if in a state of simple shear, bearing and shear stresses being computed from

$$f_b = \frac{\text{load transferred}}{\text{projected area of contact}}$$

$$f_q = \frac{\text{load transferred}}{\text{area stressed in shear}}$$

The fasteners are said to be in single shear or double shear, depending on the number of cross-sections subject to shear stress. In Fig. 13.9(a), a load is transferred by a single fastener in single shear. The bearing stress is different for the two plates, the two stresses having averages of

$$\frac{T}{t_1 d} \quad \text{and} \quad \frac{T}{t_2 d}$$

These q' otients are measures of the bearing stresses in the two plates and in the fastener where it is in contact with them.

The shear stress in the fastener is measured by

$$f_q = \frac{T}{\pi d^2/4}$$

In Fig. 13.9(b), a fastener is shown in double shear, transferring a load T. Bearing stress is measured by

$$\frac{T/2}{t_2 d} \quad \text{and} \quad \frac{T/2}{t_1 d}$$

and shear stress by

$$\frac{T/2}{\pi d^2/4} = \frac{T}{2(\pi d^2/4)}$$

Note that in joints resisting shearing loads, it is preferable that the thread should not extend into the hole. The areas available for bearing and shear are less on the threaded part of the bolt than on the unthreaded shank. The thread should be no longer than is necessary to tighten the nut.

13.3.3. Groups of rivets and bolts

As noted above, it is essential for a single pin to be used in a pinned joint in order to allow freedom of the parts to rotate. This is not so when rivets or bolts are used and most such joints contain several fasteners. Two matters arise here: one is the best arrangement of the fasteners and the other is the way the total load is shared among them.

Quantitative comparison of different arrangements of fasteners can be made if the load is applied symmetrically to the group. In that case, the load is assumed to be shared among the fasteners in proportion to their areas. While this is unlikely to be the case in service, yielding of ductile materials in the parts joined and the fasteners, as failure is approached, will make this uniform distribution of stress realistic at ultimate load. Thus, use of the assumption implies that behaviour at strains beyond yield point is the basis of design, even though working stresses and loads in service are used in the calculations.

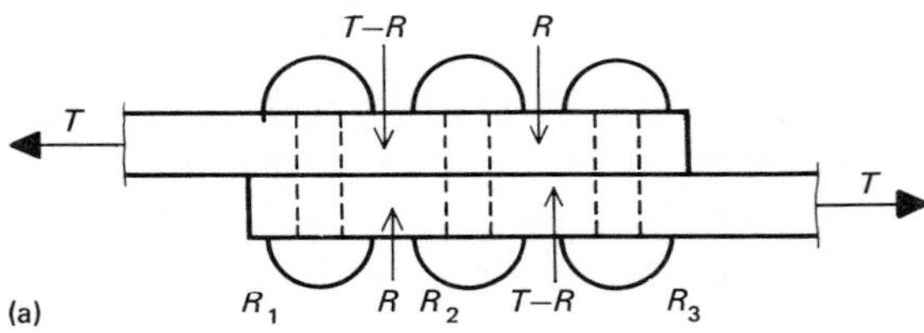

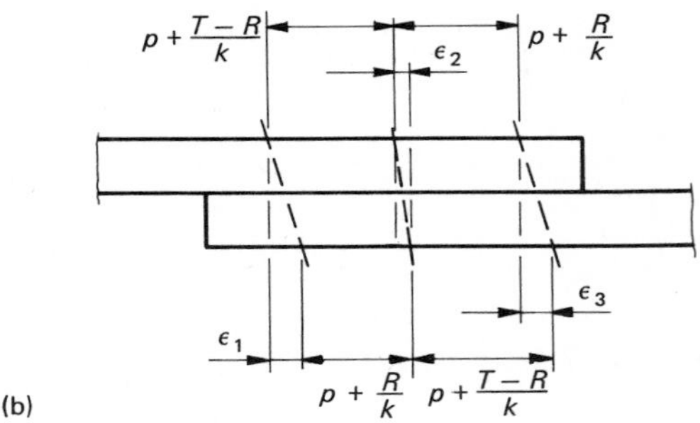

Fig. 13.10. *Forces and deflections in plates:*
(a) forces;
(b) deflections.

To make this important point clearer, consider two plates joined by three rivets and loaded as shown in Fig. 13.10(a). The shear loads carried by the three rivets are R_1, R_2 and R_3, as indicated. Clearly,

$$T = R_1 + R_2 + R_3$$

and, from the symmetry of the joint,

$$R_1 = R_3 = R, \text{ say}$$

The tensile loads in various parts of the two plates will be $T - R$ and R and these are also shown on Fig. 13.10(a).

Now if the plates are elastic and their stiffness between fasteners is k, the change in length of their various parts will be $(T - R)/k$ and R/k. Referring now to Fig. 13.10(b), where p is the pitch of the rivets and ε_1, ε_2 and ε_3 are the shear deflections in the three rivets, we can see that

$$\varepsilon_2 + p + \frac{T - R}{k} = p + \frac{R}{k} + \varepsilon_1$$

$$\varepsilon_1 = \varepsilon_2 + \frac{T}{k} - \frac{2R}{k}$$

$$= \varepsilon_2 + \frac{R_2}{k}$$

From symmetry

$$\varepsilon_3 = \varepsilon_1$$

Thus, the strain in the outer rivets is greater than the strain in the inner rivet. For elastic deformation of the rivets, the stress in the outer rivets will therefore exceed the stress in the inner one. This uneveness in the shearing of the load is caused by elastic deflection of the plates, as can be seen by putting $k = \infty$, which corresponds to zero deflection in the plates, whatever is the load. Clearly, $\varepsilon_1 = \varepsilon_2$ in that case and the rivets will be equally stressed.

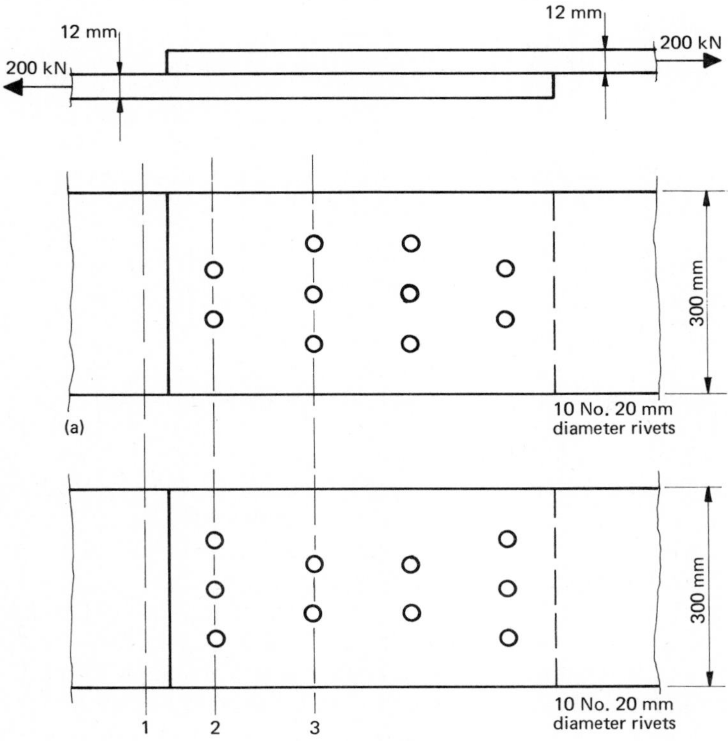

Fig. 13.11. *Alternative arrangements of rivets:*
(a) efficiency = 0.852;
(b) efficiency = 0.780.

For elastic deformation, unequal load sharing will generally occur as predicted for this example. However, if the load on the joint is increased sufficiently, the stress in the outer rivets will reach yield stress and further increased strain in them will cause no increase in stress (assuming idealised yielding). As a result, the additional load has to be carried by the inner rivets at increased stress. When the load is big enough, all the rivets will be strained beyond yield point and the stress will be the same in them all.

It should be noted that either of the following assumptions,

(a) infinitely stiff plates, and

(b) all rivets strained beyond yield point,

leads to equal stress in all rivets, but the reasons are different.

Accepting this uniform distribution of stress, a joint efficiency can be defined as the ratio of the average stress on the gross area of either plate to the highest average stress on a nett section in the joint. The arrangement of fasteners which has the highest efficiency is considered to be the best. To illustrate this, consider the alternative arrangements in Figs. 13.11(a) and (b). The load transferred across sections 1 and 2 is 200 kN in each case and the load carried by each rivet is 20 kN. Hence, the load transferred across section 3 is 160 kN in Fig. 13.11(a) and 140 kN in Fig. 13.11(b). The joint efficiency is found by calculating the average stress at sections 1, 2 and 3 in each case. This is done in the following table.

Section	Load (kN)	Area (mm^2)	Average stress (MPa)
Fig. 13.11 (a)			
1	200	3 600	55.5
2	200	3 072	65.2
3	160	2 808	57.0
		Efficiency	$= \frac{55.5}{65.2} = 0.852$
Fig. 13.11 (b)			
1	200	3 600	55.5
2	200	2 808	71.2
3	140	3 072	45.6
		Efficiency	$= \frac{55.5}{71.2} = 0.780$

On this basis, Fig. 13.11(a) is seen to be the better arrangement.

If the load is applied eccentrically, a more elaborate analysis of the load sharing is made, and the notion of joint efficiency is not relevant. For example, consider a bracket fastened to a column by means of four rivets loaded in shear, as sketched in Fig. 13.12(a). The load P is applied eccentrically, since its line of action does not pass through G, the centre of area of the sections of the four rivets. It can be replaced by an equal concentric load and a couple M equal to Pe, as shown in Fig. 13.12(b), and the concentric load P and the couple M can be dealt with separately. Thus, P is shared equally among the four rivets and causes a vertical shearing load $P/4$ in each one. The couple M also causes a shearing load in each rivet, the direction being at right angles to a line joining the centre of the rivet to G. To calculate the magnitude of this force for each

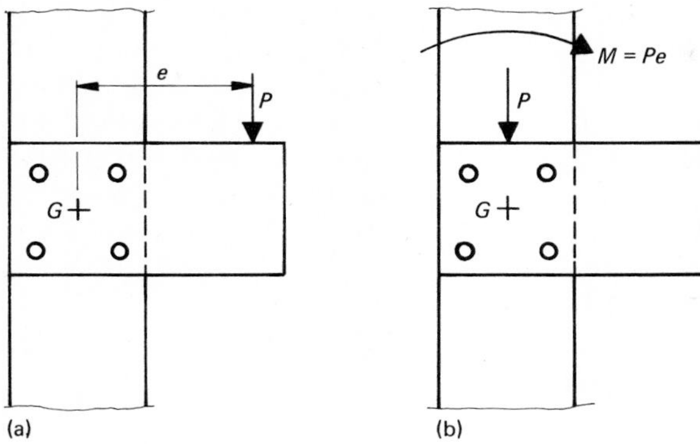

Fig. 13.12. *Bracket supporting eccentric vertical load on column.*

rivet, it is assumed that each one is proportional to the distance of the centre of the rivet from G. This is equivalent to assuming elastic deformation of the rivets in shear and infinitely rigid plates. With these assumptions, formulas can be derived for the load in any fastener.

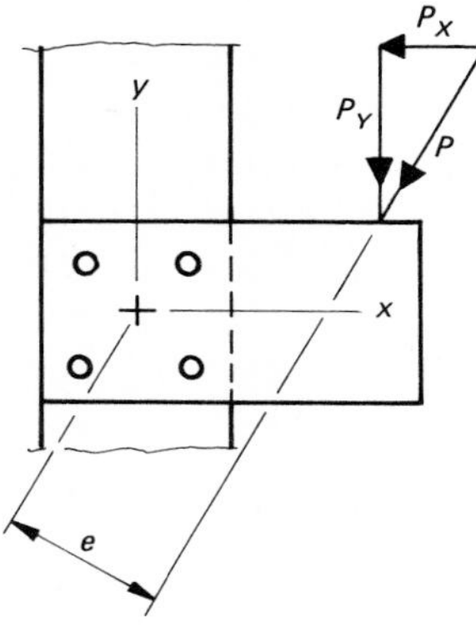

Fig. 13.13. *Bracket supporting eccentric inclined load on column.*

In a general case, with the load P inclined to the vertical, the load can be resolved into two components P_X and P_Y as in Fig. 13.13. Let R_1 be the load applied to any fastener by the load P applied concentrically; R_2 be the load applied to any fastener by the moment, Pe; x, y be the coordinates of a fastener with G as origin; and A be the cross-sectional area of one fastener.

Using the above assumptions, it can be shown that

$$R_{1X} = P_X \frac{A}{\Sigma A} \qquad R_{1Y} = P_Y \frac{A}{\Sigma A}$$

$$R_{2X} = \frac{MyA}{\Sigma A(x^2 + y^2)} \qquad R_{2Y} = \frac{MxA}{\Sigma A(x^2 + y^2)} \tag{13.8}$$

The suffixes X and Y indicate the components of the force in each case.

The total load on any fastener is the vector sum of these four loads. For n fasteners, all of the same size,

$$R_{1X} = \frac{P_X}{n} \qquad R_{1Y} = \frac{P_Y}{n}$$

$$R_{2X} = \frac{My}{\Sigma(x^2 + y^2)} \qquad R_{2Y} = \frac{Mx}{\Sigma(x^2 + y^2)} \tag{13.9}$$

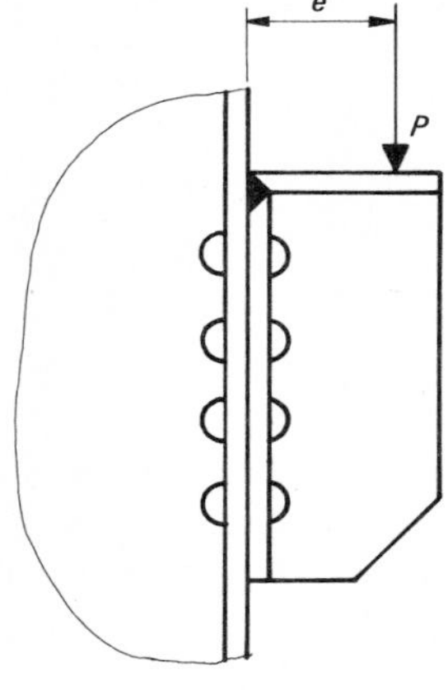

Brackets are often used to support loads on the faces of columns as sketched in the margin. At the face of the column the load P causes a shear equal to P and a moment Pe which are to be resisted by the rivets or bolts which secure the bracket to the column. The shear can be taken as being uniformly shared among the fasteners. The moment, however, causes unequal tensions in the upper fasteners and the resisting couple at the face of the column comprises the resultant of these tensions and an equal and opposite force which is the resultant of the bearing stress at the interface. An approximate analysis of the stresses in such a connection can be made as follows.

It is assumed that the effects of the load and the initial tensions in the bolts can be computed separately and the combined effect determined as described in Section 13.3.2.

The rivet or bolt material in tension is replaced by a vertical strip of equal area and width a. As sketched in the margin,

$$a = \frac{2A}{p}$$

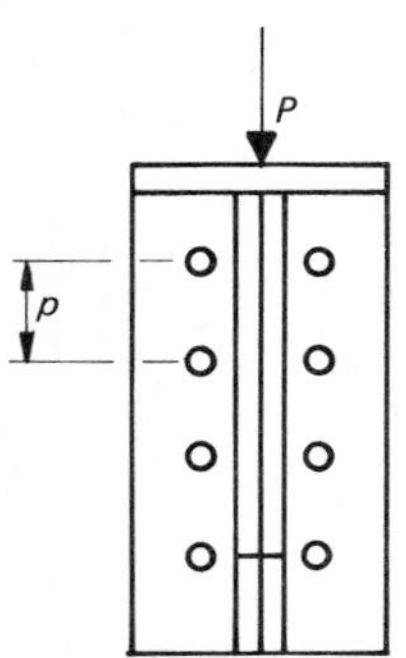

where A is the area of cross-section of one rivet or bolt, and p is the pitch of fasteners. The width of the contact area which provides the compression force in the resisting couple is b and this must be estimated, taking into account the stiffness of the bracket. In our example it would be greater than the total thickness of the projecting legs, but less than the full width of the bracket. A reasonable estimate would be twice the thickness of the projecting legs. The effective cross-section at the interface between the bracket and the column is now as sketched and it is assumed that the simple theory of bending is valid.

The neutral axis passes through the centroid, so that

$$(bkd)\frac{kd}{2} = a(d - kd)\frac{d - kd}{2}$$

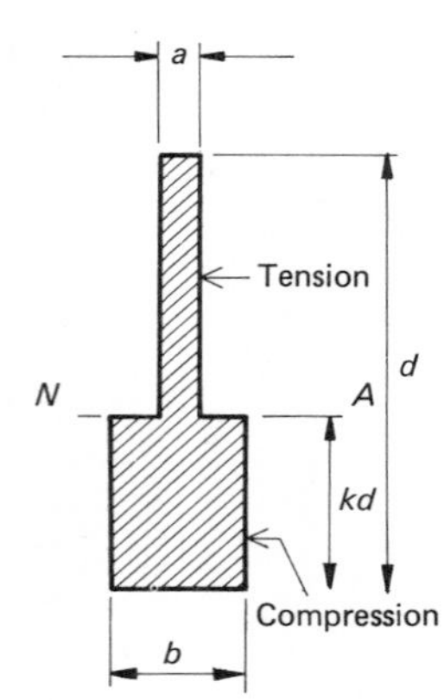

whence

$$k = \frac{1}{1 + \sqrt{b/a}}$$

It is worth noting that k is not very sensitive to inaccurate estimates of b/a. If $b/a = 1, k = 0.5$ and a change of 50 % in b/a, making it 1.5, causes k to fall to 0.45.

The second moment of area of the effective section is

$$I_{NA} = \frac{1}{3}(1 - k)^2\, ad^3$$

so that the maximum tensile stress is given by

$$f_t = \frac{3\,Pe}{(1 - k)ad^2}$$

The estimated maximum bolt tension is then given by

$$P_A = f_t\,A = \frac{3\,e\,A}{(1 - k)ad^2}P$$

This load can now be superimposed on the initial tension, P_I, with due allowance for the relative stiffness of the fastener and the plates.

13.3.4. Stresses in welds

A distinction is made between butt welds and fillet welds. As can be seen in Fig. 13.14 a butt weld is used to join two plates so that they form a continuous plate. The fillet weld is placed in a corner where two parts join. The plates may be at an angle to each other or they may be parallel, in which case the corner containing the fillet is formed by the edge of one plate and the face of the other. Such a joint might be used to transfer a tension T, as shown in Figs 13.14(c). There, the fillet welds are classified in relation to the direction of the load. Those in the left-hand sketch in Fig. 13.14(c) are end-fillet welds and those in the other view in Fig. 13.14(c) are side-fillet welds.

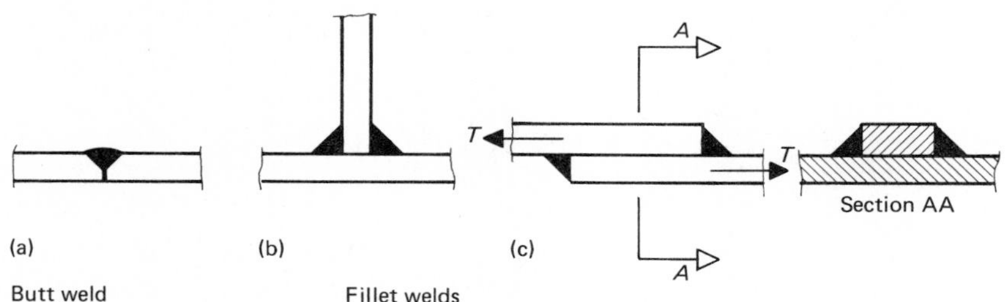

Fig. 13.14. *Classification of welds: (a) butt weld; (b) fillet welds; (c) end-fillet welds and side-fillet welds.*

The size, s, and throat thickness, t, of a fillet weld are defined in the adjacent marginal sketch. Note that a concave weld profile is bad practice and that the size in that case is determined by t rather than by the actual leg length, defined on the adjacent sketch.

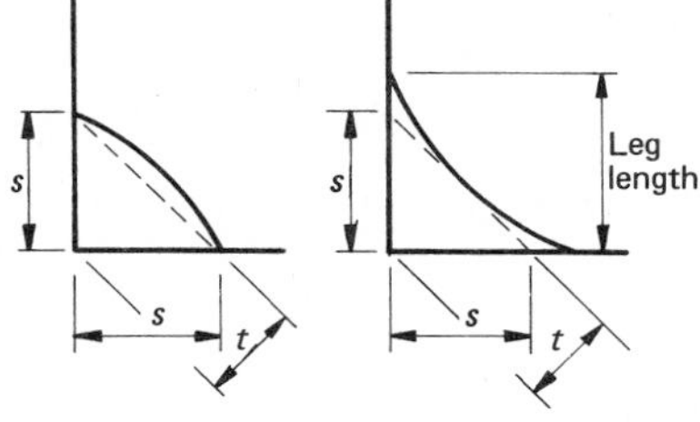

When welds are shown on engineering drawings, a standard scheme of symbols is used to describe the type, size and position of the welds. A description of some of these symbols is given in Appendix A7.

A well-constructed butt weld may be regarded as a continuation of the plates it joins. Stresses in the weld and the adjacent plates will be approximately equal. The weld acts as a stress raiser because of the discontinuity in shape due to the reinforcement. This stress-raising effect is ignored in design for static loads and it is assumed that the plates are continuous. The effect of the irregularities under fatigue conditions is described in Chapter 7.

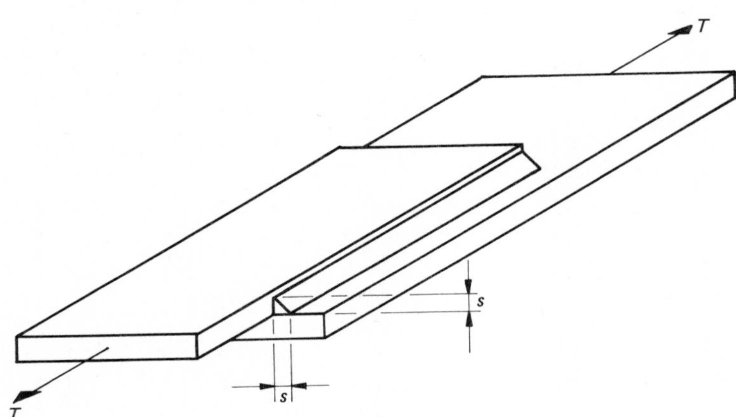

Fig. 13.15. *Dimensions of side-fillet weld. There are two side-fillet welds, each of length $\frac{1}{2}L$.*

With regard to side-fillet welds, it is assumed that failure occurs by shearing on the throat of the weld and that the shear stress is uniformly distributed. In Fig. 13.15, the total length of side-fillet weld is L and the size of the weld is s. The area of surface subject to shear stress at the throat is $(s/\sqrt{2}) \times L = tL$, so that, under a load T, the shear stress is given by

$$f_q = \frac{\sqrt{2}\,T}{sL} = \frac{T}{tL} \qquad (13.10)$$

The stress in an end-fillet weld cannot be described so simply. However, experience is that an end-fillet weld of given length can transfer a bigger load than a side-fillet weld of the same length and size. Consequently, Equation (13.10) provides a conservative basis for design of end-fillet welds.

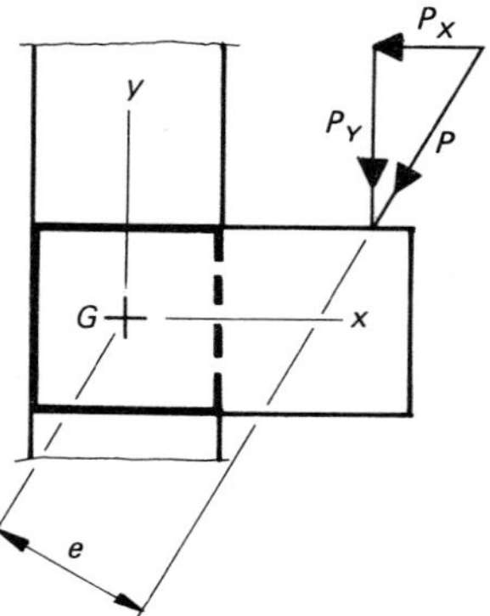

Fig. 13.16. *Welded bracket to support eccentric load on column. All welds are side-fillet welds.*

Eccentrically loaded welds may be analysed by adapting Equation (13.8). For example, in Fig. 13.16, a bracket similar to that in Fig. 13.13 is shown. However, in Fig. 13.16, the bracket is fastened to the column by means of a fillet weld all round. In this case, we have a succession of elementary lengths dl of the weld instead of a set of rivets or bolts and each element of the weld has area $dl \times (s/\sqrt{2})$, or $t\,dl$. The force on each element may be calculated from Equation (13.8):

$$R_{1X} = P_X \frac{t\,dl}{\int t\,dl} = P_X \frac{dl}{\int dl}$$

Hence, the component of force per unit area of the throat equals

$$\frac{R_{1X}}{t\,dl} = \frac{P_X}{tL}$$

where $L = \int dl$ = total length of the weld.

In similar fashion,

$$\frac{R_{1Y}}{t\,dl} = \frac{P_Y}{tL}$$

$$\frac{R_{2X}}{t\,dl} = \frac{My}{\int (x^2 + y^2) t\,dl} = \frac{My}{J}$$

and

$$\frac{R_{2Y}}{t\,dl} = \frac{Mx}{\int (x^2 + y^2) t\,dl} = \frac{Mx}{J}$$

where $$J = \int (x^2 + y^2) t \, dl$$
= polar moment of inertia of throat area of the weld.

By vector addition of these four, the resultant force per unit area of the throat can be found. The maximum value will be the critical estimated stress due to the eccentric load.

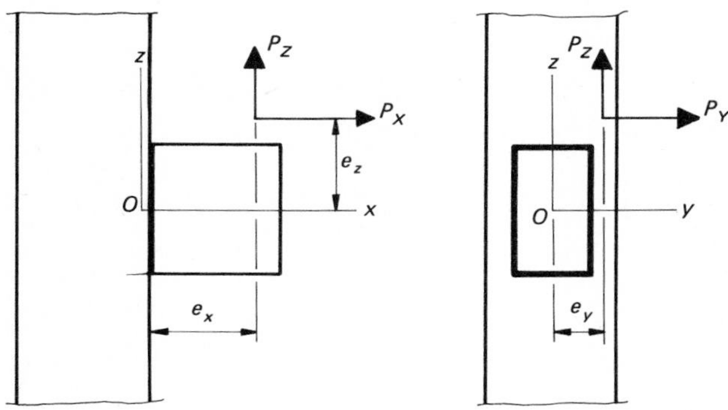

Fig. 13.17. *Welded bracket for eccentric load. Side-fillet welds and end-fillet welds.*

In a more general case, P has three components, P_X, P_Y and P_Z as shown in Fig. 13.17. The fillet weld which connects the bracket to the column is shown by the heavy lines. The centre of area of the throat of the weld is taken as origin of coordinates, O, and the point of application of P is at (e_x, e_y, e_z). The three components, P_X, P_Y and P_Z, can be replaced by equal forces through the origin and three couples M_X, M_Y and M_Z about the axes OX, OY and OZ respectively. These are given by

$$M_X = P_Y e_z - P_Z e_y$$
$$M_Y = P_Z e_x - P_X e_z$$
$$M_Z = P_X e_y - P_Y e_x$$

The resultant force per unit area of the throat of the weld is then the vector sum of six components. Three of these are due to the centrally applied forces in the equivalent system and three are due to the moments. The first three are given by

$$\frac{R_{1X}}{t\,dl} = \frac{P_X}{tL}$$

$$\frac{R_{1Y}}{t\,dl} = \frac{P_Y}{tL}$$

$$\frac{R_{1Z}}{t\,dl} = \frac{P_Z}{tL}$$

where $t = s/\sqrt{2}$, the throat thickness, and L is the total length of the weld. These forces per unit area act parallel to the axes OX, OY and OZ respectively.

The three components due to M_X, M_Y and M_Z are not uniformly distributed and, for an element of the weld with coordinates $(0, y, z)$ they may be found by means of Equation (13.10) and the simple theory of bending.

$$\frac{R_{2X}}{t\,dl} = \frac{M_Y z}{I_{YY}} + \frac{M_Z y}{I_{ZZ}}$$

$$\frac{R_{2Y}}{t\,dl} = \frac{M_X z}{J}$$

$$\frac{R_{2Z}}{t\,dl} = \frac{M_X y}{J}$$

They are parallel to OX, OY and OZ respectively.

Here $I_{ZZ} = \int y^2\,t\,dl$

= second moment of the throat area of the weld about OY

$I_{YY} = \int z^2\,t\,dl$

= second moment of the throat area about OZ

$J = \int (y^2 + z^2)\,t\,dl$

$= I_{ZZ} + I_{YY}$

= second moment of throat area about OX, or polar moment.

13.4. DESIGN OF JOINTS

Design of a joint requires decisions to be made about the material to be used, the kind of fasteners and the dimensions and arrangement of the fasteners.

Choice of material for fasteners is, in general, determined by the material used for the parts joined. For example, aluminium and its alloys are used in the construction of aircraft and the superstructures of ships. The fasteners will usually be made of similar material – if they are not, great care must be taken to keep the different metals insulated electrically from each other to prevent electrolytic corrosion.

Mild steel and cast iron are the most widely used materials in conventional metal fabrication for structures and machines. Mild steel can be welded without serious difficulty in most circumstances and suitable electrodes are normally available. Most rivets and bolts are made of mild steel, except that high tensile steel is used for high-strength friction-grip bolts. Special electrodes and techniques may be required to make welded joints in high tensile steel.

Most of this section is about the choice of size and arrangement of fasteners to transfer a known load from one part to another at a joint. We refer explicitly to mild steel rivets and bolts, high-strength friction-grip bolts made of high tensile steel, arc-welded joints in mild steel and joints in timber structures using connectors. Design of rivets and bolts using other metals is basically the same as described here. Welded joints in other metals should not be undertaken without additional careful study of the special problems that have to be solved – for example, the danger of brittle fracture in high-strength steels.

Selection of the kind of fastener depends on such matters as avoidance of bending in the members joined, cost of construction and availability of plant and skilled labour.

With regard to the first of these matters, a single pin must be used if it is necessary to avoid transferring a bending moment at the joint. Otherwise, any of the other kinds of fastener may be used. Cost will vary from place to place, depending on rates of pay for workmen, cost of material and the strength of competition among contractors. Construction which is conventional in any location should be cheapest – contractors and their workmen are familiar with all its aspects and competition is vigorous enough to keep costs down. Availability of plant acts to favour riveting and welding for shop work, and bolting (especially with black bolts) for field connections. To some extent, the validity of this comment depends on the size of the job – a large construction operation or a big plant-installation job would justify temporary welding or riveting facilities. Availability of labour favours bolting, since relatively little skill is called for. Welding has an advantage over riveting owing to its wider use in modern construction in many parts of the world. Decreasing demand for the skill in those places reduces the incentive for men to train as riveters so that workers with this skill may be hard to find.

Assuming the choice of fastener to have been made, design comprises choice of size and arrangement of the fasteners in the joint to transfer load without overstressing the material of the fasteners or of the parts. In mechanical engineering, this is usually done by selecting a factor of safety and determining a safe working stress by dividing the factor into the yield strength of the material (see Chapter 4). Sometimes codes of practice are available in which safe working stresses are specified. For example, many countries have codes of practice for construction of boilers. Codes are much more widely used in design of structures and the designer can find safe working stresses for fasteners in the relevant code.

13.4.1. Design of pins

In the design of a pinned joint, dimensions to be determined are the thickness of the plates which bear on the pin, the spacing of these plates along the pin and the diameter of the pin. When considering the location of the plates, a clearance may be necessary to avoid trapping moisture and dirt and to allow access for inspection and cleaning. It may be necessary also to allow for overturning forces, in which case it will be desirable to keep the plates widely spaced.

Design proceeds by trial. Drawing on his experience and reports of other similar structures, the designer selects suitable dimensions. The shear force and bending moment diagrams are computed and used to find the maximum bearing stress, bending stress and shear stress. Inspection of Fig. 13.5 will show that maximum compression due to bending occurs at the same place as maximum bearing stress. Maximum shear stress occurs at the same cross-section, but at the neutral axis and not at the extreme fibre. Hence, if a combination of stresses is to be taken into account, it will be the combined bearing and bending stresses. Practice varies in this respect. Some designers ignore the superposition of these stresses on the grounds that the compression due to bending and the radial and tangential compression arising from the bearing stress form a triaxial state of stress. This triaxial state causes a smaller maximum shear stress in the material than the larger compressive stress does by itself. Consequently, it is argued that it is more conservative to compare these stresses individually with the appropriate working stress. Otherwise, an equivalent combined stress is computed and compared with a safe equivalent stress, and this is the procedure adopted in BS 449, *The Use of Structural Steel in Building.*[7] In that code, a formula based on the von Mises criterion (Chapter 4) is used to calculate an equivalent stress, f_{eq}, as follows:

$$f_{eq} = \sqrt{f_{bt}^2 + f_b^2 + f_{bt}f_b + 3f_q^2}$$

or

$$f_{eq} = \sqrt{f_{bc}^2 + f_b^2 - f_{bc}f_b + 3f_q^2}$$

where f_{bt}, f_{bc}, f_q and f_b are the numerical values of bending stresses in tension and compression, shear stress and bearing stress occurring at the same point. This equivalent stress must not exceed a specified allowable equivalent stress, which is about 40% higher than the allowable tensile stress in bending.

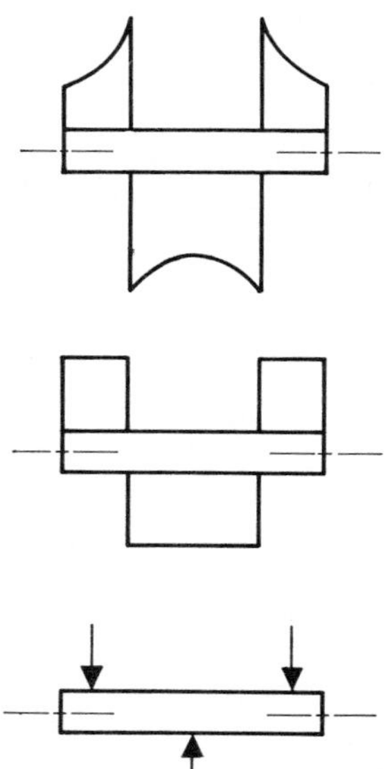

The limiting conditions in the design of a pin are bending, shear and bearing; the eye into which the pin fits must also be checked, but only the pin is considered here. It will be found that the proportions of the pin vary greatly depending on the value of the bearing stress in relation to the bending and shear stresses, and we will discuss this point briefly. If there is no relative movement between the pin and the eye in which it fits, the limiting condition is a compression or crushing failure at one of the two mating surfaces. Because of the three-dimensional nature of the stresses in this case, the allowable stress is greater than the tensile design stress. For example, in one of the codes quoted in Table 4.2, the allowable bearing stress is given as 1.25 times the design stress in tension. However, if there is to be movement between the mating surfaces, the limiting condition will be one of wear, and very much lower pressures, in the order of 10 MPa to 15 MPa, must be stipulated. This bearing pressure is suitable for pins with intermittent movement and boundary lubrication. It will usually be clear from the context which of the two meanings must be given to "bearing pressure".

Some consideration must be given to the pressure distribution on the loaded pin. Because the pin deflects under load, the distribution on a long pin will be like that shown in the first sketch. This is usually simplified to that shown in the second sketch, which would be accurate for a short well-fitting pin, or that shown in the third sketch. The last is an idealisation which simplifies calculations and is sufficiently accurate for most purposes.

The difference in proportions of a pin that results from different values of bearing pressure is illustrated in Examples 13.1(a) and (b). The layout considered is shown in the sketch. The distance L between point loads which are considered to act at the centre of each arm of the fork will be, neglecting any clearances,

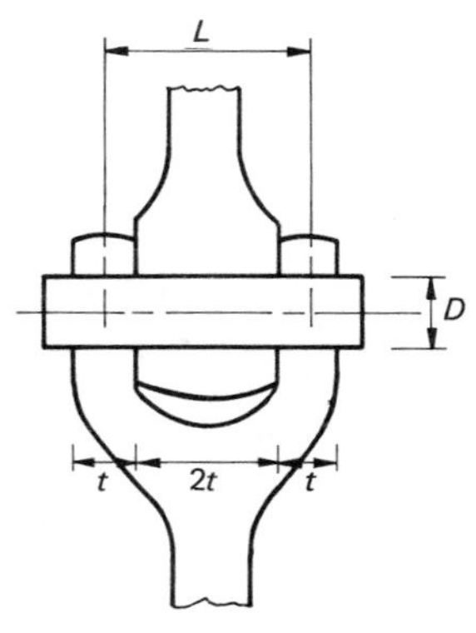

$$L = 3t$$

The equations for bending, bearing and shear stresses for the idealised point load on the pin are:

$$f_{bt} = \frac{PL}{4} \times \frac{32}{\pi D^3} = \frac{8PL}{\pi D^3} \qquad \text{(bending)}$$

$$f_b = \frac{P}{\frac{2}{3}LD} = \frac{3P}{2LD} \qquad \text{(bearing)}$$

$$f_q = \frac{2P}{\pi D^2} \qquad \text{(shear)}$$

In Example 13.1(a) the minimum value of D is 10.8, with a corresponding value of 6.6 for L. This short stubby pin would be typical of the proportions used when there is no significant moment. In Example 13.1(b) the lower bearing stress results in a larger pin, the values of D and L at minimum D being 20 and 47. These will not necessarily be the optimum dimensions; when pin deflection is taken into account (see Problem 13.13) a shorter length and larger diameter may be called for. A proportion commonly used in design is one where $D = 2t$ – that is $D = \frac{2}{3}L$. This point is shown on the $L \sim D$ plot in Example 13.1(b). A comparison between the proportions of the pins worked out in the examples is shown in the sketch.

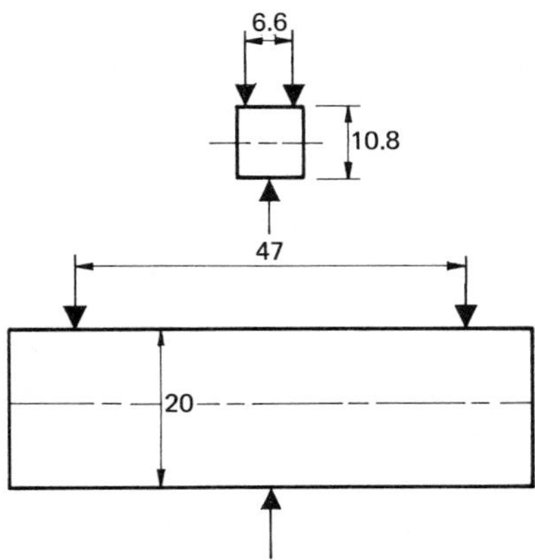

Design of a pin for a bridge bearing is illustrated in Example 13.2. In this calculation, the rules of BS 449 are applied and clearances have been allowed where they were not allowed in Examples 13.1(a) and (b).

13.4.2. Design of rivets and bolts

Joints made with rivets and bolts are most readily designed on the basis of the load-carrying capacity of a single fastener. A joint with a symmetrically applied load is then easily designed by calculating the number of fasteners required. If the load is eccentrically applied, it is necessary to choose a trial number and arrangement of the fasteners. The load on the most heavily loaded fastener can be calculated and a suitable size of fastener determined. More than one trial may be necessary.

The permitted load for a single fastener depends on the kind of fastener, the kind of load, the diameter of the fastener and the thickness of the parts joined. In practice, there is also some variation in calculating the permitted load.

For rivets, black bolts and fitted bolts in tension, the safe load or permitted load is determined by multiplying a safe stress by an area of cross-section of the fastener. We have noted that the safe stress may be specified in a code of practice or the designer may be free to decide what it should be. There is even more variety in the choice of the area to be used. For a rivet, it may be the area of the cross-section of the rivet before driving or the area after driving, which is taken to be the area of the hole. The relevant area for a black bolt or a close-tolerance bolt may be the nett area, what is known as the tensile stress area or the nominal area. The nett area is the area of cross-section at the minor diameter and the nominal area is the full area of cross-section. The stress area is between these two areas (see Appendix A9). It is generally the case that practice which allows the use of a large area also specifies a lower safe stress. Two things are clear – a consistent choice of safe stress and area must be made and the range of safe loads which can be calculated is not as wide as it appears to be at first.

British practice in structural engineering is to use the area of the hole for rivets and the nett area for black bolts and fitted bolts.

In the case of high-strength friction-grip bolts loaded in tension it is usual to specify the safe load in terms of the proof load, which depends on the size of the bolt. This is a direct and sensible way to solve the problem. In BS 3294, the maximum permissible *external* load which may be applied to a bolt is 60% of the proof load if there is no fatigue, 50% if there is.[8] Since the external, or applied, load is specified, there is no need for the designer to take into account the initial load. However, it is of interest to recall that the initial load in bolts designed to this code of practice equals the proof load.

Example 13.1 (a) Sheet 1 of 1

Design a pin for a load of 10 kN. There is no relative movement between pin and eye.

Take $p_{bt} = 140$ MPa, and $p_b = 1.5\, p_{bt}$,
$p_q = 0.5\, p_{bt}$

The equations for bending, bearing and shear stress in the pin are

Bending, $p_{bt} > \frac{8PL}{\pi D^3}$ or $\frac{L}{D^3} < \frac{\pi . 140}{8 . 10^4}$

Bearing, $1.5 p_b > \frac{3P}{2LD}$ or $\frac{1}{LD} < \frac{140}{10^4}$ or $LD > \frac{10^4}{140}$

Shear, $0.5 p_{bt} > \frac{2P}{\pi D^2}$ or $\frac{1}{D^2} < \frac{\pi . 140}{4 . 10^4}$ or $D > \sqrt{\frac{4 . 10^4}{\pi . 140}}$

These inequalities are sketched for various values of L and D.

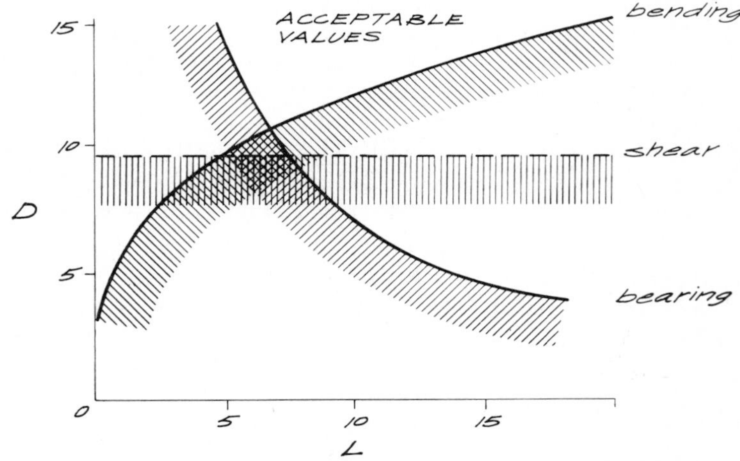

It can be seen that a range of acceptable values can be chosen, with the minimum diameter of 10.8 corresponding to a value of 6.6 mm for L.

Example 13.1 (b) Sheet 1 of 1

Repeat design for 13.1 (a), but put
f_b = 15 MPa

The limiting equations become

Bending, $\frac{L}{D^3} < \frac{\pi . 140}{8 . 10^4}$

Bearing, $\frac{1}{LD} < \frac{15.2}{3 . 10^6}$

Shear, $D > \sqrt{\frac{4 . 10^4}{\pi . 140}}$

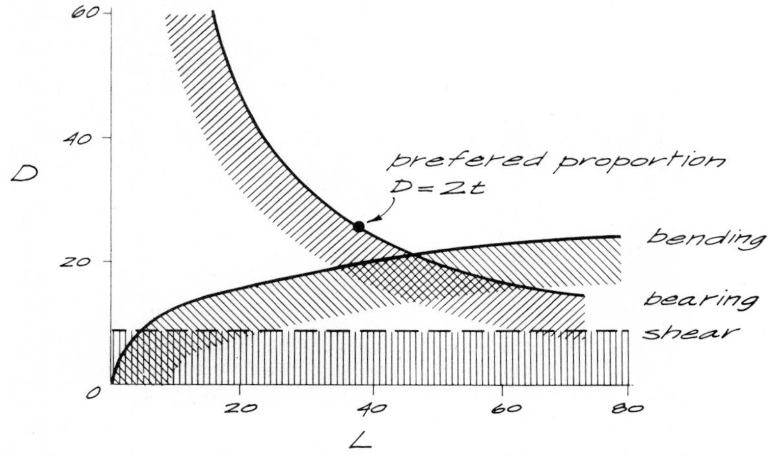

In this case the minimum value of D of 20 occurs at L = 47 mm.

Example 13.2 Sheet 1 of 2

Find the dimensions of a pin for a bridge bearing.

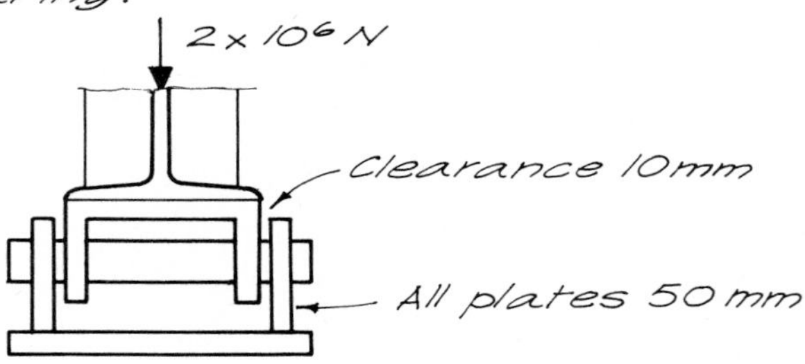

BS 449 Grade 55 steel $p_e = 360$ MPa
$p_q = 180$ MPa

Try 140 mm pin

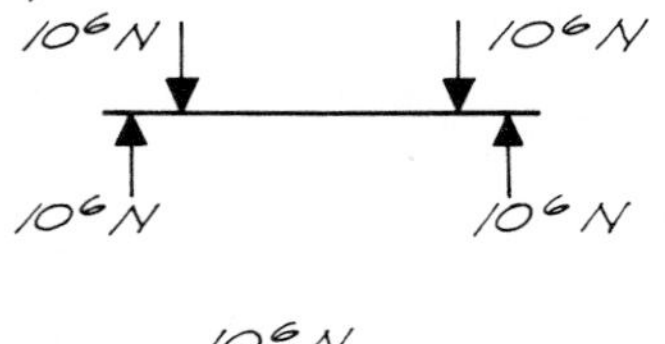

10^6 N × 60 mm
= 6×10^4 Nm

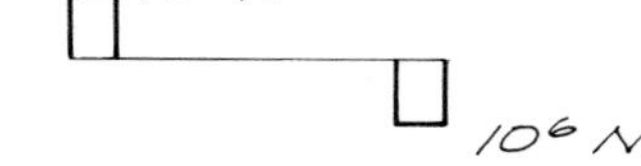

(i) Shear stress:

$$f_q = \frac{VQ}{bI} = \frac{V\left(\frac{1}{2}\pi r^2 \frac{4r}{3\pi}\right)}{2r\frac{\pi}{4}r^4} = \frac{4}{3}\frac{V}{\pi r^2}$$

$$= \frac{4}{3} \times \frac{10^6}{\pi \times 70^2 \times 10^{-6}} \text{ Pa}$$

$$= 86.6 \text{ MPa}$$ ok

(ii) Bearing stress:

$$f_b = \frac{10^6}{140 \times 50 \times 10^{-6}} \text{ Pa} = 143 \text{ MPa}$$

Example 13.2 Sheet 2 of 2

(iii) Bending stress:

$$f_{bt} = f_{bc} = \frac{M\,(r)}{\frac{\pi}{4} r^4} = \frac{4}{\pi}\frac{M}{r^3}$$

$$= \frac{4}{\pi}\,\frac{6 \times 10^4}{70^3 \times 10^{-9}}\ Pa$$

$$= 223\ MPa$$

(iv) Combined bearing and bending:

$$f_e = \sqrt{f_{bt}^2 + f_b^2 + f_{bt} f_b + 3 f_q^2}$$

$$= \sqrt{223^2 + 143^2 + 223 \times 143}$$

$$= 320\ MPa \qquad \text{ok}$$

The final load P_F is given by

$$\frac{P_F}{P_I} = 1 + \frac{1}{1 + (k_P/k_F)} 0.6$$

and

$$\frac{P_F}{P_I} = 1 + \frac{1}{1 + (k_P/k_F)} 0.5$$

as the case may be. With k_P/k_F about 10 for a joint without a gasket, the final tension is from 5% to 6% greater than the proof load.

The safe load computed for a rivet, black bolt or close tolerance bolt in tension must be compared with the final load P_F. As shown in Fig. 13.7, P_F depends on the initial tension, P_I, and the ratio of the stiffnesses of the plates and the fastener. Neither of these is readily determined. If there is no gasket, k_P/k_F can be assumed to be about 10. Approximate values for a few combinations of plate and packing as well as an approximate method of estimating k_P have been given in Section 13.3.

With regard to the initial tension, consider a rivet first. The temperature of the rivet falls from θ_1 to θ_2 after driving and the coefficient of thermal expansion, α, and the modulus of elasticity, E, both vary during cooling. For an interval $\Delta\theta$ of the total temperature drop, the unrestrained change in length would be $\alpha\Delta\theta L$, L being the length of the rivet between the rivet heads, or the grip. However, the parts joined restrain the contraction and the tension in the rivet increases by ΔP. The corresponding increase in length is $\Delta PL/AE$ where A is the gross area of cross-section of the driven rivet. Hence, the nett reduction in length during this interval of cooling is

$$\alpha\Delta\theta L - \frac{\Delta PL}{AE}$$

This equals the change in total thickness of the parts under the action of a change in clamping force equal to ΔP and this is $\Delta P/k_p$. Hence

$$\frac{\Delta P}{k_P} = \alpha\Delta\theta L - \frac{\Delta PL}{AE}$$

$$\frac{\Delta P}{A} = \frac{\alpha}{A/(L\,k_P) + 1/E}\Delta\theta = \frac{E_2\,\alpha}{k_F/k_P + E_2/E}\Delta\theta$$

where E_2 is the modulus of elasticity of the cooled rivet. Consequently

$$\int_{\theta_1}^{\theta_2} \frac{dP}{A} = \frac{P_I}{A} = \int_{\theta_1}^{\theta_2} \frac{E_2\,\alpha}{k_F/k_P + E_2/E}\,d\theta \qquad (13.11)$$

Note that it is assumed that the parts joined are at temperature θ_2 throughout the cooling of the rivet and that all stresses are within the elastic range.

To estimate this initial stress in a typical joint, let $k_P/k_F = 10$, $\theta_1 = 250$°C and $\theta_2 = 20$°C and suppose that α and E vary with temperature as shown in Fig. 13.18. The integral on the right of Equation (13.11) can be evaluated numerically and the stress in the cooled rivet is found to be 276 MPa – that is, it is approximately the yield stress.

If an external tensile load is applied to the joint, the strain in the rivet will increase while the stress remains unchanged, and the clamping force will be

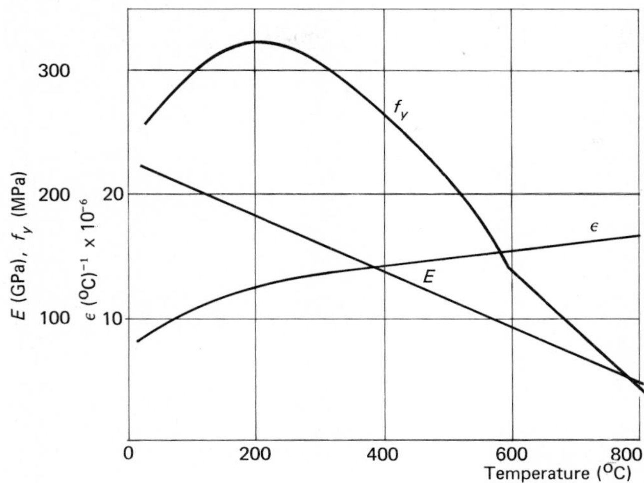

Fig. 13.18. *Modulus of elasticity, yield strength and coefficient of thermal expansion for mild steel at elevated temperatures. (Adapted from O. W. Blodgett,* Design of Weldments[14], *by permission of the James F. Lincoln Arc Welding Foundation.)*

reduced. When the external tension reaches f_yA, the clamping force will be zero, f_y being the yield stress of the rivet.

Normal practice is to limit the external tension to a permitted stress, less than yield stress, times the area of cross-section. The foregoing reasoning shows that such a load will not increase the stress in the rivet and it ensures some residual clamping force.

In the case of a black bolt or a fitted bolt also it is possible that the initial stress in the bolt will approach yield stress. By specifying the torque to be applied in tightening the nuts, the designer can ensure that the initial stress he has used in his calculations does in fact occur in the joint (see Equations (13.13) and (13.14)).

Application of these ideas to the design of a bolted joint with a gasket is illustrated in Example 13.3. The calculation follows mechanical engineering practice and is subject to the condition that there must be a residual clamping force to keep the joint sealed under load.

It will be seen that, for computation of the total load applied to the flange bolts, the internal pressure in the pressure vessel has been assumed to act on a circle with diameter equal to the inside diameter of the vessel. Practice varies in this respect, some codes requiring a larger diameter to be used. This requirement is based on the assumption of a radial pressure gradient through the joint, leading to an additional pressure force acting to open the joint.

When a shearing load is applied to a rivet, black bolt or fitted bolt, the safe load may depend on the shear stress in the fastener or the bearing stress in the fastener and the parts joined. It is therefore necessary to determine safe loads using both criteria and to use the smaller of the two to design the joint. Time can be saved in design by preparing a table of safe loads for the several kinds of fasteners of various diameters and thicknesses of parts joined. Such a table must be based on the design rules in use in the designer's office.

To illustrate the arrangement of such a table, we compute here safe loads for 12 mm rivets according to the rules in BS 449.[7] The steel in the rivets is taken to have a yield stress of 250 MPa and power-driven shop rivets are considered.

This design specification says that the area to be used is the area of the rivet hole, implying that the completed rivet fills the hole. In this example, the relevant

Example 13.3 Sheet 1 of 1

Determine the number and size of bolts for flanged joint in pressure vessel.

Data:

Vessel: I.D. 750 mm Pressure 0.5 MPa

Joint: Flanges 20 mm thick

Gasket 0.5 mm thick

Estimated $k_p = 3000$ kN/mm at each bolt

Residual clamping force required $\not< 30$ kN

Bolts: $E = 200$ GPa. Yield at 250 MPa

$p_t = 130$ MPa on nett area

ISO metric thread, coarse pitch

Try 16 mm diameter bolts

Nett area $= 144\ mm^2$

Maximum safe load $= 1.30 \times 1.44 \times 10^4$ N

$= 18.7$ kN

Estimate initial load 40% of yield

$P_I = 0.4 \times 2.5 \times 1.44 \times 10^4$ N

$= 14.4$ kN

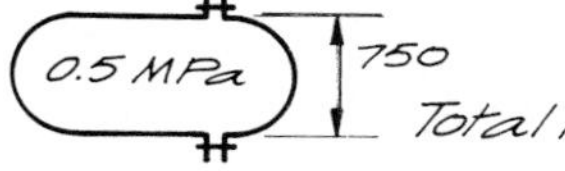

Total load $= 0.5 \times \frac{\pi}{4} \times 7.5 \times 7.5 \times 10^4$ N

$= 221$ kN

Try 15 bolts

Load applied to each bolt $= \frac{221}{15} = 14.7\ kN = P_A$

Gross area of one bolt $= 201\ mm^2$

$$k_F = \frac{AE}{L} = \frac{201 \times 200}{40.5} = 992\ kN/mm$$

$$\frac{k_P}{k_F} = 3.02$$

$$\frac{P_F}{P_I} = 1 + \frac{1}{1 + 3.02}\ \frac{14.7}{14.4} = 1.254$$

$P_F = 18.1\ kN < 18.7\ kN$ ok

20 20 0.5

Residual clamping force per bolt $= P_F - P_A$

$= 18.1 - 14.7$

$= 3.4$ kN

Total residual clamping force $= 15 \times 3.4$

$= 48.0$ kN

> 30 kN ok

15 × 16 mm ISO metric thread, coarse pitch.

diameter is 14 mm. The allowable stress in shear is 110 MPa, so that the safe load which can be transferred is

$$0.110 \times \frac{\pi}{4} \times 14^2 = 16.9 \text{ kN, if the rivet is in single shear, and}$$

$$2 \times 0.110 \times \frac{\pi}{4} \times 14^2 = 33.8 \text{ kN, if the rivet is in double shear.}$$

The safe load in bearing depends on the plate thickness and whether the fastener is in single shear or double shear. In BS 449, allowance is made for the more uniform distribution of bearing stress in the case of double shear by specifying a higher allowable stress – 315 MPa for the kind of rivets in this example. The allowable bearing stress for these rivets in single shear is less by 20% – i.e. 250 MPa.

Thus, the safe load in bearing for this 12 mm diameter rivet and plate thickness, t, is

$315 \times 14 \times t$, if the rivet is in double shear, and
$250 \times 14 \times t$, if the rivet is in single shear.

These are evaluated for a suitable range of plate thickness and all the information is assembled as follows:

Rivet diameter (mm)	Safe load in shear (kN)	Safe load in bearing for plate thickness, t (kN)					
		$t = 3$ mm	$t = 5$ mm	$t = 8$ mm	$t = 10$ mm	$t = 12$ mm	$t = 16$ mm
12	Single shear, 16.9	10.5	17.5	28.0	35.0	42.0	56.0
	Double shear, 33.8	13.2	22.0	35.3	44.1	52.9	70.5

Consider first the safe loads of a fastener in single shear. It will be seen that, for 3-mm plate, bearing limits the load which can be transferred. For all thicker plates, however, the safe load is limited to 16.9 kN by the shear stress in the rivet.

Use of the table for a rivet in double shear is a little more complicated. This is because the bearing stress at the contact with the outer plates in such a joint is approximately the same as for a fastener in single shear. Only at the centre plate can the higher stress be used. For example, the maximum load which can be transferred by the rivet in Fig. 13.19 is limited by shear in the rivet to 33.8 kN. It is limited to 22.0 kN by bearing at the contact with the centre plate and to 17.5 kN at the contact with each of the outer plates. The safe load is therefore the least of 33.8 kN, 22.0 kN and $2 \times 17.5 = 35.0$ kN.

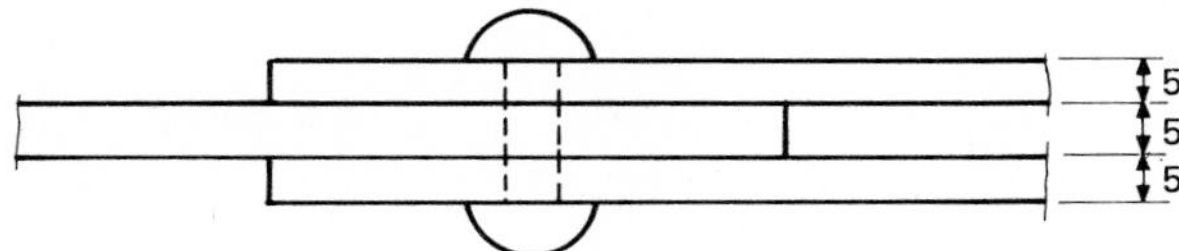

Fig. 13.19. *Single 12 mm diameter rivet in double shear, with 5 mm plates.*

It will be noted that no account has been taken of the tension induced in the fastener by the fabrication procedure. Although the fastener is subject to combined stress, the initial tension is ignored in design and the safe load is based on shear and bearing only.

This study of rivets, black bolts and fitted bolts subject to shearing loads is rounded off with the calculations in Example 13.4. There, rivets are designed for a bracket to support an eccentric load.

Example 13.4 Sheet 1 of 1

Determine the number and size of rivets for an eccentrically loaded riveted joint.

Data:

Eccentrically loaded plate — thickness 10 mm.

Design spec. BS 449

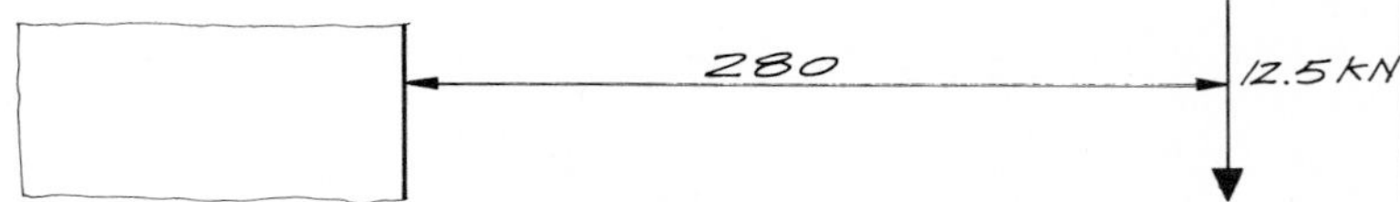

Try 8 rivets thus:

$e = 280 + 40 + 25 = 345$ mm

$M = Pe = 12.5 \times 345 = 4.31$ kNm

$\Sigma(x^2 + y^2) = 4(30^2 + 40^2) + 2 \times 30^2 + 2 \times 40^2$

$= 15000 \text{ mm}^2$

Coords of corner rivets : $x = 40,\ y = 30$

$R_{1y} = \dfrac{12.5}{8} = 1.56$ kN

$R_{2y} = \dfrac{(4.31 \times 10^3) \times 40}{15 \times 10^3} = 11.5$ kN

$R_{2x} = \dfrac{(4.31 \times 10^3) \times 30}{15 \times 10^3} = 8.62$ kN

Resultant load on corner rivet $= 15.7$ kN

$\sqrt{8.62^2 + 13.06^2} = 15.7$ kN

$1.56 + 11.5 = 13.06$

Say plate thickness for bracket = 10 mm

Safe load for one 12 mm rivet, yield 250 MPa = 16.9 kN

12 mm rivets, arranged as sketched

Yield stress, rivet steel, 250 MPa

Bracket plate thickness, 10 mm.

When high-strength friction-grip bolts are subject to shearing loads, the load which can be transferred depends on the friction between the parts joined. If no external tension is applied, the force clamping the parts together is P_I and, according to BS 3294[8],

$$P_I = \text{specified proof load.}$$

At each effective interface, the friction available to resist slipping is μP_I where μ is a coefficient of friction or slip factor. If there are j effective interfaces the total resistance is $j\mu P_I$. The safe load is determined by dividing this force by a load factor F which is greater than one. That is, the safe load which can be transferred by one bolt is $(j\mu P_I)/F$.

Typical values for μ, P_I and F are given in BS 3294: 1960.[8] The slip factor may be taken to be 0.45 provided the steel surfaces in contact are "free of paint or any other applied finish, oil, dirt, loose rust, loose scale, burrs and other defects which would prevent solid seating of the parts or would interfere with the development of friction between them". Alternatively, the slip factor to be used in design may be determined by test – the force required to cause slip in a test joint is measured. It should be noted that gaskets are not permitted in structural joints if friction is to be relied on.

We have noted that P_I is the specified proof load, provided there is no external tensile load. The load factor is given in BS 3294 as 1.4 for normal loads. For wind loads, a reduced factor of 1.2 is permitted.

The number of effective interfaces is easily determined from the layout of the joint. At each one, surfaces of two of the parts joined are in contact and the shearing load tends to make the parts slide relative to each other at this contact. In Fig. 13.20(a), there is one effective interface and the force to be transferred, T, is seen to be equal to $(\mu P_I)/F$. There are four effective interfaces in the joint in Fig. 13.20(b) and the exploded sketch shows that T in this case equals $(4\mu P_I)/F$.

Application of these ideas to design of a joint is illustrated in Example 13.5.

Fig. 13.20. *Effective interfaces with HSFG bolts:*
(a) one effective interface;
(b) four effective interfaces.

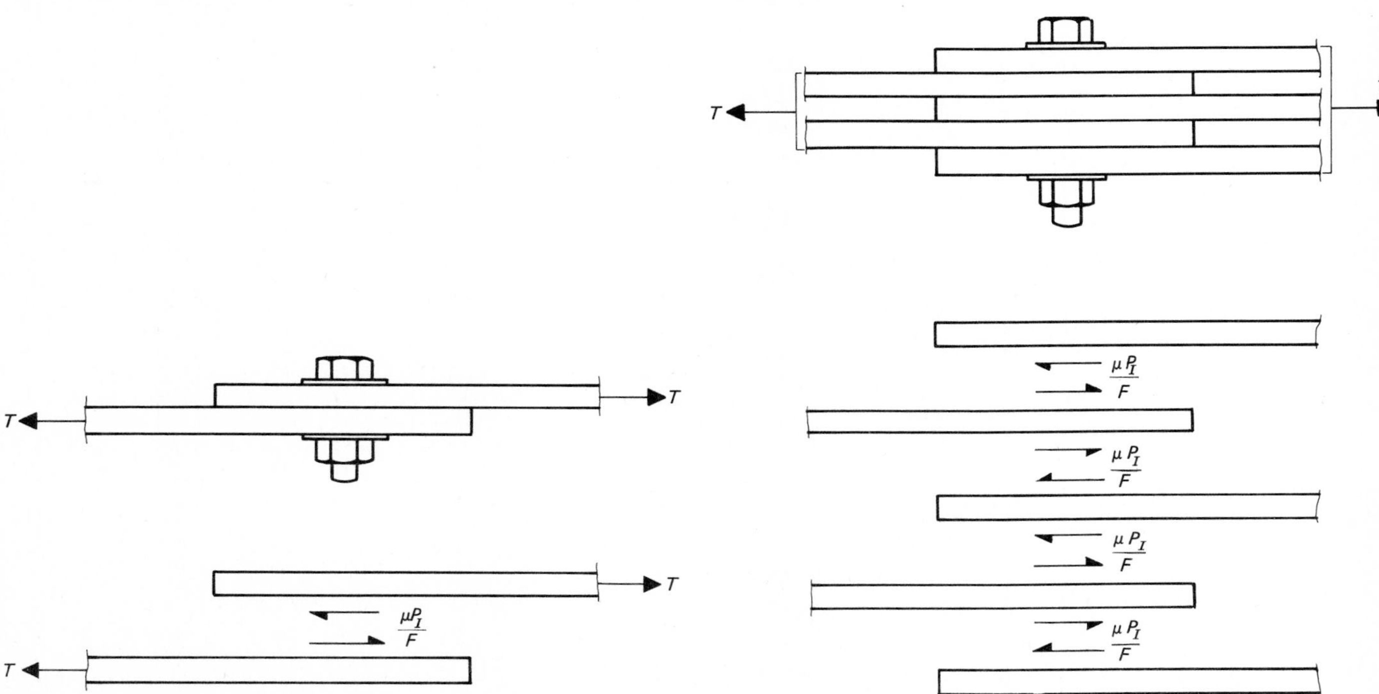

Example 13.5 Sheet 1 of 1

Find the number and size of bolts for a joint with high-strength friction grip bolts.

Design spec. BS 3294

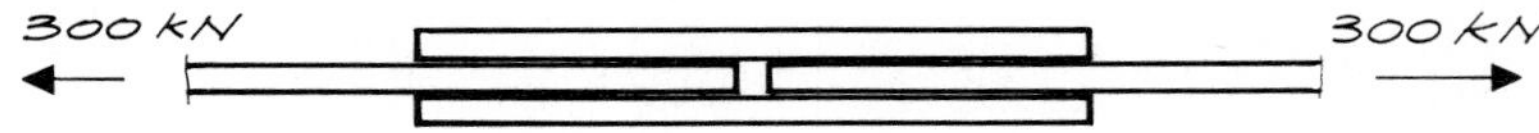

Try 3/4 in bolts - nominal diameter 19.1 mm
proof load 126 kN

No. of effective interfaces = 2

Slip factor = 0.45

Load factor = 1.4

Safe load per bolt = $\frac{2 \times 0.45 \times 126}{1.4}$

= 81 kN

No. of bolts = $\frac{300}{81}$

= 4

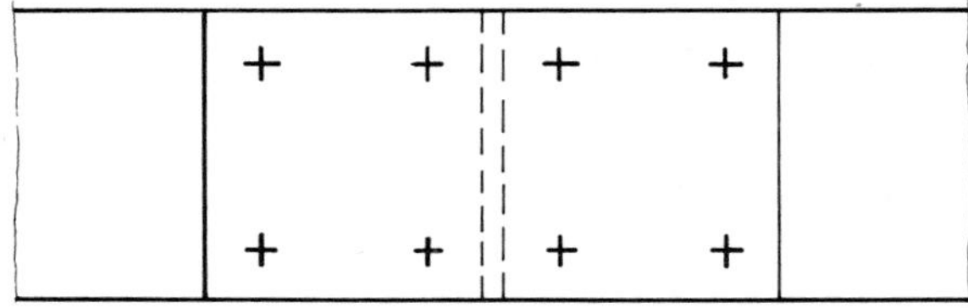

4 No. 3/4 in bolts each side

Note that use of BS 3294 requires use of Imperial dimension. Loads have been converted to SI Units.

The last point to be considered in the design of rivets and bolts is the effect of a combination of tensile and shearing loads. In structural connections, it is usually possible to assign the two loads to different groups of fasteners and to avoid the problem of combined stress. For this reason, structural codes do not give much guidance. A typical problem requires design of a beam-to-column connection able to transfer the end shear and bending moment of the beam. Such a joint is sketched in Fig. 13.21 and it is assumed that the bending moment is transferred solely through the flange connections and that the web connection transfers the shear. It can be seen that, on this basis, every loaded rivet can be designed for either a tensile or a shearing load and that none needs to be considered subject to both.

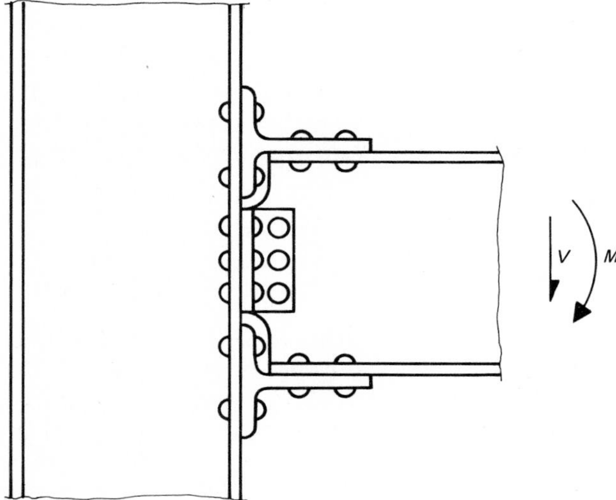

Fig. 13.21. *Riveted beam-to-column connection, capable of transferring bending moment as well as shear.*

The mechanical designer cannot always solve the problem in this fashion. The effects of the combined load on each fastener must be determined according to the principles described in Chapter 3. These stresses are then compared with yield strength divided by a factor of safety.

If an external tension is applied to a high-strength friction-grip bolt, one effect is to reduce the clamping force and, consequently, also the friction. A conservative approach to estimating the residual clamping force is required by BS 3294 – the clamping force is reduced by 1.7 times the external tension and this reduced clamping force is used instead of P_I to determine the safe load which may be transferred. That this is conservative may be seen by referring to Equations (13.2) and (13.4). From these, the residual clamping force is given by

$$C_2 = P_F - P_A = P_I - \frac{k_P/k_F}{1 + k_P/k_F} P_A$$

The factor by which the applied load is multiplied is clearly less than one. Note, however, that the factor 1.7 is consistent with a maximum external tension of 0.6 times the proof load – the rule gives a residual clamping force of zero when this maximum tension is applied. (See discussion on high-strength friction-grip bolts early in this section.)

The initial load, P_I, is an important parameter in the design of many joints – for example, when sealing a joint is required, or if loads are to be transferred by friction. The designer will want to be sure that the initial load has some value which he has used in his calculations and it is necessary for the fabricators to ensure that this load is achieved. Since the torque required to turn the nut can be related to the tension in the bolt, a ready means is available for making sure that a bolt is loaded with a specified tension. The required torque may be found directly by tests, using a load cell as described above for high-strength friction-grip bolts. For smaller jobs, such equipment will not be available and the following theoretically derived expression may be used:

$$T = P_I r_t \left\{ \frac{\tan\alpha + \mu_1 \sec\beta}{1 - \mu_1 \tan\alpha \sec\beta} + \frac{r_c}{r_t}\mu_2 \right\} \tag{13.12}$$

where T is the torque required to turn nut; P_I is the tension in bolt; r_t is the mean radius of thread; r_c is the mean collar or nut radius; α is the pitch angle of thread; β is the half the thread angle, measured in a plane normal to the thread; μ_1 is the coefficient of friction between thread of nut and thread of bolt; and μ_2 is the coefficient of friction between nut and washer or plate.

The first term is the torque necessary to advance the nut along the bolt – that is to overcome the component of P_I along the thread and friction between the thread of the nut and that of the bolt. It is easily derived from an analysis of the motion of the nut along the thread as if it were moving on an inclined plane. The second term allows for friction between the nut and the washer or plate the nut slides on as it is turned.

Equation (13.12) can be simplified considerably by noting that $\sec\beta \simeq 1$ and $\mu_1 \tan\alpha \sec\beta \ll 1$. Using these approximations, Equation (13.12) becomes

$$T = P_I r_t \{\tan\alpha + \mu_1 + \frac{r_c}{r_t}\mu_2\} \tag{13.13}$$

The three terms may be recognised as torques: one working against P_I directly, one overcoming thread friction and one overcoming seat friction. Taking representative values of 0.1 for tan α, 0.15 for μ_1 and μ_2 and 1 for r_c/r_t, the approximate formula

$$T = 0.4\ P_I\ r_t \tag{13.14}$$

is obtained. Thus, for a given desired value of P_I, the designer can specify the torque to be applied to the nuts when they are tightened during fabrication.[9]

Experimental results are in reasonable agreement with Equation (13.14) but it should be noted that, with carefully machined thread and washer surfaces that are well lubricated, tensile loads twice as big can be readily obtained for the same torque. On the other hand, if the friction in the thread is high, a torsional shear stress occurs in the bolt. Combined with the axial tensile stress, this may cause the bolt to yield before the computed tension is reached.

If the bolt yields under the action of combined torsional shear and axial tension, the tensile force will be between 50% and 90% of that calculated.[10] Since the lower figure is for a dry thread, it is desirable to specify lubrication of the thread if a tensile stress approaching yield strength is sought. Experimental checking should be considered for an important design.

13.4.3. Design of welds

Design calculations are not usually made for butt welds on the assumption that a butt weld is at least as strong as the plates it joins together. Consequently, if these are strong enough to transfer the applied loads, the butt weld will also be strong enough.

In the design of fillet welds, the load carrying capacity per unit length of weld is calculated by multiplying the throat thickness t by an allowable stress, p_q. The throat thickness is taken to be the size divided by $\sqrt{2}$, so that the allowable load per unit length is 0.7 $p_q s$ where s is the size. To save time in design, a table showing the safe load per unit length for a range of fillet-weld sizes should be prepared. For example, in BS 449, p_q depends on the grade of steel to be used and ranges from 115 MPa to 195 MPa. Safe loads per millimetre at 115 MPa, 160 MPa and 195 MPa are given in the following table.

p_q (MPa)	Safe loads on fillet welds (N/mm) — Weld size (mm)									
	3	4	5	6	8	10	12	16	20	25
115	242	322	402	483	644	805	965	1290	1610	2010
160	336	449	561	673	897	1120	1345	1790	2240	2800
195	409	546	683	819	955	1365	1640	2180	2730	3410

These loads are used in design without regard to the orientation of the weld in relation to the load and their use is illustrated in Example 13.6. The welded connection in this example carries the same load as the riveted one designed in Example 13.4.

Example 13.6 — Sheet 1 of 1

Determine the size of a fillet weld required for a welded joint.

Data:
Eccentrically loaded plate – thickness 10 mm
Design spec. BS 449
Allowable shear stress on throat 160 MPa

Try fillet weld all round, thus:

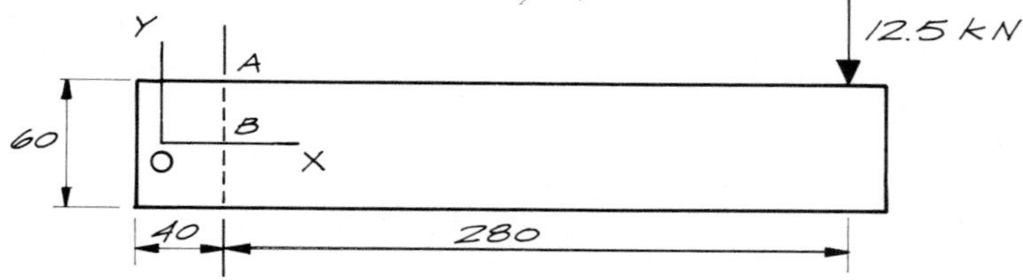

$L = 2 \times (60 + 40) = 200\,mm$

$J: \quad 4\int_0^{20} t(x^2 + 30^2)\,dx = 4t\left[1/3\,x^3 + 900x\right]_0^{20} = 82667t$

$4\int_0^{30} t(20^2 + y^2)\,dy = 4t\left[400y + 1/3\,y^3\right]_0^{30} = \frac{84000t}{166667t}$

$M = 12.5\,kN \times 300\,mm = 3.75\,kNm$

At A $\quad x = 20\,mm \quad y = 30\,mm$

$\frac{R_{1X}}{ds} = 0 \qquad \frac{R_{1Y}}{ds} = \frac{12.5 \times 10^3}{200} = 62.5\,N/mm$

$\frac{R_{2X}}{ds} = \frac{3.75 \times 10^6 \times 30}{1.67 \times 10^5} = 673\,N/mm \qquad \frac{R_{2Y}}{ds} = \frac{3.75 \times 10^6 \times 20}{1.67 \times 10^5} = 449\,N/mm$

673, 845

$62.5 + 449 = 511.5$

Resultant = 845 N/mm
8 mm fillet weld all round.

The above description shows that the quantitative aspects of the design of a welded joint may be treated quite simply. Successful design requires attention to several qualitative aspects as well and these are now discussed.

The effects to be taken into account are the microscopic and macroscopic consequences of heating and cooling steel. The most important of the microscopic or metallurgical changes are those which affect the strength and, in particular, the ductility of the metal. Built-in stresses and distortion are the principal macroscopic effects.

Although ductility does not appear specifically in design calculations, it is an important property that is often singled out for attention in materials specifications. Static strength is important, but it is often the case that the material in and around a weld will be increased in strength rather than weakened. Fatigue performance is also important, but is usually governed more by the stress concentrations that can occur at a weld than by the basic material strength (see Chapter 7). Thus it is the loss of ductility which goes with increasing strength that is likely to be the greatest cause for concern to the designer, who relies on local yielding of a ductile material to even out any high localised stresses.

Figure 13.22 shows a butt weld and it identifies schematically a zone in which molten metal from the electrode and the plates has solidified, a heat-affected zone (HAZ) in which the plates have been heated significantly by the welding and subsequently cooled and the rest of the plate, the material in which has been heated relatively little or not at all.

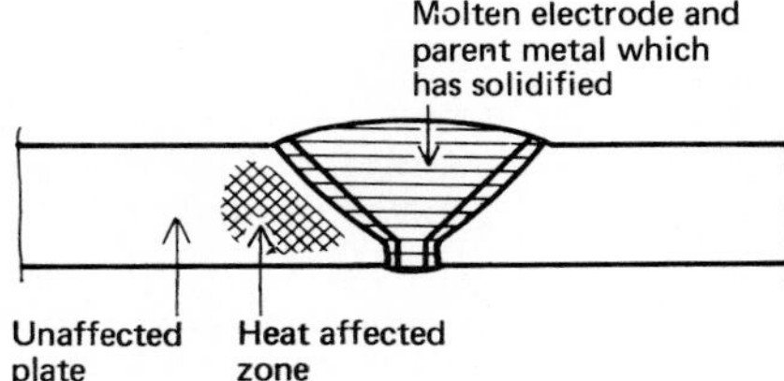

Fig. 13.22. *Butt weld, showing weld metal and heat-affected zone.*

The solidified molten metal has a coarse grain structure. Owing to the high initial temperature, cooling extends over a sufficiently long period for growth of crystals from the solid solution. For low carbon contents the intergranular material is α-iron and the steel is relatively soft, ductile and low in strength. Higher carbon contents yield cementite between the crystals with corresponding increases in hardness and strength and lower ductility. In each case, the properties of the steel reflect the properties of the intergranular material.

The heat affected zone is of particular interest. The properties in this region can be illustrated by relating the temperature profile adjacent to the weld to the iron–iron carbide phase diagram. This is done in Fig. 13.23 for a 0.3% carbon steel.[11] The following extract and Fig. 13.23 are taken from A. L. Phillips (ed.), *Introductory Welding Metallurgy*[11], by permission of the American Welding Society.

Point 1 has been heated in excess of 1300°C. The austenite that forms will be coarse-grained because of the grain growth at this temperature.

Point 2 has been heated to 1000°C and fully austenitized. Grain growth has not occurred; some grain refinement may occur.

Point 3 has been heated to just above the A_3 temperature which is not high enough to completely homogenize the austenite.

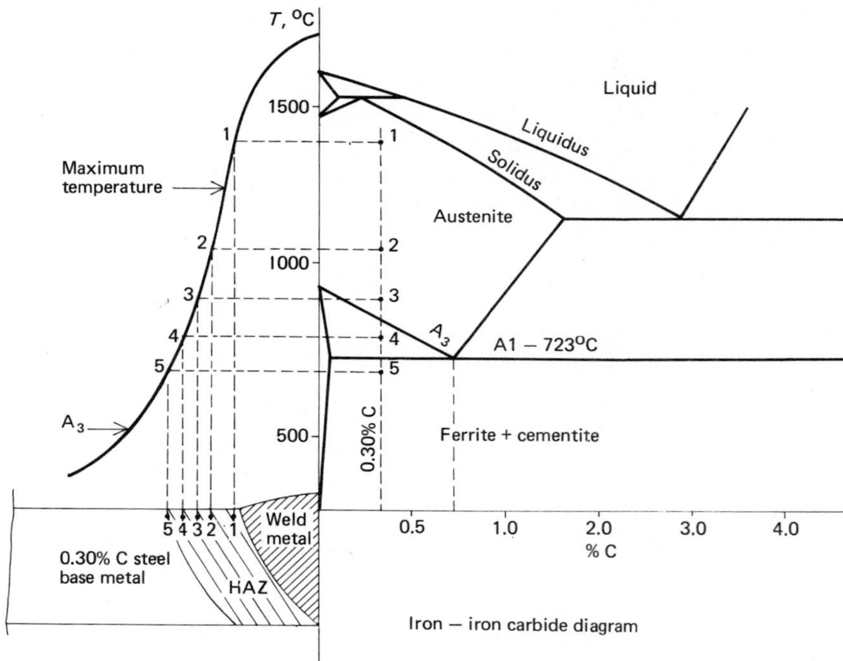

Fig. 13.23. *Iron carbide phase diagram for 0.3% carbon steel. (From A. L. Phillips (ed.),* Introductory Welding Metallurgy[11], *by permission of the American Welding Society.)*

Point 4 this area has been heated to approximately 780°C, which is between the A_1 and the A_3 temperatures. Part of the structure is converted to austenite and the resulting mixture of products during cooling can result in poor notch toughness.

Point 5 this point has been heated to 670°C, which is below the A_1 temperature, and no austenite has formed. Instead, the base metal may be spheroidized and softened.

The actual changes occurring in the HAZ will be strongly affected by alloy content and cooling rate. The latter is influenced by the heat input and the base metal thickness and temperatures. The cooling rate can be increased by using smaller size welding rods (with a corresponding increase in the number of runs), lower welding current and higher welding speeds. It is also increased by more massive base metal and is reduced by preheating the base metal. Design of welded joints must take into account such effects, and preheating may be specified with heavy plates (although it is not usually required for carbon contents below 0.2%). For a more accurate representation of the changes that occur during cooling it is essential to take into account the cooling rates; this can be done by referring to the TTT (time-temperature-transformation) or the CCT (continuous cooling transformation) diagrams for the material.[11, 12, 13]

With regard to the effect of heat on macroscopic properties, Fig. 13.18 shows typical variations of yield stress, modulus of elasticity and coefficient of thermal expansion of mild steel with temperature.[14] These can be used to show how distortion and locked-in stresses can occur in welded structures.

The effect of heat on the yield stress is to lower it, so that a plate under restraint changes shape readily on heating. Subsequent cooling will place the plate under stress if the restraint acts to prevent contraction. For example, Fig. 13.24 shows a bar held by two immovable anchors which can resist a tension or a compression in the bar. As the bar is heated, its longitudinal thermal expansion is prevented and it is subject to a compressive stress. However, this compressive stress cannot exceed the reduced yield stress and the restrained longitudinal thermal expansion will cause increased lateral dimensions by plastic flow. On cooling, longitudinal thermal contraction cannot occur and the bar is placed in tension. Since the yield stress is now increasing there is no plastic flow, and at the end of the cycle of heating and cooling the tensile stress is considerable.

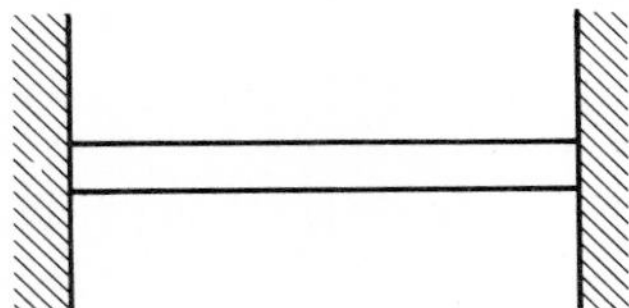

Fig. 13.24. *Bar held rigidly at its ends.*

It is not difficult to envisage a welding operation which would resemble such a cycle. Designers and fabricators need to ensure that the sequence of operations does not lead to insertion of light, easily heated members between heavy rigid sections. If such construction cannot be avoided, stress relief by heating the completed job may be necessary and this can be difficult and costly.

Distortion of welded construction is caused by thermal contraction. For example, contraction of the fillet weld in Fig. 13.25(a) will pull the upright plate over as indicated. A second fillet weld on the other side will have some corrective effect, but at the cost of locked-in stresses. Such undesirable consequences can be reduced by keeping the welding balanced. In the case illustrated in Fig. 13.25(b), with multirun fillet welds, successive runs would be made alternately on one side and the other.

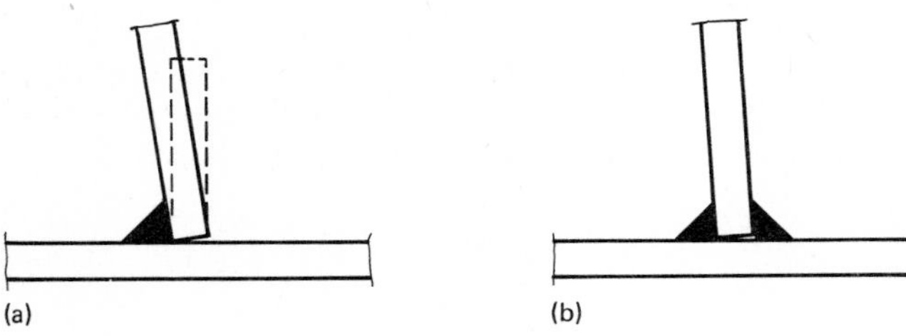

Fig. 13.25. *Distortion of joint arising from contraction of fillet welds.*

In summary, we emphasise that the designer of a welded joint must keep in mind these thermal effects. To avoid those that are adverse, he may have to specify heat treatment and welding procedure and to arrange for competent supervision of the welding. Design of welded joints in heavy members is a complex task and reference should be made to specialised texts for comprehensive treatment.[14, 15]

Example 13.7. Notes for a welding specification

When writing a specification for welded construction, standard specifications can be referred to – for example, BS 1856[16], BS 2642[17], and BS 4360.[18] The following points are significant.

(a) *Parent metal.* Alloying elements are important and difficult welding jobs can be improved by appropriate selection. Conversely, if the choice of steel is dictated by other considerations special techniques may be called for.

Mild steel used in conventional engineering practice is a low carbon steel with no other alloying elements. For special purposes there is a range of steels with varying contents of manganese (see Chapters 6, 20).

(b) *Edge preparation.* Preparation of edges to ensure full penetration needs to be described.

(c) *Cleaning procedures.* There will be clauses dealing with initial cleaning of the plate to remove impurities, and others requiring removal of slag and impurities from deposited metal between runs.

(d) *Welding procedure.* Careful specification of procedure can contribute much to successful welding. The electrodes should be described (see BS 639[19]). The welding current, the number of runs and the sequence also need to be laid down. Large butt welds are often constructed from both sides, in which case "back-gouging" to sound metal at the root of the first runs is required before starting to weld the second side. Quality of the weld metal, locked-in stresses and distortion depend strongly on procedure.

(e) *Preheating.* Preheating temperature depends on ambient temperature, metallurgical properties of plate and electrode, and the thickness of the plates. Control of cooling in relation to metallurgical properties is required.

(f) *Heat treatment.* Heat treatment of the completed job may be required to relieve locked-in stresses or to modify the metallurgical properties of the weld metal and the metal in the heat-affected zone.

(g) *Limitations on weld size.* There is a need to limit weld sizes. For example, BS 1856 sets a lower limit on the size of fillet welds in relation to plate thickness as follows:[16]

Thickness of thicker part (mm)	*Minimum size of single run fillet welds (mm)*
$10 < t \leqslant 19$	5
$10 < t \leqslant 32$	6.5
$32 < t$	8

These sizes are for manual welding using electrodes of class 2 or 3 to BS 1719: Part 1.[20]

This lower limit is necessary to avoid cracking of the weld metal as it shrinks on cooling. Similarly, when making an unduly large weld in light plate it may be difficult to control melting or burning of the plate. A rule of thumb is that weld size should not exceed plate thickness.

A conflict between minimum weld size and requirements for strength can be resolved by specifying intermittent welds. However, new problems are then introduced – effective length of the weld[7, 16], stress concentrations, and the need to seal the joint for protection from corrosion.

A conflict between maximum size and strength can be resolved by specifying a multi-run weld.

The following is an example of a detailed description of the procedure for a structural weld taken from a carefully written specification:

Material Type and Thickness – 5/8 in A149, $1\frac{1}{2}$ in A151. [These are steel types specified in an Australian manufacturing specification.]
Preheat – 200°F [93.5°C].
Welding position – Downhand.
Preparation – Double 45° bevel. Feather edge, 1/8 in gap.
Tack welding – If tack welding is required, same procedure as for welding.
Weld 1 – 5/32 in Jetweld LH70 (to be back ground before making Weld 5).
Weld 2 – 5/32 in Jetweld LH70.
Weld 3 – 3/16 in Jetweld LH70.
Weld 4 – 5/32 in Jetweld LH70.
Weld 5 – 3/16 in Jetweld LH70.

[Reprinted from *BHP House Fabrication*, September 1971, by permission of The Broken Hill Proprietary Coy Ltd.]

The weld numbers referred to are shown on the adjacent marginal sketch. Jetweld LH70 is the trade-name of the electrode required.

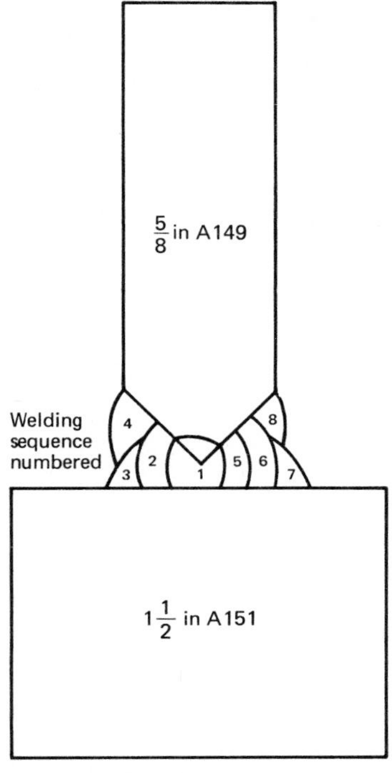

(From BHP House Fabrication, *September, 1971, by permission of The Broken Hill Proprietary Coy Ltd.)*

13.4.4. Design of timber connector joints

Design of joints in timber structures using connectors is based on test results. From these, tables of safe loads for various connector units have been developed, a connector unit being one of the following:

(a) a split ring or double-sided toothed plate and its bolt;
(b) a shear plate or a single-sided toothed plate and its bolt;
(c) two shear plates or single-sided toothed plates back-to-back and their bolt.

More than one connector may be used with the same bolt. Except when single-sided plates are used back-to-back in the same interface, connectors on the same bolt should be in different interfaces. Then, each connector, or pair of connectors back-to-back can be treated as a connector unit. The total safe load is the sum of the safe loads of the individual connector units for connectors on the same bolt or on different bolts.

The safe load per connector unit depends on the following matters:

(i) kind and size of connectors;
(ii) thickness of timber;
(iii) species of timber;
(iv) moisture content of timber;
(v) duration of load;
(vi) edge distance, end distance and spacing;
(viii) angle between load and grain.

Kind and size of connectors

Manufacture of connectors is controlled by a standard specification, for example, BS 1579: 1960, *Connectors for Timber.*[21] Commonly used sizes and types are:

Double-sided round toothed plate Single-sided round toothed plate Double-sided square toothed plate Single-sided square toothed plate	38 mm 51 mm 64 mm 76 mm

Split ring	64 mm	102 mm
Shear plate	67 mm	102 mm

(Extract from BSCP 112, reproduced by permission of the British Standards Institution, 2 Park Street, London W1A 2BS.)

For each connector there is a recommended bolt size and tabulated safe loads assume the use of these bolts.

Thickness of timber

There is a minimum thickness of timber which may be used with any connector in order to avoid having the timber cut right through at the joint by the connector. Also, load transfer may be limited by the thickness of timber remaining after bolt hole and grooves have been cut. Thus, test results show that increased loads can be transferred by a given connector as thickness increases up to a limit beyond which greater thickness has no effect, and that load capacity per connector unit is less when there are connectors in both faces of timber of given size than when the connectors are in one face only.[22]

Timber species

Load is applied by the connector to the timber by bearing where the two are in contact. The ability of different species of timber to resist such stresses varies and so, consequently, does the load which can be carried by a given connector. It is possible to group several species together with respect to connector loads, so that tables of safe loads are much simpler than they would be if every kind of timber had to be dealt with separately.

Moisture content of timber

The strength of timber decreases as its moisture content rises above about 20%, and since timber structures in highly humid atmospheres or in wet places absorb moisture readily, the safe loads for connectors in such places are reduced. For example, in BSCP 112 basic loads are for timber with moisture content less than 18%.[23] When a joint is exposed to damp conditions for long periods, safe loads are reduced by 30%.

Duration of load

Timber is a material whose strength decreases with time under permanent loads. Since basic permitted loads are usually for loads of long duration, increases over the basic loads may be allowed for temporary or transient loads. For toothed plates BSCP 112 allows increases of 12.5% for temporary loads and 25% for transient loads. Corresponding figures for split rings and shear plates are 25% and 50%.

Edge distance, and end distance and spacing

These are defined in Fig. 13.26. A distinction is made between a loaded edge or end and an unloaded edge or end. A loaded edge is such that the force applied to the member by the connector has a component directed towards it.

Provided these dimensions exceed certain values, they do not affect the safe load on the connector and it is desirable to keep them above what are called the *standard edge distance, end distance* and *spacing*. Shorter distances to edges and ends, or between connectors reduce the strength of the joint and hence the safe load per connector unit. Design specifications usually permit the use of reduced values for these dimensions, subject to a lower limit. A smaller safe load per connector unit must then be used.

The angle ϕ of the connector axis (Fig. 13.26) must be taken into account because the standard spacing sometimes depends on it.

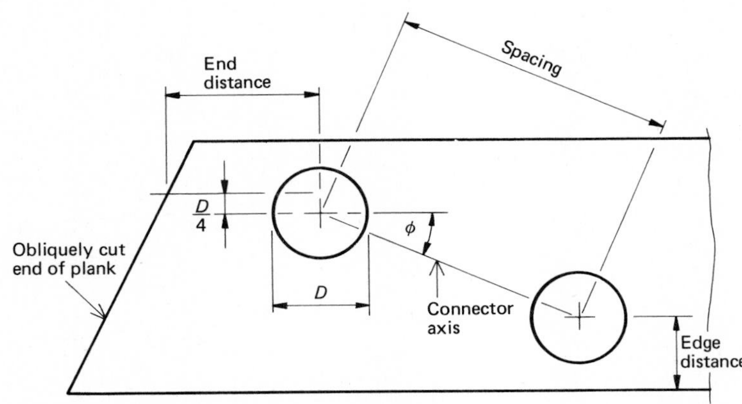

Fig. 13.26. *Definition sketch for end distance, edge distance, spacing and connector axis for timber connectors.*

Angle between load and grain

Two safe loads per connector are given in design specifications—for loads parallel to the grain and loads perpendicular to the grain. The grain of the timber is assumed to run along the length of the member. In general, it does, any deviation being limited by the grading rules in the purchasing specification.

However, it is frequently the case that connector forces are applied to the timber at some angle between 0° and 90°. The safe load which can be applied is calculated using the Hankinson formula

$$N = \frac{PQ}{P\sin^2\theta + Q\cos^2\theta}$$

where N is the safe load applied at angle θ to the grain; P is the safe load parallel to the grain; and Q is the safe load perpendicular to the grain. This empirical formula was developed for safe bearing stresses in timber and has been applied to connector loads.

Determination of the angle θ requires careful examination of the free bodies at a joint. These free bodies are
(a) lengths of the timber members being joined;
(b) the connectors.
To illustrate the point consider the heel joint of a timber truss for a pitched roof, sketched in Fig. 13.27(a). It is an important detail that the reaction of the support is applied to the edge of the bottom chord.

In Fig. 13.27(b), part of the top chord is shown. Only two forces act on it—the thrust of the rest of the top chord and the connector force. For equilibrium, they must be collinear and equal and of opposite sign. The connector force must therefore act parallel to the grain and be equal in magnitude to the thrust in the top chord.

Figure 13.27(c) shows the connectors as a free body. They also are acted on by only two forces—the reactions to the connector forces on the two chords.

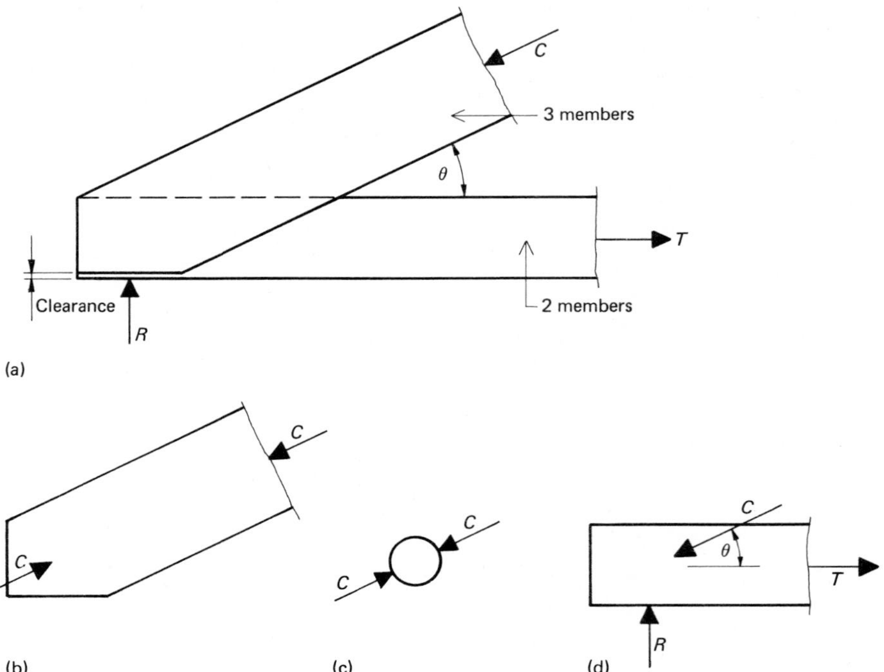

Fig. 13.27. *Transfer of forces through connectors at the heel joint of a roof truss.*

Equilibrium of the connector requires them to be collinear and since, as we have seen, one of them acts parallel to the axis of the top chord, the other does too.

Finally, in Fig. 13.27(d), a free body comprising part of the bottom chord is shown. Three forces act on it – the pull of the rest of the member, the reaction of the support and the connector force. This last is seen to be equal to the thrust in the top chord and to act at angle θ to the grain in the bottom chord.

This analysis shows that the connectors must be designed for a load equal to C acting parallel to the grain in one member (the top chord) and at an angle θ to the grain in the other (the bottom chord).

If the detail at the support were different, so that the reaction acted on the end of the top chord with the bottom chord clear, it would be found that the design force would equal T (Fig. 13.27) acting parallel to the grain in the bottom chord and at angle θ to the grain in the top chord.

Quite apart from calculation of strength, considerable skill goes into the layout of the members of a timber truss. Both chords are usually multiple members whereas web members can be single planks. A heel joint like that discussed above (Fig. 13.27) does not require splice plates, but other joints in this truss would – at the apex, for example, where the two top chords meet. Each having three members, overlapping is not possible without introducing unacceptable asymmetry and lateral bending. Also, the web to top chord joints would require splice plates if the web members were single shafts. The point we wish to emphasise is the need to plan the layout of the planks in the whole truss before starting detailed design of the joints.

Design of a truss joint using the rules in BSCP 112 is shown on Example 13.8. First, the data of the problem are listed. They include decisions already made and forces calculated from analysis of the truss. Following them is relevant information abstracted from the design specification. Calculation of the number of connectors requires examination of the forces acting on the various free bodies and identification of the member where the connector loads are critical. The number of connectors is then found by dividing the safe load per connector into the load to be transferred.

The connectors are laid out on a scale drawing. Lines parallel to the edges and ends are drawn at the appropriate edge and end distances. At critical places, lines are also drawn at half the connector radius from these. The centres of the bolts can then be located so as to satisfy the rules governing edge and end distance. A further criterion in locating the bolts is that dimensioning should be convenient for construction. Then spacing is checked. It should be noted that member sizes are frequently controlled by design of the joints and not by stresses in the members.

It will be seen in Example 13.8 that this joint has been dimensioned in relation to the splice plate. The dimensions are convenient for cutting and drilling the splice plate which can then be used as a jig for drilling the chords.

13.5. MISCELLANEOUS FASTENINGS

Apart from the standard bolts, rivets and connectors, a wide variety of other types of fasteners – ranging from the nail to the zip-fastener – are used in engineering. A selection of these is described here, along with some of the associated items used to ensure their correct operation (such as locking devices).

13.5.1. Special bolts and screws

There have been many variations in the head, shank and thread forms of screws as they have been adapted to special uses. A selection that should be wide enough to provide for most general engineering needs is described here. The main problem for industry has been one of standardisation, for the combinations that can be assembled from the variations are enormous.

Engineering and production requirements have been the cause of most variations, although aesthetic considerations have influenced some of the changes in head shapes, since fastenings are often conspicuous in an assembly or structure.

The four principal methods of turning a bolt or screw are

(a) spanner or socket on hexagon or square head;
(b) socket key in hexagon recess;
(c) screwdriver in straight slot;
(d) special screwdriver in recessed head.

The first type can be tightened rapidly with a well-fitting spanner or socket – and with accuracy if a torque wrench is used. However, with lower sizes (below, say, 10 mm), it is easy with a standard spanner to overtighten a mild steel bolt and to fracture it; since the size of the key supplied for use with hexagon recess heads makes overtightening more difficult, this type is recommended for the lower sizes. The key requires no edge space and this constitutes an important advantage, even in the larger sizes, where close spacing is necessary. An assembly of these fastenings presents a neat appearance, and they can be used to advantage where the fasteners are not concealed.

Example 13.8 Sheet 1 of 3

Design a heel joint for timber roof truss.

Data:

Both chords 2/190 mm x 45 mm (actual dimensions)
45 mm splice plate, grain // top chord
Timber species - Group J4 (CP 112)
Top chord compression = 20 kN
Bottom chord tension = 17.3 kN
Reaction = 10 kN applied to edge of bottom chord
Roof pitch = 30°

Use 64 mm round toothed plates
Standard end distance, edge distance, spacing
Moisture content < 18%
Loads permanent

From BSCP 112:
Basic connector loads:
Splice plate (connectors both sides)
$P = 3.35$ kN, $Q = 2.44$ kN
Hence, for $\theta = 30°$

$$N = \frac{3.35 \times 2.44}{3.35 \times 0.25 + 2.44 \times 0.75} = 2.15 \text{ kN}$$

Chords (connectors one side)
$P = 4.07$ kN, $Q = 3.12$ kN
Hence, for $\theta = 30°$

$$N = \frac{4.07 \times 3.12}{4.07 \times 0.25 + 3.12 \times 0.75} = 3.78 \text{ kN}$$

End distance: Loaded 95 mm ($\theta = 0°$ and $30°$)
Unloaded 44 mm ($\theta = 0°$) 52 mm ($\theta = 30°$)
Edge distance: 37 mm (Loaded and unloaded, $\theta = 0°$ and $30°$)
Spacing: $\theta = 30°$ 86 mm
$\theta = 0°$ 95 mm (Connector axis at 0° to grain)
86 mm (" " " 45° " ")
76 mm (" " " 90° " ")

Adjustments to basic loads:
None required for edge distance, end distance or spacing. For angle to grain, as above.

Example 13.8 Sheet 2 of 3

Connectors in top chord – splice plate joint

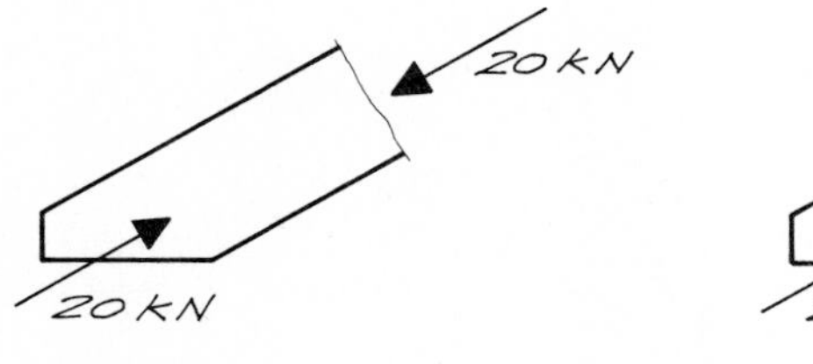

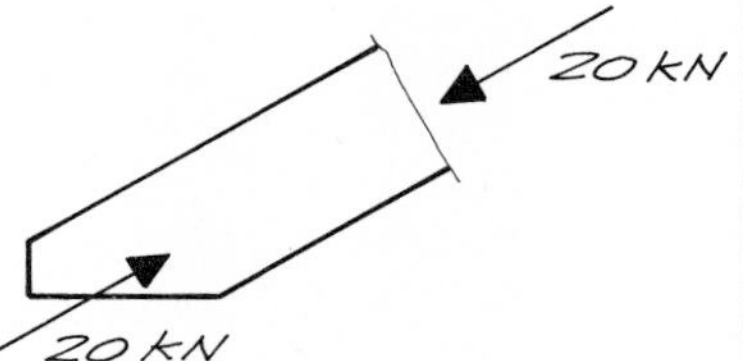

Chord
Load = 20 kN
$\theta = 0°$
$N = P = 4.07$ kN

Splice plate
Load = 20 kN
$\theta = 30°$
$N = P = 3.35$ kN

Splice plate controls — $\frac{20}{3.35} = 6$ connectors
Two connectors per bolt → 3 bolts

Connectors in bottom chord – splice plate joint

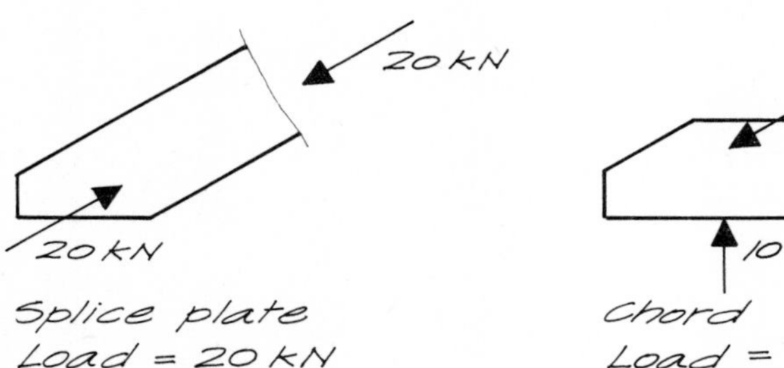

20 kN
17.3 kN
10 kN

Splice plate
Load = 20 kN
$\theta = 0°$
$N = P = 3.35$ kN

Chord
Load = 20 kN
$\theta = 30°$
$N = 3.78$ kN

Splice plate controls — $\frac{20}{3.35} = 6$ connectors
Two connectors per bolt → 3 bolts

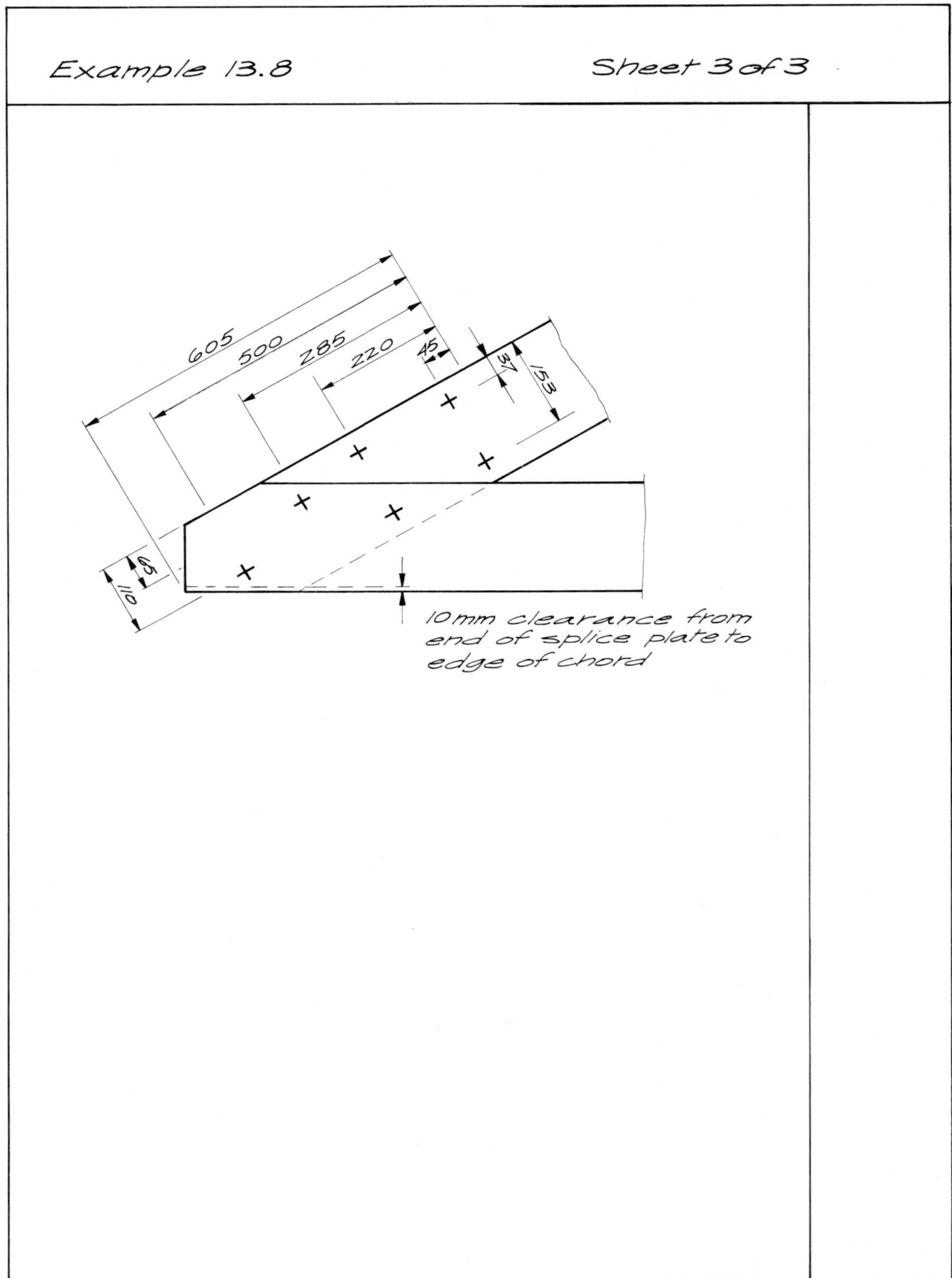
Example 13.8
Sheet 3 of 3
605
500
285
220
45
37
153
65
110
10 mm clearance from end of splice plate to edge of chord

The recessed screwhead was introduced to speed assembly, the shape of the recess helping to centralise the point of the driving tool. It also has some visual and practical advantages over the straight slot, which when damaged can spoil the appearance of a product and expose sharp edges.

There is usually a sound practical reason for the demand for special shapes of head. For example, an extra-wide flat head is useful for holding down panels of light sheet metal which often have oversize and off-centre holes.

Some standard head shapes are illustrated in Fig. 13.28; recessed heads may be used instead of the slots shown.

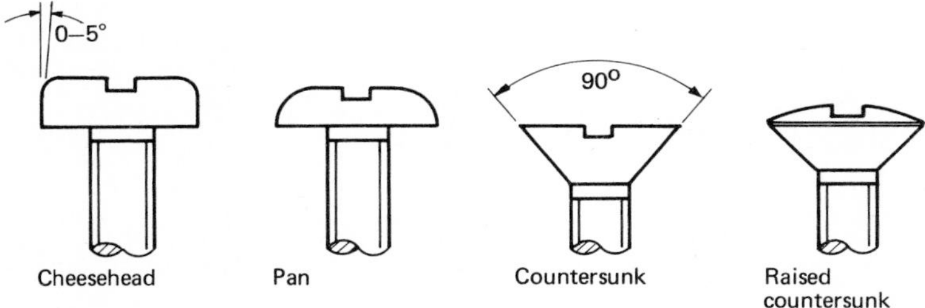

Fig. 13.28. *Some standard head shapes.*

The different terminology used in British and U.S. practices can lead to confusion; below are listed some equivalent terms.

British	*U.S.*
Countersunk	Flat
Raised countersunk	Oval
Cheesehead	Fillister

The variations in thread form from the standard metal threads were brought about mainly by demands from the plastics and sheet metal industries. The materials being used were not suited to the fine-pitch metal threads, and were often soft enough for each screw to form its own thread. Some of the thread forms developed are described in the British Standard on self-tapping screws from which Fig. 13.29 is taken.[24] These threads must be hard enough to either deform or cut the base material in which they are used. The thread-forming type have a sharp thread and large pitch; they can be driven straight into an undersize hole in light sheet metal, impregnated plywood and plastics. The thread-cutting type has some form of cut-away near the tip to aid a thread-cutting action. Both these types can be removed and replaced.

The round-headed metallic drive screw is intended for use in metals, including cast iron, and plastics. The thread angle is sufficiently steep so that the screw can be driven in directly with a hammer to form a permanent fastening.

There can also be a variation in the way in which the thread is disposed along the shank. Some of the possibilities are illustrated in the sketch. An alternative to assembling two components with a bolt is to use a stud which is located permanently in one of them. When disassembly of a component from, say, a casting is a recurring task, the wear is restricted to the nut and the threads it engages on the stud. This is advantageous since the stud can be replaced if damage occurs; damage to the thread in the casting, which could be difficult to rectify, is avoided. The studs also act in a locating and retaining role during assembly. The stud can, if necessary, have different thread forms at either end, a coarse thread being better suited for aluminium and cast iron.

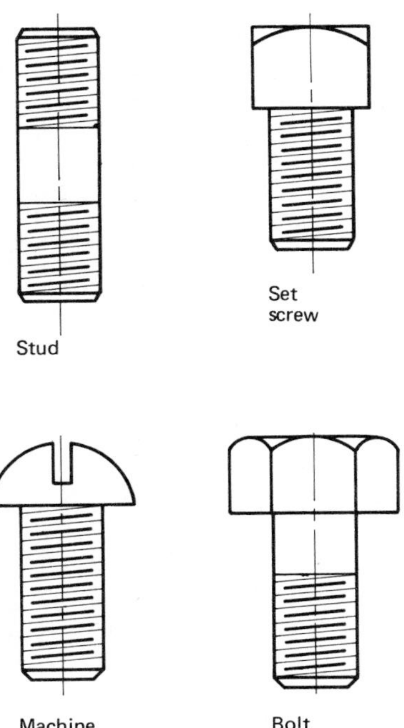

Fig. 13.29. *Tapping screws and metallic drive screws. (Adapted from BS 4174: 1967,* Self-Tapping Screws, *by permission of the British Standards Institution, 2 Park Street, London W1A 2BS.)*

Thread contour	Type of screw	Type designation for tapping screws and metallic drive screws	Purpose
	Thread-forming	A	Light sheet metal work – below 1.2 mm
	Thread-forming	B	Heavier sheet metal, non-ferrous castings, plastics
	Thread-cutting	D	Cast iron, light alloy and plastic
	Drive	U	Metal and plastic–permanent assembly

13.5.2. Locking

A nut will not work loose on a bolt when it is subjected to a steady tensile load, because the helix angle on the standard threads is low enough so that, even with lubricated surfaces, the nut will not overrun. When vibration is present the nut is still unlikely to work loose unless the load is variable and at some stage falls to zero. Under these conditions a locking device is essential. We will describe here three main types. The first uses some form of *elastic device* to ensure that some residual seating force will remain on the nut, even when the load is varying and there has been some creep in the parts that are clamped together. The *spring washer* works on this principle, and also has sharp edges which dig into the nut and the clamped surface to restrain rotation. The second type is called a *stiffnut*, and consists of a nut body containing a device which imposes friction between the nut and the screw thread with which it is engaged; the friction must be independent of the loading on the face of the nut. Finally, we come to the third type which is the *positive locking device*. Under this heading we classify the devices which will not allow the nut to rotate without the deformation or fracture of some metal restraining device.

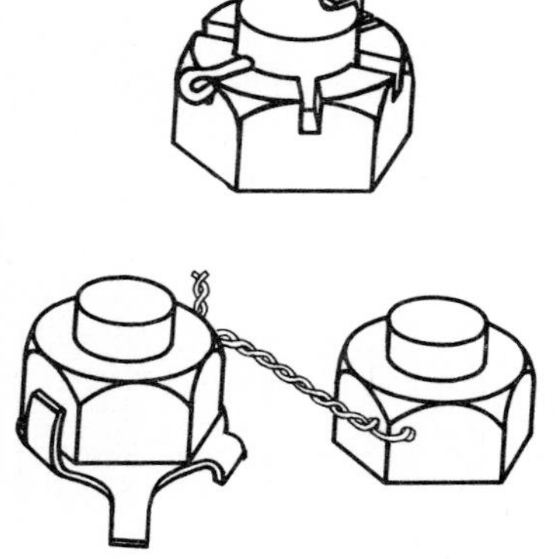

The *split pin, locking wire* and *tab washer* are examples that are shown in the sketch. When a serious structural or machine failure would result from the loosening of a nut, one of these locking methods should be used. The split pin is the most reliable and most readily checked for security. In some cases, proprietary adhesives can be substituted for these locking devices.

LIST OF SYMBOLS

A, A_1, A_2, A_g areas
a length, width of strip equivalent to fasteners
b width of gasket, effective width of contact
C compressive force
C_1, C_2 clamping forces at a joint
c clearance
D_m mean diameter of gasket
d diameter of fastener, height of bracket
E, E_1, E_2, E_g moduli of elasticity
e eccentricity of applied load
e_x, e_y, e_z coordinates of point of application of applied load

F	load factor
f_b	bearing stress
f_c	compressive stress
f_{bt}, f_{bc}	bending tensile and compressive stresses
f_{eq}	equivalent stress
f_q	shear stress
f_t	tensile stress
f_y	yield stress
I_{XX}, I_{YY}, I_{ZZ}, I_{NA}	second moments of area
J	polar second moment of area
j	number of interfaces
k	stiffness, relative height of contact area
k_F	stiffness of fastener
k_P	stiffness of plates and packing
L, l	lengths
M	moment
M_X, M_Y, M_Z	components of M
m	overpressure factor
N	permitted connector load at angle θ to grain
n	number of fasteners
P	load, permitted connector load parallel to grain
P_X, P_Y, P_Z	components of P
P_A	tensile load applied to fastener
P_F	final tension in fastener
P_I	initial tension in fastener
P_1, P_2	loads
p	internal pressure in pressure vessel, pitch of fasteners
p_q, p_t	allowable stresses
Q	permitted connector load perpendicular to grain
R, R_1, R_2, R_3	loads in fasteners
R_{1X}, R_{1Y}	components of R_1
R_{2X}, R_{2Y}	components of R_2
r_1, r_2, r_h, r_c, r_t	radii
s	size of fillet weld
T	tensile force
t	plate thickness, throat thickness of fillet weld
t_1, t_2	plate thicknesses
t_g	thickness of gasket
x, y, z	space coordinates
y	compressive stress required to seat gasket
Z	section modulus
α	coefficient of thermal expansion, pitch angle of thread
β	half angle of thread
ε, ε_1, ε_2, ε_3	shear deflections
θ	angle of load to grain in timber member
θ_1, θ_2	temperatures
μ, μ_1, μ_2	coefficients of friction
ϕ	angle of connector axis to grain in timber member.

REFERENCES

[1] *Machinery's Screw Thread Book,* 16th ed., Machinery Publishing Co., Brighton & London, 1955.

[2] BS 4190:1967, *ISO Metric Hexagon Bolts, Screws and Nuts,* British Standards Institution, London, 1967.

[3] BS 3139:Part 1:1959, "General Grade Bolts", *High-Strength Friction-Grip Bolts for Structural Engineering*, British Standards Institution, London, 1959.

[4] ABC (N.Z.), *The Gang-Nail Component System Technical Manual,* Automated Building Components (N.Z.) Limited, Auckland, 1967.

[5] V. Dobrovolsky *et al., Machine Elements,* Foreign Languages Publishing House, Moscow, undated.

[6] BS 1515:Part 1:1965, *Fusion-Welded Pressure Vessels,* British Standards Institution, London, 1965.

[7] BS 449:Part 2:1969, "Metric Units", *The Use of Structural Steel in Building,* British Standards Institution, London, 1969.

[8] BS 3294:Part 1:1960, "General Grade Bolts", *The Use of High-Strength Friction-Grip Bolts in Structural Steelwork,* British Standards Institution, London, 1960.

[9] W. C. Stewart, "Determining Bolt Tension from Torque Applied to the Nut", *Machinery Design,* **27**, 11, November 1955.

[10] BS 5380:1964, *Guide to Design Considerations on the Strength of Screw Threads,* British Standards Institution, London, 1964.

[11] A. L. Phillips (ed.), *Introductory Welding Metallurgy*, American Welding Society, New York, 1968.

[12] A. L. Phillips (ed.), "Fundamentals of Welding", *Welding Handbook*, 6th ed., American Welding Society, 1968.

[13] L. H. van Vlack, *Elements of Materials Science,* 2nd ed., Addison-Wesley, Reading, 1964.

[14] O. W. Blodgett, *Design of Weldments*, James F. Lincoln Arc-Welding Foundation, Cleveland, 1963.

[15] LEC, *Procedure Handbook of Arc-Welding Design and Practice,* Lincoln Electric Company, Cleveland, 1957.

[16] BS 1856:1964, *General Requirements for the Metal Arc-Welding of Mild Steel,* British Standards Institution, London, 1964.

[17] BS 2642:1965, *General Requirements for the Arc-Welding of Steel to BS 968 and Similar Steels,* British Standards Institution, London, 1965.
(Note that BS 968 has been superseded by BS 4360.)

[18] BS 4360:Part 2:1969, *Weldable Structural Steels,* British Standards Institution, London, 1969.

[19] BS 639:1969, *Covered Electrodes for the Manual Metal-Arc Welding of Mild Steel and Medium Tensile Steel, (Metric Units),* British Standards Institution, London, 1969.

[20] BS 1719:1969, *Classification, Coding and Marking of Covered Electrodes for Metal-Arc Welding,* British Standards Institution, London, 1969.

[21] BS 1579:1960, *Connectors for Timber,* British Standards Institution, London, 1960.

[22] H. J. Hansen, *Modern Timber Design*, 2nd ed., John Wiley, New York, 1948.

[23] BS CP 112:Part 2:1971, "Metric Units", *The Structural Use of Timber,* British Standards Institution, London, 1971.

[24] BS 4174:1967, *Self-Tapping Screws,* British Standards Institution, London, 1967.

BIBLIOGRAPHY

ALCOA, *Alcoa Structural Handbook,* Aluminium Company of America, Pittsburg, 1960.

Describes the use of aluminium in structures, including design procedures. Includes as supplements the following ASCE reports:

"Specifications for Structures of Aluminium Alloy, 2014–T6."

"Specifications for Structures of Aluminium Alloy, 6061–T6."

"Progress Reports of the Committee of the Structural Division on Design of Lightweight Structural Alloys," *ASCE Proceedings,* **82**, ST3, May 1956.

GRAY, C. S., KENT, L. E., MITCHELL, W. E. and GODFREY, C. B., *Steel Designers' Manual*, 4th ed., Crosby Lockwood, London, 1972.

In large part, the manual is supplementary to BS 449 and is intended for designers using that code.

BOWEN, L. P., *Structural Design in Aluminium*, Hutchinson, London, 1966.

Covers British practice in the use of aluminium structures and includes the following report of the Institution of Structural Engineers:

Report on the Structural Use of Aluminium, Institution of Structural Engineers, London, 1962.

BRESLER, B., LIN, T.Y. and SCALZI, J. B., *Design of Steel Structures,* John Wiley, New York, 1968.

This specialist text covers all aspects of steel design, including riveted and bolted joints. It is based on American practice.

PROBLEMS

13.1. In Section 13.4.2, Equation (13.11) and Fig. 13.18 are used to estimate the tension induced in a rivet by contraction as it cools. Do this calculation and confirm that the rivet will be stressed to its yield point. What effect will heat conducted from the rivet to the plates have on the results?

13.2. Redesign the bolted joint of Example 13.3 using Equations (13.6) and (13.7). The mean gasket diameter is 800 mm and its effective width is 22 mm. Assume that the gasket is corrugated soft copper. What is the significant difference between the two designs?

13.3. A beam-to-column joint, such as that shown in Fig. 13.21, is to be made with bolts. If the bending moment, M, is 20 kN m and the beam is a 305 mm × 165 mm × 40 kg/m universal beam, determine a suitable size and number of bolts for the flange connections.

13.4. A steel tube whose outside diameter is 75 mm and wall thickness is 5 mm has a torque of 3.75 kN m applied to it. Determine the size of fillet weld required to anchor it to a steel baseplate. Assume the permitted stress on the throat of the weld is 115 MPa.

13.5. The steel frame of a machine includes a 300 mm × 300 mm square box member built up from 15-mm plate. A force of 10 kN, inclined at 1 in 2 to this member is to be anchored to the frame by means of a bracket. Design a suitable bracket using
(a) bolts in shear,
(b) fillet welds.

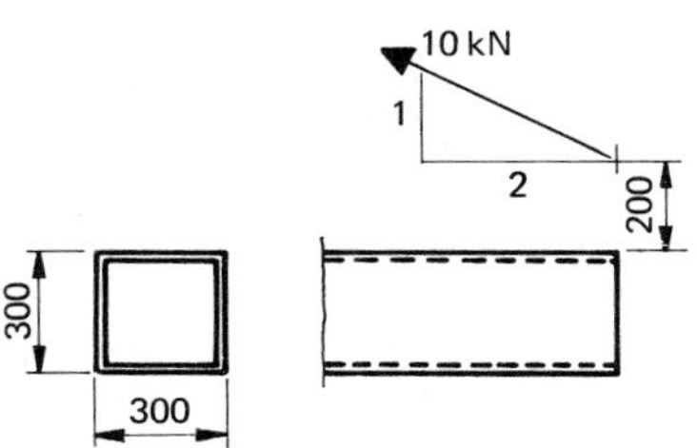

13.6. In a steel roof truss, each member is a single 51 mm × 51 mm × 8 mm angle and joints between members are made with gusset plates 10 mm in thickness. A tension of 60 kN is to be transferred from an angle to a gusset plate at one joint. Prepare alternative designs to connect the angle to the gusset plate using
(a) HSFG bolts,
(b) fillet welds.

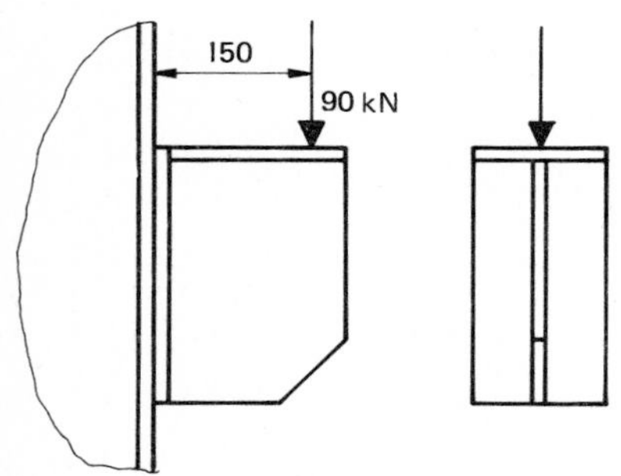

13.7. A bracket is to be fabricated from a 15-mm thick steel plate by welding and is to be bolted to the flange of a 305 mm × 305 mm × 97 kg/m universal column. The bracket has to support a load of 90 kN at a distance of 150 mm from the face of the column, as sketched. Determine suitable welding details for the bracket and the size and number of bolts to fasten it to the column. Take p_t = 130 MPa for the bolts and assume the initial stress to be 100 MPa.

13.8. Select a suitable number and arrangement of 64-mm round toothed plate connections for the apex joint of the timber roof truss of Example 13.8. Design for a compression in the top chord on each side of the apex of 15 kN and assume that a vertical member at the joint carries a tensile load of 10 kN.

13.9. A contractor is using high tensile bolts having a proof stress of 600 MPa and wishes to tighten them to 75% of this value. In the absence of manufacturer's recommendations, draw up a table of torque values for sizes M10, M20, M24 and M30. Compare the values you obtain with a manufacturer's recommendations.

13.10. Review the various methods that can be used to ensure that the correct value of pre-tension in a bolt is applied. Explain why the turn-of-the-nut method produces satisfactory results when it is apparently so crude.

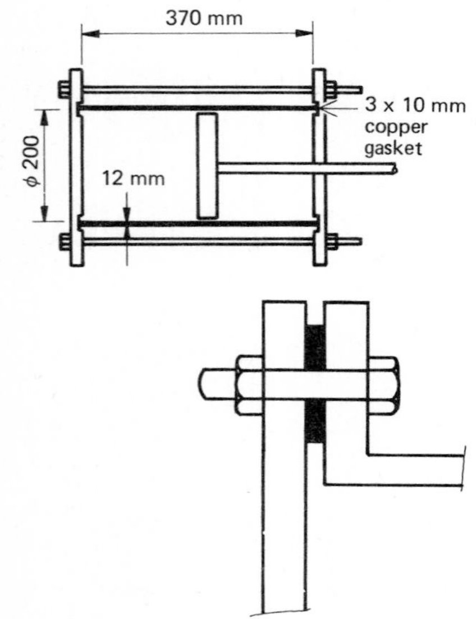

13.11. An air cylinder carrying a pressure of 1.4 MPa is shown in the sketch. The gasket is designed to be precompressed with a load of 35 kN.
(a) Calculate the diameter of the four bolts which secure the end plates. Assume that $k_P = 5\,k_F$.
(b) Specify a suitable tightening torque.
(c) Check the assumed values of k, making any reasonable assumptions.
Design for static loading, using a safety factor of 3. Yield strength for bolt material, f_y = 500 MPa.

13.12. (a) If the threads in the bolts in Problem 13.11 have a strength reduction factor of 2.5, check their safety factor in fatigue if the pressure in the cylinder varies from 0 to 1.4 MPa. Take f_e = 350 MPa.
(b) Check the safety of the bolts in fatigue for the alternative arrangement shown in the sketch, assuming that the same size of bolt is used and that $k_P = 0.2\,k_F$.

13.13. Extend the working of Example 13.1(b) by drawing lines representing various deflections on the plot of D against L. These deflections can be expressed in terms of D, for example as 0.01 D, 0.005 D, etc. What is the effect of the deflection on the operation of the pinned joint?

14 Timber Beams

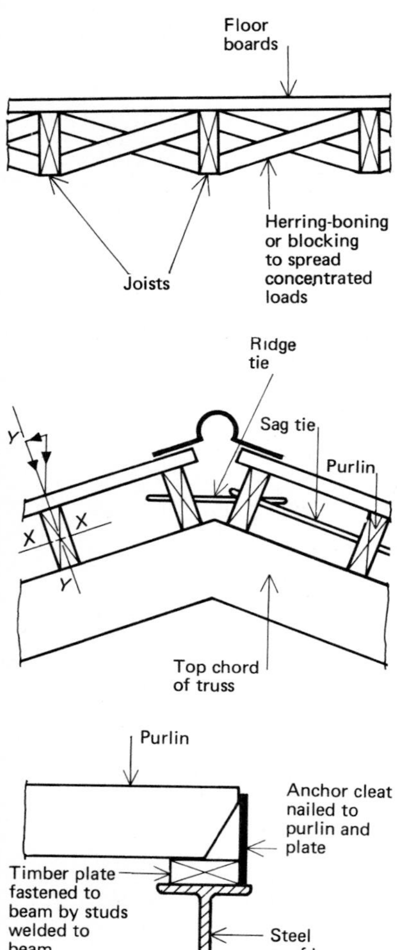

14.1. INTRODUCTION

Timber floors supported on timber or steel beams are used in many one- and two-storey buildings such as schools, halls and warehouses. In these structures, much of the timber is in the form of planks or balks known as *sawn timber*. Many larger beams are fabricated from planks glued together and these are *laminated beams*, a form of construction sometimes known as *glulam*. In addition, timber is used to support roofs and walls in halls or sheds where the main supporting frame may be built of steel or timber. If appearance is important and the cost is acceptable, timber portal frames may be used to advantage for the main structure. The construction may be solid laminated timber or a box girder with solid flanges and plywood webs.

The simplest timber beam is a plank on edge, called a *joist* in floor construction, a *purlin* in roof construction or a *girt* when used in a wall. The distance between the planks is limited by the strength or stiffness of the flooring, roofing or cladding material. In floors, for example, the joists are placed 500 mm to 600 mm apart and the floor boards span the gap between them. The joists, in turn, span from beam to beam in the main frame. Diagonal struts, called *herring-boning*, may be nailed between the joists near midspan to share concentrated loads among several joists.

In a roof, the purlins are fastened to the tops of the roof beams or the top chords of the trusses. Loads due to gravity are inclined to the principal axes of inertia of the purlins because of the pitch of the roof. Such loads can be resolved into components parallel to the roof and normal to it. The effect of the former in causing bending about the YY axis of a purlin can be eliminated if the roofing sheets are stiff enough, as diaphragms, to prevent lateral bending. The ridge purlins should be tied together to take advantage of this and it may be necessary to include sag ties in the structure as well. Wind loads act normal to the roof and often so as to lift it. There is a need for care in detailing to ensure that the roof structure is securely fastened down. Patented cleats stamped from sheet steel are available and are fastened quite simply to the purlins and the main structure by nailing.

Wall girts act as beams when they transfer horizontal forces, such as wind loads, to the main frame. The weight of the cladding can be taken directly to the foundations by timber framing between the girts (see sketch on p. 341).

14.2. DESIGN OF TIMBER BEAMS

Timber beams are designed on the basis of loads in service and permissible stresses. The simple theory of elastic bending is used to compute bending stress, maximum shear stress, bearing stress and deflection due to loads. The following formulas are used:

$$f_{bc} = f_{bt} = \frac{M}{Z} \tag{14.1}$$

$$f_q = \frac{3}{2}\frac{V}{bd} \tag{14.2}$$

$$f_b = \frac{R}{bl} \tag{14.3}$$

$$\Delta = K\frac{WL^3}{EI} \tag{14.4}$$

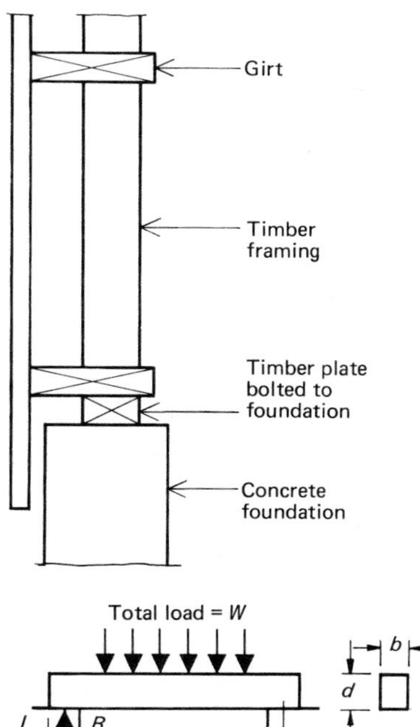

where f_{bc} is the extreme fibre compressive stress due to bending; f_{bt} is the extreme fibre tensile stress due to bending; f_q is the maximum shear stress; f_b is the bearing stress at support; Δ is the maximum deflection; M is the maximum bending moment; V is the maximum external shear; R is the reaction at support; W is the total load on beam; L is the span of beam, centre to centre of supports; l is the length of bearing at support; b is the breadth of beam; d is the depth of beam; Z is the section modulus $= \frac{1}{6}bd^2$; I is the second moment of area $= \frac{1}{12}bd^3$; E is the modulus of elasticity; and K is a factor which depends on pattern of loading.

The formulas given above for Z and I are based on rectangular cross-sections – solid beams, whether sawn or laminated, are almost invariably rectangular with their depth greater than their breadth.

The assumptions made in the derivation of these formulas are not, strictly speaking, valid for timber, so that the formulas must be regarded as qualitative, rather than quantitative. They give a measure of the strength and stiffness of timber beams shown by experience to be adequate, but they do not tell the designer exactly what the behaviour of a beam will be. For example, timber is anisotropic and highly variable in its properties from sample to sample. The stress–strain relationship is approximately linear at low loads but not for a large range of load and both strength and stiffness are time-dependent – they decrease with duration of loading.

Allowance for some of these things is made by modifying the basic stress allowable for clear timber and given in design codes such as BSCP 112, *The Structural Use of Timber.*[1] We do not intend to explore all of the details of such rules, for they are easy enough to follow. Some are fairly arbitrary, but determination of the basic stress and allowance for some natural defects in the timber can be undertaken scientifically. The procedure has considerable fundamental interest, in a wide sense, as an example of analysis of the strength of a material, as well as in clarifying the design of timber beams.

14.3. PERMISSIBLE STRESSES FOR SAWN TIMBER BEAMS

In this discussion the terminology of BSCP 112, *The Structural Use of Timber* is used and definitions of some of the terms we will use are as follows:[1]

Basic stress: The safe stress for timber without defects and carrying a permanent load.

Grade stress: The safe stress for a particular grade or quality of timber carrying a permanent load.
Permissible stress: The safe stress for a part of a structure under particular conditions of load and service.
Strength ratio: The ratio of the grade stress to the basic stress.

The permissible stress is what concerns the designer. Design of a timber beam requires the permissible stress and the stress due to the loads to be determined and the former must be the greater of the two in an acceptable design. For economy, the difference between them should be small.

Timber is a material whose strength and stiffness vary widely. One of the most significant causes of variation is the presence of defects in commercial stocks of timber. Some of these affect the appearance of the timber, a matter of considerable importance in a material valued for decoration in buildings and for construction of furniture. Such defects do not necessarily affect the strength of the timber. Of the ones that do, *knots* are most important. Knots occur where branches have grown from the stem of the tree and they affect the strength of planks sawn from the stem by distorting the grain of the timber along the plank. Other defects which affect the strength and stiffness of a piece of timber are *slope of grain*, *fissures* and *wane*. The rate of growth also has a bearing on strength.

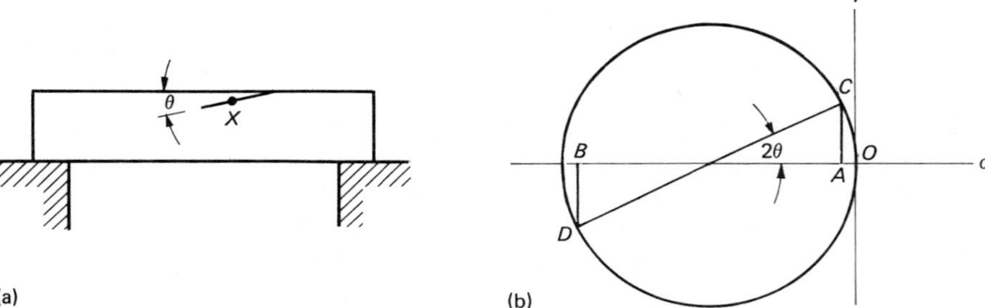

Fig. 14.1. *Effect of sloping grain.*

Slope of grain is the departure of the grain of the timber – the direction of alignment of its fibres – from a direction along a plank. It affects the strength of a plank because of the anisotropy of the material which is stronger along the grain than across it. For example, consider the stress near the top of a timber beam, as shown on Fig. 14.1(a). The bending stress is compressive and the shear stress negligible so that Mohr's circle for a point X is as shown on Fig. 14.1(b). If the grain of the timber is inclined at an angle θ to the length of the beam, components of stress on a plane parallel to the grain are OA, a compressive stress, and AC, a shear stress. Thus, the timber may well be overstressed by virtue of the compressive component across the grain or the shearing component under a load which would not overstress it if the grain were parallel to the axis of the plank.

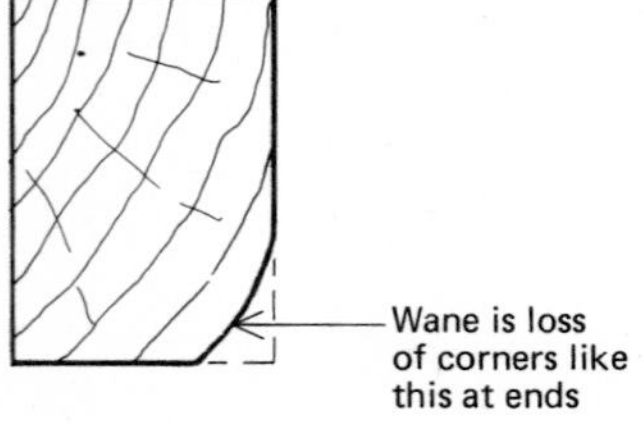

Fissures are defects which develop as the timber loses moisture and shrinks. They affect strength by modifying the geometrical properties of the cross-section. Wane is considered to reduce the bearing strength at the ends by reducing the area of contact.

Permissible stresses for commercially available timber are arrived at by determining first the basic stresses for individual species. These stresses are determined for clear timber in the green state. Clear timber is free from all defects and green timber has a moisture content in excess of the fibre saturation limit, which is about 25%. Since the strength and stiffness of timber do not vary with the moisture content if it is above the fibre saturation limit, choice of clear green timber for testing eliminates both moisture content and defects as causes of scatter in the measured strength and stiffness. Even so, results vary widely and it is necessary to test large numbers of specimens. An extensive investigation of this kind has been described by Sunley[2] whose study is a model for the determination of the strength of timber and is the basis of BSCP 112, *The Structural Use of Timber.*[1]

For bending stresses, the modulus of rupture is measured in a standard bending test. The test specimen is loaded with two point loads so as to give a uniform bending moment and no shear over a length near midspan. The modulus of rupture is computed from Equation (14.1) for the load at failure. In Sunley's experiments, the results were normally distributed. Hence, from the mean and standard deviation of the sample the modulus of rupture which would be exceeded in all but 1% of the population could be estimated. On Fig. 14.2, a histogram of some of Sunley's results is shown with the curve of a normal distribution superimposed. Thus, if $\bar{x}$ is the mean of the sample and σ its standard deviation, 1% of the population would have a modulus of rupture less than $(\bar{x} - 2.33\sigma)$. Alternatively, one can say that the probability that a specimen will have a modulus of rupture less than $(\bar{x} - 2.33\sigma)$ was 1 in 100. The basic stress was then found by dividing this modulus of rupture by a factor of safety which was taken as 2.25. Thus, 99% of the clear specimens will have a factor of safety of at least 2.25.

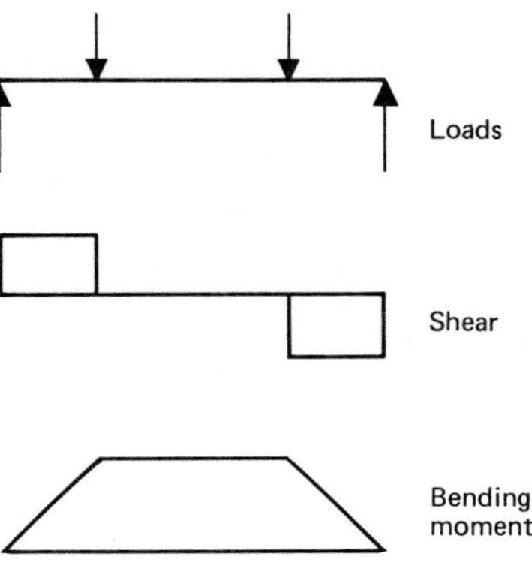

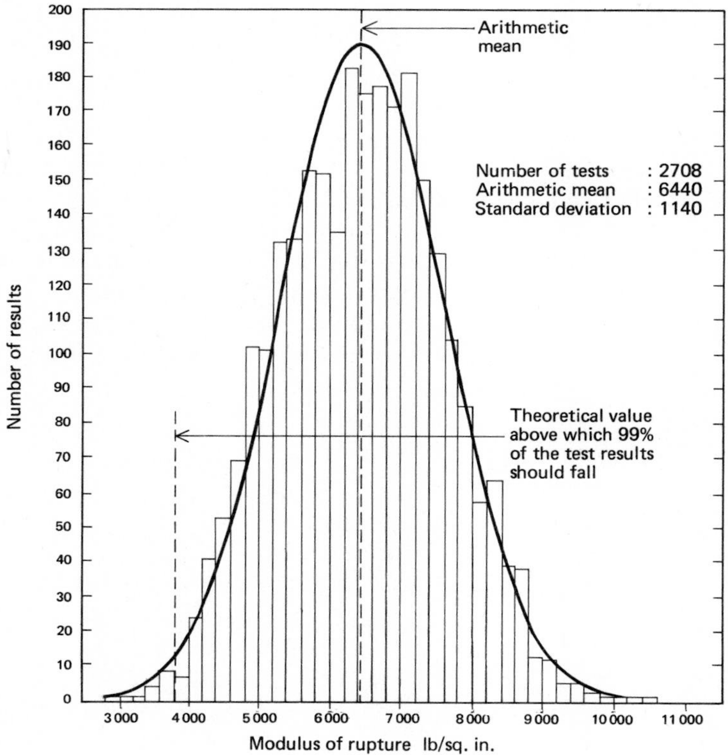

Fig. 14.2. *Histogram of bending test results with curve for normal distribution superimposed. (From J. G. Sunley, "Working Stresses for Structural Timbers"[2], with acknowledgement.)*

The modulus of rupture is used to determine basic compressive and tensile stresses in bending. The same basic stresses are used for direct compression and tension parallel to the grain. Separate crushing tests are used for compression perpendicular to the grain, which is related to bearing at the ends of beams. Since timber does not lose its strength entirely when crushed, a higher probability of failure is accepted for bearing. The limiting strength is the strength which will be exceeded in all but $2\frac{1}{2}\%$ of the population. The same factor of safety, 2.25, is used.

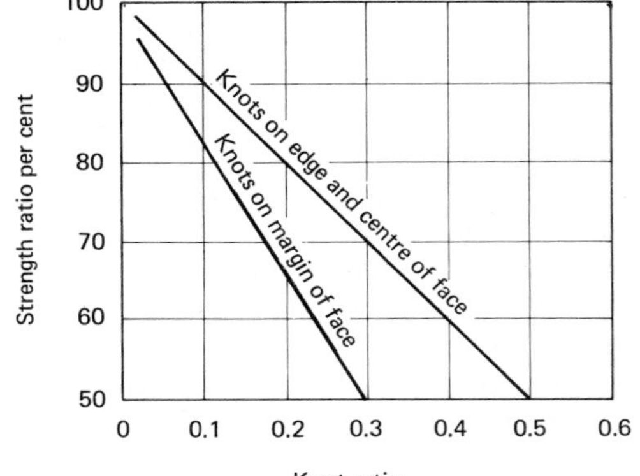

Fig. 14.3. *Effect of "knot ratio" on strength. (From J. G. Sunley, "Working Stresses for Structural Timbers"[2], with acknowledgement.)*

The effect of defects is examined next. This problem is simplified by classifying the timber into strength grades and grading rules are incorporated in BSCP 112, *The Structural Use of Timber*.[1] There are four strength grades, whose strengths are 75%, 65%, 50% and 40% of the strength of clear timber. The grade stress for Grade 65, for example, is 65% of the basic stress. If there is an experimental correlation of defects with the strengths of specimens containing them, rules for grading can be formulated, as in BSCP 112. With regard to knots, the knot ratio was related to strength by Sunley for the species he tested, the results being shown on Fig. 14.3.[2] The kind of knot ratio referred to here is the ratio of the size of the knot to the width of the face or edge in which it occurs. For a plank on edge, the position of a knot is important, those on the margin of a face (see Fig. 14.4) being more detrimental than those on an edge or in the centre of the face. The grading rules give limits for the sizes of knots in various positions, the biggest knot allowable in the grade being the one which causes the corresponding loss of strength.

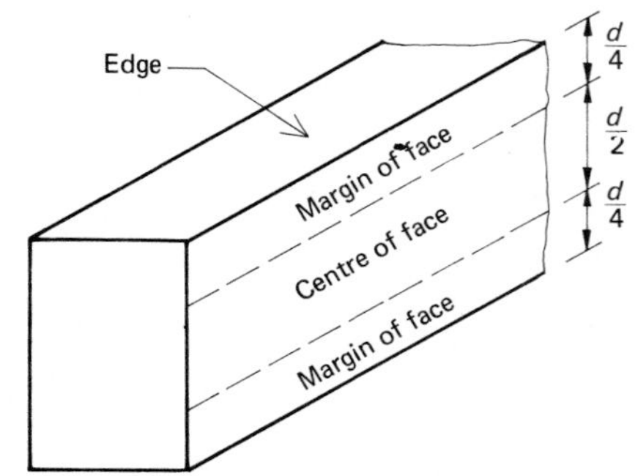

Fig. 14.4. *Definition of edge, margin of face and centre of face. (Adapted from BSCP 112,* The Structural Use of Timber[1], *by permission of the British Standards Institution, 2 Park Street, London W1A 2BS.)*

Similarly, fissures, slope of grain and wane are limited, the maximum allowable defect increasing for the lower grades. Rate of growth is measured by counting the number of growth rings per 25 mm and there is a minimum number of growth rings allowable in each grade.

In summary then the designer can find in a design specification, such as BSCP 112, basic stresses for the species of timber he intends to use, grade stresses which allow for the presence of defects and grading rules by means of which visual inspection can be used to classify commercially available timber.

The permissible stresses are obtained by multiplying the grade stresses by factors which allow for the particular circumstances under which the timber will be used. Two important features are the moisture content and the duration of loading. Green timber has a moisture content greater than the fibre saturation limit. As noted above, the strength and stiffness properties of green timber do not change much with moisture content provided it exceeds the fibre saturation limit, but as the timber loses moisture below this limit, these properties improve. The British code of practice gives basic stresses for two conditions of moisture content. The stresses for green timber are applied if the moisture content in service exceeds 18%. Otherwise what are called dry stresses are used.

The moisture content of timber is significant in the effect on the volume and distortion of a plank and the durability of the material as well as its effect on strength and stiffness. If the moisture content is less than 30%, wood shrinks as it dries and swells as it gets wet. Since wood is hygroscopic, it tends to adjust its moisture content according to its environment. To avoid distortion of completed structures, it is desirable that the moisture content of all the timber should be close to its equilibrium value at the time of construction. Table 14.1 is included here as a guide to the selection of suitable moisture contents.

Table 14.1. Moisture content of timber for various positions.

Position of timber in building	Average moisture content attained in use in a dried-out building (% of dry weight)	Moisture content which should not be exceeded at the time of erection (% of dry weight)
Framing and sheathing of timber buildings (not prefabricated)	16	22
Timber for prefabricated buildings	16	17 for precise work, otherwise 22
Rafters and roof boarding, tile battens, etc.	15	22
Ground floor joists	18	22
Upper floor joists	15	22

Source: Extract from BSCP 112, *The Structural Use of Timber,* is reproduced by permission of the British Standards Institution, 2 Park Street, London, W1A 2BS.

Duration of loading is relevant because the strength and stiffness of timber are both greater for loads of short duration than for loads of long duration. This is allowed for by multiplying the grade stress by a factor which ranges from 1 for long-term loading to 1.5 for short-term loading (see Table 14.8).

No corrections are made to allow for buckling of the compression side of a timber beam, stability being controlled by restrictions on the ratio of the depth of the beam to its breadth. Timber beams are usually rectangular in cross-section, so that the torsional stiffness is quite high. Furthermore, most beams have the compression side braced by the floor or roof supported. These things contribute strongly to the stability of a beam and provided the limits in Table 14.2 are observed, the permissible compressive stress need not be less than the tensile stress.

Table 14.2. Maximum depth-to-breadth ratios (solid and laminated members).

Degree of lateral support	Maximum depth-to-breadth ratio
No lateral support	2
Ends held in position	3
Ends held in position and member held in line, as by purlins or tie rods	4
Ends held in position and compression edge held in line, as by direct connection of sheathing, deck or joists	5
Ends held in position and compression edge held in line, as by direct connection of sheathing, deck or joists, together with adequate bridging or blocking spaced at intervals not exceeding 6 times the depth	6
Ends held in position and both edges firmly held in line	7

Source: Extract from BSCP 112, *The Structural Use of Timber*, is reproduced by permission of the British Standards Institution, 2 Park Street, London, W1A 2BS.

14.4. PERMISSIBLE STRESSES FOR LAMINATED BEAMS

Laminated timber beams are built by gluing planks together face to face or edge to edge or both. In this fashion, beams of large cross-section can be constructed and since planks can be joined end-to-end before assembly, very long beams can be built. The laminations in a beam may be horizontal or vertical. Since more is gained by horizontal lamination, it is more widely used and is the subject of this discussion.

Laminated construction is of great value to the timber industry because it permits structural use of timber from small trees. Improved quality is also obtained in comparison with beams of sawn timber. In the first place, planks of small cross-section can be dried with less risk of seasoning defects than large balks can be. Size for size, a laminated beam will be of better quality than a sawn beam for this reason alone. More important, however, is the dispersion of knots. A knot in sawn timber may continue for a considerable distance through a sawn balk. In a laminated beam, it is unlikely that a knot in one lamination will line up with another in an adjacent lamination. Lamination may be said to lead to a virtual upgrading of the timber.

Grading of timber for laminated beams follows rules somewhat different from those for sawn timber. Laminating grades described in BSCP 112, *The Structural Use of Timber* are called LA, LB and LC corresponding to knot area ratios of 0, $\frac{1}{4}$ and $\frac{1}{2}$.[1] In horizontally laminated beams, the position of a knot across the face of a lamination does not matter – since the whole face is the same distance from the neutral axis of the beam, a defect in the face causes the same loss of strength wherever it is. In vertically laminated beams, the same rules apply as for sawn timber.

With regard to other defects, slope of grain is limited so as not to cause a greater loss of strength than the knots in the grade. Wane is to be avoided altogether but there is no restriction on the rate of growth. Fissures are not considered to be detrimental, provided the angle between a fissure and a face exceeds 45°.

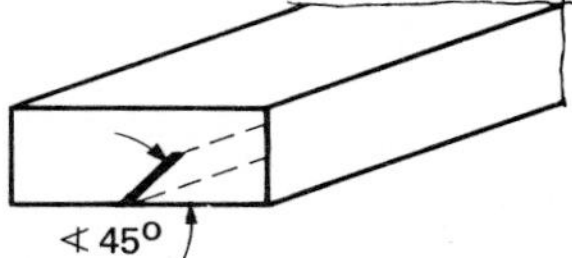

Curry has described an investigation of the strength of laminated beams undertaken to determine the strength ratios for horizontally laminated beams.[3] The tests were conducted with four species of softwood – Scots pine, Sitka spruce and larch grown in Britain, and Canadian-grown Douglas fir.

After preparation, the laminations were graded according to the rules for laminating stock. Beams were constructed with various numbers of laminations and were tested in bending to find the modulus of rupture and modulus of elasticity. The broken beams were then examined to get data on the knots in cross-sections near the site of the failure. A knot second-moment ratio is defined as $I_K : I_G$, where I_K is the second moment of area of knots in cross-section, and I_G is the second moment of area of gross cross-section.

From his experiments, Curry related modulus of rupture and modulus of elasticity to knot second-moment ratio. Knot area ratios were then found for individual laminations in cross-sections near the site of the failure, the knot area ratio being $A_K : A_G$ where A_K is the area of knots in cross-section of lamination, and A_G is the gross area of cross-section of lamination.

These data were obtained after sawing the beam through at the cross-sections of interest and sawing along the glue lines to recover short lengths of the original laminations. Randomly selected arrangements of the planks were then chosen and the knot second-moment ratio for each was computed, selection and computation being done on a digital computer, so that a large sample of knot second-moment ratios could be generated. The number of knot measurements made was about 20 000; the knot second-moment ratios were generated for 1000 samples for each of 144 constructions. Knot area ratios for the laminated constructions were also generated, but they are not relevant to this discussion.

The knot ratios were found to be normally distributed for beams of four to twenty laminations, so that the knot second-moment ratio which would be exceeded in all but 1% of the arrangements could be determined. This knot second-moment ratio and the relationship between strength ratio and knot ratio were used to get strength ratios for the various laminating grades and four to twenty laminations. Some of Curry's results have been plotted in Fig. 14.5(a), (b) and (c). These relate to beams in which all the laminations are of the same grade. Curry's results include cases where inner laminations are of lower grade than the outer ones. Such timber use has the advantage of being more efficient, since any commercial sample contains planks from all grades. However, there is the risk of getting the laminations mixed, so that careful supervision of construction of this kind is necessary.

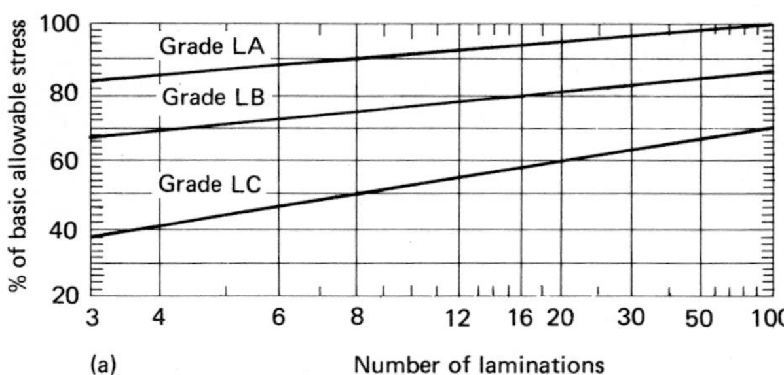

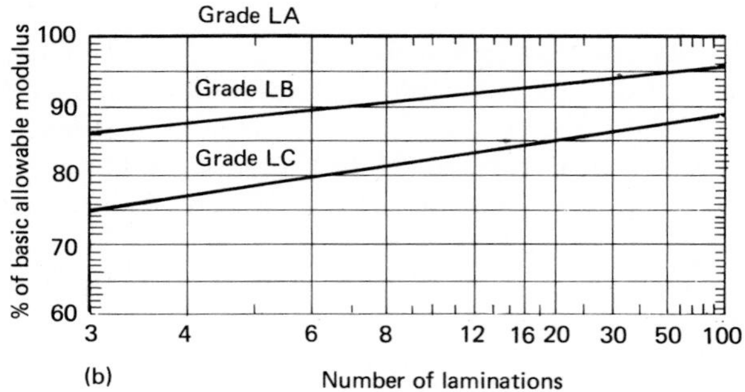

Fig. 14.5. *Effect of number of laminations on strength and stiffness: (a) ratio for bending stress; (b) ratio for modulus of elasticity in bending; (c) ratio for compressive stress parallel to grain. (Adapted from W. T. Curry, "Grade Stresses for Structural Laminated Timber"[3], with acknowledgement.)*

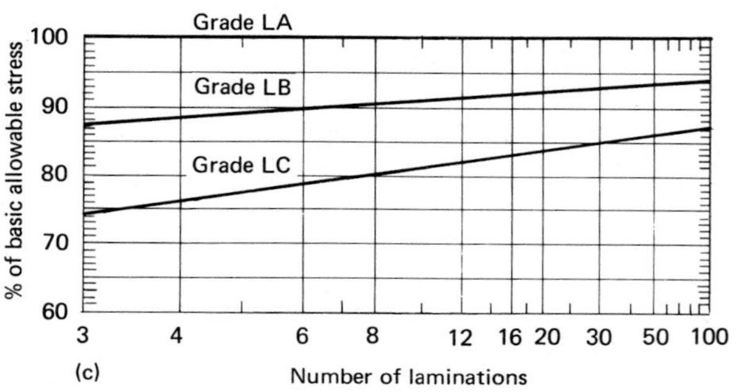

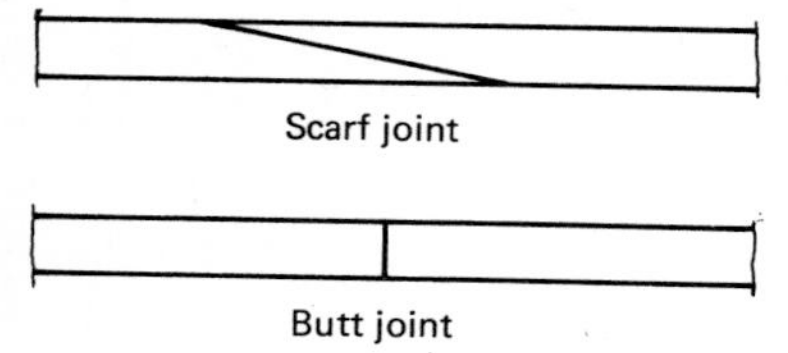

The extent to which joints interfere with the strength and stiffness of a laminated beam needs some consideration. The adhesives most widely used are synthetic resins and the designer can safely assume that well-made joints between faces and edges of laminations are at least as strong as the timber. End joints are not so easily disposed of. A good scarf joint is almost as strong as the planks it joins. On the other hand, a butt joint interrupts a lamination in tension almost completely because its strength is so low. The strength of the beam at the site of a butt joint is diminished not only because of the ineffectiveness of this lamination locally but also because of the stress-raising effect of the interruption. However, even a poorly made joint does not affect deflection much. The reduced second moment of area increases curvature locally but such local increases do not cause a significant increase in deflection, which depends on the curvature at all sections along the beam.

Efficiency ratings for plain scarf joints are given in BSCP 112, *The Structural Use of Timber*.[1] These are factors to be applied to *basic* stresses, but not to modulus of elasticity. Clearly, it is desirable to design the layout of laminations so as to avoid joints where stresses are high. Butt joints, although easy to make, should be avoided. If they are used, the lamination with the joint should be disregarded when properties of the cross-section are computed. Some authorities go further – for example, Freas and Selbo say that the nett second moment of area should be reduced by 20% to allow for butt joints in tension and that butt joints should not be allowed in the outermost laminations at all.[4]

14.5. EXAMPLES OF DESIGN OF TIMBER BEAMS

To facilitate understanding of Examples 14.1 and 14.2, some data from BSCP 112 are presented here. We emphasise that these tables are not a substitute for the local design specification and they are presented here solely to make this chapter, including the worked examples, self-contained.

Table 14.3. Softwood species groups.

Species group	Standard name	Origin
S1	Douglas fir Pitch pine	Imported
	Douglas fir Larch	Home-grown
S2	Western hemlock (unmixed) Western hemlock (commercial) Parana pine Redwood Whitewood Canadian spruce	Imported
	Scots pine	Home-grown
S3	European spruce Sitka spruce	Home-grown
	Western red cedar	Imported

Note: The terms "Home-grown" and "Imported" are relevant to U.K.

Source: Extract from BSCP 112, *The Structural Use of Timber*, is reproduced by permission of the British Standards Institution, 2 Park Street, London W1A 2BS.

In Table 14.3, fourteen species of softwood are formed into groups with similar strength and stiffness properties. Table 14.4 shows the basic and grade stresses and moduli of elasticity for the groups in a green state and Table 14.5 gives the same data for dry timber. Table 14.6 gives strength ratios for horizontally laminated members in which all of the laminations are of the same grade, and Table 14.7 gives the efficiency ratings of scarf joints. In Table 14.8 there are shown modification factors for duration of loading. The grades referred to in these tables are those described in the grading rules of BSCP 112, *The Structural Use of Timber*.[1]

Example 14.1 illustrates design of a timber floor and Example 14.2 design of a laminated beam.

Table 14.4. Green stresses and moduli of elasticity for grouped softwoods.

Species group	Grade	Bending and tension parallel to grain	Compression parallel to grain	Compression perpendicular to grain	Shear parallel to grain	Modulus of elasticity	
						Mean	Minimum
		N/mm^2	N/mm^2	N/mm^2	N/mm^2	N/mm^2	N/mm^2
	Basic	13.8	9.7	1.72	1.38		
	75	10.3	7.2	1.52	1.03		
S1	65	9.0	6.2	1.52	0.90	9000	4500
	50	6.9	4.8	1.31	0.69		
	40	5.5	3.8	1.31	0.55		
	Basic	11.0	8.3	1.38	1.38		
	75	8.3	6.2	1.17	1.03		
S2	65	6.9	5.2	1.17	0.90	6900	4100
	50	5.5	4.1	1.03	0.69		
	40	4.5	3.1	1.03	0.55		
	Basic	7.6	5.5	1.03	1.10		
	75	5.5	4.1	0.90	0.83		
S3	65	4.8	3.4	0.90	0.69	5900	3100
	50	3.8	2.8	0.76	0.55		
	40	3.1	2.1	0.76	0.41		

Note: These stresses apply to timber having a moisture content exceeding 18%.

Source: Extract from BSCP 112, *The Structural Use of Timber*, is reproduced by permission of the British Standards Institution, 2 Park Street, London, W1A 2BS.

Table 14.5. Dry stresses and moduli of elasticity for grouped softwoods.

Species group	Grade	Bending and tension parallel to grain	Compression parallel to grain	Compression perpendicular to grain	Shear parallel to grain	Modulus of elasticity	
						Mean	Minimum
		N/mm^2	N/mm^2	N/mm^2	N/mm^2	N/mm^2	N/mm^2
	Basic	17.2	13.1	2.48	1.52		
	75	12.1	9.3	2.21	1.14		
S1	65	10.3	7.6	2.21	0.97	9700	4800
	50	7.9	5.5	1.93	0.76		
	40	6.2	4.5	1.93	0.62		
	Basic	13.8	11.0	2.07	1.52		
	75	9.7	7.9	1.72	1.14		
S2	65	7.9	6.6	1.72	0.97	8300	4500
	50	6.2	4.8	1.52	0.76		
	40	5.2	3.8	1.52	0.62		
	Basic	10.3	8.3	1.52	1.24		
	75	6.6	5.2	1.31	0.90		
S3	65	5.5	4.1	1.31	0.76	6900	3800
	50	4.5	3.1	1.10	0.62		
	40	3.4	2.4	1.10	0.45		

Note: These stresses apply to timber having a moisture content not exceeding 18%.

Source: Extract from BSCP 112, *The Structural Use of Timber*, is reproduced by permission of the British Standards Institution, 2 Park Street, London W1A 2BS.

Table 14.6. Modification factors for single-grade glued laminated members and horizontally laminated beams.

Grade	No. of laminations	Values of the modification factors				
		Bending tension parallel to grain	E in bending, to be applied to the mean value of E	Compression parallel to grain	Shear parallel to grain	Compression perpendicular to grain
LA	4 or more	1.00	1.00	1.00	0.90	1.00
LB	4	0.68	0.89	0.87	0.90	1.00
	5	0.72	0.89	0.89		
	10	0.77	0.91	0.91		
	15	0.80	0.92	0.92		
	20	0.82	0.93	0.93		
	30	0.84	0.94	0.93		
	50	0.85	0.94	0.94		
	100	0.88	0.95	0.95		
LC	4	0.40	0.77	0.76	0.90	1.00
	5	0.44	0.78	0.78		
	10	0.53	0.82	0.82		
	15	0.58	0.83	0.83		
	20	0.60	0.84	0.84		
	30	0.64	0.86	0.85		
	50	0.66	0.87	0.87		
	100	0.70	0.88	0.88		

Source: Extract from BSCP 112, *The Structural Use of Timber*, is reproduced by permission of the British Standards Institution, 2 Park Street, London W1A 2BS.

Table 14.7. Efficiency ratings for plain scarf joints.

Bending and tension

Grade	Efficiency rating for a scarf slope of:			
	1 in 6	1 in 8	1 in 10	1 in 12
LA	0.69	0.77	0.84	0.88
LB	0.67	0.75	0.81	0.85
LC	0.50	0.65	0.68	0.72

Compression

Grade	Efficiency rating for a scarf slope of:			
	1 in 6	1 in 8	1 in 10	1 in 12
LA	1.00	1.00	1.00	1.00
LB	1.00	1.00	1.00	1.00
LC	0.80	0.95	1.00	1.00

Source: Extract from BSCP 112, *The Structural Use of Timber*, reproduced by permission of the British Standards Institution, 2 Park Street, London, W1A 2BS.

Table 14.8. Modification factor for duration of loading on flexural members and members in tension.

Duration of loading	Value of factor
Long-term (e.g., dead + permanent imposed)	1.00
Medium-term (e.g., dead + snow; dead + temporary loads)	1.25
Short-term (e.g., dead + imposed + wind; dead + imposed + snow + wind)	1.5

Source: Extract from BSCP 112, *The Structural Use of Timber*, is reproduced by permission of the British Standards Institution, 2 Park Street, London W1A 2BS.

Example 14.1 Sheet 1 of 1

Select joists for a timber floor for a warehouse. Live load 1000 kg/m². Main beams 2.5 m c/c. Group S1 timber, Grade 65

Try 500 mm spacing
Allow for self-weight of floor average thickness equivalent to floorboards and joists of 60 mm

At 600 kg/m³, 60 mm thickness ⟶ 36 kg/m²
Self weight = 36 × 9.81 = 353 Pa
Superimposed load = 1000 × 9.81 = 9810 Pa
10 163 Pa

For one joist $W = 10.16 \times 0.5 \times 25 = 12.7$ kN

$R = V = \frac{W}{2} = 6.35$ kN

$M = \frac{WL}{8} = \frac{12.7 \times 2.5}{8} = 3.97$ kNm

Say, moisture content 15-18% - dry conditions of service
Modification factor for duration of loading = 1.0

Permissible bending stress = 10.3 MPa

$Z = \frac{3.97 \times 10^3}{1.03 \times 10^7} = 3.86 \times 10^{-4} m^3$
$= 3.86 \times 10^5 mm^3$

Try 50 mm × 250 mm, basic size ⟶ 47 × 244 nett

Depth to breadth $= \frac{244}{47} = 5.2$
Herring-bone at midspan

$A = 1.15 \times 10^4 mm^2$, $Z = 4.66 \times 10^5 mm^3$
$I = 5.69 \times 10^7 mm^4$

$f_{bc} = f_{bt} = \frac{3.97 \times 10^3}{4.66 \times 10^{-4}} = 8.50 \times 10^6$ Pa $= 8.50$ MPa
ok

$f_q = \frac{3}{2} \times \frac{6.35 \times 10^3}{1.15 \times 10^{-2}} = 8.28 \times 10^5$ Pa $= 0.828$ MPa
< 0.97 MPa
ok

For $f_b \not> 2.21$ MPa, $\ell \not< \frac{6.35 \times 10^3}{2.21 \times 47} = 61.2$ mm

$\Delta = \frac{1}{76.8}\frac{WL^3}{EI} = \frac{1.27 \times 10^4 \times 2.5 \times 2.5 \times 2.5}{76.8 \times 9.7 \times 10^9 \times 5.69 \times 10^{-5}} = 4.69 \times 10^{-3} m$
$= 4.69$ mm

$\frac{\Delta}{L} = \frac{4.69 \times 10^{-3}}{2.5} = 0.00188$
ok

Example 14.2 Sheet 1 of 2

Design a laminated beam for the warehouse floor of Example 14.1. Group S1 timber, Grade LB. Columns spaced 6m c/c.

Total load contributed by floor to one beam
$= 10.16 \times 2.5 \times 6$
$= 152$ kN

Estimated self-weight:
Say, 600 mm × 300 mm → 0.18 m³/m
At 600 kg/m³, self-weight $= 600 \times 0.18 \times 6 \times 9.81$ N
$= 6.36$ kN

Say, $W = 160$ kN

$R = V = \frac{W}{2} = 80$ kN

$M = \frac{WL}{8} = \frac{160 \times 6}{8} = 120$ kNm

Try bending stress $= 0.8 \times 17.2 = 13.8$ MPa

$Z = \frac{1.2 \times 10^5}{1.38 \times 10^7} = 8.70 \times 10^{-3}\ m^3 = 8.70 \times 10^6\ mm^3$

Try $b = 145$ mm, $d = \sqrt{\frac{6 \times 8.70 \times 10^6}{1.45 \times 10^2}} = 601$ mm

Try 18 laminations, basic size 36 × 150 → 33 × 145 nett

$b = 145$ mm, $d = 18 \times 33 = 594$ mm, $\frac{d}{b} = 4.1$ OK

$A = 145 \times 594 = 8.60 \times 10^4\ mm^2$ $Z = 1/6 \times 145 \times 594^2 = 8.50 \times 10^6\ mm^3$ $I = 1/12 \times 145 \times 594^3 = 2.52 \times 10^9\ mm^4$

Permissible stresses:
Dry conditions, permanent load.

Bending:	0.82 × 17.2 =	14.1 MPa
Shear:	0.90 × 1.52 =	1.37 MPa
Bearing:	1.0 × 2.48 =	2.48 MPa
Modulus of elasticity:	0.93 × 9700 =	9020 MPa

$f_{bc} = f_{bt} = \frac{1.20 \times 10^5}{8.50 \times 10^{-3}} = 1.41 \times 10^7\ Pa = 14.1$ MPa OK

$f_q = \frac{3}{2} \times \frac{8 \times 10^4}{8.60 \times 10^{-2}} = 1.40 \times 10^6\ Pa = 1.40$ MPa

Example 14.2 Sheet 2 of 2

Check f_q with more accurate estimate of self-weight:

Self-weight of beam $= 600 \times 8.6 \times 10^{-2} \times 6 \times 9.81 \times 10^{-3}$ kN

$= 3.04$ kN

$$V = \frac{W}{2} = \frac{155}{2} = 77.5 \text{ kN}$$

$$f_q = 1.36 \text{ MPa} \qquad \text{OK}$$

For $f_b \not> 2.48$ MPa $\qquad l \not< \frac{8 \times 10^4}{145 \times 248} = 222$ mm

$$\Delta = \frac{1}{76.8}\frac{WL^3}{EI} = \frac{1}{76.8} \times \frac{1.6 \times 10^4 \times 6^3}{9.02 \times 10^9 \times 2.52 \times 10^{-3}}$$

$$= 1.98 \times 10^{-3} \text{ m}$$

$$= 1.98 \text{ mm}$$

$$\frac{\Delta}{L} = \frac{1.98 \times 10^{-3}}{6} = 0.00033 \qquad \text{OK}$$

Efficiency of scarf joint in tension, slope 1 in 12 $= 0.88$

Grade stress limitation more severe.

Scarf joints OK anywhere, but not in adjacent laminations at same section

SUMMARY : 145 mm × 594 mm beam
18 laminations ex 36 × 150
Scarf joints 1 in 12
$\not<$ 230 mm length of bearing.

LIST OF SYMBOLS

A_G	gross area of cross-section
A_K	area of knots in cross-section
b	breadth of beam
d	depth of beam
E	modulus of elasticity
f_b	bearing stress
f_{bc}	compressive stress due to bending
f_{bt}	tensile stress due to bending
f_q	shear stress
I, I_G	second moment of area of cross-section
I_K	second moment of area of knots in cross-section
K	factor
L	span of beam, centre-to-centre of supports
l	length of bearing at support
M	bending moment
R	reaction at support
V	external shear
W	total load on beam
$\overline{x}$	mean value of test results
Z	section modulus
Δ	deflection of beam
σ	standard deviation of test results

REFERENCES

[1] BS CP 112:Part 2:1971, "Metric Units", *The Structural use of Timber*, British Standards Institution, London, 1971.

[2] J. G. Sunley, "Working Stresses for Structural Timbers", *Forest Products Research Bulletin No. 47*, 2nd ed., H.M. Stationery Office, London, 1965.

[3] W. T. Curry, "Grade Stresses for Structural Laminated Timber", *Forest Products Research Special Report No. 15*, 2nd ed., H.M. Stationery Office, London, 1967.

[4] A. D. Freas and M. L. Selbo, "Fabrication and Design of Glued Laminated Wood Structural Members", *Technical Bulletin No. 1069*, U.S. Department of Agriculture, Washington D.C., 1954.

BIBLIOGRAPHY

Hansen, H. J., *Modern Timber Design*, John Wiley, New York, 1948.

Hinds, H. V., and Reid, J. S., *Forest Trees and Timbers of New Zealand*, Government Printer, Wellington, 1957.

An authoritative survey of the timber industry in New Zealand.

NZSR 34:1968, *Glued Laminated Timber Construction*, Standards Association of New Zealand, Wellington, 1968.

A guide to design and construction of laminated timber in New Zealand.

Pearson, R. G., Kloot, N. H. and Boyd J. D., *Timber Engineering Design Handbook*, 2nd ed., Jacaranda Press (in assn. with CSIRO), Brisbane, 1962.

A general book on timber design, written for Australian designers.

Wallis, N. K., *Australian Timber Handbook*, 3rd ed., Angus & Robertson, Sydney, 1970.

A survey of the use of timber in Australia.

PROBLEMS

14.1. What is the maximum span for 50-mm thick floorboards, for a floor load of 1000 kg m^{-2}? Assume a softwood from species group S1 and a limit on deflection of $\frac{1}{300}$ span. Allow for the floorboards being continuous over the joists.

14.2. Estimate the natural frequency of vibration of the flooring in Problem 14.1 when the span is the maximum allowed (see Chapter 15, "Steel Beams" and Reference 3 in Chapter 15). Do you think this would be a good floor?

14.3. Redesign the joists in Example 14.1 with a spacing of 750 mm. Assuming floorboards to be 50 mm thick, compare the cost per square metre of floor of the joists and floorboards in Example 14.1 and in this problem. Use cost data from Chapter 6 or locally available costs.

14.4. Repeat Problem 14.3 with joists spacings 500 mm and 750 mm, but with main beams spaced 2.8 m centre-to-centre.

14.5. Redesign the laminated beams of Example 14.2 for a spacing of 2.8 m and estimate the cost per square metre of the whole floor structure for the two designs. Use a joist spacing of 500 m.

15 Steel Beams

15.1. INTRODUCTION

The function of a beam in a building frame or a bridge has been described in Chapter 5, "Machines and Structures". Steel beams may be rolled I-sections with standard dimensions or specially built plate girders. In steel-framed buildings, the standard I-sections are predominant; both types of beams are widely used to support bridges.

There are many other places where an elevated load has to be supported and such loads range, for example, from large boilers to a hoist for unloading trucks in a cart dock. Ease of fabrication, structural efficiency and ready availability make the rolled I-section an almost automatic choice.

The I-shaped section is efficient because of the distribution of bending stress in a beam. With the usual assumptions made for the simple theory of bending of an elastic beam, highest stresses occur furthest from the neutral axis. If the area of cross-section is also concentrated there, best use can be made of it in mobilising internal forces. Furthermore, the lever arms of these parts of the section are longest. There are, then, two reasons why the flanges contribute most to the moment of resistance in elastic bending.

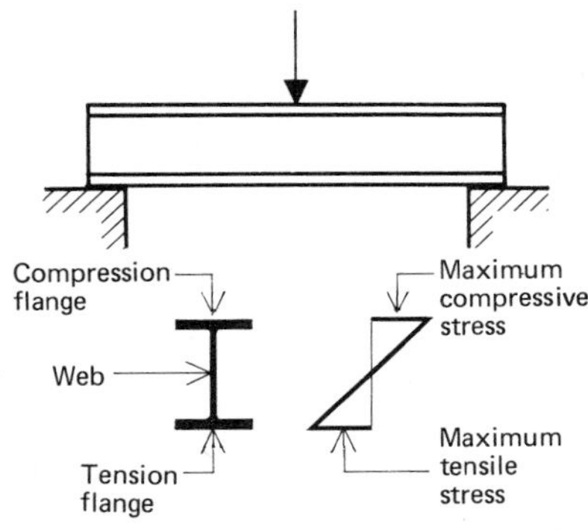

Design of steel beams may be on the basis of ultimate strength and service loads multiplied by a load factor or on the basis of working stresses and service loads. In the former, a plastic theory of bending is used; in the latter elastic behaviour is assumed. The difference between the two approaches is not important for statically determinate structures, but ultimate load analysis is considered to give a more realistic appraisal of the strength of indeterminate structures. Working stress design, based on elastic deformation, is still widely used in routine design of steel beams and this is what we will discuss here. The working stresses which must not be exceeded in service are chosen so that non-linear behaviour, such as yielding or buckling, will not occur. When it is necessary to refer to a design specification, we will use the rules from BS 449, *The Use of Structural Steel in Building.*[1] Dimensions of standard sections and their properties are taken from BS 4, *Specification for Structural Steel Sections*[2]; similar information is available in handbooks published by most steelmakers.

The term *rolled I-beam* has been used above to include both universal beams and joists, two classes of I-beams whose properties are listed in BS 4.[2] There are other I-sections that are rolled – for example, universal columns and universal bearing piles. Any of these can be used as a beam. The universal sections have flanges that are untapered or are tapered only to a small angle. Flanges of joists have a steeper taper. Universal beams are rolled in sizes 8 × $5\frac{1}{4}$ (203 mm × 133 mm) up to 36 × $16\frac{1}{2}$ (920 mm × 420 mm), while joists cover the range 8 × 4 (203 mm × 102 mm) down to 3 × 2 (76 mm × 51 mm).

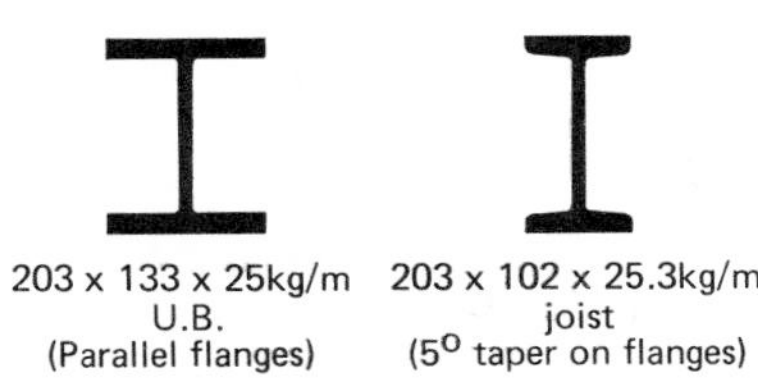

The following are the matters taken into account in design of a steel beam:

(a) strength in bending with due regard to the stability of the compression flange;
(b) stiffness in bending so as to avoid vibration and unduly large deflections;
(c) strength in shear;
(d) strength with respect to concentrated loads.

15.2. STRENGTH IN BENDING, ROLLED BEAMS

The assumptions made in the simple theory of elastic bending are that longitudinal strain is distributed linearly through the depth of the beam and that stress is proportional to strain. The distribution of stress is then given by

$$\frac{f}{y} = \frac{M}{I} = \frac{E}{R} \tag{15.1}$$

where f is the stress due to bending at distance y from the neutral axis; M is the bending moment, in general a function of distance along the beam; I is the second moment of area of the cross-section about a centroidal axis; E is the modulus of elasticity of material of beam; and R is the radius of curvature of beam under the action of the bending moment.

The maximum stress is of interest to the designer and this occurs where y is a maximum – i.e. at the extreme fibres of the beam. Thus the maximum compressive stress is given by

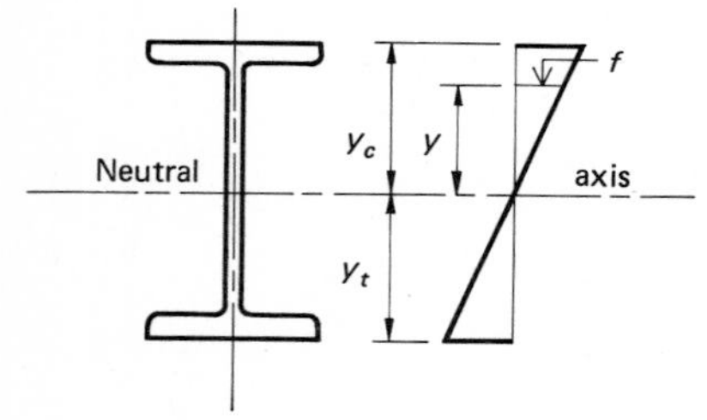

$$f_{bc} = \frac{My_c}{I} \tag{15.2}$$

and the maximum tensile stress by

$$f_{bt} = \frac{My_t}{I} \tag{15.3}$$

Since the rolled I-section is symmetrical, y_c and y_t are equal and

$$\frac{I}{y_c} = \frac{I}{y_t} = Z$$

Z, the elastic section modulus, is a property of the section tabulated in BS 4, for example.[2] Equations (15.2) and (15.3) may now be replaced by

$$f_{bc} = f_{bt} = \frac{M}{Z} \tag{15.4}$$

However, it must be noted that a rolled I-section may be made unsymmetrical by drilling holes in one flange or by plating the flanges with unequal plates. In that event, the tabulated values of I, y_c and y_t must be modified to allow for the altered distribution of area and Equations (15.2) and (15.3) must be used to calculate the stresses. Examples 15.1 and 15.2 show how these adjustments are made. The relatively large effect of the holes in Example 15.1 on the section modulus should be noted; it is clearly important to avoid holes in flanges near sections where the bending moment is large.

The tensile and compressive stresses computed from Equations (15.2) and (15.3) or Equation (15.4) must not exceed the permitted values p_{bc} and p_{bt}

Example 15.1 Sheet 1 of 1

Find y_t, y_c, I_{NA} for a 533 × 330 × 212 kg/m universal beam with two 22 mm diameter holes through the top flange.

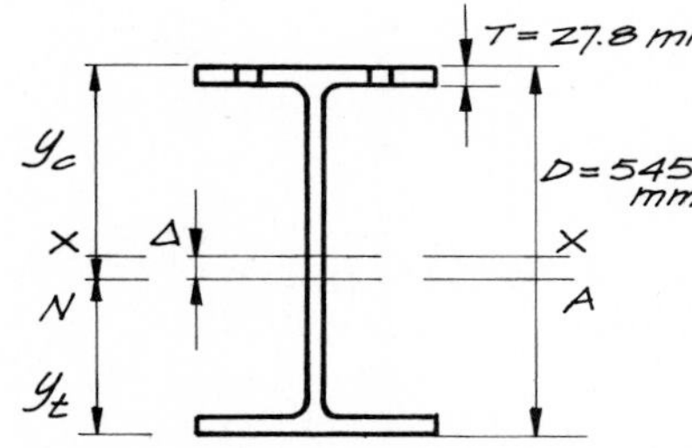

XX is the centroidal axis of gross section

NA is the centroidal axis of nett section

Area of gross section = $A = 2.696 \times 10^4$ mm²

Projected area of holes $= \Delta A = 2 \times 22 \times 27.8$

$= 1.22 \times 10^3$ mm²

Take moments about XX

$$(A - \Delta A)\Delta = A \times 0 - \Delta A\left(\frac{D}{2} - \frac{T}{2}\right)$$

$$\Delta = -\frac{\Delta A/A}{1 - \Delta A/A}\;\frac{D-T}{2}$$

$$= -\frac{0.0455}{0.9545} \times \frac{517.2}{2}$$

$$= -12.3\ \text{mm}$$ — i.e. NA is below XX

$y_t = 272.5 - 12.3 = 260.2$ mm

$y_c = 272.5 + 12.3 = 284.8$ mm

$$I_{NA} = I_{xx} + A\Delta^2 - \Delta A\left(y_c - \frac{T}{2}\right)^2$$

$$= 1.417 \times 10^9 + 2.696 \times 10^4 \times 1.23^2 \times 10^2 - 1.22 \times 10^3 \times 2.709^2 \times 10^4$$

$$= 1.417 \times 10^9 + 4.07 \times 10^6 - 8.96 \times 10^7$$

$$= 1.331 \times 10^9\ \text{mm}^4$$

$$\frac{I_{NA}}{y_t} = \frac{1.331 \times 10^9}{2.602 \times 10^2} = 5.12 \times 10^6\ \text{mm}^3$$

$$\frac{I_{NA}}{y_c} = \frac{1.331 \times 10^9}{2.848 \times 10^2} = 4.68 \times 10^6\ \text{mm}^3$$

(Compare with Z_{xx} for gross section of 5.199×10^6 mm³)

Example 15.2 Sheet 1 of 1

Find y_t, y_c, I_{NA} for a 533 × 330 × 212 kg/m universal beam with a 300 × 25 mm plate welded to the compression flange.

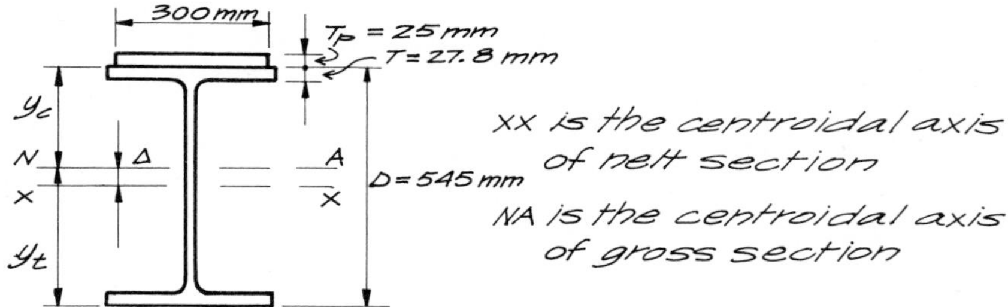

Area of nett section $= A = 2.696 \times 10^4 \text{ mm}^2$

Area of plate $= A_p = 300 \times 25$

$= 7.5 \times 10^3 \text{ mm}^2$

Take moments about XX:

$$(A + A_p)\Delta = A_p\left(\frac{D}{2} + \frac{T_p}{2}\right)$$

$$\Delta = \frac{A_p/A}{1 + A_p/A} \quad \frac{D + T_p}{2}$$

$$= \frac{0.278}{1.278} \times \frac{570}{2}$$

$$= 62.2 \text{ mm}$$

$y_t = 272.5 + 62.2 = 334.7$ mm

$y_c = 272.5 + 25 - 62.2 = 235.3$ mm

$$I_{NA} = I_{xx} + A\Delta^2 + A_p\left(y_c - \frac{T_p}{2}\right)^2$$

$$= 1.417 \times 10^9 + 2.696 \times 10^4 \times 6.22^2 \times 10^2 + 7.5 \times 10^3 \times 2.228^2 \times 10^4$$

$$= 1.417 \times 10^9 + 1.04 \times 10^8 + 3.72 \times 10^8$$

$$= 1.893 \times 10^9 \text{ mm}^4$$

$$\frac{I_{NA}}{y_t} = \frac{1.893 \times 10^9}{3.347 \times 10^2} = 5.66 \times 10^6 \text{ mm}^3$$

$$\frac{I_{NA}}{y_c} = \frac{1.893 \times 10^9}{2.353 \times 10^2} = 8.05 \times 10^6 \text{ mm}^3$$

(Compare with Z_{xx} for nett section of 5.199×10^6 mm^3)

respectively. When buckling of the compression flange does not need to be taken into account, they are equal and they depend on the grade of steel, the shape of the section and the thickness of the steel in it. For example, values of p_{bt} for rolled I-beams and universal beams are as follows:

Rolled I-beams:

Grade 43 steel,	all thicknesses	165 MPa
Grade 50 steel,	thickness $\ngtr$ 65 mm	230 MPa
	thickness > 65 mm	$f_y/1.52 \ngtr$ 230 MPa
Grade 55 steel,	thickness $\ngtr$ 40 mm	280 MPa
	thickness > 40 mm	260 MPa

(f_y is the yield strength of the steel.)

Universal beams:

Grade 43 steel,	thickness $\ngtr$ 40 mm	165 MPa
	thickness > 40 mm	150 MPa

(Extract from BS 449, *The Use of Structural Steel in Building*, is reproduced by permission of the British Standards Institution, 2 Park Street, London W1A 2BS.)

Buckling of the compression flange must be taken into account if the flange is slender enough. A slender strut is unstable under a sufficiently large axial compression and, in a similar way, the compression flange of a beam may be unstable if the transverse load on the beam is large. Consider a simply supported beam, with the ends supported in such a way that the web stays vertical at the ends. Let the load on the beam be increased until the compression flange buckles, as indicated in Fig. 15.1. Seen from above, the compression flange will be curved, while the tension flange remains straight (Fig. 15.1(b)). Furthermore, the beam is twisted about a longitudinal axis, as indicated in Fig. 15.1(c), and (d).

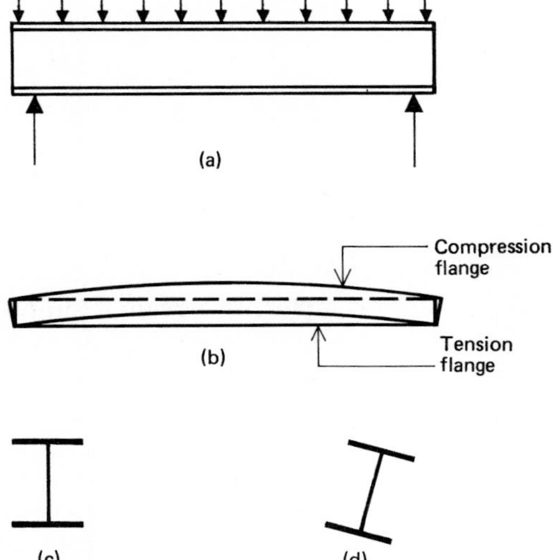

Fig. 15.1. *Buckling of compression flange of beam:*
(a) loaded beam in elevation;
(b) plan showing compression flange curved, tension flange straight;
(c) cross-section near one end;
(d) cross-section near midspan.

Although both buckling of the compression flange and buckling of a strut are fundamentally the result of instability, there are differences between them. In the first place, the stress in the strut is the same from end to end, before buckling occurs, whereas the stress in the compression flange varies with the bending moment on the beam. Secondly, buckling of the compression flange of the beam is restrained by the tension flange, which remains straight; and thirdly, buckling of the compression flange is restrained by the torsional stiffness of the beam. These factors are taken into account in deciding the safe stress in compression.

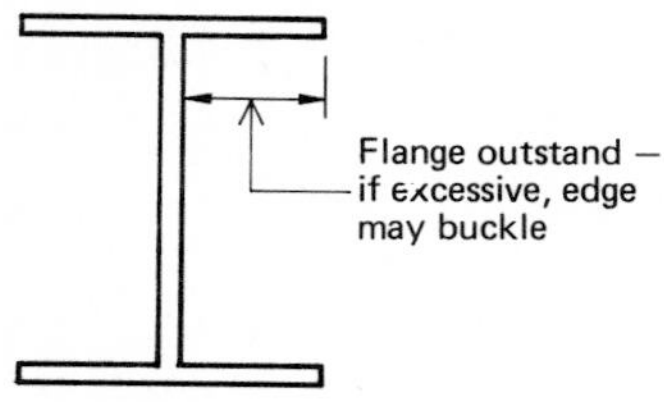

Another aspect of buckling of the compression flange is instability of the edge of the flange. Provided the flange does not project far from the web, the web can restrain this kind of buckling. Sections rolled to the shapes of BS 4 are not liable to this kind of instability[2], but there are others that are.

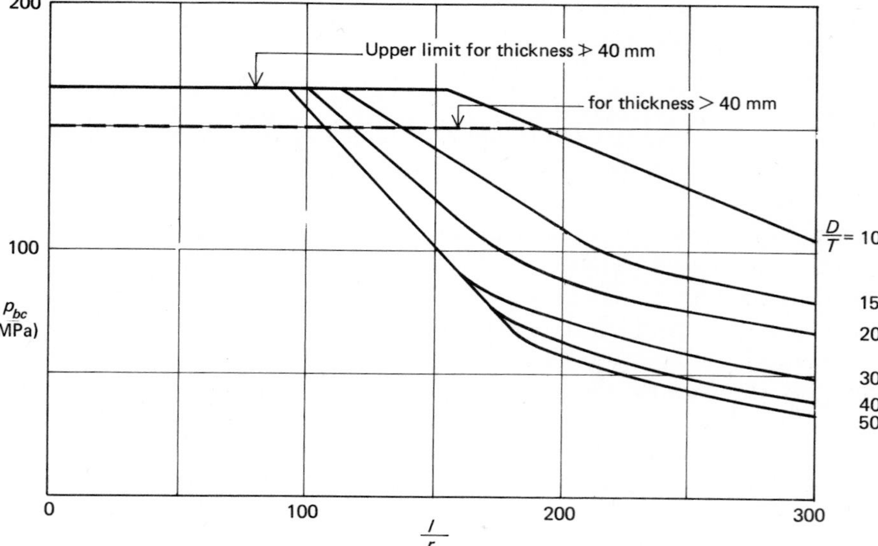

Fig. 15.2. *Permitted compressive stress in bending as a function of D/T and l/r_y for Grade 43 steel. (Based on BS 449.)*

The permitted compressive stress for rolled beams is given in BS 449 as a function of two ratios, l/r_y and D/T and is given here for Grade 43 steel in Fig. 15.2.[1] The symbols are defined as follows: D is the depth of section; T is the flange thickness; r_y is the radius of gyration of beam section about an axis lying in the plane of bending; and l is the effective length of compression flange.

Of these, D, T and r_y are geometrical properties of the section and the effective length, l, depends on the effectiveness of the restraint on lateral deflection of the compression flange. For example, if the beam is supporting a reinforced concrete floor slab, the slab entirely restrains buckling of the compression flange. At the other extreme, there may be no lateral support at all, except at the ends of the beam, and the effective length depends on torsional restraint at the ends of the beam and the resistance to rotation in plan of the end of the flange. For example, for beams adequately restrained against torsion at their ends, BS 449 specifies values of l ranging from the length of the span to 0.7 times the span.[1] If there is insufficient restraint against torsion, these effective lengths are increased by 20%. The necessary resistance to torsion can be provided by such things as web cleats, bearing stiffeners in conjunction with the seating of the beam, or building the beam into a wall. Rotation in plan of the end of the compression flange is reduced by fastening the flange to columns by cleats or welding.

Intermediate lateral restraint may be provided by secondary beams supported by main beams or girders. If buckling of the compression flange of a main beam is prevented in this way, the effective length is taken as the longest distance between such restraining members.

Application of these rules in design requires a trial and error calculation – since the permitted stress depends on the properties of the beam one can design only by testing the adequacy of trial choices. To get some idea of the order of size for a first trial, Equation (15.4) can be used with f_{bc} equal to an estimate of the expected allowable stress. This will yield a value of Z and a trial section can be chosen.

15.3. STIFFNESS OF ROLLED BEAMS

Excessive deflection of beams under service loads may cause damage to other parts of the structure or to partitions, windows or decoration in a building. Beams which are too flexible may also suffer from vibration under reciprocating loads or springiness under foot. Most design specifications put an upper limit on deflections and in the case of BS 449 this limit is related to live loads only.[1] Dead loads, such as the weight of the structure, fireproofing and floors, need not be included in the calculation of deflection. This is justified because deflections due to these loads occur before such things as partitions are constructed and before the structure has to carry any live load.

The rule is that the maximum deflection due to live load must not exceed 1/360 of the span. Thus, for a simply supported beam carrying a uniformly distributed live load W (i.e. the total live load is W), the deflection is given by

$$\Delta = \frac{1}{76.8}\,\frac{WL^3}{EI}$$

where Δ is the deflection at midspan; L is the span; E is the modulus of elasticity; and I is the second moment of area about the neutral axis. Then, since

$$\begin{aligned} \frac{\Delta}{L} &\not> \frac{1}{360} \\ \frac{WL^2}{EI} &\not> 0.214 \end{aligned} \qquad (15.5)$$

For this loading, the maximum bending moment is given by

$$M_{\max} = \frac{1}{8}WL$$

and if this and Equation (15.2) are substituted in the Inequality (15.5), the result is

$$\frac{L}{y_c} \not> 0.0268\,\frac{E}{f_{bc}}$$

With E equal to 210 GPa and f_{bc} being about 100 MPa, this yields a value of 56 for L/y_c. Thus, a rough rule of thumb is that the depth of the beam should not be less than about 1/25 of the span.

Other design specifications do not place a limit on deflection and leave this to the judgement and experience of the designer. Consideration must be given to such things as the maximum distortion that partitions, ceilings, finishing materials and glass can tolerate and, if vibrating loads are to be supported, to the natural frequency of vibration. With regard to the former, the radius of curvature of the beam may be more significant than the deflection and this can be estimated from the simple theory of elastic bending. We have

$$\frac{f}{y} = \frac{M}{I} = \frac{E}{R}$$

and

$$\frac{1}{R} = \frac{f/E}{y} = \text{gradient of strain}$$

or

$$\frac{1}{R} = \frac{M}{EI}$$

The fundamental frequency of a concentrated load can be calculated easily if the mass of the beam is negligible. For elastic deflection, the beam is like a spring and if Δ_L is the static deflection under the load W, a spring constant K can be found from

$$K = \frac{W}{\Delta_L}$$

The fundamental frequency is then given by

$$f = \frac{1}{2\pi}\sqrt{\frac{K}{W/g}} = \frac{1}{2\pi}\sqrt{\frac{g}{\Delta_L}}$$

If the mass of the beam is not negligible, or if a distributed load is supported as well as the concentrated load, an equivalent concentrated load is used. Formulas for determining this are given by Roark.[3] Undesirable resonant vibrations are possible if a natural frequency is close to the frequency of vibration of the load.

With irregular loads, it may be difficult to compute the deflection accurately. Approximate methods are justified, and the moment–area method provides a good basis for such calculations. This procedure is described in most books on structural analysis. The rule for calculating the deflection of a beam is as follows: the deflection of the point B measured from the tangent at A equals the moment about B of the area of the M/EI diagram between A and B. (See, for example, Cross and Morgan.[4]) Thus, referring to Fig. 15.3, if A is the point where the deflection is greatest, the tangent at A is horizontal, and the deflection of the end of the beam B above this tangent equals the downward displacement of A.

An estimate of the maximum deflection can be made by calculating the deflections of the two ends from the tangent at midspan and taking the mean of the two. As shown in Fig. 15.4, this mean equals the deflection at midspan and, provided the point of maximum deflection is close to midspan, it will be a good estimate of the desired maximum deflection.

To illustrate the procedure, such a calculation is made in Example 15.3. If an appropriate calculation like this shows the beam to be well within the allowable limit, no more need be done. Otherwise a more accurate calculation must be undertaken. The moment–area method will be found convenient for

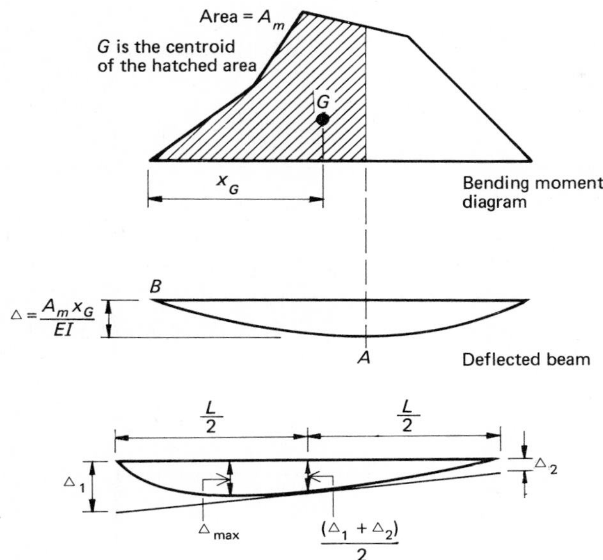

Fig. 15.3. *Moment-area method for calculation of deflection.*

Fig. 15.4. *Approximate estimation of maximum deflection.*

this also, a recommended way of using it being to search for the horizontal tangent by trial calculation of the deflections of the ends. The tangent sought is such that the upward deflections of the ends from it are equal. Alternatively, the graphical computation described in Chapter 8, "Shaft Design" may be used.

Finally, it may be desirable to camber beams before erection – that is to erect them with an upward deflection before they are loaded. Sagging beams have an undesirable appearance and camber is often used to avoid this visual offence.

15.4. STRENGTH IN SHEAR, ROLLED BEAMS

Elastic theory gives, for the distribution of shear stress in a beam

$$f_q = \frac{VQ}{bI} \tag{15.6}$$

where f_q is the shear stress at distance y from the neutral axis; V is the external shear force; Q is the first moment about the neutral axis of the part of the section further from the neutral axis than y; b is the breadth of section at distance y from the neutral axis; and I is the second moment of area of whole section about the

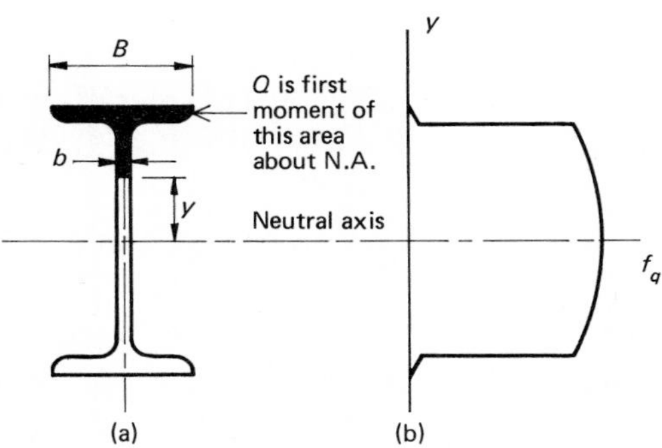

Fig. 15.5. *Shear stress in I-section: (a) definition of symbols; (b) distribution of shear stress.*

Example 15.3 Sheet 1 of 2

Estimate the deflection and natural frequency of a simply supported beam spanning 5m and carrying a load of 10 kN at a distance of 1m from one support. The beam is a 254 × 146 × 31 kg/m universal beam.

Loads and reactions:

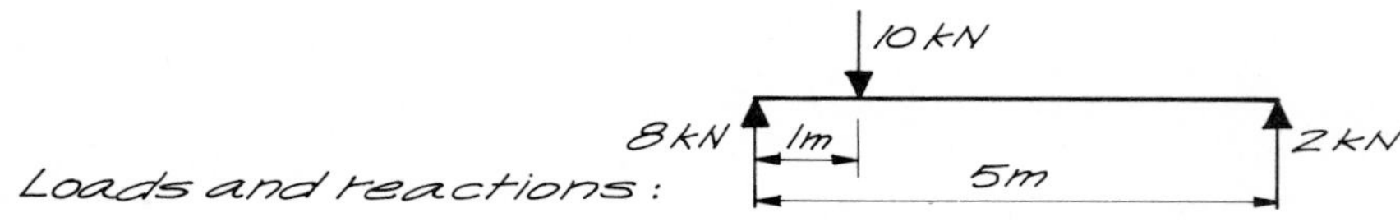

Bending moments:

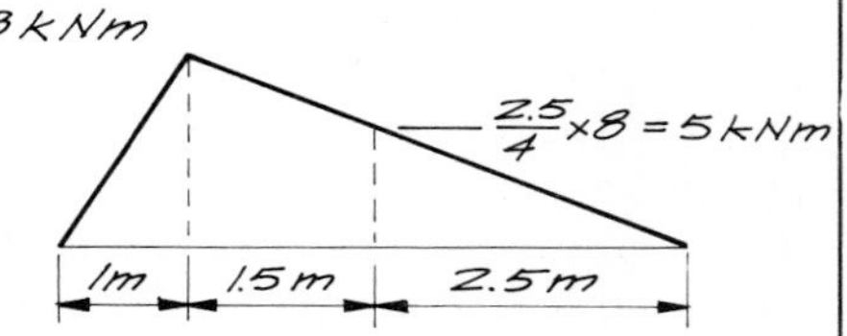

Approximate moments for calculation of deflection of ends from tangent at mid span:

6.5 kNm
4 kNm
5 kNm
0.5m
1.75m
2/3 × 2.5 = 1.67 m

For L.H. support:

$$EI\Delta = 4 \times 10^3 \times 1 \times 0.5 + 6.5 \times 10^3 \times 1.5 \times 1.75$$
$$= 1.91 \times 10^4 \text{ Nm}^3$$

For R.H. support:

$$EI\Delta = 1/2 \times 5 \times 10^3 \times 2.5 \times 1.67$$
$$= 1.04 \times 10^4 \text{ Nm}^3$$

Say, $EI\Delta = 1/2(1.91 + 1.04) \times 10^4 = 1.47 \times 10^4 \text{ Nm}^3$

$$EI = (2.1 \times 10^{11}) \times (4.427 \times 10^{-5}) \quad \text{Nm}^2$$
$$= 9.30 \times 10^6 \text{ Nm}^2$$

$$\Delta = \frac{1.47 \times 10^4}{9.30 \times 10^6} \text{ m} \quad = 1.58 \text{ mm}$$

$$\frac{\Delta}{L} = \frac{1.58 \times 10^{-3}}{5} \quad = \frac{1}{3170} \quad \text{OK}$$

Note: Accurate calculation gives Δ = 1.63 mm

Example 15.3 Sheet 2 of 2

Fundamental frequency :

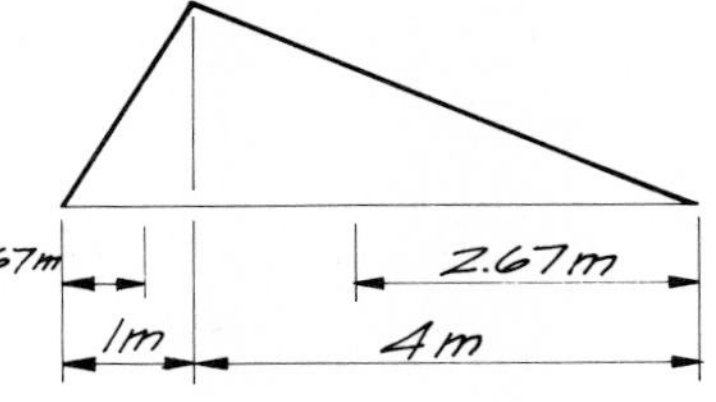

$$EI\Delta_1 = (1/2 \times 8 \times 10^3 \times 1) \times 0.667$$
$$= 2.67 \times 10^3 \ \text{Nm}^3$$

$$EI\Delta_2 = (1/2 \times 8 \times 10^3 \times 4) \times 2.67$$
$$= 4.27 \times 10^4 \ \text{Nm}^3$$

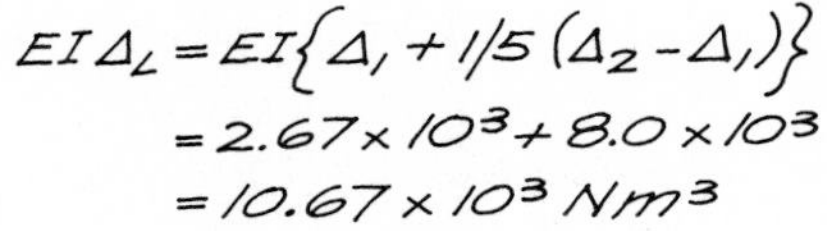

$$EI\Delta_L = EI\{\Delta_1 + 1/5(\Delta_2 - \Delta_1)\}$$
$$= 2.67 \times 10^3 + 8.0 \times 10^3$$
$$= 10.67 \times 10^3 \ \text{Nm}^3$$

$$\Delta_L = \frac{1.067 \times 10^4}{9.30 \times 10^6} = 1.15 \times 10^{-3} \text{m} = 1.15 \text{mm}$$

$$\text{Spring factor } K = \frac{W}{\Delta_L} = \frac{10^4}{1.15 \times 10^{-3}} = 8.70 \times 10^6 \text{Nm}^{-1}$$

Assuming mass of beam is negligible

$$M = \frac{10^4}{9.81} = 1020 \text{ kg}$$

$$\text{Fundamental frequency} = \frac{1}{2\pi}\sqrt{\frac{8.70 \times 10^6}{1.02 \times 10^3}}$$
$$= 14.7 \text{ Hz}$$

neutral axis. These symbols are illustrated in Fig. 15.5(a) in relation to an I-section, and the corresponding distribution of shear stress is sketched on Fig. 15.5(b). This distribution of shear stress is discontinuous at the level of the junctions of the web with the flanges where the breadth of the section is taken to change abruptly from the thickness of the web to the width of the flange. For reasons that are obvious from Fig. 15.5(b), it is often called a "top hat" distribution.

What is particularly important is that most of the resistance to shearing is provided by the web. This can be shown by considering the relationship

$$V = \int_{y_t}^{y_c} f_q \, b \, dy$$

$=$ area bounded by a curve showing $f_q b$ plotted against y.

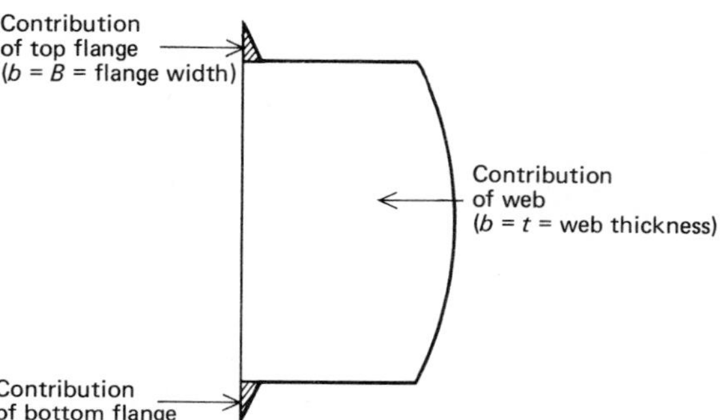

Fig. 15.6. *Product $f_q b$ as a function of y.*

In Fig. 15.6, such a curve is sketched as it would be derived from Fig. 15.5(b). These sketches make it clear that the web contributes most to the total V. Furthermore, since Q does not change much with y for points between the flanges, the distribution of f_q here is approximately uniform. As a consequence of these two facts, a reasonable estimate of the maximum value of f_q can be made by calculating the average shear stress on the web, that is

$$f_q' = \frac{V}{Dt}$$

where D is the total depth of the beam, and t is the thickness of web.

This much simpler calculation is permitted by BS 449 and the corresponding allowable stress for webs of rolled beams is as follows:[1]

Grade 43 steel,	thickness $\not>$ 40 mm	100 MPa
	thickness $>$ 40 mm	90 MPa
Grade 50 steel,	thickness $\not>$ 65 mm	140 MPa
Grade 55 steel,	thickness $\not>$ 40 mm	170 MPa
	thickness $>$ 40 mm	160 MPa

(Extract from BS 449, *The Use of Structural Steel in Building*, is reproduced by permission of the British Standards Institution, 2 Park Street, London W1A 2BS.)

We have noted in Chapter 4, "Failure and Safety Factors", that theoretically the ratio of yield strength in shear to that in tension is 0.578 and these allowable shear stresses are approximately 0.6 times the allowable stresses in tension.

The same design specification allows somewhat higher maximum stresses than these, if the maximum stress in the top hat distribution is computed. However, it is rarely that the additional work will be justified.

15.5. CONCENTRATED LOADS, ROLLED BEAMS

Any beam has to resist concentrated forces at its supports. In addition, some of the loads may be applied over short lengths of a beam. Figure 15.7 illustrates this. There, a secondary beam is shown supported by a main beam and the latter has concentrated forces applied to it at A, B and C. Under sufficiently large loading of this kind, the beam may fail by buckling of the web where the concentrated force is applied, as shown in the margin. To increase the load carrying capacity of the beam in this respect, stiffeners can be fastened to the web between the flanges. The stiffeners may be narrow strips of plate projecting from the web and welded to it or they may be angles riveted to the web. Modern practice favours the former alternative. Stiffeners that are provided to reinforce the web of a beam at concentrated load are known as *bearing stiffeners*.

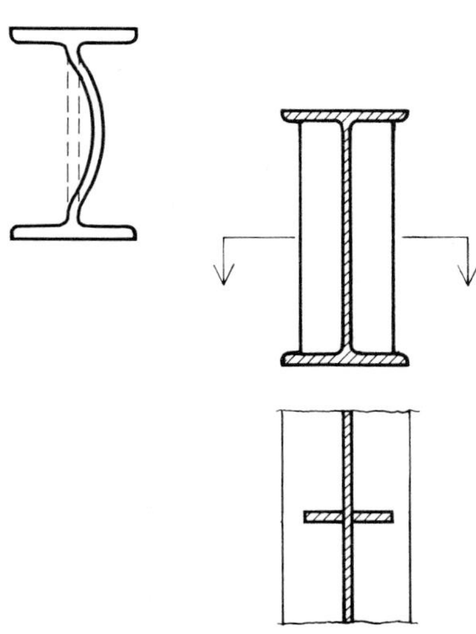

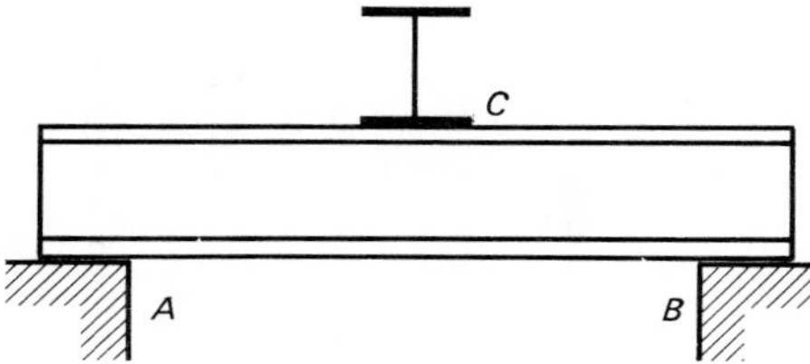

Fig. 15.7. *Concentrated load on beam.*

An approximate method of design, based on the buckling of struts, is given in BS 449.[1] The method has two steps – in the first, the safe load which can be supported by the unstiffened web is estimated. If this safe load exceeds the applied load, no more need be done. If it does not, a combination of web and stiffeners is sought which has a safe load capacity greater than the applied load.

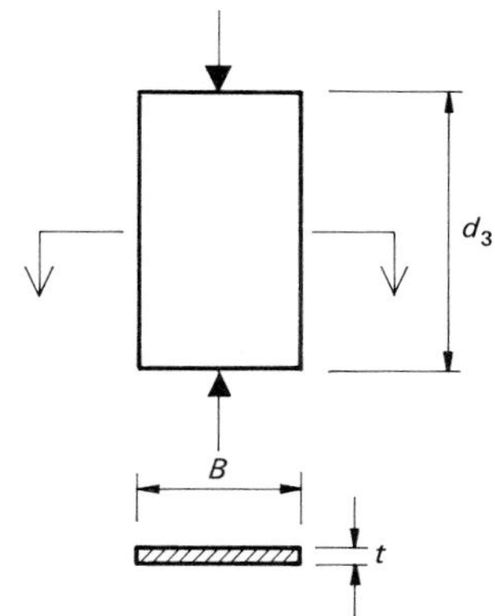

For the first step, the load is assumed to be dispersed at 45° to the neutral axis. A length B (not to be confused with flange width) is defined as shown on Fig. 15.8, the lines defining the dispersion originating at the edges of the stiff part of the contact where the force is applied to the beam. It is assumed that a length of the web equal to B acts as a strut to carry the applied force. The cross-section of this strut is a rectangle whose dimensions are B, as defined on Fig. 15.8, and t, the thickness of the web. The effective length of the strut is taken as $\frac{1}{2}d_3$, where d_3 is the clear depth of the web between the root fillets at the junctions of the flanges with the web, so that the slenderness ratio of the strut is $\sqrt{3}\,d_3/t$. Using this slenderness ratio, the permitted stress for a strut is found and hence the safe load.

This procedure is used solely to decide whether the web needs to be strengthened by stiffeners or not. If stiffeners are needed, a different rule is used to estimate the safe load that may be applied. The principle is similar, but the assumed strut is defined differently. For this purpose, it is assumed that the distance along the web contributing to the strength of the strut is 20 times the thickness of the web on either side of the stiffener plates, if this much is available. For bearing stiffeners near an end of a beam, the distance between the stiffeners and the end may be less than $20t$; in that case, only the length available can be

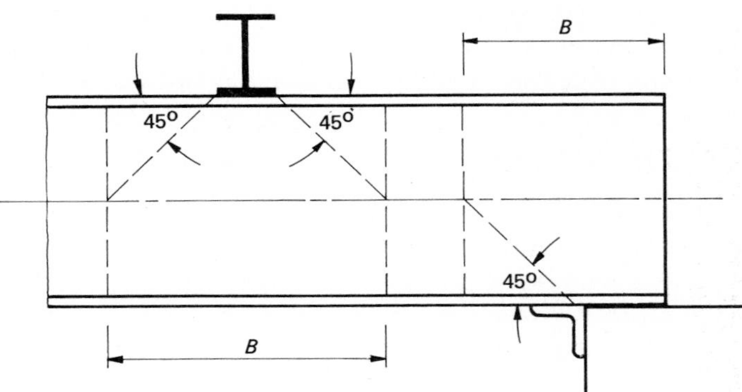

Fig. 15.8. *Dispersion of concentrated load and reaction.*

assumed effective. The cross-section of the strut is cruciform in shape, comprising this length of web and the stiffeners, as indicated on Fig. 15.9. Its effective length is taken as 0.7 times the length of the stiffener. The radius of gyration of the cruciform section is calculated for an axis in the web by working out the second moment of area about this axis and the area. The effective length divided by the radius of gyration is the slenderness ratio and the permitted stress for a strut of this slenderness is taken from the design specification. Data from the relevant table are included in Appendix A14. The permitted stress, multiplied by the area of the cross-section of the strut, gives the permitted load. For the design to be acceptable, this load must exceed the total load applied.

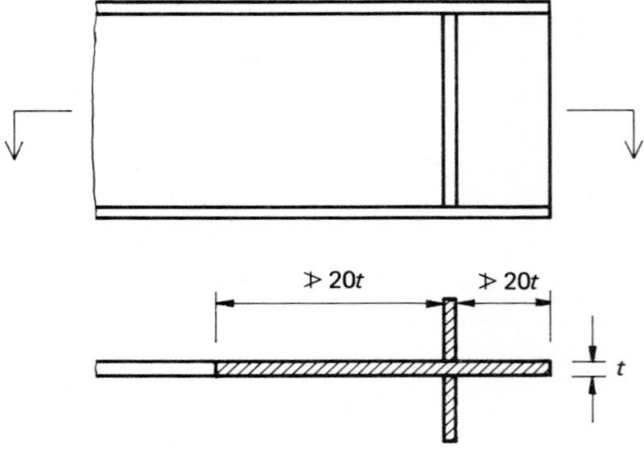

Fig. 15.9. *Bearing stiffener.*

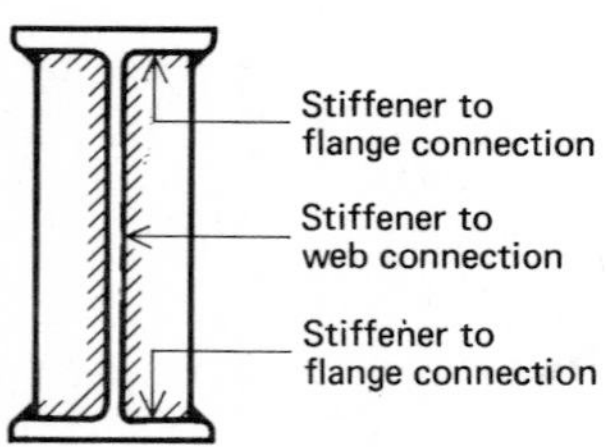

Bearing stiffeners must be fastened to the web. At the flanges, their ends have to be welded to the flanges or machined to fit against the flanges. The design specification BS 449 requires the fastenings to the web to be able to transmit the entire load and the connections of the ends of the stiffeners to the loaded flange must also be designed for this load. When stiffeners are fitted over the supports of the beam, this strength is required at both flanges. For welded construction reasonable fillet welds can usually be designed for the connection of the stiffeners to the web. However, the length available at the flanges is often too short and unduly large fillet welds would be necessary. In that case, it is better to use butt welds to connect the ends of the stiffeners to the flanges.

There is not much point in trying to refine the choice of stiffener plate. Most shops can use scrap plate for this purpose, so that extra material can be provided at negligible cost. Overdesign with respect to strength is not bad design when the extra cost incurred is small.

15.6. DESIGN OF A ROLLED BEAM

In Examples 15.4 and 15.5, calculations for the design of rolled beams are shown. These examples illustrate the application of material presented in Sections 15.2 to 15.5. The short, heavily loaded beam in Example 15.4 has high web shear stress at the ends and requires reinforcing of the web. The beam in Example 15.5 has a longer span and a lighter load. Buckling of the compression flange is more significant and the stiffness of the beam is rather low.

15.7. FLANGE PLATES ON ROLLED BEAMS

The bending strength and stiffness of a rolled beam can be increased by welding or riveting plates to its flanges. Computation of the properties of the plated beam has been illustrated in Example 15.2 and the formulas for bending stress and deflection show that increasing I will reduce both of them.

In this section we examine the strength required in the joint between the plate and the flange. Equation (15.6) gives the shear stress at any point in the beam. On an element of the beam, this stress acts on vertical and horizontal surfaces, the component on the horizontal surface being called the *complementary shear stress.* The tangential force per unit length on a horizontal surface at any distance y from the neutral axis is therefore $f_q b$ and is given by

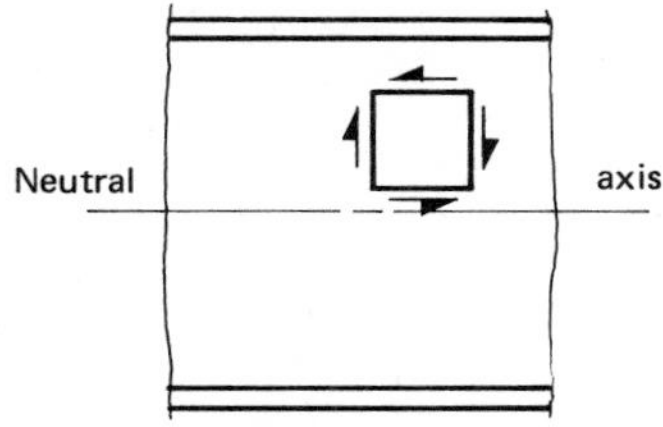

$$f_q b = \frac{VQ}{I} \tag{15.7}$$

The derivation of this formula should be studied carefully. Its important feature is the way that this traction at any level arises from the variation of bending moment along the beam. Thus, it depends on the external shear, which equals the rate of change of bending moment with respect to distance along the beam.

Then, for the joint between the plate and the flange, we have

$$Q = A_p\left(y_c - \frac{T_p}{2}\right)$$

as may be seen from Fig. 15.10. Here, A_p and T_p are respectively the area and thickness of the plate.

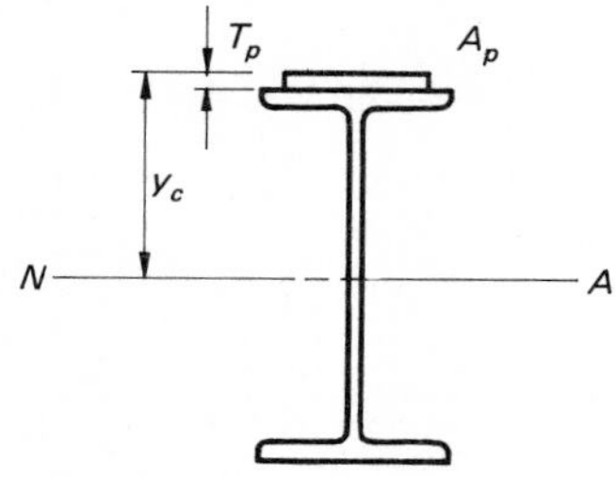

Fig. 15.10. *Flange plate on rolled beam.*

Example 15.4 Sheet 1 of 3

Select a suitable universal beam for a span of 3m centre to centre of supports, to carry two loads of 450kN at 0.5m from the centrelines of the supports. The applied loads and the reactions are each distributed over a length of 150mm of the top or bottom flange.

Design for Grade 43 steel. Assume effective lateral support for compression flange at each load.

3m; 0.5; 0.5; 0.150; 0.150; 0.150

Loads: 450 kN; 450 kN; 0.5; 0.5; 3

Live load shears: 450 kN; 450kN

Live load moments: 225 kNm

Try $p_{bc} = 165$ MPa $Z = \frac{2.25 \times 10^5}{1.65 \times 10^8}$ m^3 $= 1.36 \times 10^6 mm^3$

Try 457 × 152 × 74 kg/m, $l = 2000mm$, $r_y = 31.8$ mm

$\frac{l}{r_y} = 63.0$

$\frac{D}{T} = \frac{461}{17} = 27.2$

$p_{bc} = 165$ MPa

Dead load BM $= \frac{WL}{8} = \frac{(74 \times 9.81 \times 3) \times 3}{8}$ $= 820$ Nm

Design moment $= 225 + 0.8$ $= 225.8$ kNm

$Z_{xx} = 1.404 \times 10^6 mm^3$

$f_{bc} = \frac{2.26 \times 10^5}{1.404 \times 10^{-3}}$ Pa $= 161$ MPa OK

Check web shear stress: D.L shear $= 1/2 \times 3 \times 74 \times 9.81$ N

$= 1.09$ kN

$V = 451$ kN

$f_q' = \frac{4.51 \times 10^5}{461 \times 9.9}$ $= 98.7$ MPa

$p_q' = 100$ MPa OK

Example 15.4 — Sheet 2 of 3

Check live load deflection:

$EI\Delta = (1/2 \times 2.25 \times 10^5) \times 0.333 + (2.25 \times 10^5 \times 1.0) \times 1.0$

$= 1.87 \times 10^4 + 2.25 \times 10^5$

$= 2.44 \times 10^5\ Nm^3$

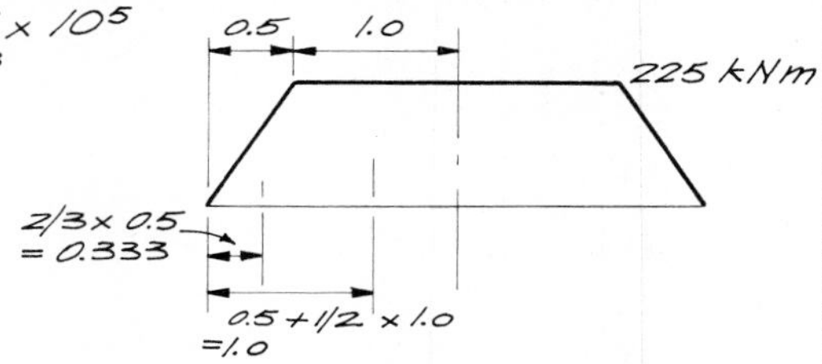

$E = 210\ GPa$

$= 2.10 \times 10^{11}\ Nm^{-2}$

$I = 3.24 \times 10^{-4}\ m^4$

$EI = 2.10 \times 10^{11} \times 3.24 \times 10^{-4} = 6.80 \times 10^7\ Nm^2$

$\Delta = \dfrac{2.44 \times 10^5}{6.80 \times 10^7} = 3.59 \times 10^{-3}\ m = 3.59\ mm$

$\dfrac{\Delta}{L} = \dfrac{3.59 \times 10^{-3}}{3} = \dfrac{1}{835}$ OK

Concentrated loads:

$d_3 = 404\ mm$ } Ref BS4
$t = 9.9\ mm$ } or Appendix A3

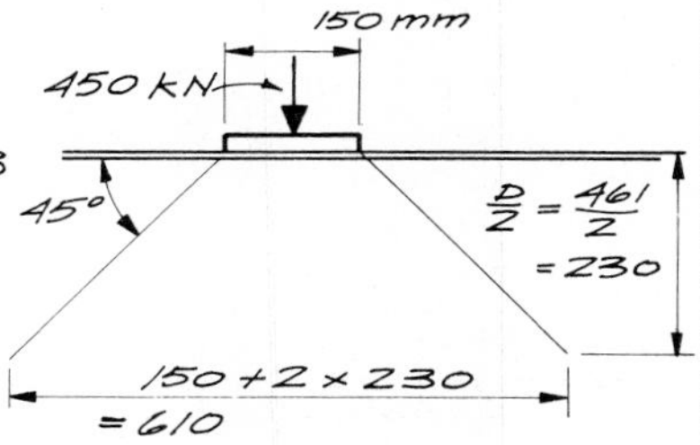

$\sqrt{3}\dfrac{d_3}{t} = \dfrac{1.732 \times 404}{9.9} = 70.8$

$p_c = 114\ MPa$
(Ref BS449 or Appendix A14)

$f_c = \dfrac{4.5 \times 10^5}{610 \times 9.9} = 74.5\ MPa$ OK

Reactions:

$p_c = 114\ MPa$ as above

$B = 150 + 230 = 380\ mm$

$f_c = \dfrac{4.51 \times 10^5}{380 \times 9.9} = 120\ MPa$

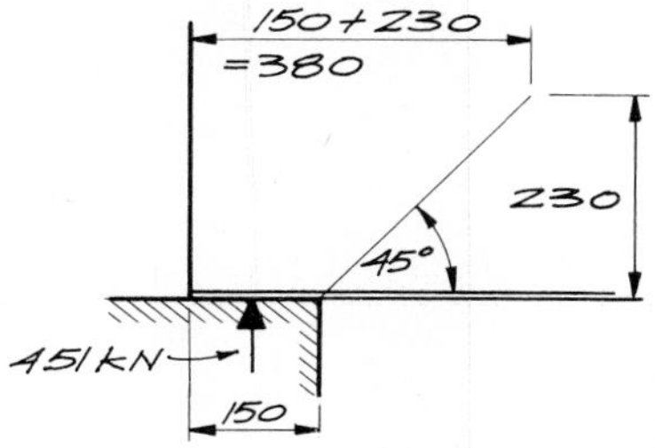

Stiffeners required:

Try 50 mm × 10 mm plates, one each side of web

$A = (278 \times 9.9) + (2 \times 50 \times 10) = 3760\ mm^2$

$I = 1/12 \times 10 \times 109.9^3 = 1.11 \times 10^6\ mm^4$

$r_y = \sqrt{\dfrac{1.11 \times 10^6}{3.76 \times 10^3}} = 17.2\ mm$

$\dfrac{\ell}{r_y} = 17.4 \qquad p_c = 148\ MPa$

$f_c = \dfrac{4.51 \times 10^5}{3.76 \times 10^3} = 120\ MPa$ OK

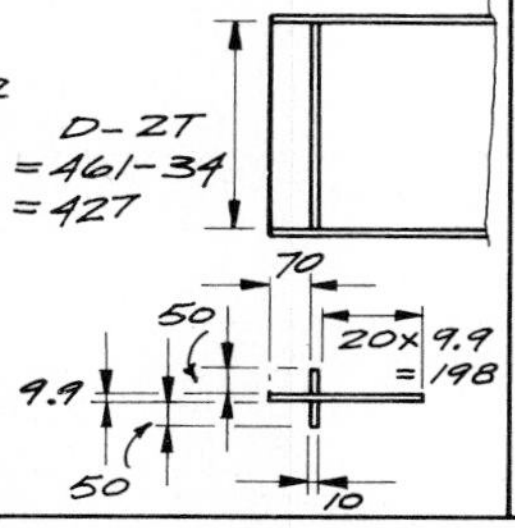

Example 15.4 Sheet 3 of 3

Welding, stiffeners to web and flanges:

Length of fillet weld, stiffener to web

$= 4 \times 427 \quad = 1708$ mm

Load per unit length $= \dfrac{4.51 \times 10^5}{1.708 \times 10^3}$

$= 264$ N/mm

4 mm fillet weld all round OK

Length of fillet weld, stiffener to flange

$= 4 \times 50 + 2 \times 10$

$= 220$ mm

Load per unit length $= \dfrac{4.51 \times 10^5}{2.20 \times 10^2}$

$= 2060$ kN/m

Too high for fillet weld

Single Vee preparation at ends of stiffener plates and butt weld to flanges

SUMMARY: 457 × 152 × 74 kg/m U.B.
2 × 50 mm × 10 mm stiffener plates each end, on ℄ bearings
4 mm fillet weld all round, stiffener to web
Single Vee butt welds, stiffener to flange, top and bottom.

Example 15.5 Sheet 1 of 3

Select a rolled beam to support a load of 86 kN centrally placed on a span of 8 m centre to centre of supports. Load and reactions spread over 150 mm of flange. Grade 43 steel.

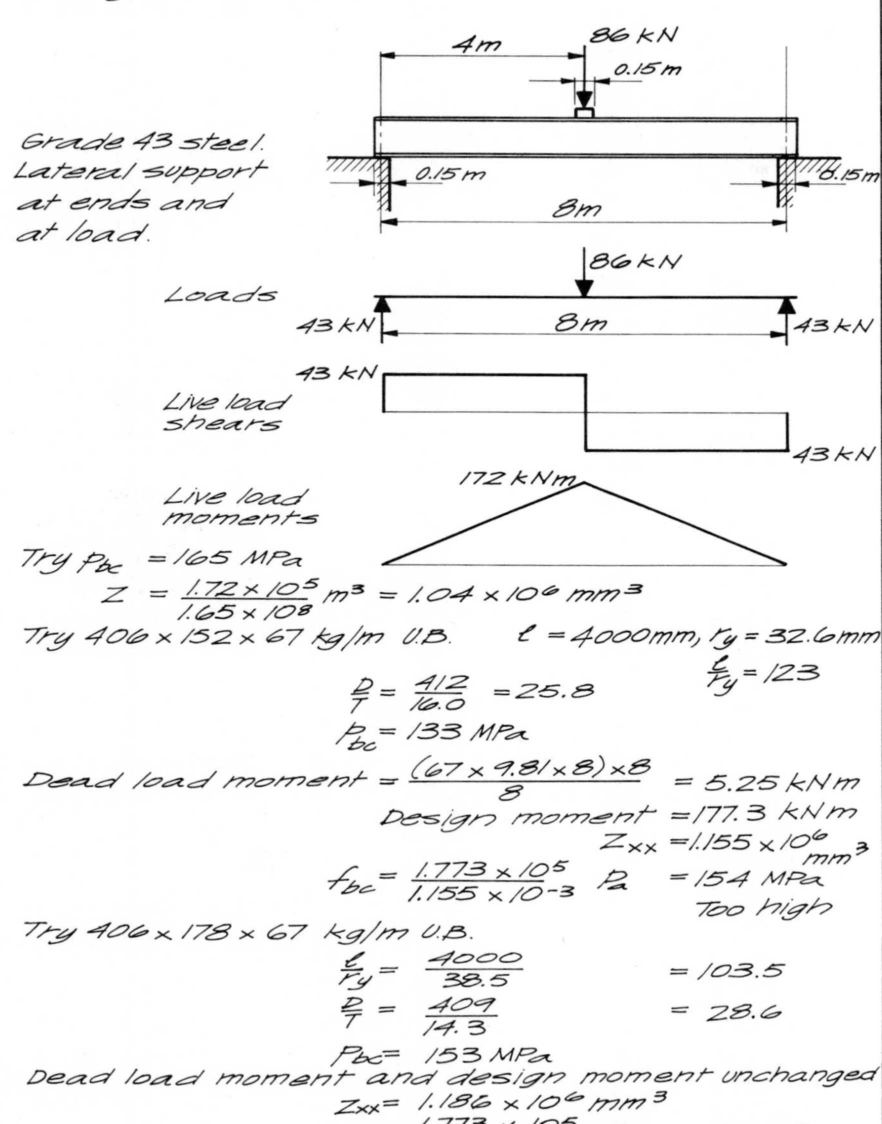

Try p_{bc} = 165 MPa

$Z = \frac{1.72 \times 10^5}{1.65 \times 10^8} m^3 = 1.04 \times 10^6 mm^3$

Try 406 × 152 × 67 kg/m U.B. ℓ = 4000mm, r_y = 32.6mm

$\frac{\ell}{r_y} = 123$

$\frac{D}{T} = \frac{412}{16.0} = 25.8$

p_{bc} = 133 MPa

Dead load moment $= \frac{(67 \times 9.81 \times 8) \times 8}{8}$ = 5.25 kNm

Design moment = 177.3 kNm

$Z_{xx} = 1.155 \times 10^6 mm^3$

$f_{bc} = \frac{1.773 \times 10^5}{1.155 \times 10^{-3}}$ Pa = 154 MPa Too high

Try 406 × 178 × 67 kg/m U.B.

$\frac{\ell}{r_y} = \frac{4000}{38.5} = 103.5$

$\frac{D}{T} = \frac{409}{14.3} = 28.6$

p_{bc} = 153 MPa

Dead load moment and design moment unchanged

$Z_{xx} = 1.186 \times 10^6 mm^3$

$f_{bc} = \frac{1.773 \times 105}{1.186 \times 10^{-3}}$ Pa = 150 MPa OK

Example 15.5 Sheet 2 of 3

Check deflection:

$EI\Delta = (1/2 \times 1.72 \times 10^5 \times 4) \times 2.67$

$= 9.20 \times 10^5 \ Nm^3$

172 kN m

$E = 2.10 \times 10^{11} \ Nm^{-2}$

2.67m

$I = 2.43 \times 10^{-4} \ m^4$

4m

$EI = 2.10 \times 2.43 \times 10^7 = 5.10 \times 10^7 \ Nm^2$

$\Delta = \dfrac{9.20 \times 10^5}{5.10 \times 10^7} = 1.80 \times 10^{-2} \ m = 18.0 \ mm$

$\dfrac{\Delta}{L} = \dfrac{1.80 \times 10^{-2}}{8} = \dfrac{1}{445}$ ok

Check web shear stress:

Dead load end shear $= 1/2 \times 67 \times 9.81 \times 8 = 2.64 \ kN$

$V = 45.6 \ kN$

$f_q' = \dfrac{4.56 \times 10^4}{409 \times 8.8} = 12.7 \ MPa$ ok

Concentrated loads:

$d_3 = 357 \ mm, \ t = 8.8 \ mm, \ \dfrac{\sqrt{3} d_3}{t} = \dfrac{1.732 \times 357}{8.8} = 70.5$

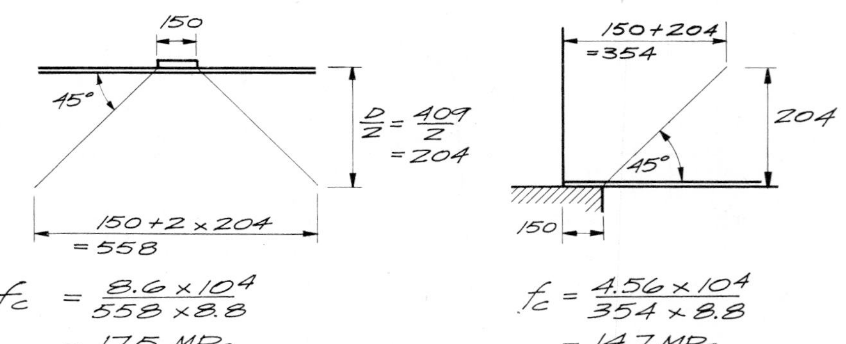

$f_c = \dfrac{8.6 \times 10^4}{558 \times 8.8}$

$= 17.5 \ MPa$

$f_c = \dfrac{4.56 \times 10^4}{354 \times 8.8}$

$= 14.7 \ MPa$

For slenderness ratio 70.5, $p_c = 114 \ MPa$ ok

SUMMARY: 406 × 178 × 67 kg/m U.B.
No stiffeners required

Example 15.5 Sheet 3 of 3

In view of large deflection, check natural frequency:

Mass of beam $= 8 \times 67 = 536$ kg

Mass of body supported $= \dfrac{8.6 \times 10^4}{9.81} = 8.77 \times 10^3$ kg

Assume mass of beam may be neglected

Spring constant $= K = \dfrac{W}{\Delta} = \dfrac{8.6 \times 10^4}{1.8 \times 10^{-2}} = 4.78 \times 10^6\ \text{Nm}^{-1}$

Fundamental frequency $= \dfrac{1}{2\pi}\sqrt{\dfrac{K}{M}}$

$= \dfrac{1}{2\pi}\sqrt{\dfrac{4.78 \times 10^6}{8.77 \times 10^3}}$

$= 3.72$ Hz

In a welded joint, the strength of two fillet welds together, one at each edge of the plate, must not be less than $f_q b$ as determined from Equation (15.7). If rivets are used in pairs, one each side of the web, with the pairs a distance s apart along the beam, the force which can be transmitted per unit length of flange is F/s. Here F is the load-carrying capacity of a pair of rivets. Since F/s is to be equal to $f_q b$, the spacing s is given by

$$s = \frac{F}{f_q b} = \frac{FI}{VQ}$$

In Example 15.6, calculation of the size of fillet weld for a flange plate is illustrated.

15.8. PLATE GIRDERS

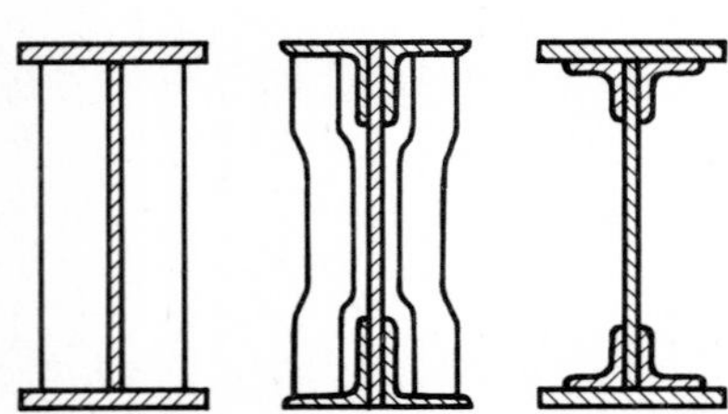

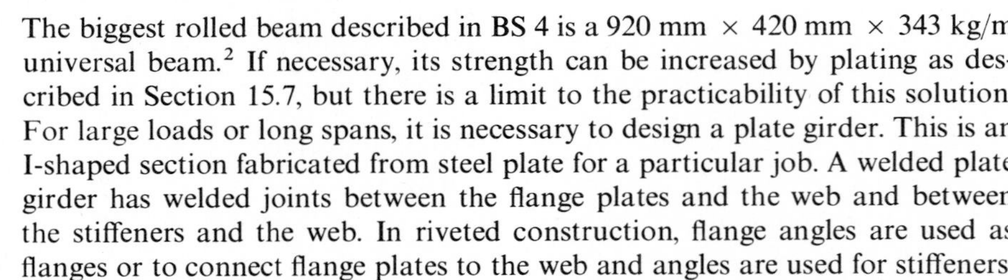

The biggest rolled beam described in BS 4 is a 920 mm × 420 mm × 343 kg/m universal beam.[2] If necessary, its strength can be increased by plating as described in Section 15.7, but there is a limit to the practicability of this solution. For large loads or long spans, it is necessary to design a plate girder. This is an I-shaped section fabricated from steel plate for a particular job. A welded plate girder has welded joints between the flange plates and the web and between the stiffeners and the web. In riveted construction, flange angles are used as flanges or to connect flange plates to the web and angles are used for stiffeners.

The designer has considerable scope for saving materials in that the strength of the beam can be varied along its length. For example, in Fig. 15.11, the bending moment and the moment of resistance of a plate girder are shown. Although the flanges have been reduced in size near the ends of the beam, the moment of resistance is everywhere greater than the bending moment. There is a clear saving in steel in the flanges, when this design is compared with one with flanges uniform from end to end. There is less scope for such saving in the web – usually, its slenderness is such that reduced thickness near midspan, where the shear is least, is not permissible.

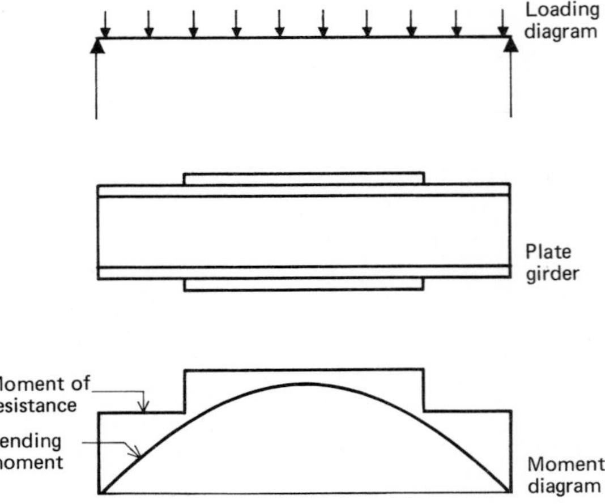

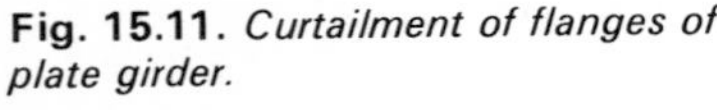

Fig. 15.11. *Curtailment of flanges of plate girder.*

Computations for beams designed to BS 449[1] are simpler if the design of the flanges is such as to keep the flanges equal. However, additional saving may be possible with unequal flanges. The dimensions of the compression flange

Example 15.6 Sheet 1 of 1

Find the size of fillet weld required at each edge of a flange plate in Example 15.2 when the external shear is 900 kN. Grade 43 steel, for which $p_q = 115$ MPa.

From Example 15.2:
For a horizontal surface between the top flange and the plate

$$Q = (7.5 \times 10^3) \times 222.8$$
$$= 1.67 \times 10^6 \text{ mm}^3$$
$$I = 1.893 \times 10^9 \text{ mm}^4$$

Hence $\dfrac{VQ}{I} = \dfrac{9 \times 10^5 \times 1.67 \times 10^6}{1.893 \times 10^9}$

$$= 794 \text{ N/mm}$$

For $p_q = 115$ MPa, one 6 mm fillet weld → 483 N/mm (Ref Chapter 13)

Hence one 6 mm fillet weld each edge of plate gives 960 N/mm OK

6 mm fillet weld each edge.

may be such that the permitted compressive stress has to be reduced to guard against buckling. In that case, the permitted stresses in tension and compression will be unequal, the former being the larger of the two. With unequal flanges and hence unequal stresses at the two extreme fibres, f_{bt} and f_{bc}, can both be kept close to the allowable stresses, p_{bt} and p_{bc}. Less steel is required than in the case where the two flange stresses are equal and limited by the lower of the two allowable stresses.

The basis of design is fundamentally similar to that for rolled beams, differences arising from the slenderness of the web and variations in flange dimensions along the beam. The problems the designer of a plate girder has to solve are to select

(a) dimensions for the web;
(b) dimensions for the web stiffeners;
(c) dimensions for the flanges;
(d) sizes for the connections of the flanges and the stiffeners to the web.

One proceeds in this order because the web can be designed independent of the flanges, whereas flange design depends to some extent on the dimensions of the web.

In the following sections, design of simple plate girders is discussed. These have equal flanges which may, however, vary along the beam, uniform web thickness and vertical stiffeners. All connections are made by welding. We do this to avoid obscuring the fundamentals of the process of design with details of more refined procedures.

15.8.1. Webs of plate girders

In most cases, the web of a plate girder has large height and length dimensions compared with its thickness. We have shown that the web of an I-section carries most of the external shear. Near the neutral axis, where bending stresses are low, the web is approximately in pure shear, with equal principal stresses in tension and compression at 45° to the length of the beam. Hence the web is prone to buckling under the action of the external shear and, to increase its resistance to buckling, it is divided into panels by stiffeners. BS 449 allows horizontal stiffeners to be used as well as vertical stiffeners[1], but we will confine ourselves to the latter, with web panels bounded by the flanges and adjacent stiffeners. The allowable average shear stress depends on the grade of steel, whether the web is stiffened or not and, if it is stiffened, on the ratio of the height of the web to its thickness and the distance between the stiffeners. There are also upper limits on the allowable slenderness in the form of the ratio of height to thickness.

To illustrate the kind of data used, the following information for Grade 43 steel from BS 449 is typical.[1]

For unstiffened webs the allowable average shear stress and maximum height to thickness ratio are as follows:

$$\left.\begin{array}{ll}\text{Web thickness} \ngtr 40\ \text{mm} & p_q' = 100\ \text{MPa} \\ \text{Web thickness} > 40\ \text{mm} & p_q' = 90\ \text{MPa}\end{array}\right\} \frac{d}{t} \ngtr 85$$

For stiffened webs, the web thickness is limited to $t \nless \frac{1}{180}$ of the smaller panel dimension and $t \nless d/200$. The allowable average shear stress is given on Fig.

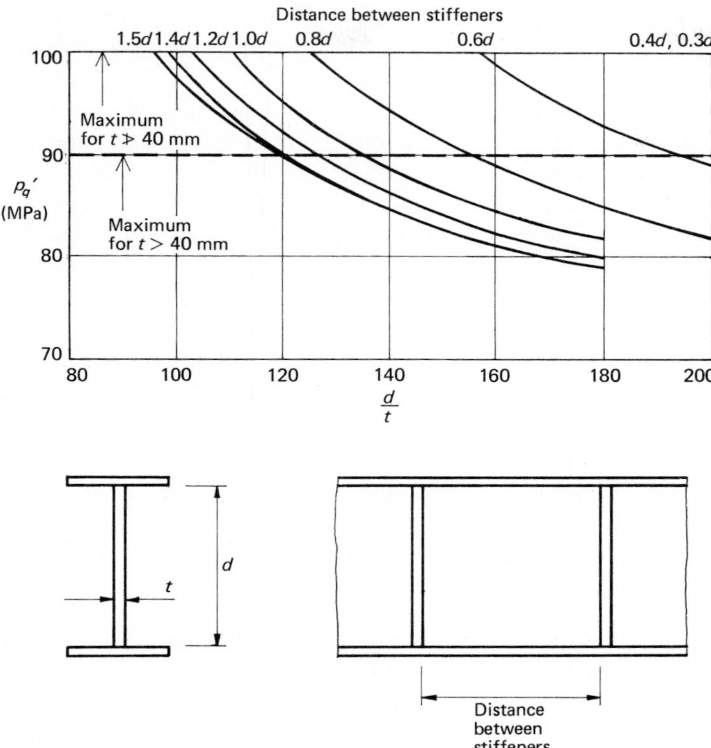

Fig. 15.12. *Permitted average shear stress on web of plate girder as a function of d/t and distance between stiffeners for Grade 43 steel. (Based on BS 449.)*[1]

Fig. 15.13. *Dimensions of web panel of plate girder.*

15.12 but it must not exceed the value allowed for an unstiffened web. The dimensions of a web panel are shown on Fig. 15.13.

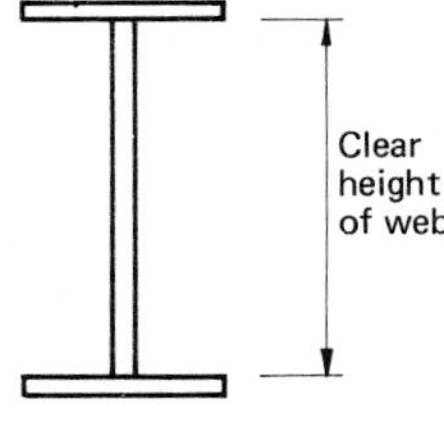

Computation of the average shear stress on the web is based on the clear height of the web – not the total depth as is the case for rolled beams. Thus, both the stress caused by the load and the allowable stress depend on the dimensions of the web plate. The designer has to work by trial and error, adjusting his choice of height, thickness and spacing of stiffeners until the allowable stress exceeds the stress caused by the loads but, for economy, by a small margin. A rule of thumb for the depth of the beam is to make it about 1/10 of the span. Greater depth will lead to smaller flanges, but at the cost of more material in the web because of increased slenderness as well as increased height. Conversely, reducing the depth of the beam reduces the lever arm of the flanges and so leads to bigger flange forces and bigger flanges.

As far as the stiffeners are concerned, BS 449 requires the spacing to be not greater than 1.5 d and the following rule to be satisfied:[1]

$$I \nless 1.5 \frac{d^3 t^3}{S^2}$$

where I is the second moment of area of a pair of stiffeners about the centre of the web; S is the maximum permitted clear distance between stiffeners; and t is the minimum required thickness of web.

If the stiffeners are made equal in thickness to the web, this rule will be satisfied if they project about $6t$ from the web, which gives some guidance in the choice of suitable dimensions. The joint between each stiffener plate and

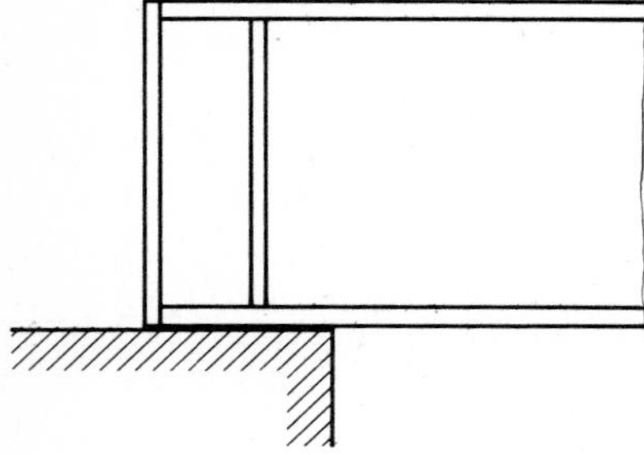

the web must be able to withstand a shearing force of $t^2/8h$ kN/mm of length where t is the web thickness in mm and h is the outstand of the stiffener, also in mm.

As in the case of rolled beams, bearing stiffeners may be required where concentrated loads are applied. They must be fitted at the reactions. The same rules are used for design of bearing stiffeners as for rolled beams. However, the loads are likely to be bigger and, because of the height of the web, the slenderness of the stiffener – web combination, greater. It may be desirable to use double stiffeners for this purpose.

15.8.2. Flanges of plate girders

The two problems in the design of the flanges are to calculate the stress caused by the bending moment and to determine the allowable stress. With regard to the former, Equations (15.2) and (15.3) are used. In the restricted context of this discussion, with the flanges equal, these equations reduce to Equation (15.4). British practice, as set out in BS 449, is to use the properties of the gross section and to adjust the computed stresses to allow for holes in the flanges.[1] The stress in a flange is adjusted by increasing the computed stress in the ratio of the gross area of the flange to the nett area.

The allowable tensile stress for flanges of plate girders depends on the quality of the steel and for Grade 43 steel it is 155 MPa if the plates are not more than 40 mm in thickness, 140 MPa if they are.

The allowable compressive stress is based on a critical stress which, in turn, depends on l/r_y for the whole cross-section and D/T for the compression flange. This critical stress is related to buckling of the flange and twisting of the girder. Consequently, it is the stiffness of some length of the compression flange and of the girder that matters. But D, T and r_y are essentially local properties and care is necessary when the properties of the cross-section vary along the beam. The critical stress for girders with equal flanges is given in BS 449[1] by

$$C_s = A = \left(\frac{1675}{l/r_y}\right)^2 \sqrt{1 + \frac{1}{20}\left(\frac{l}{r_y}\frac{T}{D}\right)^2}$$

The stress A is in MPa, and l is the effective length of the compression flange, defined the same way as for rolled beams. The radius of gyration r_y and the total depth D are measured at the section of maximum bending moment and T is the effective thickness of the compression flange given by

$$T = K_1 \times \begin{cases}\text{mean thickness of compression flange} \\ \text{at the section of maximum bending moment}\end{cases}$$

The coefficient K_1 depends on N, the ratio of the total area of both flanges at the section of least bending moment to that at the section of largest bending moment for a length of the beam between points of lateral restraint. Values of K_1 are as shown on Fig. 15.14. The breadth of a flange must not be reduced so much that N is less than 0.25. The thickness may be reduced, provided a rule given below about flange outstands is not broken.

The allowable stress depends on the grade of steel and the critical stress and, in Fig. 15.15, allowable stresses from BS 449[1] are plotted against the critical stress for Grade 43 steel. In no case may the allowable compressive stress exceed the allowable tensile stress.

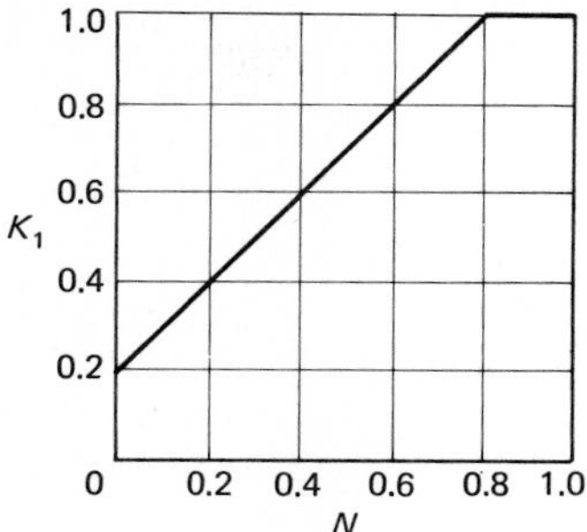

Fig. 15.14. *Determination of effective flange thickness. (Based on BS 449.)*[1]

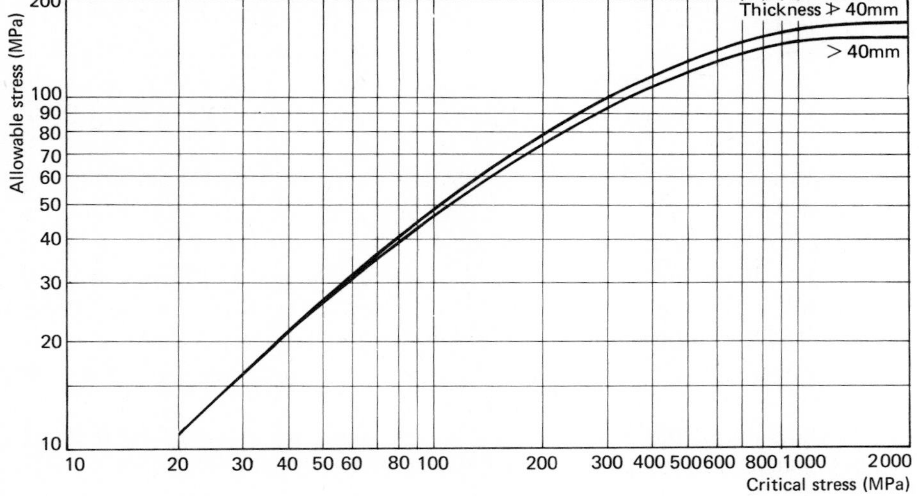

Fig. 15.15. *Permitted compressive stress for flange of plate girder, Grade 43 steel. (Based on BS 449.)*

The outstand of the flanges is controlled to avoid the risk of buckling of the edge of a flange. For this purpose, rules are given in BS 449.[1] For example, if the steel is Grade 43 and the flange plates have unstiffened edges, the outstand of the compression flange must not exceed 16 times the thickness of the thinnest plate. The factor for the tension flange is 20. The edge of the flange can be stiffened, in particular, by using a channel section as a flange. In that case, wider compression flanges are allowed.

The designer must take care with the details of flange curtailment. The change in section properties should be as gradual as possible to minimise stress raising. Tapering of flanges in breadth and thickness will achieve this, as sketched in Fig. 15.16.

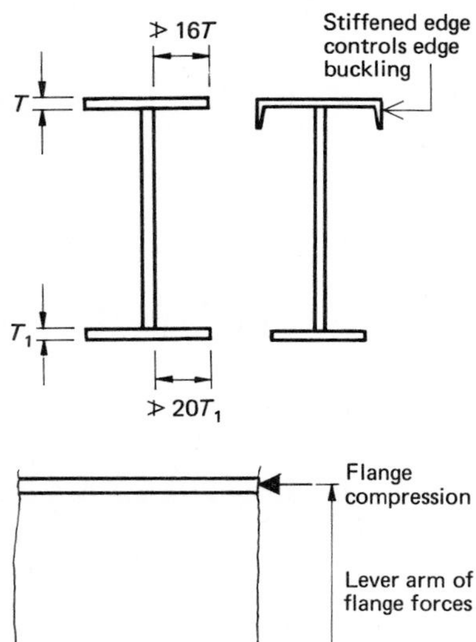

The procedure for flange design must be arranged systematically if a good solution is to be found in a reasonable time. The following steps are suggested:

(a) divide the maximum bending moment by the estimated lever arm of the forces in the flanges. This lever arm will equal the depth of the web plus the flange thickness, or the total depth of the beam minus the flange thickness. Here we assume that the contribution of the web to the moment of resistance is negligible, so that the quotient obtained in this step is an estimate of the force in each flange;

(b) divide the force in the compression flange by the estimated allowable stress in compression. This gives an area for the compression flange. For

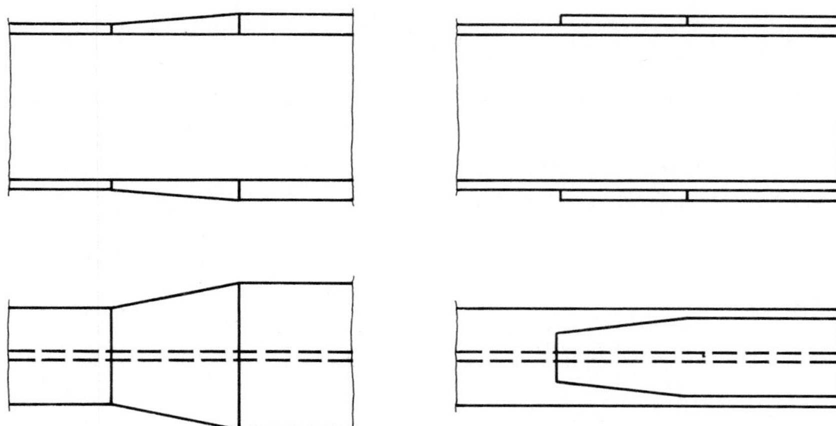

Fig. 15.16. *Tapering of flange width and thickness.*

the girders we are discussing, the tension flange is to be the same in area and shape;

(c) select suitable plates for the flanges;

(d) compute values for I_{xx}, l/r_y and D/T. Hence calculate the allowable stresses and the stresses f_{bc} and f_{bt} due to the external bending moment. These latter are then adjusted to allow for holes in the flanges;

(e) the allowable stresses and the stresses due to the loads are compared and, if necessary, further trials are made.

Curtailment of the flanges is most readily done by selecting the reduced area of flange and computing the moment of resistance with these smaller flanges. Whenever the moment of resistance exceeds the external bending moment, the reduced flanges are strong enough and the point of curtailment can be found from the bending moment diagram.

15.8.3. Connection of flanges to web

This connection has to be strong enough to resist the complementary shear stress at the junction between the web and the flange, combined with any external load applied to the flange. The force per unit length arising from shear in the web may be calculated using Equation (15.7)

$$f_q b = \frac{VQ}{I} \tag{15.7}$$

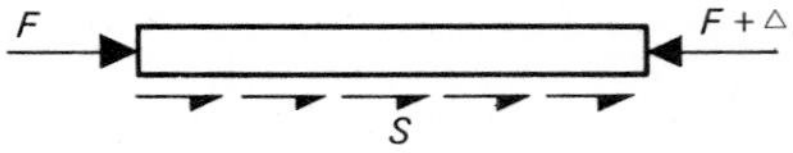

Fig. 15.17. *Forces acting on unit length of flange.*

An approximate alternative is to take the average web shear stress already computed. When this is multiplied by the web thickness, the result is the force per unit length to be transmitted across the joint. This may be seen by considering the equilibrium of a short length of the flange, as shown on Fig. 15.17. If the bending moment is not constant along the length of the beam – i.e. if the external shear is not zero – the forces applied to the ends of this element of the flange are not equal. Equilibrium is achieved with the shear force, S, exerted by the web on the flange and, for a unit length of flange

$$S = f_q t$$

15.8.4. Example of plate girder design

All of the foregoing notes are applied to the design of a plate girder in Example 15.7.

LIST OF SYMBOLS

A	area of cross-section, factor in derivation of critical stress for compression flange
A_p	area of flange plate
B	breadth of flange, effective length of web
b	breadth of cross-section at distance y from neutral axis
C_s	critical stress in compression flange
D	depth of section
d	height of web of plate girder
d_3	clear depth of web of rolled beam between fillets at junctions of flanges and web
E	modulus of elasticity
F	load-carrying capacity of rivets fastening flange plate to flange of plated beam
f	fundamental frequency of vibration, stress at distance y from neutral axis
f_{bc}, f_{bt}	bending compressive and tensile stresses at extreme fibres
f_q	shear stress at distance y from neutral axis
f_q'	average shear stress on web
f_y	yield strength
g	gravitational acceleration
h	outstand of stiffener
I	second moment of area
I_{NA}	second moment of area about neutral axis
I_{xx}	second moment of area about XX
K	spring constant for beam deflection
K_1	coefficient for calculation of effective thickness of flange of plate girder
L	span of beam
l	effective length of compression flange
M	bending moment
M_{max}	maximum bending moment
N	ratio of flange areas
p_{bc}, p_{bt}	permitted bending compressive and tensile stresses
p_q'	permitted average shear stress on web
Q	first moment about neutral axis of the part of the section further from the neutral axis than y
R	radius of curvature of deflected beam
r_y	radius of gyration about YY
S	shear force between web and flange, maximum permitted clear distance between stiffeners

Example 15.7 Sheet 1 of 4

Design a plate girder for span and loads as sketched below. Grade 43 steel. Lateral support for compression flange at loads and supports. Equal flanges.
Total depth ≯ 1.5 m.

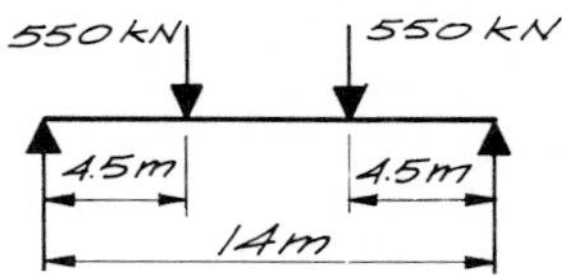

Estimated self-weight = 5 kN/m

Say, flange areas constant between loads, curtail to about 70% in one step between ends.

Shears: 550 kN 550 kN 35 kN 35 kN

Applied loads Self-weight

Moments: 2480 kN m $1/2 \times 35 \times 7 = 122$ kN m

Applied loads Self-weight

Design shear = 550 + 35 = 585 kN

Try $p_q' = 85$ MPa $\longrightarrow d \times t = \dfrac{5.85 \times 10^5}{85} = 6890$ mm²

Try $d = 1400$ mm $\rightarrow t = 4.92$ mm

Try $d = 1400$ mm $t = 6$ mm

$\dfrac{d}{t} = \dfrac{1400}{6} = 234$ – Too high

Try $d = 1400$ mm $t = 8$ mm

$\dfrac{d}{t} = \dfrac{1400}{8} = 175$ OK

Try stiffener spacing = $1.5d$

$p_q' = 79.5$ MPa

$f_q' = \dfrac{5.85 \times 10^5}{1400 \times 8} = 52.3$ MPa OK

WEB: 1.4 m high × 8 mm thick

Example 15.7 Sheet 2 of 4

Intermediate stiffeners:
Try 10mm plate, 60mm outstand

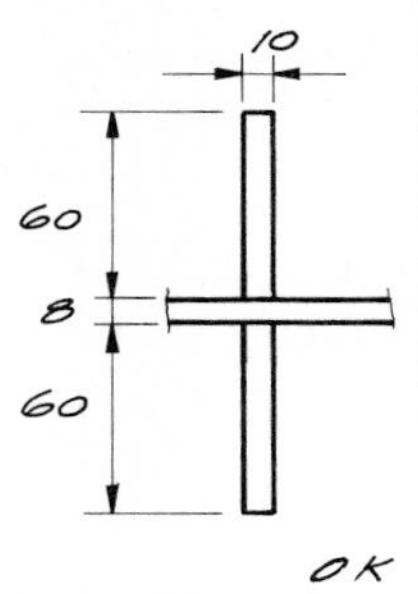

$$\frac{1.5\,d^3t^3}{s^2} = \frac{1.5 \times 1400^3 \times 8^3}{1.5^2 \times 1400^2}$$

$$= 4.77 \times 10^5 \text{ mm}^4$$

$$I = 1/12 \times 10 \times 128^3$$

$$= 1/1.2 \times 1.28 \times 1.28 \times 1.28 \times 10^6$$

$$> 10^6 > 4.77 \times 10^5 \qquad \text{OK}$$

$$\frac{t^2}{8n} = \frac{100}{8 \times 60} = 0.208 \text{ kN/mm} \text{ — 4mm fillet weld OK}$$

INTERMEDIATE STIFFENERS: 60mm × 10mm plate
2.1m centre to centre
4mm fillet weld to web

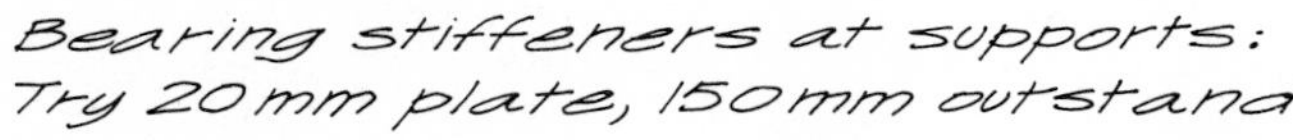

Bearing stiffeners at supports:
Try 20mm plate, 150mm outstand

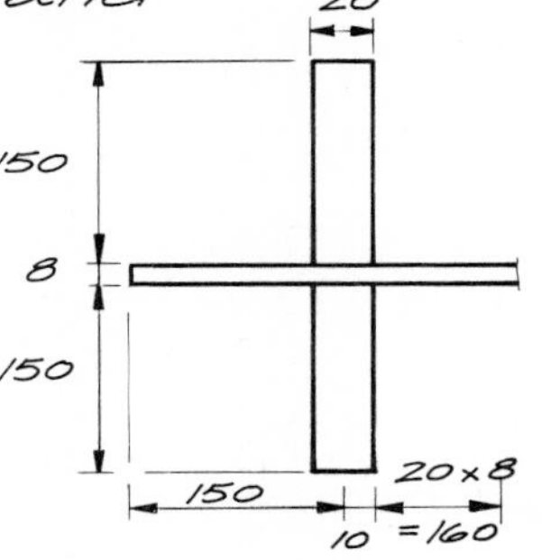

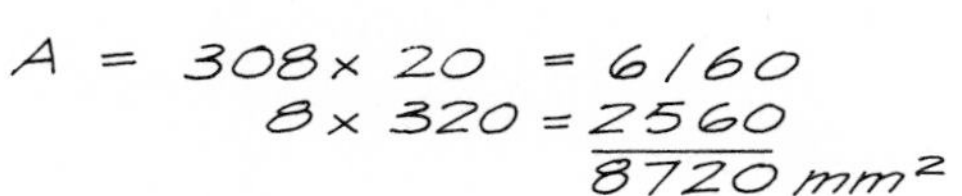

$$A = 308 \times 20 = 6160$$
$$8 \times 320 = 2560$$
$$8720 \text{ mm}^2$$

$$I = 1/12 \times 20 \times 308^3 = 4.86 \times 10^7 \text{ mm}^4$$

$$r = \sqrt{\frac{4.86 \times 10^7}{8.72 \times 10^3}} = 74.7 \text{ mm}$$

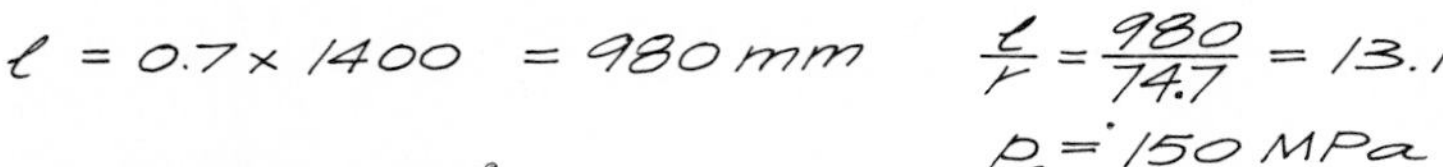

$$\ell = 0.7 \times 1400 = 980 \text{ mm} \qquad \frac{\ell}{r} = \frac{980}{74.7} = 13.1$$

$$p_c = 150 \text{ MPa}$$

$$f_c = \frac{5.85 \times 10^5}{8.64 \times 10^3} = 67.8 \text{ MPa} \qquad \text{OK}$$

Length of weld, stiffener to web = 4 × 1400 = 5600mm

$$\text{Load per unit length} = \frac{5.85 \times 10^6}{5.6 \times 10^3} = 105 \text{ N/mm}$$

4mm fillet weld OK

Length of weld, stiffener to flange = 4 × 150 + 2 × 20
= 670 mm

$$\text{Load per unit length} = \frac{5.85 \times 10^5}{670} = 875 \text{ N/mm}$$

12mm fillet weld OK

Example 15.7 Sheet 3 of 4

BEARING STIFFENERS : 2 No 150 mm × 20 mm
12 mm fillet weld to each flange
4 mm fillet weld to web

Flanges :

Design moment = 2480 + 122 = 2600 kNm

Approximate lever arm = 1400 + 25 = 1425 mm

Flange force $= \frac{2600}{1.425}$ = 1820 kN

Try p_{bc} = 160 MPa, Flange area $= \frac{1.82 \times 10^6}{1.6 \times 10^2} = 11400 \text{ mm}^2$

Try 450 mm × 25 mm flange plates

$I_{XX} = 2 \times 1/12 \times 450 \times 25^3 = 1.17 \times 10^6$

$2 \times 450 \times 25 \times 712.5^2 = 1.14 \times 10^{10}$

$1/12 \times 8 \times 1400^3 = 1.83 \times 10^9$

$1.323 \times 10^{10} \text{ mm}^4$

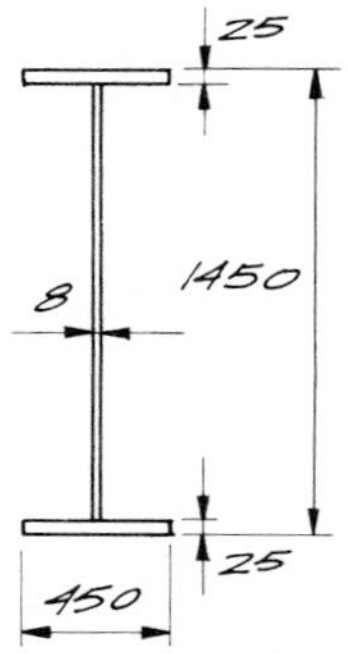

$I_{YY} = 2 \times 1/12 \times 25 \times 450^3 = 3.80 \times 10^8 \text{ mm}^4$

$A = 2 \times 450 \times 25 = 2.25 \times 10^4$

$1400 \times 8 = 1.12 \times 10^4$

$3.37 \times 10^4 \text{ mm}^2$

$r_y = \sqrt{\frac{3.80 \times 10^8}{3.37 \times 10^4}} = 106 \text{ mm}$

For maximum moment $\ell = 5000$ mm, $\frac{\ell}{r_y} = 47.1$

Flanges uniform between loads → $N = 1$, $K_1 = 1$

$\frac{D}{T} = \frac{1450}{25} = 58$

Critical stress $= \left(\frac{1675}{47.1}\right)^2 \sqrt{1 + 1/20\left(\frac{47.1}{58.0}\right)^2}$

$= 1260 \times 1.017$

$= 1280$ MPa

$p_{bc} = 165$ MPa

$f_{bc} = \frac{2.6 \times 10^6 \times 0.725}{1.323 \times 10^{-2}}$ Pa

$= 142$ MPa OK

Example 15.7 — Sheet 4 of 4

Curtailment of flanges:
Try 300 mm × 25 mm flanges

$N = \frac{300}{450} = 0.667 \qquad K_1 = 0.867$

Effective flange thickness $= 0.867 \times 25 = 21.6$ mm

$\frac{D}{T} = \frac{1450}{21.6} = 67.2 \qquad \frac{l}{r_y} = 47.1$ (same as above)

Critical stress $= \left(\frac{1675}{47.1}\right)^2 \sqrt{1 + 1/20\left(\frac{47.1}{67.2}\right)^2}$

$= 1260 \times 1.013$

$= 1280$ MPa

$p_{bc} = 165$ MPa

$I_{xx} = 2 \times 1/12 \times 300 \times 25^3 = 7.80 \times 10^5$

$2 \times 300 \times 25 \times 712.5^2 = 7.60 \times 10^9$

1.83×10^9

$9.43 \times 10^9\ mm^4$

Moment of resistance $= \frac{9.43 \times 10^{-3} \times 165 \times 10^{+6}}{712.5 \times 10^{-3}}$ Nm

$= 2180$ kNm

$x = \frac{2180}{2600} \times 4.5 = 3.78$ m

Say, change flange size at 3.5 m each side of midspan

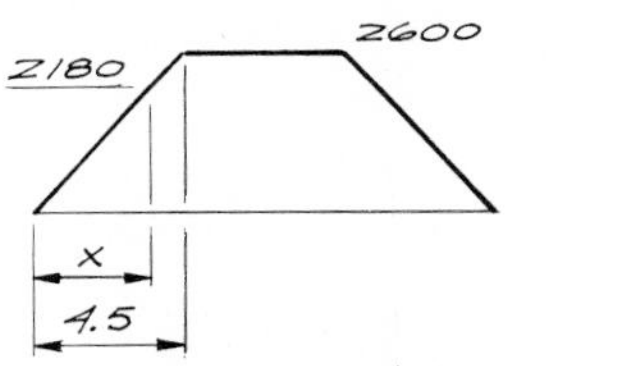

2180, 2600, x, 4.5

Approx. BM diagram

FLANGES:
Both flanges 25 mm thick throughout

6 mm fillet weld each side of web

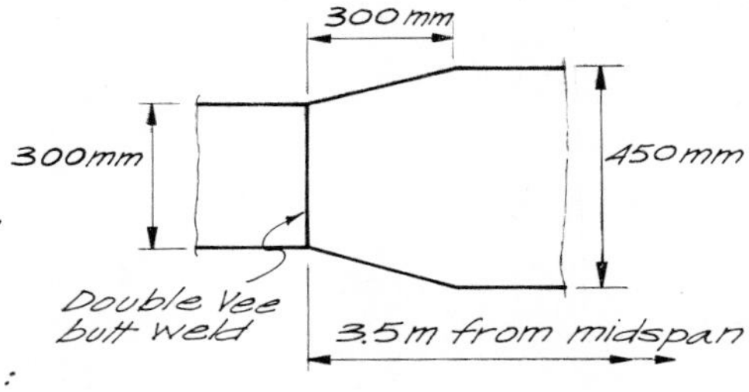

300 mm, 300 mm, 450 mm, Double Vee butt weld, 3.5 m from midspan

Web to flange welds:
Traction per unit length $= 52.3 \times 8$

$= 418.4$ N/mm

i.e. 209 N/mm in each fillet weld — 6 mm fillet weld each side

Check weight at 77 kN/m³:

$w = 77 \times 3.37 \times 10^{-2} = 2.6$ kN/m

OK

LIST OF SYMBOLS *(continued)*

s	spacing of pairs of rivets fastening flange plate to flange of plated beam
T	thickness of flange
T_p	thickness of flange plate
t	thickness of web
V	shear force
W	total load on beam
x	distance along beam
y	distance from neutral axis
y_c, y_t	distances from neutral axis to extreme fibres in compression and tension
Z	section modulus
Z_{XX}	section modulus about XX
Δ	deflection, shift of centroidal axis
Δ_L	deflection under static load
Δ_1, Δ_2	deflections of ends from tangent at midspan

REFERENCES

[1] BS 449:Part 2:1969, "Metric Units", *The Use of Structural Steel in Building*, British Standards Institution, London, 1969.

[2] BS 4:Part 1:1972, "Hot-rolled Sections", *Specification for Structural Steel Sections,* British Standards Institution, London, 1972.

[3] R. J. ROARK, *Formulas for Stress and Strain*, 4th ed., McGraw-Hill, New York, 1965.

[4] H. CROSS and N. D. MORGAN, *Continuous Frames of Reinforced Concrete,* John Wiley, New York, 1932.

BIBLIOGRAPHY

BRESLER, B., LIN, T. Y., and SCALZI, J. B., *Design of Steel Structures*, John Wiley, New York, 1968.

GERSTLE, K. H., *Basic Structural Design,* McGraw-Hill, New York, 1967.

GRAY, C. S., KENT, L. E., MITCHELL, W. A., and GODFREY, C. B., *Steel Designers' Manual*, 4th ed., Crosby Lockwood, London, 1972.

PROBLEMS

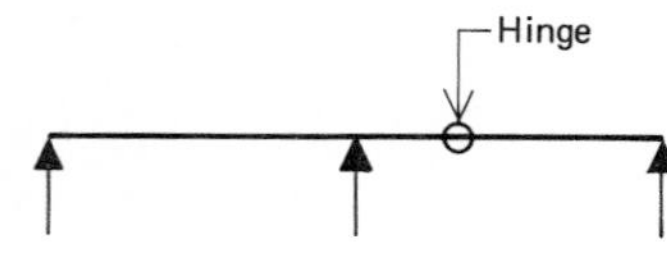

15.1. Two equal contiguous gaps are to be crossed by steel beams. Three possible structures are:

(a) one beam continuous over the central support;

(b) two simply supported beams;

(c) one beam to cross one gap and continuing over part of the other gap with a shorter, simply supported beam to complete the structure as sketched in the margin.

Sketch the bending-moment and shear-force diagrams for each case. How long should the second beam in (c) be to minimise the weight of steel? Compare the maximum bending moment for this lightest structure with the maximum bending moment in cases (a) and (b). Assume a uniformly distributed load and identify clearly any other restrictions you place on your analysis.

15.2. What effect would settlement of the central support have in the above three cases? If settlement is likely to occur, how would you allow for it in your design?

15.3. Use the graphical calculation described in Chapter 8 to estimate the deflection and natural frequency of the beam in Example 15.3.

15.4. Draw the live-load bending-moment diagram of Example 15.4 accurately to scale. Superimpose on it the bending-moment diagrams for a single central point load of 900 kN and a uniformly distributed load of 900 kN. Comment on the practicability of substituting a uniformly distributed load for a set of point loads to simplify estimates of shear force, bending moment and deflection.

15.5. Redesign the beams of Examples 15.4 and 15.5 for single rolling loads of 450 kN and 86 kN respectively. A rolling load is one that can be located anywhere along the beam.

15.6. Redesign the plate girder of Example 15.7 with a depth of web of 1.2 m and of 1.8 m and compare the total weight of steel and the cost of the three designs. Use cost data from Chapter 25 or locally available cost data.

16 Reinforced Concrete Beams

16.1. INTRODUCTION

In reinforced concrete structures, two materials, concrete and steel, are used so that the strength of each compensates for the weakness of the other. Concrete is strong in compression, but has little tensile strength; steel bars have great tensile strength but, because of their slenderness, must be restrained against buckling if loaded in compression. The basic concept in the use of reinforced concrete is to employ the concrete to resist compression and to have steel bars located in the structure where it is in tension. Implied in this use of steel is a requirement for the bars to be anchored and it is, of course, feasible to use steel bars in compression when this is appropriate. Well designed and built, reinforced concrete is tough, durable and versatile.

Construction of reinforced concrete is described in Chapter 23, "Construction Operations". In this chapter, a theoretical basis is presented for calculating dimensions of reinforced concrete beams and for design of the reinforcement.

16.2. DESIGN OF REINFORCED CONCRETE BEAMS

It is assumed that decisions have been made about the live loads (the loads in addition to the self-weight or dead load of the beam) and the span of the beam. Although these decisions may be among the most important and difficult to make in the whole design, this study is concerned with the support of known loads on a prescribed span. In this context, the problems to be solved by the designer are as follows:

(a) to determine suitable dimensions for the beam;
(b) to select the diameter and shape of the bars to reinforce the beam for bending;
(c) to make a similar selection for the shear reinforcement;
(d) to ensure that all the bars are anchored;
(e) to ensure that the beam's deflection will not be excessive.

These are related to limiting conditions, which are discussed in Chapter 4, "Failure and Safety Factors". In the theoretical development, the concept of the free body in equilibrium is used again – that is, the longitudinal stresses at any transverse section in the beam produce no resultant longitudinal force and create a couple equal to the bending moment at that section, while the resultant of the transverse stresses equals the external shear. For calculation of bending moments and shears, the span is taken from centre to centre of supports. An exception to this rule is made if the supports are very wide; then the span for calculations is the clear span plus the depth of the beam.

Design may be based on loads in service and permitted stresses which are less than yield stresses by an appropriate factor, or it may be based on service loads multiplied by a factor, and the ultimate strength of the beam. Since the trend today is firmly towards the latter, it is the only one presented. Rules for design are laid down by design codes which differ one from another. However, the same basic ideas support them all – the assumptions to be made about the behaviour of the materials, a requirement for static equilibrium and rules specifying such things as load factors and material strengths. The following analysis is made as general as possible, but the ACI *Building Code* published by the American Concrete Institute is referred to frequently.[1]

16.3. ASSUMPTIONS

Ultimate strength design of reinforced concrete is based on assumptions about the following matters:

(a) distribution of strain and strain compatibility for the two materials;
(b) relationships between stress and strain, for the two materials, concrete and steel;
(c) a definition of the ultimate strength of a reinforced concrete beam.

With regard to the first of these, it is assumed that there is a linear distribution of strain from maximum compressive strain at one extreme fibre, through zero at the neutral axis, to maximum tensile strain at the other extreme fibre. This assumption is the basis of the simple theory of bending for homogeneous beams and here it is assumed to hold at the limit of the strength of a beam made of two different materials. If strain is plotted against distance from the neutral axis, it is easy to show that the slope of the straight line equals the curvature (reciprocal of the radius of curvature) of the beam. It is assumed that the strain in any steel bar equals the strain in the concrete surrounding it – that is, there is strain compatibility. Provided the adhesion between the two materials is great enough, this assumption is acceptable on average over a short length of the beam. On the tension side, however, it must be recognised that the tensile strength of concrete is low. At large loads, the concrete cracks and the distribution of strain along a bar will become very irregular.

The distribution of stress can be determined from the strains if stress-versus-strain relationships for concrete and steel are known. Such relationships have been determined experimentally and idealised approximations are used in the design theory. Sketches of typical curves are shown in Fig. 16.1.

Loaded in compression, concrete has an approximately linear stress–strain relationship at low loads. However, as the strain increases, the curve flattens and passes through a maximum, the maximum stress being taken as the ultimate strength and given the symbol f'_c. Additional strain is accompanied by decreasing stress and failure occurs at the ultimate strain ε_u. When loaded in tension, concrete fails at a low stress, about 0.1 f'_c. Another important feature of strains in concrete is that they are time-dependent – when the stress is kept at a constant level, strain increases with time. This is not important in reinforced concrete design, but is a vital matter in relation to prestressed concrete.

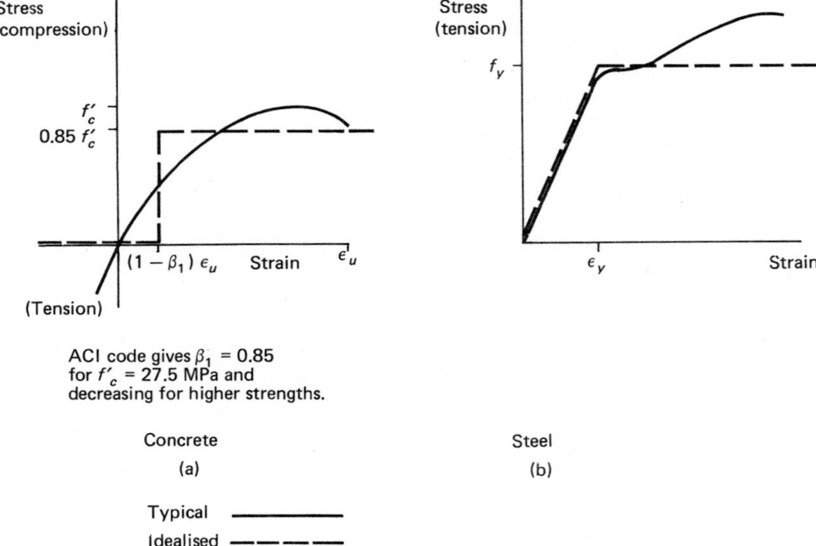

Fig. 16.1. *Typical and idealised stress–strain curves: (a) concrete; (b) mild steel.*

The rather complex behaviour of concrete under load is simplified for ultimate strength design theory, and the idealised stress–strain curve is shown dotted on Fig. 16.1(a). It has the following properties:

(a) the tensile strength of concrete is generally ignored. Exceptions to this rule are allowed in relation to inclined tension arising from shear and sometimes in prestressed concrete;
(b) failure in compression is defined by a strain of 0.003 ($= \varepsilon_u$);
(c) compressive stresses are related to strains by

$$0 < \varepsilon < (1 - \beta_1)\varepsilon_u \qquad f = 0$$
$$\varepsilon = (1 - \beta_1)\varepsilon_u \qquad 0 < f < 0.85\, f'_c$$
$$(1 - \beta_1)\varepsilon_u < \varepsilon < \varepsilon_u (= 0.003) \qquad f = 0.85\, f'_c$$

The factor $(1 - \beta_1)$ comes from the ACI Code.[1] For concrete whose strength is 27.5 MPa or less, it is 0.15 and increases with increasing strength of concrete.

The stress–strain relationship for steel in tension is less complicated, but is nevertheless simplified for the design theory. The grades of steel normally used in reinforced concrete are ductile. They show, at small strains, a linear stress–strain relationship almost up to the yield stress (f_y). Thereafter strain increases without the stress changing, through a "yield plateau". Then there is a "work hardening range" in which the stress again rises as the strain increases. When the specimen parts, the strain may be as much as 200 times the strain at yield.

The idealised approximation to this relationship is shown in Fig. 16.1(b) and comprises the elastic part of the experimental curve extended to yield stress and followed by a yield range to failure at a strain which need not be defined. The same assumption is made for steel in compression.

In addition to these assumptions about stress and strain, the theory also takes into account the warning the two materials give as they are loaded to failure. Concrete gives practically none and beams designed so as to fail by crushing of the concrete do so explosively when tested. Fragments of concrete burst out of the beam which collapses immediately. By contrast, a beam designed for tension failure will suffer large and visible deflections before it finally gives up. For this reason, it is good practice to design beams to be under-reinforced – that is, to ensure that ultimate load is defined by yielding of the steel and not by crushing of the concrete. The ultimate strength of a correctly designed beam is defined by maximum compressive strain in the concrete equal to 0.003 and strain in the tension steel greater than the strain at yield.

In the rest of this chapter, these assumptions are used to derive formulas for ultimate strength design of reinforced concrete beams. These are for beams of various kinds and use of the formulas is illustrated with worked examples.

16.4. FLEXURAL COMPUTATIONS

The requirements for static equilibrium in relation to flexure are that the total compression in any section must equal the total tension and the couple formed by these two forces must equal the external bending moment.

The external bending moment for strength design is determined by first multiplying the loads by load factors, then calculating the bending moment. Examples of load factors from the ACI code are 1.4 for dead load and 1.7 for live load. For statically determinate structures, statics alone suffice for calculation of the complete bending moment diagram, and, in this introductory text, only such structures are considered. However, it should be noted in passing that calculation of bending moments for statically indeterminate structures requires relative deflections and rotations to be taken into account to ensure continuity of the structure. These calculations are made as if the structure were behaving elastically. The admitted inconsistency in the design theory is made necessary by the difficulty of the structural calculations which describe the collapse of highly redundant structures and incomplete knowledge of their behaviour at collapse.

The flexural computations to be described below deal with the following:

(a) rectangular beams based on balanced design;
(b) oversized rectangular beams;
(c) undersized rectangular beams;
(d) T-beams;
(e) curtailment of reinforcement and anchorage.

16.4.1. Rectangular beam based on balanced design

The term *balanced design* means a design in which the concrete reaches a strain $\varepsilon_u = 0.003$ and the steel is strained to its yield point but not beyond it for the same bending moment. Although this is not a sound design, and good practice requires the steel to yield before the concrete fails, the concept is essential since it enables limiting values of certain parameters to be defined. An acceptable beam can be based on balanced design and is one in which the proportion of steel in the cross-section is a specified fraction (0.75 in the ACI Code) of what is required for simultaneous yield of the steel and crushing of the concrete.

For development of the formulas for balanced design, consider Fig. 16.2. On the left is shown a cross-section of the beam. The resistance couple at this

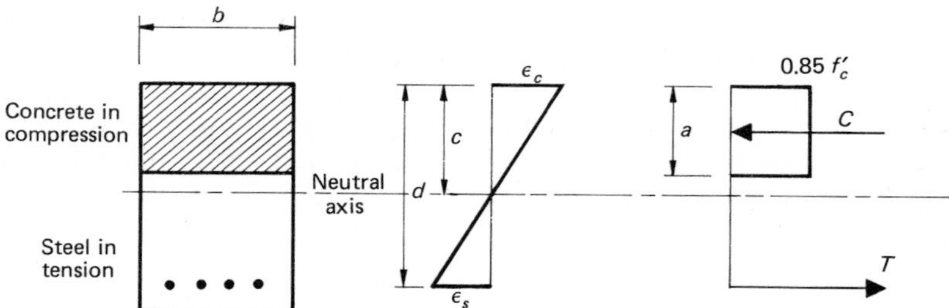

Fig. 16.2. *Rectangular beam with tension reinforcement only.*

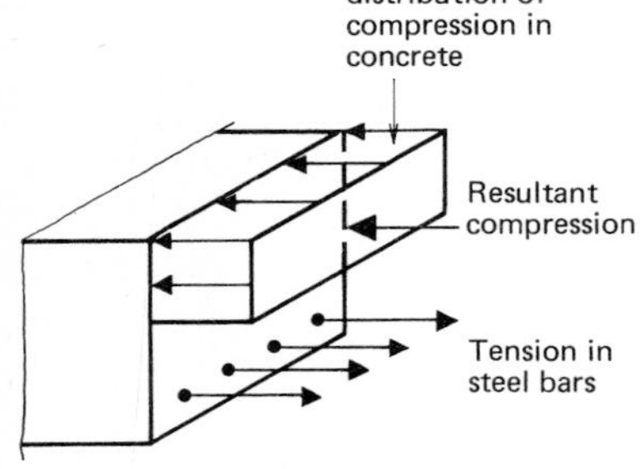

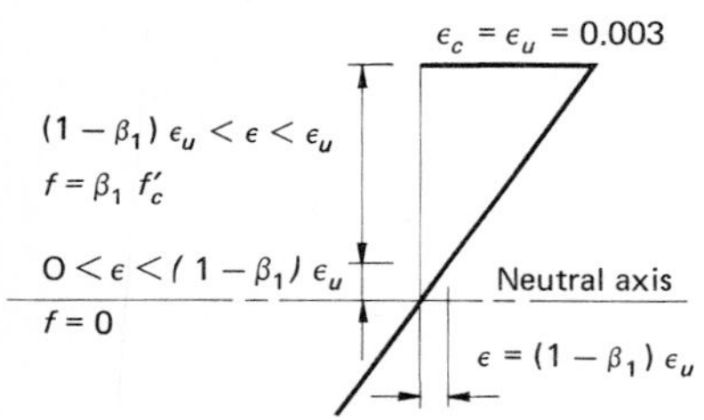

section comprises the tension, T, in the steel bars and the compression, C, which is the resultant of the stress block on the compression side of the neutral axis.

The effective depth of the beam is d, the distance from the extreme fibre in compression to the centre of tension and c is the depth of the neutral axis from the compression face. The depth of the compression stress block is a, and if $\varepsilon_c = \varepsilon_u$ $(= 0.003)$ then $a = \beta_1 c$. This follows from the idealised stress–strain relationship assumed for the concrete. The ACI Code requires the ratio $a:c$ $(= \beta_1)$ to be reduced for concrete strengths (f'_c) greater than 27.5 MPa (refer to Fig. 16.1(a)).

The design formulas may now be developed readily. From the force balance

$$A_s f_y = T = C = ab \times 0.85\ f'_c$$

where A_s is the area of steel; f_y is the yield stress in steel; a is the depth of stress block; b is the breadth of beam; and $0.85\ f'_c$ is the stress in concrete. From this

$$\frac{a}{d} = \frac{A_s}{bd}\,\frac{f_y}{0.85\ f'_c} = \rho\,\frac{f_y}{0.85\ f'_c} \tag{16.1}$$

where

$$\rho = \frac{A_s}{bd}$$

The ultimate moment of resistance is found by taking moments about the centre of tension or the centre of compression. Practice is to introduce a capacity reduction factor ϕ (< 1) at this stage, to allow for uncertainties in the assumptions, variations in the quality of construction and other imponderable factors which affect the strength of the beam. The ACI Code sets $\phi = 0.9$ in flexural computations.[1] Then, the ultimate moment of resistance, M_u, is given by

$$M_u = \phi \times C \times (d - \tfrac{1}{2}a)$$

$$= \phi \times 0.85\ f'_c\,abd\left(1 - \tfrac{1}{2}\frac{a}{d}\right)$$

$$= \phi\,(0.85\ f'_c)\frac{a}{d}\left(1 - \tfrac{1}{2}\frac{a}{d}\right)bd^2 \tag{16.2}$$

$$= K\,bd^2 \text{ where } K = \phi\,(0.85\ f'_c)\frac{a}{d}\left(1 - \tfrac{1}{2}\frac{a}{d}\right)$$

and

$$M_u = \phi \times T \times (d - \tfrac{1}{2}a)$$

$$= \phi\,A_s f_y\,d\left(1 - \tfrac{1}{2}\frac{a}{d}\right) \tag{16.3}$$

Equations (16.1) and (16.2) are the basic equations for design of a singly reinforced beam – one with reinforcement on the tension side only. To apply them, evaluation of the ratio, ρ, for balanced design (ρ_b) is now required. This will lead to the smallest allowable rectangular section.

First, the corresponding value of $c(= c_b)$ is determined from the strain distribution, namely

$$\frac{c_b}{d - c_b} = \frac{\varepsilon_c}{\varepsilon_s} = \frac{\varepsilon_u}{f_y/E_s}$$

in which $E_s = 200$ GPa is Young's modulus for the steel. Hence

$$k_b = \frac{c_b}{d} = \frac{\varepsilon_u E_s}{\varepsilon_u E_s + f_y}$$

Applying Equation (16.1) to the balanced condition,

$$\frac{\beta_1 c_b}{d} = \rho_b \frac{f_y}{0.85\ f'_c}$$

$$\rho_b = \frac{\beta_1\ (0.85\ f'_c)}{f_y} \frac{\varepsilon_u E_s}{\varepsilon_u E_s + f_y} \qquad (16.4)$$

The value of ρ to be used in design is 0.75 ρ_b as noted above so that, with β_1, E_s, f'_c and f_y known, ρ can be found. Then, a/d is computed from Equation (16.1). This value of a/d is, in turn, used to evaluate the factor $K(= \phi(0.85\ f'_c)\ a/d\ (1 - \frac{1}{2}\ a/d))$ in Equation (16.2). Then bd^2 can be calculated from Equation (16.2). Values of b and d can now be selected to provide a large enough value of bd^2 and A_s found from $\rho = A_s/bd$. The process is illustrated in Example 16.1, where calculations for design of a beam are set out as they would be in a design office.

Design data are listed at the beginning of the calculation. These include details of span and loads, concrete strength and yield stress of the steel, the factor ϕ and the load factors. Also listed are the cover and a code restriction on the spacing of bars. The significance of cover and bar spacing is described in Chapter 23, "Construction Operations". Here it need only be noted that, in the calculations for strength, d is the *effective* depth – it is not the total depth, which includes depth occupied by the bars and the concrete cover between layers of bars and between the bars and the surface of the concrete.

At the beginning of the calculation, the parameters for balanced design are calculated. These are k_b, ρ_b and the corresponding K. The maximum permitted steel ratio ($= 0.75\ \rho_b$) is also calculated. Next dead load is estimated on the basis of the anticipated size of the beam, the loads are multiplied by the load factors and shear force and bending moment diagrams are sketched.

Design of the beam then follows as described above. It should be noted that a check is made on the width required for the steel arrangement chosen and that the first choice of dimensions had to be altered to keep the steel ratio down to 0.75 ρ_b. The calculation ends with a sketch of the cross-section to summarise the design and a check on the dead load.

In routine design, no further checks need be made. Sufficient strength in the beam is ensured by bd^2 being greater than the minimum required. Yield of the steel before the concrete is crushed is ensured by having $\rho \not> 0.75\ \rho_b$.

It is, however, instructive to analyse the beam and this is done at the end of Example 16.1. The cross-section at midspan is shown on the left and near it the forces in the resistance couple. These are the tension, T, given by yield stress in the steel and the area of the bars, and the total compression, C. Since $C = T$ and the height of the stress block, $0.85\ f'_c$, and its width, b, are known, the depth of the stress block, a, can be calculated. This in turn leads to $c\ (= a/0.85)$ which gives the location of the neutral axis. With extreme fibre compression strain equal to 0.003 and c known, the strain distribution is fully determined (it is sketched on the right). It is then a simple matter to calculate the strain in the inner layer of bars.

The checks on the design are to compute the ultimate moment of resistance (including the factor $\phi = 0.9$) and to confirm that the strain in the steel exceeds yield strain.

Clearly, a table of areas for different numbers of bars of various sizes is an essential aid in a computation like this. Such a table is included in Appendix A6.

16.4.2. Oversized rectangular beam

In the preceding calculation, the size of the beam was not known in advance and design of the beam included selection of its dimensions. It is frequently the case that beam dimensions are selected for reasons other than structural. Then the structural design is limited to determination of the amount and arrangement of the steel.

The first step is to decide whether the beam is oversized or undersized relative to a beam based on balanced design. This is done very quickly by computing K for $\rho = 0.75\ \rho_b$ and comparing Kbd^2 with the design moment. (It is of first importance to remember that d is the *effective* depth – the distance from the compression face to the centre of tension – and not the total depth of the beam.) If Kbd^2 is the bigger of the two, the beam is oversized and the design proceeds as follows.

Equations (16.2) and (16.3) are valid for this case also. Equation (16.2) is rearranged with M_u equal to the design moment M to give

$$\frac{a}{d}\left(1 - \frac{1}{2}\frac{a}{d}\right) = \frac{M}{\phi\,(0.85\ f'_c)\ bd^2} = F \qquad \text{(say)}$$

whence

$$\frac{a}{d} = 1 - \sqrt{1 - 2F}$$

With all the variables which determine the value of F known, a/d can be calculated. This leads immediately to the lever arm of the ultimate moment of resistance $(d - \frac{1}{2}a)$ so that A_s can be calculated from Equation (16.3).

All of these calculations are illustrated in Example 16.2 as they would be set out in practice. Also shown is a check analysis from which it is apparent that the ultimate moment of resistance is big enough and that the steel will yield before the concrete is crushed. This detailed check would not normally be undertaken. The designer has provided more steel than the minimum required, so that he knows the ultimate moment of resistance is adequate. Furthermore, since the requirement $\rho \not> 0.75\ \rho_b$ has been met, he knows that the mode of failure would be satisfactory.

Example 16.1 Sheet 1 of 2

Determine the dimensions and reinforcement at midspan of a simply supported beam.

8.00 m span c/c supports
Live load 45 kN/m
f'_c = 20 MPa $0.85 f'_c = 17$ MPa
f_y = 270 MPa
E_s = 200 GPa
$\epsilon_u E_s$ = 600 MPa
β_1 = 0.85
Weight of concrete = 24 kN/m³

Cover = 40 mm
Bar spacing:
D or 25 mm
whichever is less
Load factors:
D.L. 1.4
L.L. 1.7
ϕ = 0.9

Balanced design: $k_b = \dfrac{600}{600 + 270} = 0.690$

$$\rho_b = \frac{0.85 \times 17}{270} \times 0.690 = 0.0369$$

$$\rho \ngtr 0.75 \times 0.0369 = 0.0277$$

For limiting value of ρ, $\dfrac{a}{d} = 0.0277 \times \dfrac{270}{17} = 0.440$

$$k = \phi(0.85 f'_c)a/d(1 - 1/2\, a/d)$$
$$= 0.9 \times 17 \times 0.44 \times 0.78$$
$$= 5.26 \text{ MPa}$$

Estimated dead load = 10 kN/m
Design load: 1.4 × 10 = 14.0
1.7 × 45 = 76.5
90.5 kN/m

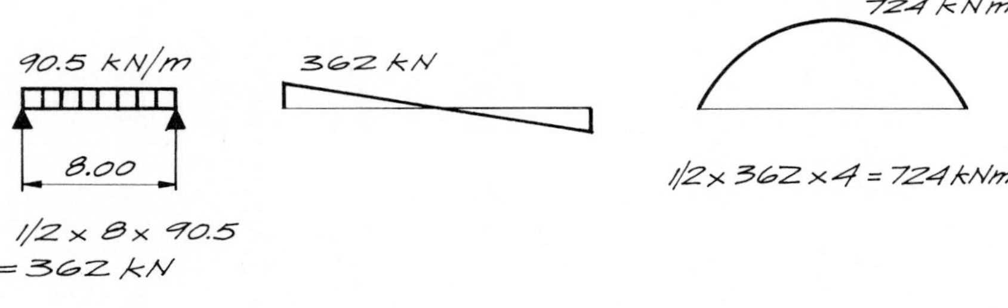

1/2 × 8 × 90.5
= 362 kN

1/2 × 362 × 4 = 724 kNm

Example 16.1 Sheet 2 of 2

$bd^2 = \frac{7.24 \times 10^5}{5.26 \times 10^6} = 0.137\ m^3$

Try $b = 410$ mm $\longrightarrow$ $d = 579$ mm

$A_s = 0.0277 \times 410 \times 579$
$= 6560\ mm^2$

14 bars × 25 mm dia $\longrightarrow$ 6870 mm^2

Two layers

Width required $= 2 \times 40 + 7 \times 25 + 6 \times 25$
$= 405$ mm OK

$\rho = \frac{A_s}{bd} \not> 0.0277$

$d \not< \frac{6870}{410 \times 0.0277} = 604$ mm

$h \not< 604 + (1/2 \times 25) + 25 + 40$
$= 682$ mm

Say, 690 mm

410

690

Check self-weight

$0.41 \times 0.69 \times 24 = 6.8$ kN/m
< allowed OK

2 × 7 × 25 mm diameter bars
25 mm clear between rows
25 mm clear between bars
40 mm clear cover

Check analysis:

$\frac{1.85 \times 10^6}{410 \times 17} = 266$ mm

C = 1850 kN

$612 - \frac{226}{2} = 479$ mm

T = 6870 × 270 N
= 1850 kN

0.003

$\frac{266}{0.85} = 313$

587

637

$M_u = 0.9 \times 1.85 \times 10^6 \times 0.479$
$= 796$ kNm
> 724 kNm OK

$\frac{587 - 313}{313} \times 0.003$
$= 0.00262$
$> \frac{f_y}{E_s} = \frac{2.7 \times 10^8}{2 \times 10^{11}} = 0.00135$ OK

Example 16.2 Sheet 1 of 1

Determine the flexural reinforcement required for a rectangular beam, given the dimensions and bending moment. The beam is oversized.
Design data as for Example 16.1.

Design moment = 724 kNm
$b = 410$ mm $h = 750$ mm
Hence $d = 673$ mm (Assuming two layers of 25 mm bars and 40 mm cover)

For $\rho = 0.75\rho_b = 0.0277$

$M_u = (5.26 \times 10^6) \times 0.410 \times 0.673^2$ Nm
$= 977$ kNm
> 724 kNm — Hence beam oversized

$$F = \frac{M}{\phi(0.85f_c')bd^2} = \frac{7.24 \times 10^5}{0.9 \times 17 \times 10^6 \times 0.41 \times 0.673^2} = 0.255$$

$$\frac{a}{d} = 1 - \sqrt{1-2F} = 1 - \sqrt{0.49} = 0.300$$

$$1 - \frac{1}{2}\frac{a}{d} = 0.850$$

Lever arm $= 0.85 \times 673 = 571$ mm

$$A_s = \frac{7.24 \times 10^5}{0.9 \times 270 \times 0.571} = 5210 \text{ mm}^2$$

Say, 12 × 25 mm bars ⟶ $A_s = 5890$ mm²

Check: $\rho = \frac{5890}{410 \times 673} = 0.0213 < 0.0277$ OK

Check analysis:

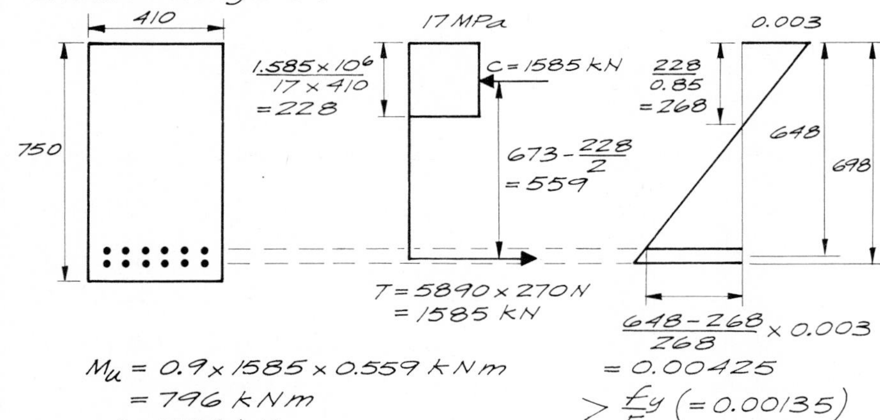

$M_u = 0.9 \times 1585 \times 0.559$ kNm
$= 796$ kNm
> 724 kNm OK

$= 0.00425$
$> \frac{f_y}{E_s} (= 0.00135)$ OK

16.4.3. Undersized rectangular beam

In this case also the dimensions of the beam are given and are tested by computing Kbd^2 for $\rho = 0.75\ \rho_b$. If this is less than the design moment, the moment of resistance has to be increased. The additional resistance is arranged by providing more steel on the tension side and bars on the compression side. Then, the total moment of resistance is made up of two couples: compression in the concrete and part of the tension in the tensile steel make one, and compression in the compression steel and the rest of the steel tension make the other. It is not sufficient to increase the tension steel only, without providing compression bars. Although the beam could be made strong enough in this way (for a modest increase in moment) the mode of failure would change from yielding of the steel with large warning deflections to crushing of the concrete.

The method of design requires the same upper limit to be placed on ρ as in the case of the singly reinforced beam – that is, the amount of tension steel is not to exceed 75% of the amount of tension steel corresponding to balanced conditions. The fundamental requirement is the same in both cases. It is that the total tension in the tension steel shall not exceed 75% of the total compression under balanced conditions.[2] Expressed algebraically, the criterion is

$$A_s f_y \ngtr 0.75 \left\{(0.85\ f_c')\ bd\left(\frac{a}{d}\right)_b + A_s'\ f_y\right\}$$

Here A_s' is the area of steel on compression side of neutral axis, and $(a/d)_b$ is the ratio of $a:d$ for balanced conditions. Other symbols are as defined previously and it is assumed that the compression bars are strained to or beyond the yield point.

Dividing both sides by $bd\ f_y$, we have

$$\rho = \frac{A_s}{bd} \ngtr 0.75\left\{\frac{(0.85\ f_c')}{f_y}\left(\frac{a}{d}\right)_b + \frac{A_s'}{bd}\right\}$$

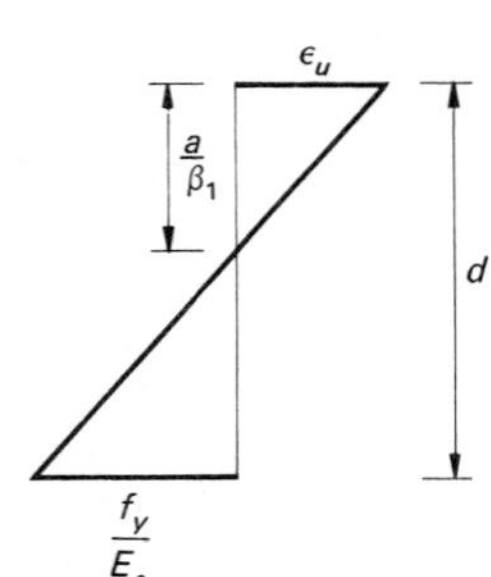

The ratio $(a/d)_b$ is obtained from the strain distribution (see marginal sketch) and is clearly the same as for the singly reinforced rectangular beam, that is

$$\left(\frac{a}{d}\right)_b = \beta_1 \frac{\varepsilon_u E_s}{\varepsilon_u E_s + f_y}$$

The constraint on the area of tension steel is then

$$\rho \ngtr 0.75\left\{\beta_1 \frac{(0.85\ f_c')}{f_y} \frac{\varepsilon_u E_s}{\varepsilon_u E_s + f_y} + \frac{A_s'}{bd}\right\}$$

The first term in brackets on the right of this inequality is the same as ρ_b as given by Equation (16.4). It is the tension steel ratio for balanced conditions in

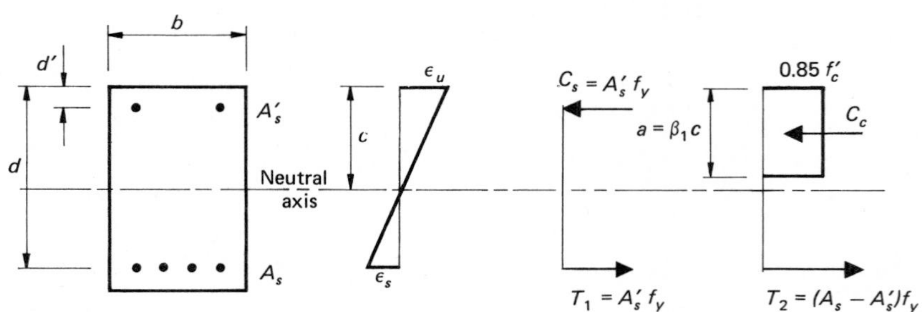

Fig. 16.3. *Rectangular beam with compression steel and tension steel.*

the given cross-section if there is no compression steel. If $\bar{\rho}_b$ is defined as this steel ratio

$$\bar{\rho}_b = \beta_1 \frac{(0.85\, f'_c)}{f_y} \frac{\varepsilon_u E_s}{\varepsilon_u E_s + f_y}$$

and

$$\rho \ngtr 0.75\,(\bar{\rho}_b + \rho')$$

where

$$\rho' = \frac{A'_s}{bd}$$

Design formulas based on these ideas can now be derived.

In Fig. 16.3 the sketches show from left to right a cross-section of the beam, the distribution of strain, a couple M_c comprising the steel compression ($C_s = A'_s\, f_y$) and a corresponding part of the steel tension ($T_1 = A'_s\, f_y$), and the rest of the resistance couple M_w – the concrete compression and the tension T_2 which is equal to $(A_s - A'_s) f_y$.

It is clear that couple M_w can be dealt with exactly as was the resistance of a singly reinforced beam, with $(A_s - A'_s)$ substituted for A_s. That is,

$$M_w = \phi(0.85\, f'_c) \frac{a}{d} \left(1 - \tfrac{1}{2} \frac{a}{d}\right) bd^2$$

$$= \phi(A_s - A'_s)\, f_y d \left(1 - \tfrac{1}{2} \frac{a}{d}\right)$$

The other couple is given by

$$M_c = \phi(A'_s\, f_y)\,(d - d')$$

provided

(a) the neutral axis is low enough to make the strain in the compression steel not less than f_y/E_s, and

(b) the compression bars are restrained against buckling.

The first condition will be satisfied if (see Fig. 16.3)

$$\frac{c - d'}{c} \varepsilon_u \nless \frac{f_y}{E_s}$$

$$\frac{c}{d} \nless \frac{\varepsilon_u E_s}{\varepsilon_u E_s - f_y} \frac{d'}{d}$$

Reference to Fig. 16.3 shows that this puts a lower limit on the compression force, C_c, and hence on T_2 :

$$(A_s - A'_s) f_y = (0.85\, f'_c)\, b \beta_1 c$$

$$\frac{c}{d} = \frac{A_s - A'_s}{bd} \frac{f_y}{\beta_1 (0.85\, f'_c)}$$

$$= (\rho - \rho') \frac{f_y}{\beta_1 (0.85\, f'_c)}$$

and

$$\rho - \rho' \nless \frac{\beta_1 (0.85\, f'_c)}{f_y} \frac{\varepsilon_u E_s}{\varepsilon_u E_s - f_y} \frac{d'}{d}$$

The second condition is satisfied by providing stirrups which enclose the compression bars and prevent them from deflecting outwards through the cover.

In summary, design of a doubly reinforced undersized beam is based on the following criteria:

$$\rho \not> 0.75\left(\frac{\beta_1(0.85\, f'_c)}{f_y}\,\frac{\varepsilon_u E_s}{\varepsilon_u E_s + f_y} + \rho'\right) \tag{16.5}$$

$$\rho - \rho' \not< \frac{\beta_1(0.85\, f'_c)}{f_y}\,\frac{\varepsilon_u E_s}{\varepsilon_u E_s - f_y}\,\frac{d'}{d} \tag{16.6}$$

Ultimate moment generated by compression in concrete

$$= M_w = \phi(0.85\, f'_c)\frac{a}{d}\left(1 - \tfrac{1}{2}\frac{a}{d}\right)bd^2 \tag{16.7}$$

Ultimate moment generated by compression in steel

$$= M_c = \phi A'_s\, f_y\,(d - d') \tag{16.8}$$

Ultimate moment generated by tension in steel

$$= M_c + M_w = \phi A'_s\, f_y\,(d - d') + \phi(A_s - A'_s)\, f_y\,(d - \tfrac{1}{2}a) \tag{16.9}$$

Application of these formulas is illustrated in Example 16.3. The limiting values of ρ and $(\rho - \rho')$ are calculated first and it will be seen that the upper limit for ρ depends on ρ', which is not known at this stage.

The first step in the design calculation is to test the need for compression steel by computing Kbd^2 with K corresponding to the maximum allowable value of ρ and $\rho' = 0$. The excess moment which is to be allowed for by adding compression steel and extra tension steel is the amount by which the design moment exceeds Kbd^2 and Equation (16.8) is used to calculate A'_s. A selection of bars for the compression steel can now be made. It is necessary to provide considerably more compression steel than is computed from the excess of the design moment over the ultimate moment when there is no compression steel. The reason is that the total tension must not exceed 75% of the total compression under balanced conditions. If provision is made for an increase in total compression of at least four-thirds of that derived from the excess moment, the desired increase in tension can be obtained without breaking this rule. However, the matter does not end there. The final selection of bars will generally yield a total A_s greater than what is computed, because it is not likely that bars can be chosen to give exactly the required area. It is sound design to provide ample compression steel and double the area computed from the excess moment will not be too much.

Suitable bars having been chosen for A'_s, M_c is computed from Equation (16.8). When this is deducted from the design moment, the design moment for the concrete is obtained and the calculation proceeds as described in Section 16.4.2.

The last step in the design is to confirm that ρ and $(\rho - \rho')$ fall within the prescribed limits. No additional checks are necessary in practice; clearly all design criteria have been satisfied. However, an analysis of the ultimate moment and the strains is included in Example 16.3.

Example 16.3 Sheet 1 of 2

Determine the flexural reinforcement required for a rectangular beam, given the dimensions and the bending moment. The beam is undersized.
Design data as for Example 16.1.

Design moment = 724 kNm
$b = 410$ mm $h = 630$ mm
Hence $d = 553$ mm, $d' = 52$ mm, assuming two layers 25 mm bars in tension, one layer 25 mm bars in compression.

$$\rho \not> 0.75\left\{\frac{\beta_1(0.85f_c')}{f_y}\frac{\epsilon_u E_s}{\epsilon_u E_s + f_y} + \rho'\right\} = 0.0277 + 0.75\rho'$$

For compression bars to yield

$$\rho - \rho' \not< \frac{\beta_1(0.85f_c')}{f_y}\frac{\epsilon_u E_s}{\epsilon_u E_s - f_y}\frac{d'}{d} = \frac{0.85\times 17}{270}\times\frac{600}{600-270}\times\frac{52}{553}$$
$$= 0.00915$$

For $\rho' = 0$ and $\rho = 0.0277$ $k = 5.26$ MPa (Example 16.1)

$$kbd^2 = 5.26\times 10^6 \times 0.41 \times 0.553^2 \text{ Nm}$$
$$= 660 \text{ kNm}$$
$$< 724 \text{ kNm}$$

Hence compression steel is required.

Excess moment = 64 kNm

$$A_s' = \frac{6.4\times 10^4}{0.9\times 0.501\times 270}$$
$$= 525 \text{ mm}^2$$

2 × 32 mm bars ⟶ $A_s' = 1610$ mm² o.k. (> 2 × 525 mm²)

$$M_c = 0.9\times 1610\times 270\times 0.501 \text{ Nm}$$
$$= 196 \text{ kNm}$$

Design moment for concrete = 724 − 196 = 528 kNm

$$F = \frac{5.28\times 10^5}{0.9\times 17\times 10^6\times 0.41\times 0.553^2}$$
$$= 0.276$$

$$\frac{a}{d} = 1 - \sqrt{1 - 0.552} = 0.330$$

$$1 - \frac{1}{2}\frac{a}{d} = 0.835$$

$$A_s - A_s' = \frac{5.28\times 10^5}{0.9\times 270\times 0.835\times 0.553}$$
$$= 4710 \text{ mm}^2$$

Example 16.3 Sheet 2 of 2

$A_s = 4710 + 1610 \qquad = 6320 \text{ mm}^2$

8×32 mm bars $\longrightarrow$ $A_s = 6435 \text{ mm}^2$

$$\rho = \frac{6435}{410 \times 553} = 0.0284$$

$$\rho' = \frac{1610}{410 \times 553} = 0.0071$$

$$0.0277 + 0.75\rho' = 0.0277 \times 0.0053 = 0.0330$$

$$\rho \not> 0.0277 + 0.75\rho' \qquad \text{OK}$$

$$\rho - \rho' = 0.0213 > 0.00915$$

OK (compression steel yields)

USE: 8×32 mm bars for tension steel
2×32 mm bars for compression steel

Check analysis:

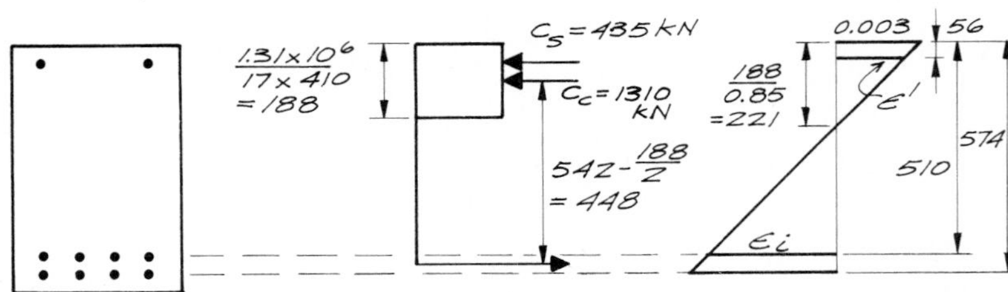

T: $1610 \times 270 = 435$
$4825 \times 270 = 1310$
$= 1745$ kN

$$\varepsilon' = \frac{221 - 56}{221} \times 0.003 = 0.00224 > \frac{f_y}{E_s} \qquad \text{OK}$$

$M_u = 0.9 \times 435 \times (0.542 - 0.056) = 191$
$0.9 \times 1310 \times 0.448 = 527$
718 kNm

$$\varepsilon_i = \frac{510 - 221}{221} \times 0.003 = 0.0392 > \frac{f_y}{E_s} \qquad \text{OK}$$

Note that M_u is smaller than the design moment by a small margin. This has occurred because the lever arm of each couple is a little smaller than assumed in the design calculations — 32 mm bars used, compared with 25 mm on which estimates of d and d' based. However, the error is insignificant — < 1%.

16.4.4. T-beam

The T-beam is a structure which arises naturally in beam and slab construction (see Chapter 5, "Machines and Structures"). The slab which spans the gap between adjacent beams is on the compression side of the beams at midspan, where bending moments are positive. It can, therefore, serve a second function as well as being a load-supporting member in its own right – that of a compression flange on the beam. How much of the slab will actually contribute to the strength of the beam is uncertain and, for this reason, codes specify an upper limit for the width of flange the designer may assume. The ACI code, for example, does not allow the flange assumed in design to be wider than a quarter of the span or to extend more than half-way to the adjacent beams on either side or to overhang the stem or rib by more than eight times the thickness of the slab.[1]

The designer must remember that the slab can add to the strength of the beam only when it is on the compression side. In monolithic construction negative bending moments occur near the supports of beams. Then, the slab, being on the tension side when the bending moment is negative, does not contribute to the strength of the beam.

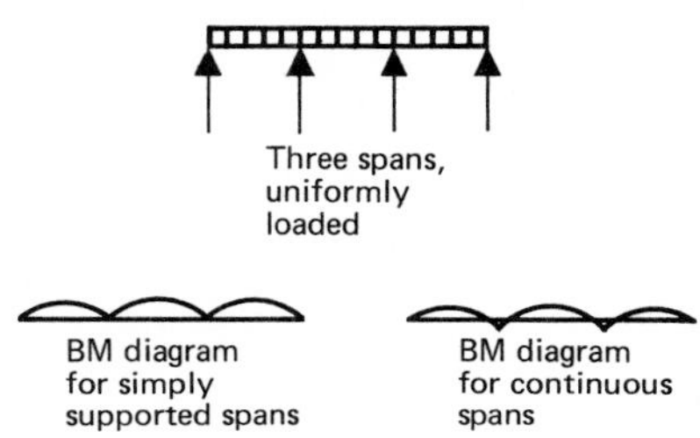

Generally the dimensions of a T-beam do not have to be fixed by flexural computations for the beam. The flange width is laid down by the code and its thickness is determined from the design of the slab, taking into account its strength and requirements for containing fires, soundproofing or other non-structural matters. The dimensions of the stem may be based on negative bending moments near the supports or on consideration of shear or they may be chosen arbitrarily. The principal item of concern to the structural designer is the reinforcement.

Two classes must be recognised depending on the location of the neutral axis. If it is high enough, none of the stem below the underside of the flange will be in compression. In this case, the beam is exactly the same as a rectangular beam of breadth b, the width of the flange. Whether the actual shape is a "Tee" or a rectangle, none of the concrete below a line $(1 - \beta_1)c$ above the neutral axis enters the flexural computations. In the second class, with a lower neutral axis, the flange and part of the stem are subject to uniform stress. The width of the compression zone is not constant and this has to be taken into account.

For conventional beam and slab floors, permitted flange widths are so large that a very big bending moment is required to push the neutral axis below the flange soffit (its underside). If the beams are close together, as they are in ribbed floors or bridge decks, the T-shape will probably have to be taken into account, but the designer will encounter many T-beams which are effectively very wide rectangular beams.

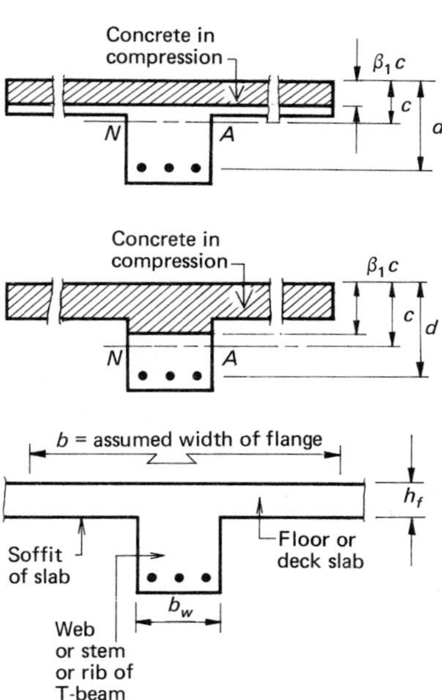

In design the first question to be asked is this: Will the compression zone extend below the slab soffit? It is very quickly answered by calculating the ultimate moment with $a = h_f$, the flange thickness. Thus,

$$C = (0.85\ f'_c)\ bh_f$$

$$\text{lever arm} = d - \tfrac{1}{2} h_f$$

$$M_u = \phi(0.85\ f'_c)\ bh_f\ (d - \tfrac{1}{2} h_f)$$

If this is greater than the design moment, less than the whole flange is required to be in compression and the procedure of Section 16.4.2 is followed. This is illustrated in Example 16.4. It is interesting to compare this design with that of the undersized rectangular beam in Example 16.2. The design moment and the effective depth are the same for both. The T-beam requires a smaller area of tension steel because the resultant compression is higher and this gives the ultimate moment a longer lever arm.

It should be noted that a possible variation in this class has not been considered – the possibility that such a beam might require compression reinforcement. For this to occur, it would be necessary to have

$$\frac{h_f}{d} > 0.75\left(\frac{a}{d}\right)_b$$

that is

$$\frac{h_f}{d} > 0.75\,\beta_1 \frac{\varepsilon_u E_s}{\varepsilon_u E_s + f_y}$$

Since $\varepsilon_u = 0.003$, $E_s = 200$ GPa, and typical values of f_y and β_1 are 270 MPa and 0.85 respectively, this inequality becomes

$$\frac{h_f}{d} > 0.44$$

Although such proportions are not unrealistic, a very large moment could be developed without compression steel and it is not likely that the design moment would exceed it. However, if it does, tension and compression reinforcement can be designed as described in Section 16.4.3.

The ultimate moment of a T-beam in the second class is considered to be the sum of three couples, as follows:

(a) compression in the overhanging parts of the flange and part of the tension steel, called A_{sf}:

$$M_f = \phi(0.85\,f'_c)(b - b_w)h_f(d - \tfrac{1}{2}h_f) = \phi A_{sf} f_y(d - \tfrac{1}{2}h_f)$$

(b) compression in any compression steel and an equal tension in another portion of the total tension steel. If the area of compression steel is A'_s and the compression steel is strained to or beyond yield point, the area of tension steel associated with this steel is also A'_s:

$$M_c = \phi A'_s f_y(d - d')$$

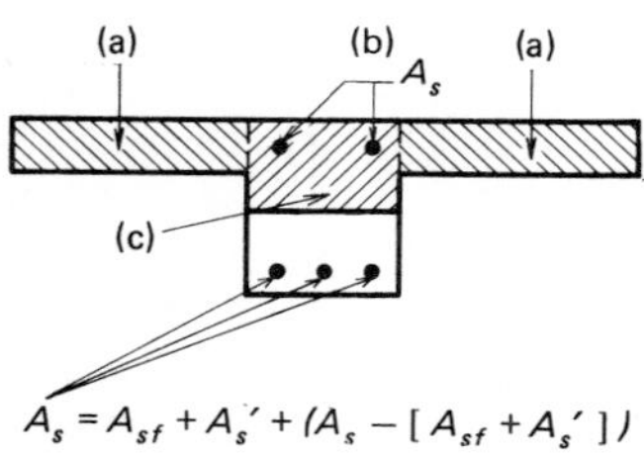

(c) compression in the remaining rectangular section of concrete and an equal tension in the remainder of the steel, $A_s - (A_{sf} + A'_s)$:

$$M_w = \phi(0.85\,f'_c)\frac{a}{d}\left(1 - \frac{1}{2}\frac{a}{d}\right)b_w d^2 = \phi(A_s - A_{sf} - A'_s)f_y(d - \tfrac{1}{2}a)$$

In order to ensure failure by yielding of the steel, the total tension is again limited to 75% of the total compression under balanced conditions, that is

$$A_s f_y \ngtr 0.75\left[(0.85\,f'_c)(b - b_w)\,h_f + A'_s\,f_y + (0.85\,f'_c)\,b_w d\left(\frac{a}{d}\right)_b\right]$$

However, since

$$(0.85\,f'_c)(b - b_w)\,h_f = A_{sf} f_y \tag{16.10}$$

and

$$\left(\frac{a}{d}\right)_b = \beta_1 \frac{\varepsilon_u E_s}{\varepsilon_u E_s + f_y}$$

Example 16.4 Sheet 1 of 1

Determine the flexural reinforcement for a T-beam, given the dimensions and the bending moment.
Design data as for Example 16.1.

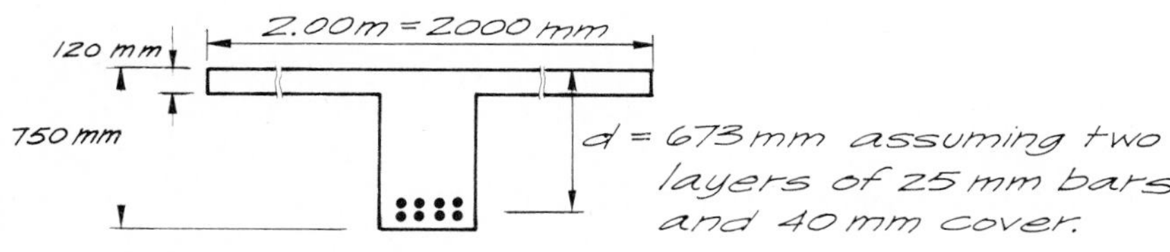

Design moment = 724 kNm

For $a = h_f$ $\quad c = 2000 \times 120 \times 17$ N $= 4080$ kN

Lever arm $= 673 - 1/2 \times 120 = 613$ mm

$M_u = 0.9 \times 4080 \times 0.613$
$= 2250$ kNm
> 724 kNm

Design as oversized rectangular beam

$$F = \frac{7.24 \times 10^5}{0.9 \times 17 \times 10^6 \times 2 \times 0.673^2} = 0.0522$$

$$\frac{a}{d} = 1 - \sqrt{1 - 0.1044} = 0.0536$$

$$1 - \frac{1}{2}\frac{a}{d} = 0.973$$

$$a = 36.1 \text{ mm}$$

Lever arm $= d - \frac{1}{2}a = 655$ mm

$$A_s = \frac{7.24 \times 10^5}{0.9 \times 0.655 \times 270} = 4550 \text{ mm}^2$$

10 × 25 mm bars → $A_s = 4910$ mm²

$$\rho = \frac{4910}{2000 \times 673} = 0.00364 < 0.0277 \quad \text{OK}$$

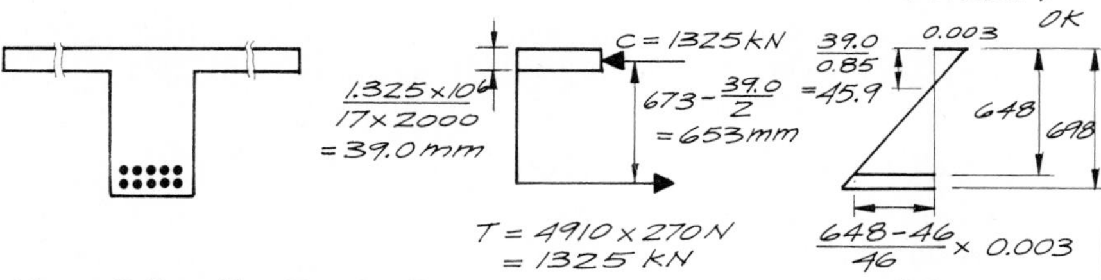

$T = 4910 \times 270$ N
$= 1325$ kN

$$\frac{648 - 46}{46} \times 0.003 = 0.0393 > \frac{f_y}{E_s}$$

$M_u = 0.9 \times 1325 \times 0.653$
$= 778$ kNm

this criterion becomes

$$A_s f_y \not> 0.75 \left[A_{sf} f_y + A'_s f_y + (0.85\ f'_c)\ b_w d\ \beta_1 \frac{\varepsilon_u E_s}{\varepsilon_u E_s + f_y}\right]$$

Hence,

$$\rho \left(= \frac{A_s}{bd}\right) \not> 0.75 \left[\frac{A_{sf} + A'_s}{bd} + \frac{b_w}{b} \frac{\beta_1(0.85\ f'_c)}{f_y} \frac{\varepsilon_u E_s}{\varepsilon_u E_s + f_y}\right] \quad (16.11)$$

The designer's problem is to select A_s and A'_s such that the total ultimate moment, M_u $(= M_f + M_c + M_w)$ is not less than the design moment M and the Inequality (16.11) is satisfied.

First, the need for compression steel can be determined by computing the maximum ultimate moment with A'_s and, hence, M_c equal to zero. Using Equation (16.10), A_{sf} is determined – this is the area of steel required on the tension side to balance the compression force which is uniformly distributed on the overhanging parts of the flange. Then the upper limit for ρ can be determined and from that the maximum allowable total area of tension steel. Since A_{sf} is known, the area of steel $(A_s - A_{sf})$ which can be assigned to the couple M_w can be determined. The stress in all of this steel is f_y, so that the tension force in each of the couples, M_f and M_w, is now known. The lever arm of the couple M_f is $(d - \frac{1}{2} h_f)$ and M_f can be evaluated. The lever arm of couple M_w depends on a and this is found by equating the tension and compression of that couple, that is

$$(A_s - A_{sf})\ f_y = (0.85\ f'_c)\ a\ b_w$$

$$a = \frac{f_y}{0.85\ f'_c} \frac{A_s - A_{sf}}{b_w}$$

With

$$M_w = \phi(A_s - A_{sf})\ f_y\ (d - \tfrac{1}{2}\ a)$$

the sum $(M_f + M_w)$ can be evaluated and this is the maximum ultimate moment that can be mobilised without compression steel.

If the design moment is less than this upper limit to $(M_f + M_w)$, then M_f, determined as described above, is deducted from the design moment. The web section can be now regarded as an oversized rectangular beam and the steel area $(A_s - A_{sf})$ can be determined by the procedure of Section 16.4.2. The calculations are illustrated in Example 16.5.

On the other hand, the design moment may exceed the maximum $(M_f + M_w)$ allowable. In that event, the design calculations are the same in principle as described in Section 16.4.3. The couple M_f is the only additional matter to be taken into account. Since its value is fixed, it introduces no real complications. Typical calculations are illustrated in Example 16.6.

16.4.5. Review of flexural computations

The basic principle common to Sections 16.4.3 and 16.4.4 is to make a decision about the amount of compression reinforcement to be used in the cross-section and then to determine the contributions to the ultimate moment made by this steel and the overhanging parts of the flange (if there is a flange). These two contributions are fixed once the compression steel and the geometry of the cross-section have been fixed. The difference between the design moment and the

Example 16.5 Sheet 1 of 2

Determine the flexural reinforcement required for a T-beam, given the dimensions and the bending moment. Design data as for Example 16.1.

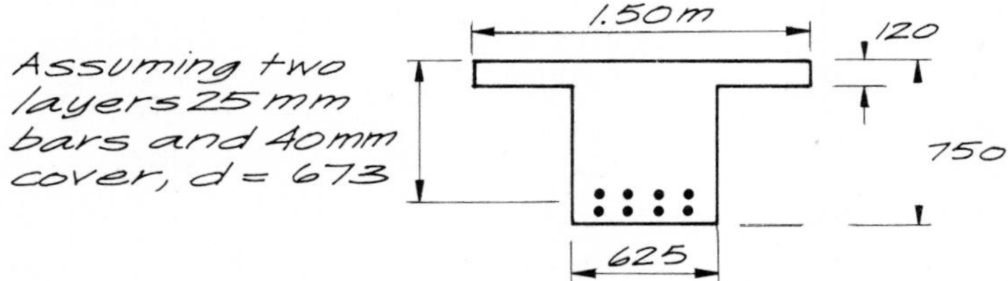

Design moment = 2 MNm

Test 1: Does compression zone extend below flange soffit?

For $a = h_f$ $\quad M_u = 0.9 \times 17 \times (1500 \times 120) \times (0.673 - 0.12/2)$

$= 1.685$ MNm

< 2 MNm

Hence compression zone must extend below flange soffit.

Test 2: Is compression reinforcement required?

For $A_s' = 0$

$$\rho \ngtr 0.75\left[\frac{A_{sf}}{bd} + \frac{b_w}{b}\,\frac{\beta_1(0.85f_c')}{f_y}\,\frac{\epsilon_u E_s}{\epsilon_u E_s + f_y}\right]$$

$$0.75\,\frac{b_w}{b}\,\frac{\beta_1(0.85f_c')}{f_y}\,\frac{\epsilon_u E_s}{\epsilon_u E_s + f_y} = \frac{625}{1500} \times 0.0277 = 0.0115$$

Since $(0.85f_c')(b - b_w)h_f = A_{sf}f_y$

$$A_{sf} = \frac{17 \times 875 \times 120}{270}$$

$= 6620\ mm^2$

$$0.75\,\frac{A_{sf}}{bd} = \frac{0.75 \times 6620}{1500 \times 673}$$

$= 0.0049$

$\rho \ngtr 0.0115 + 0.0049$

$= 0.0164$

Maximum allowable A_s, for $A_s' = 0$

$= 0.0164 \times 1500 \times 673$

$= 16\,500\ mm^2$

Example 16.5 Sheet 2 of 2

$$M_f = \phi A_{sf} f_y (d - 1/2 h_f)$$
$$= 0.9 \times 6620 \times 270 \times 0.613 \text{ Nm}$$
$$= 985 \text{ kNm}$$

$$(A_s - A_{sf}) f_y = (0.85 f_c') b_w a$$

$$a = \frac{270}{17} \times \frac{16500 - 6620}{625} = 251 \text{ mm}$$

$$M_w = \phi (A_s - A_{sf}) f_y (d - 1/2 a)$$
$$= 0.9 \times 9880 \times 270 \times 0.548 \text{ Nm}$$
$$= 1.32 \text{ MNm}$$

$$M_f + M_w = 2.30 \text{ MNm} > 2 \text{ MNm}$$

Hence compression steel not required

Design moment − flange moment = 2.00 − 0.985
= 1.015 MNm

For web section $F = \dfrac{1.015 \times 10^6}{0.9 \times 17 \times 10^6 \times 0.625 \times 0.673^2}$

$$= 0.234$$

$$\frac{a}{d} = 1 - \sqrt{1 - 0.468} = 0.268$$

$$1 - \frac{1}{2}\frac{a}{d} = 1 - 0.134 = 0.866$$

$$A - A_{sf} = \frac{1.015 \times 10^6}{0.9 \times 270 \times 0.866 \times 0.673}$$
$$= 7160 \text{ mm}^2$$

$$A_s = 13780 \text{ mm}^2$$

2 × 9 × 32 mm diameter bars → $A_s = 14500 \text{ mm}^2$

Width required $= 2 \times 40 + 9 \times 32 + 8 \times 32$
$= 624$ mm OK

$$\rho = \frac{14500}{1500 \times 673} = 0.0144$$
$$< 0.0164 \quad \text{OK}$$

Example 16.6 Sheet 1 of 2

Determine the flexural reinforcement required for a T-beam, given the dimensions and the bending moment.
Dimensions as for Example 16.5
Design data as for Example 16.1

Design moment = 2.4 MNm

Test 1: Does compression zone extend below flange soffit?
For $a = h_f$ $\quad M_u = 1.685$ MNm (Example 16.5)
Hence compression zone must extend below flange soffit.

Test 2: Is compression reinforcement required?
For $A_s' = 0$ and ρ = maximum permitted

$$M_w + M_f = 2.30 \text{ MNm} \quad \text{(Example 16.5)}$$
$$< 2.4 \text{ MNm}$$

Hence compression steel required.

Excess moment = 100 kNm

$$\phi A_s' f_y (d - d') = 10^5$$

$$A_s' = \frac{10^5}{0.9 \times 270 \times (0.673 - 0.052)}$$
$$= 662 \text{ mm}^2$$

2 x 32 mm diameter $\longrightarrow A_s' = 1610 \text{ mm}^2 \ (> 2 \times 662 \text{ mm}^2)$

$$M_c = \phi A_s' f_y (d - d')$$
$$= 0.9 \times 1610 \times 270 \times 0.621 \text{ Nm}$$
$$= 244 \text{ kNm}$$
$$M_f = 985 \text{ kNm} \quad \text{(Example 16.5)}$$
$$M_c + M_f = 1.229 \text{ MNm}$$

Balance to be provided by web = 2.4 − 1.23
= 1.17 MNm

$$F = \frac{1.17 \times 10^6}{0.9 \times 17 \times 10^6 \times 0.625 \times 0.673^2}$$
$$= 0.270$$
$$\frac{a}{d} = 1 - \sqrt{1 - 0.540} = 0.320$$
$$1 - \frac{1}{2}\frac{a}{d} = 0.840$$

Example 16.6 Sheet 2 of 2

$$\phi (A_s - A_{sf} - A_s') f_y (d - 1/2a) = 1.17 \times 10^6$$

$$A_s - A_{sf} - A_s' = \frac{1.17 \times 10^6}{0.9 \times 270 \times 0.84 \times 0.673} = 8510\ mm^2$$

From above, $A_s' = 1610\ mm^2$
From Example 16.5 $A_{sf} = 6620\ mm^2$

$$A_s = 8510 + 6620 + 1610 = 16740\ mm^2$$

14×40 mm bars $\longrightarrow$ $A_s = 17600\ mm^2$

Check: $\rho = \frac{17600}{1500 \times 673} = 0.0173$

$$0.75 \frac{A_{sf} + A_s'}{bd} = \frac{0.75 \times 8230}{1500 \times 673} = 0.0061$$

$$0.75 \frac{\beta_1 (0.85 f_c')}{f_y} \frac{\epsilon_u E_s}{\epsilon_u E_s + f_y} = \frac{0.0115}{0.0176}$$ (Example 16.5)

OK

USE 14×40 mm bars for tension steel
2×32 mm bars for compression steel

sum of these two contributions is the ultimate moment of resistance required of the rectangular section. Two degrees of freedom remain – the depth of the compression block and the area of tension steel – and there are two conditions to be satisfied. These are a balance of horizontal forces and a moment balance. In this light, the procedures described above are logical processes whereby values are assigned to certain variables until a stage is reached where the number of unknowns (two) equals the number of equations that can be written.

16.4.6. Curtailment of reinforcement and anchorage

Inspection of the bending moment diagram in Example 16.1 will show that the bending moment is not the same for the whole beam. This is generally the case and wherever the shear is different from zero the bending moment varies along the beam. Clearly, it is not necessary to provide the same strength along the whole beam and the designer can economise by reducing the area of steel in the cross-section. The ultimate moment of the cross-section is made to vary stepwise along the beam and, provided the strength exceeds the bending moment at all sections, the beam is sound.

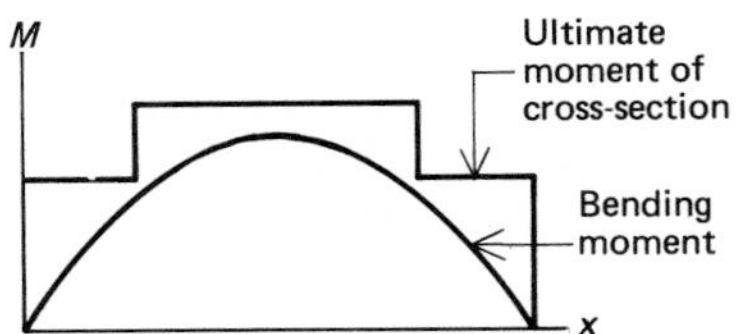

This aspect of the design can be called *curtailment* – the bars required at the section of maximum moment extend away from that section and some are curtailed where the bending moment is low enough. Curtailment means an abrupt reduction in the strength of the cross-section and the designer has to take into account the following matters:

(a) the ultimate moment of the cross-section without the curtailed bars must not be less than the bending moment;
(b) the curtailed bars must be anchored;
(c) some protection is required against the adverse effects of the abruptness of the change in properties.

Dealing with these in turn, the designer first decides how many bars he will curtail. With the geometry of the cross-section and the area of reinforcing known, the ultimate moment can be computed using the methods described above. The section where the chosen curtailment is allowable can then be found from the bending moment diagram. This may be done algebraically, but there is much to be said in favour of graphical solution on a bending-moment diagram plotted to scale.

The second point, anchorage, must now be considered. Clearly, a short distance to the left of X (see marginal sketch) the bending moment is greater than the ultimate moment without the curtailed bars, so that the strength of the beam depends on those bars being stressed. Now, the stress in them where they end is obviously zero and a tension at some distance from the end can be created only by bond between the bar and the concrete. Equilibrium of a short length of the bars shows that

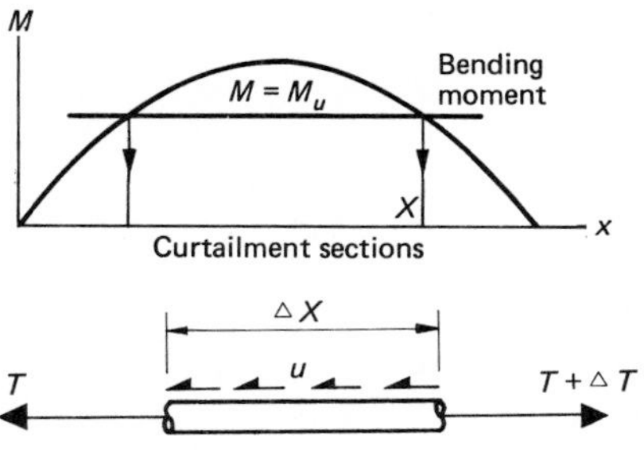

$$u\,\Sigma o\,\Delta x = \Delta T \tag{16.12}$$

where u is the bond stress or adhesion between the bars and the concrete and Σo is the total perimeter of the bars to be anchored. Bond between the bars and the concrete is improved if the bars have ribs on the surface. Such bars, known as *deformed bars*, have now displaced the older smooth bars almost entirely.

The change in tension over a finite length l from the end is

$$T = \int_0^l u \, \Sigma o \, dx$$
$$= \Sigma o \int_0^l u \, dx$$
$$= u \, \Sigma o \, l$$

assuming that the bond stress is uniformly distributed. Thus, the length required for anchorage, or the length required to develop the necessary tension in the curtailed bars is given by

$$l = \frac{T}{u \, \Sigma o} \tag{16.13}$$

The tension which has to be anchored by extending the bars beyond the section where, theoretically, they are no longer required, may be found from the details of the curtailment and the bending moment at the curtailment section.

A gradient of total tension in the bars also exists whenever the bending moment is changing – that is, whenever the shear is different from zero. Equation (16.12) can be applied and, as Δx tends to zero, it becomes

$$u = \frac{1}{\Sigma o} \frac{dT}{dx}$$

Since

$$\phi \times T \times \text{lever arm} = M$$
$$\frac{dT}{dx} = \frac{1}{\phi \times \text{lever arm}} \frac{dM}{dx}$$
$$= \frac{V}{\phi \times \text{lever arm}}$$

whence

$$u = \frac{V}{\phi \, \Sigma o \times \text{lever arm}}$$

However, such calculations for average bond at anchorage and local bond due to shear are no longer regarded as more than qualitative. Modern practice, as exemplified by the ACI *Code*[1], is to use the concept of a "development length" related more directly to experimental results. The development length is the length of bar which must be embedded in the concrete in order to develop or anchor a specified force. Any bar in a reinforced concrete structure is required to extend in both directions from any point far enough to develop the force required in the bar at that point.

Part of the requirement for development length can be met by providing a hook at the end of a bar. This device is of special value near the end of a beam – it facilitates anchorage of a bar in a short length of beam. Rules are given which enable evaluation of the length of straight bar equivalent to a standard hook – it is roughly the same as the length of bar in the hook.

As an example of the rules for development length of deformed bars, the basic development length for bars 35 mm in diameter or smaller is given by

$$l_d = \frac{0.019\ A_b f_y}{\sqrt{f'_c}} \qquad \text{(but not less than } 0.06\ d_b f_y\text{)}$$

where l_d is the basic development length in mm; A_b is the area of cross-section of bar in mm^2; f_y is the yield strength of bar in MPa; f'_c is the 28-day strength of concrete in MPa; and d_b is the diameter of bar in mm.

The development length required in particular circumstances is this basic length multiplied by factors which depend on the circumstances. Examples are as follows:

Horizontal reinforcement with more than 300 mm of concrete placed below the bars	1.4
Lateral spacing between bars less than 150 mm	0.8
More flexural reinforcement provided than required	Ratio of steel required to steel provided
Bars enclosed within a spiral of specified dimensions	0.75

In any case, l_d must not be less than 300 mm.

Compensation for the detrimental effects of the abrupt change in strength is also based on experience. Codes require increased strength in the beam if tension reinforcement is anchored in part of the beam subject to tension. The increase may be in the form of additional shear reinforcement or additional flexural reinforcement. These requirements are necessary to counter reduced ductility and increased cracking which arise when bars are curtailed in the tension zone of a beam.[2]

16.5. REINFORCEMENT FOR SHEAR

When considering the effects of external shear on a reinforced concrete beam, it must be recognised that the shear force causes inclined tensile stresses in the concrete and it is these normal stress components that are critical – not the tangential or shearing components. This fact must be emphasised because, although concrete can resist a simple shear very well, its tensile strength is low.

We have shown in Chapter 3, "Stress Analysis," how normal and tangential stress components may be combined and how principal planes across which the resultant force per unit area is normal can be located. In a reinforced concrete beam this analysis can be applied to locate the traces of surfaces across which the concrete is in tension. One such surface is sketched in Fig. 16.4, which also indicates the Mohr's circles used to define it. In general, the inclined tension per unit area acting across this surface is high near the bottom where the tension due to bending is high, and near the neutral axis where the shear stress component is large. In the compression zone it is less. The horizontal tension arising from bending is greatest where the bending moment is large and is dealt with in the flexural computations. The inclined tension near the neutral axis at places where the external shear is large has to be studied now.

With a big enough external shear a crack would open, shaped roughly as sketched in Fig. 16.4. At some place where the tensile stress is less than the strength of the concrete, the crack would peter out, although this would be much influenced by stress concentration at the end of the crack. The point is that the uncracked portion of the beam can be assumed to resist part of the

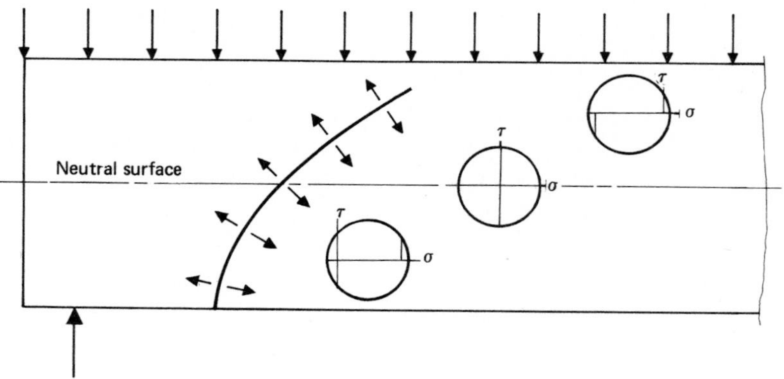

Fig. 16.4. *Inclined tension in concrete beam.*

external shear. The balance is resisted by steel bars inclined to the length of the beam. In Fig. 16.5 an elevation of the end of a beam is shown again. The potential crack is shown extending from A at the level of the tension steel to B where it is assumed to disappear. AB is projected to C at the top of the beam. The slope of ABC is assumed to be constant at 45°, the inclination it would have at the neutral axis. Also shown are bars crossing the crack, their inclination to the length of the beam being α. Each of the lines inclined at α represents one or more bars, the total area of steel on each line being A_v. The spacing of these sets of bars along the beam is s.

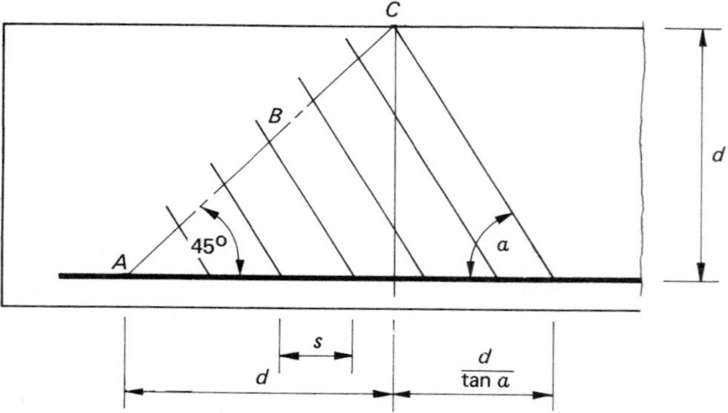

Fig. 16.5. *Design of web reinforcement.*

Vertical equilibrium of the free body to the left of AC depends on the vertical component of tension in the inclined bars being equal to that part of the shear to be resisted by them – this is designated V_s.

The number of bars crossing ABC is $(d + d/\tan\alpha)/s$ so that, at the limit of the strength of the bars

$$V_s = [A_v f_y (d + d/\tan\alpha)/s] \sin\alpha$$

whence

$$s = \frac{A_v f_y (\sin\alpha + \cos\alpha) d}{V_s} \tag{16.14}$$

Of the variables in Equation (16.14), the designer knows f_y and d. He is free to choose A_v and α so that, when he knows V_s, he can calculate the spacing.

The two independent parameters of the shear reinforcement (or preferably, by analogy with steel I-beams in which the webs resist the shear, the *web reinforcement*) are A_v and α. The inclination α is usually 45° or 90°, the former being used when main tension bars are bent up to cross the neutral axis. In the latter case, an independent set of bars called *stirrups* reinforce the web. We recommend that beginning designers use only stirrups. Then Equation (16.14) becomes

$$s = \frac{A_v f_y}{V_s} d$$

$$= \frac{A_v f_y}{v' bd} d$$

$$= \frac{A_v f_y}{v' b} \tag{16.15}$$

where $v'bd = V_s$.

Stirrups should be completely enclosed loops anchored by bending the ends of the bars through 135°. Top bars are required in the corners of the stirrups to make a self-supporting cage which can be prefabricated and lifted into place in the formwork before the concrete is placed (see Chapter 23, "Construction Operations"). Stirrups may be single stirrups or multiple stirrups and A_v is determined accordingly.

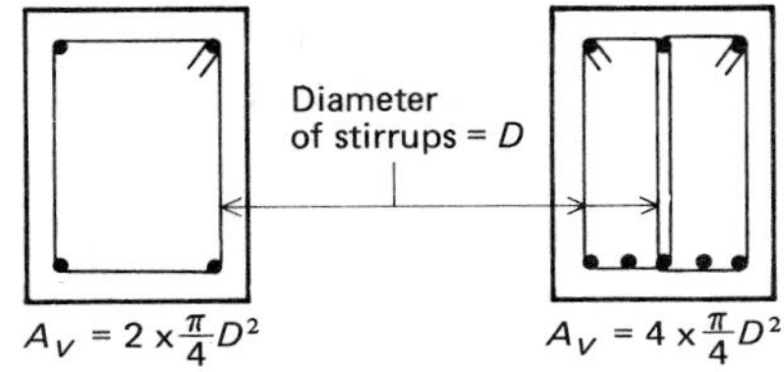

Shear calculations are based on stresses at failure and working loads multiplied by load factors. The design shear is V_u.

How much of the external shear is included in V_s is laid down by codes. The ACI Code uses the quotient $v_u = V_u/\phi bd$ as a measure of inclined tension. The capacity reduction factor, ϕ, equals 0.85 for shear. The value of v_u is the average shear stress on the effective section. It is used instead of any other measure allowing for non-uniform distribution because so little is known about the distribution. When v_u is less than an allowable stress, v_c, which depends on the tensile strength of the concrete, computations for web reinforcement need not be made. However, it is good practice to provide nominal stirrups not more than $d/2$ apart. In many codes, including the ACI Code, nominal stirrups are mandatory and we are very much in favour of this.

The limiting value, v_c, corresponding to the one in the ACI Code[1] is $0.167\sqrt{f_c'}$ where both v_c and f_c' are measured in MPa. The area of shear reinforcement should not be less than $0.345\, b_w s/f_y$ where b_w and s are in mm, f_y in MPa and the area of shear reinforcement is measured in mm^2.

If v_u exceeds the allowable measure, v_c, V_s is found by deducting $v_c bd$ from V_u. It will be seen that this method of allowing for the ability of the concrete to resist some of the shear is not consistent with the explanation given above. However, it does enable satisfactory beams to be designed without elaborate calculation and is a reasonable basis for design, bearing in mind that much has yet to be learned about this aspect of reinforced concrete.

Finally, the maximum shear for design can be taken at a distance d from the face of the beam's support. Consideration of Fig. 16.5 will show that this is consistent with the concept of failure on a crack at 45°. The whole beam must, of course, be provided with shear reinforcement.

All of the foregoing reasoning applies to T-beams, the width of the web b_w being used instead of b. Although the flange probably contributes to the shear resistance of the beam, it is not known how much it adds and it is, for this reason, ignored.

Example 16.7 Sheet 1 of 2

Determine the size and spacing of web-reinforcement for the beam of Example 16.1.

With $f_c' = 20$ MPa $\quad v_c = 0.167\sqrt{f_c'} = 0.75$ MPa

$\phi = 0.85$

$b = 410$ mm $\quad d = 612$ mm Supports 300 mm wide

On ℄ support $\quad v_u = \dfrac{V_u}{\phi bd} = \dfrac{3.62 \times 10^5}{0.85 \times 410 \times 612} = 1.70$ MPa

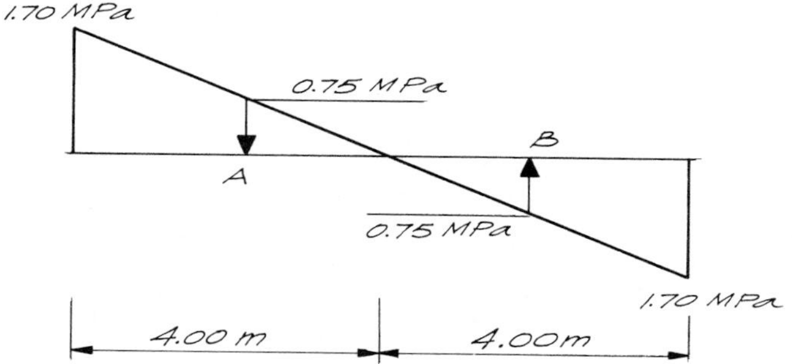

Between A and B $\quad v_u < v_c$ — nominal stirrups

$0.75/1.70 \times 4.00 = 1.77$ m either side of ℄

or 2.08 m from face of support

Say, spacing of nominal stirrups $= 300$ mm $(< \frac{d}{2})$

$A_v \nless \dfrac{0.345 \times 410 \times 300}{270} = 157 \text{ mm}^2$

10 mm diameter stirrups $\rightarrow A_v = 2 \times \pi/4 \times 10^2 \quad = 157 \text{ mm}^2$

At distance d from face of support – i.e at $150 + 612 = 762$ mm from ℄ of support

$V_u = \dfrac{4000 - 762}{4000} \times 362 = 293$ kN

$v_c bd = 0.75 \times 410 \times 612$ N $= 188$ kN

$V_s = 105$ kN

Try 10 mm diameter stirrups — $A_v = 157 \text{ mm}^2$

$s = \dfrac{(157 \times 270) \times 612}{1.05 \times 10^5}$

$= 247$ mm

Say, 10 mm diameter stirrups, 200 mm c/c

Example 16.7 Sheet 2 of 2

Location of stirrups – accumulated distance from face of support:

Distance	
0.10	To resist shear 11 No 10 mm diameter stirrups, 200 mm c/c. First stirrup 100 mm from face of support
0.30	
0.50	
0.70	
0.90	
1.10	
1.30	
1.50	
1.70	
1.90	
2.10	
2.35	Nominal stirrups 6 No 10 mm diameter stirrups, 300 mm c/c.
2.65	
2.95	
3.25	
3.55	
3.85 (℄ span)	

Same arrangement of stirrups in other half span.

In Example 16.7 calculations are shown for the web reinforcement of the beam designed in Example 16.1.

Although the foregoing discussion has been found by experience to be satisfactory as a basis for calculation of shear reinforcement, it is not put forward as a complete description of this aspect of the behaviour of reinforced concrete beams. More detailed discussions are contained in recent papers, such as those of Kani and Places and Regan.[3, 4]

16.6. DEFLECTION

Beams need to be stiff enough to avoid damage to glass, partitions and finishes. For this reason codes specify limits for deflection under live load and allowing for creep. Calculation of deflections is based on elastic theory

$$\frac{M}{E_c I} = \frac{1}{R} = \frac{d^2 y}{dx^2}$$

Difficulty arises in making these calculations because neither E_c nor I is well defined. Experience is that a linear approximation for the stress–strain curve is valid for only a small range of strains and, in any case, time-dependent strains make it difficult to rationalise the use of Young's modulus. Factors in the uncertainty of I are cracking of the concrete where tensile strains are large and the contribution of the steel, which has a large value of E, to the stiffness of the beam.

Fortunately, reinforced concrete beams are usually stiff enough to make deflection irrelevant as a design consideration. The very approximate nature of the estimated deflection does not matter too much.

We do not elaborate on this. How the calculations are made depends on the code being used and the rules therein must be followed.

LIST OF SYMBOLS

A_b	cross-sectional area of a bar (mm^2)
A_s	cross-sectional area of tension steel
A_{sf}	that part of area of tension steel required to balance compression in overhanging parts of flange of T-beam
A'_s	cross-sectional area of compression steel
A_v	cross-sectional area of web-reinforcement
a	depth of compression stress block
$\left(\frac{a}{d}\right)_b$	ratio of depth of compression stress block to effective depth for balanced conditions
b	breadth of rectangular beam, breadth of flange of T-beam
b_w	width of web of T-beam
C	compression force in resistance couple of rectangular beam
C_c	concrete compression force in rectangular beam with compression steel
C_s	steel compression force in rectangular beam with compression steel
c	depth of neutral axis from compression face
c_b	depth of neutral axis from compression face under balanced conditions

d	effective depth of beam – i.e. distance from compression face to centre of tension
d'	distance from compression face to centre of compression steel
d_b	diameter of one bar (mm)
E_c	Young's modulus for concrete
E_s	Young's modulus for steel
F	dimensionless quotient
f	stress in concrete
f'_c	28-day strength of concrete (MPa)
f_y	yield strength of steel (MPa)
h_f	thickness of flange of T-beam
I	second moment of area of cross-section
k_b	ratio of c to d for balanced conditions
K	coefficient for ultimate moment
l	a length of bar
l_d	development length (mm)
M	design moment
M_c	contribution to ultimate moment generated by compression steel
M_f	contribution to ultimate moment generated by overhanging parts of flange of T-beam
M_u	ultimate moment
M_w	contribution to ultimate moment generated by web
Σo	total perimeter of tension bars in a cross-section
R	radius of curvature of beam
s	spacing of stirrups (mm)
T	tensile force in resistance couple, tension in a bar
T_1, T_2	subdivisions of tensile force in resistance couple
u	bond stress between a bar and concrete
V_s	that part of the external shear to be resisted by web reinforcement
V_u	external shear under ultimate load
v_c	allowable nominal shear stress in concrete under ultimate load
v_u	nominal shear stress under ultimate load
v'	V_s/bd
α	inclination of web reinforcement
β_1	ratio of a to c defining depth of compression stress block
ε	strain in concrete
ε_c	strain at compression face
ε_i	strain in inner layer of tension steel
ε_s	strain in tension steel
ε_u	ultimate strain in concrete (= 0.003)
ε_y	yield strain in steel ($= f_y/E_s$)
ε'	strain in compression steel
ρ	steel ratio for tension steel ($= A_s/bd$)
ρ_b	steel ratio in rectangular beam for balanced conditions
$\bar{\rho}_b$	in the context of beams with compression reinforcement, the steel ratio for balanced conditions in a beam of the same dimensions but without compression reinforcement
ρ'	steel ratio for compression steel ($= A'_s/bd$)
ϕ	capacity reduction factor

REFERENCES

[1] ACI 318–71, *Building Code Requirements for Reinforced Concrete,* American Concrete Institute, Detroit, 1971.

[2] ACI 318–71, *Commentary on Building Code Requirements for Reinforced Concrete,* American Concrete Institute, Detroit, 1971.

[3] G. N. J. KANI, "A Rational Theory for the Function of Web Reinforcement", *Proceedings of the American Concrete Institute,* **66**, March 1969, 185–97.

[4] A. PLACES and P. E. REGAN, "Shear Failure of Reinforced Concrete Beams", *Proceedings of the American Concrete Institute,* **68**, October 1971, 763–73

BIBLIOGRAPHY

FERGUSON, P. M., *Reinforced Concrete Fundamentals,* 3rd ed., John Wiley, New York, 1973.

GERSTLE, K. H., *Basic Structural Design,* McGraw-Hill, New York, 1967.

The Proceedings of the American Concrete Institute also provide continuous access to up-to-date research and design procedures in the field of reinforced concrete.

PROBLEMS

16.1. Sketch alternative arrangements for the reinforcement for a beam continuous over two equal spans. How would you allow for the effects of settlement of
(a) the central support;
(b) one end support?

16.2. You can make the beam of Problem 16.1 statically determinate by putting a hinge in one span. Where would you put the hinge? Sketch the reinforcement in the vicinity of the hinge.

16.3. Determine the sizes and arrangement of the bars for the hinge of Problem 16.2 for spans of 8 m and the uniformly distributed load of Example 16.1. Assume the beam to be 410 mm wide by 630 mm deep overall.

16.4. The beams of Examples 16.2 and 16.3 both span gaps of 8.00 m centre-to-centre of supports and the supports are walls 300 mm thick. Prepare details of curtailment and design the shear reinforcement for both beams.

Allow for the adverse effects of the abrupt change in tensile reinforcement by using additional stirrups at the ends of curtailed bars for a length of beam equal to 0.75 times the effective depth of the beam. The additional stirrups should be such that $(A_v/bs)f_y$ computed for them alone is not less than 410 MPa.

16.5. Estimate the total cost of the beams designed in Examples 16.2 and 16.3 and Problem 16.4. Use cost data from Chapter 25 or locally available cost data.

16.6. Redesign the beam of Example 16.3 with $b = 1.00$ m and $h = 0.430$ m. Include curtailment of bending reinforcement and design of shear reinforcement. Estimate the cost of this beam and compare your estimate with the corresponding one in Problem 16.5. Examine the alternative designs from the point of view of the builder.

17 Prestressed Concrete Beams

17.1. INTRODUCTION

We recall that the value of concrete as a material for construction of beams is diminished by its low tensile strength. This deficiency is met in reinforced concrete beams by embedding mild steel bars in those parts of beams where there is tension to create an admirably versatile composite form of construction. In this chapter we will investigate prestressing, an alternative means of overcoming the low tensile strength of concrete.

The principle is to apply an axial force to the beam in such a way as to cause compressive stresses in those parts of the beam where the loads cause tensile stresses. These compressive stresses comprise the *prestress.* Combined, the *prestressing force* (the axial force which causes the prestress) and the loads cause much lower tensile stresses than the loads do by themselves. In fact, it is possible to eliminate tension entirely if this is desired. Normally, however, advantage is taken of the small tensile strength of concrete.

Stresses due to loads and a prestressing force are sketched in Fig. 17.1, where it is made clear how the prestress acts to reduce the tensile stress in service. As sketched, the distribution of the prestress is such that the maximum compressive stress is reduced by a small amount and the maximum tensile stress considerably. The designer has to apply care and skill to selection of the magnitude and line of action of the prestressing force if he is to exploit both these advantages without overloading the concrete at any time.

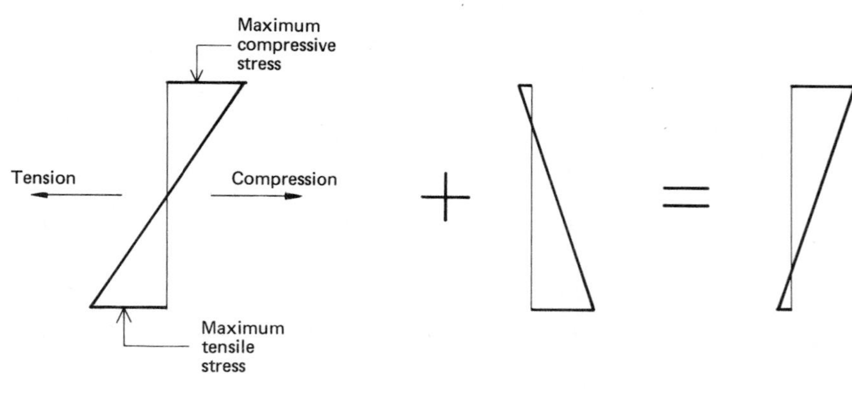

Fig. 17.1. *Stress distributions in a prestressed beam.*

At ultimate load, the condition of the beam is quite different. As the load on a prestressed concrete beam is increased, the tensile stress due to the loads also increases and, at some stage, cracks occur on the tension side of the beam. For loads greater than the cracking load the behaviour of a prestressed concrete beam is similar to that of a reinforced concrete beam. This will be examined in more detail in a later section; here we say only that the prestress does not have much effect on the ultimate strength of a beam. It is at loads below cracking load that the prestress has the most significant effect.

The stress at a given point in a prestressed concrete beam may be tensile or compressive, depending on the stage of construction and the loading. This makes it imperative for the designer to keep a careful account of the signs of stresses. In our analysis of bending, we will use the convention that compressive stresses are positive and tensile stresses are negative. We do this to follow authors who have preceded us even though it is the opposite of the convention we used in Chapter 3, "Stress Analysis". In Section 17.5.3, where the principal stress due to shear is sought, we will return to the convention of Chapter 3.

17.2. PRESTRESSING WITH STEEL TENDONS

The prestressing force can be created by stretching steel tendons from one end of the beam to the other and this is the only method we will consider. The tendons are stretched by means of a jack and, while so strained, they are anchored to the beam. When the force exerted by the jack on the tendons is released, the tendons cannot return to their unstrained condition, being anchored to the concrete. The result is a balance, with the tendons in tension and the concrete in compression. This operation is called *transfer*, implying a shift of load from the jack to the beam. More information about prestressing operations is given in Chapter 23, "Construction Operations".

A distinction is made between *pre-tensioning* and *post-tensioning*. In the former, single wires or groups of wires twisted into *strands* are used for tendons and they are stretched in the mould of the beam before the concrete is placed (see Plate 23.7). Transfer takes place after concreting, and the tendons are anchored by bond and a Poisson expansion at each end of the beam. They are also bonded to the concrete for their whole length. Before transfer, the concrete is unstrained. The balance of forces achieved after transfer requires compressive strains in the concrete, and the accompanying change in strain of the steel causes a reduction in the tensile stress in the bonded tendons. Because of this elastic shortening, the prestressing force acting on the concrete after transfer is smaller than the total tension in the tendons before transfer.

The prestress in post-tensioned beams is applied after the concrete has set. Tendons may be single wires, strands or cables comprising up to twelve wires laid alongside each other. Ducts for such cables are visible in Plate 23.8. The jack which stretches a tendon usually pushes on the end of the beam being stressed to provide a reaction for the tension in the tendon, so that the concrete is compressed at the same time as the tendon is stretched. Elastic shortening does not reduce the prestressing force in the last tendon stretched, but it does in all the others as successive tendons are stretched. There is also a reduction due to slip at the anchorage. Transfer is achieved by anchoring the tendons (with wedges, for example) and this cannot be done without slip occurring. The amount of slip is about 3 mm for wedges and it causes a corresponding reduction in the strain in the tendons. Clearly, the tension is reduced more as a result of slip if the tendons are short than it is if they are long.

Post-tensioned tendons are not bonded to the concrete immediately after transfer. It is good practice to create such a bond by pumping grout into the duct containing the tendon. The value of bonding the steel to the concrete is that cracks which occur in an overloaded beam are distributed along the beam as many fine cracks instead of being concentrated in a few wide ones. Compatibility of steel and concrete strains improves the recovery of a beam which has been overloaded and it also ensures higher ultimate strength in the beam. The analysis of prestressed concrete presented here is for beams with bonded tendons.

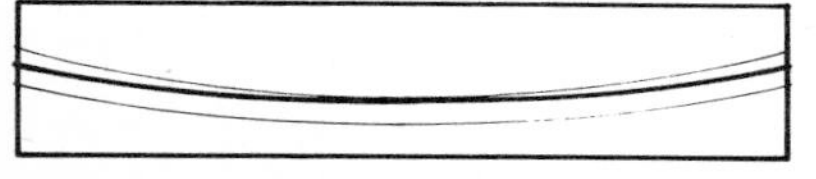

Draped tendon hugs inner wall of duct

Straight tendon – wobble causes contact

Loss of prestress is of fundamental importance to this kind of construction. It cannot be avoided and it must be allowed for. There are other losses besides those which occur at transfer because of elastic deformation of the concrete and slip at the anchorages. Elastic deformation and slip may be regarded as fundamental causes of loss in that they arise from the operations of creating the prestress. The other losses are caused by shrinkage of the concrete as it loses moisture, creep or deferred strain in the concrete under stress and creep or relaxation of the steel under stress, which are characteristics of the materials used. All of these other losses occur in pre-tensioned and post-tensioned beams. Post-tensioned beams are subject to an additional source of loss – friction against the walls of the duct. This causes the tension in a tendon to decrease with distance from the jack and is due to unavoidable contact between tendons and duct wall. It is clear how such contact occurs if a tendon is curved. In the case of a straight cable, misalignment of the duct is the cause and this is to be expected because it is impracticable to lay a duct perfectly straight. However, if the diameter of the duct is at least 25 mm greater than that of the tendon, wobble can be neglected as a cause of loss.

The early history of prestressing is a story of failure. The principle of applying a prestressing force by means of an anchored steel tendon had been recognised early, but the effects of shrinkage and creep had not. Using mild steel for their tendons, early investigators were unable to maintain a useful prestress in the beams – the total loss of strain was sufficient to erase the stress in the steel even if it was stressed to yielding. The French engineer Eugène Freyssinet was first to realise that loss of prestress can be accommodated only if the tendons are made of high tensile steel. By using high-strength steel, Freyssinet built beams in which a residual tension in the tendons remained effective after all losses had taken place.

As far as design of prestressed beams is concerned, the stress–strain function for high-strength steel has two significant features. One is that yielding does not occur. A typical stress–strain curve for a prestressing tendon is sketched in Fig. 17.2 along with the corresponding curve for mild steel and the idealised curve used in design of reinforced concrete. The high-strength steel has an elastic range, with a modulus of elasticity slightly less than that of mild steel. However, beyond the elastic range there is no yield and the function curves continuously to the point of failure. There is no valid approximation in which strain can increase without increase in stress as there is for mild steel. Furthermore, the total strain at failure is much less than that for mild steel. Large strain at constant stress between yielding and failure is the characteristic of the idealised curve for mild steel which much simplifies calculation of the ultimate strength of reinforced concrete beams. That prestressing steel lacks it, makes calculation of the strength of a prestressed concrete beam more difficult.

The other feature is the hysteresis loop. If a specimen stressed to point *a* on Fig. 17.2 is unloaded, stress and strain on unloading follow the path *ab* which is approximately parallel to the linear part of the curve followed when the load was increasing. This is what happens when the strain in a tendon decreases as losses occur and the loss of stress can be estimated by multiplying the change in strain by the modulus of elasticity.

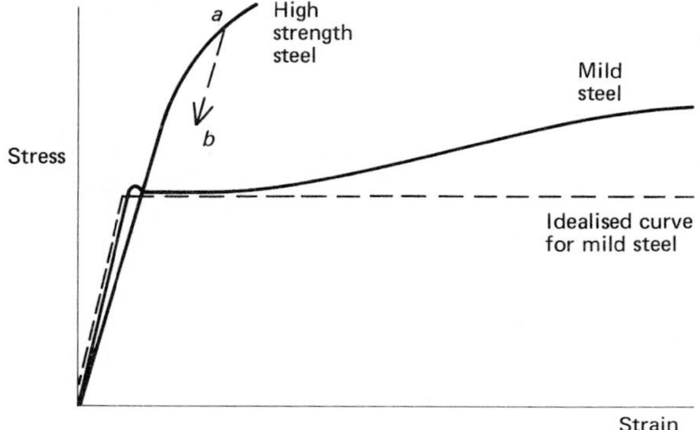

Fig. 17.2. *Stress–strain curves for mild steel and high tensile steel.*

Loss of prestress may be calculated by estimating the individual losses and adding them together. However, it is not necessary to do this and the ACI *Commentary on Building Code Requirements for Reinforced Concrete* (ACI 318–71) suggests the use of lump-sum losses – 240 MPa for pre-tensioned concrete and 170 MPa for post-tensioned concrete.[1] The source of these figures is given in ACI *Commentary* as the report of ACI-ASCE Committee 423, "Tentative Recommendations for Prestressed Concrete".[2] It is not entirely clear which items in the list of individual losses are covered by these lump sums. We have assumed that they are the losses which take place after anchorage of the tendons. In the notation that is to be used in deriving equations for design

$$P/A_{ps} = P_T/A_{ps} - 240 \text{ MPa for pre-tensioned beams}$$

$$P/A_{ps} = P_T/A_{ps} - 170 \text{ MPa for post-tensioned beams}$$

where P_T is the prestressing force immediately after transfer, and P is the prestressing force in service, after all losses have occurred. The force required in the tendons when the wires are stretched by jacking is greater than P_T by the amount of the losses that occur before anchorage – those due to slip at anchorage and elastic deformation of the concrete.

The lump-sum loss for a pre-tensioned beam is the greater of the two because the prestress occurs sooner after concreting in pre-tensioned construction than in post-tensioned construction. All losses associated with deformation of the concrete are less for concrete which is loaded at greater age.

The ACI *Building Code Requirements for Reinforced Concrete* (ACI 318–71) require the steel stress not to exceed 0.8 f_{pu} at the time of jacking and 0.7 f_{pu} after transfer, f_{pu} being the ultimate strength of the tendons.[3] The former limit is to allow some overstressing of the steel during the operation of stretching the wires on the grounds that this is a temporary state for the tendons. As we

have noted, the difference between steel stress before and after transfer is due to elastic shortening and slip and it must be estimated
(a) to enable both restrictions to be checked, and
(b) so that the jacking force required can be estimated.

If the stress in the concrete surrounding a tendon is f_c, the strain is f_c/E_c, where E_c is Young's modulus for the concrete. This is the amount by which the strain in the steel is reduced, so that the corresponding loss of stress in the steel is $(E_s/E_c)f_c$, where E_s is Young's modulus for the steel, as shown on Fig. 17.3. For simplicity, the loss is usually estimated from the average compressive stress in the concrete, rather than from that in the concrete surrounding the tendons.

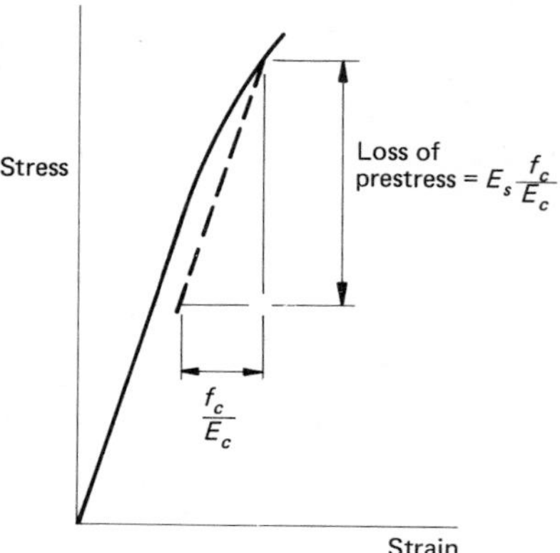

Fig. 17.3. *Estimation of loss of prestress due to elastic shortening at transfer.*

The foregoing is sufficient for a pre-tensioned beam in which the prestressing force is transferred from all of the tendons together. In a post-tensioned beam with more than one tendon, the tendons are stretched in sequence and the amount of loss is not the same for all of them. The loss is greatest for the first tendon stressed and is progressively less for the others.

The loss of prestress due to elastic deformation in a post-tensioned beam may be estimated as follows. If there are n tendons, the tension in each is approximately P_T/n and the loss of tensile stress in any anchored tendon caused by tensioning another one is $(E_s/E_c)(P_T/nA)$, where A is the area of the cross-section of the concrete. The first tendon loses this amount $(n - 1)$ times, the second $(n - 2)$ times and so on, so that the total loss of prestressing force is

$$\frac{n(n-1)}{2}\frac{E_s}{E_c}\frac{P}{nA}A_1$$

where A_1 is the area of one tendon. Since nA_1 equals the total area of prestressing steel, A_{ps}, in the cross-section we have

$$\text{total loss of tensile stress} = \frac{n-1}{2n}\frac{E_s}{E_c}\frac{P_T}{A}$$

$$= \left(1 - \frac{1}{n}\right)\frac{1}{2}\left(\frac{E_s}{E_c}\frac{P_T}{A}\right)$$

This shows the loss to be zero for a beam with one cable ($n = 1$) and that it tends to one half of what it would be if the beam were pre-tensioned for large values of n.

If the effective limit on steel stress is 0.7 f'_{pu} after transfer, addition of the loss due to elastic shortening gives the stress in the tendons at the time of jacking and the ACI *Code*[3] requires this stress to be less than 0.8 f'_{pu}. Otherwise, the effective limit is 0.8 f'_{pu} at the time of jacking and the stress after transfer is found by deducting the loss due to elastic shortening.

17.3. CONCRETE FOR PRESTRESSED CONCRETE

The strength of concrete used in prestressed construction is normally higher than that of the concrete used for reinforced concrete. This increased strength leads to smaller dimensions but, apart from this, elastic and deferred strains are reduced. Young's modulus and the shrinkage and creep characteristics all depend on the quality of the concrete. Consequently, less prestress is lost, with a corresponding saving in the amount of steel required. There is an obvious need for careful control of the manufacture of concrete for prestressed construction – a requirement that is fairly easy to satisfy in a factory manufacturing prestressed concrete products but is not so easy for construction *in situ*.

The concrete is assumed to be elastic for loads less than the cracking load. With a linear distribution of strain assumed (as is usual for the simple theory of bending) there is thus a linear distribution of stress, and the conventional formula for stress in a member subject to a bending moment (due to transverse loads) and an axial force (the prestressing force) is valid, namely

$$f = \frac{P}{A} + \frac{(Pe)y}{I} - \frac{My}{I} \tag{17.1}$$

Here f is the stress at a point a distance y from the centroidal axis of the cross-section; P is the prestressing force; M is the bending moment; A is the area of cross-section; I is the second moment of area of cross-section; and e is the eccentricity of prestressing force, measured from centroidal axis. The sign convention is such that sagging moments and compressive forces and stresses are positive. This requires e and y to be positive when measured down from the centroidal axis.

The linear relationship is assumed to persist to the limit of the tensile strength of the concrete, when

$$f = -f_r$$

f_r being the modulus of rupture. Thus, for a given beam, an estimate can be made of the bending moment which causes cracking – the *cracking moment*.

At ultimate strength of the beam, the concrete is assumed to have failed in tension, so that the entire tensile force in the resistance couple is contributed by the tendons. On the compression side of the neutral axis, the compressive stress in the concrete is assumed to be uniformly distributed over part of the area, as shown in Fig. 17.4. This stress distribution is the same as that assumed for design of reinforced concrete, but the factor β_1 is often less than 0.85 because higher strength concrete is used.

17.4. BASIS OF DESIGN

Although straightforward in essence, design of a prestressed concrete beam is made rather laborious by the need to examine the bending of the beam under

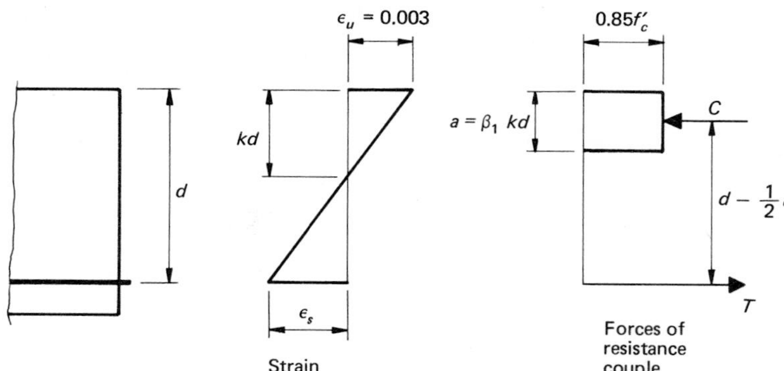

Fig. 17.4. *Strains and internal forces at ultimate strength.*

four conditions of loading instead of just one, as is the case with most materials. The four conditions are:

(a) at transfer;
(b) in service (when several distributions of live load may have to be examined);
(c) when the first crack appears;
(d) at failure, when the distribution of live load may be relevant.

For each of these conditions the forces applied to the beam are different, the strength parameters of the concrete are different and, if loss of prestress is relevant (as it is in (a), (b) and (c)) different allowances are made.

Dealing with these in turn, we examine the situation at transfer first. The load applied is the weight of the beam itself and any other parts of the structure erected before prestressing. In many cases, it will be the beam's own weight and nothing more. Other parts constructed before prestressing might include a floor.

Transfer normally occurs at a fairly early age, before the concrete has reached full strength, which acts to keep permitted stresses low. On the other hand, reduced margins of safety are allowed on the grounds that the consequences of failure at this stage are less serious than those of failure later. The nett result is that permitted stresses at transfer are often high.

The allowance for loss of prestressing force needs to cover elastic shortening and slip only.

The beam's own weight causes compression in the top of the beam and tension in the bottom of it. The prestressing force is applied eccentrically so as to cause compression in the bottom and tension in the top. The state of stress after transfer is illustrated in Fig. 17.5. Criteria to be satisfied are the limits on compressive stress at the bottom and tensile stress at the top. Elastic bending is assumed.

The limiting values for the stresses will be called p_{Tc} and p_{Tt}, the allowable stresses at transfer in compression and tension respectively. According to the ACI *Code*,[3]

$$p_{Tc} = 0.60\, f'_{ci}$$

$$p_{Tt} = 0.25 \sqrt{f'_{ci}}$$

where f'_{ci} is the compressive strength in MPa of the concrete at the time of transfer.

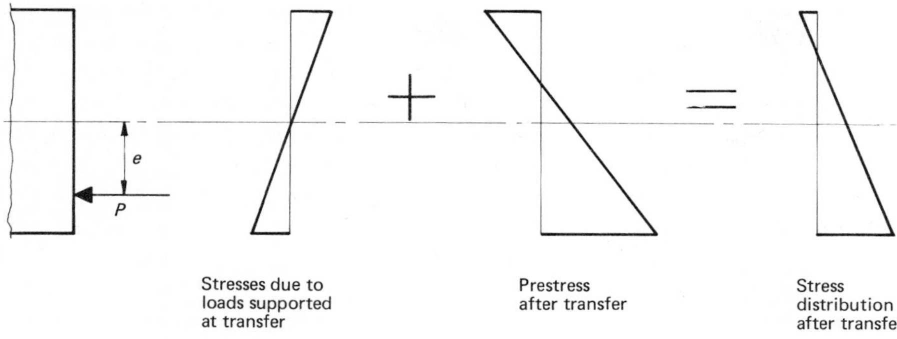

Fig. 17.5. *Bending stresses after transfer.*

Draped tendons

Bending moments due to dead load

The importance of the load supported at transfer is brought out in Fig. 17.5. When there are no stresses due to such loads, the prestressing moment must be reduced if the permitted stresses are not to be exceeded. Two points arise here. One is that there is an advantage to be won by picking up as much dead load as possible at transfer. The other is that the moment caused by such loads is zero at the ends of the beam. Thus, the beneficial effect of dead loads is reduced if the tendons are not draped so that there is full prestressing moment at mid-span but a much smaller one (possibly none at all, with the prestressing force applied axially) at the ends.

We next consider the effect of service loads or working loads. In this condition the beam is assumed to be carrying the complete dead load – its own weight and that of other structural components such as a floor or a deck – and the live load. All losses are assumed to be complete and the strength of the concrete is the 28-day strength which is the conventional basis of design. Elastic behaviour of the beam and no cracking are assumed. To obtain the stress distribution, as shown in Fig. 17.6, the prestress is reduced in proportion to the losses that have occurred since transfer and the bending stresses due to full (dead and live) load are superimposed. In this condition the limitations to be satisfied are a compressive stress at the top and a tensile stress at the bottom. These stresses will be called p_c and p_t respectively and the values allowed by the ACI *Code* are as follows:[3]

$$p_c = 0.45\, f'_c$$

$$p_t = 0.5\sqrt{f'_c}$$

where f'_c is the compressive strength in MPa of the concrete at the age of 28 days.

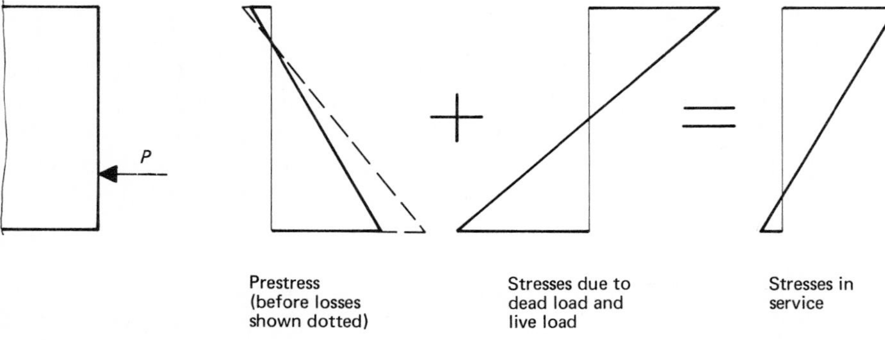

Fig. 17.6. *Bending stresses in service.*

Cracking is dealt with by computing the bending moment which would cause a stress at the bottom of the beam equal to the tensile strength of the concrete, with losses assumed to be complete. This cracking moment must exceed the bending moment due to service loads, although the codes do not always guide the designer in deciding a suitable margin. However, a margin of 8% to 10% should be enough to ensure that no visible cracks occur in service. A formal restriction on the cracking moment given in the ACI *Code* is that the ultimate moment of resistance must be at least 1.2 times as great as the cracking moment.[3] This provision ensures that an overloaded beam will give some warning before it collapses.

Finally, at ultimate load, the assumptions referred to in the preceding section are made. The load used to calculate bending moments and shears is based on service loads and load factors. For example, if the loads comprise dead load and live load, the rule in the ACI *Code* is that the ultimate strength is based on a load of $1.4\,D + 1.7\,L$ where D is the dead load and L the live load.[3] The moment of resistance computed on the basis of the distributions of stress and strain shown in Fig. 17.4 must exceed this bending moment.

By the ACI *Code*, shear stresses are required to be considered only at ultimate load.[3] Other codes require some investigations under service conditions as well but, in this case, the principal tension is affected by the prestress and it is, in fact, substantially less than it would be if there were no axial prestressing force. We will use the rules of the ACI *Code*[3] but, as a matter of interest, the effect of the prestress will be discussed.

Deflection under service load should be considered and, provided time-dependent strains are ignored, this is quite simple. Elastic behaviour can be assumed and, since tensile stresses are too low for cracks to occur, the gross cross-section of the concrete contributes to the stiffness of the beam. The eccentric prestress causes an upward deflection, or *camber*, which can be computed when the prestressing moment is known as a function of distance along the beam. It is this lifting between the ends of the beam which enables the dead load to act as a counter to the prestressing force and to allow larger prestressing moments. Because of the camber, prestressed concrete beams can sustain larger deflections under load without looking as if they are sagging than some other forms of construction can.

For a full examination of the deflection of prestressed concrete beams, reference can be made to the report of ACI Committee 435, "Deflections of Prestressed Concrete Members".[4]

17.5. FORMULAS FOR DESIGN

In this section, formulas are derived for design calculations. Those based on elastic behaviour are presented as inequalities, an important feature since it is virtually impossible to design a beam in which all allowable stresses at transfer and in service are reached. For ultimate strength in bending, the ultimate moment of resistance is derived. In consideration of shear, we examine design at ultimate load and the way that the prestress reduces the principal stress in service.

17.5.1. Elastic bending

Elastic bending is relevant at transfer, in service, and in determining the cracking moment. The general relationship to be applied has been given in Equation

(17.1). Applied to conditions at transfer, it yields the following inequalities:

$$\frac{P_T}{A} - \frac{P_T e}{Z_1} + \frac{M_T}{Z_1} \nless -p_{Tt} \tag{17.2}$$

$$\frac{P_T}{A} + \frac{P_T e}{Z_2} - \frac{M_T}{Z_2} \ngtr +p_{Tc} \tag{17.3}$$

where P_T is the prestressing force immediately after transfer; M_T is the bending moment at transfer; A is the area of cross-section of beam; Z_1 is the elastic modulus of section with respect to top of beam $= I/y_1$; Z_2 is the elastic modulus of section with respect to bottom of beam $= I/y_2$; I is the second moment of area of cross-section; y_1 is the distance from centroidal axis to top of beam; y_2 is the distance from centroidal axis to bottom of beam; e is the eccentricity of prestressing force; p_{Tt} is the allowable tensile stress in concrete at transfer; and p_{Tc} is the allowable compressive stress in concrete at transfer.

Between the time of transfer and the time when the beam is fully loaded in service, strains in the concrete change in two ways. There are the time-dependent changes which cause a loss of prestress and these are allowed for by a reduction in the prestressing force from P_T to P. There are also changes which accompany the bending of the beam under the service loads. At the level of a tendon the change is a tensile strain and it causes the tension to increase by a small amount. This change in prestressing force is disregarded.

Under working loads we have

$$\frac{P}{A} - \frac{Pe}{Z_1} + \frac{M}{Z_1} \ngtr +p_c \tag{17.4}$$

$$\frac{P}{A} + \frac{Pe}{Z_2} - \frac{M}{Z_2} \nless -p_t \tag{17.5}$$

Here P is the prestressing force after losses have occurred; M is the bending moment in service; p_c is the allowable compressive stress in concrete in service; and p_t is the allowable tensile stress in concrete in service.

Manipulation of these to get a more useful set due to Magnel,[5] yields the following:

$$e \ngtr \frac{1}{P_T}\left\{M_T + Z_1\, p_{Tt}\right\} + \frac{Z_1}{A} \tag{17.6}$$

$$e \ngtr \frac{1}{P_T}\left\{M_T + Z_2\, p_{Tc}\right\} - \frac{Z_2}{A} \tag{17.7}$$

$$e \nless \frac{1}{P_T}\left\{\frac{M}{\eta}\left(1 - \frac{Z_1 p_c}{M}\right)\right\} + \frac{Z_1}{A} \tag{17.8}$$

$$e \nless \frac{1}{P_T}\left\{\frac{M}{\eta}\left(1 - \frac{Z_2 p_t}{M}\right)\right\} - \frac{Z_2}{A} \tag{17.9}$$

The additional symbol η is defined by

$$P = \eta\, P_T$$

These inequalities can be presented graphically on coordinates e and $1/P_T$. Consider (17.6) and (17.8) first, as in Fig. 17.7. The inequalities define regions on this surface bounded by two straight lines. The slopes of the lines are indicated on Fig. 17.7 and they both meet the axis $1/P_T = 0$ at $e = +Z_1/A$. The inequalities are satisfied at all points between the lines and the criterion which establishes each boundary is noted on Fig. 17.7. We are, of course, interested in positive values of P_T. Similarly, (17.7) and (17.9) define another region bounded by two straight lines. These lines meet the axis $1/P_T = 0$ at $e = -Z_2/A$ and they are shown in Fig. 17.8. The two diagrams can be superimposed, as in Fig. 17.9, to define a region wherein pairs of P_T and e may be found to satisfy the criteria of design at transfer and in service.

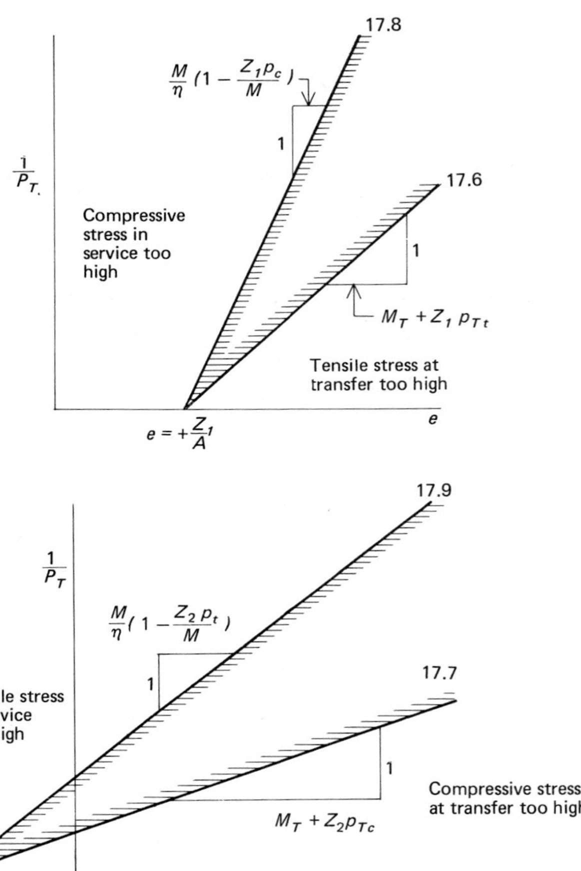

Fig. 17.7. *Inequalities for design at transfer and in service—Inequalities (17.6) and (17.8).*

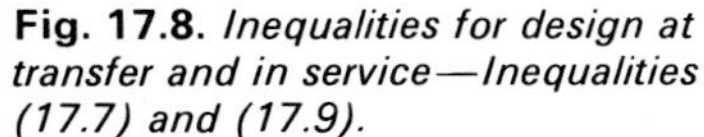

Fig. 17.8. *Inequalities for design at transfer and in service—Inequalities (17.7) and (17.9).*

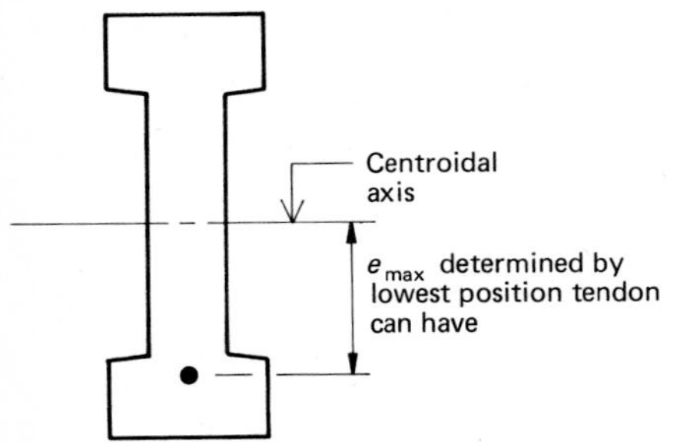

A further constraint which may also be applied has been added to Fig. 17.9 – that is, that the eccentricity cannot exceed a certain value. Physical limitations may be applied by the depth of the beam, the size and distribution of the tendons and the need for the tendons to be embedded in the concrete. Alternatively, e_{max} may be fixed by stress constraints relevant at another cross-section (at one end, for example).

This display of the constraints on the design is called a *Magnel diagram* after Gustave Magnel, the Belgian engineer who devised it.

A deduction to be made immediately is that the slopes of the lines must satisfy certain relationships if a region of permissible solutions is to exist. It is

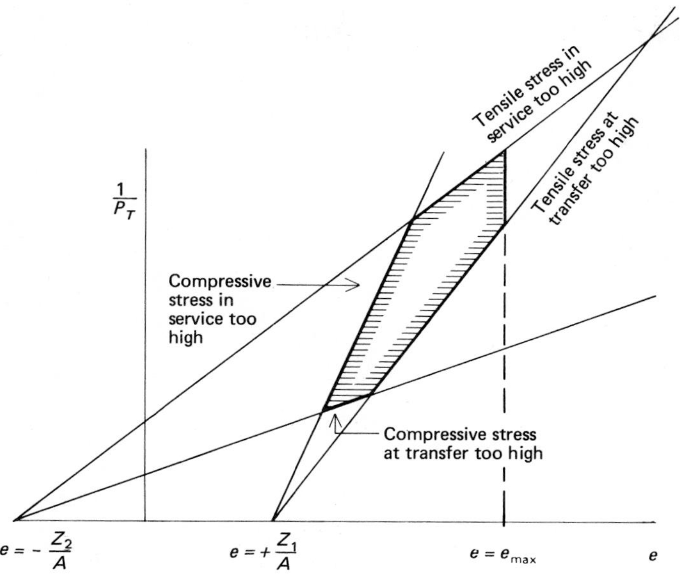

Fig. 17.9. *Magnel diagram for design at transfer and in service.*

necessary that

$$M_T + Z_1 p_{Tt} \nless \frac{M}{\eta}\left(1 - \frac{Z_1 p_c}{M}\right)$$

and

$$M_T + Z_2 p_{Tc} \nless \frac{M}{\eta}\left(1 - \frac{Z_2 p_t}{M}\right)$$

From these, the following inequalities can be determined:

$$M - \eta\, M_T \ngtr Z_1\,(p_c + \eta p_{Tt}) \tag{17.10}$$

$$M - \eta\, M_T \ngtr Z_2\,(p_{Tc} + p_t) \tag{17.11}$$

It is of some interest to examine what these inequalities would convey if $\eta = 1$ (it is usually about 0.8–0.85) and tensile stresses are not permitted, that is

$$p_{Tt} = p_t = 0$$

Then we would have

$$\frac{M - M_T}{Z_1} \ngtr p_c \tag{17.12}$$

and

$$\frac{M - M_T}{Z_2} \ngtr p_{Tc} \tag{17.13}$$

The bending moment, $M - M_T$, is the additional moment applied after transfer and p_c is the maximum compressive stress allowed in service. The conclusion from (17.12) is that the loads supported at transfer do not contribute to the compressive stress, which is limited by the design rule. This is the origin of the aphorism "the dead load carries itself". However, as we have noted, η is less than one and some tension can be allowed, so that the situation is not quite so simple. We will find a use for the Inequalities (17.10) and (17.11), if not for (17.12) and (17.13).

The significance of the intercepts Z_1/A and Z_2/A should also be noted. If, in Equation (17.1), $M = 0$, then

$$f = \frac{P}{A} + \frac{Pe\,y}{I}$$

Now if

$$e = +\frac{Z_1}{A}$$

$$f = \frac{P}{A} + \frac{P}{A}\,\frac{y Z_1}{y_1 Z_1}$$

$$= \frac{P}{A}\left(1 + \frac{y}{y_1}\right)$$

since

$$Z_1 = \frac{I}{y_1}$$

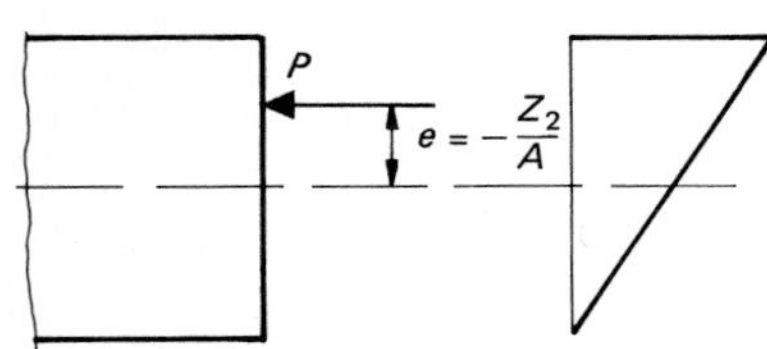

This stress distribution ranges from zero at the top of the beam where $y = -y_1$ to an extreme value in compression at the bottom. Similarly, if $e = -Z_2/A$, the stress ranges from zero at the bottom to a maximum compressive stress at the top. The eccentricities $+Z_1/A$ and $-Z_2/A$ are the limits within which the prestressing force by itself will not cause tension within the beam. The part of the cross-section between these limits is known as the *kern* of the section.

How the Magnel diagram is used in design will emerge in a following section. Here we note that the diagram can be plotted if Z_1, Z_2, A, M, M_T, η and the permitted stresses are known, which amounts to knowing the properties of the cross-section and the design data of the problem.

The linear distribution of stress is also used to calculate the cracking moment, which is given by the equation

$$-f_r = \frac{P}{A} + \frac{Pe}{Z_2} - \frac{M_{cr}}{Z_2}$$

where f_r is the modulus of rupture and is a measure of the tensile strength of the concrete, and M_{cr} is the cracking moment. Hence

$$M_{cr} = Z_2\left\{\frac{P}{A}\left(1 + \frac{e}{Z_2/A}\right) + f_r\right\} \tag{17.14}$$

Although the validity of the linear distribution of stress at the limit of the concrete's tensile strength is questionable, the moment which would cause cracking is estimated, if not forecast exactly.

17.5.2. Bending at ultimate load

A formula for the moment of resistance developed by a rectangular cross-section at ultimate load is easily derived. The internal forces are as shown in Fig. 17.10. We have

$$A_{ps} f_{ps} = T = C = 0.85\, f'_c\, ab = 0.85\, f'_c\, bd\frac{a}{d}$$

where A_{ps} is the area of prestressing steel; f_{ps} is the stress in prestressing steel at failure – i.e. when maximum strain in the concrete is 0.003; f'_c is the 28-day

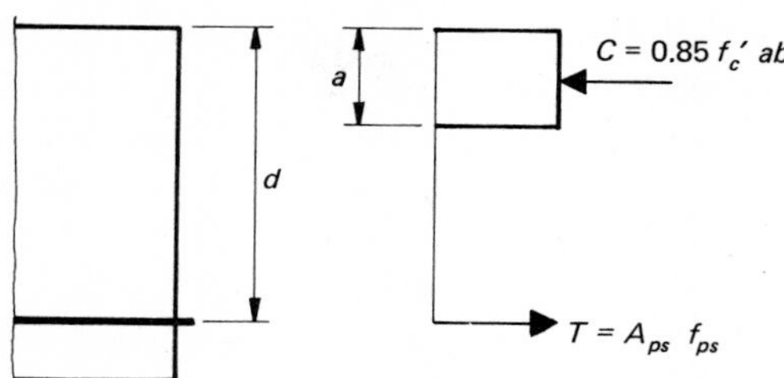

Fig. 17.10. *Internal forces of ultimate resistance couple for rectangular beam.*

strength of concrete; b is the breadth of cross-section; and a and d are as shown on Fig. 17.10. Hence

$$\frac{a}{d} = \frac{A_{ps} f_{ps}}{bd f'_c} \frac{1}{0.85} = \frac{\omega_p}{0.85} \tag{17.15}$$

where

$$\omega_p = \frac{A_{ps}}{bd} \frac{f_{ps}}{f'_c} \tag{17.16}$$

and is known as the *reinforcement index.* The moment of resistance is given by

$$M_u = \phi A_{ps} f_{ps} \left(d - \frac{a}{2}\right)$$

$$M_u = \phi A_{ps} f_{ps} d \left(1 - \tfrac{1}{2} \frac{a}{d}\right) \tag{17.17}$$

Here ϕ is the *strength reduction factor,* which is 0.9 for bending.

From Equations (17.15) and (17.16),

$$A_{ps} f_{ps} = \omega_p f'_c bd$$

$$\frac{a}{d} = \frac{\omega_p}{0.85}$$

and substituting these into Equation (17.17), we get

$$M_u = \phi\omega_p(1 - 0.59\, \omega_p) f'_c bd^2 \tag{17.18}$$

Equation (17.18) is valid for a flanged section also, provided a, the depth of the compressive stress block, is less than the thickness of the flange, h_f. Equation (17.15) shows that this will be the case if

$$\frac{h_f}{d} > \frac{1}{0.85} \omega_p$$

For a flanged section where a is greater than h_f, the same kind of reasoning leads to the following expression for the moment of resistance:

$$M_u = \phi \{\omega_{pw} (1 - 0.59\, \omega_{pw}) f'_c b_w d^2 + 0.85 f'_c (b - b_w) h_f (d - \tfrac{1}{2}h_f)\} \tag{17.19}$$

where ω_{pw} is the reinforcement index computed with the width of the web, b_w, and the area of steel required to balance the compression in the web only. This expression is the sum of two terms – the moment of resistance developed by the concrete in the web and part of the prestressing steel is one, and the moment developed by the flange and the rest of the prestressing steel is the other.

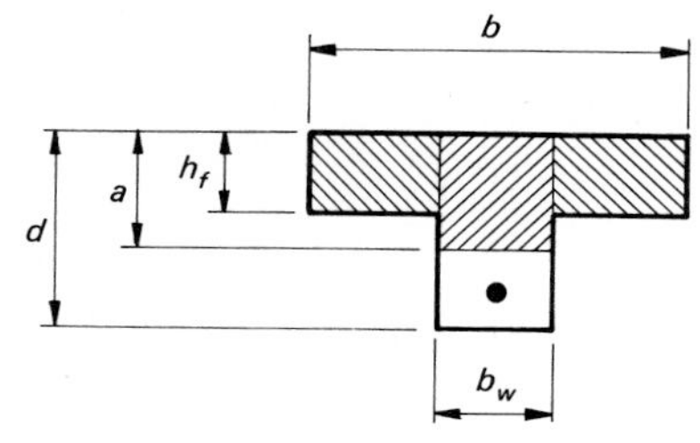

These derivations are the same as for reinforced concrete, except that the stress in the steel is f_{ps} and not f_y. This causes a significant increase in the amount of work required to compute the ultimate moment of resistance because f_{ps} varies with the strain in the steel, while f_y can be assumed to be independent of strain for the relevant range of strain. Accurate calculation of the moment of resistance is by trial and error and a complete stress–strain curve for the steel is required. One proceeds as follows.

The stress–strain curve for the steel is plotted as shown in Fig. 17.11 and, below this, the elevation of the beam is drawn to scale. If the graph is entered at the ordinate representing the stress in the steel in service, the line *ab* can be drawn on the elevation to represent a datum for strains occurring after all losses have taken place. We now consider the increases in strain that would accompany loading the beam to failure. (Note the different treatment of these strains here compared with the analysis of elastic bending.)

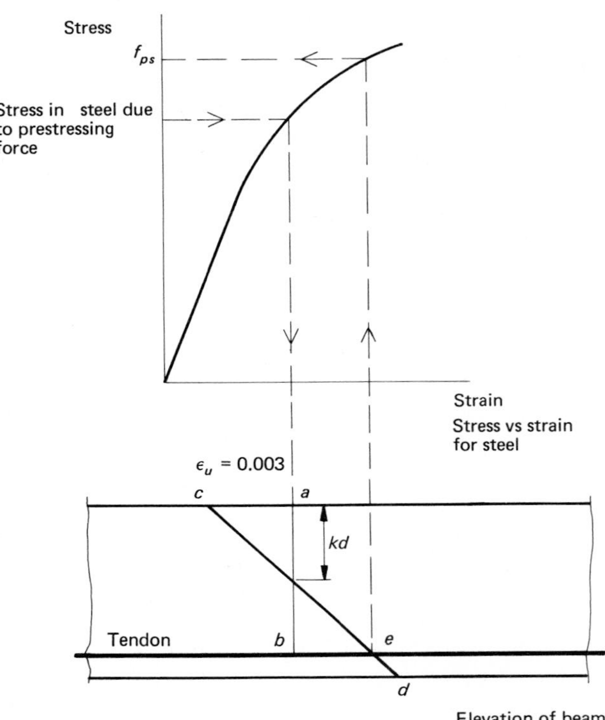

Fig. 17.11. *Calculation of steel stress at ultimate load.*

At the top of the beam, the strain in the concrete would change from a small tensile strain to the ultimate strain for concrete, which is taken to be 0.003. The former can be neglected, so that the strain at the top of the beam under ultimate load is 0.003. This is plotted as *ac* on the elevation, to the same scale as used for strain on the stress–strain diagram above. If a trial value of *kd*, the depth to the neutral axis, is now chosen, the complete distribution of increments in strain can be drawn, for we assume a linear distribution of strain. This line is *cd* and it meets the tendons on the elevation at *e*. Projecting *e* up to the stress–strain diagram gives the stress in the steel, f_{ps}. For that particular trial value of *kd*, the total compression, *C*, and the total tension, *T*, in the resistance couple can be calculated:

$$C = 0.85\, f'_c\, ab = (0.85\, f'_c)\,(\beta_1\, kd)\, b$$
$$T = A_{ps}\, f_{ps}$$

Here we have assumed a rectangular cross-section and that all the properties of the beam are known.

For equilibrium, T and C must be equal and successive trial values of kd are used until one is found for which this condition is satisfied.

The ACI *Code*[3] gives a formula for f_{ps} which enables this procedure to be avoided and leads to an approximate estimate of the ultimate moment of resistance. For bonded tendons,

$$\frac{f_{ps}}{f_{pu}} = 1 - 0.5\,\rho_p \frac{f_{pu}}{f'_c}$$

where

$$\rho_p = \frac{A_{ps}}{bd}$$

The same principles can be applied quite readily to finding the ultimate moment of resistance of a flanged section.

The reinforcement index, ω_p (or ω_{pw}) is a parameter of some importance apart from the place it occupies in Equations (17.18) and (17.19). From Equation (17.15), we have

$$\omega_p = 0.85\frac{a}{d}$$

$$= 0.85\,\beta_1 \frac{kd}{d}$$

$$= 0.85\,\beta_1 \frac{\varepsilon_u}{\varepsilon_u + \varepsilon_s}$$

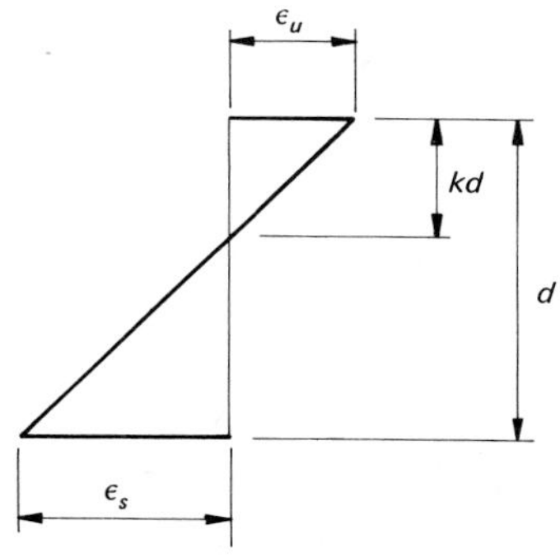

Thus, it is a measure of the strain in the steel ε_s, since ε_u has a fixed value. The ACI *Code*[3] puts an upper limit of 0.3 on ω_p (or ω_{pw}). This ensures that

$$\varepsilon_s \nless 0.003\left(\frac{2.83}{\beta_1} - 1\right)$$

and has a similar effect to the restraint on steel area in reinforced concrete design. The *Commentary* on the ACI *Code* gives as the reason for this constraint that correlation of test results with computed ultimate moments of resistance is not good if ω_p (or ω_{pw}) exceeds the limit.[1]

17.5.3. Principal tension

The principal tensile stress in the web of a prestressed concrete beam may be found by using Mohr's circle as described in Chapter 3, "Stress Analysis". An element of the web on the centroidal axis is subject to a shear stress directly associated with the external shear force, the complementary shear stress and a compressive stress due to the prestressing force. These stress components are sketched on Fig. 17.12(a). The principal tension σ_t is found by the graphical construction sketched in Fig. 17.12 (b). The stress components p_1, q_1 and q_2 are found as follows :

$$p_1 = -\frac{P}{A}$$

$$q_1 = -q_2 = \frac{VQ}{bI}$$

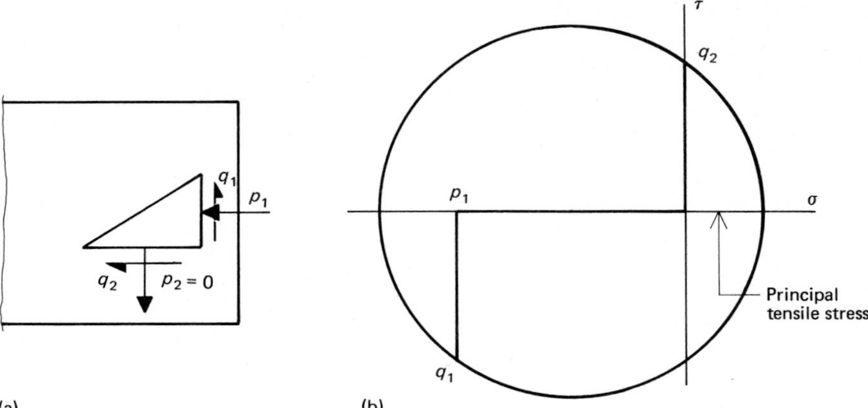

Fig. 17.12. *Computation of principal tensile stress.*

Here P is the prestressing force; V is the external shear; A is the area of cross-section; Q is the static moment about the neutral axis of the part of the cross-section beyond the surface where the shear stress is sought; b is the breadth of cross-section; and I is the second moment of area of cross-section. The sign conventions of Chapter 3 have been used – namely tensile stress is positive and positive shear stress tends to rotate the element clockwise.

If a comparative calculation is made to find the principal tension in the absence of the prestressing force, it will be found that prestressing reduces that stress. Although this is not taken into account directly in the design rules of the ACI *Code*[3], it is a feature of prestressed concrete that adds to its quality. For design according to the ACI *Code*, the average shear stress on the web at ultimate load is used[3], as discussed in the next section.

17.5.4. Shear at ultimate load

In design for shear at ultimate load, the average shear stress is used as a criterion as is the case for reinforced concrete. We have

$$v_u = \frac{V_u}{\phi\, b_w d} \tag{17.20}$$

where V_u is the external shear at ultimate load; b_w is the breadth of web; d is the depth to centre of tension. If $d < 0.8h$, h being the total depth of the beam, $0.8h$ is used instead of d in Equation (17.20); and ϕ is the strength reduction factor ($= 0.85$ in this case).

This measure of shear stress is compared with a stress v_c which is a measure of the shear strength of the beam and for prestressed concrete the ACI *Code*[3] gives

$$v_c = 0.05\sqrt{f'_c} + 4.81\zeta$$

$$\zeta = \frac{V_u d}{M_u} \quad \text{for} \quad \frac{V_u d}{M_u} \not> 1$$

$$\zeta = 1 \quad \text{for} \quad \frac{V_u d}{M_u} > 1$$

$$0.167\sqrt{f'_c} \not> v_c \not> 0.417\sqrt{f'_c}$$

The symbols f'_c, V_u, M_u and d are as defined previously.

These relationships can be displayed on a graph, for a particular value of f'_c, as shown in Fig. 17.13. At the ends of a simply supported beam, for example, $M_u = 0$ and the upper limiting value of v_c ($0.417\sqrt{f'_c}$) is relevant. Closer to midspan, V_u decreases and M_u rises, leading to a smaller value of v_c. However, v_u is also smaller and the most critical section may be at any position along the beam, except where V_u is small.

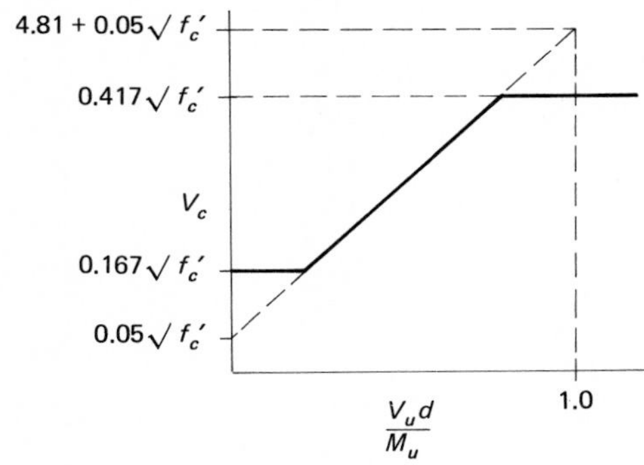

Fig. 17.13. *Rule for maximum nominal shear stress allowed in concrete at ultimate load.*

Provided $v_u < v_c$, only nominal reinforcing in the form of mild steel stirrups or their equivalent is called for. If v_u exceeds v_c, mild steel reinforcing must be designed, the rules in the ACI *Code*[3] being the same as for reinforced concrete. In no case may $v_u - v_c$ exceed $0.667\sqrt{f'_c}$.

We have not included all of the qualifications and alternatives contained in the ACI *Code*[3] – only enough for the purpose of demonstrating the design procedure and providing typical data for solution of examples and problems. If the ACI *Code* is used locally, the reader will need to become familiar with all of its provisions. Otherwise, the local code should be studied.

17.5.5. Effect of draped tendons on shear

When draped tendons are used, the eccentricity of the prestressing force can be varied from a maximum at midspan to a smaller value, or to zero or to a negative value at the ends, which allows full advantage to be taken of the load supported at transfer. Draped tendons also help to resist the external shear, leaving a reduced shearing force to be balanced by internal stresses in the concrete and the shear reinforcing.

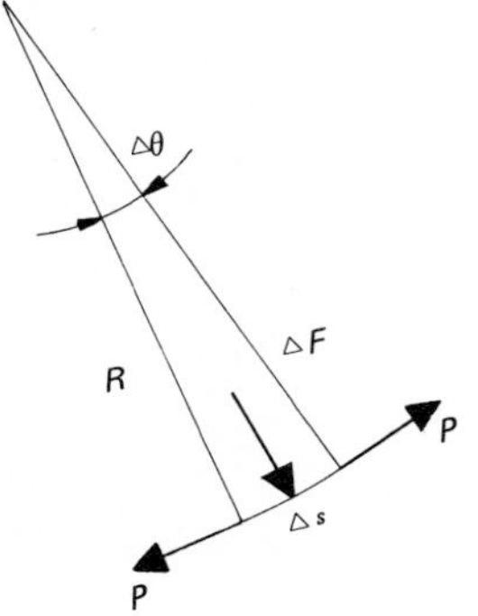

Fig. 17.14. *Radial action and reaction between tendon and beam when tendon is curved.*

Consider a place where the radius of curvature of a tendon is R, as shown in Fig. 17.14. For an element of the tendon to be in equilibrium, the concrete or the wall of the duct in contact with the tendon must exert a force ΔF as indicated and

$$\Delta F = 2\,P \sin \frac{\Delta\theta}{2} = P\,\Delta\theta$$

Now

$$\Delta\theta = \frac{\Delta s}{R}$$

so that

$$\frac{\Delta F}{\Delta s} = P\,\frac{1}{R} = P\,\frac{d^2y}{dx^2}$$

where $y = f(x)$ is the profile of the cable. The approximation

$$\frac{1}{R} = \frac{d^2y}{dx^2}$$

is valid because the slope of the profile is small. For the same reason,

$$\frac{\Delta F}{\Delta s} \approx \frac{\Delta F}{\Delta x}$$

and ΔF is almost vertical. The tendon is seen to exert an upward force on the concrete of the beam (the reaction to ΔF) and this force, per unit length of beam, is $P(d^2y/dx^2)$.

For a parabolic profile, with a total sag equal to Δe as shown in the margin,

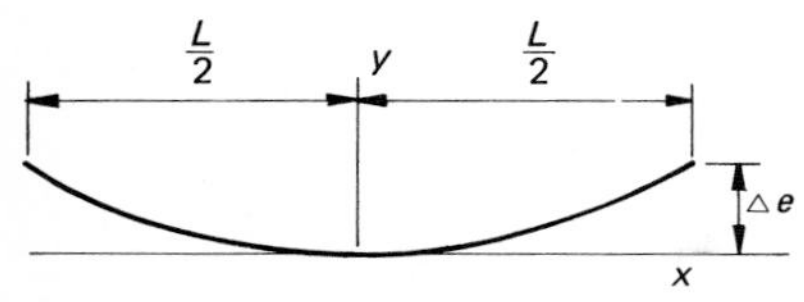

$$y = 4\Delta e\left(\frac{x}{L}\right)^2$$

where L is the span of beam. Hence,

$$\frac{d^2y}{dx^2} = \frac{8\Delta e}{L^2}$$

and is constant for the whole beam. The effect of the sag in the tendon is thus to apply a uniformly distributed upward force of $8\Delta eP/L^2$ per unit length. This internal force is balanced at the ends of the beam by the downward components of the forces tangential to the cable profile applied at the anchorages. The external reactions are clearly not affected. However, the internal force is effective in resisting the external shear.

Clearly, the gain is greatest when Δe is made as big as possible. This justifies taking the tendons above the centroidal axis at the ends. Care must be taken to ensure that the reduced eccentricity at intermediate sections and the ends does not cause a failure to satisfy the constraints on design.

17.5.6. End blocks

The tendons are anchored at the ends of a prestressed concrete beam and large forces are applied locally to the concrete. The state of stress which accompanies this is complex and exact theoretical analysis of it is not available. What is known is that substantial transverse tensile stresses exist in the end blocks. How these stresses arise is indicated on Fig. 17.15. The curved lines on Fig.

17.15(a) are tangent to the direction of the principal compressive stress at any point and equilibrium of an element such as *abcde* is shown in Figs 17.15(b) and (c) to require a tensile stress where the trajectories of principal compressive stress bend away from the edges of the beam. It is necessary to provide mild steel reinforcement to resist the tensions. The bars are required horizontally and vertically across the beam and several layers are desirable. The amount required is best estimated by means of empirical rules which are based on test results. The rules are given in Section 17.6.

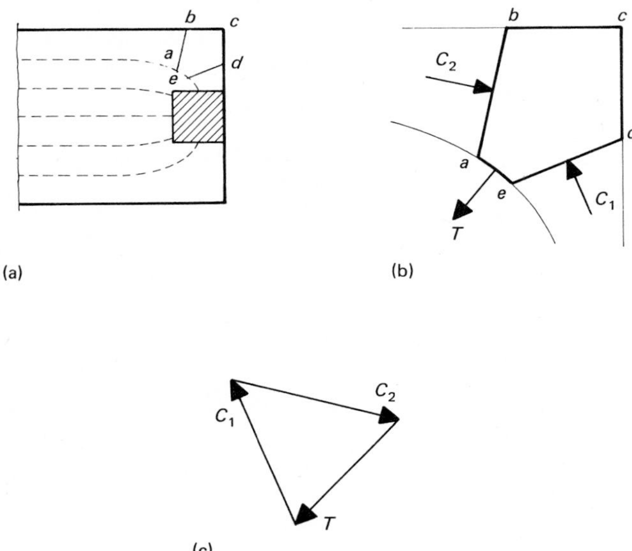

Fig. 17.15.
(a) Trajectories of principal compressive stress in end block.
(b) Forces acting on element abcde.
(c) Vector diagram for equilibrium of element abcde.

The amount of steel so computed is not large and the designer can afford to be generous in this respect. The cost of a relatively heavy-handed attitude towards the reinforcing of end blocks is not high and the additional security is worth having. As always, care is necessary to avoid putting in so much steel that concreting will be difficult, but this is not likely to be a serious problem in an end block.

Anchorage of the transverse bars in an end block has to be looked at carefully. Owing to the narrowness of the beam there may be insufficient anchorage length. If this is so, a continuous zig-zag can be used to ensure effective anchorage of all the transverse steel.

17.6. DESIGN PROCEDURE

As we have noted, design of a prestressed concrete beam is complicated by the number of conditions of loading to be examined. It must also be borne in mind that various combinations and distributions of load may require investigation as well – this aspect is dealt with in more detail in Chapter 2, "Load Analysis". For example, under the heading "working loads", a rolling load can be applied at different positions along the beam or it may be combined with a wind load. All such arrangements that are significant in the various conditions of loading must be considered.

However, some aspects of the behaviour of the material are favourable to the designer. The most important of these is that the ultimate moment of resistance is insensitive to changes in the prestressing force. Another is that the conditions of transfer and working load can be dealt with together, using the Magnel diagram to display the relevant inequalities. Finally, the cracking moment has to satisfy only two inequalities – it should be greater than the moment due to loads in service and, according to the ACI *Code*, it must be less than 1/1.2 times the ultimate moment.[3]

As far as shear is concerned, the requirements of codes vary. The ACI *Code* requires shear to be investigated at ultimate load only[3]; others may require the principal tensile stress to be investigated at transfer and in service, either instead of or in addition to the ultimate strength.

Design of a prestressed concrete beam requires decisions to be made about the following matters:

(a) whether to use pre-tensioned or post-tensioned construction;
(b) the shape and dimensions of the cross-section;
(c) the amount of prestressing steel;
(d) the location of the prestressing steel in the cross-section;
(e) the magnitude of the prestressing force;
(f) reinforcing for shear;
(g) reinforcing for tension in the end blocks.

The choice between pre-tensioned and post-tensioned construction is largely a choice between construction in a factory and construction on site. Pre-tensioning requires a stressing bed to be constructed and this is unlikely to be economical, except for a permanent installation which can be used many times. Post-tensioning can be carried out with portable equipment. Thus, long runs favour pre-tensioning, short runs and *in situ* construction favour post-tensioning.

As far as shape of cross-section is concerned, there is a wide variety of choice. A uniform rectangular cross-section has in its favour simplicity of construction and the calculations are easy. However, for large beams, much weight can be saved by using an I-section, possibly with unequal flanges. Units for the construction of floors may be precast T-sections or inverted U-sections so that the floor functions as a flange.

The shape having been chosen, dimensions can be based on strength at ultimate load. There is no assurance that the section chosen will be satisfactory under other conditions of loading, but the fact that dimensions can be decided without anticipating much about the prestressing tendons makes it logical to start with the ultimate load. Equation (17.18) is used. It is entirely valid for a rectangular section or for one with flanges if the area subject to compressive stress does not extend below the flange. For other cases, it is approximately correct and at least gives some guidance in what becomes a trial and error process.

We have

$$M_u = \phi\omega_p(1 - 0.59\,\omega_p)\,f'_c\,bd^2 \tag{17.18}$$

The external bending moment can be calculated, given the data of the design problem, and this is put equal to M_u. An estimate has to be made of the beam's own weight, which is not known at this stage. The capacity reduction factor, ϕ, is known – it is 0.9 for bending – and the quality of concrete to be used is known, so that a numerical value can be assigned to f'_c. The reinforcement index must not exceed 0.3. If an approximate value is chosen for ω_p – say, 0.2 to 0.25 – Equation (17.18) can be used to calculate bd^2 and suitable values of b and d can be chosen.

The next step is to decide on the prestressing and this is done so as to satisfy the constraints relevant to transfer and service. A Magnel diagram can be plotted using the information accumulated at this stage. First of all, losses are estimated so that a trial value of η can be found. Then the Inequalities (17.10) and (17.11) can be tested to ensure that there is a region on the Magnel diagram for permissible combinations of P_T and e. The slopes and intercepts for the four lines are computed next and a limiting value of e determined, if there is one. Stresses near the ends of the beam may be restrictive in this respect. The most favourable combination of e and P_T is then chosen – it is the one for which $1/P_T$ is greatest and corresponds to the smallest possible prestressing force. This value, divided by the stress in the steel at transfer gives the area of steel required. Some adjustments will be necessary to get a practicable area of steel, and it may be necessary to make a corresponding adjustment to the stress in the steel at transfer to keep the design inside the allowable region of the Magnel diagram. A check on the losses will also be required at this stage to make sure that η has not drifted from the value chosen initially.

The next thing to do is to calculate the cracking moment and the ultimate moment of resistance. The cracking moment is given by Equation (17.14) and is easily found. The ultimate moment of resistance is evaluated as described in Section 17.5.2.

A final check is made to ensure that ω_p or ω_{pw} is not greater than 0.3 and that the ultimate moment of resistance is at least 1.2 times as great as the cracking moment.

The objective of all the foregoing calculations is to select a cross-section and prestressing force to satisfy the criteria of bending. Shear is tested at ultimate load and the shear reinforcing designed if such reinforcing is required. When all of these have been dealt with, there remains only the matter of design of the end blocks.

Tests reported by Marshall and Mattock are the basis of design of reinforcement of end blocks of pre-tensioned beams.[6] The rule as used by the American Association of Highway Officials and quoted by Cowan and Smith[7] is

$$A_s = 0.021 \frac{PD}{f_s h}$$

where A_s is the total area of cross-section of stirrups uniformly spaced for a length at the end of the beam equal to one-fifth of its depth; P is the prestressing force; D is the depth of beam; h is the transfer length which may be taken as

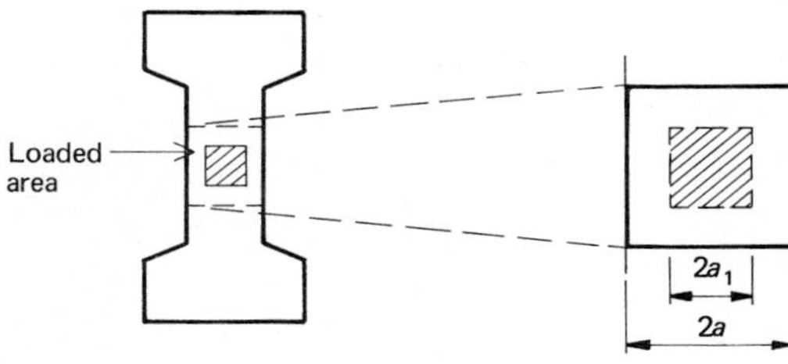

Fig. 17.16. *Loaded area for end block design in post-tensioned beam.*

50 times the diameter of a strand or 100 times the diameter of a wire; and f_s is the permissible stress in stirrups – say, 125 MPa.

For post-tensioned beams, a simple procedure is presented by Harris and Smith and attributed by them to the Cement and Concrete Association.[8] A loaded area is defined as shown in Fig. 17.16. It is the biggest square surrounding the anchorage which will fit in the cross-section. The side of the loaded area is $2a$ and a square of side $2a_1$ is assumed to represent the anchorage device, so that the area of the anchorage is $(2a_1)^2$. The average stress on the loaded area is given by

$$f_c = \frac{P_i}{(2a)^2}$$

where P_i is the part of the prestressing force anchored by the anchorage under consideration. The tensile stress in the end block is assumed to be uniformly distributed over the distance $2a$, horizontally and vertically, and its distribution along the beam is assumed to be as sketched in Fig. 17.17. The maximum tensile stress $f_{t(\max)}$ is given by

$$\frac{f_{t(\max)}}{f_c} = 0.98 - 0.825\left(\frac{a_1}{a}\right)$$

$$0.3 \ngtr \frac{a_1}{a} \ngtr 0.7$$

With these rules, the total vertical and horizontal tensions to be resisted can be estimated and a suitable area of steel calculated. It should be distributed along the end block so as to concentrate the area of steel at a distance $0.5a$ from the end, where the tensile stress is largest.

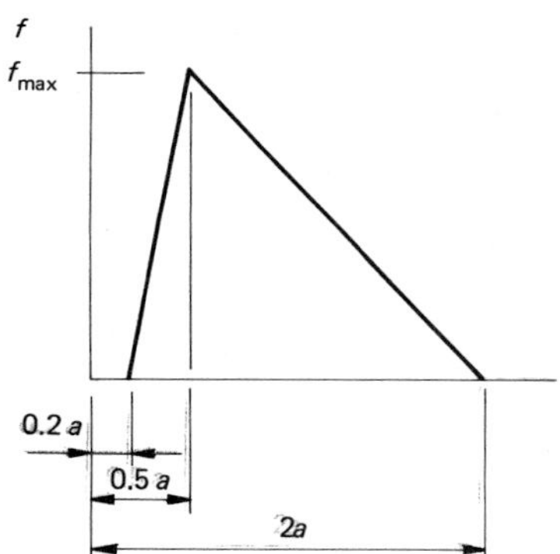

Fig. 17.17. *Empirical distribution of transverse tensile stress in end block of post-tensioned beam.*

17.7. WORKED EXAMPLES

Examples 17.1 and 17.2 show how these methods are applied to design of prestressed concrete beams. Both are comparable with the laminated timber beam of Example 14.2. In Example 17.1, a pre-tensioned beam is designed. Sheet 1 contains a list of the design data, and these are as laid down in the ACI *Code*.[3] On Sheet 2, the loads for various conditions are computed, including an estimate of the beam's own weight. From these, the relevant bending moments and shears are derived.

Sheet 2 has also the calculation for trial dimensions, using the ultimate strength equations. A rectangular cross-section has been chosen, mainly because of the small size of the beam. However, an I or a T would have been quite practicable.

On Sheet 3, the beam's own weight is checked, the properties of the cross-section are computed and the losses are examined to estimate the ratio η. Then the test is made to confirm that there is a region of permissible designs on the Magnel diagram – the Inequalities (17.10) and (17.11) are used.

Following this and carrying on to Sheet 4 is the calculation of the slopes of the lines on the Magnel diagram (Inequalities (17.6), (17.7), (17.8) and (17.9)) and the diagram is plotted on Sheet 4, showing in the shaded region where permissible pairs of e and P_T are to be found.

The limiting eccentricity shown on the Magnel diagram is based on the stresses at the ends, where

$$M = M_T = 0.$$

The inequalities that are relevant are (17.2) and (17.3) and we have

$$\frac{P_T}{A} - \frac{P_T e}{Z_1} \nless -p_{Tt}$$

$$\frac{P_T}{A} + \frac{P_T e}{Z_2} \ngtr +p_{Tc}$$

These require the stresses at transfer not to exceed the permitted tensile and compressive stresses respectively and lead to the following limits on e:

$$e \ngtr \frac{Z_1}{A} + \frac{Z_1 p_{Tt}}{P_T}$$

$$e \ngtr \frac{Z_2 p_{Tc}}{A} - \frac{Z_2}{A}$$

For a symmetrical section, with Z_1 and Z_2 equal, the first of these is the more severe, since

$$p_{Tt} \ll p_{Tc}$$

This line will be recognised as one of the restraints of a Magnel diagram plotted for the end of the beam.

The ACI *Code* allows the computed tensile stress in these parts of the beam to exceed p_{Tt}, provided mild steel reinforcing is used to resist the whole of the tensile force computed on an uncracked section.[3] We have not taken advantage of this, preferring to limit the eccentricity of the prestressing force.

The next part of the calculation deals with the eccentricity and the prestressing force. These should be chosen as high as possible in the shaded region to keep the prestressing force low. In this case, the area of prestressing steel required is 558 mm^2 and the area provided is 577 mm^2. It has been necessary to reduce the stress in the wires to compensate for this increase in area without leaving the shaded area of the Magnel diagram. This is not always the case and it is often possible to keep the tensile stress in the wires high, so providing a prestressing force greater than the minimum required. An improvement in performance in service can, in such cases, be obtained without losing anything. Another aspect of this question is brought out in Example 17.2.

Following these calculations, the losses are checked, the ultimate moment of resistance and the cracking moment are calculated, the reinforcement index is checked and the reinforcing required for shear- and end-block tension is designed.

Note that the calculation for the ultimate moment of resistance does not allow for the fact that the tendons are not concentrated at one level. We have calculated only one value of f_{ps} for each trial position of the neutral axis – that corresponding to the strain at the level of the centroid of the steel area. This must be recognised as an approximation for the mean stress in the steel because of the curvature of the stress–strain relationship.

Design of the post-tensioned beam of Example 17.2 follows the same general path. However, there are some features of special interest. First, the prestressing tendon will be a single draped cable, and we anticipate that a 12-wire Freyssinet cable will serve. This fixes minimum widths for the end block (to accommodate the anchorage cone) and the web (to contain the cable duct). The designer must have access to the dimensions of these proprietary items, from catalogues or specialised textbooks such as Cowan and Smith, Harris and Smith, or Abeles' and Turner.[7, 8, 9] A small amount of such information is included in Appendix A12.

An I-section has been chosen, with rectangular end blocks of width equal to the width of the flanges. It turns out that the neutral axis, at ultimate load, is high enough for Equation (17.18) to be valid.

When we come to investigation of conditions in service, problems about stresses at the ends do not arise because the cable is draped – at the ends, where $M_T = 0$, the prestressing force will be within the kern of the cross-section.

The region of permitted pairs of e and P_T is shaded on the Magnel diagram on sheet 3 and this time $e = e_{max}$ is not a boundary of the shaded region. Strictly speaking, this Magnel diagram is not applicable at the ends – it is based on M and M_T at midspan – but it is safe to conclude that $e = 0$ will be permissible at the ends.

Calculation of the prestressing force shows that one cable of 12 wires of 7 mm diameter is ample and the design is based on one cable fully stressed. This satisfies the criteria displayed on the Magnel diagram and leads to a cracking moment of 132 kN m. The ultimate moment of resistance, calculated on Sheet 4, is 212 kN m. The ultimate moment is high enough, and sufficiently in excess of the cracking moment which, in turn, exceeds the moment at working load by a satisfactory amount. Attention is drawn to the use of the approximate formula for f_{ps} in the calculation for ultimate moment of resistance.

The shear stress calculation on Sheets 4 and 5 includes allowance for the sag in the cable. Design of the reinforcement for the end blocks follows the empirical procedure of Section 17.6.

Example 17.3 is a straightforward application of Mohr's circle and shows how the prestressing force acts to reduce the principal tensile stress.

17.8. CONCLUSION

In this chapter we have presented the basic ideas on which design of prestressed concrete beams is based. This information, together with a locally available code, should be sufficient for the complete design of simply supported beams. For design of continuous beams and other statically indeterminate structures

Example 17.1 Sheet 1 of 7

Determine the dimensions, amount of prestressing steel and steel tension required for a pre-tensioned beam to support the timber floor of Example 14.1.

Design data: 28-day strength of concrete 35 MPa
Strength of concrete at transfer 27.5 MPa

Stress vs strain for prestressing tendons: See Sheet 5

Concrete design stresses etc:

ULTIMATE: $f_c' = 35$ MPa $0.85 f_c' = 29.8$ MPa
$\beta_1 = 0.79$

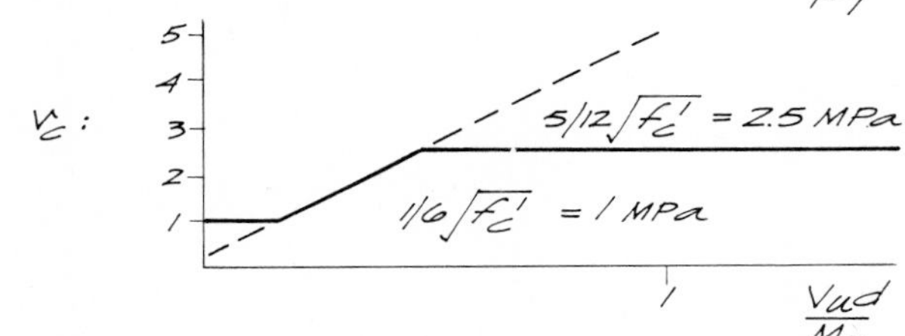

$v_u - v_c > 2/3\sqrt{f_c'}$ = 3.94 MPa

CRACKING: $f_r = 0.625\sqrt{f_c'}$ = 3.70 MPa

WORKING: $p_c = 0.45 f_c'$ = 15.7 MPa
$p_t = 0.5\sqrt{f_c'}$ = 2.96 MPa

TRANSFER: $f_{ci}' = 27.5$ MPa
$p_{Tc} = 0.6 f_{ci}'$ = 16.5 MPa
$p_{Tt} = 0.25\sqrt{f_{ci}'}$ = 1.31 MPa

Density = 2400 kg m^{-3}
Modulus of elasticity = E_c = $4750\sqrt{f_c'}$ = 28.2 GPa

Steel design stresses:

Ultimate strength = f_{pu} = 1600 MPa
Initial tensile stress $\ngtr 0.8 f_{pu}$ = 1280 MPa
Tensile stress after transfer $\ngtr 0.7 f_{pu}$ = 1120 MPa

Losses after transfer = 240 MPa
Modulus of elasticity for elastic range = 193 GPa

Example 17.1 Sheet 2 of 7

From Example 14.1.

Beam span = 6 m

Beam spacing = 2.5 m

Live load

$= 1000 \text{ kg m}^{-2} \rightarrow (1000 \times 2.5) \times 9.81 = 24600 \text{ N/m}$

Self-weight, timber floor

$= 36 \text{ kg m}^{-2} \rightarrow (36 \times 2.5) \times 9.81 = 885 \text{ N/m}$

Estimate self-weight of beam

Say,

$250 \text{ mm} \times 520 \text{ mm} \rightarrow (0.25 \times 0.52) \times 2400 \times 9.81 = 3060 \text{ N/m}$

Beam loads:

At transfer: 3060 N/m

In service:

24600
885
3060
28545 N/m

Ultimate:

$1.4D + 1.7L$

$= 1.4 \times 3945 + 1.7 \times 24600$

$= 47200 \text{ N/m}$

Bending moments and shears:

$$M_T = \frac{3060 \times 6^2}{8} = 13800 \text{ Nm at midspan}$$

$$M_T = 0 \text{ at ends}$$

$$M = \frac{28545 \times 6^2}{8} = 128000 \text{ Nm}$$

$$M_u = \frac{47200 \times 6^2}{8} = 212000 \text{ Nm}$$

$$V_u = \frac{47200 \times 6}{2} = 142000 \text{ N}$$

Estimate trial dimensions from ultimate load

$$M_u = \phi \omega_p (1 - 0.59\omega_p) f_c' bd^2$$

Try $\omega_p = 0.25 \rightarrow \phi \omega_p (1 - 0.59\omega_p) f_c'$

$$= 0.9 \times 0.25 \times 0.853 \times 35 \times 10^6$$

$$= 6.72 \text{ MPa}$$

$$bd^2 = \frac{2.12 \times 10^5}{6.72 \times 10^6} = 3.16 \times 10^{-2} \text{ m}^3$$

Try $b = 0.25 \text{ m}$

$$d = \sqrt{\frac{3.16 \times 10^{-2}}{0.25}} = 0.356 \text{ m}$$

Example 17.1 Sheet 3 of 7

Trial section:
(Centre of tension slightly below lower kern boundary)

Self-weight ok

$Z_1 = Z_2 = 1/6 \times 0.25 \times 0.52^2$
$= 1.125 \times 10^{-2}\ m^3$

$A = 0.25 \times 0.52$
$= 0.13\ m^2$

$\frac{Z_1}{A} = \frac{Z_2}{A} = 8.65 \times 10^{-2}\ m$
$= 86.5\ mm$

250
360
520
160
centre of tension

Select prestressing for service and transfer

Steel stress after transfer = 1120 MPa
Losses = 240 MPa
Final stress = 880 MPa

$\eta = \frac{880}{1120} = 0.785$

$M - \eta M_T = 128000 - 0.785 \times 13800 = 117200\ Nm$

$Z_1(p_c + \eta p_{Tt}) = 1.125 \times 10^{-2}(15.7 + 0.785 \times 1.31) \times 10^6$
$= 188000\ Nm > M - \eta M_T$ ok

$Z_2(\eta p_{Tc} + p_t) = 1.125 \times 10^{-2}(0.785 \times 16.5 \times 2.96) \times 10^6$
$= 179000\ Nm > M - \eta M_T$ ok

Slopes of lines on Magnel diagram:

17.6: $M_T + Z_1 p_{Tt} = 13800 + 1.125 \times 10^{-2} \times 1.31 \times 10^6$
$= +2.85 \times 10^4\ Nm$

17.8: $\frac{M}{\eta}\left(1 - \frac{Z_1 p_c}{M}\right) = \frac{1.28 \times 10^5}{0.785}\left(1 - \frac{1.125 \times 10^{-2} \times 15.7 \times 10^6}{1.28 \times 10^5}\right)$
$= -6.20 \times 10^4\ Nm$

17.7: $M_T + Z_2 p_{Tc} = 13800 + 1.125 \times 10^{-2} \times 16.5 \times 10^6$
$= +20.0 \times 10^4\ Nm$

17.9: $\frac{M}{\eta}\left(1 - \frac{Z_2 p_t}{M}\right) = \frac{1.28 \times 10^5}{0.785}\left(1 - \frac{1.125 \times 10^{-2} \times 2.96 \times 10^6}{1.28 \times 10^5}\right)$
$= +12.07 \times 10^4\ Nm$

Example 17.1 Sheet 4 of 7

Increments in e for increment of $10^{-6}N^{-1}$ in $\frac{1}{P_T}$:

17.6 :	$+2.85\times10^4\times10^{-6}$ m	$= +28.5$ mm
17.8 :	$-6.20\times10^4\times10^{-6}$ m	$\doteq -62.0$ mm
17.7 :	$+20.0\times10^4\times10^{-6}$ m	$= +200$ mm
17.9 :	$+12.07\times10^4\times10^{-6}$ m	$= +120.7$ mm

Magnel diagram:

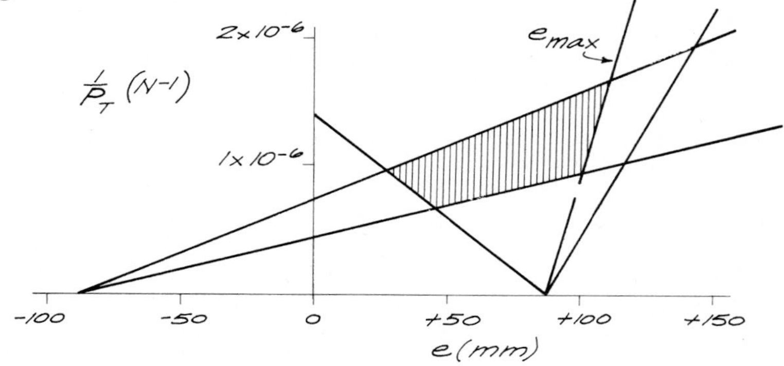

For e_{max}, as limited by stress at ends ($M_T=0$)

$$e_{max}=\frac{Z_1}{A}+Z_1p_{Tt}\frac{1}{P_T};\ Z_1p_{Tt}=1.125\times10^{-2}\times1.31\times10^6$$
$$=+1.47\times10^4\ \text{Nm}$$

Try $e = 110$ mm $= 0.110$ m

$1/P_T = 1.60\times10^{-6}N^{-1}\rightarrow P_T = 6.25\times10^5$ N

At 1120 MPa $\rightarrow A_{ps}=\dfrac{6.25\times10^5}{1.12\times10^3}=558\ \text{mm}^2$

Try 15 No 7 mm wires $\rightarrow A_{ps}=15\times\pi/4\times7^2=577\ \text{mm}^2$

At 1080 MPa $\rightarrow P_T = 6.25\times10^5$ N

$1/P_T = 1.60\times10^{-6}N^{-1}$ OK

Check elastic loss:

Average concrete stress at transfer $=\dfrac{6.25\times10^5}{0.13}$

$= 4.81$ MPa

Loss of steel tensile stress $=\dfrac{193}{28.2}\times4.81 = 33$ MPa

Initial stress $=1080+33 = 1113$ MPa < 1280 OK

Final stress $= 1080-240 = 840$ MPa

Example 17.1 Sheet 5 of 7

$$\eta = \frac{840}{1080} = 0.778 \qquad \text{OK}$$

$$P = 840 \times 577\text{N} = 485 \text{ kN}$$

Check cracking :

$$-f_r = \frac{P}{A} + \frac{Pe}{Z_2} - \frac{M_{cr}}{Z_2}$$

$$\frac{M_{cr}}{Z_2} = \frac{P}{A}\left(1 + \frac{e}{Z_2/A}\right) + f_r$$

$$= \frac{4.85 \times 10^5}{1.3 \times 10^{-1}}\left(1 + \frac{110}{86.5}\right) + 3.70 \times 10^6$$

$$= 12.20 \times 10^6$$

$$M_{cr} = 1.22 \times 10^7 \times 1.125 \times 10^{-2} = 137000 \text{ Nm}$$

(7% > M OK)

Check ultimate :

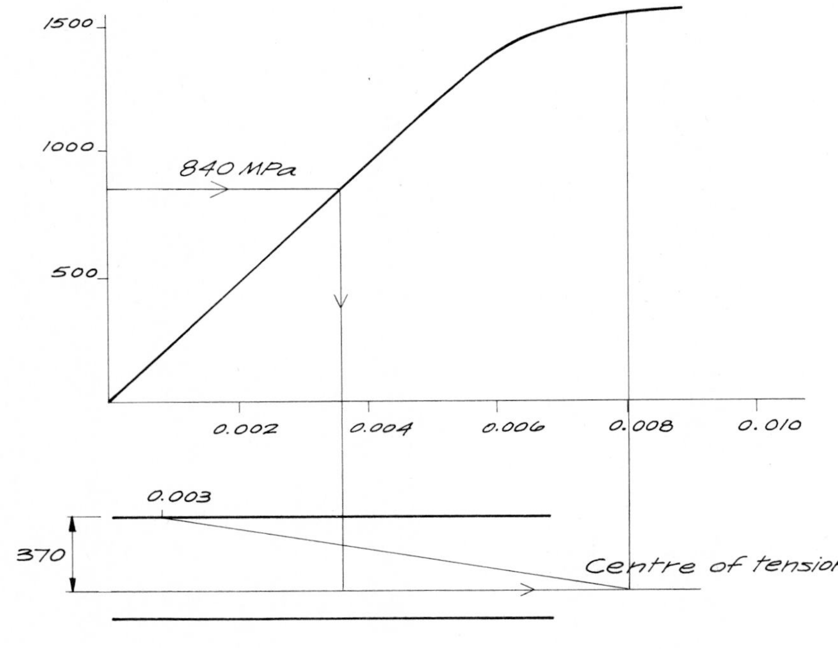

Example 17.1 Sheet 6 of 7

$C = 0.85 f'_c ab = 29.8 \times 250 \times a = 7450a$

$T = A_{ps} f_{ps} = 577 f_{ps}$

$a = 0.79\, kd$

kd	a	C	f_{ps}	T
150	118.5	8.83×10^5	1540	8.88×10^5

Further trials unnecessary

Lever arm $= 370 - 1/2 \times 118.5 = 0.311$ m

$M_u = 0.9 \times 8.83 \times 10^5 \times 0.311$

$= 248\ 000$ Nm $> 212\ 000$ Nm

$M_u > 1.2 M_{cr}$ OK

$$\omega_p = \frac{A_{ps}}{bd}\frac{f_{ps}}{f'_c}$$

$$= \frac{8.83 \times 10^5}{250 \times 370 \times 35}$$

$= 0.273 < 0.3$ OK

Check shear at ultimate:

$$v_u = \frac{V_u}{\phi bd} \qquad \phi = 0.85, \quad d \not< 0.8h = 415 \text{ mm}$$

$$= \frac{142\ 000}{0.85 \times 250 \times 415}$$

$= 1.6$ MPa

Nominal reinforcement: 6 mm stirrups, 30 mm c/c

Example 17.1 Sheet 7 of 7

End block reinforcement:

$$A_s = \frac{0.021Ph}{f_s\,(100d)} \qquad f_s = 125\ \text{MPa}$$

$$= \frac{2.1 \times 10^{-2} \times 4.85 \times 10^5 \times 5.2 \times 10^2}{1.25 \times 10^8 \times 10^2 \times 7}\ m^2$$

$$= 60.5\ mm^2$$

Two stirrups 6mm diameter (4 legs) → 113 mm^2

SUMMARY: Beam 250 mm wide × 520 mm deep
15 No 7 mm diameter wires
Initial total tension = 625000 N
Initial steel stress = 1113 MPa
Centre of tension 370 mm from top
Shear reinforcing: 6 mm stirrups 300 mm c/c
End blocks: 2 No 6 mm stirrups in 100 mm at each end.

Example 17.2 Sheet 1 of 6

As for Example 17.1, but using a post tensioned I-beam.

Design data: As for Example 17.1; except that losses after transfer = 170 MPa

Loads and bending moments as for Example 17.1

Estimate of dimensions, from Example 17.1

$bd^2 = 3.16 \times 10^{-2}\ m^3$

(Assuming compression block entirely in flange)

Try flange width 250 mm → end block wide enough for Freyssinet anchorage cone.

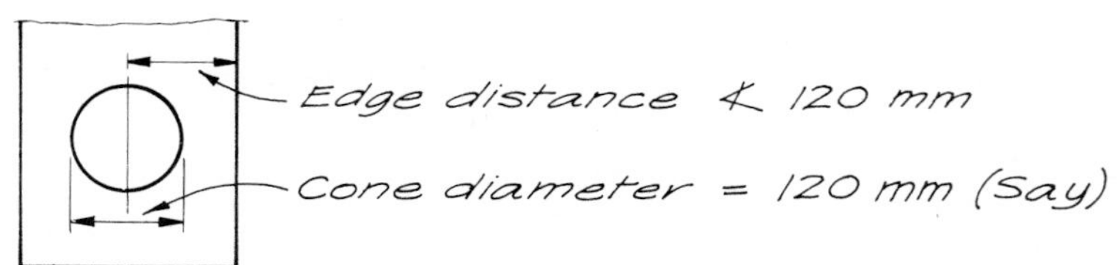

$d = \sqrt{\frac{3.16 \times 10^{-2}}{0.25}} = 356\ mm$

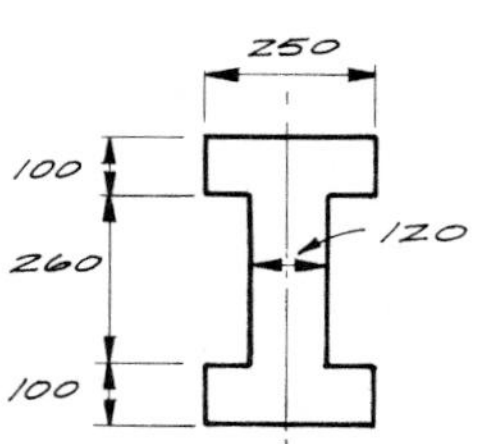

Try I-section thus

Minimum width of web:

Cable duct	45 (Say)
Cover	35
	35
Say,	120 mm

Centre of cable at least 60 mm from bottom

$e_{max} = 230 - 60 = 170\ mm$

$A = 2 \times 0.25 \times 0.1 = 0.05$
$0.12 \times 0.26 = \underline{0.0312}$
$0.0812\ m^2$

$I = 2 \times 1/12 \times 0.25 \times 0.1^3 = 0.42 \times 10^{-4}$
$2 \times 0.025 \times 0.180^2 = 16.2 \times 10^{-4}$
$1/12 \times 0.12 \times 0.26^3 = \underline{1.76 \times 10^{-4}}$
$18.38 \times 10^{-4}\ m^4$

Check self-weight:

$0.0812 \times 2400 \times 9.81 = 1920\ N/m$

$Z_1 = Z_2 = \frac{1.84 \times 10^{-3}}{0.230} = 8.00 \times 10^{-3}\ m^3$

$\frac{Z_1}{A} = \frac{Z_2}{A} = \frac{8.00 \times 10^{-3}}{8.12 \times 10^{-2}} = 0.0985\ m = 98.5\ mm$

Example 17.2 Sheet 2 of 6

Try stress at transfer $= 1120$ MPa

Allow for 3 mm slip

Loss $= 193000 \times 3/6000$ $= \underline{97}$ MPa

Jacking stress $= 1217$ MPa < 1280 OK

Allow for losses after transfer 170 MPa

Final stress $= 950$ MPa

$$\eta = \frac{950}{1120} = 0.85$$

Adjust moments and shear for reduced self-weight:

$$M_T = \frac{1920 \times 6^2}{8} = 8600 \text{ Nm}$$

$$M = \frac{27400 \times 6^2}{8} = 123000 \text{ Nm}$$

$$M_u = \frac{45700 \times 6^2}{8} = 206000 \text{ Nm}$$

$$V_u = \frac{45700 \times 6}{2} = 137000 \text{ Nm}$$

Select prestressing for service conditions:

$$M - \eta M_T = 123000 - 0.85 \times 8600 = 115700 \text{ Nm}$$

$$z_1(p_c + \eta p_{Tt}) = 8.00 \times 10^{-3}(15.7 + 0.85 \times 1.31) \times 10^6 = 134\,000 \text{ Nm} > M - \eta M_T \text{ OK}$$

$$z_2(\eta p_{Tc} + p_t) = 8.00 \times 10^{-3}(0.85 \times 16.5 + 2.96) \times 10^6 = 136\,000 \text{ Nm} > M - \eta M_T \text{ OK}$$

Slopes of lines on Magnel diagram:

17.6 $M_T + z_1 p_{Tt} = 8600 + 8.00 \times 10^{-3} \times 1.31 \times 10^6 = +1.91 \times 10^4$ Nm

17.8 $\frac{M}{\eta}\left(1 - \frac{z_1 p_c}{M}\right) = \frac{1.23 \times 10^5}{0.85}\left(1 - \frac{8.00 \times 10^{-3} \times 15.7 \times 10^6}{1.23 \times 10^5}\right) = -3.2 \times 10^4$ Nm

17.7 $M_T + z_2 p_{Tc} = 8600 + 8.00 \times 10^{-3} \times 16.5 \times 10^6 = +14.1 \times 10^4$ Nm

17.9 $\frac{M}{\eta}\left(1 - \frac{z_2 p_t}{M}\right) = \frac{1.23 \times 10^5}{0.85}\left(1 - \frac{8.00 \times 10^{-3} \times 2.96 \times 10^6}{1.23 \times 10^5}\right) = +11.7 \times 10^4$ Nm

Corresponding increments in e for increment of 10^{-6} N^{-1} in $1/P_T$

17.6 + 19.1 mm

17.8 − 32 mm

17.7 +141 mm

17.9 +117 mm

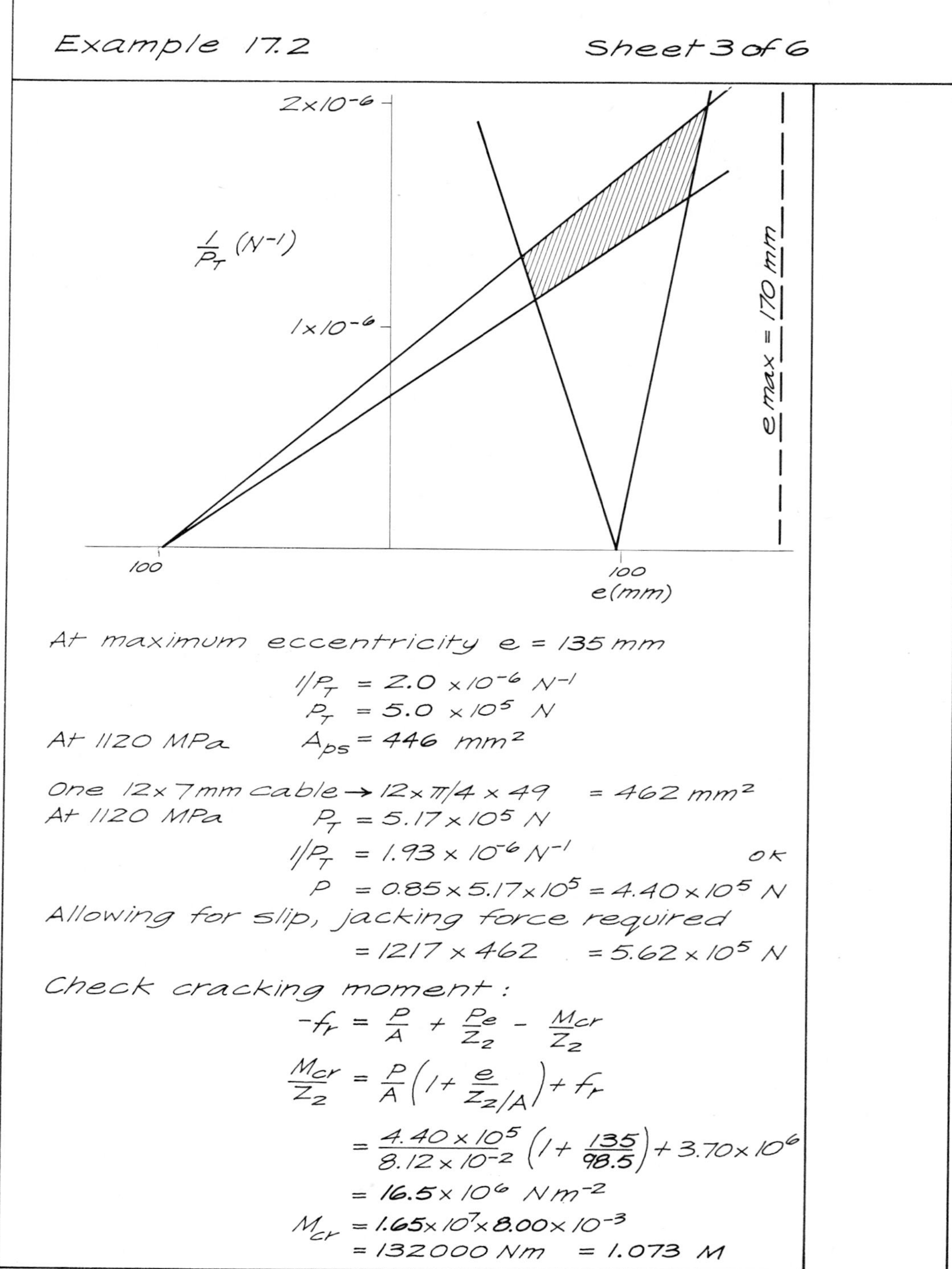

Example 17.2 Sheet 3 of 6

At maximum eccentricity $e = 135\ mm$

$$1/P_T = 2.0 \times 10^{-6}\ N^{-1}$$
$$P_T = 5.0 \times 10^{5}\ N$$

At 1120 MPa $\quad A_{ps} = 446\ mm^2$

One 12×7 mm cable → $12 \times \pi/4 \times 49 = 462\ mm^2$

At 1120 MPa $\quad P_T = 5.17 \times 10^5\ N$

$$1/P_T = 1.93 \times 10^{-6}\ N^{-1} \qquad \text{OK}$$

$$P = 0.85 \times 5.17 \times 10^5 = 4.40 \times 10^5\ N$$

Allowing for slip, jacking force required

$$= 1217 \times 462 = 5.62 \times 10^5\ N$$

Check cracking moment:

$$-f_r = \frac{P}{A} + \frac{Pe}{Z_2} - \frac{M_{cr}}{Z_2}$$

$$\frac{M_{cr}}{Z_2} = \frac{P}{A}\left(1 + \frac{e}{Z_2/A}\right) + f_r$$

$$= \frac{4.40 \times 10^5}{8.12 \times 10^{-2}}\left(1 + \frac{135}{98.5}\right) + 3.70 \times 10^6$$

$$= 16.5 \times 10^6\ N\,m^{-2}$$

$$M_{cr} = 1.65 \times 10^7 \times 8.00 \times 10^{-3}$$
$$= 132000\ Nm = 1.073\ M$$

Example 17.2 Sheet 4 of 6

Check ultimate moment:
Use alternative approximate method instead of method of Example 17.1.

$$\rho_p = \frac{A_{ps}}{bd} = \frac{462 \times 10^{-6}}{0.25 \times 0.400} = 4.62 \times 10^{-3}$$

$$\frac{f_{ps}}{f_{pu}} = 1 - \frac{0.5 \times 4.62 \times 10^{-3} \times 1600}{35}$$

$$= 0.894$$

$$f_{ps} = 1430 \text{ MPa}$$

$$T = A_{ps} f_{ps} = 6.60 \times 10^5 \text{ N} = C$$

Assuming $a <$ flange thickness

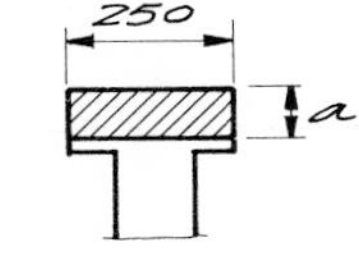

$$a = \frac{6.60 \times 10^5}{250 \times 29.8}$$

$$= 88.5 \text{ mm } (< 100 \text{ mm ok})$$

Lever arm $= 400 - 1/2 \times 88.5 = 356$ mm

$$M_u = 0.9 \times 6.60 \times 10^5 \times 0.356$$

$$= 212\,000 \text{ N m}$$

$$= 1.6 \; M_{cr} \qquad \text{OK}$$

Check ω_p:

$$\omega_p = \frac{A_{ps} f_{ps}}{bd f'_c}$$

$$= \frac{6.6 \times 10^5}{250 \times 400 \times 35}$$

$$= 0.189 < 0.3 \qquad \text{OK}$$

Example 17.2 Sheet 5 of 6

Check shear:

Assuming parabolic cable profile, with zero eccentricity at ends

UDL supported by cable

$$= \frac{8Pe}{L^2}$$

$$= \frac{8 \times 4.40 \times 10^5 \times 0.135}{6 \times 6} = 1\,32 \times 10^4 \text{ N/m}$$

At ultimate, nett shear $= 137\,000 - 3 \times 13\,000$

$= 98\,000 \text{ N}$

$$v_u = \frac{98\,000}{0.85 \times 120 \times 400}$$

$= 2.4 \text{ MPa}$

Nominal shear reinforcement:

5 mm stirrups, 300 mm c/c

End blocks:

Rectangular section at ends, 250 mm × 460 mm.

Freyssinet cone diameter = 120 mm

Side of square of equal area $= \sqrt{\pi/4} \times 120 = 106$ mm

$$\frac{a_1}{a} = \frac{106}{250} = 0.424$$

$$\frac{f_{max}}{f_c} \simeq 0.98 - 0.825 \times 0.424 = 0.63$$

$$f_c = \frac{5.62 \times 10^5}{250 \times 250} = 9.0 \text{ MPa}$$

$f_{max} = 5.67$ MPa

Example 17.2 Sheet 6 of 6

Assumed distribution of end-block tension:

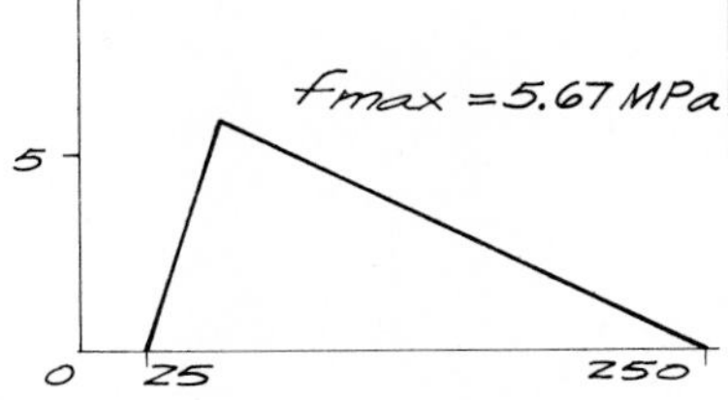

$0.2a = 0.2 \times 125 = 25$ mm

$0.5a = 0.5 \times 125 = 62.5$ mm

$2a = 2 \times 125 = 250$ mm

Total tension to be allowed for $= 1/2 \times 225 \times 5.67 \times 250$

$= 1.59 \times 10^5$ N

Allow steel tensile stress 125 MPa

→ 1270 mm²

Say, 4 vertical layers → 320 mm² each layer

3 No 12mm bars each layer → 340 mm²

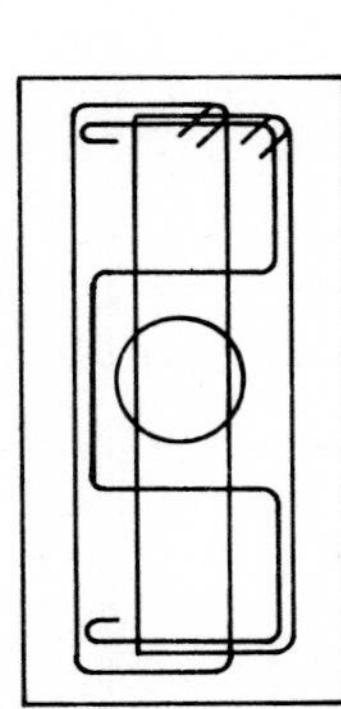

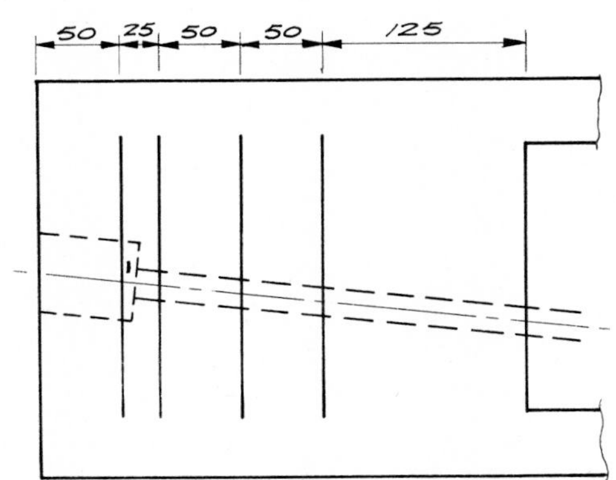

End-block reinforcement as sketched.

Example 17.3 Sheet 1 of 2

Calculate the principal tensile stress in the beams of Example 17.1 and 17.2 with and without the prestressing force.

Consider the stresses at the centre of area where the shear stress component is greatest and the prestressing compression equals P/A.

Beam of Example 17.1 Service load = 28300 N/m

End shear = 28300 × 3 = 84900 N

$$q = \frac{3}{2}\,\frac{84900}{250 \times 520} = 0.978\ \text{MPa}$$

without prestress $p_1 = 0 = p_2$

with prestress $p_1 = -\dfrac{4.80 \times 10^5}{2.5 \times 5.2 \times 10^4}$

$= -3.62$ MPa

Without prestress:

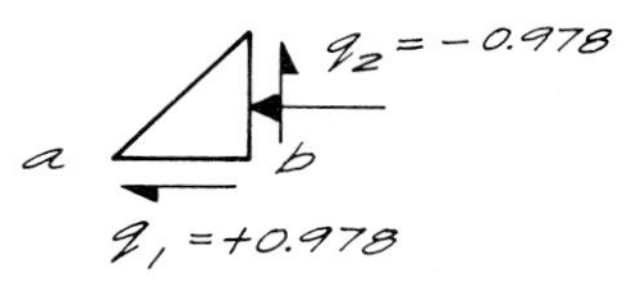

τ

$q_1 = +0.978$

$p_t = +0.978$

σ

$q_2 = -0.978$

Principal tension = 0.978 MPa on surface at −45° to ab

With prestress:

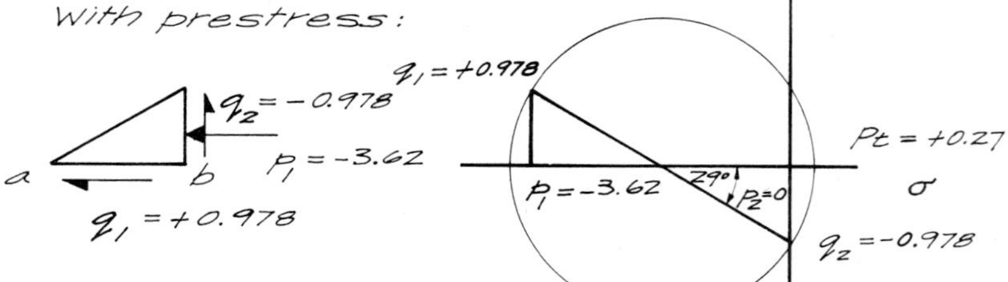

Principal tension = 0.27 MPa on surface at −14½° to ab

Example 17.3 Sheet 2 of 2

Beam of Example 17.2 Service load = 27400 N/m

Internal shear at end = (27400 − 13200) × 3 N

= 42.6 kN (Allowing for cable sag)

$b = 0.120\,m \quad I = 1.84 \times 10^{-3}\,m^4$

$$q = \frac{VQ}{bI}$$

$Q = 0.025 \times 0.180 = 0.0045$

$(0.12 \times 0.13) \times 0.065 = 0.00101$

$0.00551\,m^3$

$$q = \frac{4.26 \times 10^4 \times 5.51 \times 10^{-3}}{1.2 \times 10^{-1} \times 1.84 \times 10^{-3}}$$

$= 1.06 \times 10^6$ Pa

= 1.06 MPa

without prestress $p_1 = 0 = p_2$ $\quad q = \frac{27.4}{27.4 - 13.2} \times 1.06$

$= 2.04$ MPa

with prestress $p_1 = -\frac{4.40 \times 10^5}{8.12 \times 10^{-2}} = -5.42\,N/mm^2$

without prestress:

$q_2 = -2.04$, $p_1 = 0$, a, b, $q_2 = +2.04$

τ, $q_1 = +2.04$, $p_1 = p_2 = 0$, σ, $q_2 = -2.04$

Principal tension = 2.04 MPa

with prestress

$q_2 = -1.06$, $p_1 = -5.42$, a, b, $q_1 = +1.06$, $p_2 = 0$

τ, $q_1 = +1.06$, $p_2 = 0$, $p_t = +0.2$, 21°, $p_1 = -5.42$, σ, $q_2 = -1.06$

Principal tension

= 0.2 MPa

on surface at −10.5° to ab

or composite structures reference to specialised texts is required. Some are listed at the end of this chapter. The supports of continuous beams impose restraints on the deflection under the action of the prestressing force and these are most important. Composite construction is a very attractive application of prestressed concrete. Typical examples are the composite action of a cast *in situ* reinforced concrete slab and a precast prestressed beam. During construction the beam can support the slab so that less falsework for temporary support of the slab is necessary. In service, and at ultimate load, the slab functions as the flange of a composite beam. Such construction is well adapted to bridges and floors, but it introduces some additional problems for the designer; for example,

(a) the concrete in the slab usually has lower strength and lower modulus of elasticity than the concrete in the beam;
(b) the cross-section of the effective beam changes when the concrete in the slab has set:
(c) the horizontal component of shear stress requires steel reinforcing to ensure adequate connection of the slab to the beam;
(d) the difference in age and quality of the slab and the beam is accompanied by differences in their shrinkage, so leading to a shearing action at the interface.

For discussion of these matters, reference to the specialised texts is necessary.

LIST OF SYMBOLS

A	area of cross-section of beam
A_1	area of cross-section of one tendon
A_{ps}	area of cross-section of all tendons
A_s	area of steel for end block reinforcing
a	depth of compressive stress block, half-side of loaded area in end block
a_1	half-side of square equal in area to anchorage device
b	breadth of rectangular beam, breadth of flange of flanged beam
b_w	breadth of web of flanged beam
C	compression force in resistance couple
D	dead load, depth of beam at end block
d	depth of centre of tension from compression face of beam
E_c	modulus of elasticity of concrete
E_s	modulus of elasticity of steel
e	eccentricity of prestressing force
e_{max}	maximum eccentricity of prestressing force
Δe	sag of draped tendon
ΔF	element of force between tendon and concrete
f	stress in concrete
f_c	compressive stress in concrete
f'_c	28-day strength of concrete
f'_{ci}	strength of concrete at transfer
f_{ps}	tensile stress in prestressing steel
f_{pu}	ultimate tensile strength of prestressing steel
f_r	modulus of rupture of concrete
f_s	permissible stress in mild steel
f_t	tensile stress in concrete at the end block

$f_{t(max)}$	maximum value of f_t
f_y	yield strength of mild steel
h	depth of beam, transfer length at the end block
h_f	thickness of flange
I	second moment of area of cross-section
k	depth to neutral axis divided by depth to centre of tension
L	live load, span, length required to anchor a pre-tensioned tendon
M	bending moment due to loads in service
M_{cr}	cracking moment
M_T	bending moment due to loads supported at transfer
M_u	ultimate moment of resistance
n	number of tendons
P	prestressing force after losses, prestressing force in general
P_i	prestressing force in one tendon
P_T	prestressing force at transfer
p_1	normal component of stress
p_c	permitted compressive stress in concrete in service
p_t	permitted tensile stress in concrete in service
p_{Tc}	permitted compressive stress in concrete at transfer
p_{Tt}	permitted tensile stress in concrete at transfer
Q	first moment of part of area of cross-section
q_1, q_2	components of shear stress
R	radius of curvature of draped tendon
Δs	element of length of draped tendon
T	tensile force in resistance couple
V	external shear
V_u	external shear at ultimate load
v_c	strength of concrete in shear at ultimate load
v_u	nominal shear stress in concrete at ultimate load
x	distance along beam
Δx	increment of distance along beam
y	distance from centroidal axis of cross-section, coordinate of profile of draped tendon
y_1, y_2	distances from centroidal axis to the top and bottom faces respectively
Z_1, Z_2	elastic moduli of cross-section with respect to top and bottom faces respectively
β_1	ratio of a to kd
ε_u	ultimate strain in concrete (= 0.003)
ε_s	strain in steel
ζ	a variable
η	ratio of P to P_T
$\Delta\theta$	change in slope of draped tendon for increment of length Δs
ρ_p	ratio of area of prestressing steel to effective area of concrete ($= A_{ps}/bd$)
σ_t	principal tensile stress
ϕ	strength reduction factor
ω_p	reinforcement index for prestressing steel ($= (A_{ps}/bd)(f_{ps}/f_c)$)
ω_{pw}	reinforcement index for that part of prestressing steel acting with compression in web

REFERENCES

[1] ACI 318–71, *Commentary on Building Code Requirements for Reinforced Concrete,* American Concrete Institute, Detroit, 1971.

[2] ACI-ASCE COMMITTEE, "Tentative Recommendations for Prestressed Concrete", Report 423, *ACI Journal, Proceedings of the American Concrete Institute,* **54**, January 1958, 545–78.

[3] ACI 318–71, *Building Code Requirements for Reinforced Concrete,* American Concrete Institute, Detroit, 1971.

[4] ACI COMMITTEE, "Deflections of Prestressed Concrete Members", Report 435, *ACI Journal, Proceedings of the American Concrete Institute,* **60**, December 1963, 1697–728.

[5] G. MAGNEL, *Prestressed Concrete,* 2nd ed., Concrete Publications, London, 1950.

[6] W. T. MARSHALL and A. H. MATTOCK, "Control of Horizontal Cracking in the Ends of Pre-tensioned Concrete Girders", *Journal of the Prestressed Concrete Institute,* 7, 1962, 56–74.

[7] H. J. COWAN and P. R. SMITH, *The Design of Prestressed Concrete,* Angus & Robertson, Sydney, 1966.

[8] J. D. HARRIS and I. C. SMITH, *Basic Design and Construction in Prestressed Concrete,* Chatto & Windus, London, 1963.

[9] P. W. ABELES and F. H. TURNER, *Prestressed Concrete Designer's Handbook,* Concrete Publications, London, 1962.

PROBLEMS

17.1. The beam of Example 17.1 is, by misadventure, lifted with a single sling at midspan. What would you expect to happen? What would you do to prevent such an accident?

17.2. Estimate the cost of the beam of Example 17.2. Assume that the cost of establishing the precasting yard (cf. Fig. 25.1) is $15,000 and can be written off over 500 beams.

17.3. A cylindrical reservoir for storage of water is to be constructed of prestressed concrete, the hoop tension caused by the internal water pressure being resisted by tangential prestressing. For a reservoir 30 m in diameter select a suitable thickness of concrete and corresponding prestressing at a depth 5 m below the surface of the water. Assume that individual cables extend for one third of the circumference. Allow for friction loss using the formula

$$\frac{P_s}{P_x} = e^{\mu\alpha}$$

where P_s and P_x are the tensions at the jack and distance, x, from the jack respectively, and α is the angular distance between these points measured in radians. The coefficient of friction, μ, can be taken as 0.2. Sketch a suitable detail for anchorage.

17.4. Redesign the beam of Example 17.1 using a uniform prestress. Compare the quantities of concrete and steel required for the two designs. Is this alternative a better beam in any way?

17.5. Redesign the beam of Example 17.2 using a rectangular cross-section. Compare the quantities of concrete and steel in the two designs.

17.6. Design a floor system using pre-tensioned inverted-U floor units to support a load of 1000 kg m^{-2} on a span of 8 m. Use the same data for concrete and steel as in Examples 17.1 and 17.2.

18 Reinforced Concrete Slabs

18.1. INTRODUCTION

In Chapter 5, "Machines and Structures", we have shown the place occupied by the slab as a structural element in a bridge or a building. From the point of view of the user it is one of only a few such elements that he might be aware of. For example, to a motorist driving along a highway, the deck of a bridge is all important to his purpose. He may be, and usually is, quite unaware of the girders and foundations which support the roadway he uses and he takes them for granted. To the designer the slab is neither more important nor less important than the other elements. His concern is with the dimensions of the slab and the steel needed to reinforce it so that its strength and stiffness will be adequate.

In buildings, slabs are usually supported by beams, which may be elements of the main structural frame or secondary beams spanning from one main beam to another. Otherwise, in flat-slab construction, the slabs are supported directly by the columns. Flat-slab construction offers such advantages as reduced floor-to-floor height for given head room and simpler formwork. However, these advantages have to be offset against increased flexibility and less resistance to lateral loads. Further, transferring the load from the slab to the column poses particular problems and increased thickness in the form of a *drop panel* or a *drop* may be necessary around columns. Bridges cannot, in general, be built with flat-slab construction because of the heavy loads supported and the need for long spans.

Dimensions of slabs are rarely determined by structural calculation. Dimensions on plan depend on the arrangement of the columns and this is decided by the functional requirements of the owner and by cost. Such matters are settled early in the design process, before detailed design of slabs is undertaken. The only decision in this class likely to be of concern when the slab is being designed is whether to use secondary beams to reduce slab spans.

Thickness of a floor slab is very often determined by the fire resistance required. Safety of occupants and contents of a building is a matter of prime concern to statutory authorities and insurers who set standards for the construction of buildings. Depending on the contents and the risks involved, different fire ratings are required for the construction and this criterion often leads to slab thicknesses greater than required for strength alone. Other matters to be taken into account are deflection of the slab and problems of construction. More will be said about deflection in a later section. As far as construction is concerned, the effects of unavoidable inaccuracies in fixing the reinforcing steel are proportionately greater in a thin slab than in a deep beam. For this reason alone, slab thickness should not be less than 120 mm in normal construction. To be sure, much thinner concrete is used in shell roofs and folded plate structures, but these demand very high standards of construction and supervision.

There remain two principal problems to be solved by the structural designer. One is to determine the steel required to reinforce the slab for bending. The other occurs in flat-slab construction – the problem of transferring the load carried by the slab to the columns. For the first problem, analysis of the slab as a structure is undertaken to find the bending moment per unit width in two directions, along the slab and across the slab. With this information, the methods of Chapter 16, "Reinforced Concrete Beams", are used to proportion the reinforcement, a strip of slab of unit width being dealt with as a beam. By and large, slabs are under-reinforced, relative to design based on balanced conditions (Chapter 16), so that crushing of the concrete without yield in the steel is unusual. Except for flat slabs near the columns, shear is not often critical.

However, determination of the moments is not straightforward because most slabs are statically indeterminate structures. Unless the panel is very long in one direction its behaviour in bending is two-dimensional (compared with one-dimensional behaviour for a beam) and this cannot be ignored by the designer. Furthermore, various degrees of restraint are offered by adjacent panels, depending on variations in span and thickness and whether the panel under consideration is an interior panel or an exterior one.

Transfer of load to a column may require shear reinforcement similar to that used in beams, or a *shearhead* of structural steel. As noted above, increased thickness in the form of a drop may be desirable.

In the following sections, calculation of bending moments in slabs by means of coefficients and by what is known as the *equivalent-frame method* will be discussed. Details of these methods will not be pursued. In Section 18.4, the use of the yield-line theory will be treated in sufficient detail for slabs to be designed in practice. Rules for determining the minimum thickness required to avoid excessive deflections will be discussed briefly and the transfer of load to columns supporting flat slabs will be described. All of the structural design for strength is based on service loads multiplied by a load factor, and failure of the slab.

18.2. SLAB DESIGN BY COEFFICIENTS

The following discussion is based on the rules for slab design given in the ACI *Building Code Requirements for Reinforced Concrete*, where it is referred to as the *direct-design method.*[1] Reference should also be made to the *Commentary on Building Code Requirements for Reinforced Concrete.*[2] These rules are relevant to regular slab systems, with or without beams. In general, panels have to be rectangular, only minor departures of column positions from a rectangular grid being tolerable. In addition, there are restrictions on the ratio of length to breadth of a panel, the number of spans in each direction and the ratio of successive span lengths to each other, the ratio of live load to dead load and the relative stiffnesses of beams at the edges of a panel. Although this list of restrictions is quite long, they are not oppressive, for a majority of conventional slabs qualify for this kind of design.

The procedure is as follows:

(a) a total design moment is calculated for a span;
(b) this moment is then apportioned as a negative moment at the support and a positive moment at midspan;
(c) both these moments are then apportioned to a "column strip" and a "middle strip" as indicated in Fig. 18.1;
(d) the reinforcement necessary to provide ultimate moments equal to the negative and positive moments in these strips is then determined.

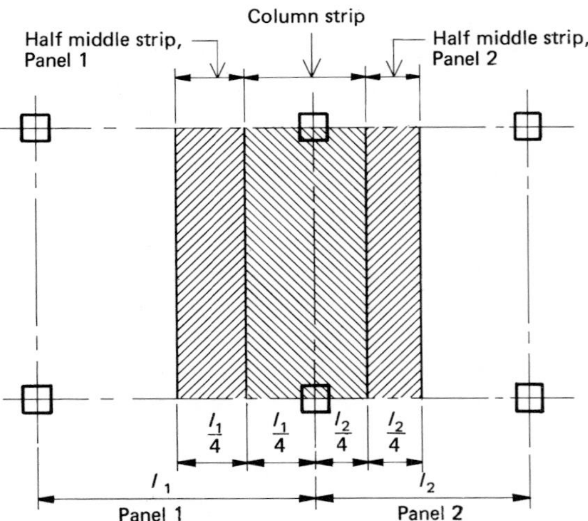

Fig. 18.1. *Subdivision of a slab floor into column strips and middle strips. Similar strips run at right angles to these.*

Rules are given for calculation of the total design moment as well as for dividing a panel into strips. The coefficients for sharing the moments in steps (b) and (c) above are also provided, the coefficients being based on theoretical analysis of bending in plates, experimental testing and experience with slabs in service.

For step (d), the column strip includes the beam (if there is one) supporting the edge of a panel. Provided construction is truly monolithic, this means that the design is applied to a T-beam at midspan, where the bending moment is positive. Over the columns, of course, the slab is on the tension side of the beam and does not act as a flange. The middle strips are designed as shallow rectangular beams.

Clearly, there is little that is difficult about designing slabs this way. One becomes familiar with the rules and applies them carefully, and this includes not applying them when they are not valid. That they are not valid for slabs of irregular shape is a disadvantage, although not a severe one. More important from our point of view is that the method offers little insight into the way a slab works. For this reason, we do not examine it more closely.

18.3. EQUIVALENT-FRAME METHOD

The equivalent-frame method is described in the ACI *Code* and *Commentary*.[1, 2] In this method, the building is assumed to comprise series of frames parallel to each other and on the centrelines of the columns. One series of frames runs along the building and the other across the building. The slabs are assigned to these frames as indicated by the hatched area in Fig. 18.1. For a frame at the edge of a building, a slab contribution is possible on only one side and it amounts to half of the first slab panel.

Analysis of the complete frame will lead to bending moments for each beam–slab combination in the equivalent frame and these moments are shared among column strips and middle strips as in the *coefficient method.* Approximate calculations, involving only part of the frame near a particular beam–slab element are permitted.

To the extent that it allows some irregularity in slab spans, this method is more versatile than the coefficient method. However, it does not contribute much to design of irregular panels. Like the coefficient method, it treats the two-dimensional character of slab bending in an arbitrary way.

18.4. YIELD-LINE THEORY

The procedure of most interest to us is the yield-line theory of the failure of slabs. Although not described in the ACI *Code*[1], it is acceptable as a design procedure; this is made clear in the *Commentary.*[2] It has the advantage of being applicable to a slab of any shape and with any condition of support at its edges. For our pedagogical purpose, another advantage is that design is based directly on the designer's understanding of the structural behaviour of the slab.

The yield-line theory was propounded by Johansen in 1943 in the Danish language. An English translation of his original work appeared in 1962.[3] An excellent brief summary of Johansen's work was presented by Hognestad in 1953[4], and the whole subject of the theory of slabs is extensively examined by Wood.[5]

Johansen's procedure is still accepted for practical design although some of its theoretical justification is now in question. Our discussion in the following sections is based largely on Hognestad's interpretation of Johansen's work.

18.4.1. Failure of a slab

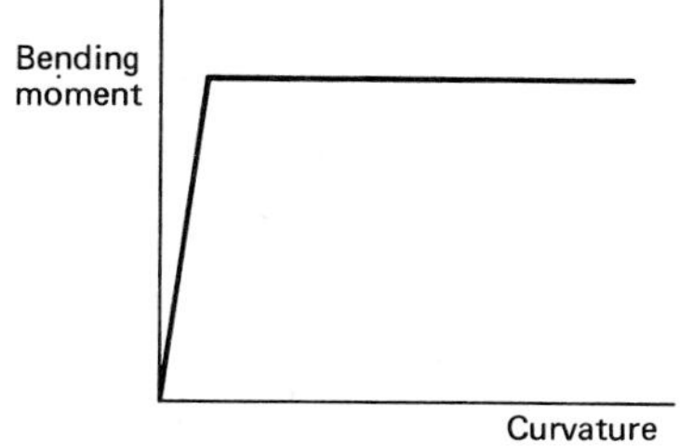

It is assumed that the material of the slab is ductile and that an idealised relationship between bending moment and curvature is valid, as shown in the margin. For curvature greater than that at yield, the bending resistance of the slab remains constant as the curvature increases. This assumption is a good one for reinforced concrete bent in a plane parallel to the reinforcement, provided there is not so much steel that failure occurs by compression in the concrete without yield in the steel. Yielding of the reinforcing steel in tension ensures that the reinforced concrete member behaves so that this is a good approximation.

When a slab conforming to this assumption is loaded to the point of failure, yielding occurs along lines where the bending moment is greatest, these lines being known as *yield lines.* Any additional deflection of the slab will occur by virtue of increasing curvature at the yield lines and elastic deflection of the parts of the slab enclosed by yield lines and edges will be negligible by comparison. Therefore, at failure the slab can be dealt with as a set of plane surfaces, supported at their edges by either the supports at the edges of the slab or the yield moment along a yield line shared with another plane surface. Neglect of elastic deflections and consideration of the constraints on rotation at the edges enable one to select arrangements of yield lines that are kinematically valid. The design

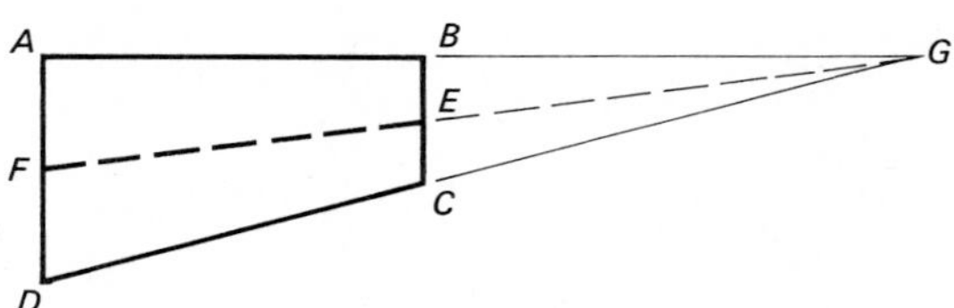

Fig. 18.2. *Kinematically valid yield line.*

process is to find the worst set of yield lines and this is best done by trial and error.

To see how kinematically valid yield lines are found, consider the simple case shown in Fig. 18.2. The slab shown in plan there is supported by walls on the edges *AB* and *CD*, the other two edges being without support. Failure of this slab would occur with a yield line such as *EF*. The panels *ABEF* and *DCEF* are, for practical purposes, planes. The area *ABEF* must rotate about *AB* and *DCEF* about *DC*. It is therefore necessary that *FE* passes through *G*, the intersection of *AB* and *DC* produced. No other yield line is compatible with these edge supports and the requirement for *ABEF* and *DCEF* to be planes. For this slab, then, different locations of the yield line would be tried, all of them passing through *G* if produced, and the one that requires the biggest yield moment for stability (given the load) is critical. Different support conditions lead to different patterns of yield lines and a few are illustrated in Fig. 18.3.

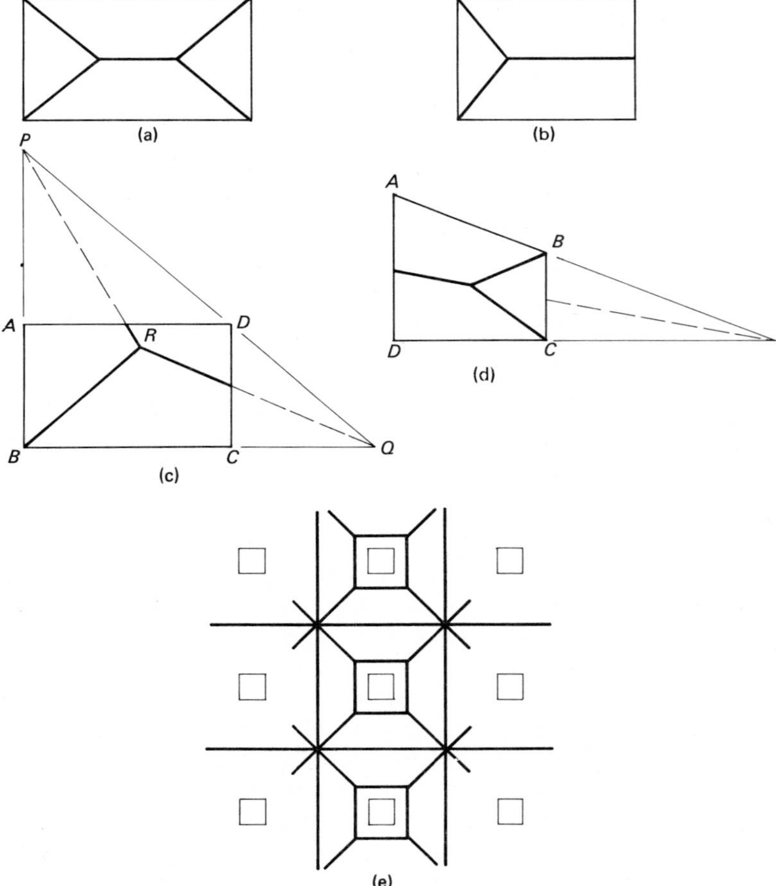

Fig. 18.3. *Examples of yield lines: (a) supported along all edges; (b) supported on three edges; (c) supported along AB and BC and at D. The axis of rotation PQ through D and the point R vary from one trial to the next; (d) supported along AB, BC and CD and unsupported along DA; (e) flat-slab floor supported on columns.*

The next matter to consider is calculation of the yield moments, given the load on the slab at failure and a pattern of yield lines. In the analysis that follows, we will assume that a beam supporting an edge of a slab panel does not fail with the slab. Such an edge is then a line about which part of the slab rotates. A beam which is not strong enough will itself yield under the load applied to the slab in somewhat the same way as an unsupported edge would.

18.4.2. Forces at a yield line

In analysis of the failure of the slab, its thickness is ignored and the slab is dealt with as a surface or a set of surfaces. This is analogous to the one-dimensional treatment of a beam or a column. Consider an element of length Δs of a line in a plane representing a slab, as in Fig. 18.4. The internal moment acting at this element will be a vector of magnitude $m\Delta s$ and, in general, in some direction different from Δs, m being the moment per unit length. It is the couple that the material on one side of the line exerts on the material on the other side. This vector can be resolved into two components, $m_b\Delta s$ parallel to Δs and $m_t\Delta s$ normal to it. Bearing in mind that the vector representing a couple is normal to the plane of the forces of the couple, we see that $m_b\Delta s$ is the bending moment in the slab at this point and $m_t\Delta s$ is the twisting moment. The orientation and magnitude of $m_b\Delta s$ and $m_t\Delta s$ depend on the orientation of the line segment Δs.

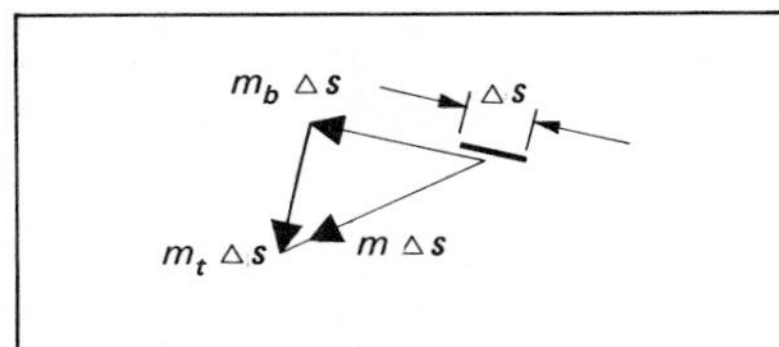

Fig. 18.4. *Bending and twisting components of moment on elementary line segment.*

If Δs is an element of a yield line, it is assumed that there is no twisting component, so that $m_b = m$. The element of length Δs represents to scale the bending moment vector at that place in the yield line.

Internal shear in the slab is taken to be zero at a yield line. This follows from the assumption that twisting components are zero and the bending component is a maximum. With no shear or twisting component acting at a yield line, the only forces which act to support a slab panel at a yield line are those that make up the internal yield moment.

This is the base on which yield-line analysis rests. One important exception to the assumptions outlined above is made when a yield line meets an edge which cannot resist twisting. This is dealt with below.

18.4.3. Isotropic slabs

If the slab is homogeneous and isotropic, the bending moment that would cause yield on a segment Δs (as in Fig. 18.4) is independent of the orientation of the segment. In a steel plate, this is clearly a good assumption. However, reinforced concrete slabs are not isotropic. Generally, slabs are reinforced in two perpendicular directions and in two planes, near the top surface for negative bending and near the bottom surface for positive bending. Provided the reinforcement is equal in the two directions, the slab can be analysed as if it were isotropic, which is shown by Wood.[5] The reinforcement does not have to be the same in the two layers, but it must be equal in the two directions in each layer.

We deal first with slabs which are equally reinforced in both directions. In Section 18.4.8 we will discuss the means whereby slabs with unequal reinforcing along and across the panel may be analysed.

18.4.4. Equilibrium analysis

The bending moment at a yield line can be found from the equilibrium of the several panels into which the slab is divided by the yield lines. Consider, for example, the slab shown in Fig. 18.5. It is reinforced equally in both directions and is simply supported on all four edges and it collapses under a uniformly distributed load w per unit area. Clearly, the pattern of yield lines must be as shown, with only one degree of freedom, the dimension x.

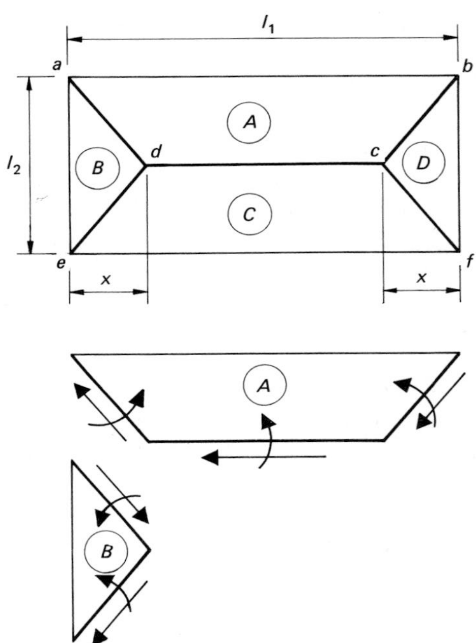

Fig. 18.5. *Bending moments at yield lines bounding panels.*

The panel A is in equilibrium under the action of

(a) the load w per unit area;
(b) an upward reaction along the edge ab;
(c) a bending moment distributed uniformly along the yield lines bc, cd and da. The lengths of these yield lines represent to scale the bending moments at the yield lines.

The panel B is similarly supported. If the equilibrium of the panels is analysed separately, two values of m (the bending moment per unit length of yield line) can be calculated – one for each panel. However, m must be the same for both panels and this is the condition which fixes x. For this problem, it is easy to find x by equating the values of m found for the two panels. In cases that are not so simple, a trial and error procedure is less laborious.

For panel A, we have, taking moments about ab:

$$\text{Moment of the load} = (\tfrac{1}{2} \times x \times \tfrac{1}{2}l_2) \times w \times (\tfrac{1}{3} \times \tfrac{1}{2}l_2) + \{(l_1 - 2x) \times \tfrac{1}{2}l_2\} \times \times w \times (\tfrac{1}{2} \times \tfrac{1}{2}l_2) + (\tfrac{1}{2} \times x \times \tfrac{1}{2}l_2) \times w \times (\tfrac{1}{3} \times \tfrac{1}{2}l_2)$$

$$= wl_1 l_2^2 \left\{ \tfrac{1}{12}\frac{x}{l_1} + \tfrac{1}{8}\left(1 - 2\frac{x}{l_1}\right) \right\}$$

Moment of the reaction $= 0$

Resultant moment on yield line $= m\{\vec{bc} + \vec{cd} + \vec{da}\} = m\,\vec{ba}$
Component parallel to ab of resultant moment $= m\,ba = m\,l_1$
For equilibrium,

$$wl_1 l_2^2 \left\{ \tfrac{1}{12}\frac{x}{l_1} + \tfrac{1}{8}\left(1 - 2\frac{x}{l_1}\right) \right\} - ml_1 = 0$$

$$m = wl_2^2 \left\{ \tfrac{1}{12}\frac{x}{l_1} + \tfrac{1}{8}\left(1 - 2\frac{x}{l_1}\right) \right\}$$

$$m = wl_2^2 \left(\tfrac{1}{8} - \tfrac{1}{6}\frac{x}{l_1} \right) \tag{18.1}$$

Similarly, from the equilibrium of panel B,

$$(\tfrac{1}{2} \times l_2 \times x) \times w \times (\tfrac{1}{3}x) - ml_2 = 0$$

$$m = \tfrac{1}{6}wx^2$$

$$m = \tfrac{1}{6}wl_2^2 \left(\frac{l_1}{l_2}\right)^2 \left(\frac{x}{l_1}\right)^2 \tag{18.2}$$

Equations (18.1) and (18.2) are sketched in Fig. 18.6 and the intersection P gives the desired value of m, which is

$$\frac{m}{wl_2^2} = \tfrac{1}{8} + \tfrac{1}{12}\left(\frac{l_2}{l_1}\right)^2 - \tfrac{1}{6}\sqrt{0.75\left(\frac{l_2}{l_1}\right)^2 + 0.25\left(\frac{l_2}{l_1}\right)^4}$$

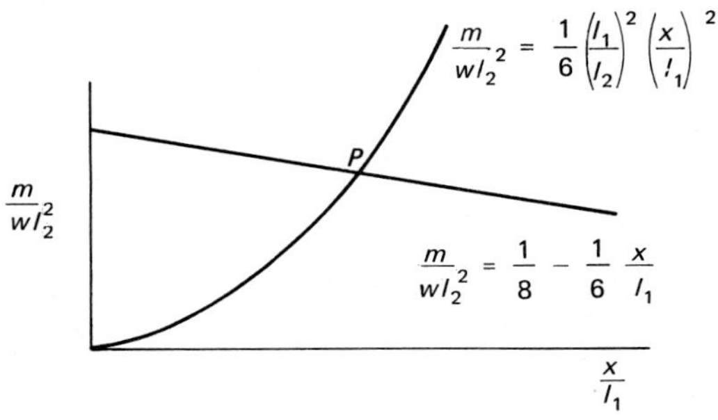

Fig. 18.6. *Solution of equations for equilibrium.*

It is clear that, as l_1/l_2 gets large, m/wl_2^2 tends to $\tfrac{1}{8}$ and this gives the bending moment in the slab spanning the short dimension. As the panel becomes increasingly elongated, two-dimensional bending decreases in importance and the slab becomes a "one-way" slab. The main reinforcement in a one-way slab lies across the panel, with nominal reinforcement parallel to the long sides.

The slab's equilibrium depends on a balance of vertical forces as well as on the moment balance described above. No shear acts at a yield line, so that the vertical forces acting on a panel of a slab are the total load downwards and the reaction of a beam or a column. It follows that this aspect of the equilibrium of a slab enables calculation of the loads applied to the supporting structure. The beams and columns can then be designed as described in the chapters on beam design and column design. If the yield lines are not known accurately, beam and column reactions can be found according to the rule in Section 18.6.

18.4.5. Energy analysis

Virtual work can also be used to calculate the moment m. By this principle, the work done by the loads for an arbitrary displacement is equal to the energy absorbed internally. For a given pattern of yield lines, an equation can be written which shows the relationship of the yield moment to the geometry of the yield lines. The pattern which corresponds to failure may be found by seeking a maximum for m. In most cases, this procedure is sufficiently involved to be unattractive in practice and the energy balance is used to check the results of the equilibrium analysis and to make the final estimate of bending moments. It has the advantage that the yield moments calculated this way are not very sensitive to the geometry of the yield lines for a pattern near the critical one. Thus, when a pattern of yield lines has been found by equilibrium analysis so that the values of m for the several panels are reasonably close to each other, a final estimate can be made by means of an energy balance.

For example, if in Fig. 18.5 $l_2/l_1 = 0.6$, then, at the point P in Fig. 18.6,

$$\frac{x}{l_1} = 0.37$$

$$\frac{m}{wl_2^{\,2}} = 0.063 \qquad (18.3)$$

Calculations for an energy balance to check Equation (18.3) are made as follows.

The arbitrary deflection chosen is such that the deflection of the line cd increases by one unit. Work will be done as the weight supported on each of the panels, A, B, C and D, descends; the work done being

$$2\left\{\tfrac{1}{4}wxl_2 \times \tfrac{1}{3} + \tfrac{1}{2}w\,(l_1 - 2x)\,l_2 \times \tfrac{1}{2} + \tfrac{1}{4}wxl_2 \times \tfrac{1}{3} + \tfrac{1}{2}wxl_2 \times \tfrac{1}{3}\right\}$$

$$= wl_1\,l_2\left(\tfrac{1}{2} - \tfrac{1}{3}\frac{x}{l_1}\right)$$

The energy absorbed at the yield lines is given by $\Sigma m \underset{\rightarrow}{\Delta s} \,.\, \underset{\rightarrow}{\Delta \theta}$, where $\underset{\rightarrow}{\Delta s}$ is an element of length of yield line and $\underset{\rightarrow}{\Delta \theta}$ is the rotation of the panel caused by the increase in deflection. Thus, for panel A,

$$\Delta\theta = \frac{1}{\tfrac{1}{2}l_2} = \frac{2}{l_2}$$

and the rotation is about ab – i.e. the direction of the vector $\underset{\rightarrow}{\Delta \theta}$ is parallel to ab. The sum $\Sigma m\, \underset{\rightarrow}{\Delta s} \,.\, \underset{\rightarrow}{\Delta \theta}$ for panel A is

$$ml_1 \times \frac{2}{l_2} = 2\frac{l_1}{l_2}\,m$$

The total energy absorbed is thus

$$4\frac{l_1}{l_2}m + 2\,ml_2\frac{1}{x} = m\frac{l_2}{l_1}\left\{4\left(\frac{l_1}{l_2}\right)^2 + 2\frac{l_1}{x}\right\}$$

The energy balance then gives

$$m\frac{l_2}{l_1}\left\{4\left(\frac{l_1}{l_2}\right)^2 + 2\frac{l_1}{x}\right\} = wl_1\,l_2\left(\tfrac{1}{2} - \tfrac{1}{3}\frac{x}{l_1}\right)$$

whence

$$\frac{m}{wl_2^{\,2}} = \frac{\left(\frac{l_1}{l_2}\right)^2\left\{\frac{1}{2}\frac{x}{l_1} - \frac{1}{3}\left(\frac{x}{l_1}\right)^2\right\}}{4\left(\frac{l_1}{l_2}\right)^2\frac{x}{l_1} + 2}$$

It will be a laborious task, although not a difficult one, to differentiate this to get the maximum value of m. However, if an approximate value of 0.4 is taken for x/l_1 in our example, with $l_2/l_1 = 0.6$, then

$$\frac{m}{wl_2^{\,2}} = \frac{\frac{1}{0.36}\{\frac{1}{2} \times 0.4 - \frac{1}{3} \times 0.16\}}{4 \times \frac{1}{0.36} \times 0.4 + 2}$$

$$= 0.063$$

This compares favourably with the value obtained from the equilibrium analysis with $x/l_1 = 0.37$ and given in Equation (18.3). The following table confirms that $m/wl_2^{\,2}$ does not change rapidly near this critical value.

$\frac{x}{l_1}$	0.30	0.35	0.40	0.45
$\frac{m}{wl_2^{\,2}}$	0.0625	0.0634	0.0635	0.0625

The point we wish to make is that an equilibrium analysis can be made to locate the optimum yield lines with sufficient accuracy to lead to a good estimate of the yield moment by means of an energy balance. What we have presented above does not prove the point, but it does demonstrate the validity of our assertion in this special case. In fact, it is valid generally and provides a basis for straightforward design calculations.

18.4.6. Knot forces

Yield lines which meet edges where there is no resistance to twisting about the edge give rise to a special problem when the angle between the edge and the yield line near the edge is not a right angle. Such edges occur where the slab is unsupported or is supported by a beam with negligible torsional and bending stiffness. Examples of yield lines which approach such edges at angles other than 90° are shown in Figs 18.3(c) and (d).

In Fig. 18.7, a yield line is shown meeting an edge incapable of resisting torsion, the angle between the yield line and the edge being α. The problem is

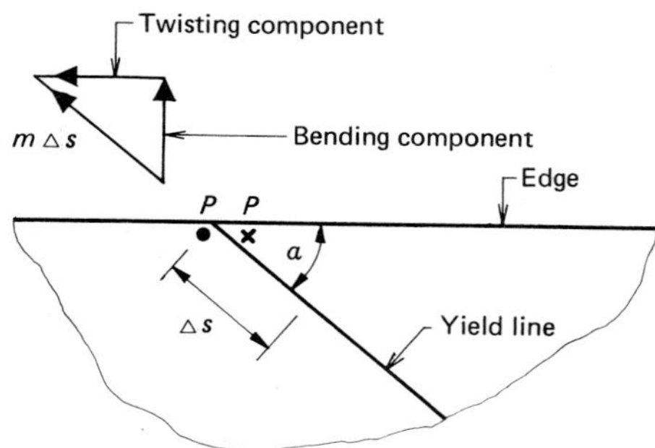

Fig. 18.7. *Equilibrium if yield line does not meet edge at right angles.*

that the bending moment in an element Δs of the yield line adjacent to the edge can be resolved into components normal to the edge and parallel to it. The latter acts to twist the slab about the edge, so that such an intersection cannot occur when the edge is unable to sustain torsion. Apparently the yield line must curve so as to give a right-angled intersection with the edge. Tests show that this does happen, the departure from the straight yield line taking place near the edge. To allow for this behaviour of the yield line adds unduly to the labour of computation and the following device is used. The yield line is assumed to be straight and to meet the edge at an angle α as shown in Fig. 18.7. The inaccuracy of alignment is confined to a short length of yield line near the edge. The twisting component which would act on the edge in such circumstances is counter-balanced by what are known as "knot forces" or "nodal forces". The names apparently arise from literal translations of Johansen's idiomatic Danish. The knot forces are a pair of equal and opposite forces which, although they do not exist, are assumed to act normal to the slab, one up and one down and are on each side of the yield line. If P is the magnitude of the knot forces,

$$P = m \cot \alpha$$

For a positive yield line – one where there is tension at the lower surface of the slab – the downward knot force is applied to the panel where the angle of intersection is acute and the upward knot force is applied to the other panel. These forces are indicated by the cross and the dot respectively in Fig. 18.7, the yield line shown being a positive one.

We do not offer any proof of the expression for P or of the validity of the concept as a means of cancelling the undesired twisting component. This is a part of yield line theory which is subject to challenge. Discussion will be found in references.[3, 4, 5, 6]

It is important to note that knot forces enter the equilibrium analysis, but they do not enter the energy analysis. Work is done by the downward-acting knot force when the intersection of the edge and the yield line descends, but an equal amount is done against the other one. The nett work done by the pair is zero.

18.4.7. Corner effects

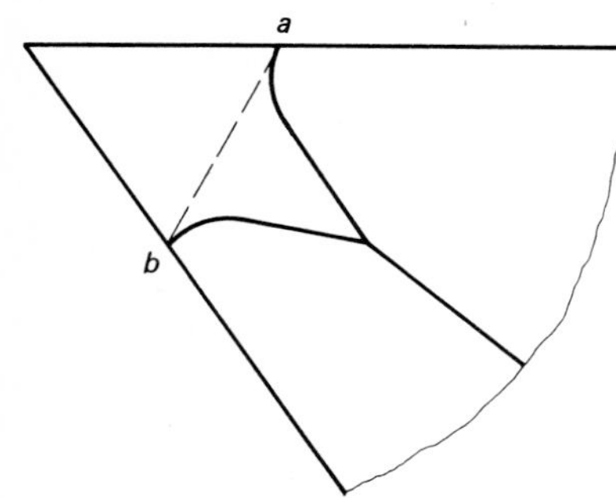

In Fig. 18.3, several instances of yield lines entering corners are shown. This is not always a valid configuration for yield lines at corners; for example, if the corner is not held down it may lift, with part of the slab rotating about a line such as *ab* in the adjacent marginal sketch. The yield line approaching the corner bifurcates to form a Y shape, with branches meeting the edges at *a* and *b*. This behaviour is more pronounced in acute angles than elsewhere and leads to a weakening of the slab – i.e. a higher yield moment is necessary for a given load.

If the corner of the slab is held down, a negative yield line may develop along *ab* and the yield strength of the slab in the corner is a matter of importance. The higher the yield strength of the slab, the more this corner effect is suppressed – that is to say, *a* and *b* get closer to the corner. Fortunately, in most corners, the yield strength required to drive *a* and *b* into the corner is not great and it is practicable to validate an array of yield lines more amenable to simple computation.

Reference to the literature will show how to allow for corner effects in design. Designers not requiring sophisticated designs for their slabs can circumvent corner effects by reinforcing the tops of the slabs in the corners with the same amount of steel as for positive bending in the centre of the same slab, this steel to be anchored in the supports to prevent the corners from lifting. Reinforcement in the top and bottom at corners is required by the ACI *Code* to extend from the corner not less than one-fifth of the longer span if the stiffness of the edge beams exceeds a specified value.[1]

18.4.8. Unequal reinforcement

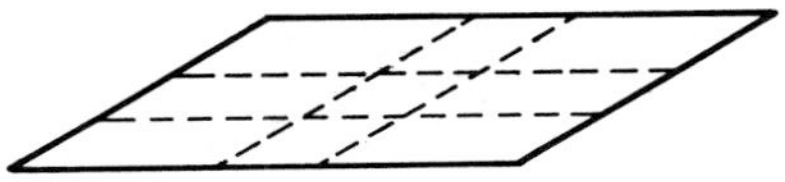

If a rectangular panel under working loads is regarded as a network of strips in two directions, some of the strips span the short way and some the long way. Continuity of the slab requires the deflections of any two strips, like those sketched in the margin, to be equal where they intersect. There are similar requirements in relation to twisting of one and bending of the other, but these do not bear on the point we wish to make here.

For equal deflections, the strip spanning the long way will be less sharply bent than the other and can therefore be reinforced with less steel. There is, then, an argument in having more steel laid across the slab than along it.

Such a slab cannot be treated as if it were isotropic and the yield moment per unit length of yield line in one depends on the orientation of the yield line. Fortunately an equivalent isotropic slab can be devised which permits the use of the analysis presented above. If the yield moment for bending in a vertical plane across the slab is m and that for bending along the slab is μm ($\mu < 1$), the equivalent isotropic slab has a yield moment equal to m and is obtained by multiplying the length of the slab by $1/\sqrt{\mu}$. The breadth of the slab and uniformly distributed loads are unchanged. Yield-line analysis of an isotropic slab with these dimensions enables m to be calculated.

Proof of this rule is given by various authors – for example, Hognestad.[4] Some care is needed in transforming line loads and point loads, and reference should be made to Hognestad or a similar source for these transformations.

18.4.9. Examples of slab design

Examples 18.1, 18.2 and 18.3 illustrate design of slabs by yield-line methods. The first is a straightforward problem and the basic computation is displayed. In Example 18.2, design of the same slab is undertaken, but with unequal amounts of reinforcement in the two directions. It is important that a significant reduction is achieved in the amount of steel required. In the third example, complications of irregular geometry and resistance to negative bending at one support are introduced. With regard to the former, scale drawings have been used to get dimensions such as sides of panels and lever arms for calculation of moments. The effect of resistance to bending at an edge is to replace the edge by a yield line when failure occurs. The bending moment acting on a panel at such an edge is negative – it causes tension near the top face – and it is designated m'. In this example, $m' = 0.5\,m$ along one edge. Note also that this bending moment in the slab acts to twist a beam supporting the edge.

18.5. DEFLECTION OF SLABS

Calculation of deflection of slabs can be avoided by satisfying rules about slab thickness. This is much the simpler process and is recommended for slabs of conventional loading and shape.

In the ACI *Code*, the following minima are laid down:[1]

Slabs without beams or drop panels	127 mm
Slabs without beams but with drop panels of specified dimensions	102 mm
Slabs having beams on all four edges with $\alpha_m \not< 2.0$	88 mm

(The ratio α_m is defined below.)

In our opinion, slabs thinner than about 120 mm pose sufficiently severe problems of construction that it is best to avoid them.

The same code states that the thickness of a slab need not be greater than that given by

$$\frac{h}{l_n} = \frac{800 + 0.725\,f_y}{36\,000}$$

where h is the slab thickness; l_n is the clear span of slab in the long direction; and f_y is the yield strength of steel reinforcement in MPa.

Clearly, an easy way to solve the problem of satisfying all the *Code*'s rules is to use this as a lower limit rather than an upper one. However, for large panels, the thickness so obtained is excessive. To avoid such excessive thicknesses, the lower limits in the code are used.

Lower limits for thickness of slabs given in the ACI *Code*[1] are

$$\frac{h}{l_n} \not< \frac{800 + 0.725\,f_y}{36\,000}\,\frac{1}{K_1}$$

$$\frac{h}{l_n} \not< \frac{800 + 0.725\,f_y}{36\,000}\,\frac{1}{K_2}$$

where

$$K_1 = 1 + \frac{5}{36}\beta\,(1 + \beta_s);$$

$$K_2 = 1 + \frac{5}{36}\beta\left\{\alpha_m - 0.5\,(1 - \beta_s)\left(1 + \frac{1}{\beta}\right)\right\};$$

α_m is the average value of ratio of flexural stiffness of beams at edges of panel to flexural stiffness of width of slab bounded by centreline of adjacent panel, if

Example 18.1 Sheet 1 of 3

Determine the reinforcement required for a slab 6m × 3m × 140 mm thick simply supported on all four edges.

Ultimate load, including self-weight = 4000 kg/m²

$f_c' = 20$ MPa $f_y = 275$ MPa $E_s = 207$ GPa

$E_s \epsilon_u = 621$ MPa

Balanced design: $k_b = \dfrac{1}{1+\dfrac{f_y}{E_s \epsilon_u}} = \dfrac{1}{1+\dfrac{275}{621}} = 0.692$

$$\rho_b = \frac{0.85 \times (0.85 \times 20)}{275} \times 0.692 = 0.0364$$

For $\rho = 0.75\rho_b = 0.0273$ $\quad \dfrac{a}{d} = \dfrac{0.0273 \times 275}{0.85 \times 20} = 0.441$

$$k = \phi(0.85 f_c')\frac{a}{d}\left(1 - \frac{1}{2}\frac{a}{d}\right) = 0.9 \times 17 \times 0.441 \times 0.781$$
$$= 5.28 \text{ MPa}$$

$w = 4000 \times 9.81$ Pa
$= 39.2$ kPa

Try yield lines as sketched.

For Panel A, take moments about ab:

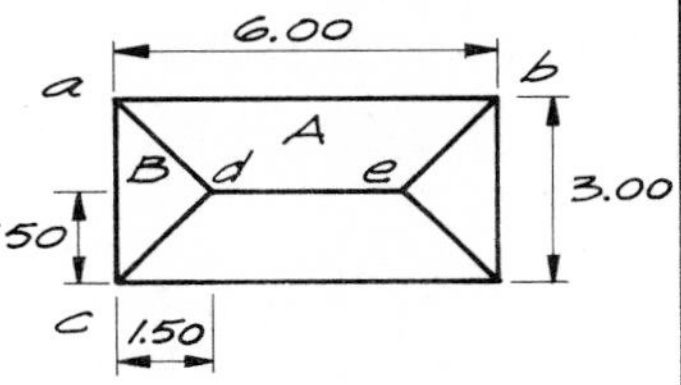

$$m \times 6 - 2 \times \frac{1.5 \times 1.5}{2} \times 39.2 \times 0.5$$
$$- 3 \times 1.5 \times 39.2 \times 0.75 = 0$$
$$m = \frac{1.5 \times 1.5 \times 3.92 \times 5}{6} + \frac{3 \times 1.5 \times 3.92 \times 7.5}{6}$$
$$= 7.36 + 22.0 = 29.36 \text{ kN m/m}$$

For Panel B, take moments about ac:

$$m \times 3 - 1/2 \times 3 \times 1.5 \times 39.2 \times 0.5 = 0$$
$$m = \frac{3 \times 1.5 \times 3.92 \times 5}{2 \times 3} = 14.7 \text{ kN m/m}$$

For unit deflection along de:

Work done by load $= (3 \times 3 \times 39.2 \times 0.5) + 8 \times \left(\dfrac{1.5 \times 1.5}{2} \times 39.2 \times \dfrac{1}{3}\right)$
$= 176 \times 118 \quad = 294$ kJ

Work absorbed at yield lines $= 2\left(6m \times \dfrac{1}{1.5} + 3m \times \dfrac{1}{1.5}\right)$
$= 12m$

Example 18.1 — Sheet 2 of 3

$12m = 294 \rightarrow m = 24.5$ kN m/m

Summary:

Panel A $\rightarrow m = 29.4$ kN m/m

Panel B $\rightarrow m = 14.7$ kN m/m

Energy balance $\rightarrow m = 24.5$ kN m/m

Adjust yield lines to make A smaller, B larger.

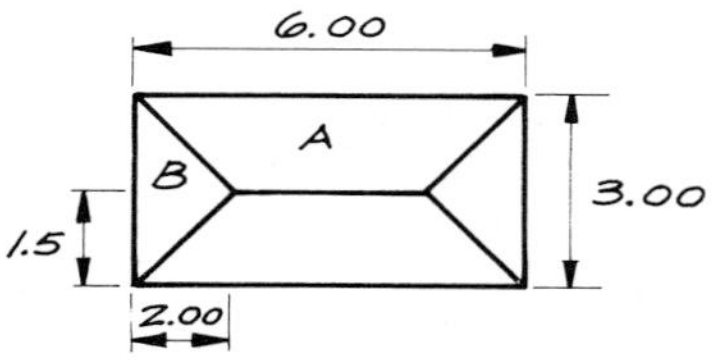

Panel A:

$$m \times 6 - 2 \times \frac{1.5 \times 2}{2} \times 39.2 \times 0.5 - 2 \times 1.5 \times 39.2 \times 0.75 = 0$$

$$m = \frac{1.5 \times 2 \times 3.92 \times 5}{6} + \frac{2 \times 1.5 \times 3.92 \times 7.5}{6}$$

$$= 9.80 + 14.7 = 24.5 \text{ kN m/m}$$

Panel B:

$$m \times 3 - 1/2 \times 3 \times 2 \times 39.2 \times 2/3 = 0$$

$$m = \frac{3 \times 3.92 \times 6.67}{3} = 26.2 \text{ kN m/m}$$

Energy balance:

$$2\left(6m \times \frac{1}{1.5} + 3m \times \frac{1}{2}\right) = (3 \times 2 \times 39.2 \times 0.5) + 8 \times \left(\frac{1.5 \times 2}{2} \times 39.2 \times \frac{1}{3}\right)$$

$$11m = (3 + 4) \times 39.2$$

$$m = \frac{7 \times 39.2}{11} = 25.0 \text{ kN m/m}$$

Design reinforcing for ultimate moment $= 25.0$ kN m/m

$$K = \frac{M_u}{bd^2} = \frac{2.50 \times 10^4}{(0.100)^2} \text{ Pa} = 2.50 \text{ MPa} < 5.28 \text{ MPa}$$

Under-reinforced

$$F = \frac{M_u}{\phi 0.85 f_c' bd^2} = \frac{2.50 \times 10^4}{0.9 \times 17 \times 10^6 \times 10^{-2}} = 0.163$$

$$\frac{a}{d} = 1 - \sqrt{1 - 2F} = 1 - \sqrt{1 - 0.326} = 0.177$$

$$1 - \frac{1}{2}\frac{a}{d} = 0.912 \quad \text{(Check)} \; \frac{a}{d}\left(1 - \frac{1}{2}\frac{a}{d}\right) = 0.162 = F \; \text{ok}$$

$$T = \frac{2.50 \times 10^4}{0.912 \times 0.100} \text{ N/m} = 274 \text{ kN/m}$$

Example 18.1 Sheet 3 of 3

$$A_s = \frac{274 \times 10^3}{275 \times 10^6} \ m^2/m = 996 \ mm^2/m$$

Try 12 mm diameter bars → 113 mm^2/bar
12 mm bars 100 mm c/c → 1130 mm^2/m

Check effective depth:

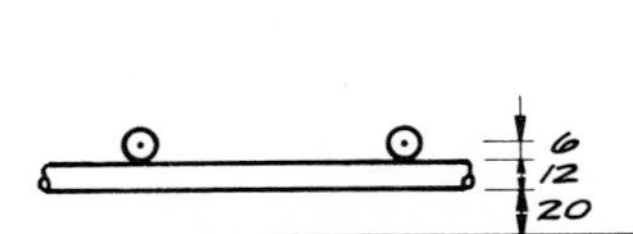

Allow 20 mm cover
12 mm bar diameter
6 mm bar radius
38 mm

Effective depth = 140 − 38 = 102 mm OK

SUMMARY: Reinforce bottom of slab with 12 mm diameter bars 100 mm c/c each way

Equal reinforcement top face in corners

Cover 20 mm.

Example 18.2 **Sheet 1 of 2**

Determine the reinforcement required for the slab of Example 18.1, but with lateral bending strength 2.5 times the longitudinal strength.

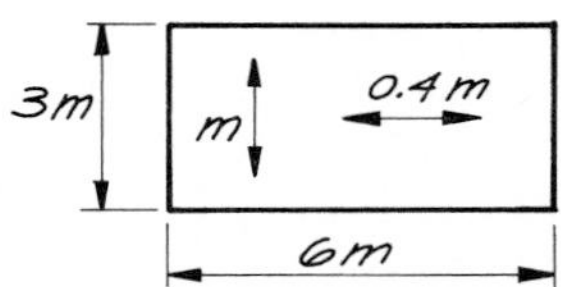

3m, m, m

$\frac{6}{\sqrt{0.4}} = 9.48\ m$

Equivalent slab

First trial:

Panel A—

moments about ab:

$$w \times \frac{1}{2} \times 3 \times 1.5 \times 0.5 - 3m = 0$$

$$m = \frac{1.5\,w}{4} = 0.375\,w$$

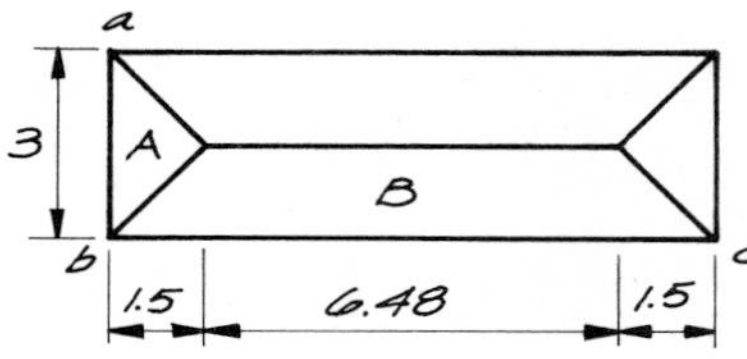

Panel B—

moments about bc:

$$2 \times w \times \frac{1}{2} \times 1.5 \times 1.5 \times 0.5 + w \times 6.48 \times 1.5 \times 0.75 - 9.48m = 0$$

$$1.125\,w + 7.30\,w - 9.48w = 0$$

$$m = \frac{8.425\,w}{9.48} = 0.888\,w$$

Second trial:

Panel A—

moments about ab:

$$w \times \frac{1}{2} \times 3 \times 2.2 \times \frac{2.2}{3} - 3m = 0$$

$$m = \frac{2.2 \times 2.2\,w}{6} = 0.805\,w$$

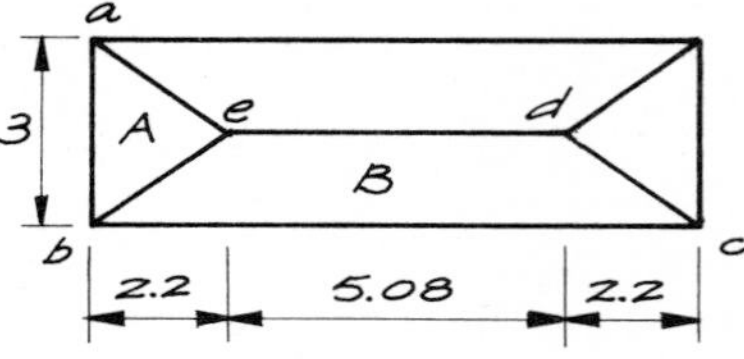

Panel B—

moments about bc:

$$2 \times w \times \frac{1}{2} \times 2.2 \times 1.5 \times 0.5 + w \times 5.08 \times 1.5 \times 0.75 - 9.48m = 0$$

$$1.65\,w + 5.71\,w - 9.48m = 0$$

$$m = \frac{7.36\,w}{9.48} = 0.777\,w$$

Check energy conservation—unit deflection along de

$$2 \times w \times \frac{1}{2} \times 3 \times 2.2 \times \frac{1}{3} + 4 \times w \times \frac{1}{2} \times 2.2 \times 1.5 \times \frac{1}{3} + 2 \times w \times 5.08 \times 1.5 \times \frac{1}{2}$$

$$= 2 \times 9.48m \times \frac{1}{1.5} + 2 \times 3m \times \frac{1}{2.2}$$

Example 18.2 Sheet 2 of 2

$$2.2w + 2.2w + 7.62w = 12.6m + 2.73m$$

$$m = \frac{12.02w}{15.33} = 0.785w$$

Design for $m = 0.785w = 30.8$ kN m/m

$0.4m = 0.314w = 12.3$ kN m/m

Longitudinal reinforcement:

$$F = \frac{1.23 \times 10^4}{0.9 \times 17 \times 10^6 \times 10^{-2}} = 0.0805 \qquad \frac{a}{d} = 1 - \sqrt{1 - 0.161} = 0.084$$

$$\left(1 - \frac{1}{2}\frac{a}{d}\right) = 0.958 \qquad T = \frac{1.23 \times 10^4}{0.958 \times 0.1} = 128.5 \text{ kN/m}$$

$$A_s = \frac{1.285 \times 10^5}{275} = 467 \text{ mm}^2/\text{m}$$

12 mm bars 240 mm c/c → 471 mm²/m

Lateral reinforcement:

$$F = \frac{3.08 \times 10^4}{0.9 \times 17 \times 10^6 \times 10^{-2}} = 0.202 \qquad \frac{a}{d} = 1 - \sqrt{1 - 0.404} = 0.227$$

$$\left(1 - \frac{1}{2}\frac{a}{d}\right) = 0.887 \qquad T = \frac{3.08 \times 10^4}{0.887 \times 0.1} = 347 \text{ kN/m}$$

$$A_s = \frac{3.47 \times 10^5}{275} = 1260 \text{ mm}^2/\text{m}$$

12 mm bars 90 mm c/c → 1260 mm²/m

(The total amount of reinforcement required for this slab should be compared with the total for the slab in Example 18.1)

Example 18.3 Sheet 1 of 4

Determine the reinforcement required for a slab with dimensions and conditions of support as sketched. Slab thickness 140 mm. Load and design data as for Example 18.1

4m
2m
6m
Unsupported
Supported $m' = 0.5m$
Supported $m' = 0$

Dimensions of panels by measurement from scale drawings:

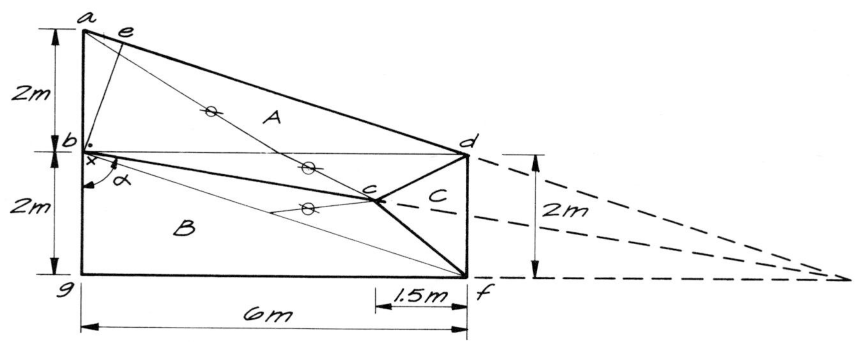

First trial – yield lines as above:

Panel A: moments about ad:

abd: Area $= 1/2 \times 6 \times 2 = 6\,m^2$ Lever arm $= 0.65\,m$

bcd: Area $= 1/2 \times 6 \times 0.77 = 2.31\,m^2$ Lever arm $= 1.07\,m$

de = component of $(\overrightarrow{dc} + \overrightarrow{cb}) = 5.68\,m$

$$\cot \alpha = \frac{2}{12} = \frac{1}{6}$$

$$w \times 6 \times 0.65 + w \times 2.31 \times 1.07 - m/6 \times 1.9 - 5.68m = 0$$

$$m = \frac{6.36w}{6} = 1.06w$$

Panel B: moments about fg:

bcf: Area $= 1/2 \times 6.3 \times 0.7 = 2.20\,m^2$ Lever arm $= 1.06\,m$

bfg: Area $= 1/2 \times 6 \times 2 = 6\,m^2$ Lever arm $= 0.67\,m$

fg = component of $(\overrightarrow{fc} + \overrightarrow{cb}) = 6\,m$

Example 18.3 Sheet 2 of 4

$w \times 2.3 \times 1.06 + w \times 6 \times 0.67 + m/6 \times 2 - 6m = 0$

$$m = \frac{6.44\,w}{5.67} = 1.13\,w$$

Panel C: moments about fd:

• Area = $1/2 \times 2 \times 1.5 = 1.5\ m^2$

Lever arm $= 0.5\ m$

fd = component of $(\overrightarrow{fc} + \overrightarrow{cf})$ $= 2\ m$

$w \times 1.5 \times 0.5 - m \times 2 - 0.5\,m \times 2 = 0$

$$m = \frac{0.75\,w}{3} = 0.25\,w.$$

Second trial:

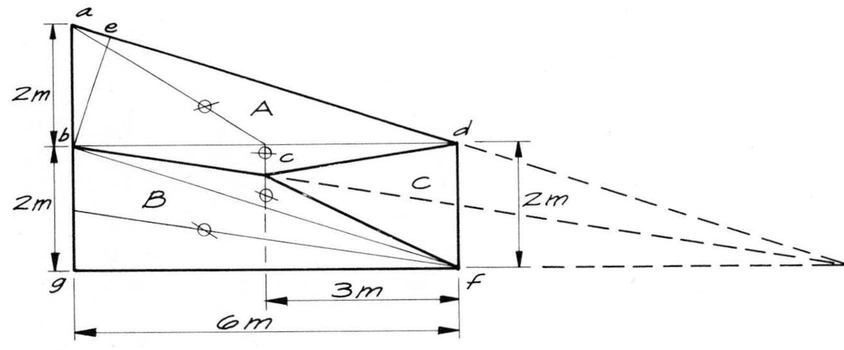

Panel A: moments about ad:

abd: Area = $1/2 \times 6 \times 2 = 6\ m^2$

Lever arm $= 0.65\ m$

bcd: Area = $1/2 \times 6 \times 0.5 = 1.5\ m^2$

Lever arm $= 1.10\ m$

$w \times 6 \times 0.65 + w \times 1.5 \times 1.10 - m/6 \times 1.9 - 5.68m = 0$

$$m = \frac{5.55\,w}{6.00} = 0.92\,w$$

Panel B:

bcf: Area = $1/2 \times 6.3 \times 0.46 = 1.45\ m^2$

Lever arm $= 1.13\ m$

bfg: Area = $1/2 \times 6 \times 2 = 6\ m^2$

Lever arm $= 0.67\ m$

$w \times 1.45 \times 1.13 + w \times 6 \times 0.67 + m/6 \times 2 - 6m = 0$

$$m = \frac{5.64\,w}{5.67} = 0.995\,w$$

Example 18.3 Sheet 3 of 4

Panel C:

Area $= 1/2 \times 2 \times 3 = 3 m^2$

Lever arm $= 1 m$

$$w \times 3 \times 1 - m \times 2 - 0.5m \times 2 = 0$$

$$m = \frac{3w}{3} = w$$

Check energy conservation — unit deflection at d

Panel A: $w \times 6 \times \frac{0.65}{1.9} + w \times 1.5 \times \frac{1.10}{1.9} = 2.92\,w$

Panel B: $w \times 1.45 \times \frac{1.13}{2} + w \times 6 \times \frac{0.67}{2} = 2.82\,w$

Panel C: $w \times 3 \times \left(\frac{1}{3} \times \frac{9}{12}\right) = 0.75\,w$

$6.49\,w$

bcd: $5.68\,m \times 1/1.9 = 2.99\,m$

bcf: $6\,m \times 1/2 = 3\,m$

fcd: $2\,m \times \left(\frac{9}{12} \times \frac{1}{3}\right) = 0.5\,m$

fd: $2 \times 0.5\,m \times \left(\frac{9}{12} \times \frac{1}{3}\right) = 0.25\,m$

$6.74\,m$

$$m = \frac{6.49\,w}{6.74} = 0.963\,w$$

Design moments: $m = 0.96\,w = 37.6$ kN m/m

$m' = 0.48\,w = 18.8$ kN m/m

Bottom steel: $k = \frac{3.76 \times 10^4}{10^{-2}} = 3.76$ MPa < 5.28 MPa

Under-reinforced

$$F = \frac{3.76 \times 10^4}{0.9 \times 17 \times 10^6 \times 10^{-2}} = 0.246$$

$$\frac{a}{d} = 1 - \sqrt{1 - 0.492} = 0.286$$

$1 - \frac{1}{2}\frac{a}{d} = 0.857$ $\quad T = \frac{3.76 \times 10^4}{0.857 \times 0.1} = 439$ kN/m

$$A_s = \frac{4.39 \times 10^5}{275} = 1600 \text{ mm}^2/\text{m}$$

12 mm bars, 70 mm c/c → 1620 mm²/m
(Both ways)

Example 18.3 Sheet 4 of 4

Top steel:

$$F = \frac{1.88 \times 10^4}{0.9 \times 17 \times 10^6 \times 10^{-2}} = 0.123$$

$$\frac{a}{d} = 1 - \sqrt{1 - 0.246} = 0.13 \qquad 1 - \frac{1}{2}\frac{a}{d} = 0.935$$

$$T = \frac{1.88 \times 10^4}{0.935 \times 0.1} = 201\ \text{kN/m}$$

$$A_s = \frac{2.01 \times 10^5}{275} = 730\ \text{mm}^2/\text{m}$$

12 mm bars, 150 mm c/c → 754 mm^2/m
(Both ways)

any, on each side of each beam; β is the ratio of clear spans in long to short direction; and β_s is the ratio of length of continuous edges (i.e. edges where there is a contiguous panel) to total perimeter.

Both of these inequalities must be satisfied. It will be seen that, in each, the right-hand side is obtained by dividing the recommended (but not mandatory) upper limit by a factor greater than one. To give some guidance about the magnitude of these factors, they are plotted in Fig. 18.8. The rules are shown to conform with one's intuitive expectations – more flexible slabs are permitted with

(a) stiffer supports at the edges (increased α_m);
(b) increased continuity at the edges (larger β_s); and
(c) panels with lower aspect ratio ($\beta \to 1$).

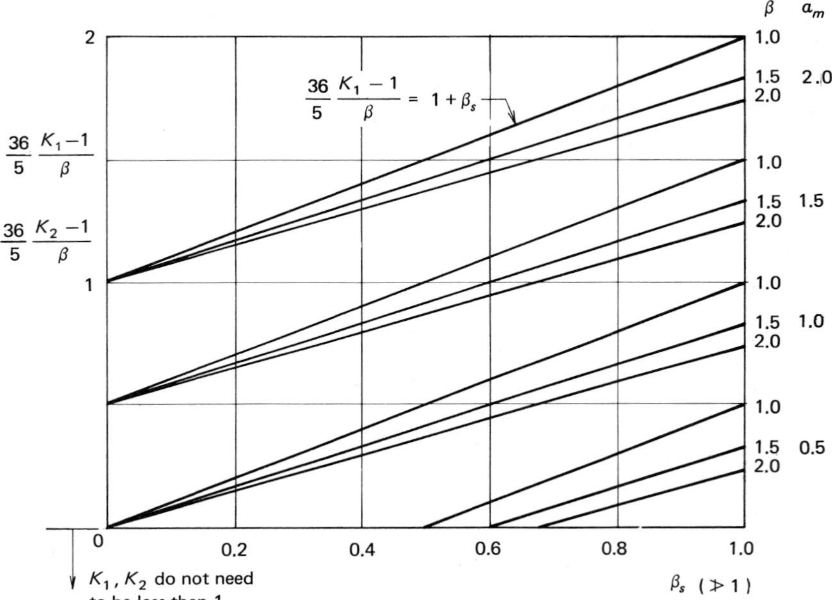

Fig. 18.8. *Coefficients to be used in rules for slab thickness. The lesser of K_1 and K_2 rules. All lines give values of K_2. Top line only gives values of K_1.*

18.6. TRANSFER OF LOAD TO COLUMNS

When the floor being designed is a beam-and-slab construction, slab panels are supported at their edges by the beams and the beams are supported by the columns. No direct transfer of load from the slab to the columns need be considered and the beam is designed to have sufficient strength in shear, as described in Chapter 16, "Reinforced Concrete Beams". The area of slab contributing load to a beam is assumed to be as shown in the margin.

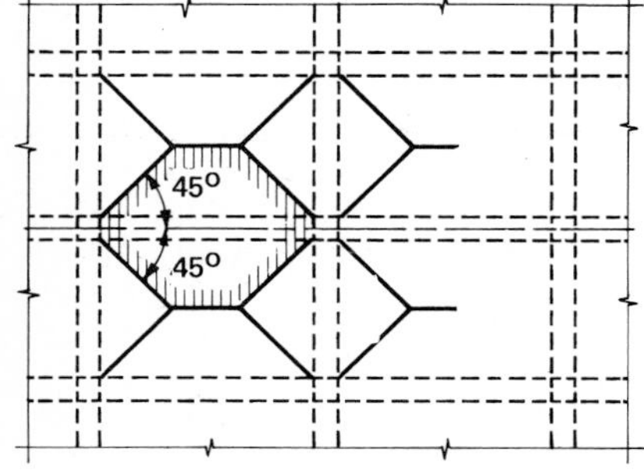

In flat-slab construction the rules for analysing shear in the slab around the head of a column are the same as for footings. These are described in Chapter 19, "Columns, Column Bases and Footings". Since the nominal shear stress in the assumed section of failure around the column is likely to be high, increased strength is required. One course of action is to build the slab with drop panels. The thickness required is determined in the same way as the thickness of a footing is found. Otherwise, the slab can be reinforced for shear and this may be done by means of vertical or inclined shear reinforcing bars. It may be necessary to test the strength of the slab at several distances from the column – not just the one at distance $\frac{1}{2}d$ from the face of the column where conditions are most severe. This is analogous to calculation of different stirrup spacings at different positions along a beam.

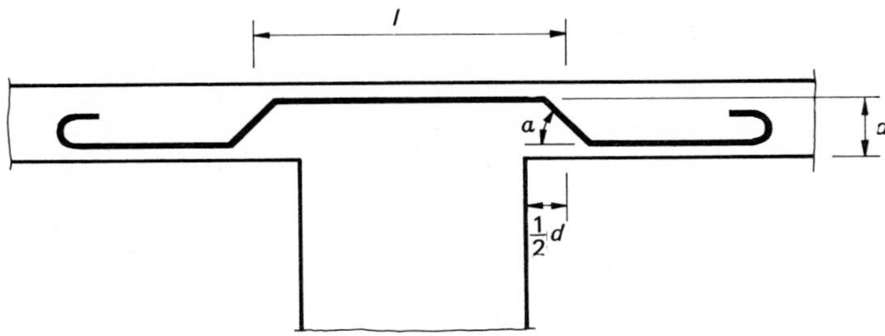

Fig. 18.9. *Reinforcement for flat slab over a column.*

The area of steel required for bars like those sketched in Fig. 18.9 may be calculated quite simply as follows. If V_u is the critical shear, the stress v_u is given by

$$v_u = \frac{V_u}{\phi\,(4l)\,d}$$

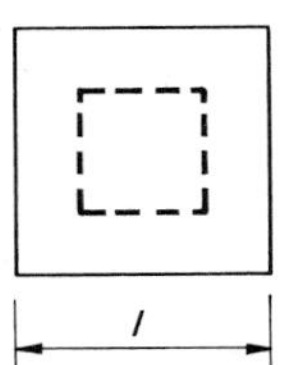

The rules in the code allow the assumption that a stress v_c is sustained by the concrete, leaving an amount $v_u - v_c$ for which steel must be provided. The external shear corresponding to this excess stress is $(v_u - v_c)(4l)\,d$ and, if T is the total tension in the bars (on all four sides of the column),

$$T \sin \alpha = (v_u - v_c)(4l)\,d$$

But

$$T = A_w f_y$$

where A_w is the area of steel crossing the critical section around the column. Thus

$$A_w = \frac{(v_u - v_c)(4l)\,d}{f_y \sin \alpha}$$

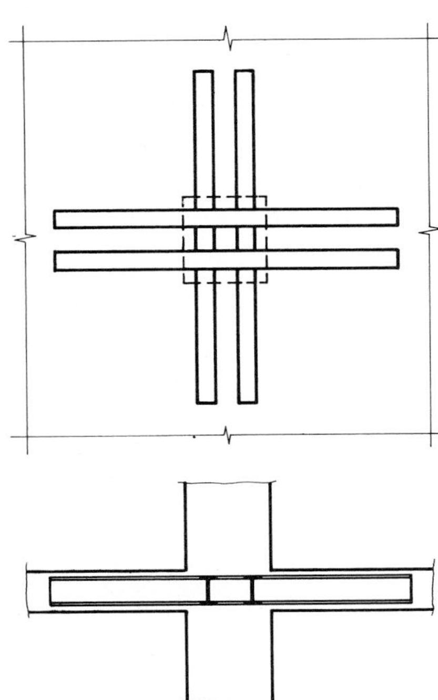

Reinforcement is required if v_u exceeds $\frac{1}{3}\sqrt{f_c'}$ and, in any case, v_u must not be greater than $\frac{1}{2}\sqrt{f_c'}$ where f_c' is the compressive strength of the concrete in MPa. The value allowed for v_c, when reinforcement is used, is $\frac{1}{6}\sqrt{f_c'}$. The stresses v_u and v_c are also measured in MPa.

Great care is necessary in detailing to ensure that shear reinforcing is adequately anchored in accordance with the rules given in Chapter 16, "Reinforced Concrete Beams". A high standard of construction is also required to ensure accurate placing of the reinforcing steel.

An alternative means of increasing the shear strength of the structure, without using a drop, is to incorporate a shearhead of steel I-sections or channel sections in the slab over the column as sketched in the margin. Rules for the design of a shearhead are given in the ACI *Code.*[1]

Example 18.4 is included to illustrate the calculations required for shear reinforcing with inclined bars.

Example 18.4 Sheet 1 of 2

Determine the dimensions of the drop or the shear reinforcement required at the head of a column supporting a flat slab.

Design data: $f_c' = 20\ MPa$ $f_y = 275\ MPa$

Allowable shear stress without reinforcement = 1.5 MPa

Allowable shear stress with reinforcement = 2.25 MPa

Sustained by concrete with reinforcement = 0.75 MPa

Ultimate load on slab (including self-weight) = 1200 kg/m^2

Columns 5 m c/c both ways

Columns 300 mm x 300 mm

Slab 125 mm thick. Effective depth 100 mm

Column load $= (5 \times 5) \times (1200 \times 9.81)$
$= 25 \times 11800\ N$
$= 295\ kN$

Effective shear at $\frac{d}{2}$ from face of column
$= 295 - 0.4 \times 0.4 \times 11.8$
$= 293\ kN$

$$v_u = \frac{2.93 \times 10^5}{0.85 \times (4 \times 0.4) \times 0.1}\ Pa$$

$= 2.15\ MPa$

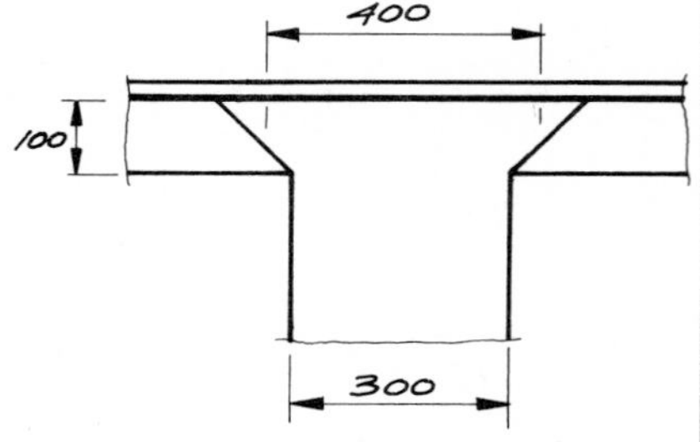

Try drop panel 50 mm below slab soffit:

$V_u = 295 - 0.45 \times 0.45 \times 11.8$
$= 293\ kN$

$$v_u = \frac{2.93 \times 10^5}{0.85 \times 4 \times 0.45 \times 0.15}\ Pa$$

$= 1.28\ MPa$ ok

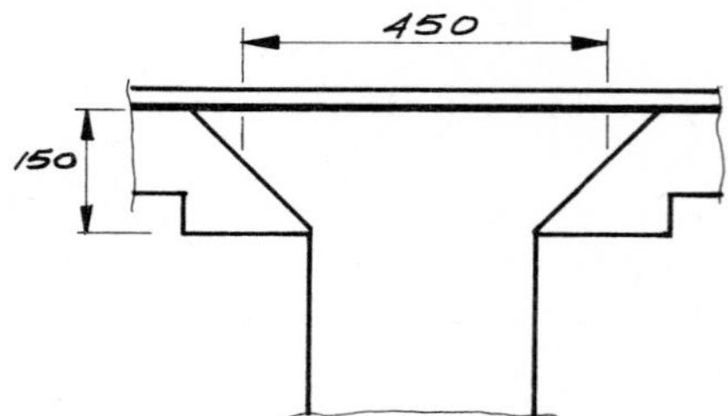

Example 18.4 Sheet 2 of 2

Try drop extending 300 mm from column

$V_u = 295 - 1 \times 1 \times 11.8$
$= 283$ kN

$v_u = \dfrac{2.83 \times 10^5}{0.85 \times 4 \times 1.00 \times 0.1}$ Pa

$= 0.83$ MPa OK

Alternatively, with reinforcing:

$v_u = 2.15$ MPa $v_c = 0.75$ MPa

$v_u - v_c = 1.40$ MPa $\alpha = 45°$

$A_v = \dfrac{1.40 \times 10^6 \times (4 \times 0.4) \times 0.1}{275 \times 10^6 \times 0.707}$ m^2

$= 1150$ mm^2

Say, 12 bars — three each way thus:

For each bar, 96 mm^2

Say, 12 mm diameter bars → total area 1360 mm^2

18.7. MINIMUM REINFORCEMENT

No slab should be built without some reinforcement in both directions. Slabs which are relatively long compared with their width may call for only small amounts of steel in the long direction; in particular, one-way slabs require none for strength. However, unsightly cracking is possible because of shrinkage of the concrete as it loses moisture and lack of longitudinal reinforcement will impair the ability of the slab to spread unexpected concentrated loads.

The ACI *Code* puts limits on the spacing of the bars at critical sections – not greater than twice the slab thickness – and on the minimum amount and the spacing of nominal reinforcement.[1] For the latter, the maximum spacing allowed is five times the thickness of the slab and not more than 457 mm and the minimum amount allowed depends on the yield strength of the steel. If the area of steel is not less than 0.2% of the gross area of concrete in a cross-section, the ACI rules will be obeyed.

18.8. CONCLUSION

Design of reinforced concrete slabs is no easy task. We have presented here the bare essentials which will enable a beginner to design serviceable slabs. Designers who require more than just serviceability – for example, greater economy in the use of steel – will have to study the topic in greater depth. The subject is a difficult one but, fortunately, the rules referred to in Sections 18.2 and 18.3 are adequate for economy and serviceability when the column grid is regular. If economy and serviceability are both sought for irregular slabs, more refined methods must be used. For a range of regular shapes, solutions based on elastic theory are available in references, such as Roark.[7]

LIST OF SYMBOLS

A_w total area of inclined shear reinforcement in slab at column
d effective depth of slab
f'_c compressive strength of concrete
f_y yield strength of reinforcing steel
h thickness of slab
K_1, K_2 coefficients
l length of side of critical section for shear in slab at junction with column
l_1, l_2 slab spans
l_n length of clear span in long direction
m moment per unit length ($= m_b$ if $m_t = 0$)
m_b bending moment per unit length of line of specified orientation
m_t twisting moment per unit length of line of specified orientation
m' negative moment per unit length, to distinguish negative moment when m is used for positive moment
P knot force
T total tension in inclined reinforcement in slab at column
V_u ultimate external shear in slab at critical section surrounding column

v_c shear stress sustained by concrete
v_u ultimate nominal shear stress
w total load per unit area on slab at collapse
x dimension in yield-line pattern
α angle between yield line and edge, inclination of inclined shear reinforcement
α_m average value of flexural stiffness of beams at edges of panel to flexural stiffness of width of slab bounded by centreline of adjacent panel, if any, on each side of each beam
β ratio of clear spans in long-to-short direction
β_s ratio of length of continuous edges to total perimeter of a panel
Δs element of length of a line
$\Delta\theta$ rotation of part of a slab
μ ratio of longitudinal yield moment to transverse yield moment
ϕ strength reduction factor

REFERENCES

[1] ACI 318–71, *Building Code Requirements for Reinforced Concrete,* American Concrete Institute, Detroit, 1971.

[2] ACI 318–71, *Commentary on Building Code Requirements for Reinforced Concrete,* American Concrete Institute, Detroit, 1971.

[3] K. W. Johansen, *Yield-line Theory*, transl. from Danish, Cement and Concrete Association, London, 1961.

[4] E. Hognestad, "Yield-line Theory for Ultimate Flexural Strength of Reinforced Concrete Slabs", *Journal of American Concrete Institute,* **49**, March 1953, 637–56.

[5] R. H. Wood, *Plastic and Elastic Design of Slabs and Plates,* Thames & Hudson, London, 1961.

[6] CCA, *Recent Developments in Yield-line Theory*, Cement and Concrete Association, London, 1965.

[A special publication containing five parts on some aspects of the theory as seen in the light of more recent research.]

[7] R. J. Roark, *Formulas for Stress and Strain,* 4th ed., McGraw-Hill, New York, 1965.

BIBLIOGRAPHY

Jones, L. L., *Ultimate Load Analysis of Reinforced Concrete Structures,* Interscience Publishers, London, 1962.

Thomas, F. E., "Load Factor Methods of Designing Reinforced Concrete", *Reinforced Concrete Review,* 540–44, **8**, 3, 1955.

PROBLEMS

18.1. Sketch the yield lines for the following slabs:

(a) the slab of Example 18.1, assuming it to be supported by columns at its corners;

(b) the slab of Example 18.3, assuming all four edges to be supported;

(c) the slab of Example 18.3, assuming it to be supported by columns at its corners.

18.2. Estimate the cost of the slabs in Examples 18.1 and 18.2.

18.3. Redesign the slab of Example 18.2, with lateral bending strength being three times the longitudinal strength and estimate the cost of this design.

18.4. Determine suitable dimensions for a drop at the head of one of the columns in Problem 18.1(a). Assume the columns are 300 mm square.

18.5. Determine the reinforcement required for the slab of Problem 18.1(a) and compare it with the reinforcement computed in Example 18.1.

19 Columns, Column Bases and Footings

19.1. INTRODUCTION

In Chapter 5, "Machines and Structures", the function performed by a column has been described. To the extent that the load carried is a compressive force parallel to the length of the member, a column resembles a strut (Chapter 12, "Ties and Struts") and the distinction between a strut and a column is somewhat arbitrary. Generally, a column is itself a complete element of the structure and supports a beam or a truss, whereas a strut is a component of some other subassembly of the structure – for example, a truss. Again in general terms, the loads carried by columns are much bigger than loads in struts and bending moments in columns may be relatively much larger than in struts. Finally, reinforced concrete presents special problems of design when used to carry compressive loads, especially when they are combined with bending moments.

Footings and bases for columns are included in this chapter. These structural elements are needed because the strength of the material supporting a steel or reinforced concrete column is usually much less than that of the column. Steel columns are often supported on concrete foundations. Since the stress in the steel may be of the order of 150 MPa, the column load must be spread over a larger area to match the bearing stress permitted on the concrete foundation – of the order of 10 MPa. This load spreading is achieved by providing a base at the bottom of the column. Similarly, the soil on which the whole structure rests may be clay with a safe bearing pressure of the order of 500 kPa and a reinforced concrete footing is necessary to spread the load over a sufficiently large area.

Bending moments frequently occur in columns. In monolithic building frames, rotation of the ends of the beams requires equal rotation of the columns and bending stresses occur in the columns even though no external load acts transversely on the column. Lateral loads on buildings, due to wind or earthquake also cause bending in columns.

When dealing with struts, we discussed the effects of eccentrically applied loads and we now show the relationship between combined axial load and bending moment on the one hand and an eccentric load on the other. Consider the effect of a longitudinal load P acting at a distance e from the axis of the column (Fig. 19.1(a)). Equilibrium will not be affected by superimposing two vertical forces P_1 and P_2, each equal and parallel to P and acting up and down on the axis, as shown in Fig. 19.1(b). Clearly P_2 and P make a couple being parallel, equal in magnitude and opposite in sense, so that this set of forces is equivalent to the axial force P_1 (also equal to P) and a moment M equal to Pe.

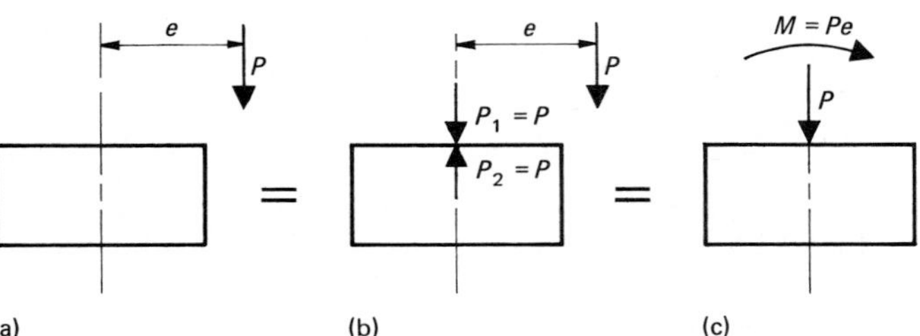

Fig. 19.1. *An axial load combined with a bending moment is equivalent to an eccentric load.*

Thus, a column supporting an eccentric load can be analysed as such or in terms of an axial load and a bending moment, whichever is more convenient.

19.2. STEEL COLUMNS

Hot-rolled sections are suitable for use as steel columns, I-sections being the most widely used. These may be the ones rolled specially for use as columns – for example, the universal column sections described in BS 4[1] – or they may be sections primarily intended for use as beams. Wide flanges are desirable to make the radius of gyration about an axis in the web as large as possible. This has the effect of reducing slenderness ratios and so making the column more resistant to buckling.

In view of the large loads involved, lighter sections such as angles are rarely suitable and only channels would normally be considered as alternatives to I-sections. However, its low radius of gyration about a centroidal axis parallel to the web makes the channel an inferior alternative.

If necessary, a rolled I-section can be strengthened by welding or riveting plates to its flanges. The required geometrical properties of the plated section are readily computed from the tabulated properties of the rolled section and those of the rectangular sections of the plates.

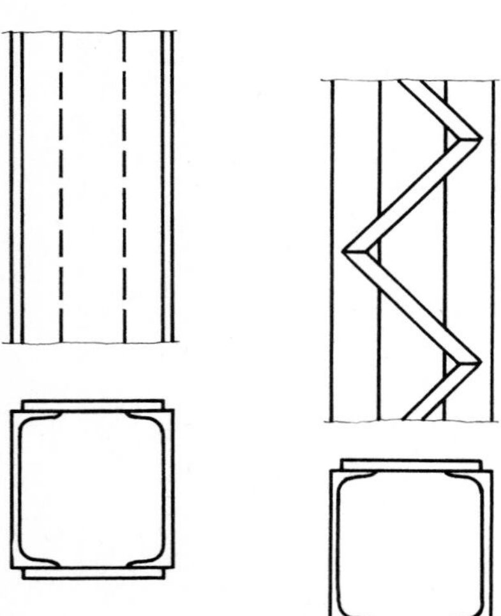

Built-up sections can also be constructed with other combinations of plate and rolled sections. For example, a strong box-section can be made by plating the flanges of two channels or by joining the flanges of two channels with battens. In the latter case, a buckling failure may involve the column as a whole or one of its components. Design codes may specify rules for the design of the battens or they may require both aspects of buckling to be investigated.

19.2.1. Design of axially loaded steel columns

The essential features of the design of axially loaded steel columns are the same as those of design of axially loaded struts (see Chapter 12, "Ties and Struts"). As pointed out above, the distinction between struts and columns is fairly arbitrary.

Design of short columns is based on failure by crushing and design of long columns takes buckling into account. In British practice, using the rules in BS 449[2], working loads and permitted stresses are used, the permitted stress being given as a function of slenderness ratio – the same function as used in the design of struts.

Design is by trial and error. A suitable section is selected, which enables the slenderness ratio to be calculated. Hence the permitted stress and allowable load can be calculated and, provided the allowable load is greater than the load in service, the choice is acceptable. Clearly, the designer wants his choice to have an allowable load only slightly greater than the load in service in the interests of economy. A typical calculation is shown in Example 19.1.

In practice, a major difficulty is to decide what the effective length is. As noted in Chapter 12, "Ties and Struts", this depends on the restraint applied to rotation and displacement of the ends of the column. Codes contain descriptions of some conditions of end fixity, but all too often the designer is faced by columns which do not fit these classifications neatly. The best that can be done is to sketch the shape the axis of the column would have if buckling were to occur. It is then usually possible to say that the curve is intermediate between two of the shapes described and conservative practice would be to use the longer of these two effective lengths.

Buckling about either principal axis should be examined, since the column may be restrained from buckling in different ways with respect to the two axes.

19.2.2. Design of eccentrically loaded steel columns

There is an additional complication in design of an eccentrically loaded column – i.e. one subject to combined axial load and bending. This complication is that permitted stresses for axial load and bending are not always equal. As described in Chapter 15, "Steel Beams", the permitted stress in bending is based on buckling of the compression flange and twisting of the beam. The permitted stress for axial loads depends on the slenderness of the whole column. Generally, it is not sufficient to compute the maximum combined stress – there is nothing to compare it with.

When the secant formula is used (Chapter 12, "Ties and Struts") some of these comments do not apply. The secant formula is based on the maximum combined stress and this is compared with the permitted stress for an axially loaded column. Implicit in this is that buckling and twisting of a beam due to bending compressive stresses depend on slenderness in the same way as buckling of a strut does.

In designing an eccentrically loaded column, a procedure of trial and error has to be used. With the properties of a trial section available, four stresses can be calculated – the compressive stresses due to the axial load and the bending moment, f_c and f_{bc} respectively, and the permitted stresses for axial compression and bending p_c and p_{bc} respectively. We have

$$f_c = \frac{P}{A}$$

and

$$f_{bc} = \frac{M}{Z} = \frac{Pe}{Z}$$

where P is the axial load; M is the bending moment; $e = M/P$ = eccentricity of equivalent eccentric load; A is the gross area of trial cross-section; and Z is the section modulus of cross-section. The permitted stresses are those allowed by the design specification. In BS 449[2], p_c is a function of the slenderness of the column and p_{bc} depends on the slenderness of the compression flange as well.

Example 19.1 Sheet 1 of 1

Design an axially loaded steel column.
Select a suitable British Standard universal column.

Design Data: Load = 600 kN
Effective length = 8 m
Design to BS 449 using Grade 43 steel.

Try 254 × 254 × 89 kg/m U.C.

$r_{yy} = 64.3$ mm $\quad \frac{\ell}{r_{yy}} = \frac{8000}{65.2} = 123$

$p_c = 57$ MPa

$A = 1.14 \times 10^4$ mm^2

$P_a = 57 \times 1.14 \times 10$ kN

$= 650$ kN

> 600 kN OK

254 × 254 × 89 kg/m U.C.

The trial section is then tested with an interaction formula – for example, the rule in BS 449 is

$$\frac{f_c}{p_c} + \frac{f_{bc}}{p_{bc}} \ngtr 1$$

or

$$\frac{P}{P_A} + \frac{M}{M_A} \ngtr 1$$

where P_A and M_A are the allowable axial load and bending moment in the absence of bending for the one and axial load for the other.[2] This rule is shown graphically in Fig. 19.2.

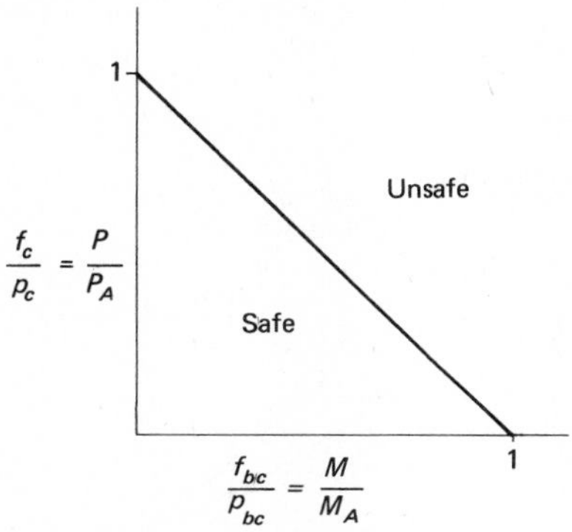

Fig. 19.2. *Linear interaction formula for eccentrically loaded column.*

The formula is consistent with the rules for axial load without bending ($f_{bc} = 0$) and for bending without axial load ($f_c = 0$). Between these extremes the formula is convenient and experience shows that reliable columns can be designed using it.

It must be realised, however, that there is nothing fundamental about this linear interaction formula – what it offers is convenience and a reasonable assessment of strength. Its weakness can be brought out by recasting the secant formula in the form of an interaction formula. We have

$$\sigma = \frac{P}{A} + \frac{Mc}{I} \sec \frac{l}{2}\sqrt{\frac{P}{EI}}$$

Since the Euler critical load is given by

$$P_c = \frac{\pi^2 EI}{l^2}$$

$$\sigma = \frac{P}{A} + \frac{Mc}{I} \sec \frac{\pi}{2}\sqrt{\frac{P}{P_c}}$$

or

$$\frac{P}{\sigma A} + \frac{Mc}{\sigma I} \sec \frac{\pi}{2}\sqrt{\frac{P}{P_c}} = 1$$

If the allowable axial load is taken to be given by

$$P_A = \sigma A$$

and the allowable bending moment by

$$M_A = \frac{\sigma I}{c}$$

this equation becomes

$$\frac{P}{P_A} + \frac{M}{M_A} \sec \frac{\pi}{2} \sqrt{\frac{P}{P_A} \frac{P_A}{P_c}} = 1 \tag{19.1}$$

This interaction equation gives the boundary between safe and unsafe designs on the P/P_A, M/M_A plane, with P_A/P_c as an additional parameter. Since $\sec \pi/2 \sqrt{(P/P_A)(P_A/P_c)}$ is greater than 1, it is clear that this boundary is lower than the linear interaction. To illustrate this, Equation (19.1) is plotted on Fig. 19.3 for $P_A/P_c = 1$ and 0.2.

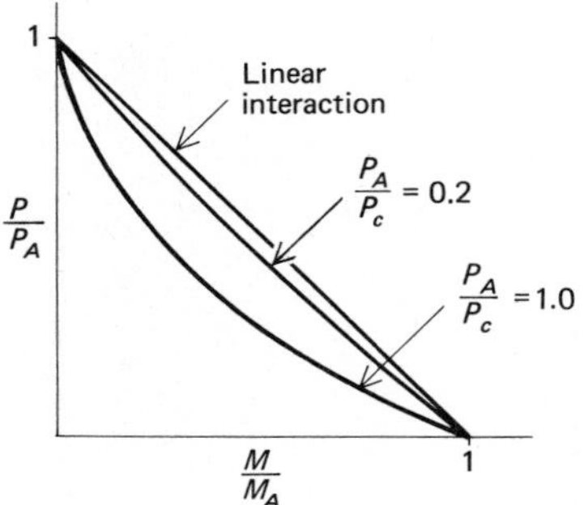

Fig. 19.3. *Secant formula recast as an interaction formula and compared with linear interaction.*

The weakness of the linear interaction is now clear – it will allow as safe some designs which would be rejected if the secant formula were used. The discrepancy between them is worse for large values of P_A/P_c than for small ones, corresponding to slender and short columns respectively. The price paid for convenience in calculation is some encroachment on the margin of safety.

This conclusion is confirmed by a more extensive analysis of the effects of combined axial load and bending. A review of the whole problem is given by Gerstle.[3]

Example 19.2 illustrates design of an eccentrically loaded steel column, using the linear interaction formula.

19.3. TIMBER COLUMNS

We do not add anything here to the discussion on design of struts in Chapter 12, "Ties and Struts". Loads carried by columns in timber buildings are not much different from those in the struts of a roof truss. Furthermore, large eccentricities are unlikely because moments are not transferred from beams or trusses to columns. Bending of columns arises from eccentricity of beam bearings, for example, and the interaction formula

$$\frac{P}{P_A} + \frac{M}{M_A} \ngtr 1$$

should be used.

Example 19.2 Sheet 1 of 2

Design an eccentrically loaded steel column.

Select a suitable plated British Standard universal column.

Design data: Load = 600 kN

Eccentricity = 550 mm

Effective length of column and compression flange = 8m

Design to BS 449 using Grade 43 steel

Try 305 x 305 x 97 kg/m U.C. plated with 305 mm x 10mm flange plates

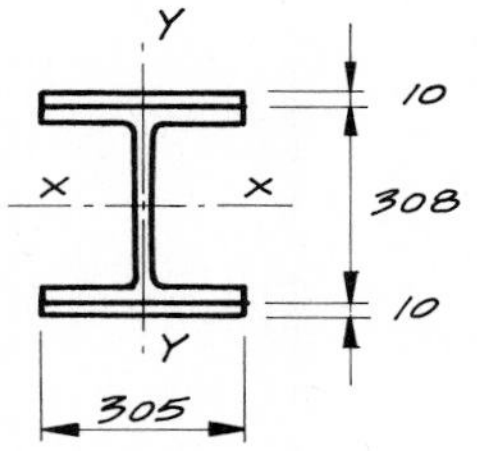

$$A = 1.233 \times 10^4 + 2 \times 10 \times 305$$
$$= 1.843 \times 10^4 \text{ mm}^2$$
$$I_{YY} = 7.268 \times 10^4 + 1/12 \times 20 \times 305^3$$
$$= 1.200 \times 10^8 \text{ mm}^4$$
$$r_{YY} = \sqrt{\frac{1.200}{1.843} \times 10^4} = 80.8 \text{ mm}$$
$$I_{XX} = 2.220 \times 10^8 + 2 \times (10 \times 305) \times 159^2$$
$$= 3.76 \times 10^8 \text{ mm}^4$$
$$Z_{XX} = \frac{3.76 \times 10^8}{1.64 \times 10^2} = 2.29 \times 10^6 \text{ mm}^3$$

$$\frac{\ell}{r_{YY}} = \frac{8000}{80.8} = 99.0 \qquad p_c = 80 \text{ MPa}$$
$$P_a = 80 \times 1.843 \times 10^4 \text{ N}$$
$$= 1.48 \text{ MN}$$
$$\frac{P}{P_a} = \frac{0.600}{1.48} = 0.405$$

$D = 328$ mm $\quad T = 25.4$ mm $\quad \frac{D}{T} = 12.9$

$\frac{\ell}{r_{YY}}$	$\frac{D}{T} = 10$	$\frac{D}{T} = 12.9$	$\frac{D}{T} = 15$
99	265	265	265

(From Fig. 15.2)

Example 19.2 Sheet 2 of 2

$$M_a = 265 \times 10^6 \times 2.29 \times 10^{-3} \text{ Nm}$$
$$= 608 \text{ kNm}$$
$$M = 600 \times 0.55 = 330 \text{ kNm}$$
$$\frac{M}{M_a} = \frac{330}{608} = 0.543$$
$$\frac{P}{P_a} + \frac{M}{M_a} = 0.949 \quad \text{ok}$$

305 × 305 × 97 kg/m U.C. with 305 × 10 flange plates

Large eccentric thrusts can occur in laminated timber portal frames, but in such structures considerable restraint against buckling is provided by the roof and walls. Also, the solid rectangular section resists twisting very well. Practice is to design these frames as beams, the combined stress being compared with a permitted stress in bending.

19.4. REINFORCED CONCRETE COLUMNS

The cross-section of a reinforced concrete column may be square, rectangular or circular. The main reinforcement comprises bars 16 mm to 40 mm in diameter and these bars are embraced by secondary reinforcement of smaller size in the form of separate ties or a continuous spiral. Reinforced concrete columns can be classified as spiral columns or tied columns, depending on the kind of secondary reinforcement. Most spiral columns are circular and tied columns are usually square or rectangular but may be circular. Sketches of typical spiral and tied columns are shown in Fig. 19.4.

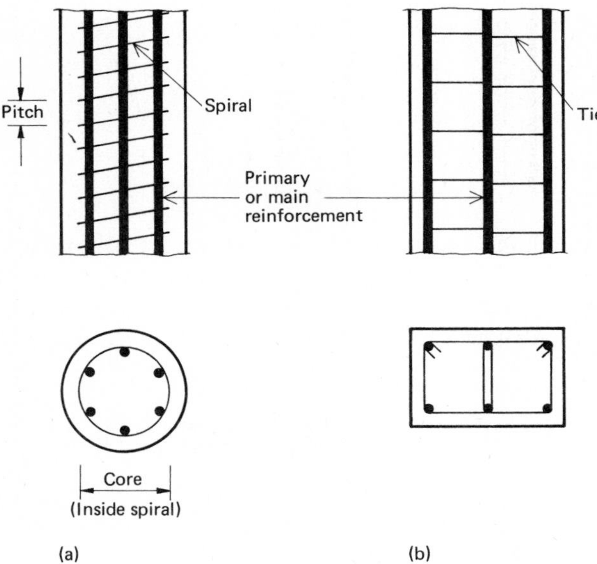

Fig. 19.4. *Reinforced concrete columns:*
(a) spiral secondary reinforcement;
(b) ties for secondary reinforcement.

The main reinforcement acts directly to contribute to the support of the load on the column – all the bars are oriented so as to be able to resist stresses parallel to the axis of the column. The contribution of the secondary reinforcement is indirect. In the first place, it helps to control buckling of the main bars. By keeping the slenderness ratio of the main bars down it enables them to resist large stresses without buckling. This is the sole effect of ties as secondary reinforcement.

Experimental evidence is that spiral binding does more. As the concrete in the core of a spiral column is loaded in compression, secondary strain makes its transverse dimension increase (the Poisson effect). Every turn of the spiral binding being well anchored, this expansion induces a tension in the spiral so that there is some resistance to expansion of the concrete in the core. The concrete is subject to triaxial stress with the result that the axial load required to cause failure is greater than it would be without the spiral binding. Provided this secondary reinforcement satisfies certain rules about the pitch of the spiral and the amount of steel in it, the gain in strength can be taken into account. In ultimate strength design, this is done by using a larger strength reduction factor, ϕ.

Tests to destruction on spiral columns and tied columns confirm that the spiral column is tougher. The behaviour described here is that of short columns, so that buckling is eliminated as a mode of failure.

Failure of a tied column begins when the concrete cover spalls off outside the main bars as sketched in Fig. 19.5. This is followed by buckling of the main reinforcement between ties and crushing of the remaining concrete. There is little reserve of strength beyond the load which causes the initial spalling.

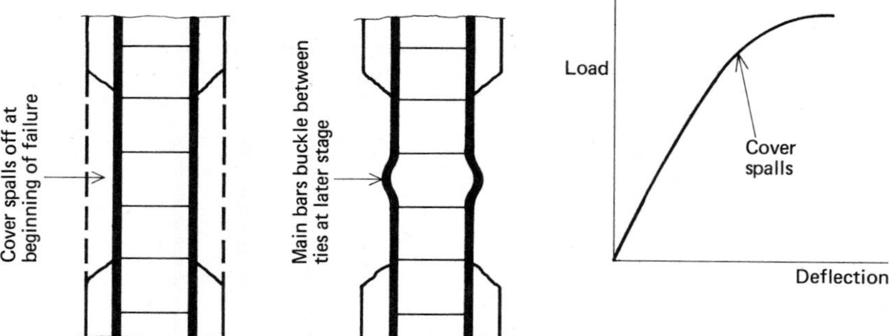

Fig. 19.5. *Failure of tied column.*

The spiral column shows similar features at the beginning of the failure, but the close pitch of the spiral prevents buckling of the main bars. Also, the confining hoop force continues to increase as the axial load increases. The result is a significant increase in load beyond the initial evidence of failure, as indicated in Fig. 19.6.

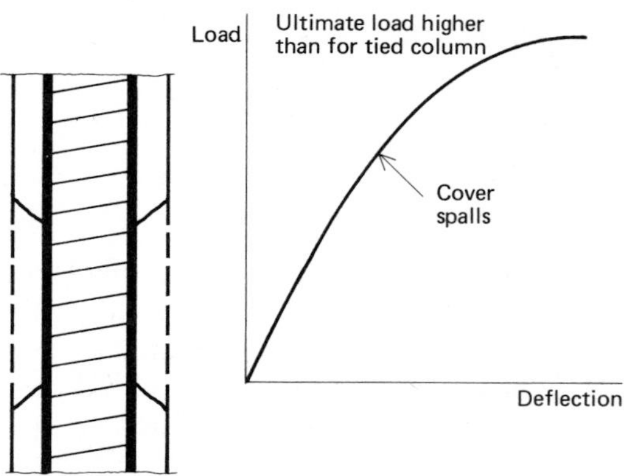

Fig. 19.6. *Failure of spiral column.*

Design of the main bars is based on the principles of mechanics and assumptions about the behaviour of reinforced concrete. The basis of design given here is the behaviour at ultimate load. It should be noted that a large-enough eccentricity can cause tensile strains over part of the cross-section, in which case some bars will be subject to tensile stress and part of the concrete in the cross-section will be ineffective.

Design of the secondary reinforcement is usually based on rules in the design specification. Such rules are based on experience in practice and testing laboratories.

19.4.1. Design of reinforced concrete columns

The assumptions used for design of reinforced concrete columns are the same as those used for beams (see Chapter 16, "Reinforced Concrete Beams"). Structural analysis leads to a load and a bending moment in the column and these will be most useful in the form of the equivalent eccentric load. When this load is multiplied by a load factor and divided by the strength reduction factor, ϕ, a design load is obtained.

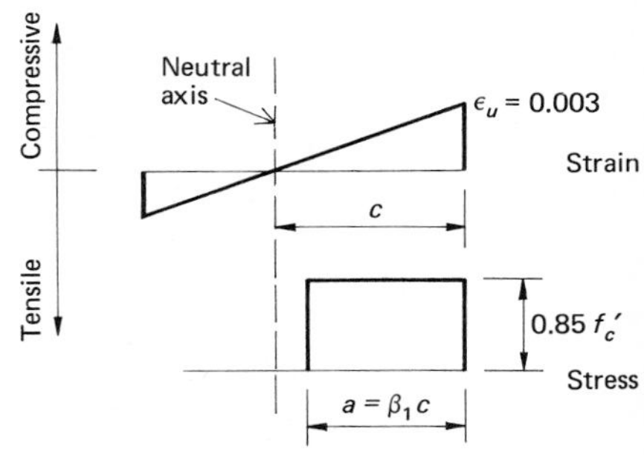

The assumptions used for computation of the strength of reinforced concrete subject to bending and axial load are repeated here for convenience:

(a) the strain in the steel and the concrete is directly proportional to distance from the neutral axis;
(b) the maximum strain in the extreme fibre in compression is 0.003;
(c) the stress in the steel equals the strain in the steel multiplied by the modulus of elasticity for strains up to the yield point and equals yield stress for larger strains;
(d) the tensile strength of concrete is negligible;
(e) the concrete compression stress block is as sketched in the margin.

The factor β_1 equals 0.85 for f_c' not greater than 26.7 MPa and reduces at the rate of 0.075 for each 10 MPa of strength in excess of 26.7 MPa.

Other rules have to be taken into account, but those above and equilibrium of the column are the basis of the formulas for design.

Equilibrium will be satisfied by a system of internal forces which is equivalent to the external load. The conditions to be satisfied are

$$\Sigma F_v = 0$$
$$\Sigma M = 0$$

describing equilibrium of the vertical forces and the moments about any convenient line. The external force is the eccentric load. For a rectangular column, the internal forces comprise:

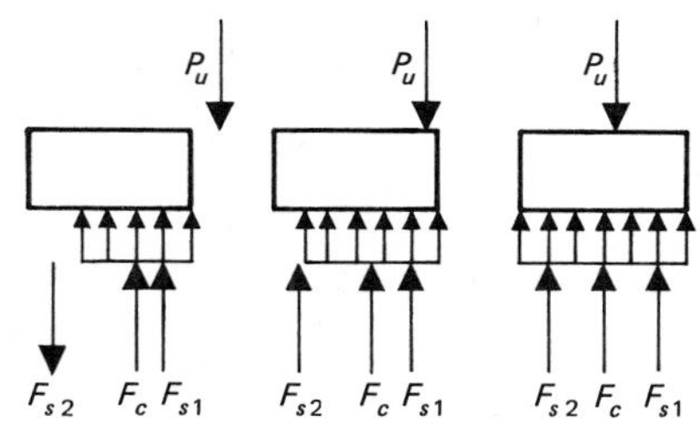

(a) F_c, the concrete compression, being the resultant of the compression stress acting over all or part of the concrete in the cross-section;
(b) F_{s1} and F_{s2}, two forces in the steel reinforcement, one near each of two opposite faces of the column. These forces may both be compressions or one may be a compression and the other a tension, depending on the eccentricity of the load.

It must be noted that the ACI *Code* requires all columns to be designed for eccentric loads.[4] Even if the structural analysis shows the column to be axially loaded, a nominal eccentricity must be assumed. This sound rule is intended to allow for inaccuracies in construction. Thus, this code gives no rules specifically for axially loaded columns; the basic assumptions could, of course, be applied to such a problem.

The equations to be solved are easily derived. Balancing the vertical forces leads to Equation (19.2) (see Fig. 19.7).

$$P_u = \phi\,(F_c + F_{s1} + F_{s2})$$

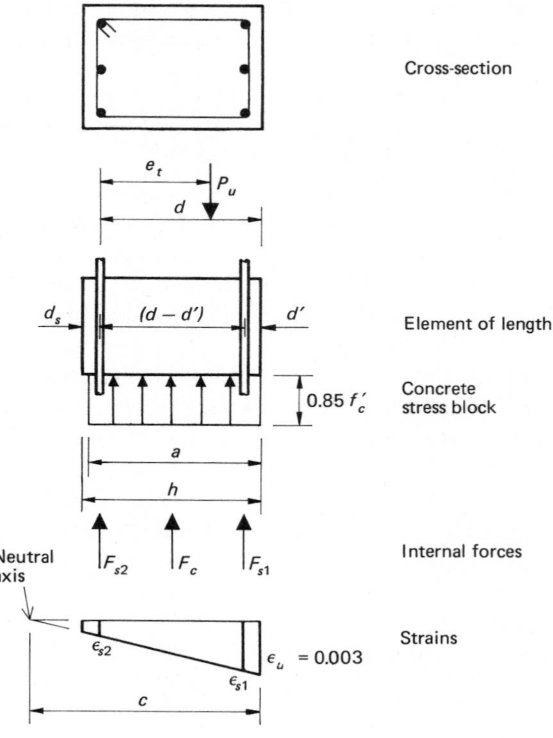

Fig. 19.7. *Distribution of stress and strain in a rectangular reinforced concrete column.*

or

$$\frac{P_u}{\phi} = F_c + F_{s1} + F_{s2} \tag{19.2}$$

Here, P_u is the ultimate load determined by multiplying the load under service conditions by the load factor.

Taking moments about the steel on the far side from the load leads to

$$P_u\, e_t = \phi \left\{ F_c\left(d - \frac{a}{2}\right) + F_{s1}\,(d - d') \right\}$$

or

$$\frac{P_u}{\phi}\, e_t = F_c\left(d - \frac{a}{2}\right) + F_{s1}\,(d - d') \tag{19.3}$$

Symbols which have not been defined are as shown in Fig. 19.7. We also have

$$a = 0.85\, c \ngtr h \tag{19.4}$$

and the strain distribution leads to

$$\varepsilon_{s1} = \varepsilon_u \frac{c - d'}{c} \tag{19.5}$$

$$\varepsilon_{s2} = \varepsilon_u \frac{c - d}{c} \tag{19.6}$$

Assuming idealised elastic-plastic deformation of the steel, these lead to

$$f_{s1} = E_s \varepsilon_u \frac{c - d'}{c} \ngtr f_y \tag{19.7}$$

and

$$f_{s2} = E_s \varepsilon_u \frac{c - d}{c} \ngtr f_y \tag{19.8}$$

where f_{s1} and f_{s2} are the steel stresses corresponding to the forces F_{s1} and F_{s2}; f_y is the yield strength of the steel; and E_s is Young's modulus.

Finally, we can express the forces F_c, F_{s1}, F_{s2} in terms of the appropriate dimensions and stresses:

$$\left.\begin{aligned} F_c &= (0.85\, f_c')\,(ab - [A_{s1} + A_{s2}]) \qquad a > d \\ F_c &= (0.85\, f_c')\,(ab - A_{s1}) \qquad d' < a < d \end{aligned}\right\} \tag{19.9}$$

$$F_{s1} = f_{s1}\, A_{s1} \tag{19.10}$$

$$F_{s2} = f_{s2}\, A_{s2} \tag{19.11}$$

How this information is best used is itself an engineering problem and we recommend careful study of the following paragraphs as an example of the organisation and manipulation of design information as well as a means to the end of designing columns.

The variables in this set of equations can be classified as follows:

Known	*Unknown, but only of passing interest*			*Unknown and required for decision making*		
$\frac{P_u}{\phi}$	a	F_c	ε_{s1}	b	f_y	A_{s1}
e	c	F_{s1}	ε_{s2}	h	f_c'	A_{s2}
ε_u	e_t	F_{s2}	f_{s1}	d		
E_s			f_{s2}	d'		

In the above list, e is the eccentricity of the applied load with respect to the axis of the column and is given by

$$M_u = P_u e$$

where M_u is the design moment. The eccentricity does not appear explicitly in the equations.

Since the number of unknowns is greater than the number of equations, the designer has to make arbitrary decisions about eight of them. Having done so, he will need an orderly procedure if he is to solve the equations for the other ten. We recommend a trial and error attack.

First, the materials to be used are chosen. This basic decision will have been made for the whole structure and fixes values of f_c' and f_y. Also, the factor in Equation (19.4) will be different from 0.85 if f_c' exceeds 26.7 MPa, as noted above. Next, the concrete dimensions and cover to the reinforcing steel are chosen and the diameter of the bars required is estimated. These decisions will give b, h, d and d'.

If a value of c is now chosen, the set of equations can be solved to yield A_{s1} and A_{s2}, the only two variables remaining in the decision-making set. The calculations are made for several values of c, each leading to steel areas compatible with the design assumptions and equilibrium. From this set of solutions, the designer chooses the one that he thinks is best. It may be one with equal areas of steel on both faces or one with the smallest area of steel in total – we favour the use of equal areas where possible, to avoid the risk of getting the steel the wrong way round when the column is built.

The calculation must be done in an orderly way, and since the same operations have to be repeated for each trial value of c, the work is best done in a table. Clearly, the method is well adapted for calculation on a computer.

Equation (19.4) is used first to get a. Then F_c can be computed from Equation (19.9). A complication here is that A_{s1} and A_{s2} are not yet known. However, the correction is small. Rules in the ACI *Code* require the total area of steel to be between 1% and 8% of the total area of cross-section.[4] The need to allow for concrete displaced by the steel is often ignored; otherwise, a nominal correction can be made. We recommend an allowance of 0.05 ab for concrete in compression displaced by steel, so that Equation (19.9) becomes

$$F_c = (0.85\ f_c')\,(0.95\ ab) \tag{19.9a}$$

The lever arm of F_c is assumed to be unchanged by this allowance for the steel area and $(d - a/2)$ is calculated from d and a which are both now available. Using Equation (19.3), $F_{s1}(d - d')$ is found:

$$F_{s1}(d - d') = \frac{P_u}{\phi} e_t - F_c\left(d - \frac{a}{2}\right)$$

The lever arm of F_{s1}, is $(d - d')$ and this is known, so that F_{s1} can be calculated and Equation (19.2) then yields F_{s2}:

$$F_{s2} = \frac{P_u}{\phi} - (F_c + F_{s1})$$

Equations (19.7) and (19.8) lead to f_{s1} and f_{s2} and one can now compute A_{s1} and A_{s2} from Equations (19.10) and (19.11).

It is obviously desirable to have some foreknowledge of the range of values of c that will be worth trying. One significant criterion is the stress in the bars on the far side from the eccentric load – is it likely to be compression or tension? The dividing criterion is given by $c = d$ and $a = 0.85d$. Then, the neutral axis coincides with these bars and there are only two internal forces – F_c and F_{s1}. We then have

$$\frac{P_u}{\phi} = F_c + F_{s1}$$

and

$$\frac{P_u}{\phi} e_t = F_c\left(d - \frac{a}{2}\right) + F_{s1}(d - d')$$

whence

$$e_t = \frac{F_c\left(d - \frac{a}{2}\right) + F_{s1}(d - d')}{F_c + F_{s1}}$$

Substituting

$$F_c = (0.85\ f_c')\,(0.95 \times 0.85d \times b)$$

$$= 0.685\ f_c'\,bd$$

$$a = 0.85d$$

and

$$F_{s1} = f_y A_{s1}$$

where it is assumed that $f_{s1} = f_y$. Thus, we get

$$\frac{e_t}{d} = \frac{0.465 + \dfrac{f_y}{0.85\ f_c'} \dfrac{A_{s1}}{bd}\left(1 - \dfrac{d'}{d}\right)}{0.807 + \dfrac{f_y}{0.85\ f_c'} \dfrac{A_{s1}}{bd}}$$

This criterion is shown in Fig. 19.8 in the form of a family of straight lines, e_t/d versus d'/d, with $(f_y/0.85\ f_c')(A_{s1}/bd)$ as a third parameter. Before starting the calculations described above, the designer can plot the point representing the column on this field. He can guess what A_{s1}/bd is likely to be and compare e_t/d as it is in his problem with the value which places the neutral axis at the steel on the far side from the load. If the eccentricity e_t is less than this critical value, the steel on the far side will be in compression and trial values of c will be greater than d. Conversely, if the eccentricity is greater than the critical value, c will be less than d.

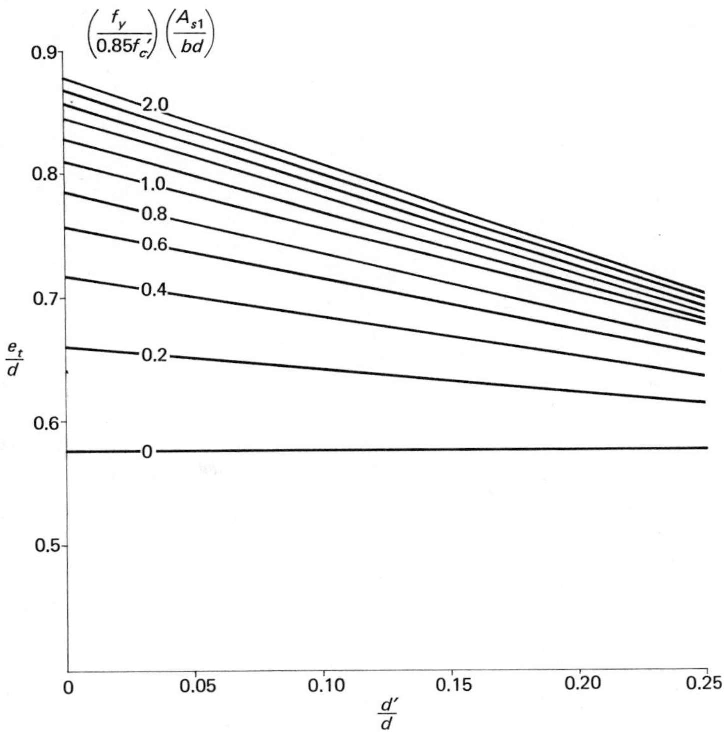

Fig. 19.8. *Relationship to be satisfied for no stress in the bars on the far side from the load (rectangular section).*

Another test which must be made is to find out whether steel in tension reaches yield stress or not. If it does, the ACI *Code* requires the proportion of steel on the tension face to be less than 0.75 ρ_b, where ρ_b is the proportion of steel (A_{s2}/bd) which would give balanced design in the absence of axial load.[4] This is the same rule as for beams (see Chapter 16, "Reinforced Concrete Beams") and the reason for it is the same.

The concept of a balanced condition enters the argument here, this being defined as simultaneous occurrence of yield in the tension steel and ultimate strain in the extreme compression fibre, i.e. 0.003. The balanced condition gives

$$a_b = 0.85 \frac{\varepsilon_u}{\varepsilon_u + \varepsilon_y} d$$

$$= 0.85\, d \Big/ \left(1 + \frac{f_y}{E_s\, \varepsilon_u}\right)$$

and

$$P_b = 0.85\, f_c'\, a_b b + A_{s1}\, f_{s1} - A_{s2}\, f_y$$

$$= 0.85\, f_c'\, a_b b$$

if $A_{s1} = A_{s2}$ and $f_{s1} = f_y$. If P_u/ϕ is greater than this balanced load, the tension steel does not yield.

Finally, if the steel areas are equal, with the steel near one face strained in tension beyond the yield point and the steel on the other side strained beyond yield in compression, calculation of the steel areas is very easy. Then, the force balance yields

$$\frac{P_u}{\phi} = (0.85\, f_c')\,(0.95ab)$$

whence a is computed. Using the moment balance,

$$\frac{P_u}{\phi} e_t = (0.85\, f_c')\,(0.95ab)\,(d - \frac{a}{2}) + F_{s1}\,(d - d')$$

whence

$$F_{s1} = 0.807\, f_c'\, ab \frac{e_t + a/2 - d}{d - d'}$$

and

$$A_{s1} = \frac{0.807\, f_c'\, ab}{f_y} \; \frac{e_t + a/2 - d}{d - d'}$$

For this calculation to be valid, Equation (19.5) shows that

$$\varepsilon_u \frac{c - d'}{c} \nless \frac{f_y}{E_s}$$

whence

$$\frac{c}{d'} \nless 1 \Big/ \left(1 - \frac{f_y}{E_s\, \varepsilon_u}\right)$$

or

$$a \nless 0.85 d' \Big/ \left(1 - \frac{f_y}{E_s\, \varepsilon_u}\right)$$

If this inequality is satisfied,

$$f_{s1} = f_y$$

The foregoing notes set out a procedure for computing the steel areas required in a column, given its dimensions and the ultimate load and moment. One other point requires consideration and that is the value of ϕ. The basic value for a column with ties for its secondary reinforcing is 0.7, and this value is relevant when the eccentricity of the applied load is small. For a given moment, the eccentricity increases as the load decreases and the column gets progressively more like a beam. It would be reasonable to allow a corresponding increase in ϕ to the value allowed for a beam (0.90) and the ACI *Code* does this.[4] If the ratio $(h - d' - d_s):h$ is less than 0.70 and P_u is less than $0.1\, f_c'\, A_g$, a larger value of ϕ is allowed, as sketched in the margin. For columns which do not meet these conditions, the linear relationship begins with $\phi = 0.7$ and $P_u = 0.1\, f_c'\, A_g$ or P_b, whichever is the less, and ends with $\phi = 0.9$ and $P_u = 0$.

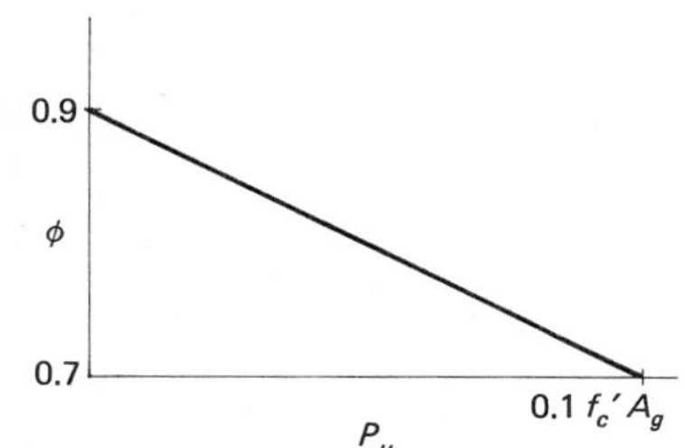

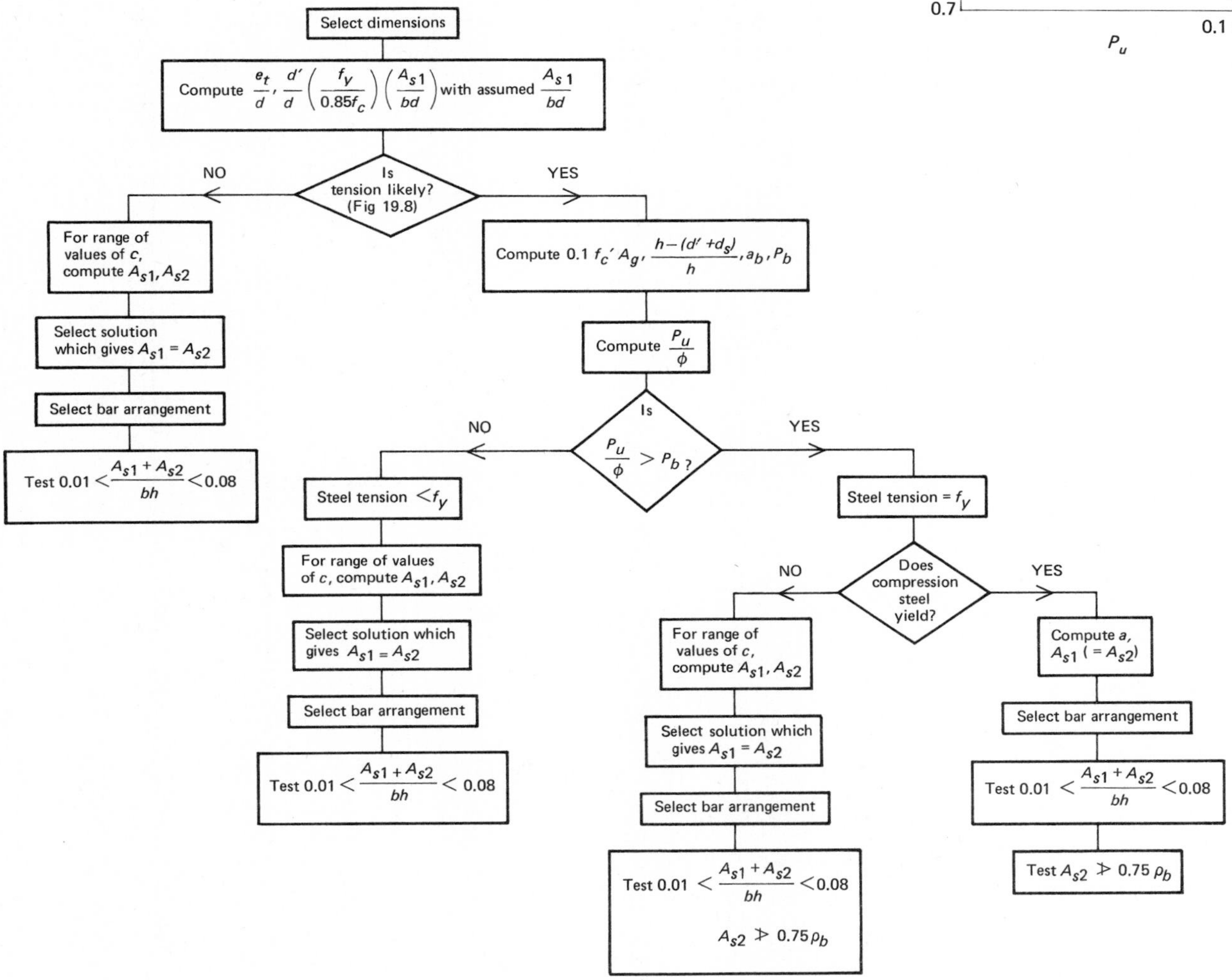

Select dimensions

Compute $\frac{e_t}{d}, \frac{d'}{d} \left(\frac{f_y}{0.85 f_c}\right)\left(\frac{A_{s1}}{bd}\right)$ with assumed $\frac{A_{s1}}{bd}$

Is tension likely? (Fig 19.8)

NO:

For range of values of c, compute A_{s1}, A_{s2}

Select solution which gives $A_{s1} = A_{s2}$

Select bar arrangement

Test $0.01 < \frac{A_{s1} + A_{s2}}{bh} < 0.08$

YES:

Compute $0.1\, f_c'\, A_g$, $\frac{h - (d' + d_s)}{h}$, a_b, P_b

Compute $\frac{P_u}{\phi}$

Is $\frac{P_u}{\phi} > P_b$?

NO:

Steel tension $< f_y$

For range of values of c, compute A_{s1}, A_{s2}

Select solution which gives $A_{s1} = A_{s2}$

Select bar arrangement

Test $0.01 < \frac{A_{s1} + A_{s2}}{bh} < 0.08$

YES:

Steel tension $= f_y$

Does compression steel yield?

NO:

For range of values of c, compute A_{s1}, A_{s2}

Select solution which gives $A_{s1} = A_{s2}$

Select bar arrangement

Test $0.01 < \frac{A_{s1} + A_{s2}}{bh} < 0.08$

$A_{s2} \not> 0.75 \rho_b$

YES:

Compute a, A_{s1} $(= A_{s2})$

Select bar arrangement

Test $0.01 < \frac{A_{s1} + A_{s2}}{bh} < 0.08$

Test $A_{s2} \not> 0.75 \rho_b$

Fig. 19.9. *Flow chart for design calculations (rectangular section).*

The flow chart in Fig. 19.9 is a diagrammatic representation of these calculations, including tests to decide the path to be followed. Examples 19.3 to 19.6 are illustrative calculations. In each case, an equal area solution is sought. The first, Example 19.3, is for a column with only nominal eccentricity. All the steel is in compression and, in the final design, a large proportion of the cross-section contributes to F_c. In Example 19.4, the eccentricity is greater and the steel near one face is in tension, but well below yield point. It will be seen that Equations (19.2) and (19.8) have given negative values for F_{s2} and f_{s2}, corresponding to the tension in these bars. In Examples 19.5 and 19.6 the steel yields in tension on one side. In Example 19.5, the compression stress in the

Example 19.3 Sheet 1 of 2

Select dimensions and reinforcing steel for reinforced concrete column given ultimate load and moment.

Design data:

$P_u = 3600$ kN Calculated eccentricity $= 0$

$f_c' = 20$ MPa Nominal eccentricity $= 0.1h$

$0.85 f_c' = 17$ MPa $f_y = 275$ MPa $E_s = 207$ GPa

$E_s \epsilon_u = 621$ MPa

Try 450 mm × 450 mm section
Allow for 50 mm cover and 40 mm diameter bars

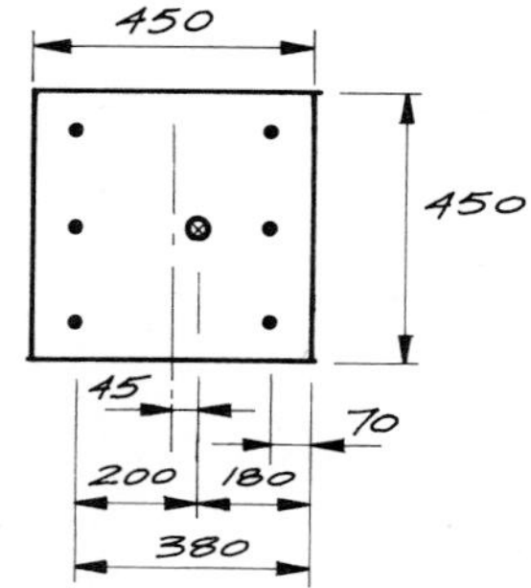

Nominal eccentricity $= 45$ mm, $e_t = 200$ mm

$$0.1 f_c' A_g = 2 \times 450 \times 450 = 405 \text{ kN} \ngtr P_u$$

$$\frac{h-(d'+d_s)}{h} = \frac{310}{450} = 0.69$$

$\phi = 0.70$ $\frac{P_u}{\phi} = 5150$ kN $\frac{P_u}{\phi} e_t = 1030$ kNm

$\frac{e_t}{d} = \frac{200}{380} = 0.527$ Anticipate $\frac{A_s}{bh} \sim 0.03$

$\frac{d'}{d} = \frac{70}{380} = 0.184$ $\frac{f_y}{0.85 f_c'} \frac{A_s}{bh} = \frac{275}{17} \times 0.03 = 0.485$

From Fig. 19.8 tension unlikely.

Example 19.3 Sheet 2 of 2

$F_c = (0.85 f_c')(0.95ab) = 17 \times 428a = 7.27a$ kN (a in mm)

$d - d' = 0.310$ m

$E_s \epsilon_u = 621$ MPa

Solution No	c	a	F_c	$d-\frac{a}{2}$	$F_c(d-\frac{a}{2})$	F_{s1}	F_{s2}
1	470	400	2900	0.180	522	1640	610
2	440	374	2720	0.193	525	1630	800

Solution No	$\frac{c-d'}{c}$	$\frac{c-d}{c}$	f_{s1}	f_{s2}	A_{s1}	A_{s2}
1	$\frac{400}{470} = 0.851$	$\frac{90}{470} = 0.192$	275	119	5960	5120
2	$\frac{370}{440} = 0.841$	$\frac{60}{440} = 0.136$	275	84.5	5930	9460

For equal areas 5 × 40 mm each face

$A_{s1} = A_{s2} = 6280\ \text{mm}^2$

$\frac{A_{s1} + A_{s2}}{bh} = 0.062$ OK

Example 19.4 Sheet 1 of 2

Select dimensions and reinforcing steel for reinforced concrete column given ultimate load and moment.

Design data:

$P_u = 1870$ kN $\quad M_u = 187$ kNm $\quad e = 0.100$ m

$f_c' = 20$ MPa $\quad f_y = 275$ MPa $\quad E_s = 207$ GPa

$$\frac{\epsilon_y}{\epsilon_u} = \frac{f_y}{E_s \epsilon_u} = \frac{275}{207 \times 10^3 \times 3 \times 10^{-3}} = 0.443$$

Try 400 mm × 400 mm section, $A_s' = A_s$

Allow for 50 mm cover and 40 mm diameter bars

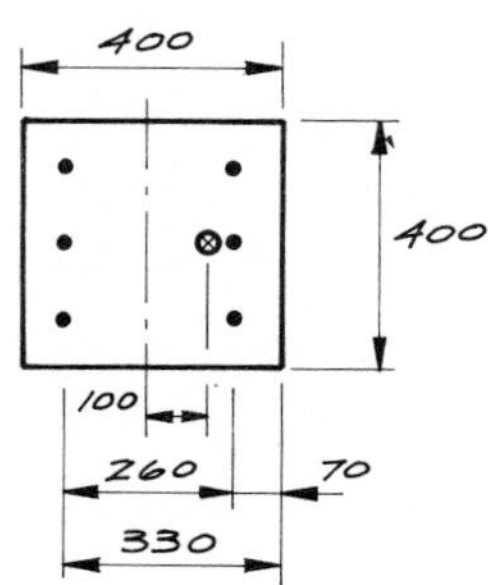

$0.1 f_c' A_g = 2 \times 400 \times 400 = 320$ kN $< P_u \quad \therefore \phi = 0.70$

$\frac{P_u}{\phi} = 2670$ kN $\quad e_t = 0.230$ m $\quad \frac{P_u}{\phi} e_t = 614$ kNm

$\frac{e_t}{d} = \frac{230}{330} = 0.697, \qquad \frac{d'}{d} = \frac{70}{330} = 0.212,$

$\frac{f_y}{0.85 f_c'} = \frac{275}{17} = 16.1.$ Say, $\frac{A_s}{bh} \sim 0.03 \rightarrow \frac{f_y}{0.85 f_c'} \frac{A_s}{bh} = 0.485$

From Fig. 19.8 tension probable $(kd < d)$

For $A_s' = A_s$: $\quad a_b = \frac{0.85 \times 330}{1.443} = 194$ mm

$P_b = 0.85 \times 20 \times 400 \times 194 = 1320$ kN $< \frac{P_u}{\phi}$

Design for compression failure

Example 19.4 Sheet 2 of 2

$F_c = (0.85 f_c')(0.95 ab) = 17 \times 0.95 \times 400 \times a = 6.45a$ kN
(a in mm)

$d - d' = 0.260$ m

$E_s \epsilon_u = 621$ MPa

Solution No	c	a	F_c	$d-\frac{a}{2}$	$F_c(d-\frac{a}{2})$	F_{s1}	F_{s2}
1	300	255	1640	0.202	331	1090	-60
2	320	272	1760	0.194	341	1050	-140

Solution No	$\frac{c-d'}{c}$	$\frac{c-d}{c}$	f_{s1}	f_{s2}	A_{s1}	A_{s2}
1	0.767	-0.100	275	-62.1	3970	966
2	0.782	-0.0313	275	-19.4	3820	7220

5 No 32 mm diameter bars each face

$$A_{s1} = A_{s2} = 4020 \text{ mm}^2$$

$$\frac{A_{s1} + A_{s2}}{bh} = \frac{8040}{160000} = 0.05 \quad \text{ok}$$

Example 19.5 Sheet 1 of 2

Select dimensions and reinforcing steel for reinforced concrete column given ultimate load and moment.

Design data:

$P_u = 200$ kN $\quad M_u = 267$ kNm $\quad e = 1.335$ m

$f_c' = 20$ MPa $\quad f_y = 275$ MPa $\quad E_s = 207$ GPa

$E_s \epsilon_u = 621$ MPa $\quad \dfrac{\epsilon_Y}{\epsilon_u} = \dfrac{275}{621} = 0.443$

Try 425 × 425 section

Allow 50 mm cover and 40 mm diameter bars

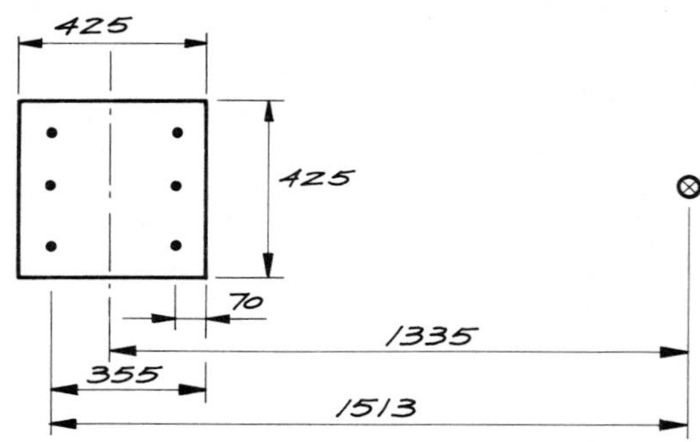

No need to test for tension.

$a_b = \dfrac{0.85 \times 355}{1.443}$

$= 209$ mm

$P_b = 17 \times 209 \times 425$

$= 1510$ kN

$P_u < P_b$ – design for tension failure

$\dfrac{h-(d'+d_s)}{h} = \dfrac{285}{425}$

$= 0.67$

$0.1 f_c' A_g = 2 \times 425 \times 425$

$= 362$ kN

$\phi = \dfrac{162}{362} \times 0.2 + 0.7$

$= 0.790$

$\rho_b = \dfrac{0.85 \times 0.85 \times 20}{275} \times \dfrac{621}{896}$

$= 0.0364$

$\rho \not> 0.75 \rho_b = 0.0273$

Example 19.5 Sheet 2 of 2

$\frac{P_u}{\phi} = \frac{200}{0.790} = 253\ kN$ $\frac{P_u}{\phi} e_t = 253 \times 1.513 = 384\ kNm$

Assuming $f_{s1} = f_Y$ — i.e. $\epsilon_{s1} > \epsilon_Y$

$a = \frac{253 \times 10^3}{17 \times 425}$ $\frac{0.85d'}{1 - \frac{\epsilon_Y}{\epsilon_u}} = \frac{0.85 \times 70}{0.557}$

$= 35.1\ mm$ $= 107\ mm$

$< 107\ mm$

Compression steel does not yield

$F_c = (0.85 f_c')(0.95ab)\ 17 \times 404a = 6.86a\ kN$ (a in mm)

$d - d' = 0.285\ m$

Solution No	c	a	F_c	$d-\frac{a}{2}$	$F_c(d-\frac{a}{2})$	F_{s1}	F_{s2}
1	120	102	700	0.304	213	600	-1047
2	100	85	583	0.312	182	709	-1039
3	95	80.7	554	0.315	174	737	-1038

Solution No	$\frac{c-d'}{c}$	$\frac{c-d}{c}$	f_{s1}	f_{s2}	A_{s1}	A_{s2}
1	$\frac{50}{120} = 0.417$	$\frac{-235}{120} = -1.96$	260	-275	2310	3800
2	$\frac{30}{100} = 0.300$	$\frac{-255}{100} = -2.55$	186	-275	3810	3770
3	$\frac{25}{95} = 0.263$	$\frac{-260}{95} = -2.74$	163	-275	4520	3770

5 No 32 mm bars each face

$A_{s1} = A_{s2} = 4020\ mm^2$

$\frac{A_{s1} + A_{s2}}{425 \times 425} = 0.0445$ OK

$\frac{A_{s2}}{bd} = \frac{4020}{425 \times 355} = 0.0266$

< 0.0273

Example 19.6 Sheet 1 of 2

Select dimensions and reinforcing steel for reinforced concrete column given ultimate load and moment.

Design data:

$P_u = 900$ kN $M_u = 240$ kNm $e = 0.267$ m

$f_c' = 20$ MPa $f_y = 275$ MPa $E_s = 207$ GPa

$E_s \epsilon_u = 621$ MPa $\dfrac{\epsilon_y}{\epsilon_u} = \dfrac{f_y}{E_s \epsilon_u} = \dfrac{275}{621} = 0.443$

Try 400 mm × 400 mm section

Allow 50 mm cover and 40 mm diameter bars

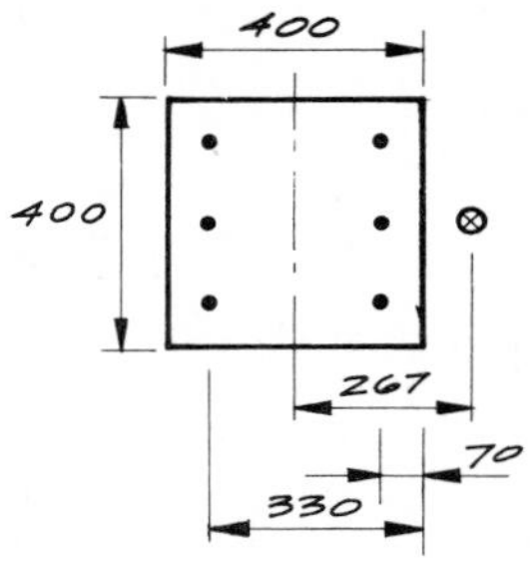

$a_b = \dfrac{0.85 \times 330}{1.443}$ $P_b = 17 \times 194 \times 400$

$= 194$ mm $= 1320$ kN

$P_u > P_b$ – design for tension failure

$\rho_b = \dfrac{0.85 \times 0.85 \times 20}{275} \times \dfrac{621}{896} = 0.0364$

$0.75 \rho_b = 0.0273$

$\dfrac{h - (d' + d_s)}{h} = \dfrac{260}{400} = 0.65$

$0.1 f_c' A_g = 2 \times 400 \times 400 = 320 \text{ kN} < P_u \rightarrow \phi = 0.70$

$\dfrac{P_u}{\phi} = 1290$ kN

Assuming compression steel yields

$a = \dfrac{1290 \times 10^3}{17 \times (0.95 \times 400)}$ $\dfrac{0.85 d'}{1 - \dfrac{\epsilon_y}{\epsilon_u}} = \dfrac{0.85 \times 70}{0.557}$

$= 200$ mm $= 107$ mm

> 107 mm compression steel yields.

Example 19.6 Sheet 2 of 2

$$e - \frac{t}{2} + \frac{a}{2} = 267 - 200 + 100 = 167 \text{ mm}$$

$$A_s = \frac{1290 \times 10^3 \times 167}{275 \times 260}$$

$$= 3020 \text{ mm}^2$$

4 No 32 mm bars each face

$$A_{s1} = A_{s2} = 3217 \text{ mm}^2$$

$$\frac{A_{s1} + A_{s2}}{bh} = \frac{6434}{400 \times 400} = 0.0402 \quad \text{OK}$$

$$\frac{A_{s2}}{bd} = \frac{3217}{400 \times 330} = 0.0244$$

$$< 0.0273 \quad \text{OK}$$

steel on the other side is less than yield stress and, in Example 19.6, this steel has yielded.

Design of a circular column with spiral secondary reinforcement is the same in principle, although made more complicated by the circular cross-section and a requirement for the longitudinal bars to be evenly spaced around the column. The ACI *Code* requires at least six bars, which means that more than two steel forces have to be calculated.[4] Considering equilibrium of the column we can write the following equations (refer to Fig. 19.10):

$$\frac{P_u}{\phi} = F_c + \Sigma F_s$$

whence

$$\Sigma F_s = \frac{P_u}{\phi} - F_c \tag{19.12}$$

and

$$\frac{P_u}{\phi} e = F_c x_c + \Sigma F_s x_s$$

whence

$$\Sigma F_s x_s = \frac{P_u}{\phi} - F_c x_c \tag{19.13}$$

Here P_u is the ultimate load; ϕ is the capacity reduction factor (= 0.75 for spirally reinforced columns); F_c is the concrete compression force; F_s is one of the steel forces; e is the eccentricity of the load with respect to the axis of the column; x_c is the distance of F_c from the axis of the column; and x_s is the distance of F_s from the axis of the column.

The resultant compression force can be written as

$$F_c = (0.85 f'_c)(0.95A)$$

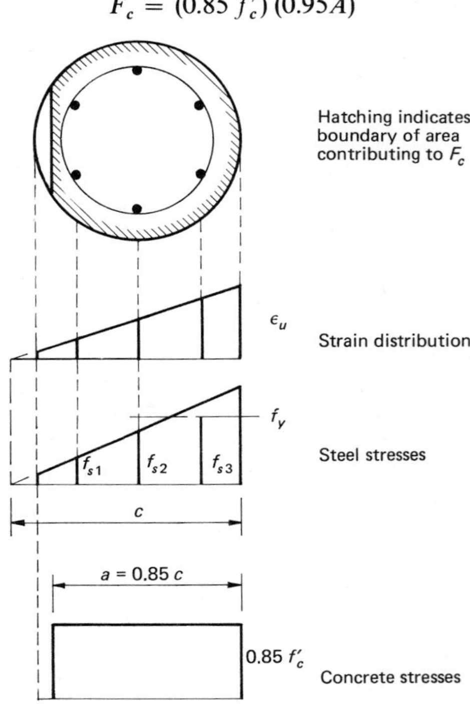

Fig. 19.10. *Distribution of stress and strain in circular reinforced concrete column.*

where A is the area of the part of the cross-section where the stress in the concrete is $0.85\, f'_c$, and an allowance of 5% has been made for the steel which displaces concrete in compression.

The total steel force is given by

$$\Sigma F_s = \Sigma f_s A_s = A_s \Sigma f_s$$

where A_s is the area of one bar or, if the arrangement is symmetrical as it is in Fig. 19.10, two bars.

Strain compatibility is used to calculate the steel stresses, bearing in mind that

$$f_s \not> f_y$$

If $\bar{x}$ is the distance of the resultant steel force from the axis of the column

$$\bar{x} = \frac{\Sigma F_s x_s}{\Sigma F_s} = \frac{\Sigma f_s x_s}{\Sigma f_s} \tag{19.14}$$

and this dimension plays a vital part in the design method we describe.

The first thing to do is to decide the size of the column, the cover to the reinforcing bars and the number and layout of the bars. The only variable to be evaluated to complete the design is the size of each bar. Again we proceed by trial.

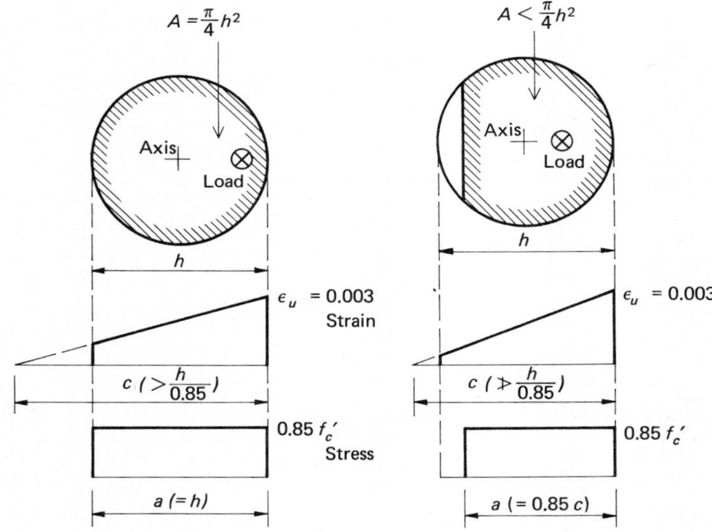

Fig. 19.11. *Examples of stressed areas of concrete in circular sections.*

If a value is assigned to c, the distance to the line of zero strain, the distribution of the concrete compression stress is defined. In Fig. 19.11 two possible distributions are sketched. Clearly, F_c and x_c can be determined, and tables can be used for rapid evaluation of A and x_c (Tables 19.1 and 19.2).

Then, using Equations (19.12) and (19.13), $\Sigma F_s x_s$ and ΣF_s are calculated and from them, $\bar{x}$ (Equation (19.14)).

The strain distribution corresponding to this value of c enables values of f_s, the steel stresses, to be computed and, since the location of each bar or pair of bars is known, $\Sigma f_s x_s$ can be computed. A second value of $\bar{x}$ is found from the quotient $\Sigma f_s x_s / \Sigma f_s$. Generally, the two values of $\bar{x}$ will not be equal and the process is repeated with a new value of c.

Table 19.1. Area of concrete in compression A as a function of $\frac{a}{h}$.
Tabulated value multiplied by h^2 gives A.

$\frac{a}{h}$	.00	.01	.02	.03	.04	.05	.06	.07	.08	.09
.0	.0000	.0013	.0037	.0069	.0105	.0147	.0192	.0242	.0294	.0350
.1	.0409	.0470	.0534	.0600	.0668	.0739	.0811	.0885	.0961	.1039
.2	.1118	.1199	.1281	.1365	.1449	.1535	.1623	.1711	.1800	.1890
.3	.1982	.2074	.2167	.2260	.2355	.2450	.2546	.2642	.2739	.2836
.4	.2934	.3032	.3130	.3229	.3328	.3428	.3527	.3627	.3727	.3827
.5	.393	.403	.413	.423	.433	.443	.453	.462	.472	.482
.6	.492	.502	.512	.521	.531	.540	.550	.559	.569	.578
.7	.587	.596	.605	.614	.623	.632	.640	.649	.657	.666
.8	.674	.681	.689	.697	.704	.712	.719	.725	.732	.738
.9	.745	.750	.756	.761	.766	.771	.775	.779	.782	.784

Source: Adapted from H. W. King, *Handbook of Hydraulics*, 3rd ed., McGraw-Hill, New York, by permission of McGraw-Hill Book Company, Inc.

Table 19.2. Location of centroid of area in compression.
Tabulated value is $\frac{x_c}{h}$ as a function of $\frac{a}{h}$.

$\frac{a}{h}$	.00	.01	.02	.03	.04	.05	.06	.07	.08	.09
.0	.500	.494	.488	.482	.476	.470	.464	.458	.452	.446
.1	.440	.434	.429	.423	.417	.411	.405	.399	.393	.387
.2	.382	.376	.370	.364	.358	.353	.347	.341	.335	.329
.3	.324	.318	.312	.307	.301	.295	.290	.284	.278	.273
.4	.267	.262	.256	.251	.245	.239	.234	.229	.223	.218
.5	.212	.207	.201	.196	.191	.185	.180	.175	.170	.164
.6	.159	.154	.149	.144	.139	.134	.129	.124	.119	.114
.7	.109	.104	.100	.095	.090	.086	.081	.077	.072	.068
.8	.063	.059	.055	.051	.047	.043	.039	.035	.031	.028
.9	.024	.021	.018	.015	.012	.009	.006	.004	.002	.001

Source: Adapted from H. W. King, *Handbook of Hydraulics*, 3rd ed., McGraw-Hill, New York, by permission of McGraw-Hill Book Company, Inc.

A set of three trials should be enough and if the two values of $\overline{x}$ – say, $\overline{x}_1$, and $\overline{x}_2$ – are plotted against each other, the intersection of the curve obtained with a straight line at 45° to the axes satisfies the condition

$$\overline{x}_1 = \overline{x}_2 \qquad \text{(See Example 19.7)}$$

The quotient $\Sigma F_s/\Sigma f_s$ is also plotted against $\overline{x}_1$ or $\overline{x}_2$ and the graph is read for the value of $\Sigma F_s/\Sigma f_s$ corresponding to the value of $\overline{x}$ necessary to solve the problem. This is the area sought.

Convenient values of c for the trial calculations are:

(a) $c = h/0.85$ which makes $a = h$ so that A is the area of the complete cross-section;

(b) $c = h$, and $a = 0.85h$;

(c) $c = d$, so that the reinforcing bars on the far side from the eccentric load are unstressed.

Example 19.7 shows the design of a circular column using the procedure set out here.

We consider that these columns are best reserved for use when the eccentricity of the load is small. Otherwise, the line of zero strain will be close to some of the bars which means that they will contribute little to the strength of the column. These columns will be most appropriate when the calculated eccentricity is zero and the nominal eccentricity for design is the minimum allowable – the $0.05h$ specified in the design code.

So far, we have not said anything about buckling in relation to reinforced concrete columns, but it must be taken into account. Some codes supply a strength reduction factor as a function of a slenderness ratio, the design philosophy being as described in Chapter 12, "Ties and Struts". The rules of the ACI *Code*[4] are discussed in more detail in the next section.

We have also not given any guidance on making the first decision on which our calculations are based – the size of the cross-section. We do not think we can. Experience in designing columns is necessary if first guesses of dimensions are to be good. The beginner in design has to rely on the experience of others – his teachers or his supervisor in the design office. Certainly, the beginner can design columns without such guidance, but he is likely to suffer many abortive trials.

19.4.2. Slenderness effects in reinforced concrete columns

The ACI *Code* has a preferred method and an alternative method for taking into account the slenderness of reinforced concrete columns.[4] In the preferred method, all loads, bending moments and deflections are computed, and allowance is made for the increase in eccentricity of axial loads caused by lateral deflection. The code makes this procedure mandatory for columns with slenderness ratios greater than 100. So much computation is required that the preferred method is practicable only when the work can be done by a computer and standard programmes are available for this kind of structural analysis. For routine design in small offices, where hand-calculation is normal, the alternative method is preferred. Fortunately, slenderness ratios greater than 100 are rare in reinforced concrete and the alternative method is usually valid.

In the alternative method, slenderness may be ignored altogether if the slenderness ratio kl_u/r is less than $(34 - 12\, M_1/M_2)$ if sidesway cannot occur, or 22 when sidesway can occur. The symbols are defined as follows: l_u is the length of the compression member, being the clear-distance between floor slabs or beams able to prevent lateral displacement or sway; k is the effective length factor to allow for rotational restraint at the ends of a column; r is the radius of gyration of cross-section (= 0.3 times appropriate lateral dimensions for a rectangular section, or 0.25 times diameter for a circular section); M_1 is the lesser end moment, positive if the column is bent in single curvature and negative if the curvature is reversed; and M_2 is the larger end moment, always positive.

For the range of slenderness ratios between the appropriate lower limit and 100, the larger of the two end moments is multiplied by a "moment magnifier", δ, to give the design moment M_c – i.e.

$$M_c = \delta M_2$$

The axial load is not changed.

Example 19.7 Sheet 1 of 3

Select dimensions and reinforcing steel for circular reinforced concrete column given ultimate load and moment.

Design data:

$P_u = 2000$ kN Computed eccentricity = 0

$f_c' = 20$ MPa Nominal eccentricity = 0.05 h

$E_s \epsilon_u = 621$ MPa $f_y = 275$ MPa $E_s = 207$ GPa

Try 400 mm diameter section. e = 20 mm
Allow 50 mm cover and 40 mm diameter bars
Say, 6 bars, radius to centres 130 mm

$\phi = 0.75$ $\frac{P_u}{\phi} = 2670$ kN

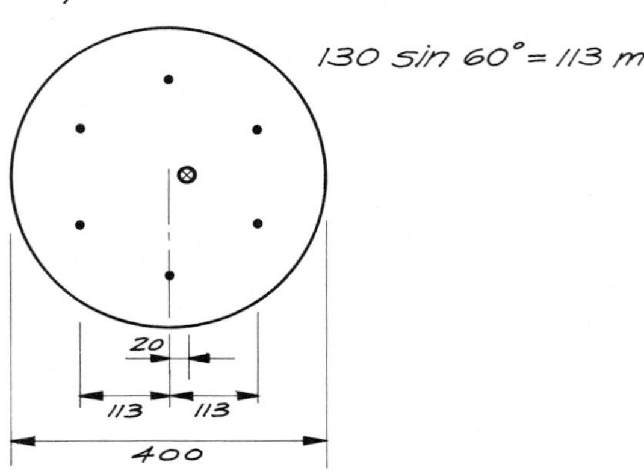

$F_c = (0.85 f_c')(0.95A) = 17 \times 0.95A = 16.2A$

$\frac{P_u}{\phi} e = 2670 \times 0.020 = 53.4$ kNm

No	$\frac{a}{h}$	A	F_c	x_c	$F_c x_c$	$\Sigma F_s x_s$	ΣF_s	$\bar{x}$
1	1.00	126000	2040	0	0	+53.4	630	+84.7
2	0.85	114000	1850	+.0172	31.8	+21.6	820	+26.4
3	0.665	88900	1440	+.0508	73.1	−19.7	1230	−16.0

Example 19.7 Sheet 2 of 3

No 1 $0.85c = h$

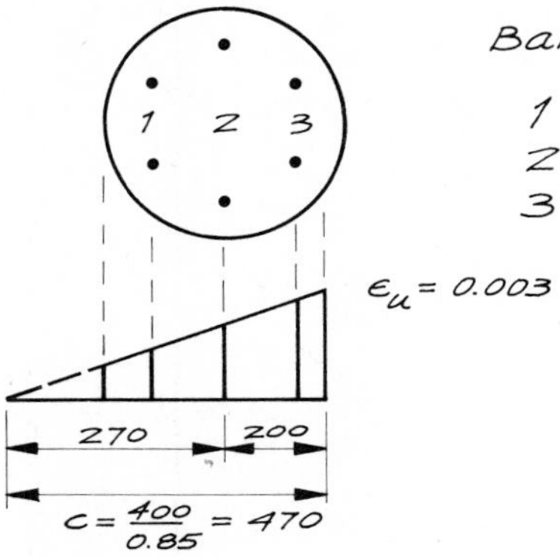
1 2 3
εu = 0.003
270
200
c = 400/0.85 = 470

Bars	Strain	f_s	x_s	$f_s x_s$
1	.00100	207	−.113	−23.4
2	.00172	275	0	0
3	.00244	275	+.113	+31.1
		757		+7.7

$\bar{x} = +9.8\ mm$

$$\frac{\Sigma F_s}{\Sigma f_s} = \frac{630}{757} = 832\ mm^2$$

No 2 $c = h$

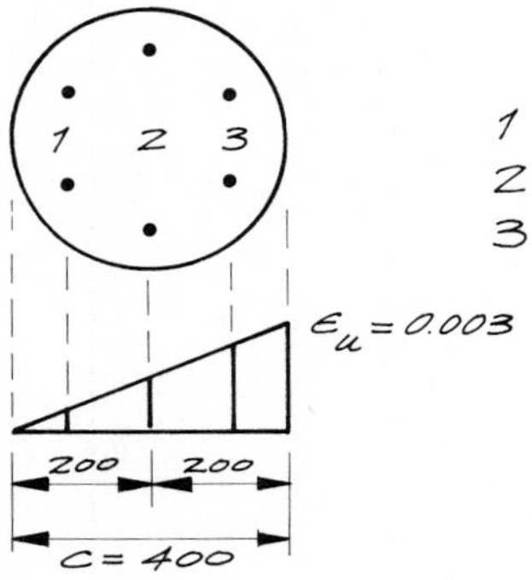
1 2 3
εu = 0.003
200
200
c = 400

Bars	Strain	f_s	x_s	$f_s x_s$
1	.0065	134	−.113	−15.2
2	.00150	275	0	0
3	.00235	275	+.113	+31.1
		684		+15.9

$\bar{x} = +23.3\ mm$

$$\frac{\Sigma F_s}{\Sigma f_s} = \frac{820}{684} = 1200\ mm^2$$

No 3 $c = d$

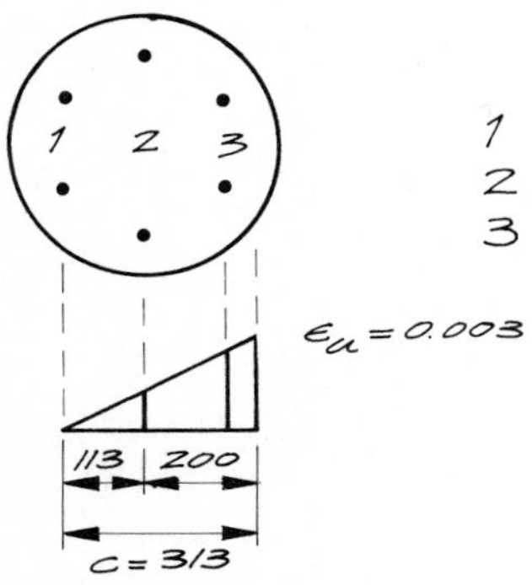
1 2 3
εu = 0.003
113
200
c = 313

Bars	Strain	f_s	x_s	$f_s x_s$
1	0	0	−.113	0
2	.00108	224	0	0
3	.00216	275	+.113	+31.1
		499		+31.1

$\bar{x} = +62.4\ mm$

$$\frac{\Sigma F_s}{\Sigma f_s} = \frac{1230}{499} = 2470\ mm^2$$

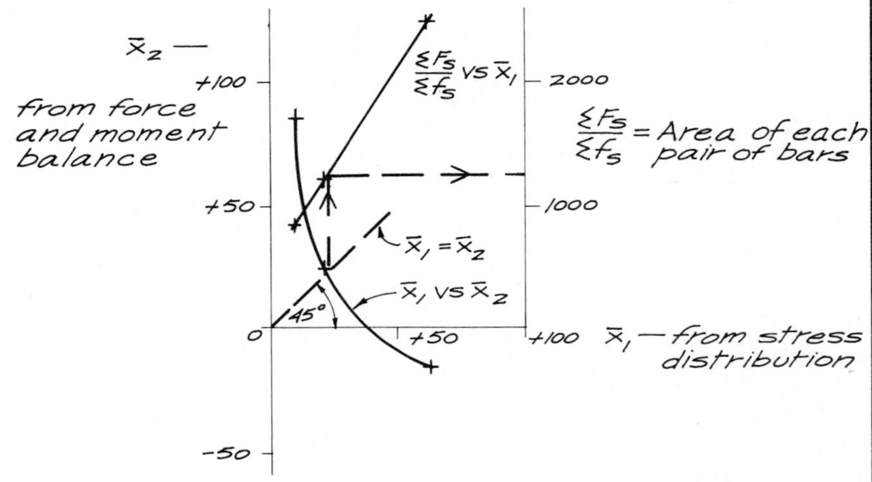

Area required for each pair = 1230 mm²

Say, 2 × 32 mm bars

(1608 mm²)

Use 6 No 32 mm bars

The first step, then, is to calculate the effective length, kl_u, and the radius of gyration, r. Of these, l_u and r are obtained from the dimensions of the structure. The factor k depends on whether the frame is braced against sidesway or not, and on the ratio of the sum of the stiffnesses of the columns to that of the beams at both ends of the column. Although the *Code* itself gives no guidance on the evaluation of k, the *Commentary on Building Code Requirements* contains nomograms for this purpose.[5]

The magnifier, δ, is given by

$$\delta = C_m \Big/ \left(1 - \frac{P_u/\phi}{P_c}\right) \nless 1$$

where P_u/ϕ is the axial design load divided by strength reduction factor; P_c is the Euler buckling load $= \pi^2 EI/(kl_u)^2$; and C_m is the moment correction factor.

If the column is braced against sidesway and has no lateral loads between its ends,

$$C_m = 0.6 + 0.4 \frac{M_1}{M_2} \nless 0.4$$

Otherwise, C_m is taken to be 1.

Computation of the Euler critical load is made difficult by uncertainty about the value of EI for a reinforced concrete section. The uncertainty arises from the non-linear relationship between stress and strain and the low tensile strength of concrete which leads to cracking on the tension side. The ACI *Code* allows the following approximations for estimation of EI:[4]

$$EI = \left(\frac{E_c I_g}{5} + E_s I_{se}\right) \Big/ \left(1 + \beta_d\right)$$

or a more conservative estimate

$$EI = \frac{E_c I_g}{2.5\,(1 + \beta_d)}$$

Here E_c is Young's modulus of the concrete; E_s is Young's modulus of the steel; I_g is the second moment of area of the gross section of the concrete about the centroidal axis, neglecting the reinforcing steel; I_{se} is the second moment of area of the steel about the centroidal axis; and β_d is the ratio of maximum dead load moment to maximum total load moment.

The procedure is supported by experimental evidence and is easy enough to apply, except for the judgement necessary to classify a frame as braced or unbraced. The *Commentary on the Building Code Requirements* offers some guidance on this aspect.[5] It is also worth noting that many reinforced concrete columns are not slender. Studies of some actual structures have shown that slenderness had no effect for about 90% of the columns in braced frames and 40% of the columns in unbraced frames.[5]

19.5. BASES FOR STEEL COLUMNS

A steel column is supported on a concrete foundation and, since the strength of concrete is a good deal less than that of steel, the column load has to be spread over an area larger than the cross-section of the column. This is readily done by welding a steel plate to the end of the column. Design of this baseplate comprises determination of the size of the weld, the plan dimensions and thickness of the plate and the size and disposition of the holding-down bolts. We will set out methods for these calculations using loads in service, a permitted bearing stress at the surface of the concrete foundation and permitted stresses in the baseplate, the welds and the bolts.

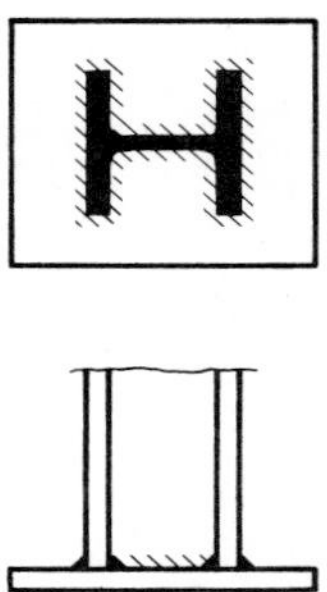

Fig. 19.12. *Baseplate for steel column.*

With regard to the weld which secures the baseplate to the end of the column, it is assumed that there is no bearing contact between the plate and the column. The whole load is assumed to be transferred by the weld. Since fillet welds can often be used, the assumption is conservative but, without machining of the surfaces in contact, it is so difficult to say how much is transferred in bearing that it is best to ignore this contribution to the strength of the joint. When the load to be transferred is large, the end of the column can be prepared for butt welding and a butt weld used to fasten the plate. No calculations need be made, on the grounds that the weld will be as strong as the flanges and web of the column and, if these are not overstressed, the weld also will not be overstressed. Many run-of-the-mill connections can be made with fillet welds. The end of the column is cut square and, if necessary, ground to ensure a reasonable match with the face of the baseplate. The design is based on a fillet weld all round both flanges and the web, as sketched in Fig. 19.12. For an axially loaded column, the load is assumed to be uniformly distributed along the whole length of the weld available. This is easily ascertained since the dimensions of the column are known, so that the strength of weld required per unit of length can be calculated. From this, the size of weld is easily determined (see Chapter 13, "Connections").

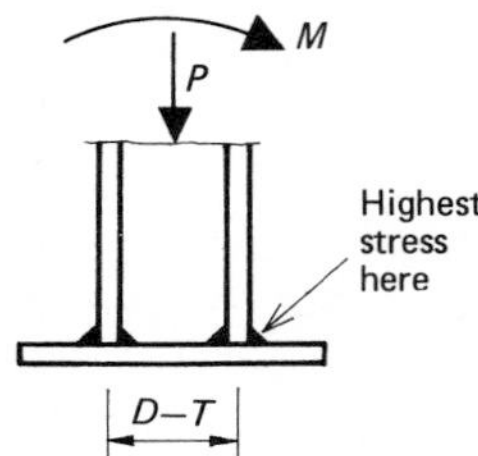

Fig. 19.13. *Fillet welds for baseplate.*

When the column supports a bending moment as well as an axial load, the stress in the weld will not be uniformly distributed and the most heavily loaded weld is along the outer edge of one flange, as indicated in Fig. 19.13. The weld size is based on an estimate of the load per unit length in this weld. Elaborate

calculations are not justified, in view of uncertainties about stress distribution and the relationship between stress and strain. A simple calculation can be made by assuming that
(a) the axial load is uniformly distributed round the total length of weld;
(b) the bending moment is transferred through the welds round the flanges only.
Using the second assumption, the bending force transmitted at the critical flange is approximately $M/(D - T)$, where M is the bending moment; D is the total depth of column section; T is the column flange thickness; and $D - T$ is the distance, centre-to-centre of flanges.

Then, if l is the total length of weld and l_1 is the length of weld round the critical flange, the load per unit length of weld round the critical flange is

$$\frac{P}{l} + \frac{1}{l_1}\frac{M}{D - T}$$

Although there is scope for varying the weld size at different parts of the joint, the gain in economy is scarcely justified and we recommend that the whole weld be the same size as required for the critical flange.

Turning now to design of the baseplate, we apply the laws of mechanics and various assumptions which depend on the size of the bending moment. The behaviour of a baseplate is quite complex, and it is only by using a simplified description of its behaviour that practicable design methods can be devised. It will be noted that a consistent set of assumptions is not used – what leads to straightforward calculation in one case is cumbersome in another. What we want is a calculation that is reasonably logical and easy to use in routine design.

There are three cases – (i), (ii) and (iii).

Case (i)
Let e be the eccentricity of applied load so that $M = Pe$. The dimensions of the baseplate are W and H, as shown in Fig. 19.14. For Case (i), $e < H/6$.

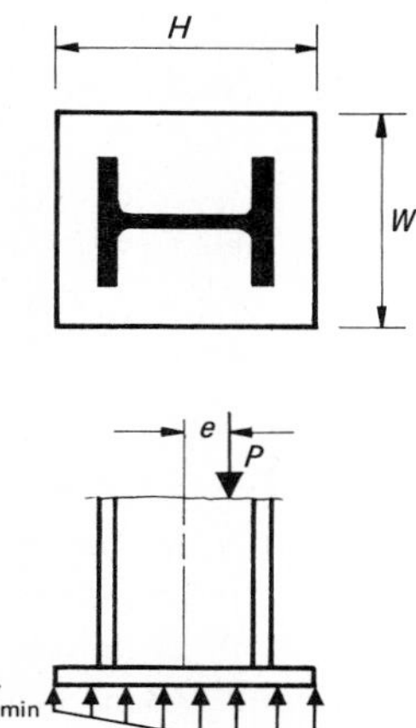

Fig. 19.14. *Design of baseplate, Case (i).*

It is assumed that the contact pressure, or bearing stress between the baseplate and the foundation is distributed linearly. This is shown in Fig. 19.14 and is the same as the stress distribution in a cross-section of a rectangular column. The maximum bearing pressure is obtained by superimposing a uniform stress due to the load P and a bending stress due to the moment M.

$$f_{max} = \frac{P}{WH} + \frac{M\left(\frac{H}{2}\right)}{\frac{1}{12}WH^3}$$

$$= \frac{P}{WH}\left(1 + 6\frac{e}{H}\right) \qquad (19.15)$$

The design criterion to be satisfied is

$$f_{max} \not> \text{permitted bearing stress for concrete}$$

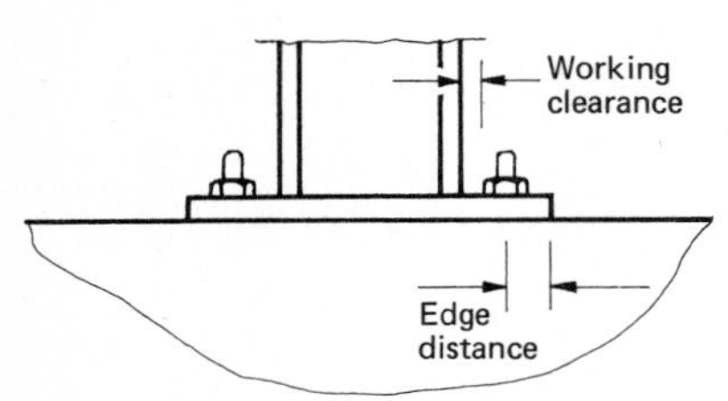

The design problem is to use known values of f_{max}, P and e to evaluate W and H. With only one criterion to satisfy, an arbitrary choice of one dimension is made and Equation (19.15) is used to calculate the other. Note that the choice of dimensions is limited by the need to leave room for holding-down bolts and a spanner to tighten the nuts. Edge distance may not be less than values laid down in the design code – typically 1.5 to 2 times the diameter of the bolt, depending on the nature of the edge of the plate.

The basis of the limiting value e/H for Case (i) (i.e. $e/H = 1/6$) is seen by considering f_{min}. We have

$$f_{min} = \frac{P}{WH} - \frac{M\left(\frac{H}{2}\right)}{\frac{1}{12}WH^3} = \frac{P}{WH}\left(1 - 6\frac{e}{H}\right)$$

and it is clear that if

$$\frac{e}{H} > \frac{1}{6}$$

a tensile stress would be required over part of the area of contact between the baseplate and the foundation. This is not possible, and design based on a continuous distribution of stress over the whole area of contact is not valid.

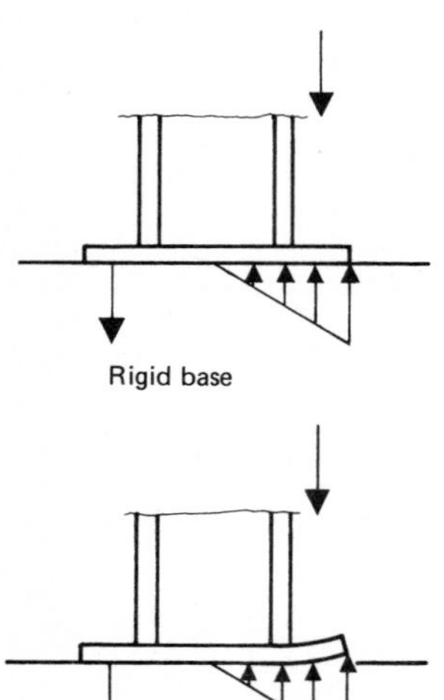

The tension necessary for equilibrium must be mobilised in some other way, and this leads to Case (ii).

Case (ii)

$$\frac{e}{D} > \frac{1}{4}\left(1 + \frac{H}{D}\right)$$

Case (ii) is such that the side of the baseplate far from the eccentric load tends to lift off the concrete foundation. Equilibrium is possible only if a tensile force in the holding-down bolts on that side acts to hold the plate down. The basic assumption made for this calculation concerns the distribution of bearing stress – it is assumed to be uniformly distributed over the part of the base outside the column section on the same side as the eccentric load and zero elsewhere. The contribution of any other part of the base in contact is ignored (see Fig. 19.15).

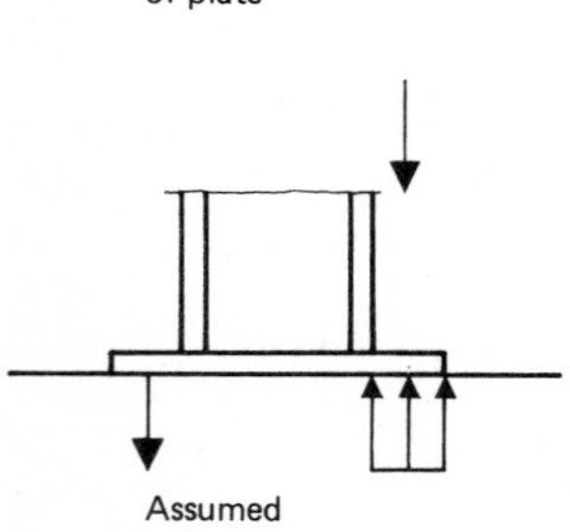

The uniform distribution is justified by upward deflection of the outer part of the baseplate. If the plate were perfectly rigid, the stress could be expected to increase with distance from the centreline. Upward deflection of a flexible plate acts to take the peak off this stress distribution and, although the result cannot be calculated, the tendency is towards a more uniform distribution.

Equilibrium of a free body comprising the lower part of the column and baseplate is described by two equations, one for vertical forces and one for

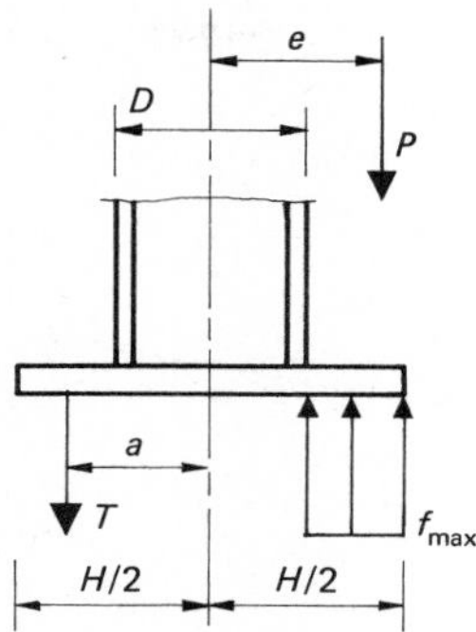

Fig. 19.15. *Design of baseplate, Case (ii).*

moments about T, the bolt tension. They are as follows:

$$T + P = f_{max}\frac{H - D}{2}W \tag{19.16}$$

$$P(e + a) = f_{max}\frac{H - D}{2}W\left(a + \frac{H + D}{4}\right) \tag{19.17}$$

When Equation (19.17) is solved for W, the result is

$$W = \frac{2P(e + a)}{f_{max}(H - D)\left(a + \dfrac{H + D}{4}\right)} \tag{19.18}$$

When Equation (19.17) is divided by Equation (19.16), we get

$$\frac{P(e + a)}{T + P} = a + \frac{H + D}{4}$$

whence

$$T = P\frac{e - \dfrac{1}{4}(H + D)}{a + \dfrac{1}{4}(H + D)} \tag{19.19}$$

These relationships for W and T can be used for design, in the following way:

(a) the stress, f_{max}, is put equal to the permitted bearing stress;
(b) a trial value of H is chosen;
(c) an estimate is made of the value of a. This depends on H, the depth, D, of the column and an allowance for working clearance;
(d) W and T are calculated;
(e) suitable holding-down bolts are selected on the basis of the computed tension and the permitted stress for tension in a bolt;
(f) if necessary, repeat the process with a better estimate of H.

The origin of the limiting value of e/H for Case (ii) can now be seen – it corresponds to $T = 0$. Clearly, $T > 0$ only if $e/H > 1/4\,\{1 + (D/H)\}$ – i.e. if $e/D > 1/4\,\{1 + (H/D)\}$.

It is now apparent that the boundary criteria for Cases (i) and (ii) can be plotted on axes e/D and H/D, as in Fig. 19.16. It will be seen that there is an area on this plane for which neither Case (i) nor Case (ii) is relevant – this is Case (iii).

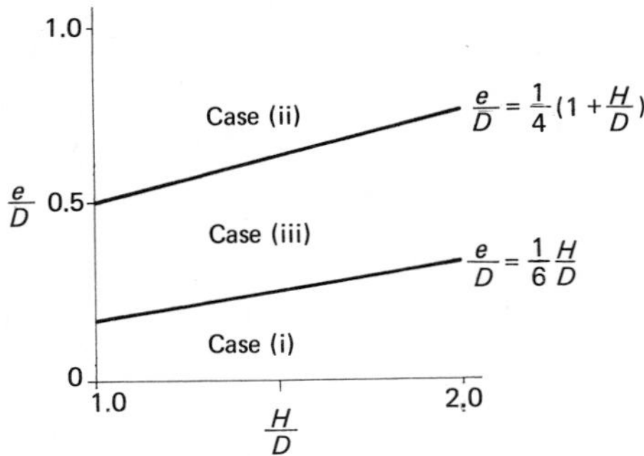

Fig. 19.16. *Boundaries for Cases (i), (ii) and (iii) of baseplate design.*

Case (iii)

$$\frac{1}{6} < \frac{e}{H} < \frac{1}{4}\left(1 + \frac{D}{H}\right)$$

$$\frac{1}{6}\frac{H}{D} < \frac{e}{D} < \frac{1}{4}\left(1 + \frac{H}{D}\right)$$

Under these circumstances, design is based on a bearing pressure distributed over part of the area of contact and without reliance on bolt tension for equilibrium. The bearing pressure is assumed to be linearly distributed, rising from zero to a maximum of f_{max} as shown in Fig. 19.17. This distribution is compatible with Case (i) at the boundary between Cases (i) and (iii), but does not match the assumption at the boundary between Cases (ii) and (iii).

For equilibrium, it is necessary for the resultant of the distributed bearing pressure to be collinear with the eccentric load – they are the only two forces acting. Hence, from Fig. 19.17,

$$e + \frac{x}{3} = \frac{H}{2}$$

$$x = 3\left(\tfrac{1}{2} H - e\right)$$

Furthermore, balancing these equal and opposite forces, we get

$$P = \tfrac{1}{2} f_{max} (Wx)$$

$$W = \frac{2P}{x f_{max}}$$

$$= \frac{2P}{3 f_{max} \left(\frac{1}{2} H - e\right)}$$

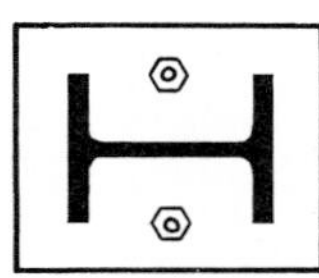

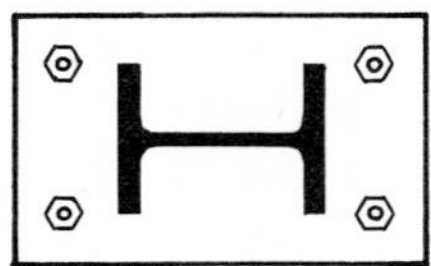

By putting f_{max} equal to the permitted bearing pressure and selecting a trial value of H, W can be calculated.

It will be seen that Cases (i) and (iii) do not require holding-down bolts for equilibrium and Case (ii) requires them on one side only. Where the calculation indicates that no bolts are necessary, nominal holding-down bolts must be provided. For a small, lightly loaded column, two will suffice and, if the eccentricity is small, they can be located between the flanges. Otherwise, there should be at least four bolts – one near each corner of the baseplate.

The effect of the different assumptions for Cases (ii) and (iii) should be noted. At the boundary between them, the stress distributions are as sketched in Fig. 19.18. Since there is no tension in the holding-down bolts, the resultant

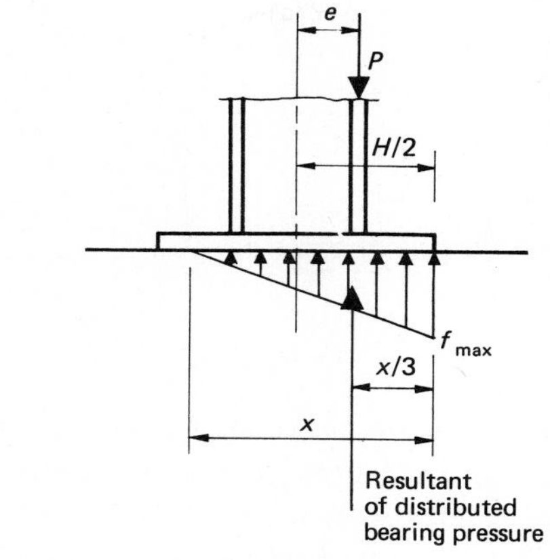

Fig. 19.17. *Design of baseplate, Case (iii).*

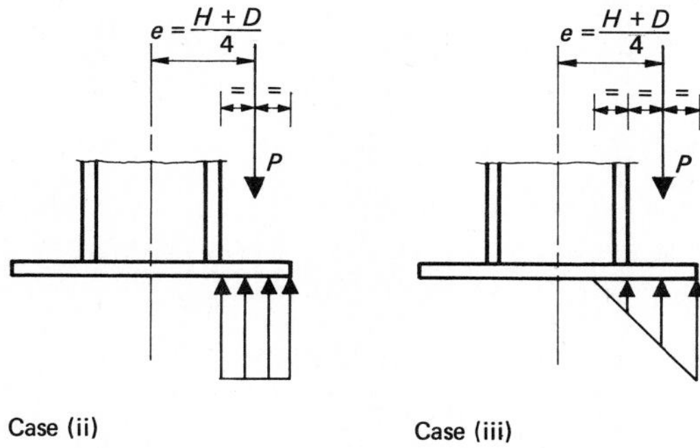

Fig. 19.18. *Comparison of Cases (ii) and (iii) at their common boundary.*

of the bearing pressure is collinear with the eccentric load in both cases. The calculated width of the baseplate will be

$$W = 2\frac{P}{f_{max}(H - D)} \tag{19.20}$$

if the rule for Case (ii) is used, and

$$W = 2.67\frac{P}{f_{max}(H - D)} \tag{19.21}$$

if the problem is considered to be Case (iii).

This discrepancy is not desirable, but can be accepted as a price paid for simplicity of calculation. Note that Cases (ii) and (iii) overlap.

The designer can keep away from this boundary by changing his choice of H. Since e and D are fixed by the loading and dimensions of the column, changing H has the effect of moving the design problem horizontally on Fig. 19.16 as indicated in the margin. Decreasing H will enable the designer to use the rules for Case (ii) with an increase in W over that given by Equation (19.20), as indicated by Equation (19.18). Increasing H makes Case (iii) relevant and reduces W from the value given by Equation (19.21). It appears that the overlap is not as wide as Equations (19.20) and (19.21) suggest at first sight.

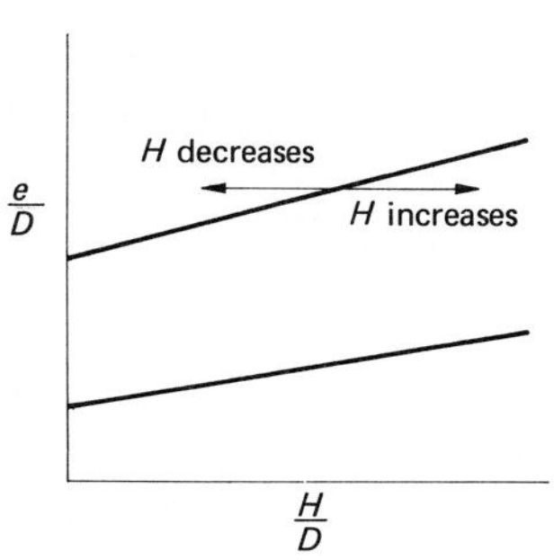

The remaining decision to be taken concerns the thickness of the baseplate. A design formula can be derived by considering the part of the baseplate projecting beyond the column as a cantilever. This cantilever is loaded by the

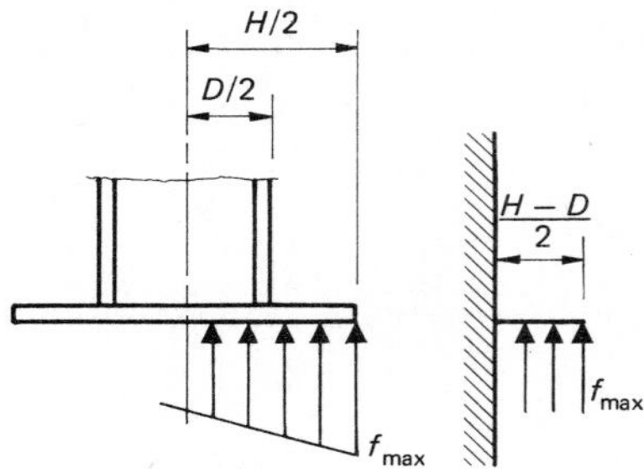

Fig. 19.19. *Outstanding part of baseplate is designed as a cantilever.*

upward contact pressure of the base, as shown in Fig. 19.19. The bearing stress on this projecting part can be taken as being uniformly distributed, so that we have to design a cantilever with a span of $\frac{1}{2}(H - D)$ and carrying a total load of $f_{max}\, W\, \frac{1}{2}(H - D)$ uniformly distributed. The bending moment in the baseplate, M_B is given by

$$M_B = f_{max}\, W\, \tfrac{1}{2}(H - D) \times \frac{H - D}{4}$$

$$= \tfrac{1}{8} f_{max}\, W\, (H - D)^2$$

The maximum permitted bending moment, for an allowable bending stress of p_{bt} in the baseplate is

$$p_{bt} \times \tfrac{1}{6} W t^2$$

where t is the thickness of the baseplate. Hence,

$$t = \sqrt{\frac{6\,\dfrac{MB}{W}}{p_{bt}}} = \sqrt{\frac{3\, f_{max}}{p_{bt}}\left(\frac{H - D}{2}\right)^2} \qquad (19.22)$$

In deriving this formula, we have assumed two-dimensional bending. For large eccentricities, this is a reasonable assumption, but when the load is axial, or nearly so, three-dimensional bending will occur as the plate tends to deflect to the shape of a dish around the column. Equation (19.22) would then give a conservative estimate of the thickness required. The British design code (BS 449) gives a formula which allows for this dishing as follows:

$$t = \sqrt{\frac{3\, f_{max}}{p_{bt}}\left(A^2 - \frac{B^2}{4}\right)} \qquad (19.23)$$

where A is the greater projection of the baseplate beyond the column, and B is the lesser projection of the baseplate beyond the column. If Equation (19.22) were used for an axially loaded column, we would replace $\frac{1}{2}(H - D)$ by A and the relationship between the two is clear. A limiting case for Equation (19.23) arises when A and B are equal. Then Equation (19.23) gives a thickness which is about 87% of that given by Equation (19.22).

Design of a baseplate as described here is illustrated in Example 19.8.

19.6. BASES FOR TIMBER COLUMNS

Bases for timber columns are not required to spread loads but rather to provide for fastening a column to its base and to ensure that the timber is kept dry and

Example 19.8 Sheet 1 of 2

Select dimensions for column baseplate and holding down bolts, given column loading and dimensions.
Refer to Example 19.2 for design of column.

Design data:

$P = 700$ kN $e = 550$ mm $D = 328$ mm
Allowable bearing stress on concrete = 7.5 MPa
Allowable stress in baseplate = 150 MPa
Allowable tensile stress on nett area of bolts = 130 MPa

$\frac{e}{D} = \frac{550}{328} = 1.68$ Ref. Fig. 19.16 : Case II

Try $H = 750$ mm

Allowing edge distance = 50 mm
$a = 325$ mm

$e = 550$
$a = 325$
$e + a = 875$

$H + D = 1078$ $\frac{H+D}{4} = 269.5$
$H = 750$ $a = 325$
$D = 328$ $a + \frac{H+D}{4} = 594.5$
$H - D = 422$

50 | 325 | 375 | 750

$$W = \frac{2 \times 7 \times 10^5 \times 0.875}{7.5 \times 10^6 \times 0.422 \times 0.595} \text{ m} = 650 \text{ mm} \quad \text{OK}$$

$$t = \sqrt{\frac{3 \times 7.5 \times 10^6 \times 0.211 \times 0.211}{150 \times 10^6}} \text{ m} = 82 \text{ mm} \quad \text{Too big?}$$

$$T = 700 \times \frac{280}{595} = 330 \text{ kN}$$

Try 4 bolts @ 130 MPa $A = \frac{330 \times 10^3}{4 \times 130} = 635 \text{ mm}^2$

4 No 36 mm diameter bolts

Example 19.8 Sheet 2 of 2

To reduce thickness of baseplate, weld stiffeners to base of column:

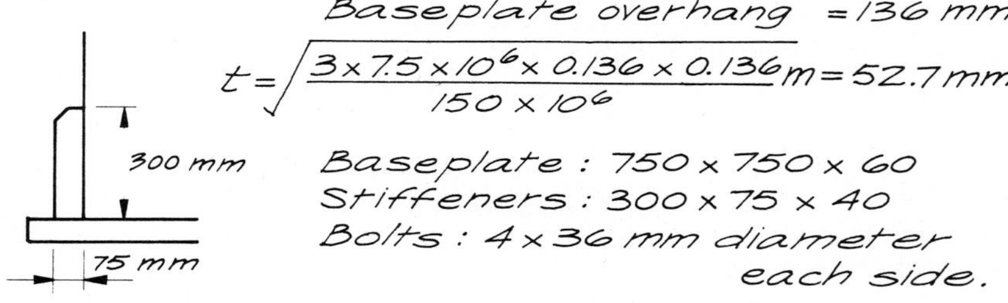

Baseplate overhang $= 136$ mm

$$t = \sqrt{\frac{3 \times 7.5 \times 10^6 \times 0.136 \times 0.136}{150 \times 10^6}}\ \text{m} = 52.7\ \text{mm}$$

Baseplate : $750 \times 750 \times 60$
Stiffeners : $300 \times 75 \times 40$
Bolts : 4×36 mm diameter each side.

For welding of baseplate to column try fillet welds at 115 MPa

$D = 328$ mm $\quad T = 10 + 15 = 25$ mm $\quad D - T = 303$ mm

$$\frac{M}{D-T} = \frac{700 \times 550}{303} = 1270\ \text{kN}$$

ℓ : 328
305
295
928
×2
1856 mm

ℓ_1 : 305
50
295
650 mm

$$\frac{P}{\ell} + \frac{1}{\ell_1}\ \frac{M}{D-T} = \frac{700}{1856} + \frac{1270}{650} = 2330\ \text{N/mm}$$

Too much for fillet welds

Double – V edge preparation, web and flanges of column. Butt weld all round to baseplate.

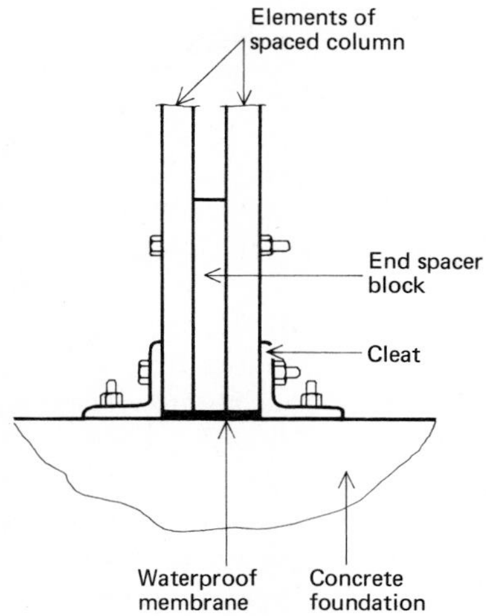

Fig. 19.20. *Base for timber column.*

well ventilated. Steel cleats bolted to the sides of the column and to the foundation, and a membrane of waterproof fabric under the timber meet these requirements (Fig. 19.20).

19.7. REINFORCED CONCRETE FOOTINGS

We have shown how, because of the difference in strength of steel and concrete, the load carried by a steel column must be spread over an enlarged area at the base where the column is in contact with the concrete foundation. In the same way, the concrete foundation must be larger than the column (if it is of reinforced concrete) or the baseplate of a steel column, because of the limited strength of the clay or rock beneath. This enlarged concrete base is the *footing* and here we examine the calculations required for design of a square footing to support an axially loaded reinforced concrete column.

The objective of the designer is to determine the plan dimensions, thickness and reinforcing steel required. Criteria to be satisfied are that a safe bearing pressure on the subsoil is not exceeded and that the ultimate strength of the footing allows a suitable margin of safety under working conditions.

The plan dimensions are determined from consideration of a safe bearing pressure on the subsoil. This safe pressure may be found from tests on samples of the soil or it may be based on visual classification of the type of soil. The former procedure is more reliable but it is also more costly and is therefore justified only for large and important foundations. Many lightly loaded footings can be designed using safe bearing pressures taken from handbooks or a local authority's by-laws.

The total load applied to the subsoil comprises the column load (including its own weight) and the weight of the footing, so that the design criterion is

$$\frac{P + wL^2 t}{L^2} \not> p_b$$

where P is the total column load; L is the side of square footing; t is the thickness of footing; w is the weight per unit volume of concrete; and p_b is the safe bearing stress for subsoil. Hence,

$$L^2 \nless \frac{P}{p_b - wt}$$

To find the size of footing required, the thickness is estimated and the weight per unit area of the footing (i.e. wt) is deducted from p_b. The contact area required is calculated by dividing this nett bearing pressure into the total load on the column.

The thickness of a footing usually depends on shear or diagonal tension. Considered as a cantilever beam bending upwards under the action of the upward pressure of the subsoil, the footing has a short span and high intensity of loading. Short spans and large loads make shear more critical than bending. The thickness of a footing can certainly be reduced by including inclined tension reinforcement, and the ACI *Code* has rules to cover this.[4] We do not think it is worth while. Footings are easy to build; they are quite free from congested reinforcement and the concrete does not have to be lifted into place. These things are reflected in the low cost of concrete in footings so that it is usually better to increase the amount of concrete in the footing and keep the amount of steel down.

According to the ACI *Code* two modes of failure have to be investigated.[4] In one, the projecting part of the footing shears like a two-dimensional cantilevered slab, as sketched in Fig. 19.21. Failure by inclined tension is assumed to occur on a plane inclined at 45° to the top surface and extending right across the footing from the edge of the column to the top layer of reinforcing steel at a depth d below the top. The shear which causes this failure is taken to be the resultant of the forces outside a vertical plane at a distance d from the face of the column.

Under working conditions, this external shear is given by

$$V = pL\,\frac{L - h - 2d}{2} - wt\,L\,\frac{L - h - 2d}{2}$$

$$= (p - wt)\,L\,\frac{L - h - 2d}{2}$$

$$= \frac{P}{L^2}\,L\,\frac{L - h - 2d}{2}$$

where p is the actual bearing pressure due to total load on subsoil, h is the side of column, and the other symbols are as previously defined. Note that the weight of the footing itself does not contribute to this shear force – the weight of the footing is balanced by part of the upward soil pressure and only the remainder has to be resisted by internal stresses in the footing.

Multiplying both sides by a load factor gives the external shear V_u for ultimate strength design of the footing

$$V_u = \frac{P_u}{L^2}\,L\,\frac{L - (h + 2d)}{2}$$

From consideration of the strength of the footing, we have

$$V_u \ngtr \phi\, v_{u1}\, Ld$$

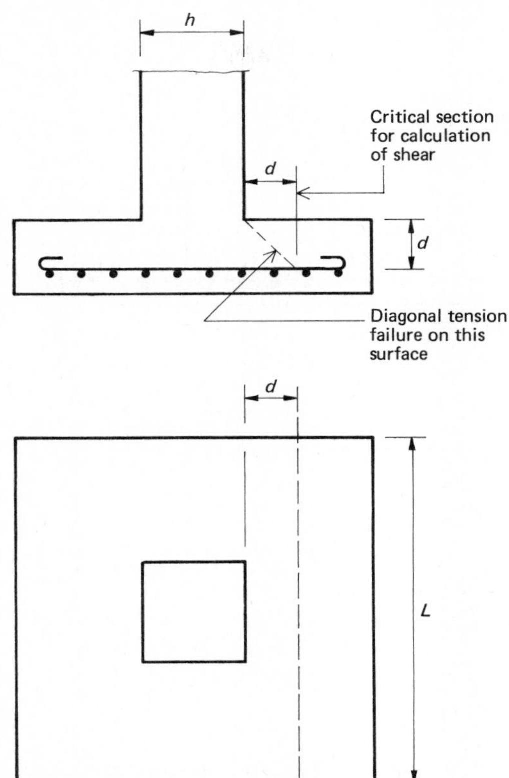

Fig. 19.21. *Two-dimensional case of diagonal tension in footing.*

where v_{u1} is the strength of concrete in diagonal tension, and ϕ is the strength reduction factor (= 0.85 for shear).
Hence, the effective depth is given by

$$d \nless \tfrac{1}{2}(L - h) \Big/ \left(1 + \frac{\phi v_{u1} L^2}{P_u}\right)$$

The other mode of failure which must be taken into account assumes that inclined cracks form on all four sides of the column, extending at 45° from the horizontal from the faces of the column at the top of the footing to the top layer of reinforcing steel. They are the side surfaces of a frustum of a pyramid, as sketched in Fig. 19.22. The critical section for computation of the external shear and the resistance is a vertical surface surrounding the column, as seen in plan, and at a distance $d/2$ from the column.

In this case, we have for the ultimate external shear

$$V_u = \frac{P_u}{L^2}\{L^2 - (h + d)^2\} \tag{19.24}$$

From the internal stresses in the footing, we obtain

$$V_u = \phi\, v_{u2} 4\,(h + d)\, d \tag{19.25}$$

and the condition to be satisfied is

$$\frac{P_u}{L^2}\{L^2 - (h + d)^2\} \ngtr \phi\, v_{u2} 4\,(h + d)\, d$$

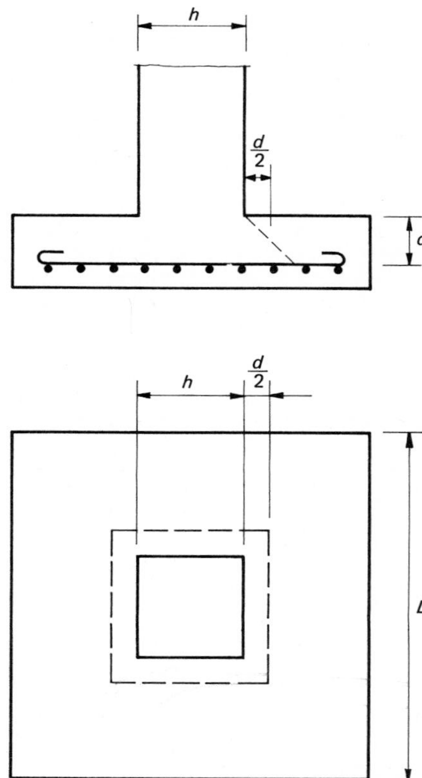

Fig. 19.22. *Three-dimensional case of diagonal tension in footing.*

The limiting case can be solved as a quadratic equation, but it is almost as easy, and much more revealing to plot V_u against d from Equations (19.24) and (19.25). Equation (19.24) gives the external shear and Equation (19.25) gives the strength. Acceptable values of d are found where the strength exceeds the load, as indicated in Fig. 19.23.

It is important to note that the ACI *Code*[4] allows a larger strength in diagonal tension for the three-dimensional case – i.e. $v_{u2} > v_{u1}$.

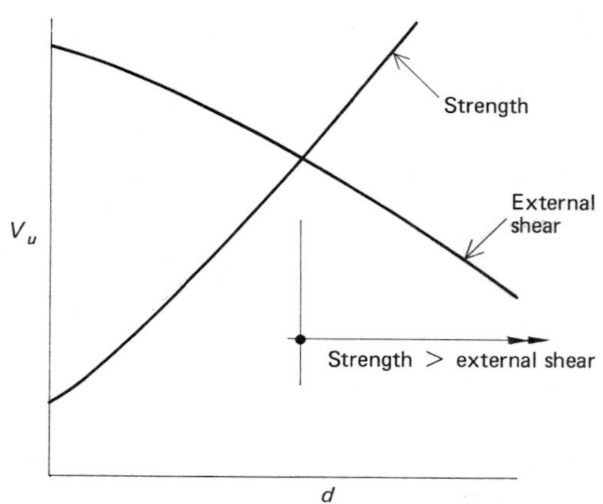

Fig. 19.23. *Strength of footing and external load plotted against effective depth (three-dimensional case).*

Example 19.9 Sheet 1 of 3

Select dimensions and reinforcing steel for reinforced concrete footing for column of Example 19.3.

Design data:

Column dimensions: 450 mm × 450 mm

Column load: $P = 2300$ kN $P_u = 3600$ kN

Safe bearing load on soil $= 500$ kPa

$f_c' = 20$ MPa $f_y = 275$ MPa

$v_{u1} = 0.75$ MPa $v_{u2} = 1.5$ MPa

$w = 22.5$ kN/m²

$\phi = 0.9$ for bending, 0.85 for diagonal tension

$E_s \epsilon_u = 621$ MPa

Balanced design — bending:

$$\rho_b = \frac{0.85 \times 17}{275} \times \frac{1}{1 + \frac{275}{621}} = \frac{0.85 \times 17}{275 \times 1.443} = 0.0364$$

$0.75 \rho_b = 0.273$

$$\frac{a}{d} = \frac{0.0273 \times 275}{17} = 0.441$$

$$k = \phi (0.85 f_c') \frac{a}{d} \left(1 - \frac{1}{2}\frac{a}{d}\right) = 0.9 \times 17 \times 0.441 \times 0.780$$
$$= 5.27 \text{ MPa}$$

Estimated thickness $= 0.6$ m

Estimated weight of footing $= 22.5 \times 0.6 = 13.5$ kPa

Nett bearing pressure $= 500 - 14 = 486$ kPa

$$L^2 = \frac{2300}{486} = 4.73 \text{ m}^2$$

$$L = 2.18 \text{ m}$$

Make $L = 2.20$ m $\qquad \frac{P_u}{L^2} = 744$ kPa

For two-dimensional shearing of footing:

$$\frac{\phi v_{u1} L^2}{P_u} = \frac{0.85 \times 0.75}{0.744} = 0.858$$

$$L - h = 2.20 - 0.45 = 1.75 \text{ m}$$

$$d \not< \frac{1.75}{2 \times 1.858} = 0.471 \text{ m}$$

For three-dimensional shearing:

External shear $= 744\{4.84 - (0.45 + d)^2\}$ kN

Strength $= 0.85 \times 1500 \times 4(0.45 + d)d$

$= 5100(0.45 + d)d$ kN

Example 19.9 Sheet 2 of 3

d	$0.45+d$	$5100(0.45+d)d$	$4.84-(0.45+d)^2 = A$	$744A$
0.4	0.85	1730	4.12	3060
0.5	0.95	2420	3.94	2920
0.6	1.05	3210	3.74	2780

With $b = L = 2.2$ m

$$kbd^2 = 5.27 \times 10^6 \times 2.2 \times 0.577 \times 0.577 \text{ Nm} = 3860 \text{ kNm}$$

$$M_u = 1/8 \times 744 \times 10^3 \times 2.2 \times 1.75 \times 1.75 \text{ Nm} = 626 \text{ kNm}$$

Design as oversize beam

$$F = \frac{M}{(0.85 f_c')\,bd^2} = \frac{6.26 \times 10^5}{17 \times 10^6 \times 2.2 \times 0.577 \times 0.577} = 0.0502$$

$$\frac{a}{d} = 1 - \sqrt{1 - 2F} = 1 - \sqrt{1 - 0.1004} = 0.052$$

$$1 - \frac{1}{2}\frac{a}{d} = 0.974 \qquad d - \frac{1}{2}a = 0.562 \text{ m}$$

$$T = \frac{626}{0.562} = 1120 \text{ kN} \qquad A_s = \frac{1120 \times 10^3}{275} = 4070 \text{ mm}^2$$

Say, 36 No 12 mm diameter bars

Thickness of footing: d = 577
Bar radius = 6
Bar diameter = 12
Cover = 75
670 mm

Example 19.9 Sheet 3 of 3

Check total load :

Weight of footing $= 0.67 \times 22.5 = 15.1$ kPa

$$\frac{P}{L^2} = \frac{2300}{4.84} = 475 \text{ kPa}$$

490 kPa

< 500 kPa OK

Check development length :

$$\ell_d \not< \frac{0.019\, A_b f_y}{\sqrt{f_c'}} = \frac{0.019 \times 113 \times 275}{\sqrt{20}}$$

$= 132$ mm

$$\ell_d \not< 0.06\, d_b f_y = 0.06 \times 12 \times 275$$

$= 198$ mm

$\ell_d \not< 300$ mm.

Bars to cross footing completely, both ways;

gives $\ell_d \simeq (2.20 - 0.45) \times 1/2 = 0.875$ m OK

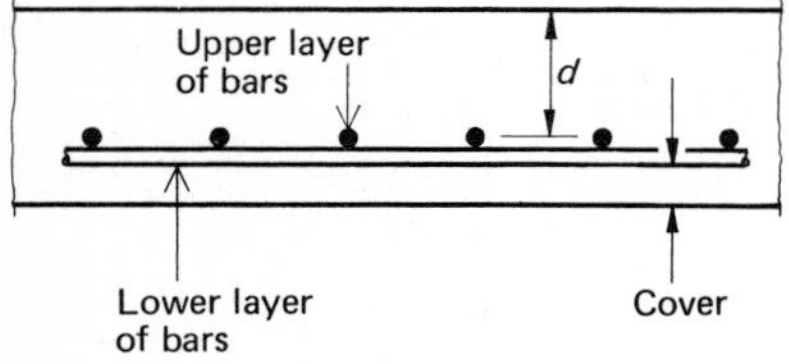

These calculations lead to an acceptable value for d, the depth to the centre of tension in the top layer of reinforcement. To this must be added

(a) the radius of the bars in that layer;
(b) the diameter of similar bars in another layer at right angles, because the footing has to be reinforced both ways;
(c) an allowance for concrete cover under the bars.

With these additions the thickness of the footing is obtained.

The last part of the design is to calculate the amount and disposition of the reinforcing steel required. For this, the projecting part of the footing is treated like a two-dimensional cantilever bending upwards. The critical section is taken at the face of the column, so that the design moment is given by

$$M_u = \underbrace{\frac{P_u}{L^2} \times L\left(\frac{L-h}{2}\right)}_{\text{Nett upward force}} \times \underbrace{\tfrac{1}{2}\left(\frac{L-h}{2}\right)}_{\text{Lever arm}}$$

that is $$M_u = \tfrac{1}{8}\frac{P_u}{L^2} L\,(L-h)^2$$

Reinforcement for a cross-section of breadth L and effective depth d and this ultimate moment is designed as described in Chapter 16, "Reinforced Concrete Beams". We draw attention particularly to the importance of development length and anchorage of the reinforcing steel. There is not overmuch length for development and anchorage in most footings which may make this a critical aspect of the design.

Example 19.9 shows the application of these methods to design of a footing for the column of Example 19.3.

LIST OF SYMBOLS

A	area, larger projection of baseplate beyond column
A_g	gross area of concrete in cross-section
A_s, A_{s1}, A_{s2}	areas of steel in cross-section
a	depth of compression stress block, lever arm of tension in bolts
a_b	depth of compression stress block for balanced condition
B	lesser projection of baseplate beyond column
b	breadth of rectangular reinforced concrete section
C_m	moment correction factor for eccentrically loaded reinforced concrete columns
c	distance from neutral axis to extreme fibre in compression
D	depth of steel section
d	effective depth of reinforced concrete section – i.e. depth from compression face to steel near the opposite face
d_s, d'	distance of centre of steel from adjacent concrete face
E	modulus of elasticity
E_c, E_s	modulus of elasticity for concrete and steel respectively
e	eccentricity of applied load, measured from axis of column
e_t	eccentricity of load in reinforced concrete column, measured from steel on far side from load

F_c	resultant internal concrete force
F_s, F_{s1}, F_{s2}	resultant internal steel forces
F_y	vertical force or component
f_{bc}	compressive stress due to bending
f_c	compressive stress due to axial load
f_c'	strength of concrete in compression
f_{max}	maximum stress between baseplate and foundation
f_{min}	minimum stress between baseplate and foundation
f_s, f_{s1}, f_{s2}, f_{s3}	stresses in steel reinforcement
f_y	yield strength of steel reinforcement
H	length of side of baseplate
h	total depth of reinforced concrete section
I	second moment of area
I_g	second moment of area of gross concrete section about centroidal axis, neglecting reinforcement
I_{se}	second moment of area of steel reinforcement about centroidal axis of section
k	effective length factor for reinforced concrete columns
L	length of side of square footing
l, l_u	lengths of columns
l, l_1	lengths of welds
M	bending moment
M_A	allowable bending moment
M_B	bending moment in baseplate
M_c	moment to be used for design of reinforced concrete column
M_u	ultimate moment
M_1	lesser end moment in reinforced concrete column, positive for single curvature, negative if curvature reversed
M_2	larger end moment in reinforced concrete column, always positive
P	load parallel to axis of column
P_A	allowable axial load
P_b	balanced load
P_c	Euler critical load
P_u	ultimate load
P_1, P_2	longitudinal forces
p	bearing pressure on foundation
p_b	safe bearing pressure on foundation
p_{bc}	allowable compressive stress due to bending
p_{bt}	allowable tensile stress due to bending
p_c	allowable compressive stress due to axial load
r	radius of gyration of section
T	flange thickness, tension in holding-down bolts
t	thickness of baseplate or footing
V	external shear
V_u	ultimate external shear
v_u, v_{u1}, v_{u2}	ultimate inclined tensile stresses
W	side of baseplate
w	weight per unit volume of concrete
x	width of stressed area under baseplate

LIST OF SYMBOLS *(continued)*

$\overline{x}$	distance of resultant steel force from axis of column
x_c	distance of F_c from axis of column
x_s	distance of F_s from axis of column
Z	section modulus
β_1	ratio of depth of compression stress block to distance from extreme fibre in compression to neutral axis
β_d	ratio of dead-load moment to total moment in reinforced concrete column
δ	moment magnifier
$\varepsilon_{s1}, \varepsilon_{s2}$	strains in steel reinforcement
ε_u	ultimate strain in concrete
ε_y	yield strain in steel
ρ	ratio of area of steel on tension side of column to effective area
ρ_b	steel ratio for balanced conditions
σ	compressive stress due to combined axial load and bending
ϕ	strength reduction factor

REFERENCES

[1] BS 4:Part 1:1972, "Hot Rolled Sections", *Specifications for Structural Steel Sections,* British Standards Institution, London, 1972.

[2] BS 449:Part 2:1969, "Metric Units", *The Use of Structural Steel in Building,* British Standards Institution, London, 1969.

[3] K. H. GERSTLE, *Basic Structural Design*, McGraw-Hill, New York, 1967.

[4] ACI 318–71, *Building Code Requirements for Reinforced Concrete,* American Concrete Institute, Detroit, 1971.

[5] ACI 318–71, *Commentary on Building Code Requirements for Reinforced Concrete,* American Concrete Institute, Detroit, 1971.

BIBLIOGRAPHY

BSCP 112:Part 2:1971, "Metric Units", *The Structural Use of Timber in Building,* British Standards Institution, London, 1971.

FERGUSON, P. M., *Reinforced Concrete Fundamentals,* 3rd ed., John Wiley, New York, 1973.

A comprehensive text on reinforced concrete design, The rules in the ACI *Code* are explained in detail.

GRAY, C. S., LEWIS, E. K., MITCHELL, W. E. and GODFREY G. B., *Steel Designers' Manual*, 4th ed., Crosby Lockwood, London, 1972.

A manual based on BS 449[2]

HANSEN, H. J., *Modern Timber Design*, 2nd ed., John Wiley, New York, 1948.

PROBLEMS

19.1. Obtain copies of the codes of practice in use locally and prepare brief summaries of the principal rules relating to steel, timber and reinforced concrete columns.

19.2. Redesign the column of Example 19.1 using a box section made up of two plated channels. Compare the area of cross-section with that obtained in Example 19.1.

19.3. A steel column comprises a 305 × 305 × 97 universal column (refer to BS 4) and its effective length is 6 m. Calculate the safe load for the column if the eccentricity is 250 mm and bending is about the strong axis. Use locally relevant design data or the data of Example 19.2.

19.4. A spaced timber column comprises three 200 mm × 75 mm (nominal) planks of Douglas fir. The column has an effective length of 3 m. Calculate the safe working load using the design rules applicable locally or those in BSCP 112, *The Structural Use of Timber*.

19.5. Take the column section designed in Example 19.3 and calculate the ultimate load which could be supported at eccentricities of 0, 50 mm, 100 mm, 500 mm, 1 m and 1.5 m. Refer to Chapter 16, "Reinforced Concrete Beams" and calculate the ultimate moment without an axial load. Plot your results in the form of an interaction diagram. Use the design data of Example 19.3, except for load and eccentricity.

19.6. Design a base for the column of Problem 19.2. Use your local code, or take data from Example 19.8.

19.7. Determine the dimensions and reinforcement required for a wall footing to support a wall whose thickness is 250 mm. The total load applied to the footing is 450 kN/m. Use design data relevant locally or, if these are not accessible, the data of Example 19.9.

19.8. Determine the dimensions and reinforcement required for a square footing of a 400 mm × 400 mm reinforced concrete column, the load applied to the footing being 2000 kN and the safe bearing stress on the soil 400 kPa. For other data, take the same values as in Example 19.9.

19.9. Redesign the footing of Problem 19.8 using a safe bearing stress of 600 kPa. Calculate the quantities of concrete, formwork and reinforcing steel in each case and, using local unit rates, estimate the cost of each footing.

19.10. Two columns 2 m apart are both 400 mm × 400 mm in cross-section. The total load at the bottom of one is 2000 kN and at the bottom of the other 2500 kN. Design a rectangular footing to support both columns, using the design data of Example 19.9.

PART III

20 Properties of Materials

20.1. INTRODUCTION

In Chapter 6, some of the more important factors that must be taken into account when selecting a material were discussed. We now consider the properties and behaviour of some of the common engineering materials in more detail. Representative values for a selection of these are given in Appendix A4. In many cases, a range of values can be produced, for example, by varying the heat treatment of metals, and such detailed information should be sought in the references that are given. Some typical mechanical and physical properties for the materials discussed here are given in Table 6.1. The properties of timber have been considered in Chapter 14.

It must be understood that strength values specified in standards are usually minimum values, and that most materials supplied by manufacturers will surpass them. A distinction should be made between minimum values and "typical" values. When a material is to be purchased to a manufacturer's specification, the values given in the specification can be used for design.

It is a relatively simple matter to select from the very large number of available types of materials one that will meet any reasonable requirements as regards properties and finish. To make the selection with regard to economy, availability and conservation of the scarcer elements is less simple. For important applications involving large amounts of material it may be advisable to use the consultant services provided by manufacturers.

In the selection process, the various limiting conditions that are considered at the outset of a design provide a useful starting point. Although concentration of design simply on the avoidance of failure may appear to be a somewhat negative approach, the list of possible failure modes gives a positive guide when determining the qualities required in a material.

20.2. CONCRETE

Concrete has the lowest cost per unit mass or volume of any of the materials we discuss. However, it is used in very large quantities on individual jobs, and optimum efficiency and economy require that it be correctly specified.

The cost of ready mixed concrete in bulk quantities is about $25 per cubic metre, which is about $10 per tonne. The range of strength when the age is 28 days in normal building construction is from 15 MPa to 25 MPa. For prestressed concrete the 28-day strength is usually about 35 MPa and special mixes can be designed with higher strength than this.

The factor which affects the strength of concrete most is the ratio of water to cement in the plastic concrete. As this ratio rises, the 28-day strength falls.

Design of a concrete mix requires suitable weights of water and cement for each batch to be determined, usually by making trial mixes before construction starts. A central mixing plant would not, of course, be required to make such trial mixes, records of its performance in regular production being available for inspection.

However, the matter does not end here, for the workability of the plastic concrete must be taken into account. For example, a concrete mix of suitable strength may have so little water in it that it cannot be compacted in the formwork. Although laboratory tests would indicate suitable strength, the final result in the field may well be quite unsatisfactory because of voids in the concrete and around the reinforcing. Workability can be improved without sacrificing strength by increasing the amounts of water and of cement.

Without careful attention to grading of the mineral aggregates – that is, the frequency distribution of various sizes of particles – additional water is detrimental. It leads to segregation of the various grain sizes in the formwork, and consequent defects in the finished construction. Also, excess water in the plastic concrete will cause increased shrinkage after the concrete has set and, if this volume change is restrained, unsightly cracks occur.

In selecting the grade of concrete to use, the designer wants to economise by keeping down the amount of the most expensive ingredient – the cement. To minimise the volume of concrete, he wants to keep the 28-day strength as high as possible. How far he can go is limited by the need to ensure that the plastic concrete can be compacted in the formwork. Well-graded aggregates and sufficient water must be used, the latter having its effect on cement content and water-to-cement ratio. Further information on the strength and behaviour of concrete can be found in *Concrete Manual* of the US Bureau of Reclamation and in Murdock and Blackledge.[1, 2]

20.3. CAST IRON

Grey cast iron will be the first choice for a ferrous material if good casting qualities are sought and strength is not of first importance. The cast metal consists of a ferrite and pearlite matrix in which graphite flakes are embedded. As the graphite content increases, the strength and hardness decrease; in the most commonly used grades the strength increases from 150 MPa to 400 MPa as the carbon content ranges from about 3.5% to 2.5%. The lower-strength material should not be regarded as a low-grade material – it has superior casting and machining qualities and it is uneconomic to specify higher strength than is necessary. The high compressive strength of cast iron should be made use of where possible.

Cast iron in all the forms discussed here has excellent machinability – up to 25% better than free-machining steel. The good wear and abrasion qualities can be further improved by surface hardening, but when enhanced properties are required, some alloying materials are generally used. Ordinary grey iron has reasonable corrosion resistance and can be used up to 350°C, while alloying with small amounts of chromium increases this upper temperature to 700°C. Under these conditions the creep strength is low. Other alloys are used where special corrosion or magnetic properties are required. Design data for several grades of cast iron are given in Table A4.1.[3] More detailed information can be found in reference books.[4, 5, 6, 7]

The greatest deficiency of grey iron for many applications is its brittleness and low resistance to shock loading. Greatly improved toughness is obtained in malleable iron. This is formed by casting white iron, with a total carbon content of 2% to 2.5%, and subjecting it to prolonged annealing. This treatment, which may last for 100 or more hours, adds to the cost, but results in a ductile material consisting of ferrite interspersed with flake graphite aggregates. The most common form is known as *blackheart* iron, and has a yield strength ranging from 220 MPa to 250 MPa, with elongation of 10% to 14% at failure. This material remains ductile down to low temperatures and can be used up to 400°C. The modulus of elasticity is high – 170 GPa – compared with that of grey iron which varies from 70 to 150 as the carbon content decreases. The properties are not sensitive to section thickness since the heat treatment gives uniformity throughout. Specifications for ferritic malleable castings are given in BS 310[8], while BS 3333 gives properties for a pearlitic form that has substantially higher strength.

Spheroidal graphite (SG) or nodular iron is another form that has high ductility, strength and toughness. By treating a melt, which is similar to that for grey iron, with cerium or magnesium, the graphite in the casting separates out in the form of nodules. In the as-cast form the pearlite matrix is not very ductile and the castings are usually heat-treated by annealing or normalising.

The grades described in BS 2789 range from SNG 24/17, a ferritic iron with high impact resistance, to SNG 47/2, which is a high-strength iron with pearlite structure.[9] Mechanical properties for these two and an intermediate grade are given below.

BS grade	Minimum tensile strength (MPa)	Elongation %	0.5% permanent set stress (MPa)
SNG 24/17	370	17	230
SNG 32/7	495	7	340
SNG 47/2	725	2	460

Modulus of elasticity, 145 to 170 GPa.

Further information on the use and heat treatment of nodular iron can be found in *Casting Design Handbook*.[6]

20.4. STEEL AND ITS ALLOYS

Steels can be classified under the headings of low-, medium- and high-carbon and low- and high-alloy steels.[10] The properties of one or more types in each category are given in Tables A4.2 and A4.3. The selected steels are described in terms of mechanical properties, alloying materials and applications; typical specifications have also been listed, but these are given mainly as a general guide and to help when more detailed information is sought.

A difficulty encountered here is in the varied nomenclature that is used for steels. The numbering system developed by the Society of Automotive Engineers (SAE) and American Iron and Steel Institute (AISI) is based on chemical composition and has been widely used all over the world. The basis of the AISI system, which uses four or five digits, is shown in Table 20.1. British steels have been widely described by an arbitrary number allocated to each grade and prefixed by the letters "En". Preliminary information has been issued by the British Standards Institution of a new designation to replace the "En" system, which has been found inadequate to describe the number and variety of new steels.[11] In this six-digit system, the first three numbers, from 000 to 299, represent 100 times the mean manganese content for carbon steels. The fourth symbol, a letter, indicates whether the steel is supplied to analysis (A), mechanical property (M) or hardenability requirements (H). The last two digits represent 100 times the mean carbon content. In the case of alloy steels, the first three numbers, from 500 to 999, denote the alloy type, the remaining digits being used as for carbon steels.

Table 20.1. AISI–SAE numbering system for steels.

Symbol	Description
10xx	Plain carbon
11xx	Free-cutting
13xx	Manganese
2xxx	Nickel
3xxx	Nickel-chromium
30xxx	Corrosion- and scale-resistant
40xx	Molybdenum
41xx	Chromium-molybdenum
50xx	Low chromium
51xx	Medium chromium
6xxx	Chromium-vanadium
86xx 87xx	Chromium-nickel-molybdenum
9xxx	Silicon-manganese

Second digit denotes variations in composition.
Last two digits denote (percentage carbon content) x 100.

The letter S is used for the fourth symbol for stainless steels, the first three digits, taken in the range 300 to 499, being arbitrary and corresponding, where possible to the AISI nomenclature. The last two digits denote the specific alloy.

This new system conforms to ISO proposals, and will no doubt come into wide use; however, the AISI system will continue to be used by manufacturers and engineers in many countries.

Before discussing each type, it is useful to describe briefly the way in which the cost of a particular type of steel is determined. Structural steels are taken as an example.

Steel is supplied in the form of bars and billets for forging, structural sections, plates, tubes for general construction work and merchant bar – rounds, hexagons and other shapes – for machining. The surface may be hot-rolled, cold-drawn or ground. All these factors will affect the price, and the economy in using the steel.

The cost of a particular type and finish is built up from a basic cost per tonne. In Australia, in the middle of 1974, this cost was about $150 per tonne for structural steel. Mild-steel plate (yield strength 250 MPa) up to 50 mm thick would be supplied at this price, while thicker plate would cost $10 over the basic price. For a higher yield-strength material (yield strength 350 MPa) the extra cost would be $24. Steel supplied to a specified impact value would be subject to an additional charge. The same price would hold for smaller structural sections, but an extra charge would be added for larger sizes. At the same time the basic price for free-cutting steels in merchant-bar form was $135 per tonne, and the extra costs for leaded and resulphurised steels were $45 and $30 per tonne respectively for semi-killed steels. Fully killed steels would have a higher price. Apart from the various technical aspects, the price of steel is complicated by factors such as the size of order and freight charges, and this summary only gives a simplified picture of the cost structure for the material supplied by the manufacturer.

The designer may not specify or even be particularly aware of metallurgical aspects such as the content of minor alloying elements or deoxidation procedures. The steelmaking practices for particular products are well established and will only be varied in special circumstances. For example, rimmed steels are well suited to the manufacture of sheet or strip for applications where surface finish is important. However, the fatigue properties at the surface of this steel are not good, and the designer should be aware of this and other such limitations connected with the metallurgy of steelmaking. These steels are specifically excluded from use in some specifications of structural steels. Fully killed steels are specified where uniform properties are required for highly stressed and heat-treated components.

One reason for the lack of attention given by the designer to metallurgical aspects is that it is common practice to specify steels in terms of their strength and end use and leave the manufacturer to select the most suitable composition to satisfy these requirements. It nevertheless behoves the designer to be aware of the effects of the principal alloying elements, and some of these are listed in Table 20.2.

Steels with a carbon content less than 0.15% have low strength and poor machining qualities. Their surface hardness can be improved by case-hardening, and free-machining qualities imparted by the addition of 0.25% sulphur, but their usual application is for cold-forming operations, such as pressing or deep-drawing, where their high ductility is an advantage.

Much of the steel used for construction and structural work falls in the range of 0.15% to 0.3% carbon content. The four grades shown in Table A4.2 are classified as weldable structural steels, the grade numbers 40, 43, 50 and 55 denoting their minimum tensile strength in 10^7 Pa.[12]

Within each of the grades there are several sub-grades distinguished by increasing stringency of yield stress and notch ductility requirements. For example, there are no Charpy V-notch or yield requirements for grade 50A, while for grade 50D the yield stress is specified as 350 MPa and impact test values are given down to −10°C. These improved properties are required to avoid the possibility of brittle fracture. The yield stress is specified for sections varying from 16 mm to 100 mm, and other properties specified in the standard are also dependent on section thickness. No endurance limit is quoted, and if required for design, a value of about 0.48 f_t (tensile strength) would be used (see Chapter 7).

Table 20.2. Effect of alloying elements on properties of steel.

Alloying element	Effect on strength and toughness	Other effects
Manganese	Marked strengthening effect; for same strength toughness is improved	Neutralises sulphur, increases hardenability
Silicon	Improves strength, reduces toughness	Improves oxidation resistance
Nickel	Strengthens and toughens	With high chromium gives stainless steel.
Chromium	Slightly strengthens	Improves corrosion resistance, hardenability and high temperature properties. Used over 12% for stainless steel
Sulphur	Lowers toughness	Up to 0.25% improves machinability; lowers weldability
Phosphorus	Strengthens	Improves hardenability and machinability
Copper	Little effect in small quantities	Improves resistance to atmospheric corrosion
Molybdenum	Improves toughness	Improves high temperature properties

These steels are weldable, provided the procedures detailed in BS 1856 are followed for grades 40 and 43 and those in BS 2642 for Grades 50 and 55.[13, 14] They are available in the form of sections, bars and plates, usually in the as-rolled condition, although the higher grades may be normalised.

The first steel (Grade 40) given in Table A4.2, with a yield strength of about 210 MPa, is typical of what would be used if the designer specified simply a "mild steel".

Design information for a small selection of steels with yield strengths ranging from 470 MPa to over 1000 MPa is given in Table A4.3.[4, 15, 16, 17] Each of these will have its own range of applications and its special problems of fabrication and heat treatment; only very general comments are made about these in the table. In particular, nothing has been said of the need for the designer, when specifying a heat-treatable steel, to be satisfied that the required hardness and strength can be attained. The steel must be cooled during quenching faster than a critical rate which varies with alloy content. For plain carbon steels this rate is so high that sections larger than about 12 mm cross-section cannot be fully hardened. Some of the alloy steels have low critical cooling velocities, so that they can be through-hardened without danger of distortion or cracking. Some handbooks describe the hardenability of various steels for different sections, stating what type of quenching is necessary to achieve a certain hardness. However, if an accurate assessment of hardenability is required, information based on the Jominy test can be used.[18] In this test, a jet of water cools the end of a test bar at a controlled rate. After quenching, hardness measurements along the bar give a measure of the hardenability of the material.

Reference should be made to some of the textbooks and handbooks on materials for further information on this and other topics such as the selection of steels for use at high temperatures.[4, 10, 15, 16, 17]

We have described steels in rather general terms. However, there are standards and manufacturers' catalogues which specify steels for particular purposes – for example, for pressure vessels, for dies and so on.

No attempt has been made to deal with any of the very high-strength steels, such as the maraging and ausforming types, which can be produced with yield strengths up to 2000 MPa and 2800 MPa respectively. Information on these materials can be found in V. F. Zackay, *High Strength Materials.*[19]

20.5. ALUMINIUM ALLOYS

Aluminium has a density of 2.7, about one-third that of steel and, when alloyed and heat-treated, can have a strength (0.2% proof stress) in the same range as that of structural steel. The alloys have good resistance to weathering but information must be sought from manufacturers' catalogues to determine whether surface protection is needed for specific cases.

The mechanical properties of aluminium do not deteriorate at low temperatures – they may improve slightly – and there is no brittle fracture problem; for many alloys there is a marked reduction in strength above about 150°C, although some retain good strength up to 250°C.

The choice of alloying elements depends on the properties that must be enhanced and on whether the finished product is to be in wrought or cast form.[4, 5, 20] Some of the alloying elements give improved strength qualities without heat treatment; others form alloys which require heat-treatment for optimum properties. An important subdivision is therefore between those alloys that require heat treatment and those that do not. Alloys in the latter category are capable of being cold-worked to improve their yield strengths, but the heat-treatable alloys generally give the highest strengths.

In the British Standard system of nomenclature each alloy is given a number, preceded by the letter **H** or **N** to denote "heat-treatable" or "non-heat-treatable".[20] The condition or temper of the alloy is described by a letter following the number. Casting alloys in the British Standard system are given a prefix **LM**. The nomenclature of the Aluminium Association is also widely used. The meaning of some of the symbols is described in Table 20.3.

Alloys that respond to heat treatment have the alloying element retained in solution when quenched after being held at about 500°C. The metal is soft for a short period after this solution heat treatment, during which time it may be cold-formed, but it hardens or ages spontaneously with time. If necessary, this process can be accelerated by heating in what is termed an *artificial ageing* or *precipitation treatment.* The precipitation of the alloying element as a fine dispersal through the parent metal causes a distortion of the crystal lattice. The strengthening effect in this case comes largely from the resistance to the movement of dislocations that this distortion causes.

Some aluminium alloys are more suitable for welding than others, and manufacturers' recommendations should be sought. For the weldable alloys, the processes recommended for high quality welds are the ones in which an inert gas (argon or helium) is used to shield the molten weld from contamination. The electrode may be non-consumable tungsten or a consumable metal wire,

Table 20.3. Summary of nomenclature for aluminium alloys.

Metal condition or heat treatment	BS and ISO designation	Nearest equivalent Aluminium Association designation
Material in fully annealed condition	**O**	**O**
As manufactured	**M**	**F**
Non-heat-treatable alloys		Letter **H** followed by two digits (first indicating process and second, the degree of strain hardening)
Material subjected to varying degrees of cold work. The designations are in ascending order of strain hardening and tensile strength	**H1** **H2** **H3** **H4** **H5** **H6** **H7** **H8**	
Heat-treatable alloys		
Solution heat-treated and naturally aged	**TB**	T4
Cast metal, solution treated and stabilised	**TB7**	
Solution treated, cold-worked and naturally aged	**TD**	T3
Cooled from hot-working process and precipitation-treated	**TE**	T5
Solution treated and precipitation-treated	**TF**	T6
Cast metal, fully heat-treated and stabilised	**TF7**	
Solution treated, cold-worked then precipitation-treated	**TH**	T8

Wrought alloys in the BS system are distinguished by the letter **H** (heat-treatable) or **N** (non-heat-treatable) and a second letter denoting the form of the product. The most common are

sheet, **S**; plate, **P**; tube, **T**; bars and sections, **E**.

The alloy number follows the two letters.
The BS casting alloys have a number prefixed by the letters **LM**.

the two processes being labelled TIG and MIG respectively. The non-heat-treatable alloys are favoured for welding; the welds will have about 90% of the strength of the parent metal. The loss with the **H**-type alloys is much greater unless welding can be followed by heat treatment.[21] Electric resistance welding – for example, spot and seam welding – is also used.

The machinability of aluminium alloys is high – at least four or five times that of steel. There is, in addition, a wide variety of drawn and extruded shapes, and complicated shapes and thin sections can be cast. These factors go some way to compensating for the price difference between aluminium and steel.

The mechanical properties of a small selection of aluminium alloys are given in Table A4.4. Specifications for the wrought materials vary slightly, depending on the product form – e.g., forged, extruded or rolled sheet – and markedly depending on the amount of cold work or the heat treatment. Typical values have been given here. The properties of the casting alloys also vary with casting process and heat treatment.

20.6. OTHER LIGHT ALLOYS

Where low weight is important, alloys of magnesium (density 1.8) and titanium (density 4.5) have increasing use. Their application in design is somewhat specialised, but for completeness we will discuss briefly some of their characteristics.[22, 23]

High production rates can be achieved with magnesium alloys; they have very high machinability and can be extruded and cast by any of the usual methods. Some alloys can be welded, under good conditions, using the shielded-arc process.

Typical strength properties for a heat-treatable magnesium casting alloy ($7\frac{1}{2}\%$ aluminium + zinc and manganese) are

Tensile strength	270 MPa
Elongation	5%
Yield strength	125 MPa

The low modulus of elasticity (45 GPa) makes it important to pay particular attention to stiffness in design.

Titanium has good strength at room temperature and, for short term loads, up to about 500°C. The ratio of strength to weight is amongst the highest attainable. Tensile strengths are in the range 700 MPa to 900 MPa with elongation greater than 10%, and 0.1% proof stresses from 600 MPa to 800 MPa. These properties give it increasing application in the aerospace industries.

20.7. COPPER ALLOYS

High purity high-conductivity copper is widely used for electrical conductors of all types. For some purposes its strength and fabrication properties can be improved by the addition of small quantities (less than 1%) of elements such as chromium or tellurium. However, the most widely used alloys are the brasses, which contain up to 40% of zinc, and the tin-bronzes.[6, 24]

Brasses containing zinc in the range 30% to 37% (the *alpha* brasses) are very ductile and are suitable for cold-forming operations such as deep drawing. They can be joined by soldering. Work hardening will produce a proof stress of 400 MPa with some remaining ductility. Brass in this composition range is sometimes termed *cartridge brass.*

With zinc content greater than 37%, the *beta* phase appears in the structure, strengthening the brass but lowering its ductility. The 60/40 alloy, sometimes called *Muntz metal,* is widely used for hot-worked products. In this class of material it is relatively cheap and gives moderate mechanical properties and corrosion resistance. The yield strength ranges from 200 MPa to 350 MPa, depending on the amount of work hardening. Machinability of the brasses can be improved by the addition of up to 3% of lead.

The bronzes are alloys of copper and tin, the tin content varying from 3% up to 10%. They have higher strength and better corrosion resistance than the brasses.

Some of the more commonly used copper alloys are listed with design applications in Table 20.4. Quite small additions of some alloying elements give marked improvements in properties. The more expensive materials will usually be selected where special problems of strength and corrosion must be solved.

Table 20.4. Summary of some commonly used copper alloys.

Material	Alloying elements	Properties and design applications
High-strength brass	Aluminium, manganese, iron and other elements added to brass; sometimes misleadingly called "manganese bronze"	Highly stressed components at temperatures below 150°C. Not recommended for fatigue loads or high rubbing speeds. Yield strength ranges from 200 MPa to 300 MPa
Leaded bronze	Bronze with up to 23% lead	A material that is softer than phosphor bronze, suitable for bearings and bushes where misalignment and marginal lubrication occur
Phosphor bronze	A tin-bronze with small additions of phosphorus	Heavily loaded gears and bearings where sliding wear occurs
Gunmetal	A tin-bronze with up to 5% zinc	Good casting characteristics and low coefficient of friction. Used where medium strength and corrosion resistance is required.
Aluminium bronze	Copper-aluminium alloys with iron, nickel or manganese	Gives high strength and shock resistance, with good corrosion and oxidation properties up to 400°C. Suitable for gravity diecasting
Silicon bronze	Copper-silicon alloy with manganese and iron in small quantities	Used in cast and wrought condition in chemical plant where good corrosion resistance is required

20.8. PLASTICS AND RUBBERS

Plastics are available with a wide range of properties and they can be considered as possible replacements for many other materials; however, cost considerations will not always uphold the place they gain by their versatility. This versatility comes largely from the way in which the organic molecules that are the major constituents can be tailored for specific applications.

Plastics can be broadly subdivided into two categories based on their behaviour on heating. The first are the *thermoplastic* materials which soften and then melt on heating. The second are the *thermosetting* plastics that undergo a reaction in which a three-dimensional molecular network is formed, causing them to harden and set; further heating will result in chemical breakdown. The thermosetting plastics are usually used with fillers or reinforcing materials to give the mechanical properties needed for moulded parts.

Tensile strengths of plastics vary widely, but typical values fall in the range 20 MPa to 60 MPa at room temperature. The strength can be improved by the use of fibre reinforcement. Values of the modulus of elasticity quoted in the literature range from 0.5 GPa to 10.0 GPa. It must be understood that such values are the results of a standard test, and that the deflection obtained in a component will be a function of time and temperature. Plastics at room temperature are close to their softening points, and the problem of determining deflection under load resembles that encountered when designing metals in the creep range. Although some plastics can be used up to 200°C, others will soften and distort under load at 60°C.

The designer will often be as much concerned with problems of stiffness and resistance to impact as with strength, and the performance of a component will depend on the shape of the part and the production process used as well as on the mechanical properties. It is clearly important for the designer to ensure that the service conditions are well defined when selecting a plastic for a particular application. Information such as that given in Tables 20.5 and 20.6, taken from a paper by D. W. Saunders and J. M. Stuart in *Engineering Materials and Methods*[25], is helpful in making a preliminary selection. Only comparative properties are given, but it can be taken that for tensile strength "excellent" corresponds to more than 55 MPa, while "poor" is less than 20 MPa.

As far as strength design goes, much of the handbook information on mechanical properties is not well suited for direct use in design, since service conditions are likely to be very different from those specified for the standard tests. High safety factors, in the range 6 to 10, have been used with such data. New design methods, which take into account the viscoelastic behaviour of these materials, are being developed. References to such methods are given in the paper by Saunders and Stuart.

The most widely used and cheapest of the thermosetting plastics are those based on phenol formaldehyde resins. Depending on the fillers used, strengths up to 60 MPa can be obtained. Phenolics are easy to mould and withstand temperatures up to 200°C. Their colour range is limited. Epoxy resins when reinforced with glass fibres give strengths equal to those obtained in light alloys, and can be used when the need for a high-strength material justifies the considerably higher cost. Other thermosetting plastics that are used for domestic ware and electrical accessories are the urea and melamine formaldehydes and the alkyds.

Thermoplastic materials do not generally require the extensive use of fillers and reinforcing materials. Polyethylene is used on its own in a wide variety of forms in the packaging industry. It has excellent properties as an electrical insulator, and in its high density form can be extruded to give pipes that have good resistance to chemical attack. Polyvinyl chloride (PVC) is used in an equally wide range of products and is produced in soft or rigid form. Nylon is a valuable engineering material with a yield strength of about 70 MPa. It has properties of toughness and wear resistance that are useful in design. It can be formed by extrusion or injection moulding into gears, cams, bearings and other components which will require little or no lubrication. Among the many other thermoplastic materials, such as polystyrene and methylmethacrylate (Perspex), ABS polymers stand out as having useful engineering properties. These plastics are based on acrylonitrile, butadiene and styrene resins, and have excellent strength and impact properties. They are readily moulded

Table 20.5. Properties of thermoplastics.

Polymer	*Specific gravity*	*Tensile strength*	*Compressive strength*	*Impact strength*	*Heat resistance*	*Machining qualities*	*Chemical resistance*
Acetal	1.41–1.43	excellent	good	good	good	excellent	good
Methylmethacrylate	1.17–1.20	excellent	good	poor	good	good	fair
Cellulose acetate	1.23–1.34	excellent	excellent	fair	good	excellent	poor
Cellulose acetate butyrate	1.15–1.22	good	good	good	good	excellent	poor
Cellulose nitrite	1.34–1.40	good	excellent	good	fair	excellent	poor
Chlorinated polyether	1.4	fair	—	poor	excellent	excellent	excellent
Nylon 6.6	1.10–1.14	excellent	good	fair	excellent	excellent	good
Nylon 6	1.13–1.16	excellent	fair	good	good	excellent	good
High density polythene	0.94–0.96	fair	poor	excellent	good	excellent	excellent
Low density polythene	0.91–0.93	poor	—	excellent	fair	good	excellent
Polypropylene	0.90–0.91	fair	fair	good	excellent	excellent	excellent
P.T.F.C.E.	2.10–2.20	fair	excellent	good	excellent	excellent	outstanding
P.T.F.E.	2.13–2.22	fair	poor	good	outstanding	excellent	outstanding
F.E.P.	2.14–2.17	poor	—	excellent	excellent	excellent	outstanding
Polystyrene (G.P.)	1.04–1.06	excellent	good	poor	fair	fair	fair
Polystyrene (high impact)	0.98–1.10	good	fair	good	fair	good	fair
Acrilonitrile-butadiene-styrene	1.01–1.15	excellent	fair	excellent	good	good	fair
Polycarbonate	1.2	excellent	fair	excellent	good	excellent	fair
P.V.C. (rigid)	1.35–1.45	excellent	good	excellent	poor	excellent	good
P.V.C. (flexible)	1.16–1.35	fair	poor	excellent	fair	poor	good
P.V.D.C.	1.65–1.72	fair	poor	fair	fair	good	good

Source: Reproduced by permission of the Council of the Institution of Mechanical Engineers from *Engineering Materials and Methods.*[25]

Table 20.6. Properties of thermosetting plastics.

Polymer	*Specific gravity*	*Tensile strength*	*Compressive strength*	*Impact strength*	*Heat resistance*	*Machining qualities*	*Chemical resistance*
Melamine formaldehyde							
Unfilled	1.48	—	excellent	—	good	—	good
Cellulose filler	1.47–1.52	excellent	excellent	poor	excellent	fair	good
Asbestos filler	1.70–2.00	good	good	fair	good	fair	good
Glass fibre filler	1.80–2.00	excellent	excellent	good	excellent	good	good
Phenol formaldehyde							
Unfilled	1.25–1.30	good	good	poor	good	fair	fair
Woodflour filler	1.32–1.45	excellent	excellent	fair	excellent	fair	fair
Mica filler	1.65–1.92	good	good	poor	excellent	poor	fair
Glass fibre filler	1.75–1.95	excellent	good	excellent	outstanding	—	fair
Polyester							
Glass fibre filled moulding compound	1.8 –2.3	excellent	good	excellent	excellent	good	fair
Mineral filled	1.60–2.30	good	good	fair	excellent	fair	fair
Silicone							
Asbestos filled	1.60–1.90	outstanding	good	—	—	fair	fair
Glass fibre filled	1.68–2.0	fair	fair	excellent	outstanding	fair	fair
Urea formaldehyde cellulose filled	1.47–1.52	excellent	excellent	fair	fair	fair	fair
Epoxy moulding compound							
Glass fibre filled	1.8 –2.0	outstanding	excellent	excellent	outstanding	good	excellent
Mineral filled	1.60–2.06	good	good	fair	outstanding	good	excellent

Source: Reproduced by permission of the Council of the Institution of Mechanical Engineers from *Engineering Materials and Methods*.[25]

and are widely used for housings, covers and containers in engineering and domestic products.

A range of natural rubber materials can be made by mixing raw rubber with fillers and protective agents and vulcanising it with chemicals such as sulphur and stearic acid. Vulcanised rubber will have a tensile strength of 14 MPa to 28 MPa, calculated on the original cross-section, but design stresses will usually be in the range of 2 MPa to 3 MPa. It can be readily bonded to metallic and non-metallic surfaces, and makes an excellent compression spring. Design data can be found in Lindley.[27] Natural rubber is attacked by oils, some gases and organic solvents. Where resistance to these chemicals is required, one of the synthetic rubbers can be selected. These have similar strength properties, and there is a wide range from which to select hoses, seals, packings and tank linings that will be resistant to specific fluids.[25, 28, 29]

REFERENCES

[1] U.S. BUREAU OF RECLAMATION, *Concrete Manual,* 7th ed., U.S. Government Printing Office, Washington, 1963.

[2] L. J. MURDOCK and G. F. BLACKLEDGE, *Concrete Materials and Practice,* 4th ed., Arnold, London, 1968.

[3] BS 1452:1961, *Grey Iron Castings,* British Standards Institution, London, 1961.

[4] ASM, *Metal Progress Databook*, American Society for Metals, Ohio, 1968.

[5] A. DAVIDSON, "Materials", vol. 2 in *Handbook of Precision Engineering,* Philips Technical Library, Macmillan, London, 1970.

[6] ASM, *Casting Design Handbook*, American Society for Metals, Ohio, 1962.

[7] H. T. ANGUS, *Physical and Engineering Properties of Cast Iron,* The British Cast Iron Research Association, Birmingham, 1960.

[8] BS 310:1958, *Blackheart Malleable Iron Castings,* British Standards Institution, London, 1958.

[9] BS 2789:1961, *Iron Castings with Spheroidal or Nodular Graphite,* British Standards Institution, London, 1961.

[10] BS 971:1950, *Commentary on British Standard Wrought Steels (En series)*, British Standards Institution, London, 1950.

[11] PD 6423:1968, *New Designation System for Carbon Steels,* British Standards Institution, London, 1968.

[12] BS 4360:1972, *Weldable Structural Steels,* British Standards Institution, London, 1972.

[13] BS 1856:1964, *General Requirements for the Metal-arc Welding of Mild Steel,* British Standards Institution, London, 1964.

[14] BS 2642:1965, *General Requirements for the Arc Welding of Carbon Manganese Steels,* British Standards Institution, London, 1965.

[15] C. J. SMITHELLS, *Metals Reference Book*, vols 1 and 2, 3rd ed., Butterworths, London, 1962.

[16] J. WOOLMAN and R. A. MOTTRAM, *The Mechanical and Physical Properties of the BS En Steels,* vol. 1, 1964, vol. 2, 1966, vol. 3, 1969, Pergamon Press, Oxford.

[17] BS 1515:Part 1:1965, "Carbon and Ferritic Alloy Steels", *Fusion Welded Pressure Vessels,* British Standards Institution, London, 1965.

[18] ASTM A255, *End Quench Test for Hardenability of Steels,* American Society for Testing and Materials, Philadelphia, 1967.
[19] V. F. Zackay (ed.), *High Strength Materials*, John Wiley, New York, 1965.
[20] P. C. Varley, *The Technology of Aluminium and its Alloys,* Newnes-Butterworths, London, 1970.
[21] ASM, "Welding and Brazing", vol. 6 in *Metals Handbook*, 8th ed., American Society for Metals, Ohio, 1971.
[22] P. Greenfield, *Magnesium*, M & B Monograph, ME/11, Mills and Boon, London, 1972.
[23] V. Weiss, "Non-Ferrous Alloys", vol. 2 in *Aerospace Structural Metals Handbook*, Syracuse University Press, Syracuse, 1963.
[24] H. J. Sharp (ed.), *Engineering Materials*, Heywood Books, London, 1966.
[25] E. G. Semler (ed.), *Engineering Materials and Methods,* The Institution of Mechanical Engineers, London, 1971.
[26] The Society of the Plastics Industry, *Plastic Engineering Handbook,* 3rd ed., Reinhold, New York, 1960.
[27] P. B. Lindley, *Engineering Design with Natural Rubber,* 3rd ed., The Natural Rubber Producers' Research Association, London, 1970.
[28] A. T. McPherson and A. Klemm, *Engineering Uses of Rubber,* Reinhold, New York, 1956.
[29] E. Baer (ed.), *Engineering Design for Plastics,* Reinhold, New York, 1964.

21 Industrial Design

21.1. INTRODUCTION

The engineering designer does not work in isolation from other trades and professions, and it is important that he should have a knowledge of the working methods and capabilities of those with whom he may be associated. In this chapter, the work of industrial designers will be discussed with this in mind and, in particular, with the intention of describing some of the methods and techniques that are likely to be of interest and use to the engineer.

Industrial design (ID) is as diverse an activity as engineering and it is even harder to find a definition which covers the widening scope of the field, which now ranges from product design to typography and includes graphic and interior design.

We will describe the way in which industrial design has grown over the past 100 years to be a distinct professional activity, and then look at the skills and training that designers bring to bear on the problems they tackle. Finally, some of the methods used by the product designer and his approach to a problem will be discussed.

The emphasis that is placed on the visual aspects of design in this chapter does not mean that this is the province solely of the industrial designer; indeed it is hoped that it is by now clear to the reader that this is an aspect that must increasingly be the concern of the engineer engaged in creative design.

21.2. EARLY HISTORY

Over a period of about 100 years centred at the beginning of the nineteenth century, the Industrial Revolution altered a traditional way of working that had existed for hundreds of years. Before this the craftsman shaped his product from start to finish according to a design which had been handed down to him. This design had evolved slowly to the stage where it was well-suited to the tools and materials available and to the needs of the user.

With the coming of iron and then steel in quantities sufficient for use on a large scale – Bessemer invented his steelmaking process in 1856 – buildings and machines appeared in new forms. This time of energetic development of industrial production was marked by mediocrity and confusion in design. This mediocrity was seen by William Morris (1834–1896) who sought to make goods having artistic merit available to the ordinary man. If he fostered excellence in design, Morris did little to dispel confusion, since he eschewed modern production methods and advocated a return to handcrafts. Despite this, his art has been the starting point of much commercial production, and this sets him as one of the initiators of the modern movement.

The story of the progression of design during the formative period makes fascinating reading. Pevsner traces the Modern Movement in architecture and design back to William Morris, to *Art Nouveau* and to the work of the nineteenth century engineers.[1] Work from the *Art Nouveau* period is familiar today, since the work of some of the artists of the period has enjoyed a comeback. While it may be difficult for engineers to grasp immediately the significance of this work, they will readily appreciate the impact of the work of the early builders, architects and engineers. The skylines were changing and tall buildings and structures, such as the Eiffel Tower, loudly proclaimed the new power of the engineer, and the opportunities that lay in the new technology.

So the way opened for goods that could be produced in quantities large enough to satisfy the demands of the masses. Unfortunately, in many cases little thought went to the problems of suiting the design to the changed production methods and the changing social climate. The tendency was to reproduce the ornamentation which in earlier times had added prestige and value to handcrafted products, and thoughtless embellishment became a hallmark of the times. The artist, who had been brought into the manufacturing process in the potteries of Wedgwood as early as 1775, now stood aloof.

At the beginning of the twentieth century, this debasement of the new technology was being countered by the efforts of many individual artists and designers, but it was not until 1919 that the first strong movement appeared. The *Staatsliches Bauhaus* was set up by Walter Gropius as a centre and school of arts and of arts and crafts. The artists and craftsmen who worked and trained there were responsible for a great surge of progressive activity in a wide-ranging field of design, and the *Bauhaus* is generally regarded as having been the cradle of modern design. The American contribution built up rapidly from 1927 onwards, and many designers had become well-established by the late 1930s; notable among these were Loewy, Teague and Dreyfuss.

If at first the search for an aestheticism of form and function was a purist crusade, the economic implications were soon grasped by manufacturers who sought to improve the marketability of their products. It is interesting to note that in an early textbook on industrial design by Harold van Doren, first published in 1940, the province of industrial design is described wholly in terms of manufacturing and marketing concepts. It is defined there as "the practice of analysing, creating and developing products for mass manufacture. Its goal is to achieve forms which are ensured of acceptance before extensive capital investment has been made, and which can be manufactured at a price permitting wide distribution and reasonable profits." [From H. van Doren, *Industrial Design*, 2nd ed., McGraw-Hill, New York, 1954[2], by permission of McGraw-Hill Book Company Inc.]

21.3. THE INDUSTRIAL DESIGNER

While the early practitioners may have been drawn from a variety of other occupations, such as stage and fashion design, designers are now being trained to diploma and degree level in their specialist fields. The courses are often associated with schools of art but, in addition to subjects which relate to the visual arts, they may involve studies in marketing, psychology, production processes and materials, along with design and shop experience. Generally, no attempt is made to equip the designer to carry out detailed stress or structural analysis and, where needed, a specialist is employed.

When tackling a project, the first requirement is that the designer should have a thorough understanding of the problem and all the circumstances that will affect the manufacture and marketing of the product. These factors are, of course, applicable to the approach an engineer would make to the same problem, and the solutions could be equally well within his competence. The question arises then: Why call in an industrial designer? Two points can be made here which will describe some of his activities and help to define his field of action.

First, the designer will nearly always be consulted because a sales or marketing problem is to be solved. Even an organisation which employs its own ID staff may call on an outside consultant to provide a fresh and original approach; this may be done to meet a fall in sales or, in cases of farsighted management, to ensure the continuation of a successful venture. The designer must, therefore, be in touch with market trends and public preference – or be able to initiate surveys to determine these. And in this sort of activity, the consultant will have an advantage over the engineer or designer employed in an industry in that he is free from day-to-day pressures and problems and is not in any way tied to existing methods and ideas.

Second, the designer must bring a high standard of aesthetic perception to the design of the product. He must seek a balance between the undoubted attractions of innovative design and the reluctance of some of the public to accept new ideas. He will have to meet the often-conflicting ideas of the salesman and the engineer without compromising his vision of the product. It is clear that he must have a special competence in this area which the engineer does not usually have an opportunity to develop.

The designer must be able to work with his client's personnel and obtain its goodwill and cooperation; as far as the engineer is concerned, the association can be pleasant and profitable and will give him the opportunity to extend his knowledge while working as one of a team. He may well find that he can adopt something of the approach of the industrial designer to problems that come his way.

The ID consultant may work on his own or may come from an office which has a number of supporting personnel – engineers, model-makers and graphic designers – and all the other talents that can be brought to bear on marketing, graphic presentation and engineering problems. Apart from product design, large offices may undertake specialised architectural and engineering development work; interior design and exhibition design are other fields of activity.

21.4. DESIGN CONCEPTS

When he examines critically a structure or machine, an engineer will comment on its "looks" – the visual stimulus it arouses – as readily as he will on technical aspects, such as structural efficiency or economy. In doing this he is expressing an aesthetic judgment – a judgment on the visual merit or beauty of the object. It is possible that his comment will be based largely on the looks of the object in relation to its function and in relation to others of its type that he has encountered. A designer, when commenting in a similar manner, will usually discuss his views in aesthetic terminology: he will talk of balance, contrast, proportion, colour and style as well as fitness for a purpose. These expressions are familiar and are used by designers in much the same way as they are by engineers.

This is not the place for a discussion of the concepts and values that underlie aesthetic appreciation. In such a discussion, the main difficulty is that, although there are accepted values, there are no absolute standards that hold irrespective of time or place. It will, however, be useful to discuss style at this stage, since it plays an important part in the visual impact of an object and is the keynote of a designer's work.

21.4.1. Style

When the total expression of line, form and colour of several products shows some similarity, this is ascribed to a similarity in style. It may run through all the products of a company, being assiduously fostered as a house style in the design of products and packaging and in advertising.

The elements that go to create a style are hard to pinpoint, though in some cases they are simple and easily recognised. The radiator of the Rolls Royce car has changed over the years, but the basic form remains the same and expresses a style which is known throughout the world. In other cases, a subtle use of colour, materials, form and lettering produces a visual impression which, unless it is copied, marks the style of an individual designer. Lovers of painting and sculpture, devotees of cars and yachts, followers of fashion, all look for and come to recognise the style of their favourite artist or designer.

As a particular style achieves success, it may be widely copied and become popular and fashionable. However, fashions change and move on to leave a style to mark and date an era. The reasons for these changes in fashion have social as well as technical facets. It would be pleasing to think that the changes, when they occur, come as a result of some fundamental new need or of some development in material or manufacturing capabilities. However, the real basis for these changes, which may stem from a search for aesthetic merit, is very largely an economic one; and so styling may become a tool which is used deliberately to label a commodity as being up with the times – or outdated.

Styling is not a word that finds great favour among designers, mainly because it has (probably with justification) some connotation of superficiality about it. A good deal of discussion concerning design has as a starting point the relation between form and function. And perhaps the extent of the superficiality of styling is that some of it has little or nothing to do with either of these. Thus there is some debate on the merits of styling as compared with what has come to be called "good design".

These two orientations have been the principal ones in product design since it became established as a professional activity forty years ago. In broad terms, "styling" is held to derive some of its forms from considerations outside the nature of the product and to be concerned with projecting such abstract notions as luxury, speed or prestige by means of motifs currently popular in certain social strata. The "good design" philosophy, on the other hand, maintains that the attributes of the product itself, such as its function, its manufacturing constraints and its materials, all expressed simply and honestly, should provide the basis for its aesthetic form.

Stepform (C 1925-1935)
architecturally oriented
ziggurat-like stepping
rounded cascading
symmetry, vertical stress
fluting, reeding
flared forms
sunburst motifs
zig-zags, triangles

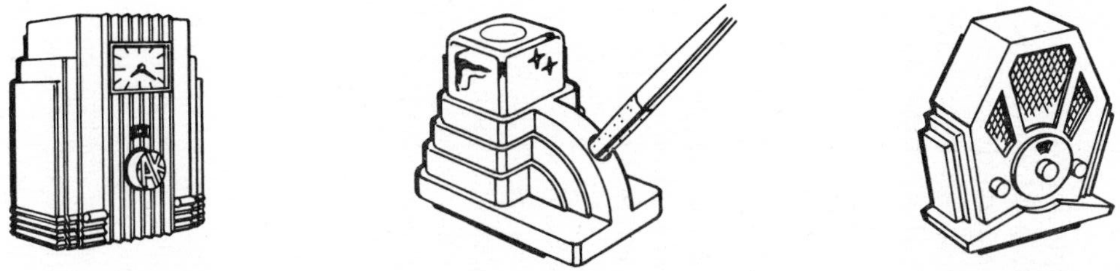

Streamlining (C 1930-1945)
false streamlining
tear-drop forms
asymmetry, horizontal stress
wrap-around banding
jelly-mould, soft forms
three-line accents
slátting, reeding
ball corners

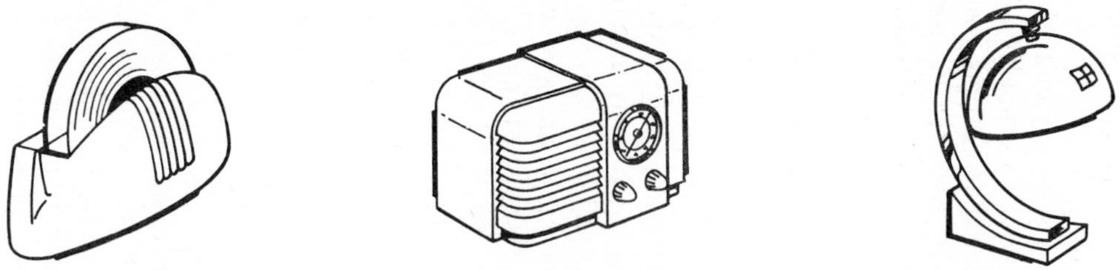

Taperform (C 1945-1957)
trapezoidal solids
crisp corners
tapered, frenched panels
textured, patterned panels
boomerang, saddle and jog accents
facetted accent jewellery
conical knob caps
jet aviation allusions

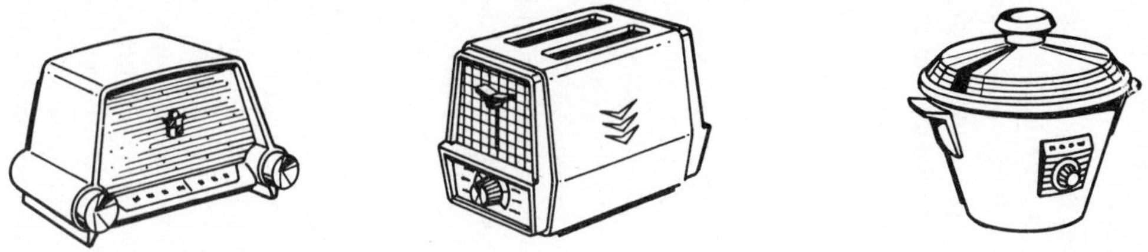

(a)

The Sheer Look (1957-)
box-like solids
thin-edged bezels and trim
asymmetrical panel division
small-scale patterns
spin-brushed knobs and panels
rectangular escutcheons
butterfly and slant panels
refinement in detail

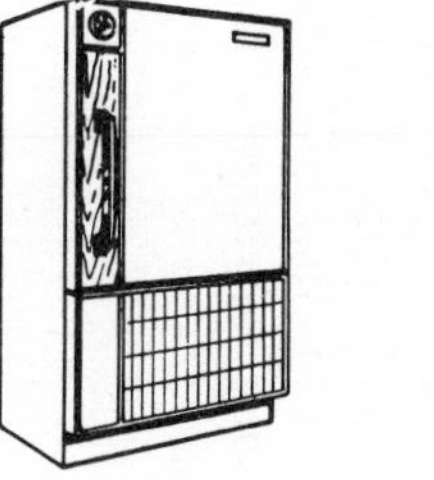
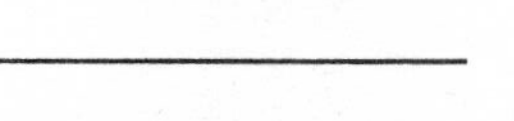

The Sculptured Look (C 1960-)
curved, carved solids
lens forms, conic sections
flying wedge forms
peaked edges
pulled-out, chewing-gum forms
scooped minor forms
trumpet forms
influence of clay-modelling

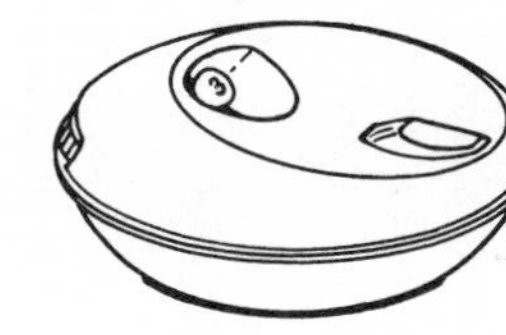
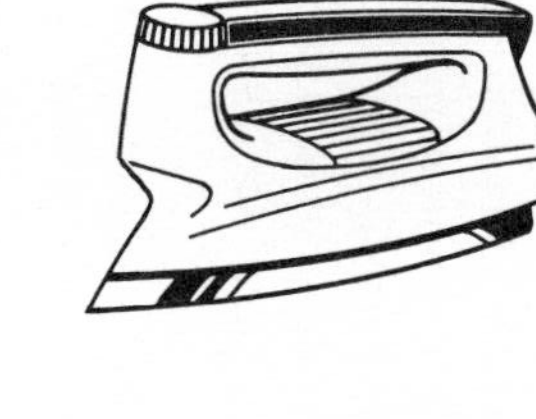

New Trends Emerging (late 1960s)
(here only for engineered products)
1920s, 1930s revivalism
wrap-around motifs
rounded cascading, stepping
radiused corners in a plane
blow-up, pneumatic forms
transparency
(this style uncertain yet)

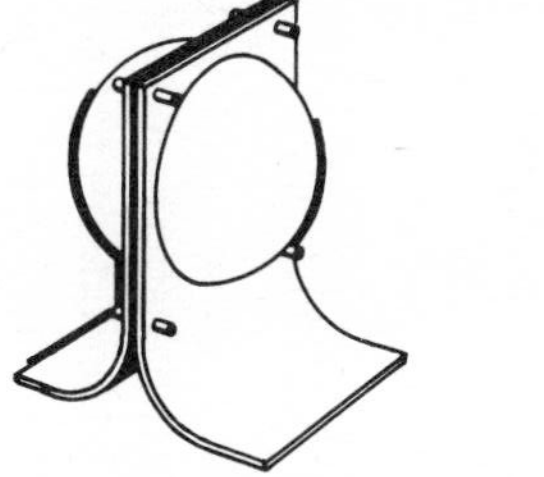
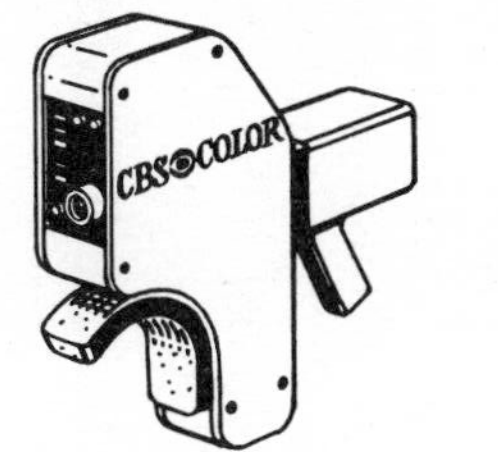

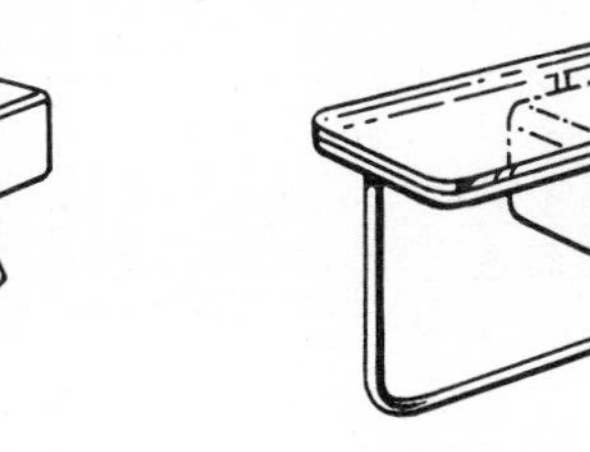

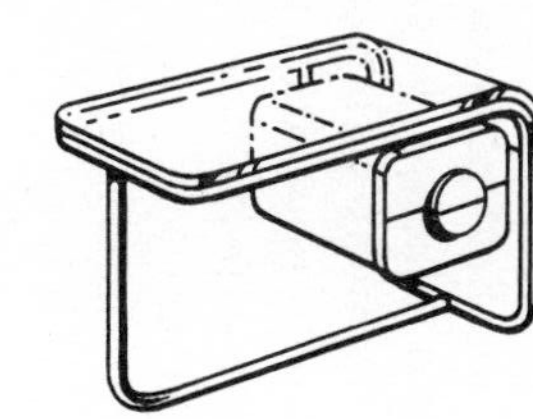
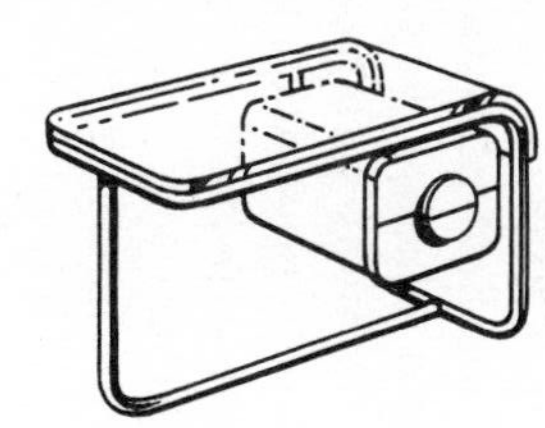
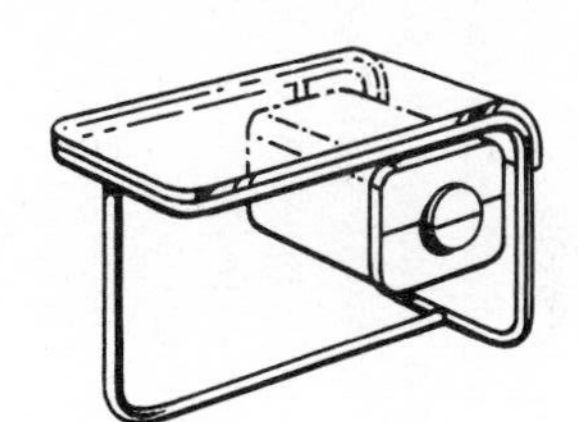

(b)

Fig. 21.1. *(a) and (b). Cavalcade of sub-styles in American product styling over forty years. (From G. Jackson, "The High Style"[3], by permission of the author.)*

Whatever may be the virtues of these two approaches – and many designs end up as a compromise between them – it is unlikely that ventures by engineers into the field of styling will be successful; generally speaking they are well advised to adopt the good design philosophy, the simplest expression of which is in harmony with basic engineering objectives. Engineering designers who do have the opportunity to develop a style within this framework will find that it can be expressed in many design details as well as in more extensive constructions.

An interesting classification of sub-styles for American engineered products has been proposed by Jackson and is shown in Fig. 21.1.[3] The chart begins at the time of the appearance of modern professional industrial design in America in the late 1920s. Jackson points out that designs in each sub-style may tend either towards restraint or high styling; this variation is illustrated in Fig. 21.2, also taken from Jackson's work, in which four shadings of style for a typical consumer product are shown. The first three are familiar; the fourth, labelled "New trends emerging" accounts for several new movements in the fashion end of the spectrum, such as "Pop", "Mod" and revivals of past styles.

21.5. THE DESIGNER AT WORK

Apart from visual aspects, the industrial designer is much concerned with functional requirements and with the human factors that are involved in the use of machines. He has been quick to use the techniques and stored knowledge of other disciplines, many of which are equally useful to engineers. Two of these are described briefly here. The first is the technique of model-making, the second the discipline of "ergonomics".

21.5.1. Models

The engineer, often adept at model-making, frequently handicaps himself by attempting to represent his thinking in two dimensions. Although the final mode of communicating his ideas often will be an orthographic projection drawing, there may be much to be gained by using models at some stage in design. These can range from simple cardboard and wire constructions to elaborate plant layouts built up from commercially available components.

When a sketch or drawing – even a perspective drawing – is made of a body, the choice of viewpoint, subtleties of light and shade – these and other factors will combine to make the visual effect different from that achieved by the final construction. A model will avoid some of these difficulties, and will provide an opportunity to experiment with the form and massing. Where possible, a full-size representation should be used; in the intermediate scale range – say, half-scale – a false impression of size may be given. When smaller scale factors are used, the viewer is more likely to make adequate mental adjustment for the effects of scaling.

Two basic types will be described; first, the one used to study the line and form of the body-shape of a product; and, second, the one used to settle problems in the arrangement of components whose spatial relationships are too complicated to be readily represented in two-dimensional drawings.

The first type is one in which the external form is the principal interest; it can be built-up from any convenient plastic material – modelling clay or

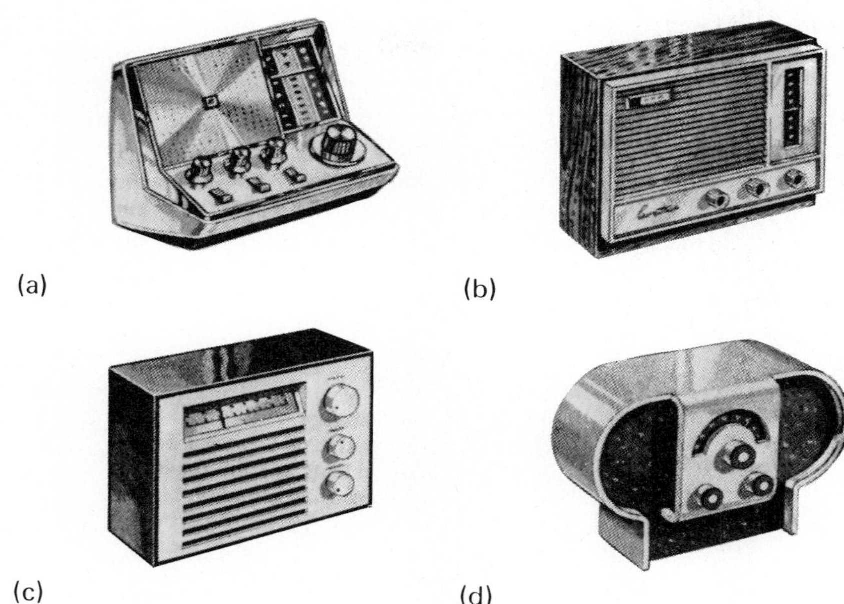

(a) (b) (c) (d)

Fig. 21.2. *Four shadings of product style:*
(a) "High Style";
(b) "Good Design";
(c) "The New Rationalism";
(d) "New Trends Emerging".
(From G. Jackson, "The High Style"[3]*, by permission of the author.)*

plasticine are suitable. Plaster of Paris can also be used, if need be in conjunction with timber formers, but gives less freedom to experiment with form. These materials may be painted after suitable priming of the surface. This type lies mainly in the province of the industrial designer and may be used to study the massing of anything from a toaster to a car, scale or full size.

The second type, which is more likely to be of use to the engineer, is the system-layout model which will show, full size or to scale, the actual locations, proportions and spacing of the structure and components. Where clearances are critical, some moving parts may be modelled to ensure that correct functioning is obtained with the minimum waste of space. These models may be built from paper, cardboard, balsa wood, aluminium sheet, perspex and so on; glues are available for all these materials – for example, epoxy glues are suitable for aluminium.

Not only does this approach provide a clearer picture of complicated mechanisms than do large numbers of drawings, but in building the model the engineer will come to grips with the forming and assembly problems; some feel for the rigidity of the structure can also be gained from the model, although, of course, this analogy should not be taken too far. Engineers should not hesitate to build such models for their own enlightenment and, if need be, to use them when passing design instructions on to draughtsmen and other engineers.

The construction of elaborate models of large plant need not involve any special skills in model-making, since scale-moulded plastic representations of many standard parts – pipes, valves, pressure vessels, channels, beams etc. – can be purchased. The setting up of the individual items may, in fact, be treated as part of the design process in which the design engineer takes part. If the locations of various pipe junctions and bends can be determined from the completed model, the exact dimensions of the pipe runs can be computed, and the drawings of complicated pipe arrangements replaced by a schematic representation. The model can be kept as a guide for the construction process and thereafter may well be used for planning or training operations. The scales usually adopted for such work are 1 in 25 and 1 in 50. Plate 21.1 illustrates a model of a section of a power station made up in this way. In this deaerator bay, the layout for many complicated pipe runs was almost completely designed on the model. Note that the pipes are represented by wires which have discs spaced along their length to show their actual diameter. The cost of such a model is high, and must be justified by a saving in drawing time and by the further uses to which it can be put. Some savings may be ascribed to the possibility of checking assembly problems; for example, ensuring that a large component can be manoeuvred into position at a certain stage of construction.

Plate 21.1. *Model of deaerator bay at Fawley Power Station. (Photograph by courtesy of Central Electricity Generating Board and Industrial Models Limited.)*

21.5.2. Ergonomics

Over the years since the Industrial Revolution, improvements in productivity and in working conditions have come from many causes, and the principal one of these has been the vast increase in the power that has replaced manual labour. Initially, the application of power did little to bring improvements in conditions. More recently, given an awareness of the needs, improvements have come in the main from a study of the relationship between man and his working environment; this last phrase can serve to define *ergonomics*. The application of these studies is directed at improving the efficiency and minimising the hazards and discomforts of a wide range of operations; the term used in North America – *human engineering* – is descriptive of the work involved.

During the Second World War, there was an urgent need to train men and women in large numbers to operate complex machinery. This brought about an upsurge in the study of the performance of human beings, and an increasing awareness of the gains to be made by designing machines that were matched to the capabilities of the operators. Workers in the medical sciences, psychologists, physicists, architects and engineers, found themselves working on various aspects of the same problem; these were some of the disciplines that were brought together when, in 1949, a society was formed to advance the interests of workers in this field.[4] It was at this time that the word *ergonomics* was coined from the Greek *ergon* (work) and *nomos* (law).

The industrial designer and the engineer find considerable use for this approach. Many data are available on such topics as lighting and noise levels, layout of controls and instrumental displays and anthropometric measurements, and some of these are described below. It is important that the designer should not just call on such information when the finishing touches are being put to a machine. Consideration must be given to these factors at the preliminary design stage – and this "ergonomic approach" is, of course, particularly important where the efficiency and accuracy of the human operator are important to the safe functioning of the equipment. For example, the mechanical and structural engineers working on the design of a crane should attach great importance to the location of the control cabin; attention must be paid to visibility, lighting, layout of controls, ventilation and other factors affecting the comfort of the operator, if he is to realise his full working potential.

Some of the important aspects of ergonomics which are dealt with below are given only brief coverage, the intention being to indicate what sort of information is available and where further data can be found. In some cases, considerable latitude can be tolerated in the data presented; this latitude arises from the very great adaptability of the human body to a wide range of conditions. It is this adaptability which has been one of the causes of delaying strong action to make improvements. However, it is now recognised that the penalty of bad design – in terms of added fatigue and strain – is often high and can be avoided.

Equipment should be matched to the size of operators and operating levers and pedals positioned so that they do not impose unreasonable demands on strength, stamina and reach. In seeking to meet this requirement, the designer can use measurements of body size that have been collected for various populations. A set of measurements of this type is reproduced from K. F. Murrell's *Ergonomics*[4] in Table 21.1. A great deal of information of this type is available in chart and tabular form[5]; the designer must ensure, of course, that it is relevant to the population he is dealing with. It will be noted that these anthropometric measurements are tabulated in terms of the mean and the 5th and 95th percentiles. Thus, in the case of head clearance in the sitting position, the use of the 95th percentile for height above seat (960 mm) will ensure that this percentage of the population is catered for. When selecting a seat height that will allow the feet to reach the ground, the designer would usually select the 5th percentile value of the popliteal height (lower leg under knee) – in this case, 370 mm. Those with greater leg lengths can accommodate themselves by stretching out their legs. Of course, in this and many other cases the complete, but usually costly, solution is to provide some form of adjustment.

In laying out work areas and control systems, considerable use is made of information of the type set out in Fig. 21.3, taken from F. T. Kellerman *et al.*,

Table 21.1. Anthropometric measurements for a civilian working population (mm).

	Males proportion of stature*	Mean	Percentiles 5th	Percentiles 95th	Females proportion of stature*	Mean	Percentiles 5th	Percentiles 95th
*Standing in Shoes**								
Stature	—	1750	1650	1850	—	1650	1540	1750
Eye level	0.936	1640	1550	1740	0.938	1550	1440	1640
Elbow height	0.608	1070	990	1140	0.603	1000	930	1090
Shoulder height (maximum height for controls)	0.811	1420	1330	1500	0.805	1330	1250	1440
Symphysis height (minimum height for controls)	0.507	890	810	950	0.490	810	760	890
Gluteal furrow	0.468	810	740	880	0.437	740	690	810
Lower leg length (crotch height)	0.485	850	770	910	0.462	760	710	840
Sitting								
Height above seat	0.522	910	850	960	0.532	850	750	880
Eye level above seat	0.477	800	740	850	0.468	750	650	770
Elbow height	—	240	190	270	—	230	220	250
Buttock to back of knee	0.280	480	430	520	0.286	460	410	500
Buttock to patella (knee cap)	0.346	600	550	630	0.350	560	550	610
Bitrochanteric (buttock) width	—	350	330	390	—	370	320	410
Minimum height of controls above seat	—	25	—	—	—	25	—	—
Shoulder height (max. height of controls above seat)	0.338	580	550	630	0.333	530	480	570
Popliteal (lower leg under knee) height (no shoes)	0.242	420	390	440	0.246	390	370	420
Patella height (no shoes)	0.316	550	510	600	0.310	500	430	550
Breadth across both knees	—	200	180	230	—	190	160	230
Standing or Sitting								
Reach for operation of controls from c.l. of body at shoulder height	—	(690)	610	(750)	—	(610)	530	(670)
Maximum span at working level	—	(1520)	1400	(1660)	—	(1400)	1270	(1510)
Normal span at working level	—	(1220)	1090	(1330)	—	(1120)	990	(1230)
Maximum forward reach from front edge of bench	—	(510)	470	(550)	—	(430)	380	470
Normal forward reach from front edge of bench	—	(330)	300	(360)	—	(280)	270	(320)
Shoulder width	—	460	420	500	—	410	380	460
Shoulder to elbow	0.213	370	340	410	0.210	330	300	370
Elbow to finger tips	0.272	470	430	510	0.270	430	410	470
Foot length	—	270	250	290	—	240	220	250
Elbow width	—	440	380	500	—	410	360	470
Dimensions relating to equipment								
Clearance between seat and work surface	—	165	130	190	—	170	130	200
Minimum leg room	—	—	—	840	—	—	—	760
Back of seat to front edge of bench	—	—	—	360	—	—	—	360
Clearance between floor and work surface	0.338	580	550	630	0.350	560	500	610

*Standing dimensions and proportions are based on stature to which has been added 25 mm for shoes in males and 50 mm in females. When using proportions for other statures heel height must be added before calculating standing dimensions.

The dimensions given in this table are the author's "best estimate" from available anthropometric surveys. Dimensions are given to the nearest 10 mm; to do otherwise would imply an accuracy which the data do not possess; they are not necessarily accurate to the nearest 10 mm but they are probably reasonably consistent within themselves.

Source: Adapted from K. F. Murrell, *Ergonomics*[4], by permission of the Associated Book Publishers Ltd.

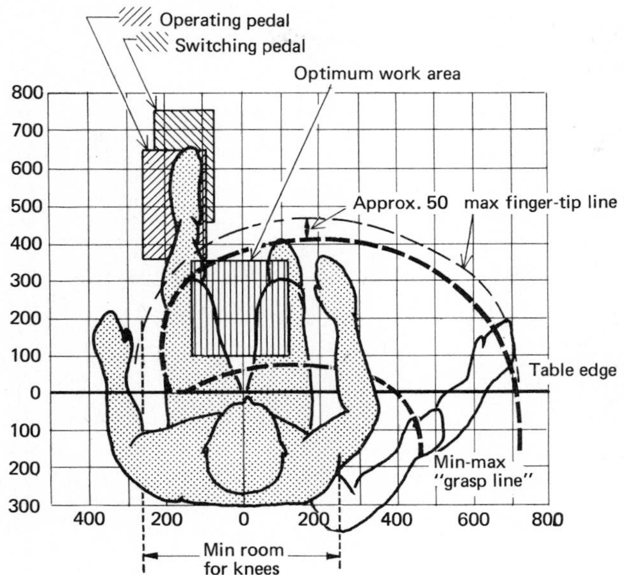

Fig. 21.3. *Optimum working areas for hands and feet. (From F. T. Kellerman* et al., Ergonomics in Industry[6], *by permission of the Centrex Publishing Co.)*

Ergonomics in Industry.[6] The working area can, of course, be larger than that illustrated, but a penalty in operator fatigue would be paid, resulting perhaps in slower and substandard work.

It is often necessary to know what forces can be applied by different parts of the body and Figs 21.4 and 21.5 illustrate typical information of this type for various arm and leg positions.[6] The forces shown are maximum values and, when designing equipment where repeated applications of force are required, values greater than 10% of those given will quickly cause fatigue. As a general principle, muscles are best loaded when they are in movement, contracting or relaxing; static loads on muscles – for example, those incurred when holding an object at arm's length – are tiring and unpleasant. When determining total loads on limbs, the weight of the limb itself must be taken into account.

The fact that many machines are now being automated does not mean that all loads should be taken off the operator. If an operator is required at all, it may well be advantageous to the operation, as well as to the man, to avoid turning him into a mere button-pusher. The operator must not be overloaded, and the rate of work demanded must be well below the maximum capabilities that are indicated in Fig. 21.6 taken from T. R. Nonweiler, *The Man-powered Aircraft.*[7]

Lighting and vision is another topic which is within the field of ergonomics. On reading recommendations made in the literature stressing the need for adequate lighting and clear marking of instruments and controls, many engineers are inclined to feel that the problem is straightforward and easily solved by the common-sense application of a few simple principles. This may be so, but there are still many installations to be seen where these principles are neglected. Here again, the adaptability of the human sense involved makes up for many deficiencies in design.

The lighting problem comes down to supplying adequate levels of light in the right places, without shadow or glare. After this requirement has been satisfied, the visual efficiency with which a task can be carried out will depend

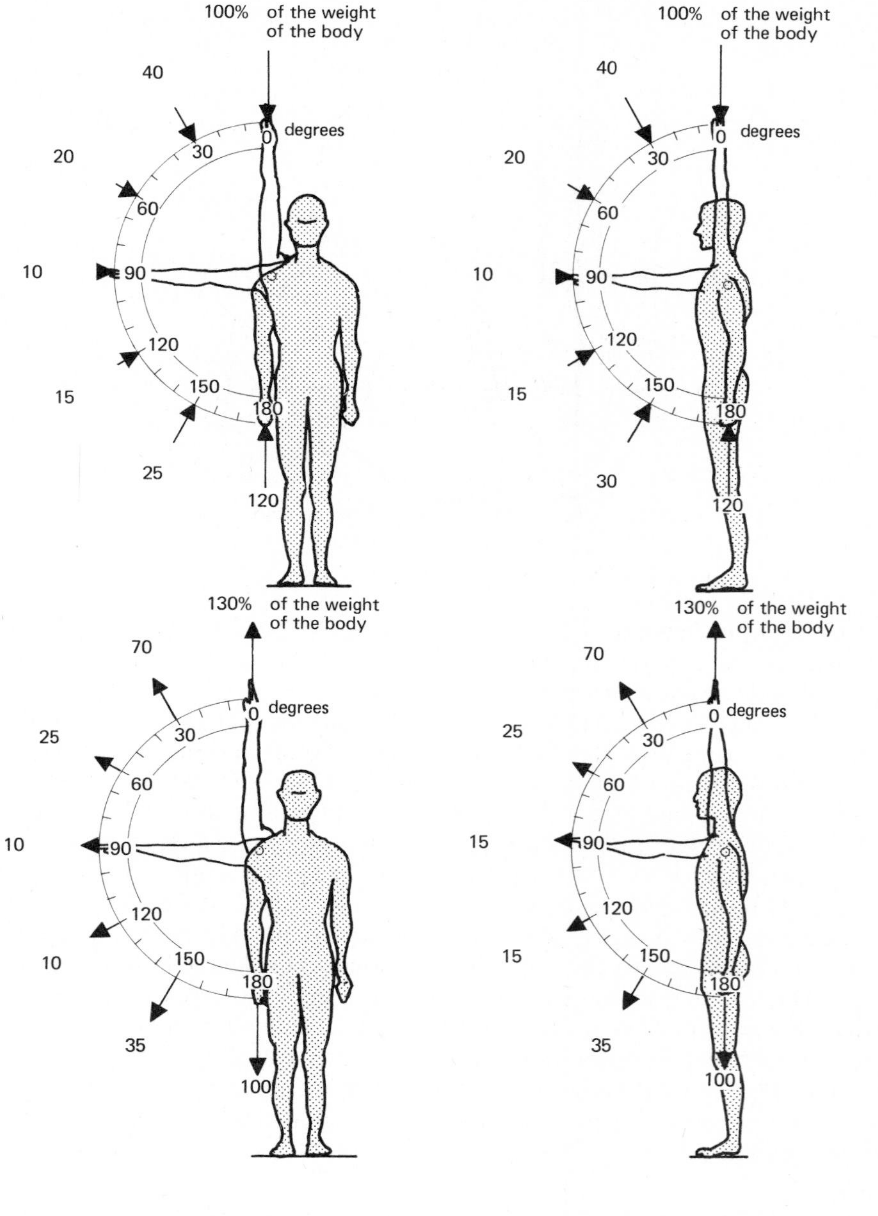

Fig. 21.4. *Maximum pushing and pulling forces for the arm. (From F. T. Kellerman* et al., Ergonomics in Industry[6], *by permission of the Centrex Publishing Co.)*

on colour, contrast and size of the working area and such other details as shape and layout of controls and of scales and lettering.

Typical values of illumination levels suitable for various locations and tasks can be found in handbooks on the subject.[4] Quite large variations may be found between one reference and another; these arise from several causes. First, values in the past were influenced by what was economically attainable. Secondly, the ability to see increases as the logarithm of the illumination, and there is disagreement among authorities as to how far it is economical to pursue small gains which can be achieved only at the expense of high energy consumption. Although lighting equipment is becoming more efficient, capital and

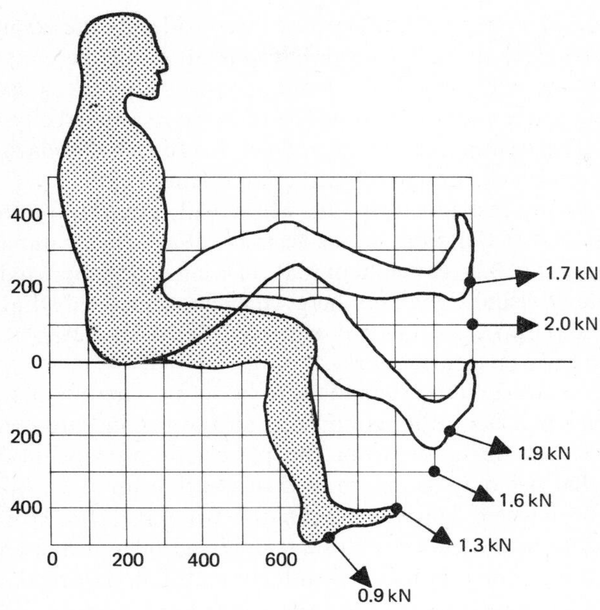

Fig. 21.5. *Maximum thrust forces for the leg. (From F. T. Kellerman* et al., Ergonomics in Industry[6], *by permission of the Centrex publishing Co.)*

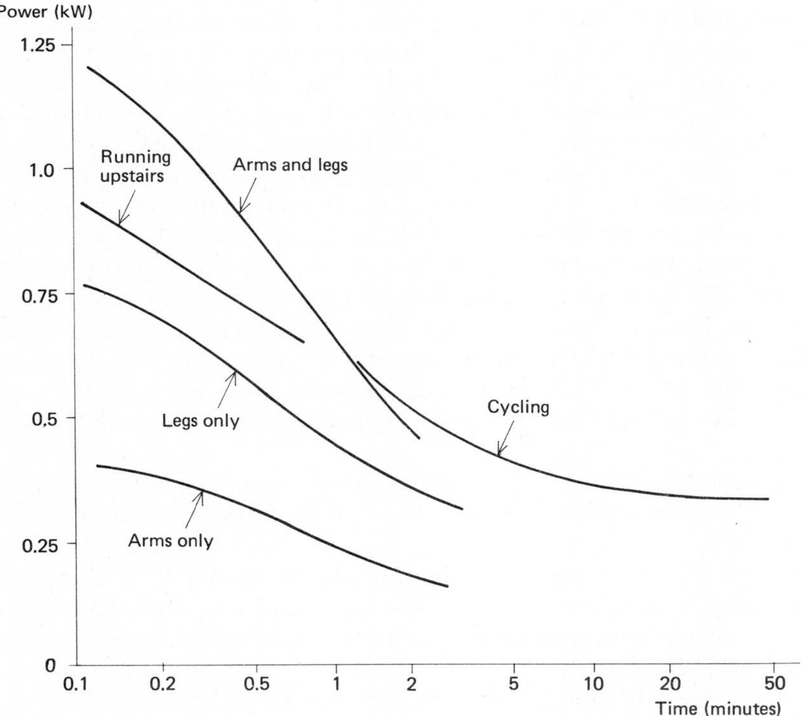

Fig. 21.6. *Estimates of man's maximum power output, compiled by Nonweiler from various sources. (From T. R. Nonweiler, "The Man-Powered Aircraft",* Journal of the Royal Aeronautical Society, **62**, *October 1958, by permission of the* Journal of the Royal Aeronautical Society.*)*

running costs are high and the overall planning of lighting in a large project will usually be a job for specialists.

If some particular element of a machine is to be accentuated for some reason (such as safety) lighting, contrast of colour or form – or some combination of these – may be used.

Some of these points will be illustrated by considering the arrangement and marking of instrument dials. Two types of indicators will be considered: first, a type which, along with others of a similar type in a panel, is used for check reading; and second, an instrument which must be read correctly to the limits of its accuracy. These two types are the subject of a British Standard which gives useful information on the design of such components.[8]

A check-reading instrument is one which will require no action from an operator unless it deviates from some desired value – for example, an engine oil-pressure gauge. Such an instrument may, of course, be supported or replaced by some audible or visual signal, but here we will consider only dials of the type shown. For this purpose a simple uncluttered dial is preferable; the safe or desired range should be clearly marked.

Where a bank of such instruments is to be set up, they should be arranged so that under normal operating conditions all the needles are oriented in the same direction. Any deviation from the pattern is quickly detected by the operator. The dial size may be determined by considering the minimum movement that can be detected from the expected viewing distance. Murrell suggests that a pointer will be clearly seen to have deviated from a marked zone if its movement from the zone is at least 1.3 mm per metre of viewing distance.[4] This criterion makes it possible to calculate the minimum dial size.

Consider the case of a temperature gauge calibrated from 0 to 800 with a safe limit from 650–700. If the allowable deviation from one range is 20° and the instrument is to be read from 3 m, the movement for 20° must be 3×1.3 mm. Thus, the scale length for 800° must be $(800/20) \times 3.9$ or 160 mm – the dial size is thus fixed. An instrument of this type should have a pointer which stands out clearly from its background and care should be taken to avoid highly reflective bezels or other distracting objects in the immediate surround.

The second class of instruments, used for quantitative readings, requires careful design if it is to be read to its potential accuracy. There are two main types of error: the gross error caused by, say, reading a temperature of 1800 as 800, and the error of not interpolating between adjacent scale markings accurately. The chances of such errors occurring are lessened by correct choice of shape, distribution and contrast of scale and numerals. This is a question which cannot be dealt with here in depth, and readers are referred to a text, such as Murrell.[4] However, the following points can be made to illustrate the approach:

(a) match the number of scale markings to the instrument accuracy, allowing for visual interpolation of not more than fifths between minor scale markings;

(b) number the major scale marks in ones, twos or fives and adjust the minor markings so as not to require the observer to carry out mental arithmetic – for example, avoid an interval of 20 with four minor divisions (giving intermediate values of 4, 8, 12, 16);

(c) avoid more than three digits in the scale numbers.

A final point can be made regarding the shape and proportions of numerals: BS 3693 gives recommendations on the shape and proportions of digits, and these provide an excellent basis on which to work. Figures 21.7 and 21.8 are taken from this Standard. Figure 21.7 shows the recommended series of figures, while Fig. 21.8 illustrates how these can be dispersed around a part-circular scale to produce a clear and uncluttered dial.

12345
67890

Fig. 21.7. *Recommended series of figures. (Extract from BS 3693 is reproduced by permission of the British Standards Institution, 2 Park Street, London W1A 2BS.)*

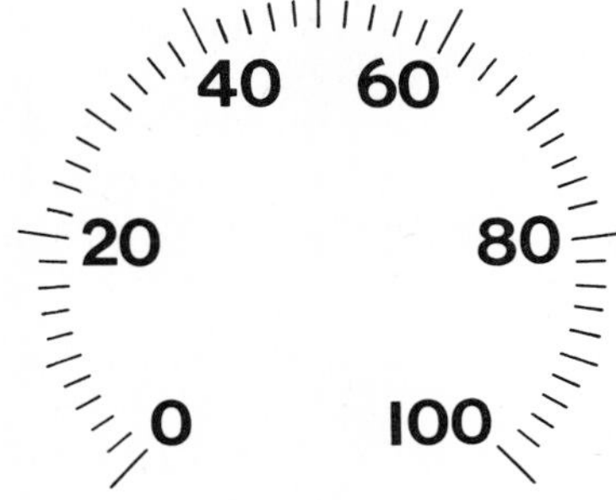

Part circular scale

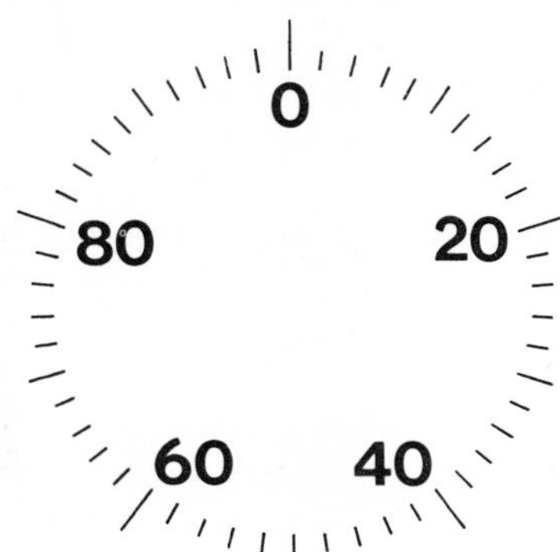

Full circular scale

Fig. 21.8. *Examples of types of scales. (Extract from BS 3693 is reproduced by permission of the British Standards Institution, 2 Park Street, London W1A 2BS.)*

REFERENCES

[1] N. Pevsner, *Pioneers of Modern Design*, Penguin, Harmondsworth, 1960.
[2] H. van Doren, *Industrial Design*, 2nd ed., McGraw-Hill, New York, 1954.
[3] G. Jackson, "The High Style", *Designscape 12/1*, New Zealand Industrial Design Council, Wellington, 1970.
[4] K. F. H. Murrell, *Ergonomics: Man in his Working Environment*, Chapman & Hall, London, 1965.
[5] H. Dreyfuss, *Designing for People*, Simon & Schuster, New York, 1955.
[6] F. T. L. Kellerman, P. A. van Wely, and P. J. Willems, *Ergonomics in Industry*, Philips Technical Library, Centrex Publishing Co., Eindhoven, 1963.

[7] T. R. F. Nonweiler, "The Man-Powered Aircraft", *Journal of The Royal Aeronautical Society*, **62**, 1958.
[8] BS 3693:Parts 1 and 2:1964, *The Design of Scales and Indexes*, The British Standards Institution, London, 1964.

BIBLIOGRAPHY

Beresford-Evans, J., *Form in Engineering Design*, Oxford University Press, London, 1954.

Blake, J., and Blake, A., *The Practical Idealist*, Lund Humphries, London, 1969.

Chaparis, A., *Man-Machine Engineering*, Tavistock Publications, London, 1965.

Ferebee, Ann, *A History of Design from the Victorian Era to the Present*, Van Nostrand Reinhold, New York, 1970.

Mayall, W. H., *Industrial Design for Engineers*, Iliffe Books, London, 1967.

Middleton, M., *Group Practice in Design*, Architectural Press, London, 1967.

Pilditch, J., and Scott, D., *The Business of Product Design*, Business Publications, London, 1965.

PROBLEMS

21.1. This exercise is intended to introduce students to simple methods of increasing the usefulness and visual appeal of a drawing, and to enable them to discover the medium best suited to their abilities.

Engineering drawings showing the external outlines of an object do not usually need shading or colouring, but we are considering the case where information on shape, massing or colour will be conveyed by such treatment.

Select one of the objects listed below, and make an orthographic elevation, an isometric and a perspective drawing on tracing paper. Make several prints of these drawings on white paper and treat them in the following ways:

(a) Use pencil shading of object and background to accentuate shape;
(b) Use colour pencils to show a colour scheme;
(c) Use pastels to show a colour scheme;
(d) Use the opaque qualities of pastel to experiment with changes in shape of the original drawing.

Suitable objects are: control levers and timing knobs (try different shapes), a simple electrical instrument, a hydraulic valve, an instrument panel, a laboratory balance, or one of the items illustrated in Fig. 21.1.

21.2. A small firm with capacity to manufacture light sheet-metal and wire products requires ideas for items that could be produced during slack periods. Provide one or more designs for this purpose, illustrating your ideas with suitable sketches.

21.3. Your employer has asked you to design a baseplate on which to mount a small, electrically driven air compressor which will be used to provide a portable air supply in the factory. You have decided that you will try to interest him in a more elaborate design which would be suitable for larger-scale production and could be marketed for garage and home workshop use. Write a design brief for the industrial designer you intend to call in as a consultant. Select either a very small compressor which could be carried, or a larger one which would be mounted on wheels.

21.4. Produce a design to the brief of Problem 21.3. Make a colour drawing suitable for presentation to management or to a customer.

21.5. Investigate some of the new shapes that have been proposed for numerals to improve their legibility under conditions of adverse visibility.

22 Manufacturing Operations

22.1. INTRODUCTION

The designer as he works makes decisions which determine the machines and processes to be used in manufacture. These decisions often relate to functional requirements and may be concerned with shape, size, surface finish and material; nevertheless each influences the production methods to be used and hence the economy with which the part can be made. The designer must always keep this interrelationship in mind and it is with the object of highlighting it, rather than describing actual machines and processes in detail, that we approach this chapter.

A classification of the more important processes is given in Table 22.1; we will describe some of these briefly before going on to discuss how their use is influenced by design decisions.

Table 22.1. Classification of some of the principal manufacturing operations.

Casting	Plastic moulding	Hot forming	Cold forming	Machining	Miscellaneous
Sand	Injection	Hammer forging	Bending	Turning	Gas cutting
Shell-mould	Pressure	Drop forging	Stretch forming	Grinding	Welding
Gravity die	Blow	Press forging	Deep drawing	Drilling	Riveting
Pressure die		Upset forging	Rolling	Boring	Surface finishing
Investment		Extrusion	Spinning	Reaming	Heat treating
Centrifugal		Rolling		Milling	
				Planing	
				Shaping	
				Broaching	

Although some of the knowledge of processes required by the designer must come from experience and observation in the workshop, a lot can be gleaned from books on the subject. Some can be quantified, and Figs 11.2 and 22.1 show information of the type that the designer must be conversant with. In the first chart (Chapter 11) accuracy is related to machining operations; once a tolerance quality has been fixed the choice of operations is limited to those which will give that tolerance, or better.[1] In the same way the second chart relates the surface finish to the production process.[2] We will refer to these again.

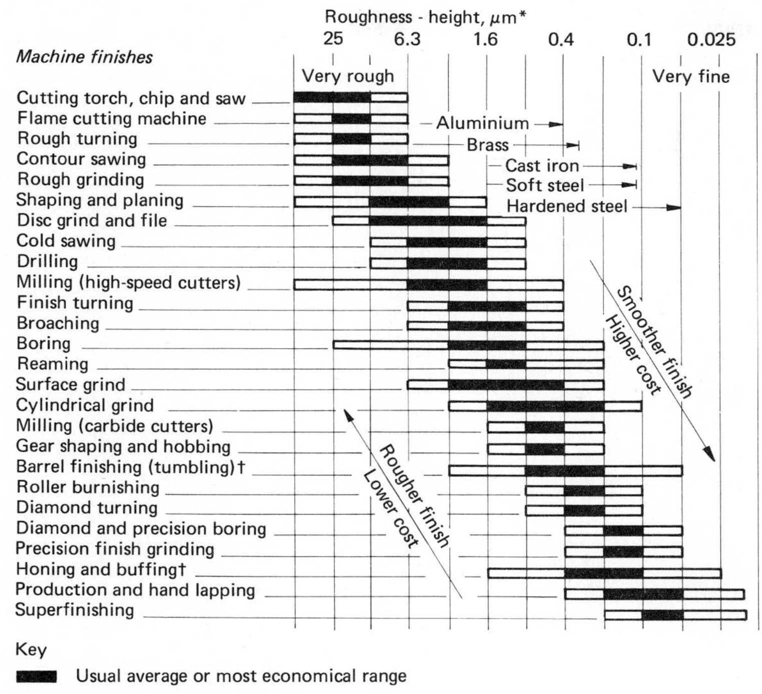

Fig. 22.1. *Roughness–height ratings for various production processes. (After Table 6 in J. A. Broadston, "Surface Finish Requirements"[2], by permission of The American Society of Mechanical Engineers.)*

The designer will also be influenced by factors not considered here. For example, where local industries are comparatively underdeveloped he will perforce specify processes for which equipment and skills are readily available; in any case he will be inclined to favour processes with which he has previously had some successful experience.

There are many useful text and reference books that can be consulted. Some, like *Engineering Design* by Matousek[3], adopt a systematic approach to form design. Others discuss a wide range of materials and the processes that are used to shape them.[4, 5, 6]

The values used in this chapter to set limits to such factors as accuracy or minimum production quantities are typical ones; they can be varied in special cases. For example, where the market is limited, smaller production runs – and the resulting higher costs – may have to be accepted.

22.2. PRODUCTION PROCESSES

22.2.1. Casting

There are several different types of casting processes, and the one selected for a particular job will depend on the material, size, and tolerances involved and especially on the number of pieces to be produced. Once these factors have determined the process to be used, the designer must respect the various limitations – such as those on accuracy and minimum wall thickness – that it imposes. Some of the principal features of the processes described here are

Table 22.2. Design aspects of some casting processes.

			Permanent mould		
	Sand casting	Shell moulding	Gravity diecasting	Pressure diecasting	Investment casting
Material	Wide selection of ferrous and non-ferrous	Wide selection of ferrous and non-ferrous	Mostly aluminium alloys	Aluminium, magnesium and some copper alloys	Wide selection of ferrous and non-ferrous
Form	Complex shapes	Complex shapes	Simple shapes	Simple shapes	Complex shapes
Size	Up to 2×10^5 kg	Up to 10 kg	Up to 250 kg	Up to 45 kg	Up to 2 kg
Quantity	One, upwards	500 minimum	1000 minimum	5000 minimum	Usually large
Finish	Poor	Good	Good	Very good	Excellent
Draft	1° or more	$\frac{1}{2}$° sufficient	2° per side minimum	2° per side minimum	No limitation
Dimensional accuracy* (mm in 25 mm)	Grey iron ±1 Steel ±4 Light alloy ±2 Brass ±3	±0.15	±0.4	±0.15	±0.05
Minimum section† (mm)	Grey iron 4 Steel 8 Light alloy 4 Brass 4	2	4	2	0.1

*Values of accuracy vary with size and type of casting.
† These minimum values should only be demanded when essential.

summarised in Table 22.2; the accuracies and surface finishes can be compared with other methods of manufacture in Figs 11.2 and 22.1.

In *sand casting*, a cavity in the sand packed into a mould is produced by means of a wooden pattern; this pattern may be built up from several pieces to simplify removal from the mould. Holes and cavities in the casting can be obtained by the use of cores, also made of sand, which are broken up and removed from the casting after it has solidified. The only permanent items specific to each different shape of casting are the pattern and the equipment to make the cores; the process is therefore suitable for small runs.

Shell-mould casting requires a moulding sand that has a plastic binder mixed with it; this is dropped or blown onto a metal pattern heated to about 230°C. After a short period, the shell that has set round the pattern – about 5 mm to 10 mm in thickness – is pulled off and hardened by baking. The inner surface of the mould in this process will be much smoother than that of a sand mould. Two half-shells are joined together to form the mould and cores may be used as in sand casting. After pouring, the remains of the shell are stripped from the casting. Because of the cost of the metal pattern, the process is best suited to long production runs; the cost of the moulding material makes the cost of the process higher than that of sand moulding. However, fine detail and complex shapes can be cast with good surface finish and close tolerances, and the smaller allowances for machining help to offset the higher costs of casting.

Shell moulding lends itself to automation and is widely used in large scale production.

The permanent-mould method gives economy in large scale production by using a mould manufactured from a heat-resistant steel which can be used repeatedly. In the simpler form of the method, the molten metal flows into the mould under gravity. This process is termed *gravity diecasting*, although in the United States the term *permanent-mould casting* is used. It is suitable for quantities of 1000 or more.

Higher production rates and castings having thinner and more complex sections may be obtained if the molten metal is fed to the die under pressure. *Pressure diecasting* requires elaborate production equipment, but production rates as high as 200 cycles per hour may be attained. It is used where large quantities – 5000 or more – are involved.

Investment casting starts with a wax replica of the shape to be cast; refractory materials are consolidated around the wax, which is then melted out leaving a jointless mould. There are no problems of coring and intricate shapes can be obtained using any of the casting metals. Good accuracy can be achieved, and the method is particularly suitable for high-temperature alloys which are hard to machine. Although the process can be used for small numbers of special parts, it is mainly used in industry for volume production.

When parts are hollow and have radial symmetry *centrifugal casting* may be used. In this process a sand or metal mould rotates about an axis of symmetry and the molten metal is forced out against the wall; the outer surface of the casting takes up the shape of the mould. When the axis of rotation is horizontal the inner surface is cylindrical. Good quality castings for pipes, cylinder liners and similar shapes can be produced.

The operations involved in producing a simple sand casting are illustrated in Fig. 22.2. The part is shown with the pattern and core in Fig. 22.2(a). The internal hole has two annular grooves which are reproduced on the core. The pattern has a cylindrical extension at each end – the core print – which forms a cavity in the sand into which the end of the core will sit. In Fig. 22.2(b), the way in which the sand is built-up round the pattern is shown; in the last step, the boxes have been separated to remove the pattern and to put the core in place. The gate system through which the molten metal is poured is cut in the sand before the boxes are finally assembled. When the metal has been cast in the cavity, the sand mould – including the core – is broken up and the sprue (which is formed from metal which solidifies in the gate) is cut away from the casting.

22.2.2. Plastic moulding

Hot compression moulding is suitable for thermosetting plastics; a measured amount of loose material, or a shape that has been roughly preformed, is placed in a heated die and compressed by a plunger. The material sets in the shape of the die, and can be pushed out when it is cured. The flash that is shown in the sketch must be removed, but it can be eliminated by using a specially shaped plunger.

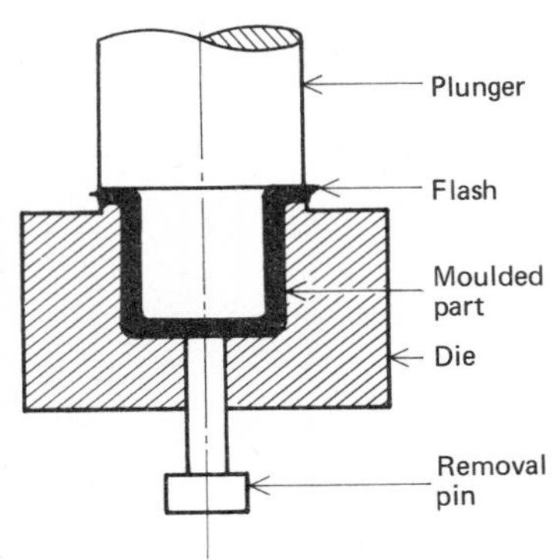

Injection moulding is used for thermoplastic materials. Plastic granules are fed through a heater, where they are melted, into a cool die. When the plastic has solidified the die opens to eject the part; the cycle may take only a few seconds.

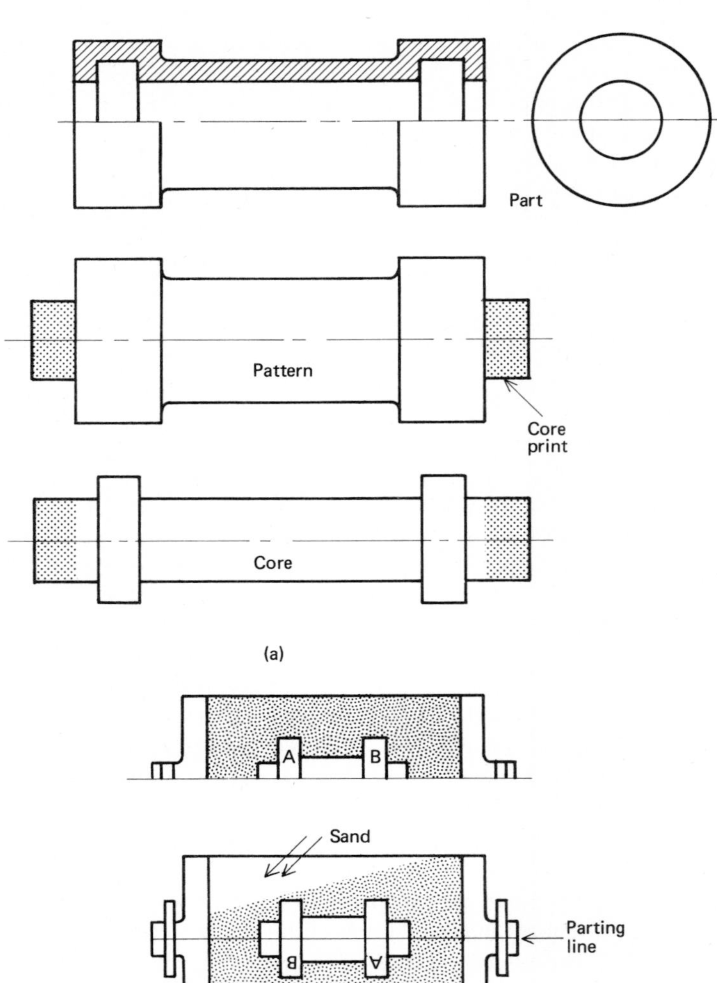

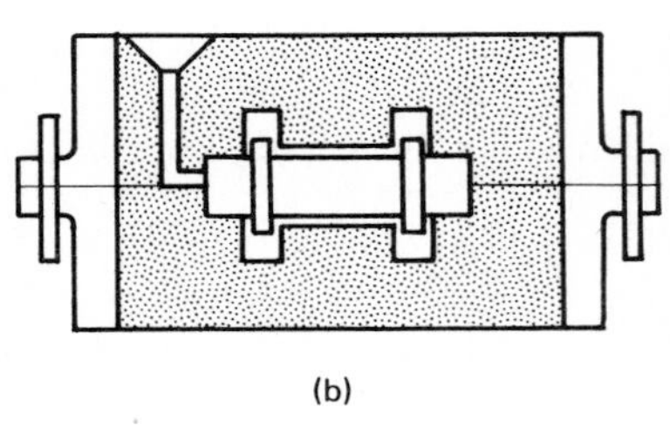

Fig. 22.2. *Casting a simple part: (a) part, pattern and core, (b) three stages in preparation of mould.*

Hollow shapes, such as bottles, can be made by *blow moulding*. A two-part mould is closed around a heated thermoplastic tube, and low pressure air is connected to the inside of the tube to blow it into contact with the mould; the mould opens to release the product. In other similar processes sheets of plastic are softened and pulled into contact with a die by a vacuum. In such operations, the thickness of the part is determined by the initial tube or sheet thickness, but local thinning will occur where the material deforms to take up the shape of the die.

The processes described for moulding plastics are suitable for large-scale production; and in these and similar methods for which expenditure on tooling is high, it is important that the designer consult the manufacturer at an early stage in design.

Table 22.3. Standard moulding tolerances for ABS plastic moulding material.

Drawing code	Dimension (mm)	Tolerance, plus or minus (μm)	
A = diameter B = depth C = height	Refer graph		
		Commercial	Fine
D = bottom wall		100	50
E = side wall		75	50
F = hole diameter	0 to 3	50	25
	3 to 6	50	25
	6 to 12	75	50
	12 and over	100	50

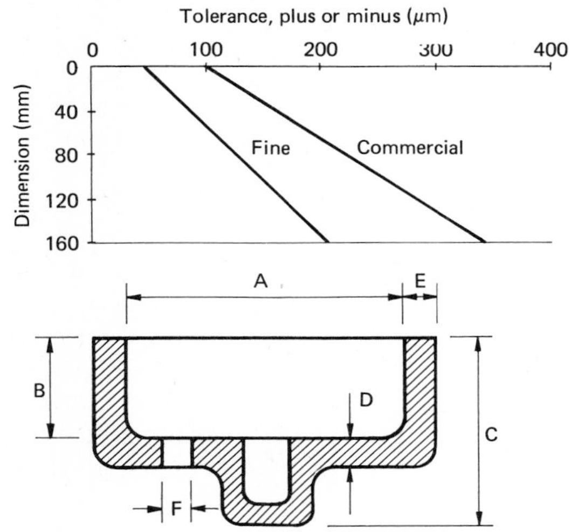

Note: The "Commercial" values represent common production tolerances. The "Fine" values represent closer tolerances that can be held but at a higher cost.

Extensive information is available on the accuracy that can be expected in moulding various plastic materials. Examples of such data are published by the Society of the Plastics Industry.[7] The way in which the tolerances are presented is illustrated in abridged form in Table 22.3 for the plastic moulding material ABS. The dimensions are based on the hypothetical article shown in the table. For more complete information and for other materials and processes, the reference quoted above and Beck, *Plastic Product Design*[8] should be consulted.

22.2.3. Hot and cold forming

A wide range of materials and shapes are produced in wrought form, to the extent that the designer looks on plates, bars, tubes, channels and rails as the raw materials for his work. These shapes will have been made by rolling, or sometimes extruding hot metals; along with drawing and forging, these are the principal *hot-forming processes.*

From the producer's viewpoint, the advantages of hot forming are:

(a) for a given change in shape, power requirements are minimised;
(b) large plastic deformations can be obtained without changes in ductility;
(c) grain size can be controlled;
(d) major defects in the raw material will usually be revealed by the large deformations involved, while minor ones due to slag and other inclusions will be spread into a fibrous structure.

The advantages will also hold for further hot processing of the basic shapes listed above. The main disadvantage of hot forming seen by the designer is the poor surface finish that results; some cold working as a final operation will improve the finish and accuracy, and will also give an increase in the yield strength, at the expense of ductility.

The hot-forming processes, except forging, usually result in standardised products and we will discuss only the latter.

When a simple shape is required this can often be produced by *hammer forging* using shaped tools. This is a mechanised development of the blacksmith's work, and is used on small production runs. The forging operation may save a considerable amount of machining, and can also give a favourable grain orientation. Accuracies will be about ± 3 mm or ± 4 mm.

Drop forging using a die must be restricted to long runs since the die is expensive. It is made of high-strength steel and carries an impression of the surface to be formed. Very large forces are needed to make the hot billet flow to conform to the shape of the closed die and, where large shape changes are needed, a progression of die shapes may be used. Adequate machining allowances are left on the final forged form.

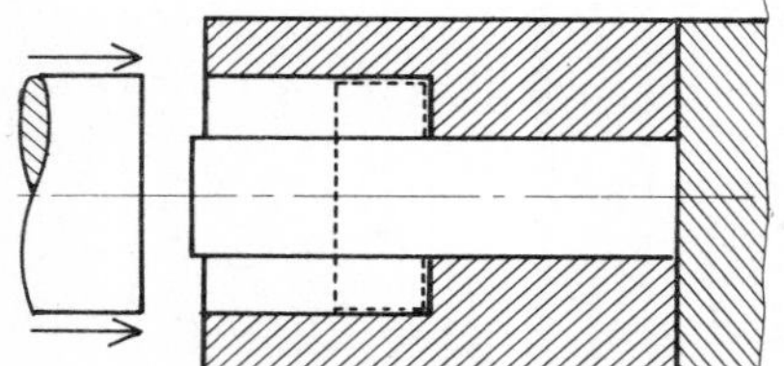

Upset forging can be used to produce local changes in thickness along the length of a bar; an axial force squeezes an unsupported part of the bar into a cavity in the die. The unsupported length must not be greater than about three times the bar diameter, so that the increase in diameter that can be achieved is limited.

Cold-forming operations can be classified under the headings of *squeezing*, *bending*, *drawing* and *shearing*.[4] The hot-forming operations just described can be carried out cold and fall in the first of these classes, along with *riveting* which can be performed hot or cold, and *thread rolling*. The *bending* operations produce parts having simple curvature, while those classed as *drawing* are capable of producing shapes with compound curvature; *spinning* and *stretch-forming* are typical of these. *Shearing* is usually done to give a straight, clean edge to a flat plate, but *blanking* and *piercing* are also included under this heading. Figure 22.3 shows diagramatically some of the common forming processes.

22.2.4. Machining

We classify *machining operations* as ones in which material is removed in chip form by means of a cutting tool or an abrasive wheel or block. More will be said later about the extent to which the designer will predetermine which process should be used. It is sufficient to say here that he differentiates between them mainly on the basis of the cost to achieve a certain shape, accuracy and surface finish.

The first point to be made about machining operations is that they are costly and produce scrap material as well as finished parts. If they can be avoided, they should be. The second point is that a high price must be paid for high accuracy. The accuracy to be expected from different processes is shown in Fig. 11.2; as the accuracy improves, the rate of removal of material decreases sharply. The overlap in Fig. 11.2 between the best casting and the least precise machining finishes indicates that there will be opportunities to eliminate machining on better quality castings. The surface texture of shell- and die-cast products is often as acceptable as a rough machined finish.

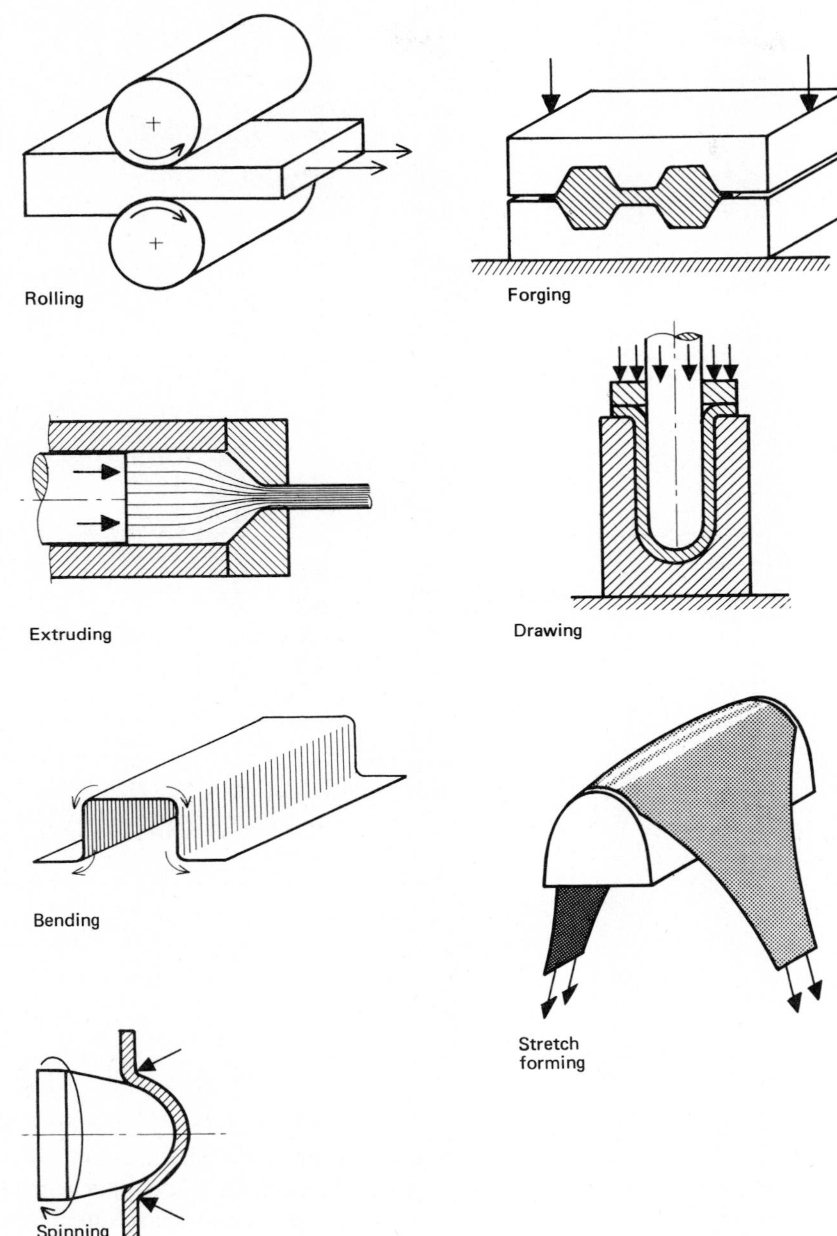

Fig. 22.3. *Schematic representation of some common forming and shearing processes.*

Where machining is necessary and several possible alternatives have been fixed by accuracy requirements, the shape of the machined surface will dictate the most economical process.

The basic machining processes are illustrated schematically in Fig. 22.4. Some of these – for example, *broaching* – can be used to produce internal or external shapes. *Drilling*, *boring* and *reaming* are restricted to the production of holes, and will be selected on the basis of the accuracy, size and shape required.

Most of the machining operations done with a cutting tool can be performed by similar grinding operations; the main differences are that the latter

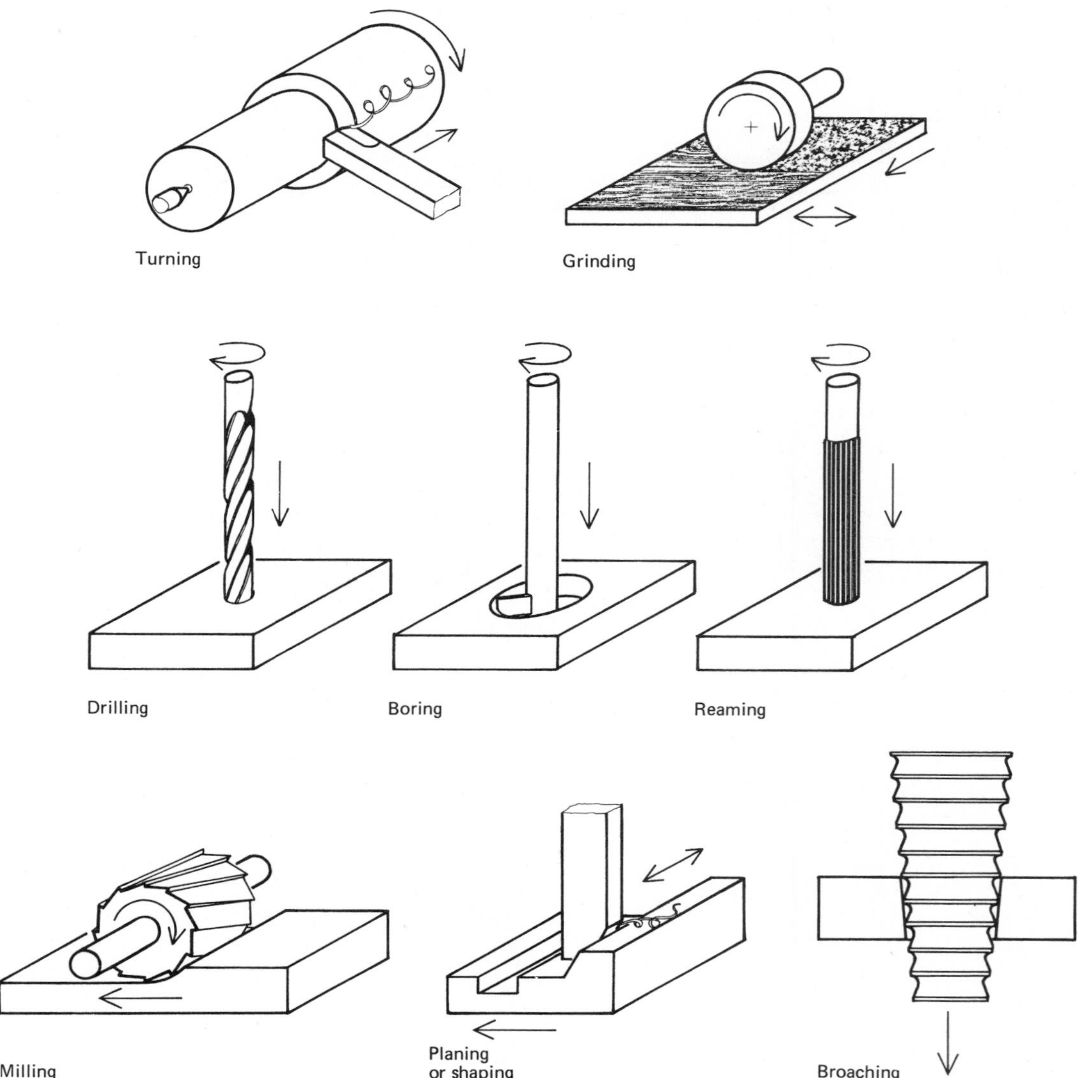

Fig. 22.4. *Schematic representation of some common machining processes.*

can produce a better surface finish and that they generally remove far less material. The designer will specify these abrasive methods when he seeks high accuracy and surface finish or when the material is too hard for other cutting tools. When high-strength heat-treated steels are used, preliminary machining can be done while the material is in the annealed state. A small machining allowance can be left for removal by grinding after heat treatment, eliminating inaccuracies that might be caused by distortion.

22.3. THE INFLUENCE OF SHAPE

Having reviewed some commonly encountered production processes we will look at the way in which the shape of a product influences their selection; the relationship between process and surface finish and accuracy should be borne in mind.

22.3.1. The shape of a casting

For a complicated shape, the choice of process in practice is limited. Apart from cutting it out of the solid – and this can be accomplished at a cost – one of the various casting methods presents the most likely choice. The casting will usually require some machining where mating and running surfaces are involved, but with the more accurate methods this is reduced, since machining allowances are cut down.

While the designer has great freedom in the choice of shape, he must exercise care if the task of the foundry is to be simplified and the best economy obtained. When large-scale production is planned, foundry engineers will be consulted before the shape is finalised. In any case, the designer is well advised to check the suitability of a design for casting before it is too late to make changes.

Some design aspects of the various casting processes are shown in Table 22.2. Each material is subject to different limitations; for example, the greater fluidity and lower shrinkage of cast iron as compared to steel make it an easier material to cast and give greater dimensional accuracy.

Some general points regarding the design of casting are summarised below:

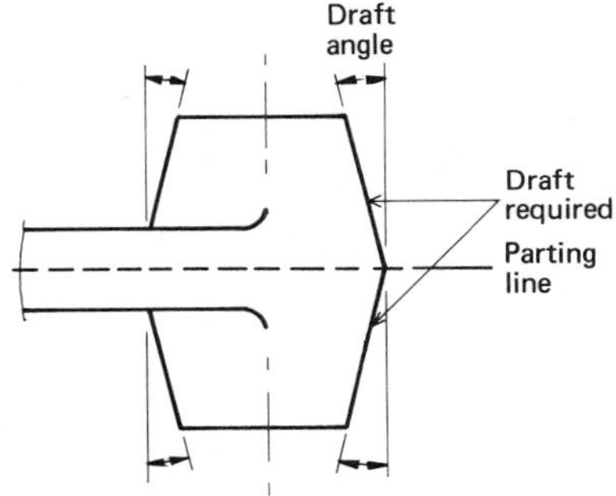

(a) flat surfaces perpendicular to the parting line must be given a taper or draft so that the pattern can be removed from the mould without causing damage. Typical values of the draft required for different methods are given in Table 22.2. The designer will usually show the draft on the drawing, but may simply indicate the surfaces on which draft is, and is not, acceptable;
(b) no special allowances for casting need be shown on the original drawing; the patternmaker will allow for shrinkage and for additional material (typically about 3 mm or 4 mm) where machined surfaces are specified;
(c) where possible, the wall thickness should be kept constant. If they are necessary, changes in section should be gradual. Because of shrinkage, the last sections to solidify (usually the most massive) may be starved of molten metal and voids will result. The foundryman can minimise this problem by using metal inserts in the sand, called *chills*, to absorb heat from the slower-cooled sections, and by providing a reservoir of molten metal, called a *riser*, adjacent to critical sections. The designer will minimise the problem by correctly proportioning the parts where shape changes occur.

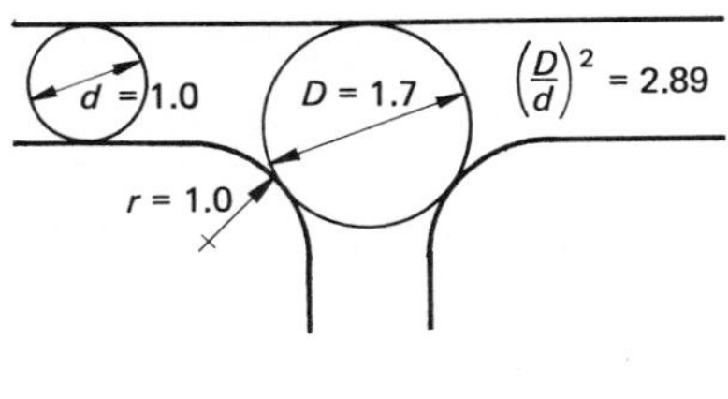

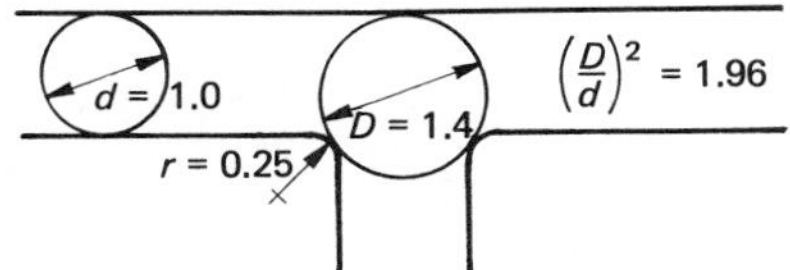

The "inscribed circle" method enables a comparison to be made between the masses of metal at two different locations in a section – the mass being proportional to the area of the circle. In the sketch, the junction between three walls is drawn for two different values of fillet radius. It will be seen that the ratio $(D/d)^2$ is 2.89 in one case and 1.96 in the other. Therefore, the larger fillet radius gives an increase in mass at the section of nearly 50%. Clearly, in the design of a casting, the need to avoid the hot spots associated with large local masses must be balanced by the need to avoid sharp notches that act as stress-raisers. No hard and fast recommendations can be made as to dimensions, but a variety of preferred methods and proportions are described in Broadston[2] and Matousek.[4] Two typical junctions, with an improved layout, are shown in Fig. 22.5(a) and (b). The base in the first drawing can be improved from the viewpoint of castability by either of the modifications shown on the right – the dotted version being easier to mould. In the second drawing, the intersection between a wall and flange is shown. The flange is commonly made thicker than the wall since it must have adequate strength and rigidity even after machining. A sudden change in section is avoided by the tapered junction shown on the right-hand side;

(d) small variations in wall thickness may occur because of inaccurate location of cores, and this possibility must be recognised where strength is a ruling factor. Dimensional variations between two features will be greatest when they lie on different sides of the parting line;

(e) large flat surfaces should be avoided since they are not rigid and are often unsightly in the as-cast condition because of inevitable imperfections. A conical surface is more rigid and less subject to built-in stress. A flat surface can be broken up by webs which will stiffen it and, if machining is required, pads can be provided to decrease the area that must be machined. The provision of such pads on a baseplate will simplify lining up the base on an uneven surface. In the same way, the inner surface of the supporting bracket shown in the sketch need not be machined along its full length.

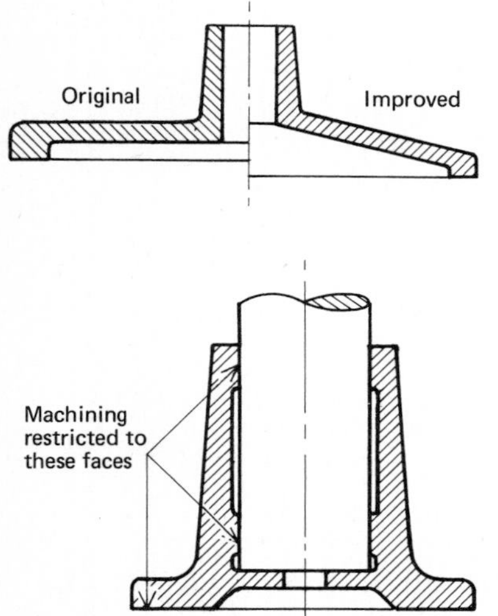

It is noted in Table 22.2 that the permanent-mould method is suitable for simple shapes. It must be borne in mind that economy in large-scale production can be achieved by moulding into one casting as many as possible of the various bosses, flanges and cavities that make up the whole component. The resulting die will be complicated and costly; the shape of the casting must nevertheless still conform to the constraints of having cores that can be retracted to permit the casting to be removed from the die.

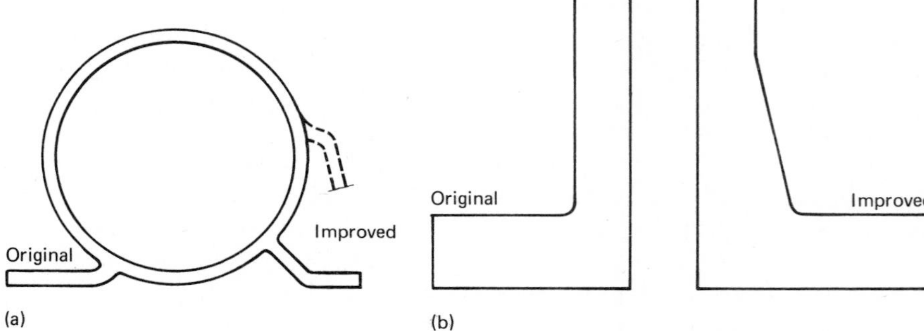

Fig. 22.5. *Shape of a casting: (a) base-to-body junction, (b) wall-to-flange junction.*

22.3.2. Hot- and cold-formed shapes

Many structural components are built up from rolled shapes, such as channels and angles. Circular, square and rectangular steel and light alloy sections are readily available, and extruded light alloys are produced in a great variety of shapes. These, in their turn, can be cut, riveted, welded and bonded to give new shapes, and the designer should be aware of the availability of these standard sections before he calls for a shape that requires special forming operations.

Some of the most commonly encountered standard sections are illustrated in Fig. 22.6, but there are many others that are available, particularly those extruded from light alloys or plastics.

Some of the forming and shearing methods have been illustrated in Fig. 22.4. Rolling, forging and drawing can be used to produce massive parts, while

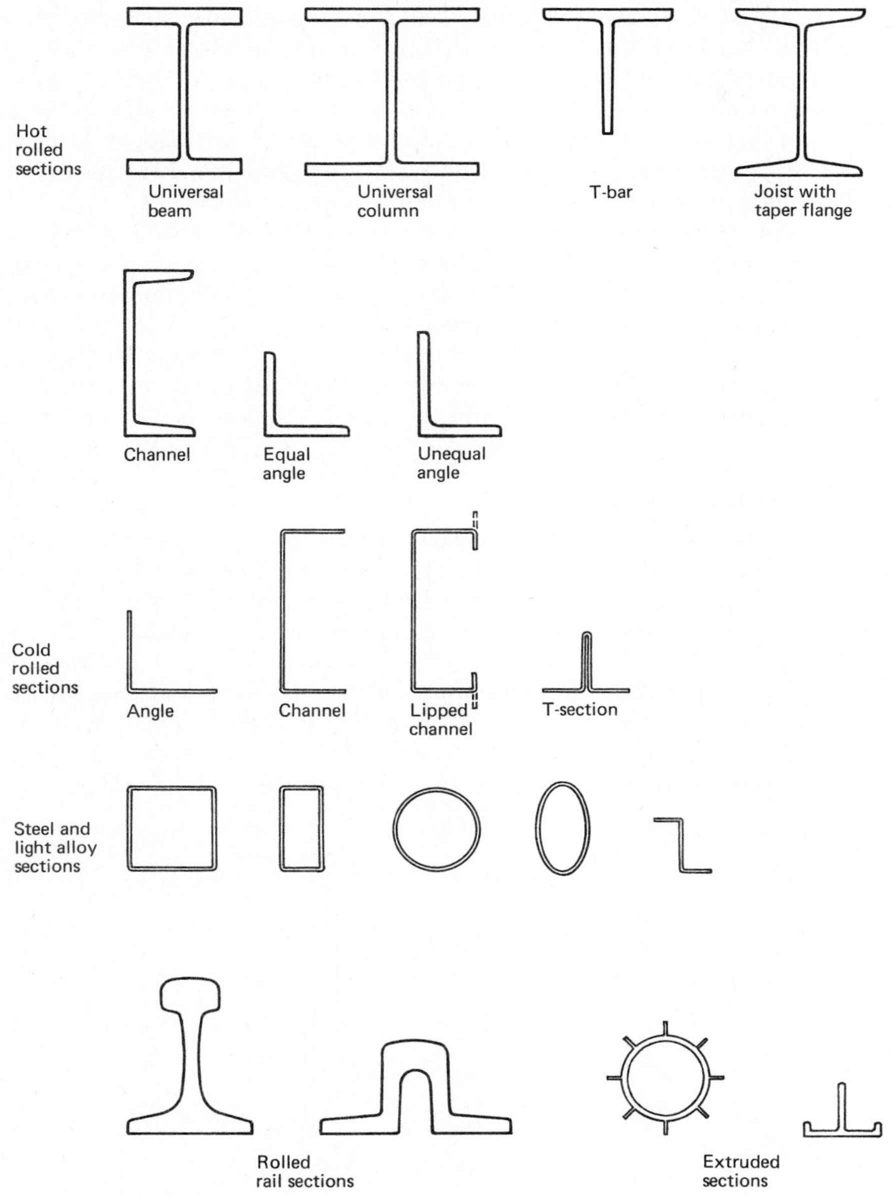

Hot rolled sections

Universal beam

Universal column

T-bar

Joist with taper flange

Channel

Equal angle

Unequal angle

Cold rolled sections

Angle

Channel

Lipped channel

T-section

Steel and light alloy sections

Rolled rail sections

Extruded sections

Fig. 22.6. *Some standard rolled and extruded shapes.*

the other processes of stretching, bending, spinning and shearing are normally used to produce components from plate or from relatively light sheet metal.

Flat metal sheets have little rigidity and buckle readily. By giving them single or double curvature, the designer shapes them for a particular application and improves their rigidity. The shapes that are produced in most of the cold-forming processes are basically simple, although the operations themselves may demand considerable skill.

We will look briefly at the forging process which is used to prepare materials for further machining. Some design aspects of the closed-die drop-forging process are illustrated in the sketch of a forged connecting rod. In the first place,

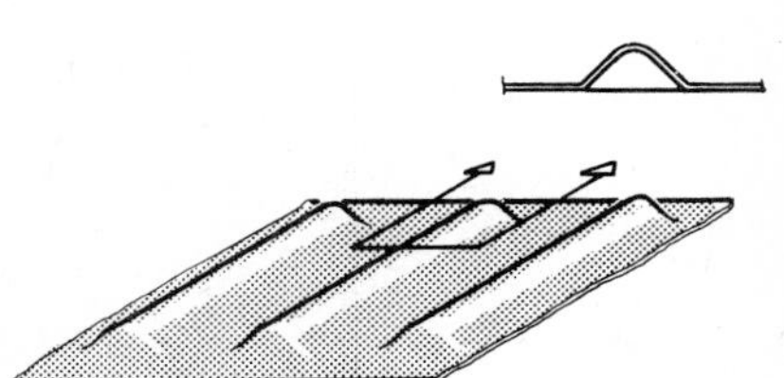

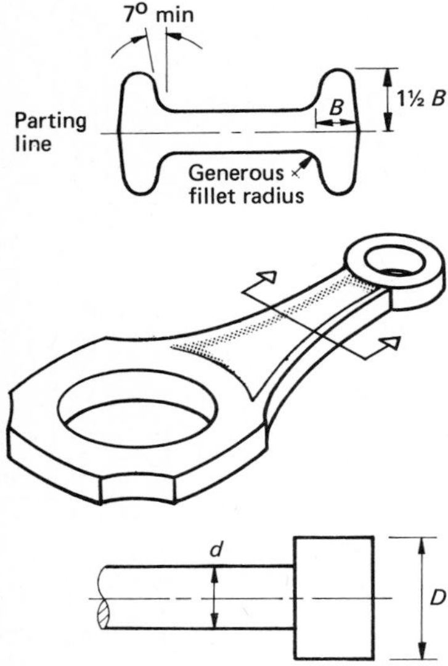

a parting line which lies in a single plane and goes through the centre of the part is preferred. The height of the rib that can be forged is limited and depends on the fillet radius and the draft; a value of about $1\frac{1}{2}$ times the base thickness of the rib would be satisfactory. The draft should be not less than 7° and the fillet radius as generous as possible – 5 mm would be a typical value on a small forging. Web thicknesses less than 4 or 5 mm should be avoided if possible. More detailed recommendations can be found in *Metals Handbook*.[9]

A simple shape frequently encountered in machine design is a shaft with a flange or end plate of various proportions. If the ratio D/d is small (say, less than 2) and the production number small, the part can be economically manufactured from a solid bar. For larger values of D/d, a good deal of machining can be saved by upset-forging the end of the bar. When D/d is large, this operation becomes impractical and the shaft and bar would then have to be manufactured separately and fastened together.

22.3.3. Machined shapes

We discuss first the turning and boring processes suitable for producing cylindrical surfaces, and then look at the processes used to produce flat surfaces. Since the designer distinguishes between cutting and abrasive methods mainly in terms of surface finish and accuracy, grinding operations will not be described specifically.

Turning on a lathe is a continuous cutting operation and is a highly developed and efficient process of its kind. Many different types of machines are available to give a wide range of production rates and accuracies. On some machines, several cutting tools may work simultaneously.

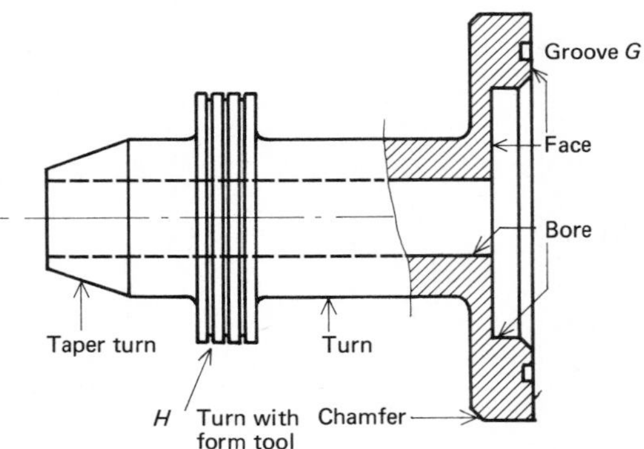

Fig. 22.7. *Typical turned part.*

The workpiece is gripped in such a way that very little setting-up is required. If a chuck is used, the three gripping jaws move together and automatically centre the cylindrical object being held. When bar stock is being machined, it can be gripped accurately and quickly in a *collet* – a split bush which is sprung inwards to grip the bar.[4] Once a part has been centred, several operations can be carried out with the assurance of accuracy. For example, in Fig. 22.7, which shows a typical turned part, groove G will be concentric with the bored hole, while the end face will be perpendicular to the axis, within the limits of accuracy of the machine. These advantages show up best in parts that are symmetrical about the centreline, but they can also be obtained, with more setting-up time, by fastening an unsymmetrical part to a faceplate used in place of a chuck. This guarantee of accurate location of planes and holes is very important to the designer, and provides a strong incentive for him to select cylindrical shapes which lend themselves to rapid production on a lathe.

In quantity production, it is possible to use a contoured tool to cut shapes, such as that at *H* in Fig. 22.7, in one operation. Quite complex shapes can be produced with such form tools, but they are expensive to make.

The production of holes by drilling does not give high accuracy of size or location, the drill usually finishing a hole oversize. An undersize drilled hole can be reamed to give better size accuracy. More precise location and size may be obtained on a jig-boring machine, while high quality surface finish can be achieved in holes by honing.

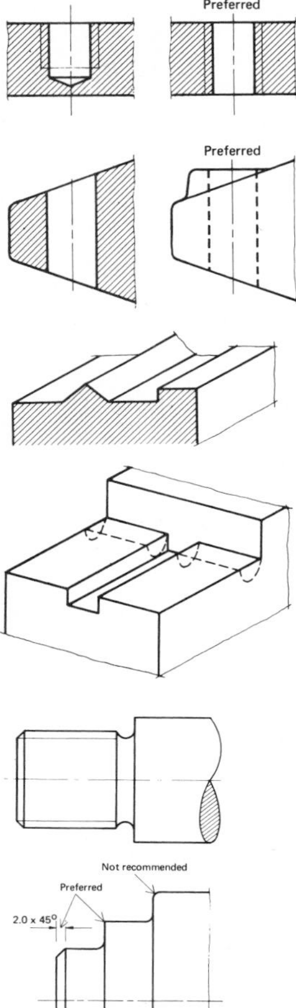

It is more economical to specify a through hole than a blind hole if it is to be tapped, since only one taper tap is needed and it can be driven right through the hole. At the location where a drilled hole enters a shape, the surface should be perpendicular to the axis for ease of starting the drill.

The flat surfaces shown in the sketch can be formed on shaping or planing machines (see Fig. 22.4). In the former machine, the tool moves across the workpiece, while in the latter, the workpiece is moved. Differentiation between the machines is based mainly on the fact that the planer can handle very large parts.

Flat surfaces can also be produced on a milling machine, on which a circular cutting head with a number of teeth produces a continuous cutting action that gives high rates of metal removal. These machines are very versatile and a great variety of shapes can be produced using special cutters. By combining a gang of cutters on one spindle, a stepped surface can be produced in a single pass.

It is usually difficult to carry a cutting tool right up to an adjoining face, and care should be taken when detailing the lines of intersection of two planes. For example, it would be difficult to cut the groove shown in the sketch right up to the adjacent face, and the machining operation is simplified if a gap (shown dotted) is left on the inside corner. For the same reason, screw threads on a turned part must terminate before a step in diameter is reached, and it is often convenient to allow the threads to run out in a groove, as illustrated.

External corners can usually be left sharp – in good shop practice any rough edges are removed – but if required, a 45° chamfer can be specified as shown. Rounded corners are not usually required, and detract from the clean look of a part.

22.4. PROCESS SPECIFICATION

The control exerted by the designer on the selected manufacturing process varies, and may at times seem tenuous. For example, he may hold a clear concept of a shaft to be shaped by rough machining on a lathe with a final grinding operation to provide an accurate finish at the location of bearings and seals. These operations may not be called for explicitly on the drawings, yet the instructions for tolerance and surface finish at the different locations will leave little room for doubt about the machines to be used.

Such understanding between designer and manufacturer is necessary and useful, but must not be presumed on. If the designer cannot accept any other than a ground surface he must specify this on the drawing with the tolerance and surface finish. Responsibility for making technical decisions must not be left to the man operating the machine. He may have the technical knowledge, but he will seldom have the background information needed for a sound decision.

The designer must provide adequate information to ensure his minimum requirements are met without going to extremes which may hamper production. He should not specify a particular machine or process if there are suitable alternatives; production planning is simplified if he states his requirements as simply and as precisely as possible – this will often be in terms of tolerance and surface finish – and leaves it to the shop supervisor to take the responsibility for meeting them.

If the designer is uncertain as to the processes that are available, or is unable to choose between the merits of several, he will do well to consult the shop supervisor. He must not get round his difficulty by resorting to vague instructions which throw what should be the design decisions onto shop personnel. The results may be satisfactory but he learns nothing and will have to take responsibility for the mistakes that do occur. An example can be taken from the specification for heat treatment of a shaft. If he calls for "quench at 820°C and temper" the designer can only expect the supervisor to select the quenching fluid and tempering temperature in accordance with accepted practice. However, these decisions affect the strength of the part and must be taken by the designer; it may be that sufficient strength can be obtained by quenching in oil; in addition, it may be essential to lower the shaft into the quenching bath axially to avoid distortion. It is also important to specify the tempering temperature. The selection of a heat-treatable steel must involve consideration of all these factors and the information must be given to shop personnel.

REFERENCES

[1] BS 1916:Part 2:1953, *Guide to the Selection of Fits in BS 1916:Part 1,* British Standards Institution, London, 1953.

[2] J. A. Broadston, "Surface Finish Requirements in Design", in O. J. Horger (ed.), *Metals Engineering—Design*, ASME Handbook, 2nd ed., McGraw-Hill, New York,

[3] R. Matousek, *Engineering Design*, Blackie, London, 1966.

[4] E. P. de Garmo, *Materials and Processes in Manufacturing,* 3rd ed., Macmillan, New York, 1969.

[5] H. J. Sharp (ed.), *Engineering Materials,* Heywood Books, London, 1966.

[6] L. E. Doyle, *Manufacturing Processes and Materials for Engineers,* 2nd ed., Prentice-Hall, Englewood Cliffs, 1969.

[7] The Society of The Plastics Industry, *Plastics Engineering Handbook*, 3rd ed., Van Nostrand Reinhold, New York, 1967.

[8] R. D. Beck, *Plastic Product Design*, Van Nostrand Reinhold, New York, 1970.

[9] ASM, "Properties and Selection of Metals", vol. 1 in *Metals Handbook*, 8th ed., American Society for Metals, Ohio, 1961.

BIBLIOGRAPHY

ASM, "Machining", vol. 3 in *Metals Handbook*, 8th ed., American Society for Metals, Ohio, 1967.

A comprehensive description of machining related to processes and materials.

WRIGHT BAKER, H. (ed.), *Modern Workshop Technology*, Part 1, "Materials and Processes", 2nd ed., 1956; Part 2, "Machine Tools and Manufacturing Processes", 2nd ed., 1960; Cleaver-Hulme, London.

23 Construction Operations

23.1. INTRODUCTION

The objective of this chapter is to describe briefly some of the work required for construction of a building, a bridge or some other structure. It is not intended to be a guide for the engineer about to undertake such construction. Successful practice can be learned only in the field. However, the student of engineering design has not, in general, had time to acquire such experience and, to the extent that good design takes into account construction practice, he is at a disadvantage. The descriptions that follow contain enough information for a student who is careful, though inexperienced, to ensure that the structures he designs could be built.

23.2. EXCAVATION

Excavation for foundations and basements is normally the first operation undertaken. Even though no useful space may be provided, excavation is necessary because the uppermost layers of soil are rarely suitable for the support of loads. They are deficient in strength and stability. The kind of excavation required varies widely, depending on the depth of a suitable bearing stratum, its strength and the total load to be applied. There is a corresponding range in excavation techniques. At one extreme, lightly loaded columns can be supported on individual footings on clay at, say, 0.5 m to 1.0 m below ground level. Then, excavation by hand is appropriate. In another case, a multistorey building may be supported on a *raft* (essentially one large footing under the whole building) at a depth of 5 m. In that case, bulk excavation using power shovels or front-end loaders would be required for economy. Plate 23.1 shows bulk excavation for an underground parking building with additional deeper excavations required for column foundations.

Matters of concern to the designer are

(a) access to the excavation for plant and retrieval of the plant on completion;
(b) the need for temporary support of the sides of the excavation; and
(c) drainage.

In locating the works on the site, the designer may be able to leave enough space between them and the boundaries for whatever is needed for access, temporary support and drainage sumps or pumping equipment or both. However, in urban areas land values are so high that buildings often extend to the boundaries of the site. Special procedures will then be required. These include *underpinning* of neighbouring buildings whose foundations are at a higher level and specifying a sequence of excavation and construction in stages so that only parts of the whole site need be opened up at any one time.

Plate 23.1. *Bulk excavation. Note temporary support of the face in the background and the use of heavy plant for excavation and haulage of spoil. (Photo: C. Collins. Reproduced by permission of the University of Auckland.)*

It is worth taking care to keep the bottom of an excavation clean. Succeeding operations require men to work in the excavation and, if this is clay and rain falls (it usually does), the foundation area becomes a sea of mud. Good protection can be provided by placing a thin layer of concrete, 30 mm to 50 mm thick over the bottom of the excavation. This concrete provides an accurate surface for fixing reinforcing steel and formwork for footings. The layer of concrete should be shown on the drawings, described in the specification and allowed for in the bill of quantities.

23.3. ERECTION OF STRUCTURAL STEELWORK

Structural steelwork is fabricated in a workshop, the operations required there being cutting, drilling, riveting, bolting and welding. Machining is rarely required. All of these shop operations are described elsewhere in this book. The extent to which steelwork is made into sub-assemblies varies, depending largely on facilities for transport. Legal restrictions, obstructions (such as tunnels and underpasses) and the size of vehicles available for transport are matters that the designer needs to bear in mind. In general, columns for full-storey height, complete beams and complete roof trusses can be transported. Delivered to the site, they are hoisted into place by cranes. In modern construction and under normal circumstances, crane capacity is unlikely to be a limitation. An exception to this generalisation is construction on a remote site. Then limited capacity for transport and hoisting may make it necessary to deliver individual members of trussed frames – for example, the towers of a transmission line in mountainous country.

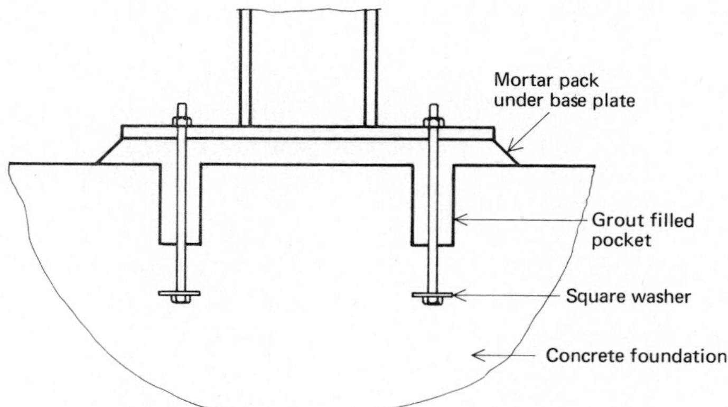

Fig. 23.1. *Detail for bolting a steel baseplate to a concrete foundation.*

The size of units to be fabricated is not the only aspect of design controlled by construction operations. There are important matters of detail in matching the steelwork and its concrete foundation, and in supporting beams temporarily. With regard to the first of these, holes in baseplates of columns can be located much more accurately than the bolts in the concrete can be. Also, the top surface of the concrete foundation cannot be made smooth enough to match the steel baseplate or with the required accuracy of elevation. A good detail to allow for these is shown in Fig. 23.1. When the foundation is concreted, the holding-down bolts are located as accurately as possible and formwork for pockets is provided to surround them. This formwork may well be in the form of industrial cardboard tubes and is hung, along with the bolts, from a temporary beam spanning across the formwork of the foundation. The completed foundation has the heads of the bolts and part of their shanks anchored in the concrete, while the rest of the shank is free. The advantage is that adjustments can be made by bending the free part of the bolts, so that they can be positioned accurately before arrival of the steelwork. Adjustment for level is made by setting the column baseplate on temporary steel packing. To complete erection of a column, the nuts are tightened on the holding-down bolts, *grout* is poured into the pockets and *mortar* is packed under the baseplate. It is necessary to use grout (a fluid mixture of cement and water) to fill the pockets because the filling must be poured into place. Since it has the undesirable characteristic of relatively high shrinkage it should not be used under the baseplate. Shrinkage can be reduced by using less water and adding sand to the mix – this is mortar, and a fairly stiff mortar can be pushed under a baseplate. Other details can, of course, be devised, the essential characteristic being to permit relatively generous tolerances in the concreting operation.

Temporary support of beams is necessary when a lengthy period is required to make the permanent connections. It is clearly undesirable to have a crane immobilised while providing temporary support. *Erection cleats* are used to support the ends of beams temporarily. Such connections need to be made quickly and do not have to carry large loads. With some ingenuity it may be possible to incorporate them in the permanent joint, but there is no compelling reason to strive for this.

Detailing of permanent connections should take into account the practical difficulties of working under arduous conditions of height and exposure to the weather.

23.4. CAST *IN SITU* REINFORCED CONCRETE

Freshly mixed concrete is a plastic mixture of Portland cement, mineral aggregates and water. It flows more or less readily and can thus be moulded into a variety of shapes. After some hours, it sets and its strength as a solid material continues to increase over a period of months. These are characteristics which make concrete a very versatile material for construction.

The ingredients are mixed in batches in a concrete mixer. This may be done on the construction site, in which case storage for cement and aggregate is required. The mixed concrete is transported directly from the mixer to the job. Alternatively, the concrete may be mixed in a central mixing plant and delivered to the construction site in *transit mixers*. These are truck-mounted mixers in which the fresh concrete is agitated slowly while *en route* to the site.

In concrete construction, the first stage is to build the moulds, known as the *boxing* or *formwork*. Formwork has to be supported on a temporary structure called *falsework* (Plate 23.2). Timber is widely used for formwork and falsework, and sheets of plywood are very useful for forming large areas. Unless great care is taken in handling formwork and falsework, re-use of timber is limited. Thus, although the initial cost is low, the cost per square metre of completed concrete surface may be relatively high. The greater durability of steel makes it competitive economically, especially when standardised shapes lead to repeated use. Another material sometimes used for circular columns is cardboard. Cardboard tubes of considerable strength are available in some places and they can be used as moulds with minimal site work.

The completed formwork is a mould in the shape of the concrete outlines. The designer of the structure is not concerned directly with design of the falsework and formwork but he needs to know what is done and how much it costs. For example, there is scope for varying the cross-section of a beam along the length as the bending moment varies, with a clear economy in materials. However, there is a compensating increase in cost because of the higher labour content in the more elaborate formwork and falsework. Unless wages are low, a prismatic beam is likely to be cheaper.

It is impossible to complete most structures by placing concrete continuously. Furthermore, certain things make it undesirable to place large quantities without pausing. For example, concrete shrinks on setting. If a long length is placed, and is subject to restraint, shrinkage causes tensile stresses which may exceed the tensile strength of the concrete. Many unsightly cracks are caused this way. Another aspect concerns escape of the heat released as hydration of the cement proceeds. Substantial temperature rises occur in large masses of concrete and significant thermal strains may be caused by expansion and contraction. For such reasons, *construction joints* are required to divide the whole structure into units suitable for a day's concreting. These joints are located in relation to the internal stresses in the loaded structure so as to avoid highly stressed places, and they are made as strong as possible. Reinforcing steel is continuous across such joints and before concrete is placed in adjacent sections, the concrete surface at the joint is roughened to improve the bond between the two pours. In beams and slabs, joints are usually placed at about quarter span. Formed joints with vertical faces are required. Construction joints in columns are located just below and just above floors to suit construction operations. The designer is responsible for deciding where construction

Plate 23.2. *Falsework and formwork for a beam-and-slab floor. Left foreground, steel props for support of formwork. Background, timber falsework. Right foreground, completed formwork with steel fixers placing reinforcing steel. (Photo: A. Estie. Reproduced by permission of the University of Auckland.)*

joints will be permitted and where they will not. He may do this by describing these places in general terms in the specification or he may require the joints to be in specific locations. The first is preferable, since it permits more flexibility in arranging the concreting.

When the formwork for a pour has been completed the *steel fixers* move in. The reinforcing steel is supplied by *steel benders* who may work in a shop remote from the site. They use bar bending machines and cutting equipment to cut and bend the steel to the shapes and dimensions required. The steel is then sent to the site in labelled bundles of bars. The steel fixers erect the bars inside the formwork (Plate 23.3). To maintain the required thickness of concrete between a bar and an adjacent surface (called the cover), they use *biscuits*. These are blocks of mortar about 50 mm × 50 mm and of thickness equal to the cover. A piece of soft iron wire is embedded in the mortar, with about 100 mm of each end clear of the biscuit. The wire is used to fasten a biscuit to the steel in such a fashion as to pack the steel off the formwork. The biscuits are embedded in the permanent concrete and become part of the permanent structure.

Where bars intersect, as, for example, main bars and stirrups do in a beam, they are tied together with soft iron wire. The completed reinforcement fixed in place thus forms a rigid cage. It should be strong enough to support its own weight and not be distorted by men walking on it during construction. This strength and rigidity can be exploited by assembling beam or column

Plate 23.3. *Reinforcing steel in place, ready for placing concrete in a beam-and-slab floor. (Photo: A. Estie. Reproduced by permission of the University of Auckland.)*

reinforcement into cages outside the formwork—even away from the site—and lifting complete cages into the formwork.

Long runs of reinforcing steel require *splices* to be made, depending on the length of bar available. It is necessary for the designer to know what lengths can be supplied locally. Splices may be made by welding, but this is expensive under site conditions and is reserved for very large bars. In conventional construction, splices are made by overlapping. The end of one bar in the splice is offset and the offset end overlaps the end of the other bar. Bar forces are transferred by bond between the steel and the concrete and shear in the concrete. Ideally, the offset should be sufficient to allow both bars in the splice to be completely surrounded by concrete. This is rarely done, and the offset is usually one bar diameter, so that the overlapping bars are in contact. Construction is much simplified and the loss of strength is unimportant if the splice is long enough.

The next operation is placing the concrete in the moulds (Plate 23.4). It is of first importance that the formwork should be completely filled and the steel completely surrounded by dense concrete of uniform quality. While good workmanship on site is a major factor in achieving this, there is no doubt that much defective construction has its origin on the drawing board. The design must be such as to facilitate placing and compaction of the concrete. Too often designers, particularly inexperienced designers, try to keep concrete dimensions down.

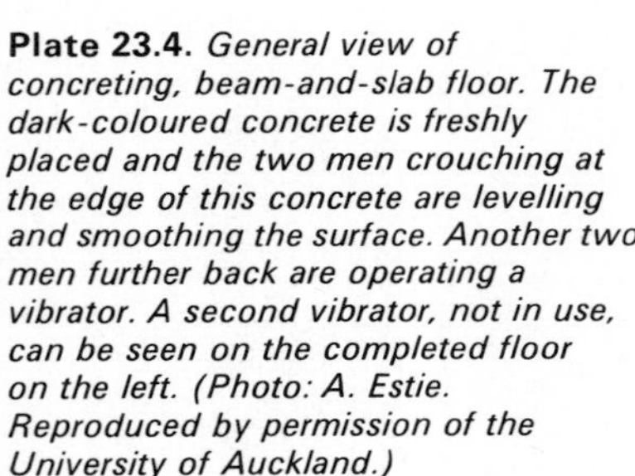

Plate 23.4. *General view of concreting, beam-and-slab floor. The dark-coloured concrete is freshly placed and the two men crouching at the edge of this concrete are levelling and smoothing the surface. Another two men further back are operating a vibrator. A second vibrator, not in use, can be seen on the completed floor on the left. (Photo: A. Estie. Reproduced by permission of the University of Auckland.)*

This bad practice leads to congestion of steel in the formwork. Codes of practice contain rules about limits on the percentage of steel area in the cross-section and spacing between bars. It must be remembered that these are limits and, wherever possible, the designer should be more generous in the interests of good concrete in place.

In modern construction, *vibrators* are used to aid concrete placing (Plate 23.5). A vibrator is a hollow steel cylinder with a shaft inside carrying an eccentric mass. The shaft is driven by an air motor in the vibrator or by means of a flexible drive from a remote motor, and causes a high-frequency vibration. When the vibrator is immersed in the concrete the agitation causes compaction of the concrete in the mould and around the steel.

The concrete must be cured while setting. The hardening and growth of strength result from a chemical process, hydration of the cement. If there is

Plate 23.5. *Close-up view of concreting, bridge deck. A skip of concrete hangs from the crane hook in the background. There is a vibrator in the foreground. Note the workmen walking on the reinforcement. Although this is usually forbidden, it always happens and must be borne in mind in design. (Photo: G. J. N. Blake. Reproduced by permission of Mr Blake.)*

inadequate water present, the process ceases early and the gain in strength is curtailed. *Curing* prevents loss of water from the setting concrete. The mass of concrete may be kept wet by sprinkling or by covering it with fabric kept moist. Curing compounds are available which, when sprayed onto the surfaces of the concrete, seal them and prevent loss of water by evaporation. Such compounds are advantageous when it is undesirable to use water.

The final operation is removal of formwork and falsework. We have noted that economy of construction is improved if they can be used again. Care is taken to control removal or *striking* of falsework, so that the green concrete is not left without support before it is strong enough to carry its own weight and any construction loads. Clauses in the specification usually require concrete to be supported for from 14 to 28 days after placing.

23.5. PRECAST CONCRETE

Precast concrete enables the designer to exploit the advantages of factory production, such as better working conditions, better control of quality and, for standardised items, economy in the use of moulds. On site, cost is reduced by eliminating falsework or, at least, minimising it (Plate 23.6). For reasons like this, standard bridge beams have been designed. These are for a range of spans and for suitable standard highway loading. Under these circumstances factories can be built for production of such beams at maximum efficiency.

Beams like this are made in reinforced concrete and prestressed concrete. The completed beams are transported to the site, usually by road, and hoisted into place. They are widely used in construction of bridges. Most such beams are simply supported because of the difficulty of making moment-resisting field connections at their ends.

To take full advantage of precasting, the designer has to ensure long runs for the manufacturer. This means reducing the number of different kinds of beams on the job, even if some beams are thereby bigger than is strictly necessary.

Plate 23.6. *Column for precast concrete building frame being hoisted into place. Note reinforcing bars for joint to beam. (Photo: A. Estie. Reproduced by permission of the University of Auckland.)*

At the same time, it should be noted that long runs justify more elaborate shapes to save material. In a large number of beams, only a little need be saved on each one to amount to a substantial quantity in total. Good judgement and knowledge of costs is required to balance these considerations. Precasting also offers the advantage of close control of manufacture. Higher working stresses are justified by the better quality of the completed product.

23.6. PRESTRESSED CONCRETE

There are two broad classes of prestressed concrete construction – *pre-tensioned* and *post-tensioned*. In both, the prestressing force is applied to the concrete after it has set, but before it is loaded, except for its self-weight. In pre-tensioning, the prestressing wires are tensioned before the concrete is cast and permanent anchorage of the prestressing force is mainly by bond near the ends of the wires. In post-tensioned prestressed concrete, the wires are tensioned after the concrete has set and mechanical anchorages are used.

Plate 23.7. *Pre-tensioning beds in a factory. Far left, two poles. Centre, mould for T-beam being erected. Right, completed beams. (Photo: A.C.L. Photography. Reproduced by permission of Stresscrete Limited, Auckland.)*

Pre-tensioning is done on a *stressing bed*, usually in a factory (Plate 23.7). This comprises a long bed on which the moulds for several items – for example, beams – are assembled end to end. There is a strong anchor block at each end of the bed and one end has facilities for jacking. High tensile steel wires are strung between the anchor blocks and placed under the specified tension by jacking. Any mild-steel reinforcement is also fixed at this stage. Concrete is then placed in the moulds so as to fill them completely and to surround the bars and the tensioned wires. When the concrete is strong enough to sustain the prestress, the wires are released from the anchors. A balance is achieved between compression strain in the concrete surrounding the wires and a reduction in tensile strain in the wires, the nett result being the desired distribution of compressive stress. The wires are cut between adjacent units, leaving short lengths protruding at the ends of the units. The consequent increase of tension in the wires from zero at their ends to some high value a short distance from the end is possible because of two things. One is the bond between the wire and the concrete surrounding it. The other is the small increase in lateral dimension corresponding to the reduction of axial tension when the prestressing force is transferred. This Poisson effect enables the end of each wire to act like a wedge in preventing slip.

There is an obvious need for accelerated strength gain to keep the rate of production of a stressing bed high. Steam curing is used extensively. By covering the bed and supplying steam under the covers, the atmosphere is kept warm and moist and under these circumstances concrete gains strength rapidly. The time at which the prestressing force is transferred is controlled by crushing test cylinders cured under similar conditions and is often only 24 hours after concreting.

Typical products manufactured on pre-tensioning beds are beams for bridges and buildings (including T-beams and inverted U-beams for floor construction), poles, planks, railway sleepers.

Plate 23.8. *Cable ducts, anchorages and reinforcement for post-tensioned beam. Note the heavy reinforcement required in the end block. Other features of interest are the ribs on the deformed bars and the mortar biscuits to ensure a cover of concrete over the reinforcing in the finished beam. (Photo: C. Collins. Reproduced by permission of the University of Auckland.)*

In post-tensioning, groups of wires in the form of *cables*, or bars are used instead of individual wires (Plate 23.8). The mould is built as for reinforced concrete. Cable *ducts* and any mild steel reinforcement are fixed in place, and this includes anchor cones or anchor plates at the ends of the ducts. These last are special devices required to resist and spread the large force to be anchored at each end of a cable or bar. After concreting, the cables or bars are threaded through the ducts and when the concrete is strong enough the beam is stressed.

When bars are used, they are usually anchored by means of nuts running on threads cut in the ends of the bar. One nut is run on and the bar is stretched by means of a special jack pulling the other end of the rod and pushing on the beam. When the prescribed strain has been induced in the bar, the nut on this end is tightened and the jack is removed.

The procedure is similar when cables are used, except that the wires of the cable are secured by means of wedges. Cables are used much more extensively than bars are. In either case, specially designed jacks are required to ensure that the required load can be jacked into the cable or bar and later transferred to the concrete without slip causing unacceptable loss of tension (Plate 23.9).

Plate 23.9. *Freyssinet jack stretching a tendon. Individual wires are secured to the cylinder of the jack by wedges. A workman kneeling on the scaffold planks overhead measures the extension of the jack with a rule graduated in inches. (Photo: A.C.L. Photography. Reproduced by permission of Stresscrete Ltd., Auckland.)*

Finally, grout is pumped into the duct for protection against corrosion and to bond the cable or bar to the beam.

Post-tensioning is used on large beams when precasting is impossible because of limited facilities for transport or hoisting or both. Thus, its field is, primarily, the construction of large bridges.

Advantages of post-tensioning in comparison with pre-tensioning are as follows:

(a) as noted above, there is virtually no limit to the size of beam that can be prestressed;
(b) draped cables can be used, so that the eccentricity and distribution of the prestress can be varied along the length of the beam;
(c) beams continuous over several supports can be prestressed;
(d) portable equipment is used, so that prestressing on the construction site is practicable.

There are the following disadvantages:

(a) ducts may be displaced or damaged during construction. Quite apart from a consequent error in locating the prestressing force, threading of the bars or cables may be impossible if a duct is crushed or contains grout which has leaked in from the concrete;

(b) some prestress is lost to friction against the duct and in slip at the anchorages;
(c) there is a considerable content of skilled labour in post-tensioning and this makes it relatively expensive.

When prestressed beams are also precast, our comments at the end of Section 23.5 are relevant. For all prestressed construction the designer must not forget the fundamental need to avoid congestion in the mould and, in post-tensioning, he must ensure that there is room at the ends of a beam for jacking. Special attention may be required to avoid congestion, since some dimensions of a beam can be quite small – e.g. the thickness of the web of an I-beam.

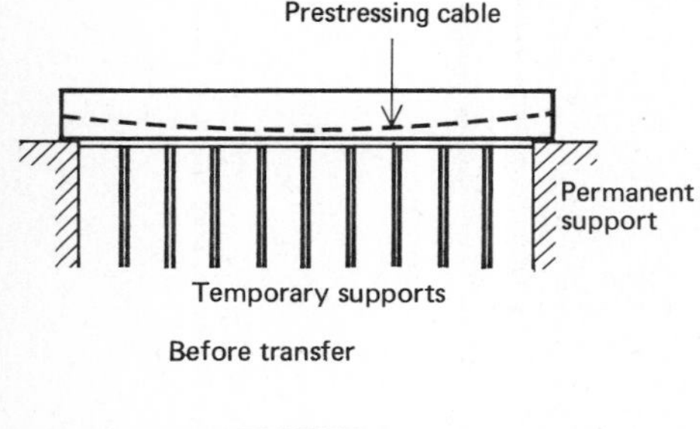

Before transfer

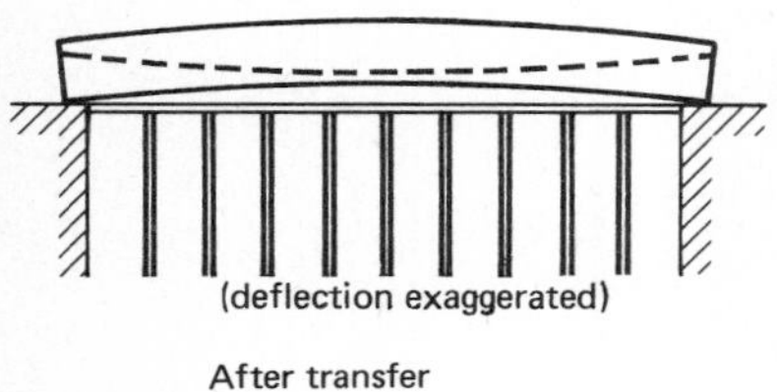

After transfer

It is of some importance that the weight of post-tensioned construction can be used to advantage. In Fig. 17.5, the stress distributions due to dead load, the eccentric prestressing force and the two combined are shown. The point is that the beam is never loaded by the prestress by itself. As the tension in the cables increases, the beam deflects upwards and is lifted off the temporary supports causing the stresses due to dead load to be imposed simultaneously with the prestress. Consequently, the compressive stress at the top of the beam due to dead load cancels part of the tensile stress due to prestressing as the latter is applied. The same thing happens at the bottom of the beam with the result that a larger residual of prestressing compression is left in the lower fibres to counter the tensions caused by loads applied later. To some extent at least, the greater the load at the time of prestressing the greater the advantage. However, the gain is offset by the increased cost of temporary support and here, too, judgement and familiarity with costs are required.

This use of a structure's own weight in design is reserved for post-tensioned construction because there is little to be gained from the self-weight of a beam by itself and, more important, the dangerous consequences of a precast beam being accidently overturned cannot be ignored.

23.7. TIMBER

In timber construction, the operations affecting the designer are primarily those of manufacturing lumber from trees. In large mills, trees are cut into slices by passing them longitudinally through saws (Plate 23.10). Additional longitudinal cuts transform these slices into planks with standard nominal cross-sectional dimensions and the planks are then cut to length. The sawn timber is classified into grades depending on the presence of visible defects.

In further processing, the sawn surfaces are made smooth, or dressed, by passing the planks through a machine which planes the surface with knives mounted on rapidly rotating spindles. The edges only, or all four sides may be dressed. In the former case, planks of uniform width are produced; in the latter thickness also is consistent.

As cut, timber has a relatively high moisture content – about 30% – and since most of its structural properties improve with reduced moisture, manufacture of lumber includes drying or seasoning. In a suitable climate, the planks may be stacked in a yard in such a way as to permit ventilation of the stack. Otherwise, batches of lumber may be dried at elevated temperatures with forced ventilation in kilns.

Finally, treatment with preservative chemicals may be undertaken. Many species of timber are subject to fungus attack if exposed to moisture or a highly

Plate 23.10. *Manufacture of lumber. A balk of timber is driven by rollers through a gang saw. (Photograph by courtesy of New Zealand Forest Products Ltd.)*

humid atmosphere. Also, certain insects consume some species. These attacks can be countered most effectively by impregnation under pressure with solutions of toxic salts. Sufficient of the chemicals is retained in the cells of the timber to increase remarkably the durability of some species – for example, New Zealand-grown *radiata* pine.

As far as the designer is concerned, the following points are relevant to his calculations and details. First, grading rules describe the defects which are permitted in various grades. Working stresses corresponding to these grades are found in the design specification and the two together enable the designer to ensure that he gets what he wants. Sometimes the design specification includes grading rules. Second, it should be noted that lumber is purchased on the basis of nominal size. The actual size of plank supplied is smaller, the difference being the timber lost in sawing and dressing. (Refer to Chapter 12, "Ties and Struts" and Chapter 14, "Timber Beams".) Table A8.1 in Appendix A8 shows typical dimensions and geometrical properties for a range of nominal sizes. This aspect affects not only the strength of a member but also the matching of its dimensions with other components, especially fittings or parts made of steel (e.g., steel gusset plates in a timber truss).

Other points concern preservative treatment. This is best done after dressing because the highest retention of salts is in the outermost cells. Consequently, it is technically and economically unsound to remove this timber in dressing. As noted below, compromise may be necessary when glued joints are to be made. In that event, compatibility of preservative and adhesive must be checked. Some preservatives inhibit setting of certain adhesives, either absolutely or if present in sufficient concentration. In the latter case, a compromise can be reached by dressing the timber after preservative treatment. Otherwise, it may be possible to assemble a glulam structure (for example, a beam) before treatment. However, there are disadvantages, such as the size of the units to be handled and placed in pressure treatment vessels and less effective penetration into the big beam than into the planks from which it is built because of the larger surface area exposed by the latter.

As far as construction of timber structures is concerned, we have described the use of connections of various kinds in Chapter 13, "Connections". When adhesives are used, construction requirements are for jigs to hold components in place and clamping facilities to force them into intimate contact while the adhesive is setting. Details of practice vary depending on the adhesive and the best source of them is technical literature supplied by manufacturers.

23.8. CONCLUSION

We conclude this chapter by repeating that our objective is not to write a construction manual. These brief descriptions serve to introduce the student of design to some fundamental aspects of construction which affect the work of engineering designers. Some of the descriptions are necessarily vague, because details and dimensions vary from place to place. It is imperative for the student to acquire this local detail as quickly as he can. Visits to works during his courses, practical experience during vacations and discriminating collection of trade literature are the essential features of this process.

BIBLIOGRAPHY

ABELES, P. W. and TURNER, F. H., *Prestressed Concrete Designers' Handbook*, Concrete Publications, London, 1964.

Basic principles, design methods and data.

ANTILL, J. M. and RYAN, P. W. S., *Civil Engineering Construction,* Angus & Robertson, Sydney, 1957.

A comprehensive discussion of construction.

BARRON, T. B., *Erection of Constructional Steelwork,* Iliffe & Sons, London, 1956.

A textbook for junior site engineers and students.

CHUGG, W. A., *Glulam*, Ernest Benn, London, 1964.

Theory and practice of glued laminated timber structures.

COWAN, H. J. and SMITH, P. R., *The Design of Prestressed Concrete,* Angus & Robertson, Sydney, 1966.

Basic principles and design to satisfy the Australian code.

FREAS, A. D. and SELBO, M. L., *Fabrication and Design of Glued Laminated Wood Structural Members*, U.S. Department of Agriculture, Washington, 1954.

GEDDES, S., *Building and Civil Engineering Plant*, Crosby Lockwood, London, 1951.

Data for purchase, application and operation of construction plant.

GLOVER, C. W., *Structural Precast Concrete*, CR Books, London, 1964.

Design and construction in precast concrete.

HALPERIN, D. A., *Building with Steel*, Technical Press, London, 1960.

Covers design, detailing, fabrication and erection of steelwork.

HARRIS, J. D., and SMITH, I. C., *Basic Design and Construction in Prestressed Concrete*, Chatto & Windus, London, 1963.

HINDS, H. V. and REID, J. S., *Forest Trees and Timbers of New Zealand*, Government Printer, Wellington, 1957.

ISE, *Industrialized Building and the Structural Engineer*, Institution of Structural Engineers, London, 1966.

Papers presented at a symposium in May 1966. Application of industrialised methods to construction in steel, concrete and other materials.

REYNOLDS, C. E., *Concrete Construction*, 2nd ed., Concrete Publications, London, 1945.

SMITH, R. C., *Principles and Practices of Heavy Construction*, Prentice-Hall, Englewood Cliffs, 1967.

Covers the whole range of construction operations. Contains data for design of temporary works and for other operations.

WALLIS, N. K., *Australian Timber Handbook*, Angus & Robertson, Sydney, 1970.

Properties of timber at various stages of manufacture.

WYNN, A. E., *Design and Construction of Formwork for Concrete Structures*, Concrete Publications, London, 1947.

24 Communications and Organisation

24.1. INTRODUCTION

This chapter is largely concerned with the way in which designers and engineers communicate among themselves and with workmen, contractors and management. While we are primarily interested in communications in their various forms, a clearer understanding will come if we consider also the organisation of the people who generate them and are linked by them. The variety of organisations encountered in engineering is very great and the modes of communication are correspondingly varied. We will simplify our discussion by considering communications in only one kind of organisation for getting work done, this being the formal cooperation of client, consultant and contractor. The basic principles are quite relevant in other situations and what we describe can be readily adapted when less formality is required.

In its simplest form, the problem is to convey information from engineer to draughtsman and to the workshop or construction site; there is a corresponding progression from rough sketch to finished drawing which requires little detailed description. Nevertheless, there is scope for misunderstanding if the organisation is not flexible enough to cope with priorities, progress checks and changes in design, or if the designer is not familiar with basic shop and site methods.

At the other end of the scale lie the projects which may run through the stages of design brief, consulting engineer's studies and recommendations, detailed drawings and specifications, tendering and finally construction. We will study these stages from the viewpoint, at all times, of the communications required. The communications that we will discuss can be represented as in Fig. 24.1 where the more important relationships and stages and forms of communication are set out.

24.2. THE ENGINEERING AND CONSTRUCTION INDUSTRIES

Before studying the details of engineering communication, it is necessary to understand the structure of these industries. This is best done by considering the parts played by the three parties principally involved in the design and construction of a structure, machine or plant – the client, the designer and the constructor. We will consider them in terms of civil and structural engineering, when the designer is a consulting engineer or architect and an independent contractor undertakes the construction. Much design and construction also takes place wholly within a single organisation, especially in mechanical and

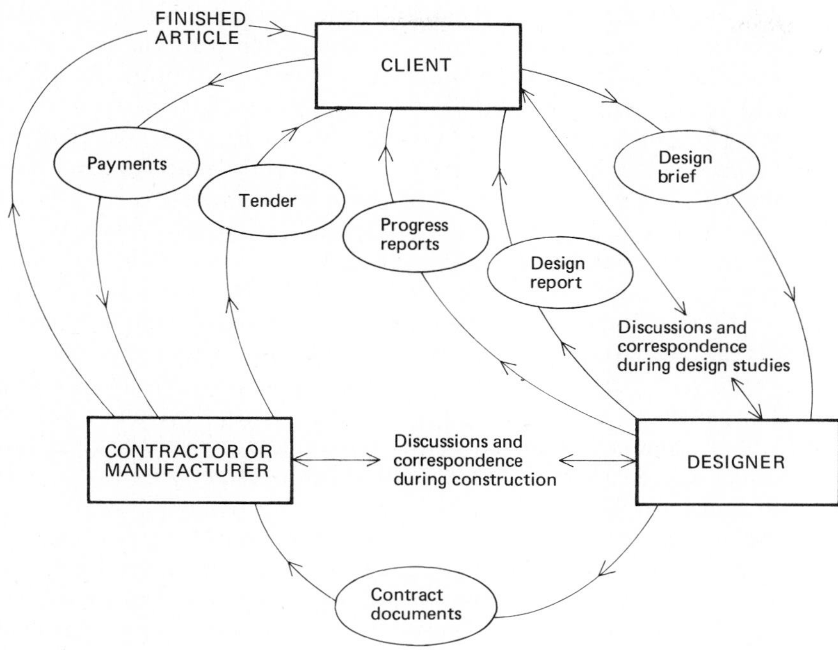

Fig. 24.1. *Communications among client, contractor and designer.*

production engineering; in such cases, three parties can still be identified – in the form of management, design office and production shop – and the communications used are similar to those described below. However, since formal legal and contractual liabilities are less stringent, there tends to be more scope for variation in the disposition of work and responsibility. Also, the accounting procedures are different.

From our point of view, the client is the promoter of a scheme requiring buildings and machines to be built and purchased. He has decided that these are necessary for his operations to make a profit or provide a service. He knows what functions the buildings and machines have to perform. He has the responsibility of conveying these facts to his advisers – in the form of a design brief, to be described in Section 24.3 – and of ensuring that money is available to pay for them. Although we have referred to the client as an individual, this party may be a company or an element of local or central government. Thus, the client may be an individual who wants a small factory for the manufacture of knitted garments or a company expanding its production of newsprint from 120 000 tonnes per year to 180 000 tonnes per year. In each of these examples, the client intends to make a profit from the new operations. Another example would be a city council which requires a garage for its street cleaning machines. In this case, the client requires a building and vehicle maintenance plant so that it can perform a service.

The *design brief* is a communication from the client to the second party referred to above, the consulting engineer or architect. As in the case of the client, the role of consulting engineer or architect may be filled by an individual or a group of people, usually a partnership. The individual or partnership will employ assistants if the size of the practice warrants it. The consulting engineer or architect is retained by the client as his technical adviser and the engagement is usually in two phases. In the first, before any construction starts, the consultant is wholly engaged in serving his client's interests. He examines all aspects of the proposal, in collaboration with the client, and prepares alternative schemes for achieving the client's objectives. In the early stages, only outline designs are prepared for comparison and the best of these are studied in more detail. Economic aspects of this process are discussed in Chapter 25, "Economics in Design". When the consultant has completed this work he sends his conclusions and recommendations to the client in a *design report*, which is described in more detail in Section 24.4. Ultimately, a scheme is accepted by client and consultant as suitable in that it enables the client to carry out the operations he has specified and to make the profit or provide the service he considers necessary. The consultant then prepares a complete design for this scheme and contractors are invited to bid for its construction. Information is supplied to them in the form of *contract documents* which comprise *drawings* and *specifications, bill of quantities* and *conditions of contract*. These documents describe the work to be done, materials and parts to be supplied and the rights and obligations of all the parties. The conditions of contract comprise a legal document, which is usually standardised within a given country; the others are engineering or architectural documents generally prepared for particular jobs.

When one of the bids, or *tenders*, is accepted by the client, a contract comes into existence. The contractor agrees to provide all the materials, plant and labour required to complete the construction described in the documents prepared by the consultant; the client agrees to pay the contractor the amount of his tender or some other sum determined in accordance with the contract documents and the law of contract.

At this stage, the role of the consultant changes to some extent. His engagement by the client normally continues and he supervises the work of the contractor to ensure that all construction and all manufactured items supplied comply with the contract documents. He also certifies to the client the value of work completed from time to time, usually monthly, and the client makes progress payments accordingly. In addition to these responsibilities in the interests of the client, the consultant also has to interpret the contract documents and he must do this without bias. In this respect he is legally required to act independently and must be quite fearless in doing so, even if his decision is adverse to his client. Such occasions should not arise and will do so only when there had been an earlier misunderstanding between client and consultant – for example, if the client thinks some item is included in the contract and the consultant is unaware of the need for it, it may in fact be omitted from the documents which are the basis of the contract.

In the descriptions above, the term *consultant* has been used deliberately because of the need for a name which describes several kinds of professional advisers – architect, mechanical engineer, building services engineer, structural engineer, civil engineer. It is necessary to do this because a large scheme requires the services of more than one of these and a small one may require the services of any of them. However, only one will, as a rule, have the direct responsibility to the client for ensuring that a contract is let and completed correctly. Who this is will depend on the kind of work to be done. In a building contract, the principal adviser is usually the architect and the structural engineer and building services engineer are subordinate to him. On the other hand, a consulting mechanical engineer would be the leader in arranging construction of an industrial plant, with the architect and structural engineer assisting him. Although there are no set rules, it is very much in the client's interest to have only one professional adviser who can see to the effective collaboration of the whole professional team. An interesting development in this respect is the growth of professional partnerships of architects, town planners and engineers from the various disciplines. One consultant can in this way provide the whole range of professional service.

The last of our three parties is the manufacturer, or contractor, who supplies or builds everything required for the physical realisation of the client's dream and the consultant's design. In civil engineering contracts, the structures are usually unique and construction on site to the specific requirements is the rule; however, at least some of the components used will be standard ones, and these will be drawn from a manufacturer's range of standard products. These components, and some of the service equipment used in such contracts, will be produced by a manufacturer who, with his own design, construction and sales staff, endeavours to produce and stock equipment that will satisfy a predictable market need.

For specialised supply and construction, contractors often engage subcontractors. Although responsibility for performance generally remains with the contractor, some of the work is frequently done by such specialists. Subletting often includes design to satisfy a specified performance. For example, in reinforced concrete construction, the supply, bending and fixing of reinforcing steel may be sublet. Heating and ventilating plant for a building is usually supplied and often designed as well by a subcontractor, who may be nominated by the client – that is, the main contractor is not necessarily free to select the subcontractor. However, in this case the subcontractor is directly responsible to the client for the work he does.

We have noted above that contracts are let after bids or tenders have been invited. Thus there is a strong element of competition for jobs among contractors. This has the advantage to the client of keeping costs down. However, for every contract let, several unsuccessful tenders are prepared and the cost of this work adds to contractors' indirect costs and thus to the cost of construction. Another disadvantage of the system of competitive tendering is that special skills possessed by certain contractors may not be fully utilised. For these reasons variations of the organisation described are not unusual. For example, a building contract may be let before detailed designs are prepared. The designer can then work in collaboration with the contractor to take advantage of whatever specialised plant or labour he has. The price paid to the contractor in such a case may be a negotiated lump sum or it may be based on a schedule of unit rates to be applied to the various items of construction. Another similar variation has the client dealing directly with a contractor who does both design and construction; however, the loss of an independent adviser may be to the disadvantage of the client. Again, the client may employ salaried staff to design the works or to construct them or to do both. This kind of organisation is frequent among local bodies and central government departments for, at least, part of their work.

In this description we have outlined briefly the structure of the industries which are served by engineering designers. Understanding of this structure clarifies the kinds of communication that are needed and the various people and organisations who have to communicate with each other. The process relies heavily on the use of drawings and in Section 24.5.1 we will describe the various types that are used. It will be accepted that internal communications – say, between members of a design team – will be less formal than the external communications between parties who are not in contact daily, or in situations where there are contractual obligations. Most of what we have to say is about the more formal procedure.

24.3. THE DESIGN BRIEF

Coming as it does right at the beginning of the design process, the brief plays an important part in the success of the enterprise it launches. It is usually in writing, but small jobs may be initiated verbally, or the written brief supplemented by discussion. In any case, the designer is well advised to confirm in writing any part of the original brief conveyed verbally or any changes agreed to in conversation with the client.

The design brief is a statement of the client's requirements and wishes, as they are known when the project begins. It incorporates details of special requirements and information that the designer cannot be expected to know or find out without the client's help. The following is a list of some typical items:

(a) a statement of the purpose of the project;
(b) sizes and capacities of buildings and machinery, including information about possible extensions in the future;
(c) information about the flow of materials and components through a manufacturing process;
(d) location and characteristics of the sites of plants and buildings;
(e) financial details – for example, the amount of capital available, expected returns from manufacturing;
(f) target dates for completion of design and construction;
(g) details of special legal restraints in addition to those in recognised codes and standards;
(h) demands for special or unusual items that would not normally be provided:
(i) items which may require special attention – for example, aesthetic requirements in a building, pollution hazards and any special preferences or dislikes the client may have.

There is no set form for a design brief. Fundamentally, it has to satisfy only one criterion – comprehensive description of the client's plans. The initiative in this is with the client, but the designer must realise that he may be able to contribute to the client's understanding of his own problem; he should be willing to suggest changes in the brief if he thinks he can make improvements. Clearly, mutual respect and shared confidence will contribute much to a successful brief. This aspect of the design brief is emphasised because a communication must be received as well as sent to be effective. The designer must attempt to foresee all requirements and continue his investigations until he is certain that he has all the necessary background information. It will be recalled that, in Chapter 1, this was described with good reason as the "divergent" stage of design. Although it is the client's responsibility to make sure that the brief is adequate, a good designer will put time and effort into helping him do this.

24.4. THE DESIGN REPORT

The design report is a formal method of presenting the results of a design investigation. It may be prepared by a junior engineer for his chief or be the result of a substantial team effort in the design office for presentation to a client. Its make-up will vary, depending on the size and scope of the subject matter, but there are principles fundamental to the preparation of all reports which are described here.

A common feature of reports is that they are read by people with a variety of backgrounds and interests – for example, managers seeking the conclusions reached and the justification for them, or specialists who want to know the technical basis of a decision. Part of the art of writing a good report is the ability to satisfy the requirements of different classes of readers without requiring any of them to read what does not interest them or what they cannot understand because of its technical complexity. Any reader of a report wants to know first what the report is about, second, what results have been obtained from the investigations and, third, how the results have been obtained – in short, *object, results, method.*[1] However, as noted above, since not all readers want all the detail, the report should be in two main divisions – a *summary* of object and results and a *detailed account* of the object, method and results. Note that only the detailed account includes description of the methods. Non-technical readers need read only the first part and even the reader with an interest in the technical aspects of the investigation can get the information he needs for immediate action before reading the whole report.

The first division, or *summary*, should include reference to the design brief and should review agreed alterations. For the report to be useful the author must make a firm recommendation and give his reasons for making it – basically, that it is the best solution to the problem posed in the brief.

Judgement is necessary in the presentation of the *detailed account*, for even here too much detail can be included. Complete calculations are not normally given although they must, of course, be carefully filed for reference. It is important that calculations should be described and the assumptions which have been made must be acknowledged. If numerical detail (such as data or calculations for estimates or design) is included, it should be relegated to appendixes.

Estimates of cost will be an important section in the detailed account and here the basis of the estimates must be shown clearly. Circumstances beyond the control both of client and of designer can change rapidly. If, for example, rates of taxation are changed after the design report has been presented, it may be necessary to determine very quickly how the recommended plan will be affected. As noted above, numerical detail of estimates is properly placed in appendixes.

The guiding principle in deciding on levels of subdivision is to avoid inclusion of detail which obstructs understanding.

Illustrations should be chosen with the same principles in mind. In the summary of object and results, pictorial illustrations may be more useful than conventional engineering drawings. The latter have a place in the detailed account, but careful selection is necessary. To avoid making the report too bulky, reduced-size drawings can be used. These are quite suitable for showing the general features of the design and the fact that fine detail cannot be presented is no disadvantage at this stage in the process.

24.5. SKETCHES AND DRAWINGS

We have noted above that drawings are used in construction and manufacture as a primary means of conveying information. Heavy emphasis must be placed on good drawing practice and, although senior designers do little draughting themselves, no engineer who aspires to eminence in design can succeed without knowing what is required in a good engineering drawing and how to prepare one. This knowledge demands draughting experience for a secure foundation; students and junior engineers must be prepared to put some time to acquiring this experience.

Our emphasis on this must not be misunderstood. In practice, the drawing is a means to an end which is the manufacture or construction of a product. The drawing is not an end in itself, to be prepared with loving care as a work of art. But its importance as a means is such that instruction in the skills and conventions of engineering draughting should precede or be part of the course in design. Much of a designer's success – and the success of any engineer – rests on his ability to judge the merits of a scheme at the stage of the sketch plan or general arrangement drawing. The mechanical skills of draughting – which develop rapidly with practice – are not needed for their own sake, but to facilitate the use of this medium of communication.

24.5.1. Conventions of engineering drawing

Detailed description of draughting practice is outside the scope of this text. For this information, reference can be made to BS 308, *Engineering Drawing Practice*[2] and AS CZ1 – 1973, *Australian Standard Engineering Drawing Practice.*[3] However, a short review of the basic principles is relevant.

In general, engineering drawings represent to scale three-dimensional objects in two dimensions. Conventions in the use of projection of solids on planes and of section planes enable this to be done completely and with clarity when both draughtsman and reader are familiar with the conventions. These conventions are widely used and there are others specific to different branches of engineering. For example, in drawings of reinforced concrete the reinforcing steel is shown in plans and elevations as if it were visible. Strict conformity with the rules of projection and sectioning would require an inordinate number of sections. To this extent the conventional portrayal of the reinforcing is encoded. Such use of coding is taken further in electrical and hydraulic drawings where standard symbols are used to signify various components and the runs of conductors and pipes may be highly schematic.

Other conventions relate to the kind of drawings used for communication. For informal communication, *engineering sketches* are appropriate. The need for the designer to keep a clear pictorial record of layout in the formative stages of a design has been emphasised before. The drawings used may be informal and only a very minor component may be involved. Sometimes sketches may be schematic and not to scale. But in many cases, and particularly when any layout or arrangement of mating parts is involved, it is important that they be drawn to scale. In this way a check is kept on the proportioning and balance of the various elements. In addition, a clear record of the designer's intent is available for checking or to be passed on to a draughtsman.

Apart from reduced accuracy of line and formality of arrangement, an engineering sketch should retain the essential qualities of an engineering drawing. It should not be confused with a freehand sketch, perspective sketch or an artist's impression; these can be useful in the communication process but they have different qualities from an engineering sketch. They can all play a part in conjunction with the written word in conveying information.

Formal drawings for a big job are arranged in sets which may be very large. The arrangement of a set of drawings depends to some extent on the nature of the job.

In mechanical design, the *general arrangement* (or GA) is the basic drawing which describes the complete machine. It is supplemented by other drawings which describe the components of the machine – one or more drawings for each component, but preferably not more than one component on each sheet.

The basic functions of the GA drawing are, broadly, as follows:

(a) to illustrate the disposition of the components, arranged together to make up the final assembly;
(b) to provide a guide for the assembly and mode of operation of these elements;
(c) to provide a basis for a parts list which details all the various components that make up the whole;
(d) to provide a basis for further subdividing the assembly into subassemblies (if required) and individual parts.

When design work is in progress, the GA provides a skeleton on which to hang the developing forms of the design. When the design is complete, the work of manufacture or purchase is initiated; the parts list on the GA has an important function in this operation. Each part, or one representative of each group of identical parts, is identified in the list along with information which will aid in its procurement – the material specification and the quantity required for each assembly. Where parts are to be bought in – standard nuts, bolts, bearings, or other proprietary equipment – suitable identification must be provided on the list.

The GA should also provide the basis for a rational and flexible scheme for numbering and filing drawings. A scheme for this purpose is described in AS CZ1 – 1973[3], and every design office will have some such system for allocating numbers. The principal requirements of a numbering system are that it should provide means of relating the individual part drawings to the subassembly or main assembly to which it belongs and that it should identify drawings that have been modified.

It should be noted that parts are usually identified during manufacture by their drawing number. This number is not necessarily retained to describe the part in service manuals and in spare parts' lists.

In civil or structural engineering, a plan usually shows the location of the works in relation to survey pegs in the ground. There may also be drawings of excavation and filling required to be done on the site before erection of structures begins.

Drawings of the structure itself should include some representation of the complete structure just as the GA represents a complete machine. However, the purpose of this representation of a structure is less comprehensive than the purpose of the GA in mechanical practice. The size of civil engineering structures is often such that the complete project can be shown only to a very small scale. Such a drawing is useful in conveying an overall impression and in showing how various parts are related to each other, but it cannot do much more than that.

Framing plans show the location of the principal structural elements. A frame of reference that is essentially Cartesian is used to identify them and *detailed drawings* describe them. The framing plans show where the elements are to be and the details show how they are to be built. To make use of a Cartesian frame of reference, *setting-out lines* are defined. In general, these form a rectangular grid in plan and they pass through the centrelines of the columns in both directions. If one set is numbered and the other identified by a sequence of letters, any column can be identified by a letter and a number. The third, vertical, coordinate can be given as a level – either as the number of a storey above ground or as a height above a specified datum.

Arrangement of a set of drawings is determined by the operations in construction (see Chapter 23, "Construction Operations"). Thus, structural steel drawings are kept separate from reinforced concrete drawings because the structural steel will be fabricated in a workshop remote from the site. In reinforced concrete construction three sets of drawings are required – concrete outlines, details of reinforcing steel and steel-bending schedule. The first set is required by the carpenters who construct the boxing, or formwork. Reinforcing details are required by the steel fixers who erect the reinforcing steel in the boxing and the schedule is for the steel benders. A description of the work of these men is given in Chapter 23. Here, we are concerned with the kind of information to be presented to them on the drawings.

Dimensions on the drawings of concrete outlines should be related to the setting-out lines. These are realised physically on the site by stretching string lines and dimensions from them can be used directly. Vertical dimensions can be referred to a significant part of the element under construction – for example, the bottom or soffit of a beam. When this has been established with the aid of a surveyor's level and staff, the carpenter's task is straightforward. Our point here is that the draughtsman should choose a significant and convenient part of the element and refer dimensions to it. A good knowledge of construction procedures will ensure that the dimensioning does not require workmen to do a lot of arithmetic under adverse conditions to get the distances they will use.

Dimensions on drawings of reinforcement should be referred to concrete outlines. The point is that by the time the steel fixers do their work, the completed boxing is the most convenient frame of reference they have. Also, they receive from the benders bundles of bent bars and their task is to fix these bars in place in the boxing. However, it is necessary to provide complete information about the bending on these drawings since they are often the documents used to prepare the bending schedule.

The bending schedule is a list of bars required for the job. It shows the number, diameter and length of steel bar to be drawn from stock to provide the various classes of bars in the job. Each class is given a coded description and the schedule shows how the bars are to be bent. Sketches, usually not to scale, are used to convey bending information concisely and the schedule also contains notes describing the location of the bars in the job. The same coding is used to identify the bars on the reinforcing details and the final link in this chain of communication is the label fastened to a bundle of bars by the steel benders. This label shows the number of bars in the bundle and their code number and is an obvious necessity for the steel fixers.

Preparation of the steel schedule is often left to the subcontractor supplying and fixing the steel. In that case, he takes the information he requires from the details of the reinforcing steel.

24.6. CONTRACT DOCUMENTS

Of the formal contract documents, the *conditions of contract* and the *bill of quantities* are of little direct interest to the designer. The former are based on the law of contract and they are contained in a legal document which is usually standardised, at least nationally. One set of standard conditions widely used in civil engineering is that of the Institution of Civil Engineers.[4] The conditions of contract define the legal obligations of the client ("the owner" in the terminology of the conditions) and the contractor. The powers of the engineer are also described. The bill of quantities is a list of all the items of construction, the quantity of each and, when completed by the contractor, the unit rate and the amount for each item (see Chapter 25, "Economics and Costing"). The bill of quantities enables the estimated total cost of the project to be determined. It is also used for progress payments – the quantity of work completed under each item is measured and the unit rate applied to the quantity – and to determine the value of any alterations made after the contract has been let. In civil engineering, the bill of quantities is often prepared by the engineer. In building contracts, quantity surveyors are usually engaged to prepare bills.

We have discussed drawings fully above and there remains one other formal contract document – the *specification.* Here the designer is directly and intimately concerned. Specifications can be considered to fall into two types. The first is related to the kind of contract we have described above where the detailed design is the responsibility of an independent consultant or the client. In this context, the specification is a description in words of the quality of materials and workmanship and of the methods to be used in completing the project. With regard to materials, physical properties are of fundamental importance. The designer bases his design on certain numerical values for properties such as yield strength, ultimate strength, elasticity, hardness, etc. The specification is the document wherein he defines the quality of material needed to guarantee the validity of his calculations. It is possible to use different methods for materials that the contractor will purchase and those that he will manufacture on site. For the former, standard specifications can be used. There is an obvious economic advantage in having all manufacturers of a given material supply a product which satisfies the same specification. Steel is a commodity in point. It is widely used in manufacturing and construction and there are well-tried standard specifications for steelmaking. Clearly, a designer need do no more than refer to the appropriate standard. This principle extends to many manufactured articles, for example, nuts and bolts.

Standards are published by national organisations, examples of which are British Standards Institution, *Deutsche Industrie-Norm* (German Industrial Standards), Standards Association of Australia, Standards Association of New Zealand, American Society for Testing Materials. Collaboration among such bodies leads to uniformity on an international scale and results in the standard specifications of the International Organization for Standardization (ISO).

Of the materials likely to be manufactured on site by the contractor, concrete is by far the most important. There are standard specifications for the ingredients of concrete – cement and aggregates. Cement being manufactured as an item of merchandise, the standard specification generally provides the necessary control over its production. Sometimes, for very large jobs, contracts are let for manufacture of cement to a special specification; this may be done if unusual characteristics are required in the cement. Aggregates are produced by mining operations with more or less processing depending on the need for it. There is a wide range of source materials and of competence of quarry operators and, to some extent, a tolerance on quality, depending on the job. Some standards allow quite wide tolerances on various aspects of aggregate quality and engineers often prefer their own specifications. Restraints on the procedure for mixing, placing and curing concrete are desirable, in addition to limiting values of relevant physical properties.

With regard to workmanship, standard specifications cover some operations, for example, welding.[5] In any case, the qualifications and level of skill of workmen are defined and, where a variety of procedures would apparently satisfy requirements, it is necessary to say clearly which one is demanded. For example, when holes are to be made in steel it must be clear in the specification whether gas-cut holes are permitted, whether punching is permitted, if reaming is required, whether burrs are to be removed. The point is that different ways of doing the job differ in cost and in the quality of the finished article. The contractor must know which he is required to use and what he will be paid for.

In the other type of specification, the client or the consultant deals only with the broad aspects of the design, responsibility for the details being included in the contract. This sharing of the work is common in contracts for large items of plant – for example, a power station or a dockside crane. The specification in a contract of this kind is a description of the performance to be achieved and the constraints under which the performance is required. Performance and constraints are determined by the client or the consultant and often call for a substantial effort in design and calculation. The crane referred to above, for example, would be required to handle cargo at a prescribed rate in units of specified size. This performance would be required subject to certain overall dimensions, requirements for braking and other control, protection against overload and so on. The details of a suitable crane would be left to the contractor.

This second kind of specification is open to negotiation up to the time of signing a contract. Although negotiation on the specification complicates comparison of tenders, it may be in the interests of the client. Specifications of the first type are usually accepted without alteration by tenderers. What are known as "tagged tenders" in this context are not regarded with favour by consultants and they are usually not accepted as valid bids.

Writing of specifications requires experience in performing or supervising shop operations and site operations. A good knowledge of the alternative ways of carrying out various kinds of work is required as well as the ability to assess their relative merits and costs. Careful attention to detail is necessary to ensure that no item is inadequately or ambiguously described or omitted from the specification. Some assistance in writing specifications may be obtained from a publication of the British Standards Institution entitled *Guide to the Preparation of Specifications.*[6]

REFERENCES

[1] C. W. Knight, "The Writing of Technical Papers and Reports", in *The Presentation of Engineering Evidence,* The Institution of Civil Engineers, London, 1946.

[2] BS 308:1972, *Engineering Drawing Practice*, Part 1, "General Principles"; Part 2, "Dimensioning and Tolerancing of Size"; Part 3, "Geometrical Tolerancing"; British Standards Institution, London, 1972.
[3] AS CZ1—1973, *Australian Standard Engineering Drawing Practice*, The Institution of Engineers, Australia, Sydney, 1973.
[4] ICE, *General Conditions of Contract*, 4th ed., amended, The Institution of Civil Engineers, London, 1969.
[5] BS 1856:1964, *General Requirements for the Metal-Arc Welding of Mild Steel*, British Standards Institution, London, 1964.
[6] PD 6112, *Guide to the Preparation of Specifications*, British Standards Institution, London, 1967.

BIBLIOGRAPHY

BANKER, J. A., *Reinforced Concrete Detailing*, Oxford University Press, London, 1965.

BOOKER, P. J., *A History of Engineering Drawing*, Chatto & Windus, London, 1963.

DAWE, J. and LORD, W. J., *Functional Business Communication*, Prentice-Hall, Englewood Cliffs, 1968.

HARWELL, G. C., *Technical Communication*, Macmillan, New York, 1960.

MORRIS, J. E., *Principles of Scientific and Technical Writing*, McGraw-Hill, New York, 1966.

TWORT, A. C., *Civil Engineering Supervision and Management*, Arnold, London, 1966.

WALLACE, I. N. D., *Hudson's Building and Civil Engineering Contracts*, Street & Maxwell, London, 1970.

PROBLEMS

24.1. Sketch an organisation chart for two partners and a staff of about six in a small consulting practice. Indicate the vertical and horizontal communications that would be exchanged among them. What happens if the typist resigns and a replacement cannot be found?

24.2. You are writing the specification for a building. Prepare clauses covering excavation for the footings and preparation for placing the reinforcement and for concreting.

24.3. You are a client who wishes to build a small general engineering workshop and equip it with a range of machine tools. List the headings and sub-headings of the design brief.

24.4. The brief of Problem 24.3 has been transmitted to the consultant, and an assistant engineer has been sent to inspect alternative sites and to report on the sites and the necessary investigations. You are the assistant engineer. List the headings and subheadings of your report to the partner who is your chief.

24.5. Think about the role of photography in communications at all stages in a project for construction of a building.

24.6. Write a check-list for use when checking the drawings for an assembly such as a gearbox. Build it up under different headings such as:

Bearing – dimensions, location, lubrication, sealing . . .

Finish – plating, painting . . .

24.7. A client of the consulting firm for which you work has asked you to design a small storage shed for frozen foodstuffs. What information would you expect him to give you? Suggest a form for the design brief if you are to have full responsibility for deciding on the storage volume and layout as well as the cooling plant.

25 Economics in Design

25.1. INTRODUCTION

In addition to technical factors, economic factors must be studied by the designer if he is to reach a solution to a problem that is financially advantageous as well as technically sound. The methods described in this chapter are aimed primarily at helping the designer with some of the economic decisions he will have to make; some of the problems that arise when attempting to establish the profitability of a project are also described.

The final responsibility for establishing the value or profitability of a venture will usually rest with management, and marketing prospects as well as production costs will be studied before the decision is made to proceed with a venture. The designer will be involved in providing some of the information needed for this and he should know how such decisions are reached.

Of more direct interest to the designer are the cost comparisons that can be made during the design process – especially at the stage where alternative solutions are being considered. In many cases, this will involve comparison between the direct costs of two schemes, and here the problem is simply one of estimating costs. In other cases, the comparison cannot be made directly and the costs must be processed, using some simple economic methods, before they become meaningful. For example, the high initial cost of a quality product cannot be compared with the cost of a lower quality item which will require a greater future expenditure on maintenance, unless allowance is made for the difference, with time, of the value of money.

We are concerned then with two aspects – the estimation of costs, and their processing. The latter is examined first, so that we can establish the cost figures that are important and how they are best broken down. When considering the cost estimation, a few simple methods of determining costs will be described, and some typical cost figures given. It must be stressed that these should be checked and brought up to date before being used in practice; they are provided here mainly for illustrative purposes.

25.2. ECONOMIC FACTORS

The engineer's concern with costs is generally related to a need to show a profit on some capital investment, and we will now consider briefly some of the factors involved in predicting or calculating these profits.

The profits considered here will be the direct ones. The indirect and intangible profits – and perhaps losses – that come from some types of engineering works will not be accounted for at this stage. It is clear that the profit from the installation of an exhaust-purification system would be negative unless some allowance were made for the benefits of cleaner air. The development of water resources and transportation systems are other cases where a full "cost-benefit" analysis must be made if the expenditure is to be justified objectively. A brief description of this type of analysis is given in Section 25.4.

The total investment, I, is the total capital that must be committed and risked in the proposed project. In some cases, the investment is centred on a single construction – for example a commercial building intended for rental – and no further breakdown need be considered. In other cases, it is useful to subdivide it into portions which carry different risks; thus the fixed investment, I_F, which provides the specialised equipment for a production process is subject to a greater risk than auxiliary investments, I_A, in items which, like a steam supply or a concrete-batching plant, will often serve a number of separate operations. The costs for such auxiliary items are often covered by allowing a unit cost for the material or service it provides – for example, the cost of concrete per unit volume, delivered to the site, is easily obtained; this procedure is convenient in many costing calculations.

The revenue, R, from the investment is the direct return that comes from sales or services provided. To determine the profit, the cost incurred in producing the revenue must be deducted from R.

These costs will usually involve labour, materials and expenses, and the way they are broken down will depend on the use that is to be made of the information. For our purposes they can be divided into *direct* and *indirect* costs. The direct costs are those that can be clearly traced to a particular unit of output – for example, the material and labour required from one specific component. These direct costs are proportional to output quantity. The indirect costs cannot be associated with any particular unit of output, and comprise such expenses as administration, maintenance, rent and insurance.

We can write

$$\text{Gross profit } GP = R - \text{direct costs}$$
$$\text{Net profit } NP = GP - \text{indirect costs.}$$

To determine the *net profit*, allowance must be made in the indirect costs for tax and for the need to spread the original cost of the fixed investment over its economic life. We assume here that a sum is provided each year, so that at the end of the expected life of the plant the investment I_F has been recovered; for a life of n years the annual sum will be $(1/n)I_F$. This is the simplest method of allowing for depreciation and is termed *straight-line depreciation*. If the asset has a scrap value at the end of its depreciated life, this is allowed for by deducting it from I_F.

Tax is levied on an assessable income, and the details of how this income is calculated will vary from one country to another, depending on the tax regulations. We will not consider these details, but will assume that the tax rate, t, is actually applied to an amount $(GP - f\,I_F)$, where f is the yearly fractional loss of value on the fixed equipment allowed by tax authorities. Thus the net profit NP after tax is given by

$$NP = GP - \frac{I_F}{n} - t\,(GP - f\,I_F) \qquad (25.1)$$

The importance of the tax term is that t will often have a value as great as 0.5; clearly, the final term in Equation (25.1) cannot be neglected when the absolute value of profit is important, although it may not be significant when the engineer is making direct comparisons between alternative solutions.

Labour costs may be a significant fraction of the total costs. They are determined by multiplying the man-hours required by the unit rate (per hour) of labour. In some simplified cost estimates it is convenient to absorb in this term the administration, supervision and other overhead charges that comprise part of the indirect costs. In this case, the unit rate for labour will be appreciably higher than the standard labour rate.

When the profit has been determined, some criterion must be adopted by which the success of the venture can be gauged. The simplest approach is to calculate the return on the original investment; this is the ratio of the average yearly profit during the earning life to the capital invested. The return will be expected to reach a percentage which should be comparable with other investments that are subject to a similar degree of uncertainty. Where the risks are small, 10% would be accepted as a return, whereas the limiting condition might be set at double this figure, or higher, where the design has novel features or the market is unpredictable. Other methods of assessing economic worth are available, which are more suitable if an investigation to maximise the worth of the project is to be made. These are described by Rudd and Watson.[1] More detailed methods of investment appraisal are given in *An Introduction to Engineering Economics*[2] which presents many examples based on engineering practice. A guide to capital cost estimation is given in a publication of The Institution of Chemical Engineers.[3]

When considering improvements to existing plant, a simple criterion sometimes used is the "pay-off" time. This is the number of years the improvement will take to pay for itself. Some industries may use a pay-off time as low as one or two years.

Example 25.1

The investment in operating an aircraft in agricultural aviation work (aircraft, loading equipment, etc.) is \$45,000, and its operating time is charged out at \$60 per hour of flying time. If costs – pilot, fuel, insurance, maintenance, etc. – are \$35 per hour and the annual operating time is 700 hours, determine the net profit. Assume that depreciation is spread over five years, and the tax rate, t, is 0.5. Take $f = 0.25$.

		\$
Revenue	700 × \$60	42,000
Costs	700 × \$35	24,500
Gross profit		17,500
Depreciation	\$45,000/5	9,000
Operating profit		8,500
Tax	0.5 (17,500 − 0.25 × 45,000)	3,125
Net profit		5,375

Thus the year's profit after tax on the \$45,000 investment is \$5,375, or about 12%. Whether this would be regarded as a satisfactory return on capital for this type of operation is questionable. For example, certain costs – maintenance crew, insurance, rent – will accrue irrespective of the flying time, and unforeseen events such as a spell of bad weather, reducing the flying time to 600 or 500 hours, would cut down the profits drastically.

In Example 25.1, the basis of the calculations is a cost of $35 per hour of flying time. Such figures must be determined by an analysis of the costs – estimated ones in the case of a new aircraft, but ones which can be modified in the light of operating experience. Example 25.2 gives details of an estimate of this type in which a cost per hour is worked out for operating two Britten Norman aircraft in an undeveloped area.[4]

Example 25.2

Estimated cost of operating two BN2 Islanders in an undeveloped area, based on a total utilisation of 1500 flying hours/year. (From an article by D. G. Videan in *Shell Aviation News*, 388, 1970[4], by permission of the author.)

Basic cost of 2 aircraft	£68.000
Instruments, radio and navigation equipment for 2 aircraft	8.000
Spares holding and special items (See Note 1)	12.000
TOTAL	£88.000
Fixed costs per annum	
Depreciation 10% to Nil (See Note 2)	8.800
Interest on capital 10% (See Note 3)	8.800
Insurance 5% (See Note 4)	4.400
Salaries, leave and fringe benefits: (See Note 5)	
1 Manager/Chief Pilot @ £8.000	8.000
3 Pilots (See Note 6) @ £5.000	15.000
4 Engineers @ £3.000	12.000
Accommodation and messing:	
8 — £100 per month	9.600
6 Local labour, cleaning, loading, etc., — £500 per annum	3.000
Hangar and facility maintenance and sundry expenses	2.000
	£71.600
Variable costs based on 1500 flying hours/year	
Fuel (22 Imp gal/hour @ 3/–)	4.950
Oil (at 10% of fuel cost)	495
Engine overhaul (based on 1800 hours TBO and £1.250 per engine and propeller)	2.084
Consumable spares @ £2 per hour	3.000
Total variable costs per annum	£10.529

Total costs per annum = £82.129
Cost per flying hour = £54.75
(See Note 7)

Note 1: Because of the remoteness of the area in which the aircraft will be operating, a larger sum than usual has been allocated under this heading. Experience may well show that the spares holding is too lavish, in which case it can be run down gradually with a resultant decrease in costs.

Note 2: The depreciation rate to be used is either a matter of company policy or an inspired guess. With a simple aircraft such as the Islander, there seems no reason why for this particular operation it should not perform adequately for ten years. If after that time it continues in service, extra benefits accrue.

The most difficult thing is to try to anticipate a market value ten years ahead. It is thought that such a figure can be of academic interest only, and of little practical value. It is preferable therefore to depreciate to zero value even though this may be somewhat pessimistic.

Notes 3 and 4: Both figures are mean values which are considered to be fair for calculation purposes. Actual figures may vary considerably between operators.

Note 5: Salaries paid to the personnel under this heading may be approximately half the figures shown. Yet it is especially necessary in remote areas to ensure that welfare arrangements and amenities are as good as it is possible to make them. Adequate staff amenities and welfare under adverse conditions is the best way to ensure an efficient operation.

Note 6: This will be a single pilot operation. The flying staff specified may seem extravagant, but it is in line with the safeguards considered necessary for an operation of this type.

Note 7: Comparative costings for a typical operation in the UK return a figure less than 50% of this. However, the operation considered above is to be carried out under difficult tropical conditions, and provisioning and allowance in all sectors of expenditure must therefore be much more generous.

One difficulty that arises when comparing payments or returns is that they do not necessarily fall in the same year. A cash return of $100 at the present time is worth more than an identical payment five years from now. In broad terms, a fulfilled immediate need carries a greater value than the promise of some future fulfilment. A direct measure of the difference is the interest that the present payment can earn over the five-year period. In practice, the difficulty can be resolved by comparing all moneys on the basis of their *present value* at some specified compound interest rate, i.

Let S be the amount in a fund at time θ; the change dS in the fund in time $d\theta$ will be given by

$$dS = i\, S\, d\theta$$

so that

$$i\,(\theta_2 - \theta_1) = \int_1^2 \frac{dS}{S} = \ln \frac{S_2}{S_1}$$

where S_1 and S_2 are the amounts at times θ_1 and θ_2. Thus,

$$\frac{S_2}{S_1} = e^{i(\theta_2 - \theta_1)}$$

Putting $\theta_1 = 0$ gives

$$S_0 = S_\theta e^{-i\theta} \tag{25.2}$$

where S_0 is the present value of S_θ dollars at some time θ in the future. Tables of S_0 for various values of i and θ are available;[1] however, the computation is simple on slide rules which have log log scales.

We have assumed here that the interest on S is compounded continuously with time. In nearly all practical cases, the interest is payable at discrete time intervals – annually, half-yearly or quarterly. The equivalent formula for the present value is

$$S_0 = \frac{S_n}{(1 + i)^n}$$

where i is the interest rate per period and n is the number of periods. The continuous compounding method is used here since it is convenient for slide-rule work and for mathematical manipulation; the numerical differences between the two methods are small and will serve as a reminder that our approach does not have the accuracy and completeness of that used by accountants. It can be noted that in the limit, as n becomes large and θ becomes small, the differences are negligible. Compound interest tables are given in *An Introduction to Engineering Economics.*[2]

Example 25.3 illustrates the way in which the present value concept is used.

Example 25.3

If the prevailing interest rate, i, is 8 %, compare the merits of two possible ways of providing a pumping service for 10 years.

Scheme (*a*) High grade pump	*Scheme* (*b*) Lower quality pump
Initial cost: \$1200	Initial cost: \$500
Life: 10 years	Life: 5 years
Annual power and maintenance costs: \$90	Annual power and maintenance costs: \$120
Salvage value after 10 years: \$200	Salvage value: nil

It is assumed in Scheme (b) that the pump is replaced after five years, and that the price has risen at the rate of 2 % per annum.

Scheme (*a*): present \$ values

Equipment cost: \$1200 initial cost

Annual costs: $S_0 = \int_0^{10} 90\, e^{-0.08\theta}\, d\theta = \left[-1125\, e^{-0.08\theta}\right]_0^{10}$

$= 620$

Salvage: $S_0 = 200\, e^{-i\theta} = 200\, e^{-0.08 \times 10} = 89.8$

Total present value of costs: \$1910

Scheme (*b*): present \$ values

Equipment cost: $500 + 500\,e^{-(0.08-0.02)5} = 870$

Annual costs: $S_0 = \int_0^{10} 120\,e^{-0.08\theta}\,d\theta = 826$

Salvage: $S_0 = 0$

Total present value of costs: \$1696

On the basis of present values, Scheme (*b*) *is more economical.*

25.3. PREPARATION OF ESTIMATES

There are two general approaches possible when the cost of an engineering project is to be estimated. The first involves a full and detailed accounting of each and every separate piece of material and equipment and man-hour of labour. This detailed cost estimate can be made to a high degree of accuracy but requires the completion of at least the preliminary drawings and specifications. While the designer himself may well use this method to estimate the cost of some minor items, the detailed costing of major projects will usually be handled by specialist personnel. Since it is carried out when the design has reached an advanced stage, it is of limited use at the time when several alternative courses of action are being considered by the designer.

The second approach can be used early in the design operation to give approximate costs on which design decisions can be based, and is therefore of more relevance here. It relies heavily on past experience, and ranges from the knowledge that "this type of structure will cost \$200 per m^2" to the similar but more sophisticated extensions of this method that are described in Section 25.3.2.

25.3.1. Detailed estimates

The question of detailed estimates needs to be considered from two points of view – that of the engineer and that of the contractor or manufacturer. Both use the records and knowledge they have acquired with experience but their reasons for estimating are different and their records consequently contain different information. The engineer's estimate is for the information of his clients or employers and it is a forecast of the money required for construction. The contractor's estimate is the sum for which he is prepared to do the work. The contractor's estimate (or tender) is prepared in greater detail than that of the engineer and is based on the operations of construction rather than the results of those operations.

The engineer bases his detailed estimate on the *bill of quantities* (Chapter 24). This shows the quantities of all the items that make up the complete construction and each of these gives the total cost for that item when multiplied by the appropriate unit rate. The engineer can usually find an appropriate unit rate in his records of previous contracts. Table 25.1 shows a typical sheet from a bill of quantities with the column for unit rate and total for each item filled in as they would be by the estimating engineer.

The contractor applies more fundamental reasoning to the question of unit rates. Whereas the engineer's only concern is with the cost of the item in

Table 25.1. Sheet from a bill of quantities.

Project: . . . . Sheet No.: . . . .

Item	Description	Unit	Quantity	Rate	Amount, $
	Brought forward				212,800
9	Establish precasting yard, fabricate and erect formwork for precast beams	LS*			52,000
10	Reinforcing steel in precast beams	tonne	70	480	33,600
11	Concrete in precast beams	m^3	1100	100	110,000
12	Travelling falsework for superstructure	LS*			38,000
13	Formwork for superstructure	m^2	2000	17	34,000
14	Reinforcing steel in superstructure	tonne	100	480	48,000
15	Concrete in superstructure	m^3	850	70	59,500
	Carried forward				587,900

*Lump sum.

place, the contractor is concerned with the quantity and cost of material he will have to buy and the number of man-hours of labour and machine-hours of plant that will be required for construction. These require detailed knowledge of the local market for materials and the performance rates of labour and machines. Such data are available in the contractor's records and in the literature – e.g., in Geddes[5] and in such journals as *Civil Engineering and Public Works Review*. Although generally relevant to contracting, this more fundamental approach will be required of the engineer if he has to estimate the cost of a job for which he does not have the unit rates.

A great deal of this kind of work is required for a big job and all operations will be covered – from the preliminary preparation of the site to clearing-up and landscaping. Much of the routine computation can be programmed for a computer.

Some allowance is usually made for unforeseen contingencies, and expected or predictable cost rises are taken into account. Although this method can, at its best, produce good accuracy, many cases of gross inaccuracies (nearly always underestimates!) are publicised each year. Small errors will arise in the estimating process owing to incorrect cost information, or can be caused by unexpected cost changes; however, the major errors can usually be traced to some design assumption that was incorrect or was based on insufficient information. For example, an inadequate investigation of sub-surface conditions for a building may lead to unjustified assumptions regarding the foundations needed. Excavation during construction may reveal conditions that require extensive and costly treatment; this is clearly a design error and not an estimating error.

25.3.2. Approximate cost estimation

Engineers who work in a particular field collect a good deal of information on equipment and construction costs and use this to predict the cost of new work; provided no new techniques or processes are involved and excessive extrapolation is not attempted, good approximations can be arrived at. While experience will count for a great deal, there are many published data and methods which can be used for such estimates; some typical information of this type will be considered here.

Mention has already been made of the idea of estimating the cost of a building on the basis of a cost per unit area. Obviously this method will be inaccurate when it is used for structures of unusual shape, size or finish or where novel methods are introduced; however, quite accurate estimates can be made in a limited range.

An example of this approach is given in a publication prepared by Bertlin and Partners.[6] The cost per square metre of enclosed area for a number of port transit sheds in the United Kingdom is tabulated, the cost varying from £11/m^2

Table 25.2. Typical cost figures for use in preliminary cost estimation.

Item	Unit	Rate $	Comment
Excavation	m^3		Excavation rates vary widely depending on the material excavated, cartage to point of disposal and method of payment. These rates are typical ones for clay, with no haulage, and payment for actual quantity excavated
Bulk		1.0	
In trenches		4.0	
Formwork	m^2		
Simple		15	
Elaborate		23	
Reinforcing steel	tonne		
Placed		480	
Concrete	m^3		
Precast		100	
Foundations		60	
Superstructure		70	
Post-tensioning	each		Includes the cost of cables, anchorage, fixing in place and jacking
10 m long		240	
20 m long		300	
Labour	h		Includes overheads. Wide variation can be expected, depending in part on how overheads are taken into account
Fitter or welder		4.50	
Machinist		5.00	
Sandcastings	kg		Cost depends on size, complexity and production quantity. These are approximate values for reasonably high production rates on medium size castings
Grey iron		0.33–0.55	
S.G. iron		0.55–0.75	
Steel		0.55–0.75	
Aluminium		1.10–1.30	
Bronze		2.20–2.60	
Utilities			
Steam, 0.7 MPa	tonne	1.2	
Water, unprocessed	m^3	0.005	
Water, city supply	m^3	0.050	
Electrical energy	kWh	0.015	Depends on total consumption
Road transport	tonne km	0.04–0.08	Depends on location and on load and distance
Rail transport	tonne km	0.04–0.08	

This table of approximate costs is given for use in student projects. Costs are subject to considerable variation with time and place.

to £35/m^2. These cost figures are further analysed on the basis of the clear width spanned by the structure and the type of construction. In this case, the cost range is considerably smaller than the range quoted above, and useful preliminary cost figures can be obtained. The cost figures are, of course, specific to Britain, and their use elsewhere would require careful consideration of differences in methods and in the costs of labour and materials. It is noted that the cost figures are adjusted to 1968 values: they can be brought up to date by the application of a suitable cost index which takes into account the variation of labour and material costs over the years (see, for example, *Chemical and Process Engineering*).

After the preparation of general-arrangement drawings, approximate estimates can be made using cost figures of the type shown in Table 25.2. The first requirement when using such unit cost figures is that the job is broken down into items and stages which conform to the data that are available. A typical example of such an estimate is given in Example 25.4; the data and methods are similar to those used in detailed estimates but the individual tasks are not broken down into such detail.

Example 25.4

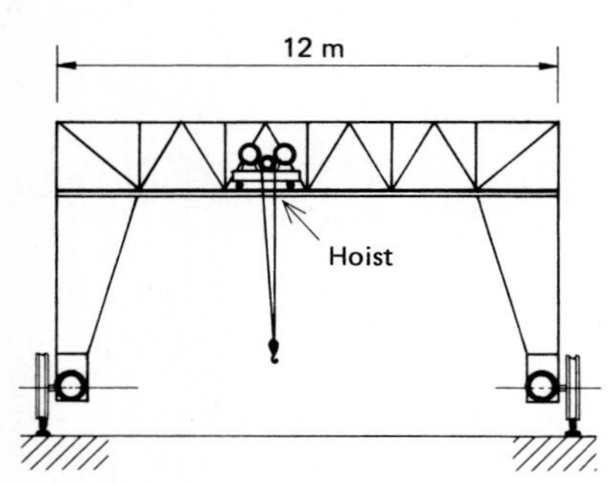

Estimate the cost of a 12 m rail-mounted gantry of the type shown in the sketch. From an existing design, the quantity of steel is estimated at 5000 kg. No site work – electrical or structural – needs be allowed for. In this type of structure, it is known that about 75 hours of labour per tonne of steel is required for fabrication.

The cost estimate is summarised in the table below, and notes are given on the makeup of each item. No allowance has been made for design costs, but on the assumption that some redesign of an existing gantry is required, 60 hours drawing-office time could be added. Some allowance could be made for labour "down-time" to account for cleaning-up and other short breaks. This would involve adding, say, 10% of the items with a large labour content – items (a), (b), (c), (d) and (h).

JOB DESCRIPTION: 12 m RAIL MOUNTED GANTRY

ENQUIRY No.................
CUSTOMER..................
DATE........................

SUMMARY OF COSTS

Item	Description	Costing details	Cost
(a)	Structure	Fabrication labour: allow 75 hours/tonne of steel, i.e. 375 hours at $4.50 per hour.	1680
(b)	Machine shop	Estimate machining time for wheels, drive pinions, axles, etc. as 75 hours. Charge at $5.50 per hour.	413
(c)	Fitting	Assembly of wheel drive: allow 12 hours at $4.50 per hour.	54
(d)	Electrical	Cost of electrical equipment, plus installation, estimated at $700.	700

SUMMARY OF COSTS (*continued*)

Item	Description	Costing details	Cost
(e)	Shop materials	Structural steel charged at 20c per kg.	1000
(f)	Bought out items	Wheel castings—4 x 45 kg. castings at 35c per kg (standard pattern available). 2 motors for wheel drive ("down shop drive") at $240. Bearings and other parts $100. Total $403.	403
(g)	Subcontract	Gearcutting for drive pinions and drive wheels. Subcontract $50.	50
(h)	Handling	Handling time for steelwork in shop. Allow 10% of fabrication time. Charge at $4.10 per hour.	154
(i)	Finishing	Painting charged at $18 per tonne. Standard paint job.	90
(j)	Transport	Transport to point of installation (local). Quote from transport operator: $80.	80
(k)	Craneage	Crane required for 4 hours for installation. $15 per hour hire charge.	60
(l)	Installation labour	Installation at site. 20 man-hours at $5 per hour.	100
(m)	Sundries	Travel and travelling time: $40. Test and certification of crane: $80.	120
(n)	Contingencies	To allow for contingencies, add 10% of items (a), (b), (c), (d) and (h).	300
(p)	Cost of hoist	Hoist bought out at $3500.	3500
		TOTAL	8704
	PROFIT	A mark-up of from 15% to 30%, depending on local conditions, may be made. Here take 25% on shop items, 15% on item (p).	1826
	ROYALTIES	NIL	
	FINAL ESTIMATED SELLING COST:		$10,530

The labour involved in using this technique on works of any magnitude is very great; another approach known as the factored estimate method[1] enables the total investment cost of a major scheme to be extrapolated from the delivered cost of the major items of equipment. This method was developed in the chemical engineering industry and can be used for large process plant.

In the first place, the estimated initial cost of the major items of processing equipment – boilers, pressure vessels, compressors and so on – is summarised to give the investment cost of the major equipment, I_E. On the basis of data

Table 25.3. Factors for estimating total plant cost.

Installed cost of process equipment	I_E
Experience factors as fraction of I_E	
Process piping	f_1
Solids processing	0.07–0.10
Mixed processing	0.10–0.30
Fluids processing	0.30–0.60
Instrumentation	f_2
Little automatic control	0.02–0.05
Some automatic control	0.05–0.10
Complex, centralized control	0.10–0.15
Manufacturing buildings	f_3
Outdoor construction	0.05–0.20
Indoor-outdoor construction	0.20–0.60
Indoor construction	0.60–1.00
Auxiliary facilities	f_4
Minor additions at existing site	0.00–0.05
Major additions at existing site	0.05–0.25
Complete at new site	0.25–1.00
Outside lines	f_5
Among existing facilities	0.00–0.05
Separate processing units	0.05–0.15
Scattered processing units	0.15–0.25
Total physical cost	$I_E\left(1 + \sum_i f_i\right)$
Experience factors as fraction of physical cost	
Engineering and construction	f_{I_1}
Straightforward engineering	0.20–0.35
Complex engineering	0.35–0.50
Size factor	f_{I_2}
Large commercial unit	0.00–0.05
Small commercial unit	0.05–0.15
Experimental unit	0.15–0.35
Contingencies	f_{I_3}
Firm process	0.10–0.20
Subject to change	0.20–0.30
Tentative process	0.30–0.50
Indirect costs factor $f_I = \sum_i f_{I_i}$	
Total plant cost $I_F = I_E\left(1 + \sum_i f_i\right) f_I$	

Source: Rudd and Watson, *Strategy of Process Engineering,* John Wiley, New York, 1968, reprinted by permission of John Wiley & Sons Inc.

accumulated from many cost studies, a relationship has been determined between I_E and the costs of other items – piping, instrumentation and other facilities – needed to complete the plant. Thus the direct plant cost becomes

$$I_E + \sum_i f_i I_E$$

where f_i are the experience factors for the additional items. A multiplying factor, f_I, is used to convert this direct cost to I_F, the fixed investment in the complete system. Thus

$$I_F = f_I\,(I_E + \sum_i f_i\, I_E)$$

where f_I accounts for indirect expenses such as engineering fees, contractor's costs and profit, etc. This method was initiated by Lang[7], and the factor relating I_F and I_E is called the *Lang factor*, f_L.

$$I_F = f_L \times I_E$$

where

$$f_L = f_I\,(1 + \Sigma f_i) \tag{25.3}$$

Table 25.3 gives values of the experience factors, f_i, that are relevant for processing systems. Other data of this type can be found in Happel's *Chemical Process Economics.*[8] The value of the Lang factor, f_L, is typically of the order of 3 or 4 for a process that has no unusual features; this is confirmed by the illustrative figures given in Table 25.4. The implication of this value should be noted – the finished plant will cost between three and four times the cost of the major items of equipment.

Table 25.4. Typical experience factors for fluids-processing systems.

Delivered costs of major equipment		I_E
Additional direct costs as fraction of I_E		
Labor for installing major equipment		0.10–0.20
Insulation		0.10–0.25
Piping (carbon steel)		0.50–1.00
Foundations		0.03–0.13
Buildings		0.07
Structures		0.05
Fireproofing		0.06–0.10
Electrical		0.07–0.15
Painting and clean-up		0.06–0.10
	$\sum f_i$	1.09–2.05
Total direct cost $(1 + \sum f_i)I_E$		
Indirect costs as fraction of direct costs		
Overhead, contractors costs, and profit		0.30
Engineering fee		0.13
Contingency		0.13
	$f_I = 1 + 0.56 = 1.56$	
Total cost $I_F = (1 + \sum f_i)f_I I_E = (3.1 \text{ to } 4.8)\, I_E$		

Source: Rudd and Watson, *Strategy of Process Engineering*, John Wiley, New York, 1968. (Adapted from J. Happel, *Chemical Process Economics*, Wiley, New York, 1958.) Reprinted by permission of John Wiley & Sons Inc.

25.4. COST—BENEFIT ANALYSIS

As suggested in the introduction to Section 25.2, some projects can only be evaluated by taking into account benefits which provide no direct economic return. Much government finance is channeled into work of this type; where capital is scarce and some investment criterion must be used to fix priorities, it is important that the indirect and intangible benefits, both financial and social, should be given due allowance. If this is not done, projects which have high intangible but few direct benefits will fare badly in the competition for scarce funds. This type of evaluation is known as cost-benefit analysis; it is a comparatively recent development which is finding increasing application in many fields.

Take as an example the development of a hydroelectric scheme. It will be seen that there are clear-cut direct costs and benefits that accrue directly to the sponsoring authority. The direct costs can be attributed to the construction of dams, generating sets and transmission lines; the benefits come from the sale of electrical energy and perhaps also of irrigation water. The indirect costs and benefits are usually more difficult to evaluate and may not result in any return to the authority. In this case the benefit to the surrounding areas, in terms of development in industry and agriculture and the additional recreational facilities provided by the large body of stored water, could be taken into account. Another aspect that should be considered is the improved flood control resulting from construction of the dam. It is likely that, in providing these indirect benefits, additional direct costs will be incurred. For example, if the recreational possibilities are to be fully developed, it will be necessary to clear all vegetation from the flooded land. In any case, the neglect of such work may trigger off the hostility of those interested in the preservation of natural resources.

It is clear that the estimation of some of the social benefits in money terms will require the cooperation of people in a whole range of disciplines and that the difficulties of ascribing values to some of the benefits will be very great. In spite of these difficulties and the necessity to rely in many cases on judgement, some economists feel that the method provides a rational basis for choice from a range of projects. For more detailed information on the methods used in this type of analysis the reader is referred to R. McKean and Prest and Turvey.[9, 10]

LIST OF SYMBOLS

f	fractional loss of value of I_F for tax purposes
f_i	experience factor
f_I	multiplying factor
f_L	Lang factor, defined by Equation (25.3)
GP	gross profit
i	interest rate
I	total investment
I_A	auxiliary investment
I_E	cost of major equipment
I_F	fixed investment
n	economic life
NP	net profit
R	revenue
S	amount in a fund
t	tax rate
θ	time

REFERENCES

[1] D. F. Rudd and C. C. Watson, *Strategy of Process Engineering*, John Wiley, New York, 1968.

[2] ICE, *An Introduction to Engineering Economics*, The Institution of Civil Engineers, London, 1969.

[3] I. Chem. E., *A Guide to Capital Cost Estimation*, The Institution of Chemical Engineers, London, 1969.

[4] D. G. Videan, "An Oil Company's Representative Costings for Exploration Support", *Shell Aviation News*, 388, 1970.

[5] S. Geddes, *Estimating for Builders, Public Works Contractors, Architects, Civil Engineers and Surveyors*, 4th ed., Newnes, London, 1966.

[6] Bertlin & Partners, *Port Structures*, vols 1 and 2, National Ports Council, London, 1969.

[7] H. J. Lang in *Chemical Engineering*, **54**, 10, Octobėr, 1947, 117–21.

[8] J. Happel, *Chemical Process Economics*, John Wiley, New York, 1958.

[9] R. N. McKean, *Efficiency in Government Through Systems Analysis*, John Wiley, New York, 1958.

[10] A. R. Prest, and R. Turvey, *Cost-Benefit Analysis: A Survey*, Macmillan, London, 1967.

PROBLEMS

25.1. Which of the two projects, A or B, is the most profitable? Is the profit adequate for a high-risk venture?

Project	A	B
Total investment I_F($)	49,000	95,000
Production costs ($/year)	45,000	21,000
Sales income ($/year)	36,000	43,000
Life of project (years)	7	10

Take $t = 0.5, f = 0.15$.

25.2. The major equipment for a dairy factory is expected to cost $89,000. Use the factored estimate method to estimate the upper and lower bounds for the total cost, taking data from Tables 25.3 and 25.4. Include labour, piping for mixed processing, instrumentation with little automatic control, indoor construction, electrical and paint work. Take the indirect cost factor, f_I, as 1.56.

25.3. A welded structure is required to take a steam pipeline over a road. A clear span of 15 m is required, and the pipeline weighs 3.5 kN/m. Estimate the cost of the complete finished structure. Make any assumptions you need regarding footings, finish, etc.

25.4. The planning for a process plant for a 10-year period indicates that additional steam supplies will be required. The alternatives are to upgrade an existing geothermal supply, at a cost of $55,000, or to purchase an oil fired plant for $32,000. Maintenance costs for the geothermal pipelines and plant are high – \$1,800 per annum, while those for the conventional supply are \$200 p.a. The annual running costs for the boiler (mainly fuel) are \$3,800 for the first year, increasing at 3% p.a. Select the most favourable alternative, on the basis of present values, if the company expects a 12% return on capital.

APPENDIXES

APPENDIX A1

Values of Various Units in Terms of SI Units

Although SI units are used in this book and are preferred in many countries, designers will sometimes be forced to use data given in other units. A short list of quantities is given below with the equivalent in SI units; the values have been rounded off to the fourth significant figure.

LENGTH		
	1 in	25.4 mm*
	1 ft	304.8 mm*
MASS		
	1 lb_m	0.4536 kg
	1 ton	1016 kg
	1 slug	14.59 kg
FORCE		
	1 lb_f	4.448 N
	1 ton	9.964 kN
DENSITY		
	1 lb_m/ft^3	16.02 kg/m^3
AREA		
	1 acre	0.4047 ha
STRESS		
	1 lb_f/in^2	6895 Pa
	1 lb_f/ft^2	47.88 Pa
	1 ton_f/in^2	15.44 MPa
ENERGY		
	1 ft lb_f	1.356 J
	1 BTU	1.055 kJ
	1 calorie	4.187 J
POWER		
	1 HP	745.7 W

*exact values

APPENDIX A2
Formulas for Shear Force, Bending Moment and Deflection of Beams

Bending moment and shear force diagrams for horizontal beams with a number of different loading conditions are drawn in Table A2.1. In each case, formulas are given for maximum values of shear force, bending moment and deflection.

This information has been taken from the *AISC Manual of Steel Construction*, 7th Edition, 1970, by permission of the American Institute of Steel Construction Inc.

The following nomenclature is used:

E	modulus of elasticity of steel
I	moment of inertia of beam
M_{max}	maximum moment
M_1	maximum moment in left section of beam
M_2	maximum moment in right section of beam
M_x	moment at distance x from end of beam
P	concentrated load
P_1	concentrated load nearest left reaction
P_2	concentrated load nearest right reaction, and of different magnitude than P_1
R	end beam reaction for any condition of symmetrical loading
R_1	left end beam reaction
R_2	right end or intermediate beam reaction
V	maximum vertical shear for any condition of symmetrical loading
V_1	maximum vertical shear in left section of beam
V_2	vertical shear at right reaction point, or to left of intermediate reaction point of beam
V_x	vertical shear at distance x from end of beam
W	total load on beam
a	measured distance along beam
b	measured distance along beam which may be greater or less than a
l	total length of beam between reaction points
w	uniformly distributed load per unit of length
w_1	uniformly distributed load per unit of length nearest left reaction
w_2	uniformly distributed load per unit of length nearest right reaction, and of different magnitude than w_1
x	any distance measured along beam from left reaction
x_1	any distance measured along overhang section of beam from nearest reaction point
Δ_{max}	maximum deflection
Δ_a	deflection at point of load
Δ_x	deflection at any point x distance from left reaction

Table A2.1. Beam diagrams and formulas for various static loading conditions.

1. SIMPLE BEAM—UNIFORMLY DISTRIBUTED LOAD

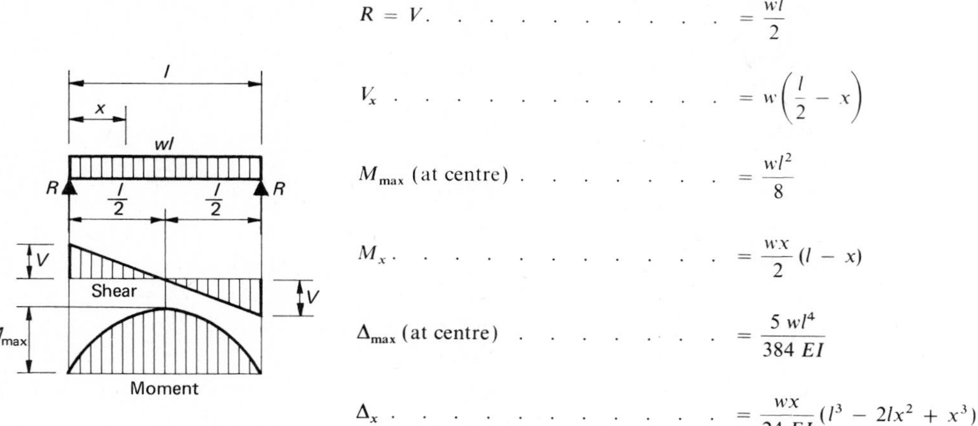

$R = V \ldots = \frac{wl}{2}$

$V_x \ldots = w\left(\frac{l}{2} - x\right)$

$M_{\max}$ (at centre) $\ldots = \frac{wl^2}{8}$

$M_x \ldots = \frac{wx}{2}(l - x)$

$\Delta_{\max}$ (at centre) $\ldots = \frac{5\,wl^4}{384\,EI}$

$\Delta_x \ldots = \frac{wx}{24\,EI}(l^3 - 2lx^2 + x^3)$

2. SIMPLE BEAM—LOAD INCREASING UNIFORMLY TO ONE END

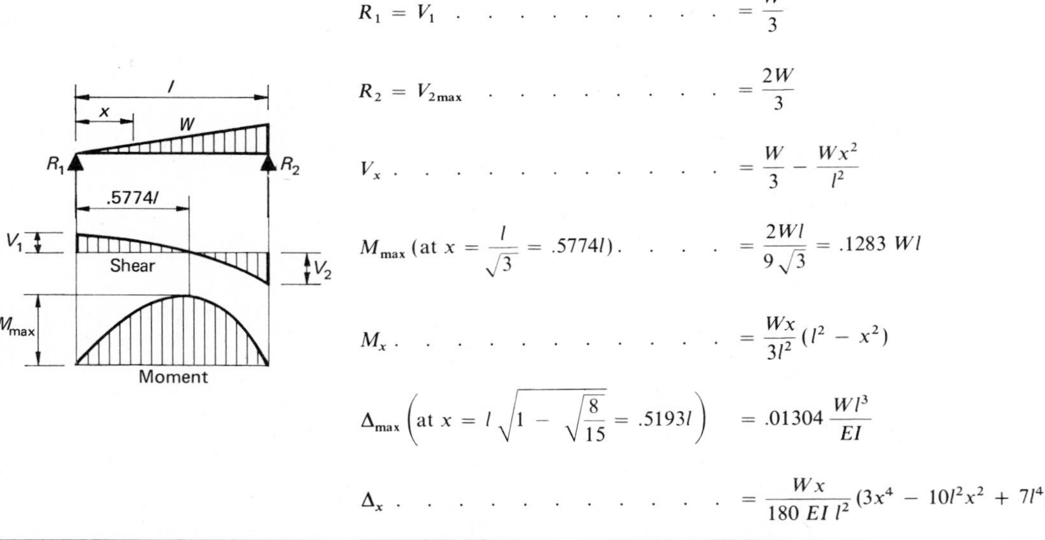

$R_1 = V_1 \ldots = \frac{W}{3}$

$R_2 = V_{2\max} \ldots = \frac{2W}{3}$

$V_x \ldots = \frac{W}{3} - \frac{Wx^2}{l^2}$

$M_{\max}$ $\left(\text{at } x = \frac{l}{\sqrt{3}} = .5774l\right) \ldots = \frac{2Wl}{9\sqrt{3}} = .1283\,Wl$

$M_x \ldots = \frac{Wx}{3l^2}(l^2 - x^2)$

$\Delta_{\max}$ $\left(\text{at } x = l\sqrt{1 - \sqrt{\frac{8}{15}}} = .5193l\right) = .01304\,\frac{Wl^3}{EI}$

$\Delta_x \ldots = \frac{Wx}{180\,EI\,l^2}(3x^4 - 10l^2x^2 + 7l^4)$

3. SIMPLE BEAM—LOAD INCREASING UNIFORMLY TO CENTRE

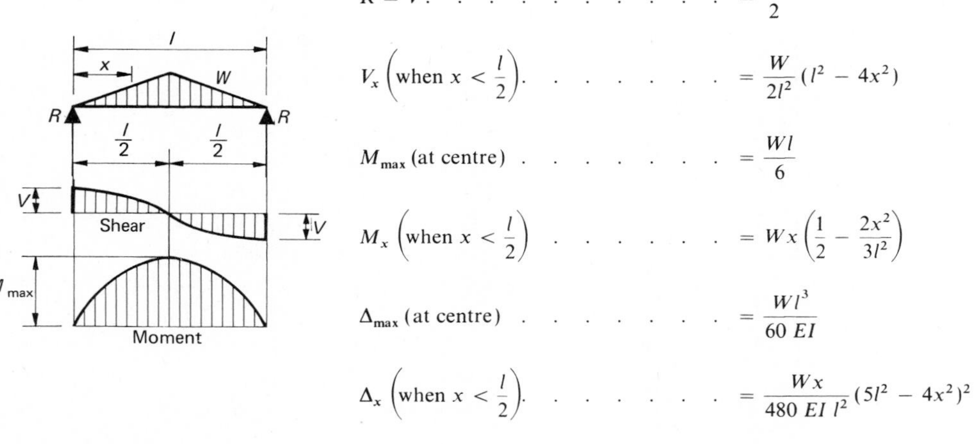

$R = V \ldots = \frac{W}{2}$

$V_x \left(\text{when } x < \frac{l}{2}\right) \ldots = \frac{W}{2l^2}(l^2 - 4x^2)$

$M_{\max}$ (at centre) $\ldots = \frac{Wl}{6}$

$M_x \left(\text{when } x < \frac{l}{2}\right) \ldots = Wx\left(\frac{1}{2} - \frac{2x^2}{3l^2}\right)$

$\Delta_{\max}$ (at centre) $\ldots = \frac{Wl^3}{60\,EI}$

$\Delta_x \left(\text{when } x < \frac{l}{2}\right) \ldots = \frac{Wx}{480\,EI\,l^2}(5l^2 - 4x^2)^2$

Table A2.1. Beam diagrams and formulas for various static loading conditions.

4. SIMPLE BEAM—UNIFORM LOAD PARTIALLY DISTRIBUTED

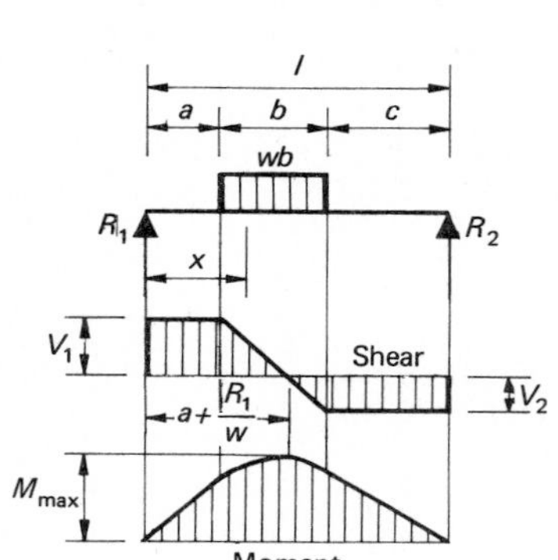

$R_1 = V_1$ (max when $a < c$) . . . . $= \dfrac{wb}{2l}(2c + b)$

$R_2 = V_2$ (max when $a > c$) . . . . $= \dfrac{wb}{2l}(2a + b)$

V_x (when $x > a$ and $< (a + b)$) . . $= R_1 - w(x - a)$

$M_{max}\left(\text{at } x = a + \dfrac{R_1}{w}\right)$ . . . . . $= R_1\left(a + \dfrac{R_1}{2w}\right)$

M_x (when $x < a$) . . . . . . $= R_1 x$

M_x (when $x > a$ and $< (a + b)$) . . $= R_1 x - \dfrac{w}{2}(x - a)^2$

M_x (when $x > (a + b)$). . . . . $= R_2(l - x)$

5. SIMPLE BEAM—CONCENTRATED LOAD AT ANY POINT

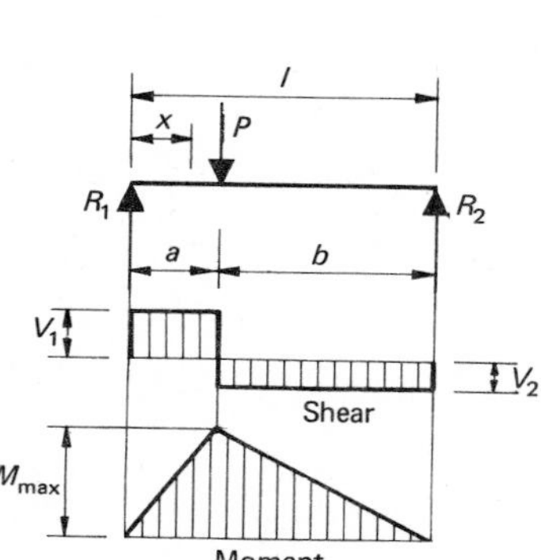

$R_1 = V_1$ (max when $a < b$). . . . $= \dfrac{Pb}{l}$

$R_2 = V_2$ (max when $a > b$). . . . $= \dfrac{Pa}{l}$

M_{max} (at point of load) . . . . . $= \dfrac{Pab}{l}$

M_x (when $x < a$). . . . . . . $= \dfrac{Pbx}{l}$

$\Delta_{max}\left(\text{at } x = \sqrt{\dfrac{a(a + 2b)}{3}} \text{ when } a > b\right) = \dfrac{Pab(a + 2b)\sqrt{3a(a + 2b)}}{27\,EI\,l}$

Δ_a (at point of load) . . . . . . $= \dfrac{Pa^2b^2}{3\,EI\,l}$

Δ_x (when $x < a$) . . . . $= \dfrac{Pbx}{6\,EI\,l}(l^2 - b^2 - x^2)$

6. BEAM FIXED AT ONE END, SUPPORTED AT OTHER—UNIFORMLY DISTRIBUTED LOAD

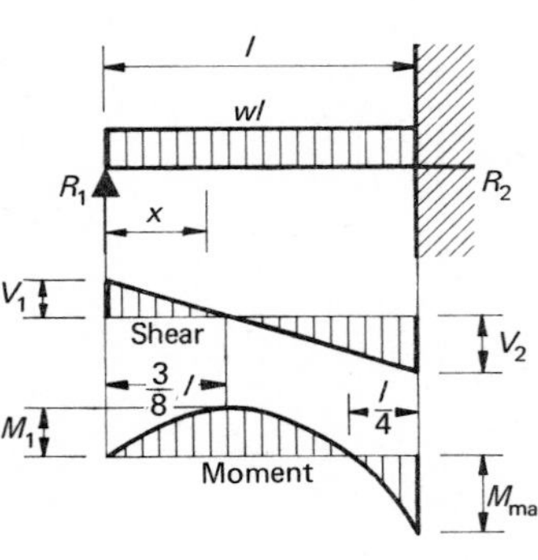

$R_1 = V_1$ . . . . $= \dfrac{3wl}{8}$

$R_2 = V_{2\,max}$ . . . . $= \dfrac{5wl}{8}$

V_x . . . . $= R_1 - wx$

M_{max} . . . . $= \dfrac{wl^2}{8}$

$M_1\left(\text{at } x = \dfrac{3}{8}l\right)$ . . . . $= \dfrac{9}{128}wl^2$

M_x . . . . $= R_1 x - \dfrac{wx^2}{2}$

$\Delta_{max}\left(\text{at } x = \dfrac{l}{16}(1 + \sqrt{33}) = .4215l\right)$. $= \dfrac{wl^4}{185\,EI}$

Δ_x . . . . $= \dfrac{wx}{48\,EI}(l^3 - 3lx^2 + 2x^3)$

Table A2.1. Beam diagrams and formulas for various static loading conditions.

7. BEAM FIXED AT ONE END, SUPPORTED AT OTHER—
CONCENTRATED LOAD AT ANY POINT

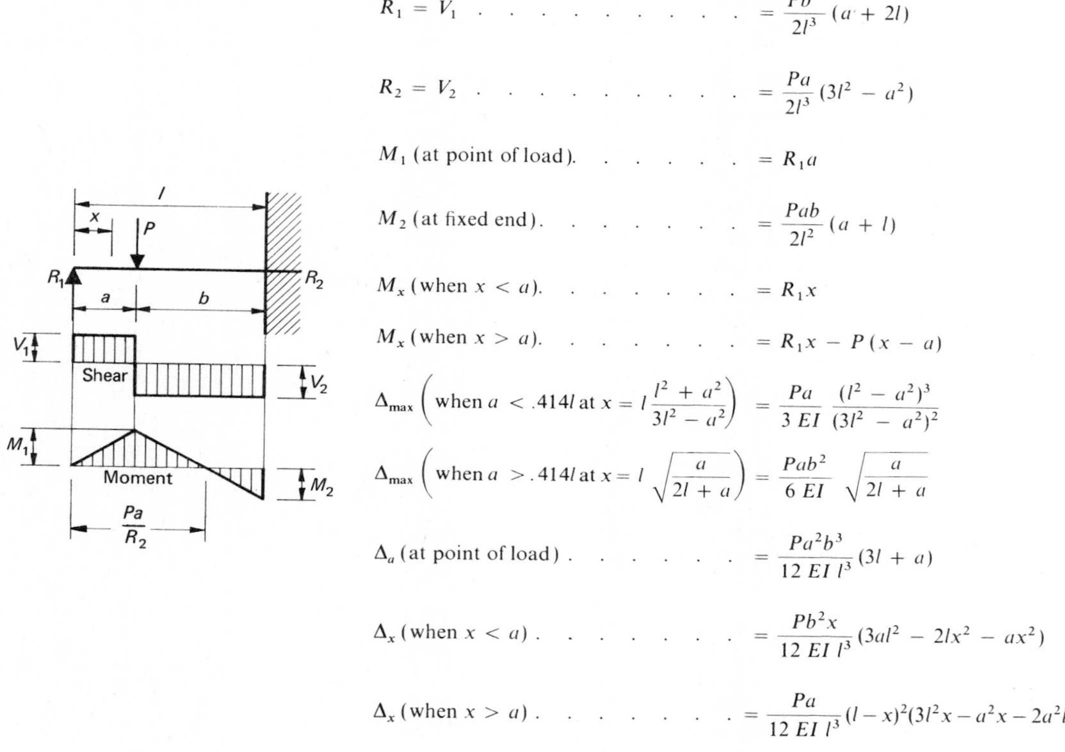

$R_1 = V_1$ . . . . $= \frac{Pb^2}{2l^3}(a + 2l)$

$R_2 = V_2$ . . . . $= \frac{Pa}{2l^3}(3l^2 - a^2)$

M_1 (at point of load). . . . . . . $= R_1 a$

M_2 (at fixed end). . . . . $= \frac{Pab}{2l^2}(a + l)$

M_x (when $x < a$). . . . . $= R_1 x$

M_x (when $x > a$). . . . . $= R_1 x - P(x - a)$

$\Delta_{max}\left(\text{when } a < .414l \text{ at } x = l\frac{l^2 + a^2}{3l^2 - a^2}\right) = \frac{Pa}{3\,EI}\frac{(l^2 - a^2)^3}{(3l^2 - a^2)^2}$

$\Delta_{max}\left(\text{when } a > .414l \text{ at } x = l\sqrt{\frac{a}{2l + a}}\right) = \frac{Pab^2}{6\,EI}\sqrt{\frac{a}{2l + a}}$

Δ_a (at point of load) . . . . $= \frac{Pa^2b^3}{12\,EI\,l^3}(3l + a)$

Δ_x (when $x < a$) . . . . $= \frac{Pb^2x}{12\,EI\,l^3}(3al^2 - 2lx^2 - ax^2)$

Δ_x (when $x > a$) . . . . $= \frac{Pa}{12\,EI\,l^3}(l - x)^2(3l^2x - a^2x - 2a^2l)$

8. BEAM FIXED AT BOTH ENDS—UNIFORMLY DISTRIBUTED
LOADS

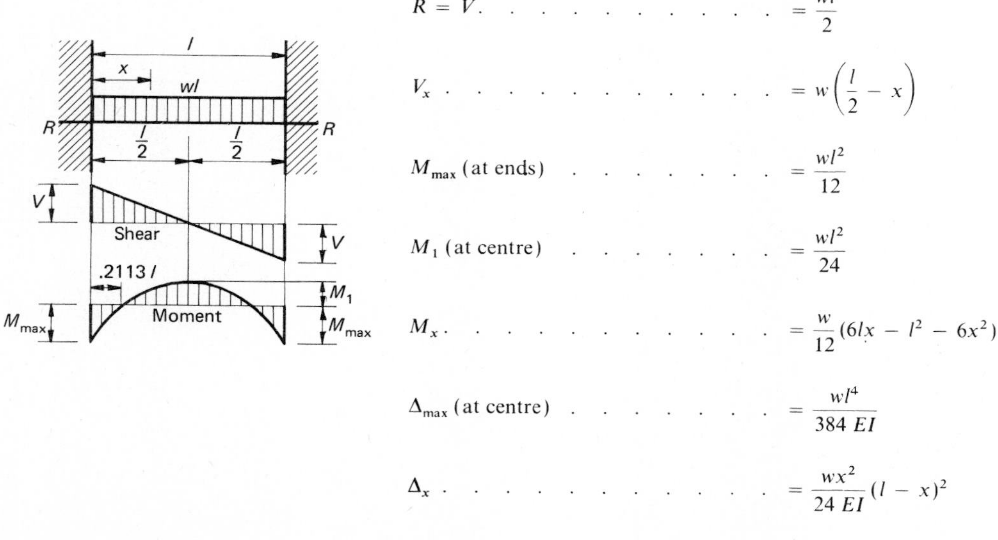

$R = V$. . . . . $= \frac{wl}{2}$

V_x . . . . $= w\left(\frac{l}{2} - x\right)$

M_{max} (at ends) . . . . $= \frac{wl^2}{12}$

M_1 (at centre) . . . . $= \frac{wl^2}{24}$

M_x . . . . $= \frac{w}{12}(6lx - l^2 - 6x^2)$

Δ_{max} (at centre) . . . . $= \frac{wl^4}{384\,EI}$

Δ_x . . . . $= \frac{wx^2}{24\,EI}(l - x)^2$

Table A2.1. Beam diagrams and formulas for various static loading conditions.

9. BEAM FIXED AT BOTH ENDS—CONCENTRATED LOAD AT ANY POINT

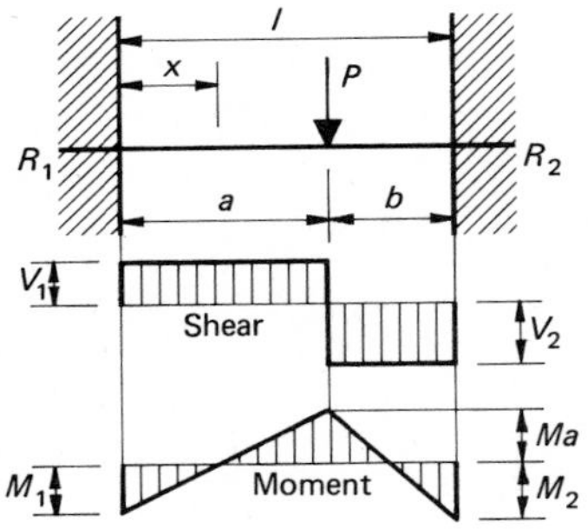

$R_1 = V_1$ (max when $a < b$) . . . . $= \dfrac{Pb^2}{l^3}(3a + b)$

$R_2 = V_2$ (max when $a > b$) . . . . $= \dfrac{Pa^2}{l^3}(a + 3b)$

M_1 (max when $a < b$) . . . . . . $= \dfrac{Pab^2}{l^2}$

M_2 (max when $a > b$) . . . . . . $= \dfrac{Pa^2b}{l^2}$

M_a (at point of load) . . . . . . $= \dfrac{2Pa^2b^2}{l^3}$

M_x (when $x < a$) . . . . $= R_1x - \dfrac{Pab^2}{l^2}$

$\Delta_{max}\left(\text{when } a > b \text{ at } x = \dfrac{2al}{3a + b}\right)$ . . . $= \dfrac{2Pa^3b^2}{3\,EI\,(3a + b)^2}$

Δ_a (at point of load) . . . . . . $= \dfrac{Pa^3b^3}{3\,EI\,l^3}$

Δ_x (when $x < a$) . . . . $= \dfrac{Pb^2x^2}{6\,EI\,l^3}(3al - 3ax - bx)$

10. CANTILEVER BEAM—LOAD INCREASING UNIFORMLY TO FIXED END

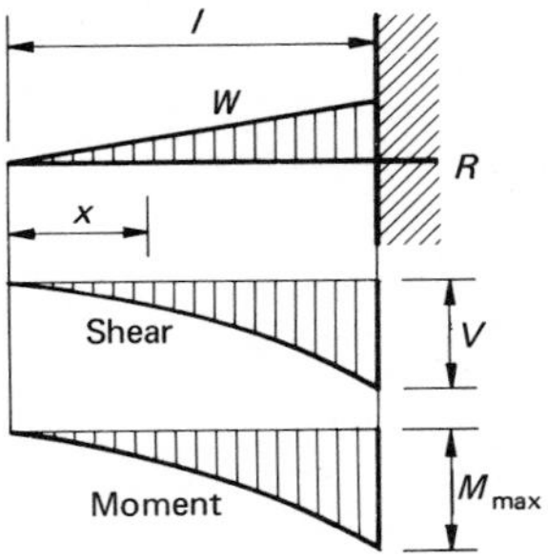

$R = V$ . . . . $= W$

V_x . . . . $= W\dfrac{x^2}{l^2}$

M_{max} (at fixed end) . . . . $= \dfrac{Wl}{3}$

M_x . . . . $= \dfrac{Wx^3}{3l^2}$

Δ_{max} (at free end) . . . . $= \dfrac{Wl^3}{15\,EI}$

Δ_x . . . . $= \dfrac{W}{60\,EI\,l^2}(x^5 - 5l^4x + 4l^5)$

11. CANTILEVER BEAM—UNIFORMLY DISTRIBUTED LOAD

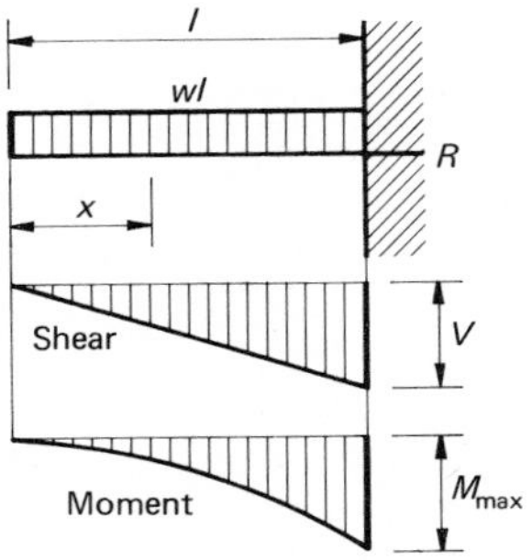

$R = V$ . . . . $= wl$

V_x . . . . $= wx$

M_{max} (at fixed end) . . . . $= \dfrac{wl^2}{2}$

M_x . . . . $= \dfrac{wx^2}{2}$

Δ_{max} (at free end) . . . . $= \dfrac{wl^4}{8\,EI}$

Δ_x . . . . $= \dfrac{w}{24\,EI}(x^4 - 4l^3x + 3l^4)$

Table A2.1. Beam diagrams and formulas for various static loading conditions.

12. BEAM FIXED AT ONE END, FREE TO DEFLECT VERTICALLY BUT NOT ROTATE AT OTHER—UNIFORMLY DISTRIBUTED LOAD

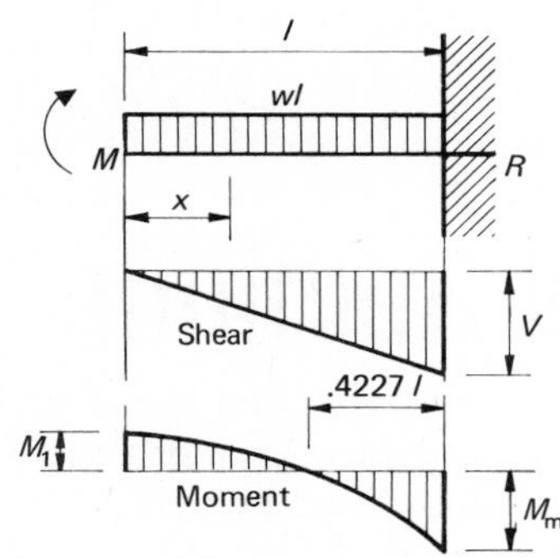

$R = V$ . . . . $= wl$

V_x . . . . $= wx$

M_{max} (at fixed end) . . . . $= \dfrac{wl^2}{3}$

M_1 (at deflected end) . . . . . . $= \dfrac{wl^2}{6}$

M_x . . . . $= \dfrac{w}{6}(l^2 - 3x^2)$

Δ_{max} (at deflected end) . . . . . . $= \dfrac{wl^4}{24\,EI}$

Δ_x . . . . $= \dfrac{w(l^2 - x^2)^2}{24\,EI}$

13. CANTILEVER BEAM—CONCENTRATED LOAD AT ANY POINT

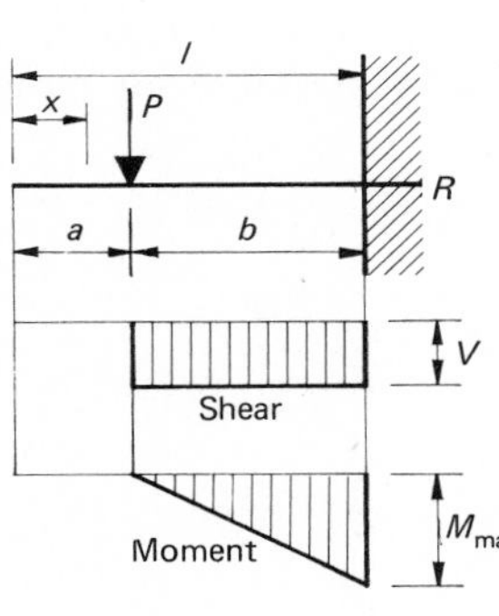

$R = V$ . . . . $= P$

M_{max} (at fixed end) . . . . $= Pb$

M_x (when $x > a$). . . . . $= P(x - a)$

Δ_{max} (at free end). . . . . $= \dfrac{Pb^2}{6\,EI}(3l - b)$

Δ_a (at point of load) . . . . $= \dfrac{Pb^3}{3\,EI}$

Δ_x (when $x < a$) . . . . $= \dfrac{Pb^2}{6\,EI}(3l - 3x - b)$

Δ_x (when $x > a$) . . . . $= \dfrac{P(l - x)^2}{6\,EI}(3b - l + x)$

14. BEAM FIXED AT ONE END, FREE TO DEFLECT VERTICALLY BUT NOT ROTATE AT OTHER—CONCENTRATED LOAD AT DEFLECTED END

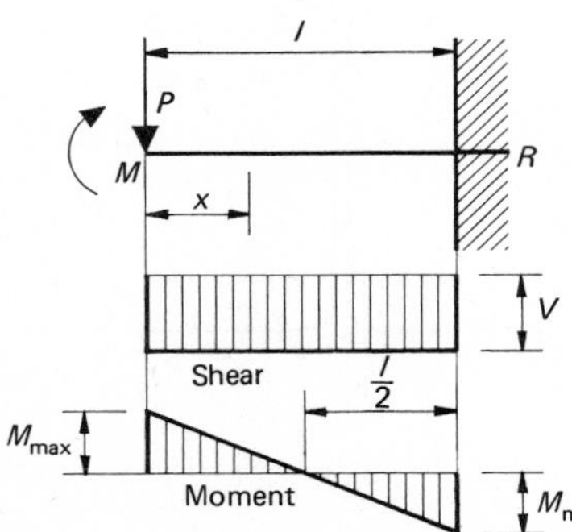

$R = V$ . . . . $= P$

M_{max} (at both ends) . . . . $= \dfrac{Pl}{2}$

M_x . . . . $= P\left(\dfrac{l}{2} - x\right)$

Δ_{max} (at deflected end) . . . . . . $= \dfrac{Pl^3}{12\,EI}$

Δ_x . . . . $= \dfrac{P(l - x)^2}{12\,EI}(l + 2x)$

APPENDIX A3

Properties of Steel Sections

The following tables give the properties of a selection of structural steel sections. These properties and accompanying illustrations are extracts from BS 4: Part 1: 1972, *Specification for Structural Steel Sections*, reproduced by permission of the British Standards Institution, 2 Park Street, London W1A 2BS.

At the present time (1974) "hard conversion" of steel sections has not taken place – that is, sections are still rolled to dimensions in Imperial units. What is presented here is known as a "soft conversion" – dimensions and properties are presented in metric units. The first two columns in the following table are headed "Designation" and this information can be regarded as a description for identifying sections. Alternative designations are given, based on significant dimensions in Imperial units and in metric units.

In practice, the first two columns contain information required for purchase. The remaining columns contain information required for design calculations in SI units.

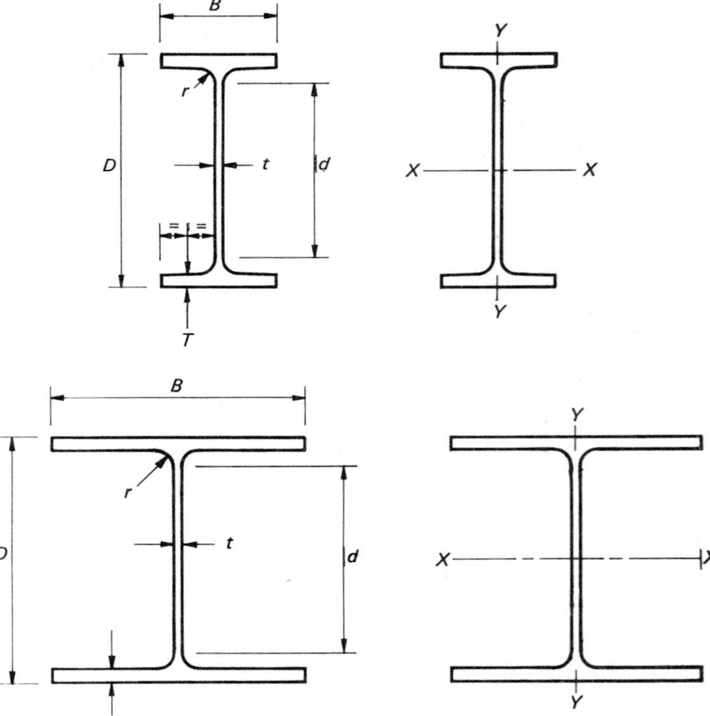

Fig. A3.1. *Definition of symbols, universal beams for Table A3.1.*

Fig. A3.2. *Definition of symbols, universal columns for Table A3.2.*

Table A3.1. Properties of universal beams.

DESIGNATION				Depth of section D	Width of section B	Thickness		Root radius r	Depth between fillets d	Area of section	Moment of inertia		Radius of gyration		Elastic modulus	
Serial size		Mass/unit length				Web t	Flange T				About X–X	About Y–Y	About X–X	About Y–Y	About X–X	About Y–Y
mm	(in)	kg/m	(lb/ft)	mm	mm	mm	mm	mm	mm	cm^2	cm^4	cm^4	cm	cm	cm^3	cm^3
533 × 330	(21 × 13)	212	(142)	**545.1**	**333.6**	16.7	27.8	16.5	450.1	269.6	141 682	16 064	22.9	7.72	5199	963.2
533 × 210	(21 × 8¼)	109	(73)	**539.5**	**210.7**	11.6	18.8	12.7	472.7	138.4	66 610	2 755	21.9	4.46	2469	261.5
		92	(62)	**533.1**	**209.3**	10.2	15.6	12.7	472.7	117.6	55 225	2 212	21.7	4.34	2072	211.3
457 × 191	(18 × 7½)	82	(55)	**460.2**	**191.3**	9.9	16.0	10.2	404.4	104.4	37 039	1 746	18.8	4.09	1610	182.6
457 × 152	(18 × 6)	74	(50)	**461.3**	**152.7**	9.9	17.0	10.2	404.4	94.9	32 380	963	18.5	3.18	1404	126.1
		67	(45)	**457.2**	**151.9**	9.1	15.0	10.2	404.4	85.3	28 522	829	18.3	3.12	1248	109.1
406 × 178	(16 × 7)	67	(45)	**409.4**	**178.8**	8.8	14.3	10.2	357.4	85.4	24 279	1 269	16.9	3.85	1186	141.9
		54	(36)	**402.6**	**177.6**	7.6	10.9	10.2	357.4	68.3	18 576	922	16.5	3.67	922.8	103.8
406 × 152	(16 × 6)	67	(45)	**412.2**	**152.9**	9.3	16.0	10.2	357.4	85.3	23 798	908	16.7	3.26	1155	118.8
381 × 152	(15 × 6)	60	(40)	**384.8**	**153.4**	8.7	14.4	10.2	333.2	75.9	18 632	814	15.7	3.27	968.4	106.2
356 × 171	(14 × 6¾)	57	(38)	**358.6**	**172.1**	8.0	13.0	10.2	309.1	72.1	16 038	1 026	14.9	3.77	894.3	119.2
		45	(30)	**352.0**	**171.0**	6.9	9.7	10.2	309.1	56.9	12 052	730	14.6	3.58	684.7	85.4
305 × 165	(12 × 6½)	40	(27)	**303.8**	**165.1**	6.1	10.2	8.9	262.6	51.4	8 500	691	12.9	3.67	559.6	83.71
305 × 127	(12 × 5)	37	(25)	**303.8**	**123.5**	7.2	10.7	8.9	262.6	47.4	7 143	316	12.3	2.58	470.3	51.11
254 × 146	(10 × 5¾)	31	(21)	**251.5**	**146.1**	6.1	8.6	7.6	216.2	39.9	4 427	406	10.5	3.19	352.1	55.53
203 × 133	(8 × 5¼)	25	(17)	**203.2**	**133.4**	5.8	7.8	7.6	169.9	32.3	2 348	280	8.53	2.94	231.1	41.92

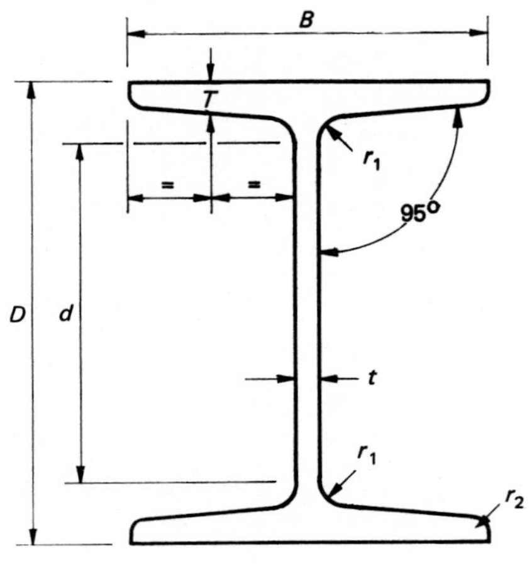

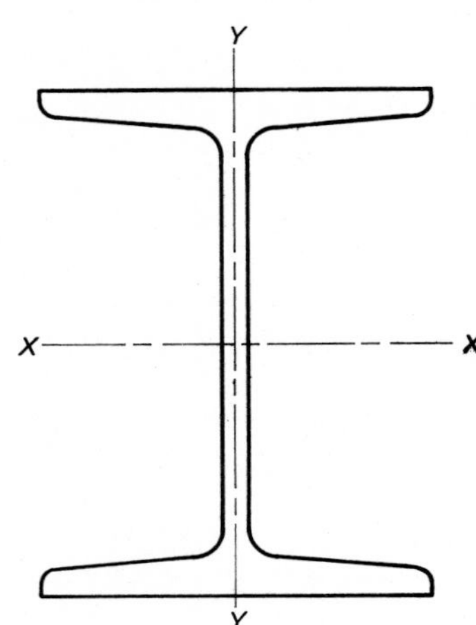

Fig. A3.3. *Definition of symbols, joists with 5° taper flanges for Table A3.3.*

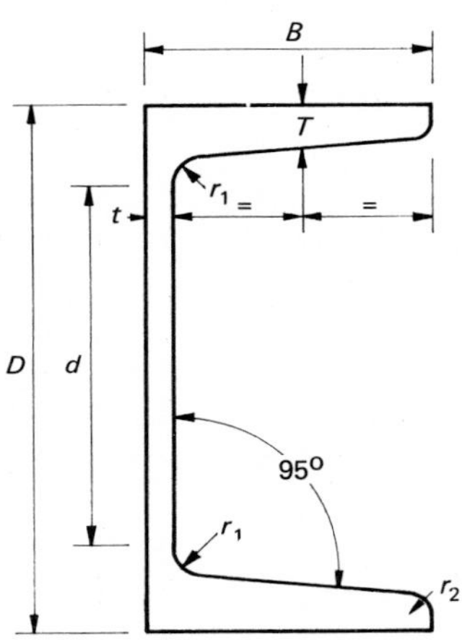

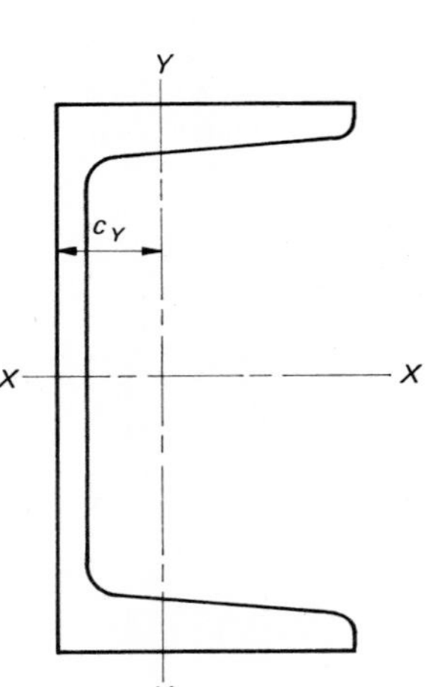

Fig. A3.4. *Definition of symbols, channels for Table A3.4.*

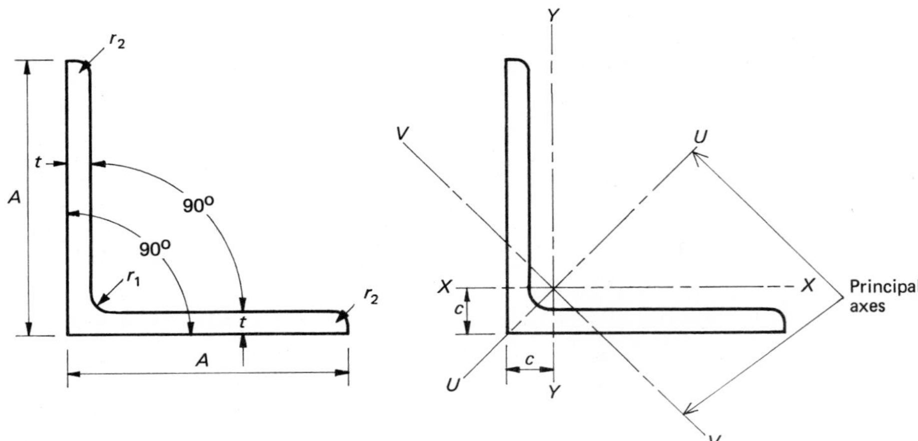

Fig. A3.5. *Definition of symbols, equal angles for Table A3.5.*

Table A3.2. Properties of universal columns.

DESIGNATION				Depth of section	Width of section	Thickness		Root radius	Depth between fillets	Area of section	Moment of inertia		Radius of gyration		Elastic modulus	
Serial size		Mass/unit length		D	B	Web t	Flange T	r	d		About X–X	About Y–Y	About X–X	About Y–Y	About X–X	About Y–Y
mm	(in)	kg/m	(lb/ft)	mm	mm	mm	mm	mm	mm	cm^2	cm^4	cm^4	cm	cm	cm^3	cm^3
305 × 305	(12 × 12)	97	(65)	**307.8**	**304.8**	9.9	15.4	15.2	246.6	123.3	22 202	7268	13.4	7.68	1442	476.9
254 × 254	(10 × 10)	89	(60)	**260.4**	**255.9**	10.5	17.3	12.7	200.2	114.0	14 307	4849	11.2	6.52	1099	378.9
		73	(49)	**254.0**	**254.0**	8.6	14.2	12.7	200.2	92.9	11 360	3873	11.1	6.46	894.5	305.0
203 × 203	(8 × 8)	60	(40)	**209.6**	**205.2**	9.3	14.2	10.2	160.8	75.8	6 088	2041	8.96	5.19	581.1	199.0
		46	(31)	**203.2**	**203.2**	7.3	11.0	10.2	160.8	58.8	4 564	1539	8.81	5.11	449.2	151.5
152 × 152	(6 × 6)	30	(20)	**157.5**	**152.9**	6.6	9.4	7.6	123.4	38.2	1 742	558	6.75	3.82	221.2	73.06
		23	(16.0)	**152.4**	**152.4**	6.1	6.8	7.6	123.4	29.8	1 263	403	6.51	3.68	165.7	52.95

Table A3.3. Properties of joists with 5° taper flanges.

DESIGNATION				Depth of section	Width of section	Thickness		Radii		Depth between fillets	Area of section	Moment of inertia		Radius of gyration		Elastic modulus	
Nominal size		Mass/unit length		D	B	Web t	Flange T	Root r_2	Toe r_2	d		About X–X	About Y–Y	About X–X	About Y–Y	About X–X	About Y–Y
mm	(in)	kg/m	(lb/ft)	mm	mm	mm	mm	mm	mm	mm	cm^2	cm^4	cm^4	cm	cm	cm^3	cm^3
203 × 102	(8 × 4)	25.33	(17.00)	**203.2**	**101.6**	5.8	10.4	9.4	3.2	161.0	32.26	2294	162.6	8.43	2.25	225.8	32.02
178 × 102	(7 × 4)	21.54	(14.50)	**177.8**	**101.6**	5.3	9.0	9.4	3.2	138.2	27.44	1519	139.2	7.44	2.25	170.9	27.41
152 × 89	(6 × 3½)	17.09	(11.50)	**152.4**	**88.9**	4.9	8.3	7.9	2.4	117.9	21.77	881.1	85.98	6.36	1.99	115.6	19.34
127 × 76	(5 × 3)	13.36	(9.00)	**127.0**	**76.2**	4.5	7.6	7.9	2.4	94.2	17.02	475.9	50.18	5.29	1.72	74.94	13.17
102 × 64	(4 × 2½)	9.65	(6.50)	**101.6**	**63.5**	4.1	6.6	6.9	2.4	73.2	12.29	217.6	25.30	4.21	1.43	42.84	7.97
76 × 51	(3 × 2)	6.67	(4.50)	**76.2**	**50.8**	3.8	5.6	6.9	2.4	50.3	8.49	82.58	11.11	3.12	1.14	21.67	4.37

Table A3.4. Properties of channels

DESIGNATION				Depth of section D	Width of section B	Thickness		Radius		Depth between fillets d	Area of section	Distance of centre of gravity c_y	Moment of inertia		Radius of gyration		Elastic modulus	
Nominal size		Mass/unit length				Web t	Flange T	Root r_1	Toe r_2				About X–X	About Y–Y	About X–X	About Y–Y	About X–X	About Y–Y
mm	(in)	kg/m	(lb/ft)	mm	mm	mm	mm	mm	mm	mm	cm^2	cm	cm^4	cm^4	cm^2	cm^2	cm^3	cm^3
381 × 102	(15 × 4)	55.10	(37.0)	**381.0**	**101.6**	10.4	16.3	15.2	4.8	312.4	70.19	2.52	14 894	579.8	14.6	2.87	781.8	75.87
305 × 89	(12 × 3½)	41.69	(28.0)	**304.8**	**88.9**	10.2	13.7	13.7	3.2	245.4	53.11	2.18	7 061	325.4	11.5	2.47	463.3	48.49
254 × 89	(10 × 3½)	35.74	(24.0)	**254.0**	**88.9**	9.1	13.6	13.7	3.2	194.8	45.52	2.42	4 448	302.4	9.88	2.58	350.2	46.71
229 × 76	(9 × 3)	26.06	(17.5)	**228.6**	**76.2**	7.6	11.2	12.2	3.2	178.1	33.20	2.00	2 610	158.7	8.87	2.19	228.3	28.22
203 × 76	(8 × 3)	23.82	(16.0)	**203.2**	**76.2**	7.1	11.2	12.2	3.2	152.4	30.34	2.13	1 950	151.4	8.02	2.23	192.0	27.59
178 × 76	(7 × 3)	20.84	(14.0)	**177.8**	**76.2**	6.6	10.3	12.2	3.2	128.8	26.54	2.20	1 337	134.0	7.10	2.25	150.4	24.73
152 × 76	(6 × 3)	17.88	(12.0)	**152.4**	**76.2**	6.4	9.0	12.2	2.4	105.9	22.77	2.21	851.6	113.8	6.11	2.24	111.8	21.05
127 × 64	(5 × 2½)	14.90	(10.0)	**127.0**	**63.5**	6.4	9.2	10.7	2.4	84.1	18.98	1.94	482.6	67.24	5.04	1.88	75.99	15.25
102 × 51	(4 × 2)	10.42	(7.0)	**101.6**	**50.8**	6.1	7.6	9.1	2.4	65.8	13.28	1.51	207.7	29.10	3.96	1.48	40.89	8.16
76 × 38	(3 × 1½)	6.70	(4.5)	**76.2**	**38.1**	5.1	6.8	7.6	2.4	45.7	8.53	1.19	74.14	10.66	2.95	1.12	19.46	4.07

Table A3.5. Properties of equal angles.

DESIGNATION				Leg lengths	Thick-ness	Radius		Mass/unit length		Area of section	Distance of centre of gravity	Moment of inertia			Radius of gyration			Elastic modulus
Nominal size		Nominal thickness		A	t	Root r_1	Toe r_2				c	About X–X, Y–Y	About U–U	About V–V	About X–X, Y–Y	About U–U	About V–V	About X–X, Y–Y
mm	(in)	mm	(in)	mm	mm	mm	mm	kg/m	(lb/ft)	cm^2	cm	cm^4	cm^4	cm^4	cm	cm	cm	cm^3
		13	($\frac{1}{2}$)		12.6			29.07	(19.52)	37.03	4.24	819	1303	335	4.70	5.93	3.01	74.5
152 × 152	(6 × 6)	11	($\frac{7}{16}$)	**152.4**	11.0	12.2	4.8	25.60	(17.18)	32.61	4.18	727	1156	297	4.72	5.96	3.02	65.7
		9	($\frac{3}{8}$)		9.4			22.02	(14.79)	28.06	4.11	631	1003	258	4.74	5.98	3.03	56.7
127 × 127	(5 × 5)	10	($\frac{3}{8}$)	**127.0**	9.5	10.7	4.8	18.30	(12.29)	23.31	3.49	359	571	147	3.93	4.95	2.51	39.0
		13	($\frac{1}{2}$)		12.6			18.91	(12.70)	24.09	2.98	228	361	94.3	3.07	3.87	1.98	31.7
102 × 102	(4 × 4)	11	($\frac{7}{16}$)	**101.6**	11.0	9.1	4.8	16.69	(11.21)	21.27	2.92	203	323	83.8	3.09	3.90	1.99	28.1
		9	($\frac{3}{8}$)		9.4			14.44	(9.69)	18.39	2.86	178	283	73.1	3.11	3.92	1.99	24.4
89 × 89	($3\frac{1}{2}$ × $3\frac{1}{2}$)	8	($\frac{5}{16}$)	**88.9**	7.9	8.4	4.8	10.58	(7.10)	13.47	2.48	99.8	159	41.0	2.72	3.43	1.74	15.6
		13	($\frac{1}{2}$)		12.6			13.85	(9.30)	17.64	2.35	90.4	143	38.2	2.26	2.84	1.47	17.1
		11	($\frac{7}{16}$)		11.0			12.20	(8.19)	15.55	2.29	80.9	128	33.8	2.28	2.87	1.47	15.2
76 × 76	(3 × 3)	9	($\frac{3}{8}$)	**76.2**	9.4	7.6	4.8	10.57	(7.10)	13.47	2.23	71.1	113	29.5	2.30	2.89	1.48	13.2
		8	($\frac{5}{16}$)		7.8			8.93	(5.99)	11.37	2.16	60.9	96.8	25.1	2.31	2.92	1.48	11.2
		6	($\frac{1}{4}$)		6.2			7.16	(4.81)	9.12	2.10	49.6	78.8	20.3	2.33	2.94	1.49	8.97
		9	($\frac{3}{8}$)		9.4			8.78	(5.89)	11.18	1.92	40.5	64.0	17.0	1.90	2.39	1.23	9.15
64 × 64	($2\frac{1}{2}$ × $2\frac{1}{2}$)	8	($\frac{5}{16}$)	**63.5**	7.9	6.9	2.4	7.45	(5.00)	9.48	1.86	35.0	55.5	14.6	1.92	2.42	1.24	7.80
		6	($\frac{1}{4}$)		6.2			5.96	(4.00)	7.59	1.80	28.6	45.4	11.8	1.94	2.45	1.25	6.28
		9	($\frac{3}{8}$)		9.4			6.85	(4.60)	8.72	1.60	19.6	30.8	8.42	1.50	1.88	0.98	5.64
51 × 51	(2 × 2)	8	($\frac{5}{16}$)	**50.8**	7.8	6.1	2.4	5.80	(3.90)	7.39	1.54	17.0	26.8	7.17	1.52	1.91	0.98	4.81
		6	($\frac{1}{4}$)		6.3			4.77	(3.20)	6.08	1.49	14.3	22.7	5.95	1.53	1.93	0.99	3.98
		5	($\frac{3}{16}$)		4.6			3.58	(2.40)	4.56	1.42	11.0	17.4	4.54	1.55	1.95	1.00	3.00

Table A3.6. Properties of unequal angles.

DESIGNATION Nominal size (mm)	(in)	Nominal thickness (mm)	(in)	Leg lengths A	Leg lengths B	Thickness t	Radius Root r_1	Radius Toe r_2	Mass/unit length		Area of section	Distances of centre of gravity c_x	c_y	Moment of inertia About X–X	About Y–Y	About U–U	About V–V	Radius of gyration About X–X	About Y–Y	About U–U	About V–V	Angle α	Elastic modulus About X–X	About Y–Y
mm	(in)	mm	(in)	mm	mm	mm	mm	mm	kg/m	(lb/ft)	cm²	cm	cm	cm⁴	cm⁴	cm⁴	cm⁴	cm	cm	cm	cm	tan α	cm³	cm³
152 × 102	(6 × 4)	13	($\frac{1}{2}$)	**152.4**	**101.6**	12.6	10.7	4.8	23.99	(16.10)	30.56	5.00	2.48	716	257	825	148	4.84	2.90	5.20	2.20	0.437	70.0	33.4
		11	($\frac{7}{16}$)			11.0			21.14	(14.19)	26.93	4.94	2.42	637	229	734	132	4.86	2.92	5.22	2.21	0.439	61.8	29.6
		9	($\frac{3}{8}$)			9.5			18.30	(12.29)	23.31	4.88	2.36	555	201	641	115	4.88	2.93	5.24	2.22	0.441	53.6	25.7
152 × 76	(6 × 3)	13	($\frac{1}{2}$)	**152.4**	**76.2**	12.6	9.9	4.8	21.45	(14.40)	27.33	5.52	1.74	647	110	685	71.5	4.87	2.01	5.01	1.62	0.259	66.6	18.7
		11	($\frac{7}{16}$)			11.0			18.92	(12.70)	24.10	5.46	1.68	575	98.5	610	63.7	4.89	2.02	5.03	1.63	0.261	58.8	16.6
		9	($\frac{3}{8}$)			9.5			16.39	(11.00)	20.87	5.39	1.62	503	86.7	534	55.7	4.91	2.04	5.06	1.63	0.263	51.0	14.4
127 × 76	(5 × 3)	13	($\frac{1}{2}$)	**127.0**	**76.2**	12.6	9.1	4.8	18.91	(12.70)	24.09	4.41	1.89	389	105	429	64.4	4.02	2.09	4.22	1.63	0.354	46.9	18.3
		11	($\frac{7}{16}$)			11.0			16.69	(11.21)	21.27	4.35	1.83	346	94.2	383	57.3	4.04	2.10	4.25	1.64	0.358	41.5	16.3
		9	($\frac{3}{8}$)			9.4			14.44	(9.69)	18.39	4.28	1.77	302	82.8	335	50.0	4.06	2.12	4.27	1.65	0.360	35.9	14.1
		8	($\frac{5}{16}$)			7.8			12.06	(8.10)	15.37	4.21	1.70	255	70.2	283	42.3	4.07	2.14	4.29	1.66	0.362	30.1	11.9
102 × 76	(4 × 3)	13	($\frac{1}{2}$)	**101.6**	**76.2**	12.6	8.4	4.8	16.38	(11.00)	20.87	3.34	2.08	207	98.8	251	54.3	3.15	2.18	3.47	1.61	0.540	30.3	17.8
		11	($\frac{7}{16}$)			11.0			14.44	(9.70)	18.40	3.28	2.02	185	88.5	225	48.2	3.17	2.19	3.50	1.62	0.544	26.8	15.8
		9	($\frac{3}{8}$)			9.4			12.50	(8.39)	15.92	3.22	1.96	162	77.8	197	42.0	3.19	2.21	3.52	1.62	0.547	23.3	13.7
		8	($\frac{5}{16}$)			7.9			10.58	(7.10)	13.47	3.16	1.90	138	66.8	169	35.9	3.20	2.23	3.54	1.63	0.549	19.7	11.7
89 × 64	($3\frac{1}{2}$ × $2\frac{1}{2}$)	9	($\frac{3}{8}$)	**88.9**	**63.5**	9.4	7.6	4.8	10.57	(7.10)	13.47	2.91	1.65	104	43.9	123	24.6	2.78	1.80	3.02	1.35	0.493	17.4	9.34
		8	($\frac{5}{16}$)			7.8			8.93	(5.99)	11.37	2.85	1.59	88.8	37.7	106	21.0	2.79	1.82	3.05	1.36	0.496	14.7	7.92
		6	($\frac{1}{4}$)			6.2			7.16	(4.81)	9.12	2.78	1.53	72.1	30.7	85.8	17.0	2.81	1.83	3.07	1.37	0.498	11.8	6.37
76 × 51	(3 × 2)	9	($\frac{3}{8}$)	**76.2**	**50.8**	9.4	6.9	2.4	8.78	(5.89)	11.18	2.62	1.36	63.2	22.3	72.3	13.2	2.38	1.41	2.54	1.09	0.426	12.6	5.99
		8	($\frac{5}{16}$)			7.9			7.45	(5.00)	9.48	2.56	1.30	54.5	19.3	62.5	11.3	2.40	1.43	2.57	1.09	0.431	10.8	5.12
		6	($\frac{1}{4}$)			6.2			5.96	(4.00)	7.59	2.49	1.24	44.4	15.9	51.1	9.20	2.42	1.45	2.59	1.10	0.436	8.66	4.13
64 × 51	($2\frac{1}{2}$ × 2)	8	($\frac{5}{16}$)	**63.5**	**50.8**	7.8	6.6	2.4	6.55	(4.40)	8.35	2.03	1.40	32.1	18.1	40.7	9.50	1.96	1.47	2.21	1.07	0.618	7.44	4.93
		6	($\frac{1}{4}$)			6.2			5.35	(3.59)	6.82	1.97	1.35	26.7	15.1	34.0	7.86	1.98	1.49	2.23	1.07	0.622	6.10	4.06
64 × 38	($2\frac{1}{2}$ × $1\frac{1}{2}$)	8	($\frac{5}{16}$)	**63.5**	**38.1**	7.8	6.1	2.4	5.80	(3.90)	7.39	2.26	1.00	29.2	7.79	32.1	4.87	1.99	1.03	2.08	0.81	0.347	7.13	2.77
		6	($\frac{1}{4}$)			6.3			4.77	(3.20)	6.08	2.20	0.94	24.4	6.59	27.0	4.05	2.00	1.04	2.11	0.82	0.353	5.89	2.30

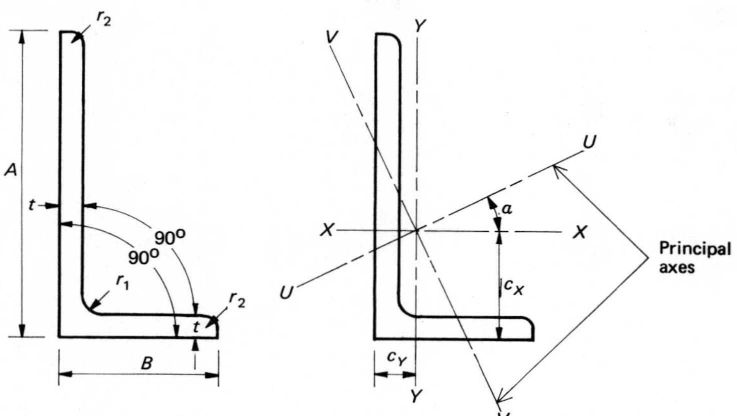

Fig. A3.6. *Definition of symbols, unequal angles for Table A3.6.*

APPENDIX A4
Properties of Materials

The tables in this appendix give the mechanical properties of some commonly used engineering materials. The following tables are included:

Table A4.1. Minimum tensile strength of grey cast iron.
Table A4.2. Properties for structural steel sections and flat bars.
Table A4.3. Typical properties of medium-carbon and alloy steels.
Table A4.4. Properties of wrought- and cast-aluminium alloys.

The strengths given for cast iron and structural steel are the minimum values quoted in the relevant standards. The properties of medium-carbon and alloy steels and wrought aluminium are typical values. In the case of cast aluminium, the minimum values from BS 1490: 1970 are given. It is noted in this standard that the specified properties apply to separately cast test bars, and may not be realised in all parts of the castings. In all cases the strength values are sensitive to the section thickness.

Extracts from BS 1452: 1961, *Specification for Grey Iron Castings* (Table A4.1), BS 4360: 1972, *Specification for Weldable Structural Steels* (Table A4.2), BS 1474: 1972, *Specification for Wrought Aluminium and Aluminium Alloys* (Table A4.4) and BS 1490: 1970, *Specification for Aluminium and Alloy Ingots and Castings* (Table A4.4) are reproduced by permission of the British Standards Institution, 2 Park Street, London W1A 2BS. Table A4.3 is adapted from various sources.

Table A4.1. Properties of grey cast iron.

Cross-sectional thickness of casting (mm) Over	Up to and including	Nominal diameter of test bar as cast (mm)	Gauge dia. (mm)	Tensile strength, minimum (MPa) Grade 10	12	14	17	20	23	26
—	9.5	15.2	10.1	170	200	247	293	340	386	432
9.5	19.0	22.2	14.3	162	193	231	278	324	371	417
19.0	29.6	30.5	20.3	155	185	216	263	309	355	402
29.6	41.3	40.6	28.6	147	178	208	247	293	340	386
41.3	—	53.3	37.9	140	170	200	232	278	325	371

Table A4.2. Properties of structural steel sections and flat bars. Summary of values given in BS 4360: 1972.

Grade	Tensile strength f_t (MPa)	Yield stress (minimum) f_y (MPa) varies with section size	Elongation % on 200 mm gauge length	Charpy V-notch Temp. °C	Charpy V-notch Energy J	Equivalent specification in ISO R 630
40 A		—	22		—	
40 B		240/210	22	RT	27	Fe 42 B
40 C	400/480	240/210	22	0	27	Fe 42 C
40 D		240/210	22	−15	27	Fe 42 D
40 E		255/225	22	−30	27	—
43 A		255/225	20	—	—	Fe 44 A
43 B		255/225	20	RT	27	Fe 44 B
43 C	430/510	255/225	20	0	27	Fe 44 C
43 D		255/225	20	−15	27	Fe 44 D
43 E		270/240	20	−30	27	
50 A		—	18	—	—	
50 B	490/620	355/325	18	—	—	Fe 52 B
50 C		355/325	18	0	27	Fe 52 C
50 D		355/340	18	−10	27	Fe 52 D
55 C	550/700	450/415	17	0	27	—
55 E		450/400	17	−50	27	—

Table A4.3. Typical properties of medium carbon and alloy steels.

Steel type and alloy content (%)	Typical En and AISI classification*	Tensile stress f_t (MPa)	Elongation %	Yield stress or 0.2% proof stress, f_y (MPa)	Endurance limit f_e (MPa)	Brinell hardness	Charpy V-notch, (J)	Machinability, percent of mild steel	Weldability	Comments
0.4% carbon steel 0.4 C, 0.2 Si, 0.8 Mn	En 8 C 1038 C 1039	Normalised 540 H and T 700	 20 20	275 470	270 340	150–200 200–255	— 38	70	With special precautions	Gives better strength than mild steel, is cheaper than alloy steel. Used for machine parts requiring strength and wear resistance without specially high-impact strength. Free-cutting grade also available
Carbon-manganese steel 0.2 C, 0.2 Si, 1.5 Mn 0.4 Ni max, 0.25 Cr max	En 14A 1320 C 1024	Normalised 540 H and T 700	 20 20	325 465	250 325	150–205 200–255	—	65	Preheating required	Weldable steel used in preference to carbon steel where improved strength required. En 14B, C 1027, have higher carbon content
1% chrome-molybdenum steel 0.4 C, 0.2 Si, 0.6 Mn 1.2 Cr, 0.3 Mo	En 19 4137 4140	H and T 1000	 16	800	480	290–340	35	40	Special precautions followed by heat treatment	Good ductility and resistance to shock for heavily loaded machine parts
3½% nickel steel 0.4 C, 0.2 Si, 0.6 Mn 3.5 Ni, 0.3 Cr max	En 22 2340	H and T 850	 18	680	440	250–300	38	50	Special precautions followed by heat treatment	Used for high strength machine parts where toughness and low temperature properties are important
1½% nickel-chrome-molybdenum 0.4 C, 0.2 Si, 0.6 Mn 1.6 Ni, 1.2 Cr, 0.3 Mo	En 24 4340	H and T 1250	 14	1050	650	360–415	25	50 soft, 35 hard	With special precautions followed by heat treatment	Good hardenability. Used where toughness and wear are important
Spring steel 0.5 C, 0.5 Si max, 0.6 Mn 1.0 Cr, 0.2 V	En 47 6150	H and T 1500	 10	1400	650	450–500	—	—	—	Oil quenched steel for leaf and coil springs
18–8 stainless steel Austenitic rust- and heat-resistant steel 0.16 C max, 18 Cr 8 Ni	En 58 A 302	Cold drawn 1200	 22	650	490	—	80	30 (75 for free cutting)	Use welding grade 58 E or 304	High resistance to corrosion; good strength and scaling properties up to 650°C. Welding grade (58 E or 304) available, also free-cutting grade
Nickel-chromium alloy for high temperatures	Nimonic 80 A	Room temperature properties 1100	 35	 610	—	—	90	—	—	For highly stressed parts at high temperatures. Good scaling and creep properties. Gas turbine components up to 750°C. 75 MPa gives 0.2% creep in 10 000 hours at 740°C

*For other comparable specifications see BS 3179, *Comparison of British and Overseas Standards for Steels:* Part 1: 1967, "Chemical Composition of Wrought Carbon Steels"; Part 2: 1962, "Chemical Composition of Wrought Alloy Steels".

Table A4.4. Properties of aluminium alloys.

Alloy type	BS designation	Other designation	Principal alloying elements %	Condition	Proof stress f_y (MPa)	Tensile strength f_t (MPa)	Elongation %	Fatigue strength at 5×10^7 cycles f_e (MPa)	Applications
Non-heat-treatable wrought alloys. Weldable (BS 1474: 1972)	N3	3203	1.3 Mn	0 H2 H6	59 120 150	110 130 175	40 17 8	48 55 70	Ductile corrosion-resistant alloy used for panelling and general sheet metal work
	N4	5152	2.0 Mg	0 H2 H6	85 155 215	195 230 280	22 14 10	95 110 125	Sheet metal work and marine applications
	N5	5154	3.5 Mg	0 H2 H6	110 205 245	240 270 310	27 15 12	120 125 140	Structural, marine and pressure-vessel applications
	N6	5056	5.0 Mg	0 H2	145 270	285 325	25 14	125 140	General structural and marine work
	N8	5083	4.5 Mg 0.7 Mn	0 H2 H4	120 205 255	260 290 325	22 12 10	130 140 145	Preferred alloy for shipbuilding and marine use. Structural and cryogenic applications.
Heat-treatable wrought alloys (BS 1474: 1972)	H9	6063	0.5 Si, 0.6 Mg	TB TF	110 190	155 240	20 15	80 85	General purpose extrusion alloy. Used for architectural sections. Weldable
	H15	2014	4.5 Cu + Si, Mg, Mn	TB TF	280 450	420 500	9 7	170 170	Aircraft and heavy-duty structural applications. Forgings
	H20	6061	1.0 Mg, 0.6 Si + Cu	TB TF	140 250	230 290	18 12	120 120	Marine and automotive uses. Corrosion-resistant; weldable
	H30	6351	1.0 Mg, 1.0 Si + Mn	TB TF	110 250	200 295	15 7	95 100	Heavy-duty structural work, where corrosion resistance required. Weldable
Casting alloys (BS 1490: 1970)	LM2	307	1.5 Cu, 10 Si	M chill cast	—	150	—	—	Preferred alloy for pressure diecastings. Not heat-treatable
	LM4	303	3 Cu, 5 Si, 0.4 Mn	M TF sand cast	—	140 230	2 —	— —	Widely used for sand and permanent mould castings. General purpose alloy with moderate strength. Heat-treatable
	LM6	401	11 Si	M sand cast M chill cast	—	160 190	5 7	— —	Used for sand and permanent mould casting where corrosion resistance required. Chemical plant and marine applications. Weldable
	LM21		4 Cu, 6 Si + Mg, Mn	M sand cast M chill cast	—	150 170	1 1	— —	Sand and gravity diecastings. High proof stress and hardness. Not heat-treatable
	LM24	313	3.5 Cu, 8 Si	M chill cast	—	180	1.5	—	Similar to LM2. Widely used for domestic and automotive components. Not heat-treatable
	LM25	601	7 Si, 0.3 Mg	TB7 TF chill cast	—	230 280	5 2	— —	High strength casting alloy used for aircraft and automotive applications. Heat-treatable

APPENDIX A5

Limits and Fits

Values of upper and lower deviations are given here for a range of shafts and holes. Tables A5.1 and A5.2 are for the selected fits recommended in BS 4500. The additional values given in Table A5.3 extend the range of fits that can be obtained.

The information has been taken from BS 4500: 1969, *ISO Limits and Fits*. Reproduced by permission of the British Standards Institution, 2 Park Street, London W1A 2BS.

Table A5.1. Limits of tolerances for selected holes. (Upper and lower deviations.)

ES = Upper deviation
EI = Lower deviation

Unit = μm

Nominal sizes		**H7**		**H8**		**H9**		**H11**	
Over	**Up to and including**	**ES** +	**EI**	**ES** +	**EI**	**ES** +	**EI**	**ES** +	**EI**
mm	mm								
—	3	10	0	14	0	25	0	60	0
3	6	12	0	18	0	30	0	75	0
6	10	15	0	22	0	36	0	90	0
10	18	18	0	27	0	43	0	110	0
18	30	21	0	33	0	52	0	130	0
30	50	25	0	39	0	62	0	160	0
50	80	30	0	46	0	74	0	190	0
80	120	35	0	54	0	87	0	220	0
120	180	40	0	63	0	100	0	250	0
180	250	46	0	72	0	115	0	290	0
250	315	52	0	81	0	130	0	320	0
315	400	57	0	89	0	140	0	360	0
400	500	63	0	97	0	155	0	400	0

Table A5.2. Limits of tolerances for selected shafts.
(Upper and lower deviations.)

es = upper deviation
ei = lower deviation

Unit = μm

Nominal sizes		c11		d10		e9		f7		g6		h6		k6		n6		p6		s6	
Over	To	es −	ei −	es −	ei −	es −	ei −	es −	ei −	es −	ei −	es −	ei −	es +	ei +	es +	ei +	es +	ei +	es +	ei +
mm —	mm 3	60	120	20	60	14	39	6	16	2	8	0	6	6	0	10	4	12	6	20	14
3	6	70	145	30	78	20	50	10	22	4	12	0	8	9	1	16	8	20	12	27	19
6	10	80	170	40	98	25	61	13	28	5	14	0	9	10	1	19	10	24	15	32	23
10	18	95	205	50	120	32	75	16	34	6	17	0	11	12	1	23	12	29	18	39	28
18	30	110	240	65	149	40	92	20	41	7	20	0	13	15	2	28	15	35	22	48	35
30	40	120	280	80	180	50	112	25	50	9	25	0	16	18	2	33	17	42	26	59	43
40	50	130	290																		
50	65	140	330	100	220	60	134	30	60	10	29	0	19	21	2	39	20	51	32	72	53
65	80	150	340																	78	59
80	100	170	390	120	260	72	159	36	71	12	34	0	22	25	3	45	23	59	37	93	71
100	120	180	400																	101	79
120	140	200	450	145	305	85	185	43	83	14	39	0	25	28	3	52	27	68	43	117	92
140	160	210	460																	125	100
160	180	230	480																	133	108
180	200	240	530	170	355	100	215	50	96	15	44	0	29	33	4	60	31	79	50	151	122
200	225	260	550																	159	130
225	250	280	570																	169	140
250	280	300	620	190	400	110	240	56	108	17	49	0	32	36	4	66	34	88	56	190	158
280	315	330	650																	202	170
315	355	360	720	210	440	125	265	62	119	18	54	0	36	40	4	73	37	98	62	226	190
355	400	400	760																	244	208
400	450	440	840	230	480	135	290	68	131	20	60	0	40	45	5	80	40	108	68	272	232
450	500	480	880																	292	252

Table A5.3. Limits of tolerance for commonly used holes.
(Upper and lower deviations.)

Nominal sizes		J				K				M				N				P			
		6		7		6		7		6		7		6		7		6		7	
Over	To	ES +	EI −	ES +	EI −	ES +	EI −	ES +	EI −	ES −	EI −	ES	EI −	ES −	EI −	ES −	EI −	ES −	EI −	ES −	EI −
mm —	mm 3	2	4	4	6	0	6	0	10	2	8	0	12	4	10	4	14	6	12	6	16
3	6	5	3	6	6	2	6	3	9	1	9	0	12	5	13	4	16	9	17	8	20
6	10	5	4	8	7	2	7	5	10	3	12	0	15	7	16	4	19	12	21	9	24
10 14	14 18	6	5	10	8	2	9	6	12	4	15	0	18	9	20	5	23	15	26	11	29
18 24	24 30	8	5	12	9	2	11	6	15	4	17	0	21	11	24	7	28	18	31	14	35
30 40	40 50	10	6	14	11	3	13	7	18	4	20	0	25	12	28	8	33	21	37	17	42
50 65	65 80	13	6	18	12	4	15	9	21	5	24	0	30	14	33	9	39	26	45	21	51
80 100	100 120	16	6	22	13	4	18	10	25	6	28	0	35	16	38	10	45	30	52	24	59
120 140 160	140 160 180	18	7	26	14	4	21	12	28	8	33	0	40	20	45	12	52	36	61	28	68
180 200 225	200 225 250	22	7	30	16	5	24	13	33	8	37	0	46	22	51	14	60	41	70	33	79
250 280	280 315	25	7	36	16	5	27	16	36	9	41	0	52	25	57	14	66	47	79	36	88
315 355	355 400	29	7	39	18	7	29	17	40	10	46	0	57	26	62	16	73	51	87	41	98
400 450	450 500	33	7	43	20	8	32	18	45	10	50	0	63	27	67	17	80	55	95	45	108

APPENDIX A6

Areas of Reinforcing Bars

Table A6.1 (below) shows the cross-sectional area in mm^2 of various numbers of bars. Two series of bar sizes are covered:

(a) preferred metric sizes of bars as designated in BS 4449: 1969, *Hot Rolled Steel Bars for the Reinforcement of Concrete*;

(b) sizes currently rolled by Broken Hill Proprietary Ltd., Melbourne, Australia.

Table A6.1. Cross-sectional area of groups of round bars.

BS preferred sizes (mm)	Sizes rolled by BHP (mm)	Number of bars									
		1	2	3	4	5	6	7	8	9	10
6		28	57	85	113	141	170	198	226	255	283
8		50	101	151	201	251	302	352	402	452	503
10	10	79	157	236	314	393	471	550	628	707	785
12	12	113	226	339	452	566	679	792	905	1 018	1 131
16	16	201	402	603	804	1005	1 206	1 407	1 608	1 809	2 010
20	20	314	628	942	1257	1571	1 885	2 199	2 513	2 827	3 142
	24	452	905	1357	1810	2262	2 714	3 167	3 619	4 072	4 524
25		491	982	1473	1964	2455	2 945	3 436	3 927	4 418	4 909
	28	616	1232	1847	2463	3079	3 695	4 311	4 926	5 542	6 158
32	32	804	1608	2413	3217	4021	4 825	5 629	6 434	7 238	8 042
	36	1018	2036	3054	4071	5089	6 107	7 125	8 143	9 161	10 179
40	40	1257	2513	3770	5027	6283	7 540	8 796	10 053	11 310	12 566
	50	1964	3927	5891	7854	9818	11 781	13 745	15 708	17 672	19 635

APPENDIX A7

Welding Symbols

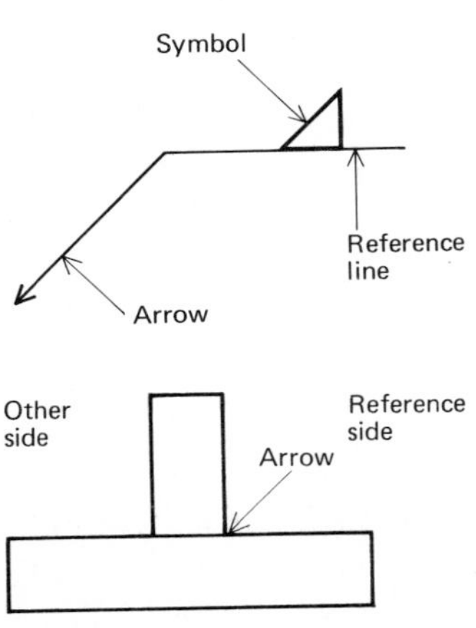

A standard scheme of symbols is used on drawings to describe the type, size and position of welds. A summary of the more important symbols is given here. Reference should be made to BS 499: Part 2: 1965, *Symbols for Welding*, for more complete information.

The scheme requires the use of a weld symbol, an arrow and a reference line and other notation pertaining to dimensions and procedure. The welds may also be shown on the drawing by thickened lines or hatching, but this is optional.

The location of the arrow on the drawing fixes the reference side of the joint. If the weld symbol is placed below the reference line, the weld is to be made on the reference- or arrow-side of the joint. If the symbol is placed above the line, the weld is made on the other side of the joint. If the symbol is placed on both sides of the line, the weld is to be made on both sides of the joint.

The standard weld symbols are given in Fig. A7.1 and examples of the scheme are shown in Fig. A7.2. The use of the symbols described here conforms to British and American practice. Continental practice is to place the symbol above the reference line for welds on the arrow side.

Additional information is conveyed by symbols at the junction of the arrow and reference line. A circle indicates "weld all round" and a filled-in circle denotes a field weld.

The leg size of the weld can be shown on one side of the weld symbol, while the spacing of intermittent welds – with the gap dimensions in brackets – can be added on the other side.

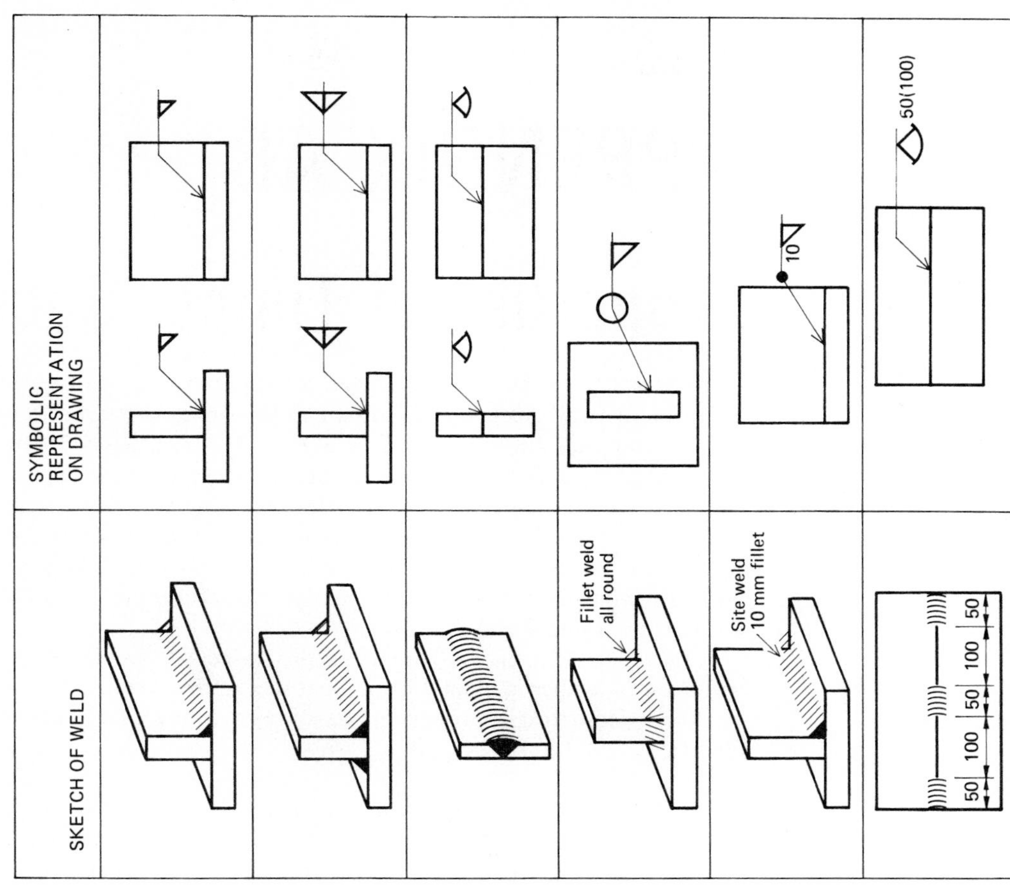

Fig. A7.2. *Examples of the use of welding symbols.*

FORM OF WELD	SECTIONAL REPRESENTATION (Need not be shown on drawing)	SYMBOL
Fillet		
Square butt		
Single-V and double-V butt		
Single-U and double-U butt		
Single and double bevel butt		
Single-J and double-J butt		
Stud		
Seal		

Fig. A7.1. *Standard welding symbols.*

APPENDIX A8

Properties of Timber Sections

Geometrical properties of rectangles corresponding to a range of timber sizes are given in Table A8.1. These sizes are approximately the same as the following sizes in Imperial units: 6 × 2, 6 × 3, 8 × 2, 8 × 3, 10 × 2, 10 × 3, 10 × 4, 12 × 2, 12 × 3 and 12 × 4. Although it is not certain that supplies will be standardised in these sizes in countries changing to metric measurement, the sizes selected are realistic. Properties of a much wider range are available in BSCP 112, *The Structural Use of Timber*. In any case, good practice will be to ascertain what sizes are available locally, and what dimensions can be expected for a given nominal size.

In Table A8.1 gauged timber is processed to give dimensional accuracy and dressed timber is planed for appearance and accuracy.

The geometrical properties have been computed from the following formulas:

$$A = ht \qquad Z_{XX} = \frac{1}{6}h^2t \qquad Z_{YY} = \frac{1}{6}ht^2$$

$$I_{XX} = \frac{1}{12}h^3t \qquad I_{YY} = \frac{1}{12}ht^3$$

$$r_{XX} = \frac{h}{\sqrt{12}} \qquad r_{YY} = \frac{t}{\sqrt{12}}$$

h = height of rectangular section,
t = width of rectangular section.

Both h and t are for the finished size.

Table A8.1. Geometrical properties of timber sections.

Nominal size	Gauged, height and thickness								Dressed four sides							
	Finished size	A 10^3 mm^2	Z_{XX} 10^3 mm^3	Z_{YY} 10^3 mm^3	I_{XX} 10^6 mm^4	I_{YY} 10^6 mm^4	r_{XX} mm	r_{YY} mm	Finished size	A 10^3 mm^2	Z_{XX} 10^3 mm^3	Z_{YY} 10^3 mm^3	I_{XX} 10^6 mm^4	I_{YY} 10^6 mm^4	r_{XX} mm	r_{YY} mm
150 × 50	144 × 47	6.768	162.4	53.02	11.69	1.246	41.6	13.6	140 × 45	6.300	147.0	47.25	10.29	1.063	40.4	13.0
150 × 75	144 × 69	9.936	238.5	114.3	17.17	3.942	41.6	19.9	140 × 65	9.100	212.3	98.58	14.86	3.204	40.4	18.8
200 × 50	194 × 47	9.118	294.8	71.42	28.60	1.678	56.0	13.6	180 × 45	8.100	243.0	60.75	21.87	1.367	52.0	13.0
200 × 75	194 × 69	13.39	432.8	153.9	41.98	5.311	56.0	19.9	180 × 65	11.70	351.0	126.8	31.59	4.119	52.0	18.8
250 × 50	244 × 47	11.47	466.4	89.83	56.90	2.111	70.4	13.6	230 × 45	10.35	396.8	77.63	45.63	1.747	66.4	13.0
250 × 75	244 × 69	16.84	684.7	193.6	83.53	6.680	70.4	19.9	230 × 65	14.95	573.1	162.0	65.90	5.264	66.4	18.8
250 × 100	244 × 94	22.94	932.7	359.3	113.8	16.89	70.4	27.1	230 × 90	20.70	793.5	310.5	91.25	13.97	66.4	26.0
300 × 50	294 × 47	13.82	677.1	108.2	99.53	2.544	84.9	13.6	280 × 45	12.60	588.0	94.50	82.32	2.126	80.8	13.0
300 × 75	294 × 69	20.29	994.0	233.3	146.1	8.048	84.9	19.9	280 × 65	18.20	849.3	197.2	118.9	6.408	80.8	18.8
300 × 100	294 × 94	27.64	1354	433.0	199.1	20.35	84.9	27.1	280 × 90	25.20	1176	378.0	164.6	17.01	80.8	26.0

APPENDIX A9

ISO Hexagon Bolts and Screws

Design information for bolts and screws is given in Table A9.1. The nominal sizes tabulated here represent the preferred sizes. Other sizes are manufactured and their properties are given in the literature. The information in this table has been extracted from BS 3692: 1967, *ISO Metric Precision Hexagonal Bolts, Screws and Nuts*, and BS 3643: Part 2: 1966, "Limits and Tolerances for Coarse Pitch Series Threads in *ISO Metric Screw Threads*.

Table A9.1. Design information for ISO metric precision bolts and screws.

Nominal size and thread diameter *d* (mm)	Thread pitch (coarse series) (mm)	Width across flats (max) *s* (mm)	Height of head *k* (mm)	Tensile stress area* (mm^2)	Area at minor diameter
M1.6	0.35	3.2	1.1	1.27	1.08
M2	0.4	4	1.4	2.07	1.79
M2.5	0.45	5	1.7	3.39	2.98
M3	0.5	5.5	2.0	5.03	4.47
M4	0.7	7	2.8	8.78	7.75
M5	0.8	8	3.5	14.2	12.7
M6	1.0	10	4.0	20.1	17.9
M8	1.25	13	5.5	36.6	32.8
M10	1.5	17	7	58.0	52.3
M12	1.75	19	8	84.3	76.2
M16	2.0	24	10	157	144
M20	2.5	30	13	245	225
M24	3.0	36	15	353	324
M30	3.5	46	19	561	519
M36	4.0	55	23	817	759
M42	4.5	65	26	1120	1050
M48	5.0	75	30	1470	1380
M56	5.5	85	35	2030	1910
M64	6.0	95	40	2680	2520

*Based on mean of effective and minor diameter.

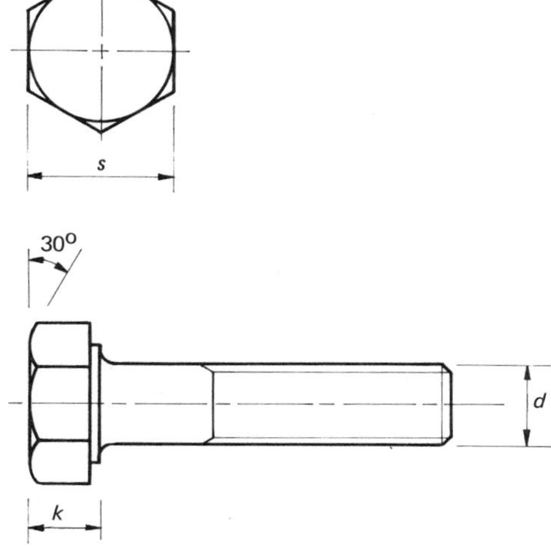

APPENDIX A10

Bearing Tables

The information given in these tables is intended for use in student exercises and for preliminary design purposes. It has been extracted from the *SKF General Catalogue* 2800 (1970) by permission of the SKF Bearing Company.

Four bearing types have been selected, as follows:

Table A10.1. Deep-groove ball bearings, Series 62 and 63
Table A10.2. Self-aligning ball bearings, Series 12 and 13
Table A10.3. Angular-contact ball bearings, Series 72 and 73
Table A10.4. Cylindrical roller bearings, Series 2 and 3

For each shaft size tabulated here, two bearing sizes have been selected. Most manufacturers produce a wider selection than this. For example, in the catalogue quoted above there are six deep-groove ball bearing sizes listed for a 50 mm shaft, with dynamic load ratings ranging from 4.8 to 67 kN. Clearly, it is necessary to refer to a comprehensive catalogue to obtain the most efficient installation. Such a publication will also give more information on abutment and fillet dimensions, and will illustrate the alternative arrangements that are available, for example, for the lip on the cylindrical roller bearing races.

Table A10.1. Deep-groove ball bearings, Series 62 and 63.

Equivalent dynamic bearing load
$P = XF_r + YF_a$

F_a/C_0	e	$F_a/F_r \leqslant e$ X	Y	$F_a/F_r > e$ X	Y
0.025	0.22	1	0	0.56	2
0.04	0.24	1	0	0.56	1.8
0.07	0.27	1	0	0.56	1.6
0.13	0.31	1	0	0.56	1.4
0.25	0.37	1	0	0.56	1.2
0.5	0.44	1	0	0.56	1

Boundary dimensions (mm)			Basic load ratings dynamic static (kN)		Limiting speeds rpm × 10^{-3}		Designation (SKF)	Abutment and fillet dimensions (mm)		
d	D	B	C	C_0	lubrication grease	oil		d_a min	D_a max	r_a max
10	30	9	3.91	2.22	24	30	6200	14.0	26	0.6
	35	11	6.09	3.78	20	26	6300	14.0	31	0.6
15	35	11	5.87	3.56	19	24	6202	19.0	31	0.6
	42	13	8.72	5.34	17	20	6302	20.0	37	1
20	47	14	9.79	6.09	15	18	6204	25.0	42	1
	52	15	12.2	7.83	13	16	6304	26.5	45.5	1
25	52	15	10.7	6.94	12	15	6205	30.0	47	1
	62	17	16.9	11.3	11	14	6305	31.5	55.5	1
30	62	16	14.9	10.0	10	13	6206	35.0	57	1
	72	19	21.4	14.5	9	11	6306	36.5	65.5	1
35	72	17	19.6	13.8	9	11	6207	41.5	65.5	1
	80	21	25.4	17.8	8.5	10	6307	43.0	72	1.5
40	80	18	23.6	16.2	8.5	10	6208	46.5	73.5	1
	90	23	31.6	22.2	7.5	9	6308	48.0	82	1.5
45	85	19	25.4	18.5	7.5	9	6209	51.5	78.5	1
	100	25	40.7	29.8	6.7	8	6309	53.0	92	1.5
50	90	20	27.1	19.6	7.0	8.5	6210	56.5	83.5	1
	110	27	47.1	36.3	6.3	7.5	6310	59.0	101	2
60	110	22	36.9	28.0	6.0	7.0	6212	68.0	102	1.5
	130	31	62.3	48.0	5.0	6.0	6312	71.0	119	2
70	125	24	47.1	37.8	5.0	6.0	6214	78.0	117	1.5
	150	35	80.1	62.3	4.5	5.3	6314	81.0	139	2
80	140	26	55.6	44.5	4.5	5.3	6216	89.0	131	2
	170	39	94.3	80.1	3.8	4.5	6316	91.0	159	2
90	160	30	73.8	59.6	3.8	4.5	6218	99	151	2
	190	43	109	97.9	3.4	4.0	6318	103	177	2.5
100	180	34	94.3	78.3	3.4	4.0	6220	111	169	2
	215	47	136	133	3.0	3.6	6320	113	202	2.5

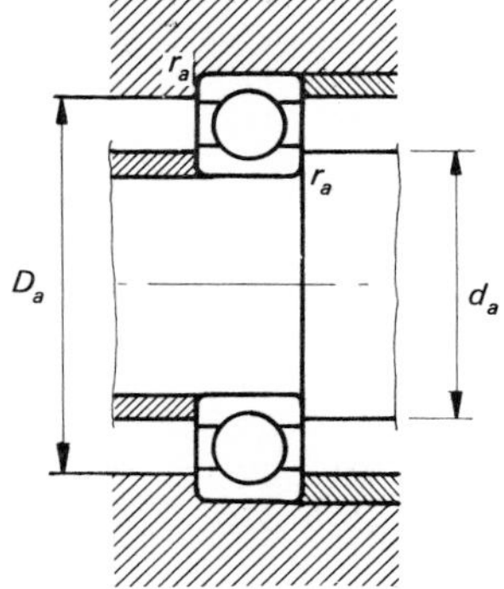

Table A10.2. Self-aligning ball bearings, Series 12, 13.

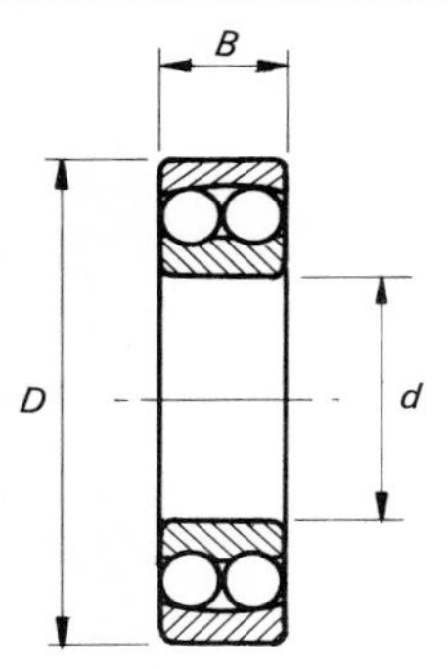

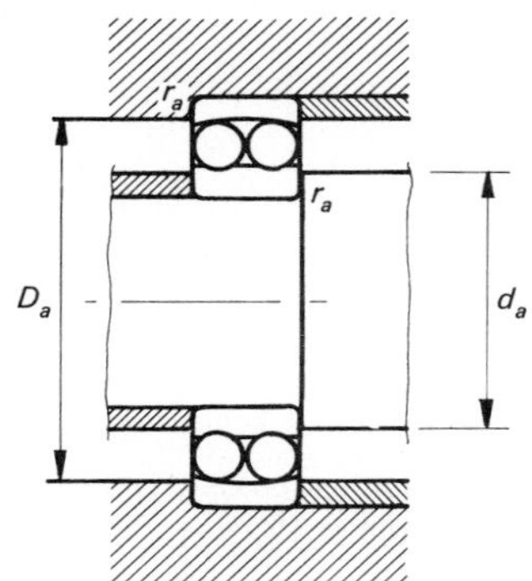

Equivalent dynamic bearing load
$P = XF_r + YF_a$

Boundary dimensions (mm)			Basic load ratings dynamic static (kN)		Limiting speeds rpm × 10^{-3}		Designation (SKF)	Abutment and fillet dimensions (mm)			Calculation factors				
					lubrication							$F_a/F_r \leqslant e$		$F_a/F_r > e$	
d	D	B	C	C_0	grease	oil		d_a min	D_a max	r_a max	e	X	Y	X	Y
15	35	11	5.65	2.02	19	24	1202	19	31	0.6	0.33	1	1.9	0.65	3.0
	42	13	7.38	2.67	17	20	1302	20	37	1	0.33	1	1.9	0.65	3.0
20	47	14	7.70	3.16	15	18	1204	25	42	1	0.27	1	2.3	0.65	3.6
	52	15	9.61	3.91	12	15	1304	26.5	45.5	1	0.30	1	2.1	0.65	3.3
25	52	15	9.25	4.00	13	16	1205	30	47	1	0.27	1	2.3	0.65	3.6
	62	17	13.8	5.74	9.5	12	1305	31.5	55.5	1	0.28	1	2.2	0.65	3.5
30	62	16	12.0	5.56	10	13	1206	35	57	1	0.25	1	2.5	0.65	3.9
	72	19	16.0	7.56	9	11	1306	36.5	65.5	1	0.26	1	2.4	0.65	3.7
35	72	17	12.0	6.23	9	11	1207	41.5	65.5	1	0.23	1	2.7	0.65	4.2
	80	21	19.1	9.44	7.5	9	1307	43	72	1.5	0.25	1	2.5	0.65	3.9
40	80	18	14.5	8.01	8.5	10	1208	46.5	73.5	1	0.22	1	2.9	0.65	4.5
	90	23	22.7	11.8	6.7	8	1308	48	82	1.5	0.24	1	2.6	0.65	4.1
45	85	19	16.2	8.90	7.5	9	1209	51.5	78.5	1	0.21	1	3.0	0.65	4.6
	100	25	29.1	15.1	6.3	7.5	1309	53	92	1.5	0.25	1	2.5	0.65	3.9
50	90	20	16.9	9.96	7	8.5	1210	56.5	83.5	1	0.20	1	3.2	0.65	4.9
	110	27	33.4	16.7	5.6	6.7	1310	59	101	2	0.24	1	2.6	0.65	4.1
60	110	22	23.1	14.2	5.6	6.7	1212	68	102	1.5	0.18	1	3.5	0.65	5.4
	130	31	43.6	25.4	4.5	5.3	1312	71	119	2	0.23	1	2.7	0.65	4.2
70	125	24	26.7	16.9	5	6	1214	78	117	1.5	0.18	1	3.5	0.65	5.4
	150	35	56.5	34.0	4	4.8	1314	81	139	2	0.22	1	2.9	0.65	4.5
80	140	26	30.2	21.4	4.5	5.3	1216	89	131	2	0.16	1	3.9	0.65	6.1
	170	39	68.1	40.7	3.6	4.3	1316	91	159	2	0.22	1	2.9	0.65	4.5
90	160	30	42.9	29.1	3.8	4.5	1218	99	151	2	0.17	1	3.7	0.65	5.7
	190	43	89.0	53.4	3.2	3.8	1318	103	177	2.5	0.22	1	2.9	0.65	4.5

Table A10.3. Angular contact ball bearings, Series 72 and 73.

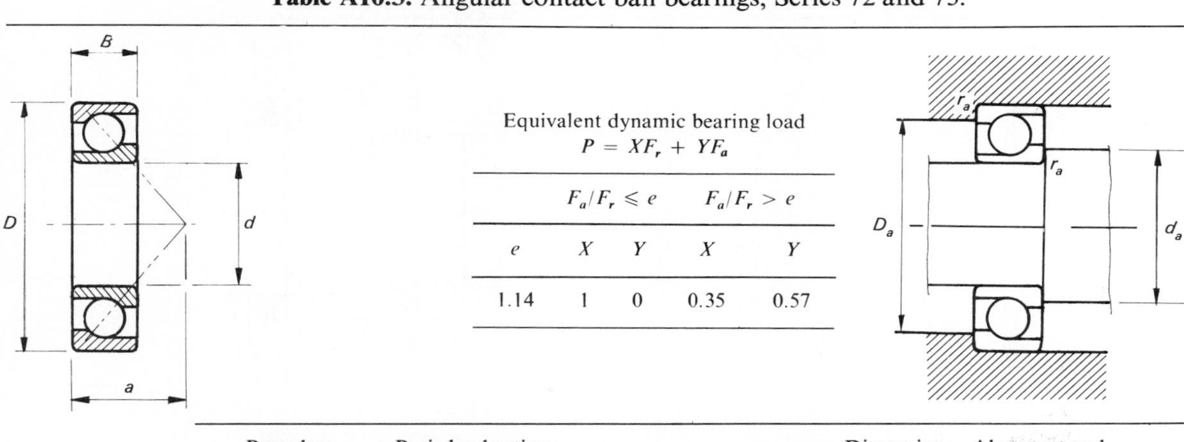

Equivalent dynamic bearing load
$P = XF_r + YF_a$

	$F_a/F_r \leqslant e$		$F_a/F_r > e$	
e	X	Y	X	Y
1.14	1	0	0.35	0.57

Boundary dimensions (mm)			Basic load ratings dynamic static (kN)		Limiting speeds rpm × 10^{-3}		Designation (SKF)	Dimension a (mm)	Abutment and fillet dimensions (mm)		
d	D	B	C	C_0	lubrication grease	oil			d_a min	D_a max	r_a max
15	35	11	6.09	3.69	16	22	7202B	16	20	30	0.6
	42	13	8.90	5.25	14	19	7302B	19	21	36	1
20	47	14	10.1	6.36	11	16	7204B	21	26	41	1
	52	15	13.6	8.14	10	15	7304B	23	27	45	1
25	52	15	11.3	7.70	9.5	14	7205B	24	31	46	1
	62	17	18.9	12.2	8.5	12	7305B	27	32	55	1
30	62	16	15.3	10.9	8.5	12	7206B	27	36	56	1
	72	19	24.0	16.2	7.5	10	7306B	31	37	65	1
35	72	17	20.7	14.9	7.5	10	7207B	31	42	65	1
	80	21	28.0	20.0	7	9.5	7307B	35	44	71	1.5
40	80	18	24.5	18.5	6.7	9	7208B	34	47	73	1
	90	23	34.7	24.9	6.3	8.5	7308B	39	49	81	1.5
45	85	19	27.6	21.1	6.3	8.5	7209B	37	52	78	1
	100	25	44.5	33.4	5.6	7.5	7309B	43	54	91	1.5
50	90	20	28.5	23.1	5.6	7.5	7210B	39	57	83	1
	110	27	51.6	40.0	5	6.7	7310B	47	60	100	2
60	110	22	42.9	36.3	4.8	6.3	7212B	47	69	101	1.5
	130	31	69.4	53.4	4.3	5.6	7312B	55	72	118	2
70	125	24	52.5	46.3	4.3	5.6	7214B	53	79	116	1.5
	150	35	87.2	72.5	3.6	4.8	7314B	64	82	138	2
80	140	26	60.9	55.6	3.6	4.8	7216B	59	90	130	2
	170	39	103	89.0	3.2	4.3	7316B	72	92	158	2
90	160	30	81.4	75.6	3.2	4.3	7218B	67	100	150	2
	190	43	120	111	2.8	3.8	7318B	80	104	176	2.5

Table A10.4. Cylindrical roller bearings, Series 2 and 3.

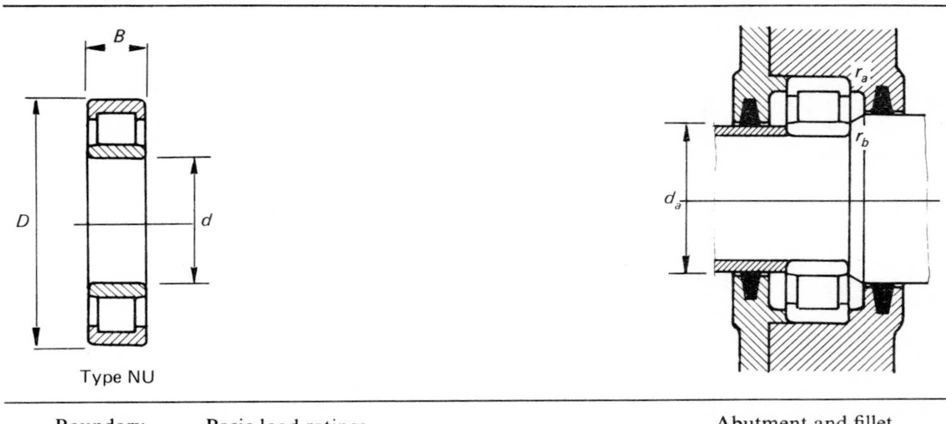

Boundary dimensions (mm)			Basic load ratings dynamic static (kN)		Limiting speeds rpm × 10^{-3}		Designation (SKF)	Abutment and fillet dimensions (mm)				
d	D	B	C	C_0	lubrication grease	oil		d_a min	d_a max	D_a max	r_a max	r_b max
20	47	14	13.6	7.38	15	18	NU204	25	26	42	1	0.6
	52	15	20.2	11.6	12	15	NU304	26.5	27	45.5	1	0.6
25	52	15	15.1	8.72	12	15	NU205	30	31	47	1	0.6
	62	17	26.0	14.9	9.5	12	NU305	31.5	33	55.5	1	1
30	62	16	20.2	12.0	10	13	NU206	35	37	57	1	0.6
	72	19	34.0	20.0	8.5	10	NU306	36.5	40	65.5	1	1
35	72	17	29.1	17.3	9	11	NU207	41.5	43	65.5	1	0.6
	80	21	42.9	27.1	8	9.5	NU307	43	45	72	1.5	1
40	80	18	38.5	24.0	8.5	10	NU208	46.5	49	73.5	1	1
	90	23	50.7	32.7	6.7	8	NU308	48	51	82	1.5	1.5
45	85	19	40.0	25.4	7.5	9	NU209	51.5	54	78.5	1	1
	100	25	69.4	45.4	6.3	7.5	NU309	53	57	92	1.5	1.5
50	90	20	42.3	27.6	7	8.5	NU210	56.5	58	83.5	1	1
	110	27	80.1	51.6	5.6	6.7	NU310	59	63	101	2	2
60	110	22	60.9	42.9	5.6	6.7	NU212	68	71	102	1.5	1.5
	130	31	111	77.0	4.8	5.6	NU312	71	75	119	2	2
70	125	24	72.5	50.7	5	6	NU214	78	82	117	1.5	1.5
	150	35	145	101	4	4.8	NU314	81	87	139	2	2
80	140	26	96.1	68.1	4.5	5.3	NU216	89	94	131	2	2
	170	39	173	125	3.6	4.3	NU316	91	99	159	2	2
90	160	30	136	99.6	3.8	4.5	NU218	99	105	151	2	2
	190	43	218	158	3.2	3.8	NU318	103	111	177	2.5	2.5
100	180	34	162	125	3.4	4	NU220	111	117	169	2	2
	215	47	291	218	2.8	3.4	NU320	113	125	202	2.5	2.5

APPENDIX A11

Theoretical Stress Concentration Factors

The theoretical stress concentration factor, K_t, is defined as

$$K_t = \frac{\text{maximum stress at a stress raiser}}{\text{nominal or engineering stress}}$$

Values of K_t are given for the following commonly encountered cases:

Fig. A11.1. Notched flat bar in tension
Fig. A11.2. Notched flat bar in bending
Fig. A11.3. Grooved shaft in tension
Fig. A11.4. Grooved shaft in bending
Fig. A11.5. Grooved shaft in torsion
Fig. A11.6. Shaft with a shoulder fillet – tension case
Fig. A11.7. Shaft with a shoulder fillet – bending case
Fig. A11.8. Shaft with a shoulder fillet – torsion case
Fig. A11.9. Straight portion of a keyway in a shaft in torsion

Figures A11.1 to A11.9 have been taken from R. E. Peterson, *Stress Concentration Design Factors*, John Wiley, New York, 1953, by permission of John Wiley & Sons, Inc.

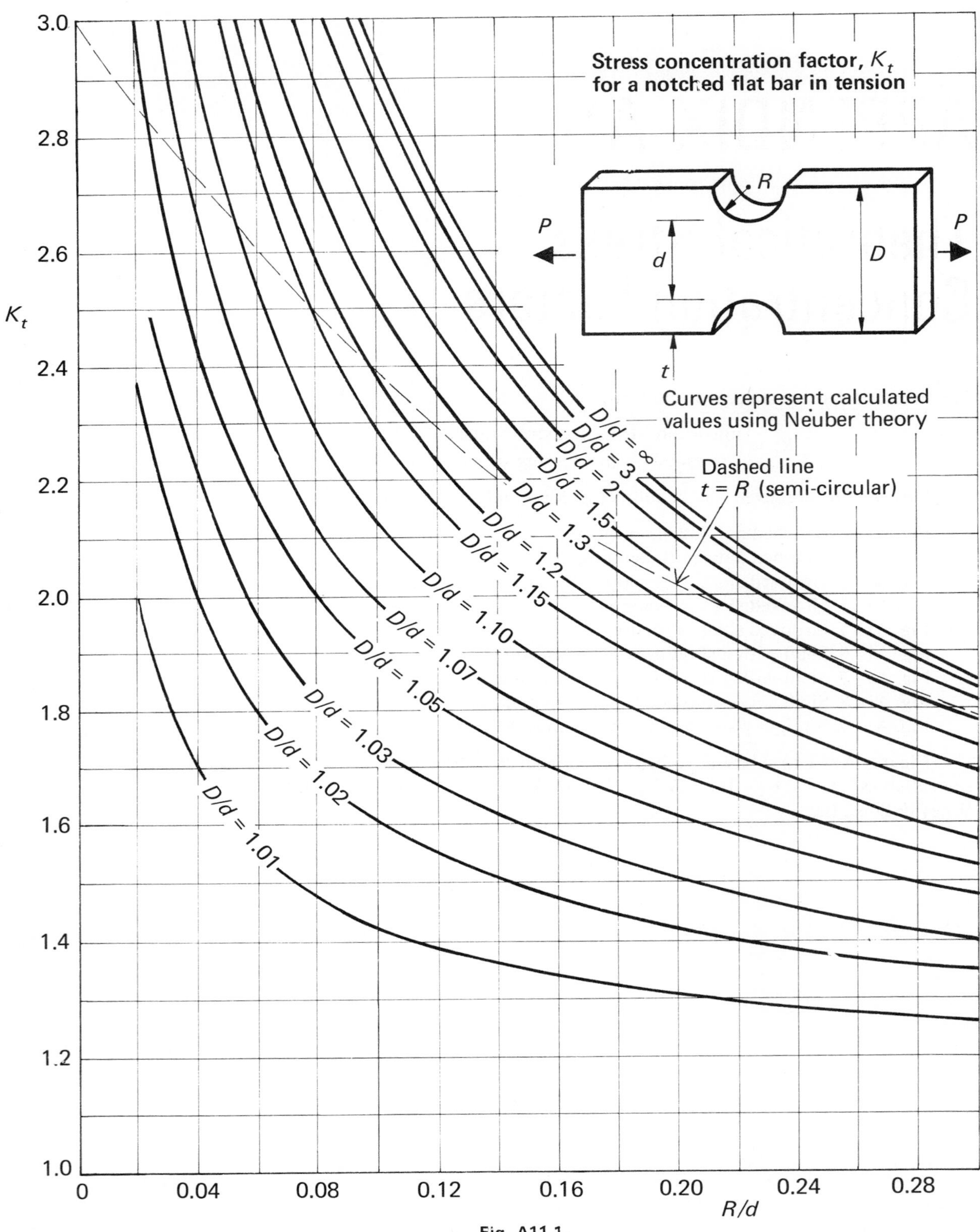
Stress concentration factor, K_t for a notched flat bar in tension
Curves represent calculated values using Neuber theory
Dashed line $t = R$ (semi-circular)
P
P
R
d
D
t
K_t
R/d
3.0
2.8
2.6
2.4
2.2
2.0
1.8
1.6
1.4
1.2
1.0
0
0.04
0.08
0.12
0.16
0.20
0.24
0.28
D/d = ∞
D/d = 3
D/d = 2
D/d = 1.5
D/d = 1.3
D/d = 1.2
D/d = 1.15
D/d = 1.10
D/d = 1.07
D/d = 1.05
D/d = 1.03
D/d = 1.02
D/d = 1.01

Fig. A11.1.

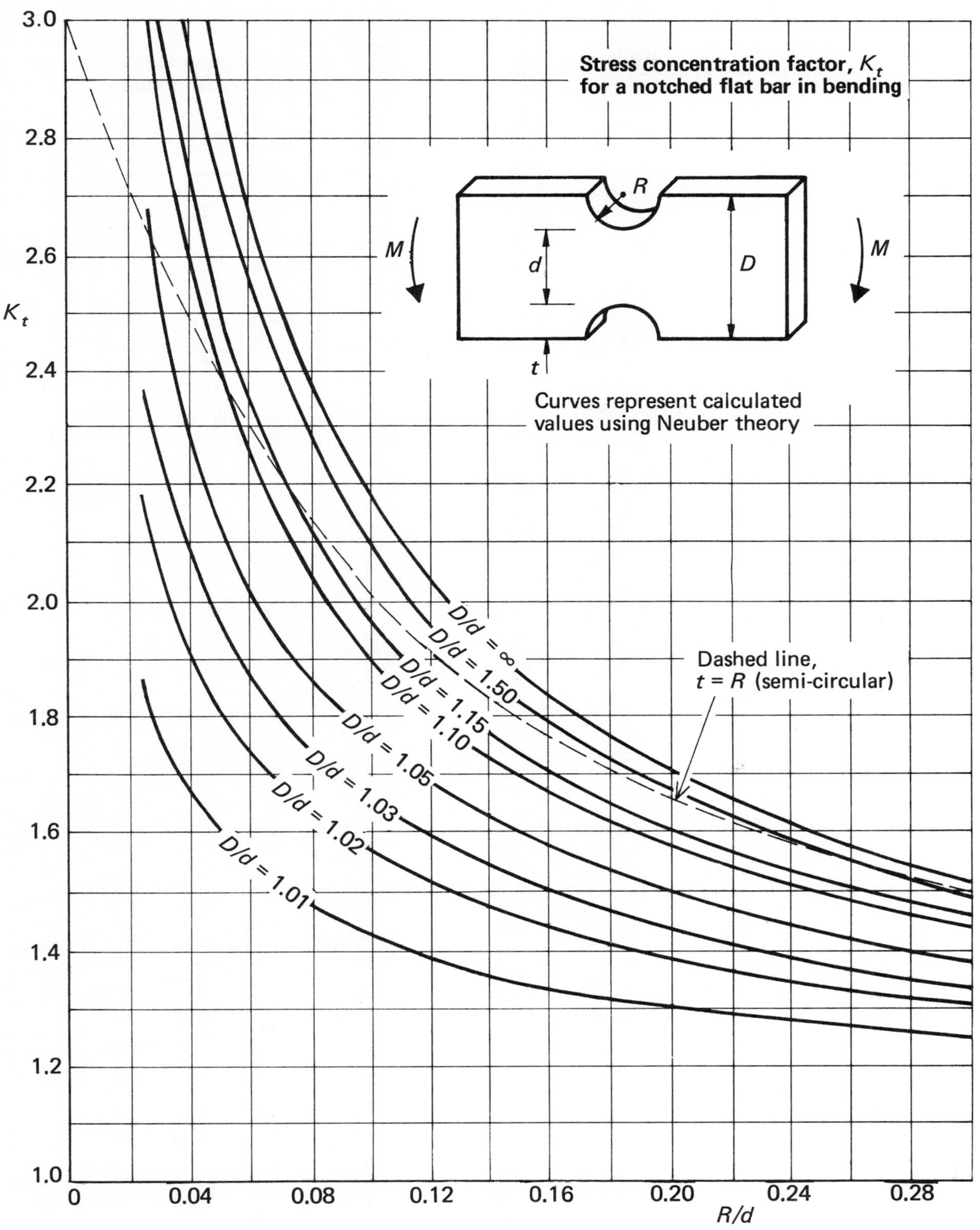
Stress concentration factor, K_t for a notched flat bar in bending
R
M
d
D
M
t
Curves represent calculated values using Neuber theory
Dashed line, $t = R$ (semi-circular)
$D/d = \infty$
$D/d = 1.50$
$D/d = 1.15$
$D/d = 1.10$
$D/d = 1.05$
$D/d = 1.03$
$D/d = 1.02$
$D/d = 1.01$
K_t
3.0
2.8
2.6
2.4
2.2
2.0
1.8
1.6
1.4
1.2
1.0
0
0.04
0.08
0.12
0.16
0.20
0.24
0.28
R/d

Fig. A11.2.

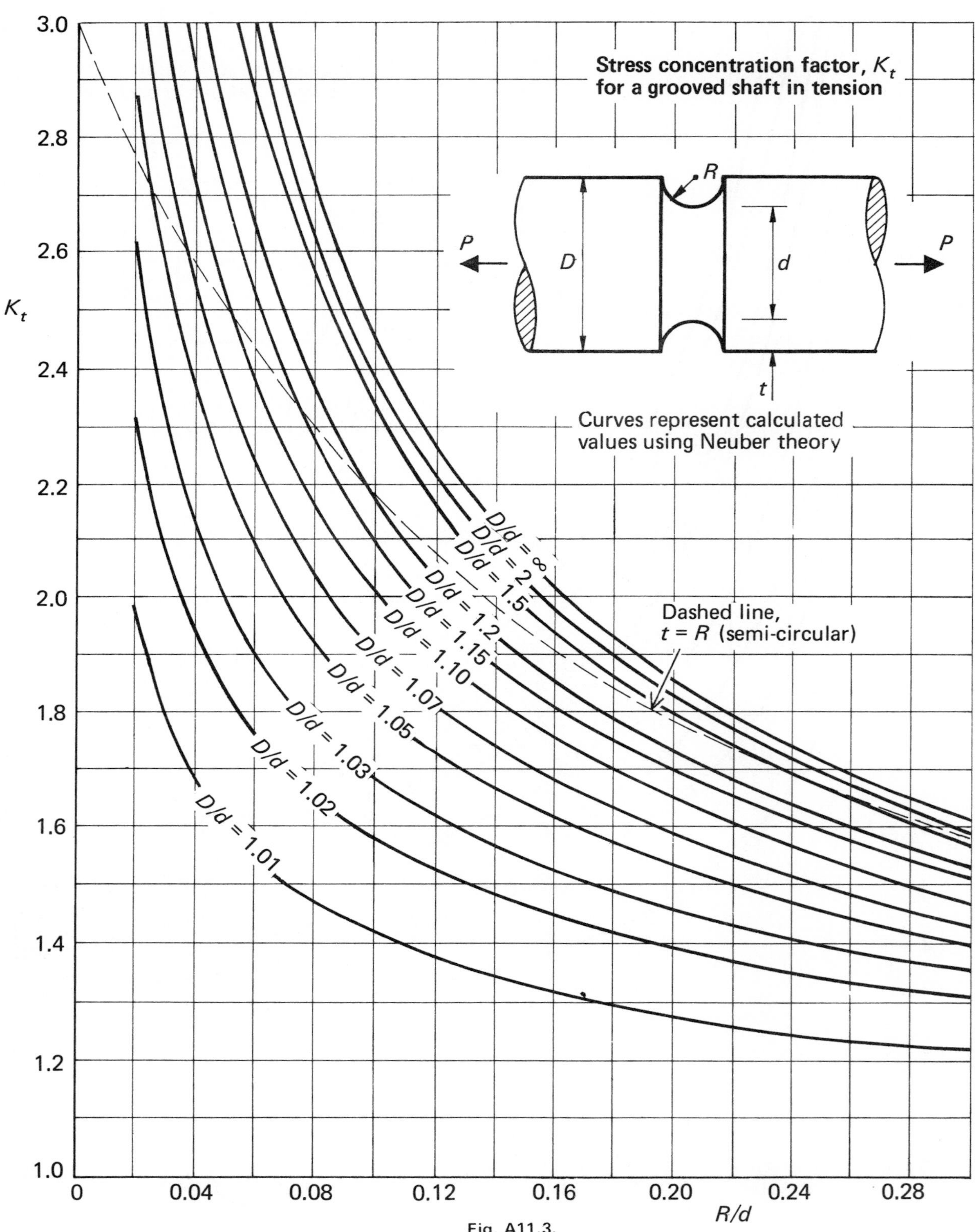
Stress concentration factor, K_t
for a grooved shaft in tension
R
P
D
d
P
t
Curves represent calculated
values using Neuber theory
Dashed line,
$t = R$ (semi-circular)
D/d = ∞
D/d = 2
D/d = 1.5
D/d = 1.2
D/d = 1.15
D/d = 1.10
D/d = 1.07
D/d = 1.05
D/d = 1.03
D/d = 1.02
D/d = 1.01
K_t
3.0
2.8
2.6
2.4
2.2
2.0
1.8
1.6
1.4
1.2
1.0
0
0.04
0.08
0.12
0.16
0.20
0.24
0.28
R/d

Fig. A11.3.

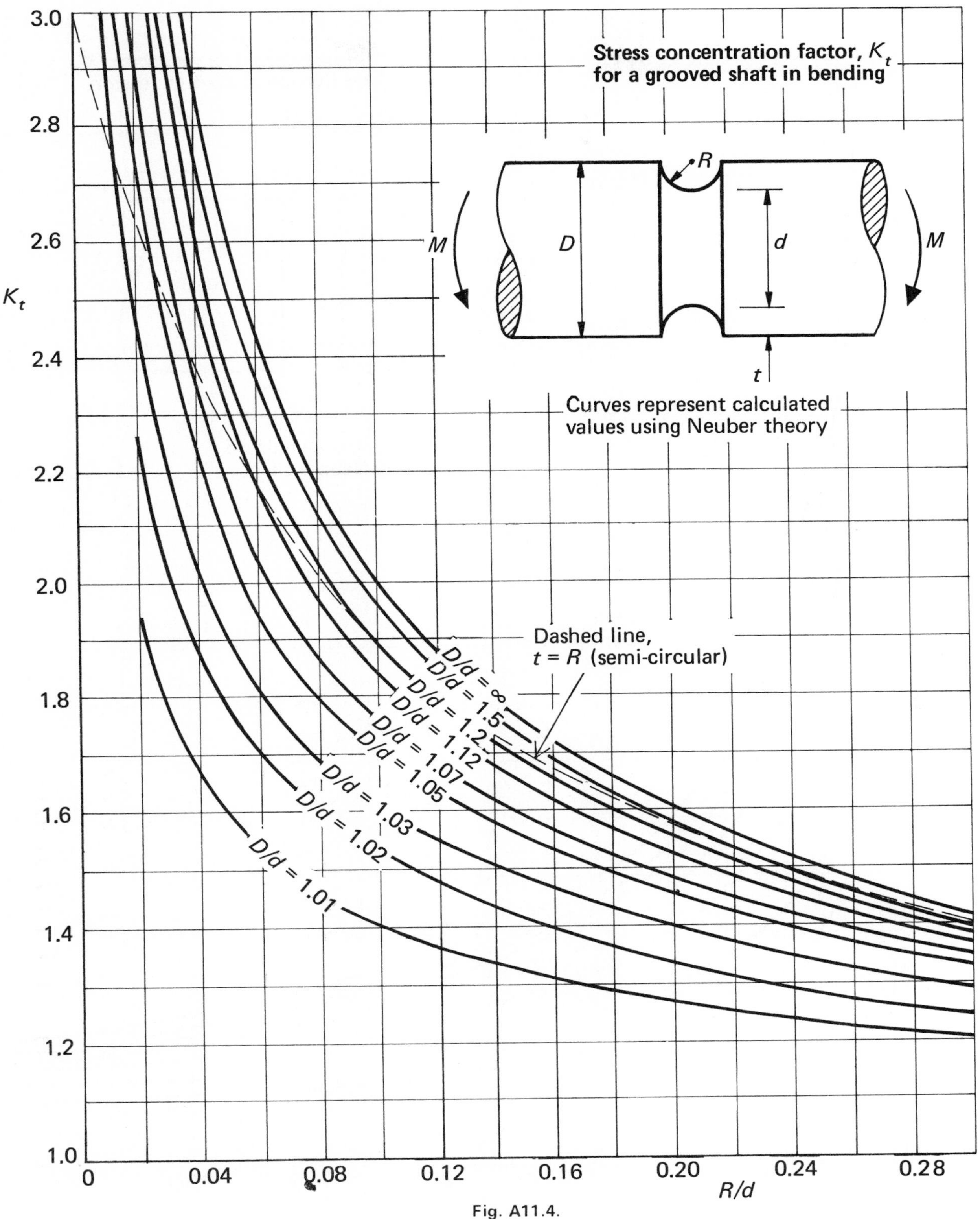
Stress concentration factor, K_t for a grooved shaft in bending
R
M
D
d
M
t
Curves represent calculated values using Neuber theory
Dashed line, $t = R$ (semi-circular)
D/d = ∞
D/d = 1.5
D/d = 1.2
D/d = 1.12
D/d = 1.07
D/d = 1.05
D/d = 1.03
D/d = 1.02
D/d = 1.01
K_t
3.0
2.8
2.6
2.4
2.2
2.0
1.8
1.6
1.4
1.2
1.0
0
0.04
0.08
0.12
0.16
0.20
0.24
0.28
R/d

Fig. A11.4.

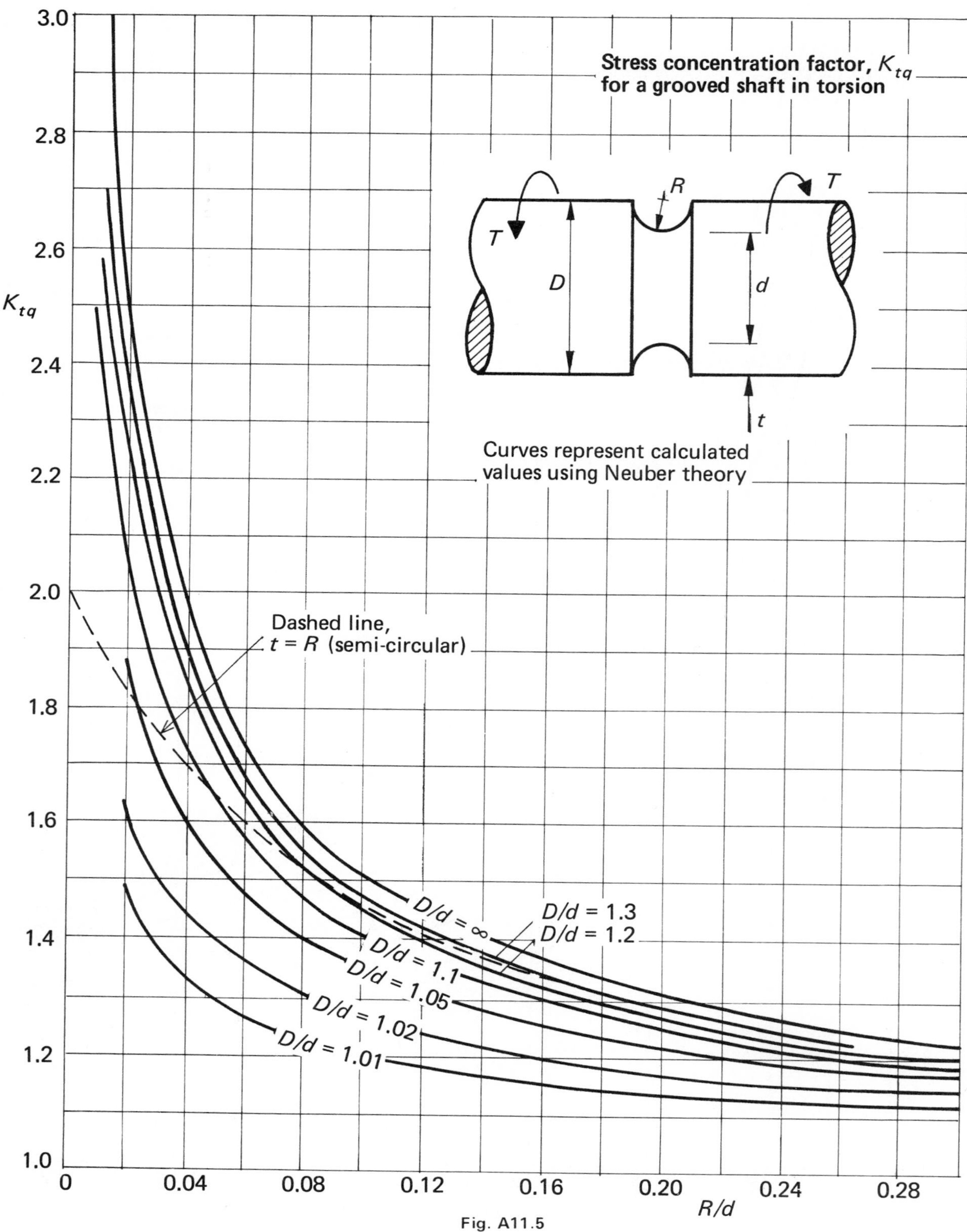
Stress concentration factor, K_{tq}
for a grooved shaft in torsion
T
R
T
D
d
t
Curves represent calculated
values using Neuber theory
Dashed line,
$t = R$ (semi-circular)
$D/d = \infty$
$D/d = 1.3$
$D/d = 1.2$
$D/d = 1.1$
$D/d = 1.05$
$D/d = 1.02$
$D/d = 1.01$
K_{tq}
3.0
2.8
2.6
2.4
2.2
2.0
1.8
1.6
1.4
1.2
1.0
0
0.04
0.08
0.12
0.16
0.20
0.24
0.28
R/d

Fig. A11.5

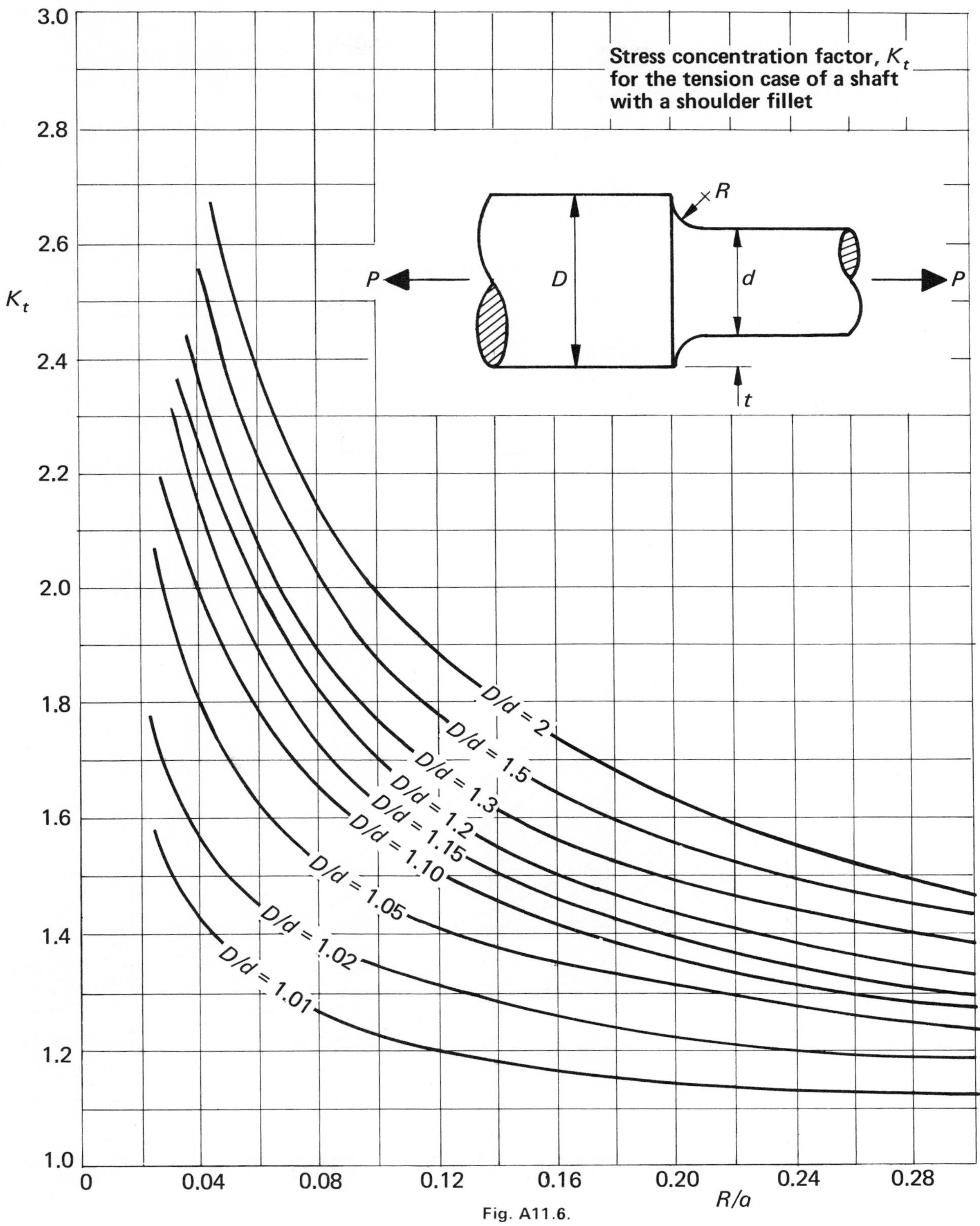
Stress concentration factor, K_t for the tension case of a shaft with a shoulder fillet
R
P
D
d
P
t
K_t
3.0
2.8
2.6
2.4
2.2
2.0
1.8
1.6
1.4
1.2
1.0
D/d = 2
D/d = 1.5
D/d = 1.3
D/d = 1.2
D/d = 1.15
D/d = 1.10
D/d = 1.05
D/d = 1.02
D/d = 1.01
0
0.04
0.08
0.12
0.16
0.20
0.24
0.28
R/a

Fig. A11.6.

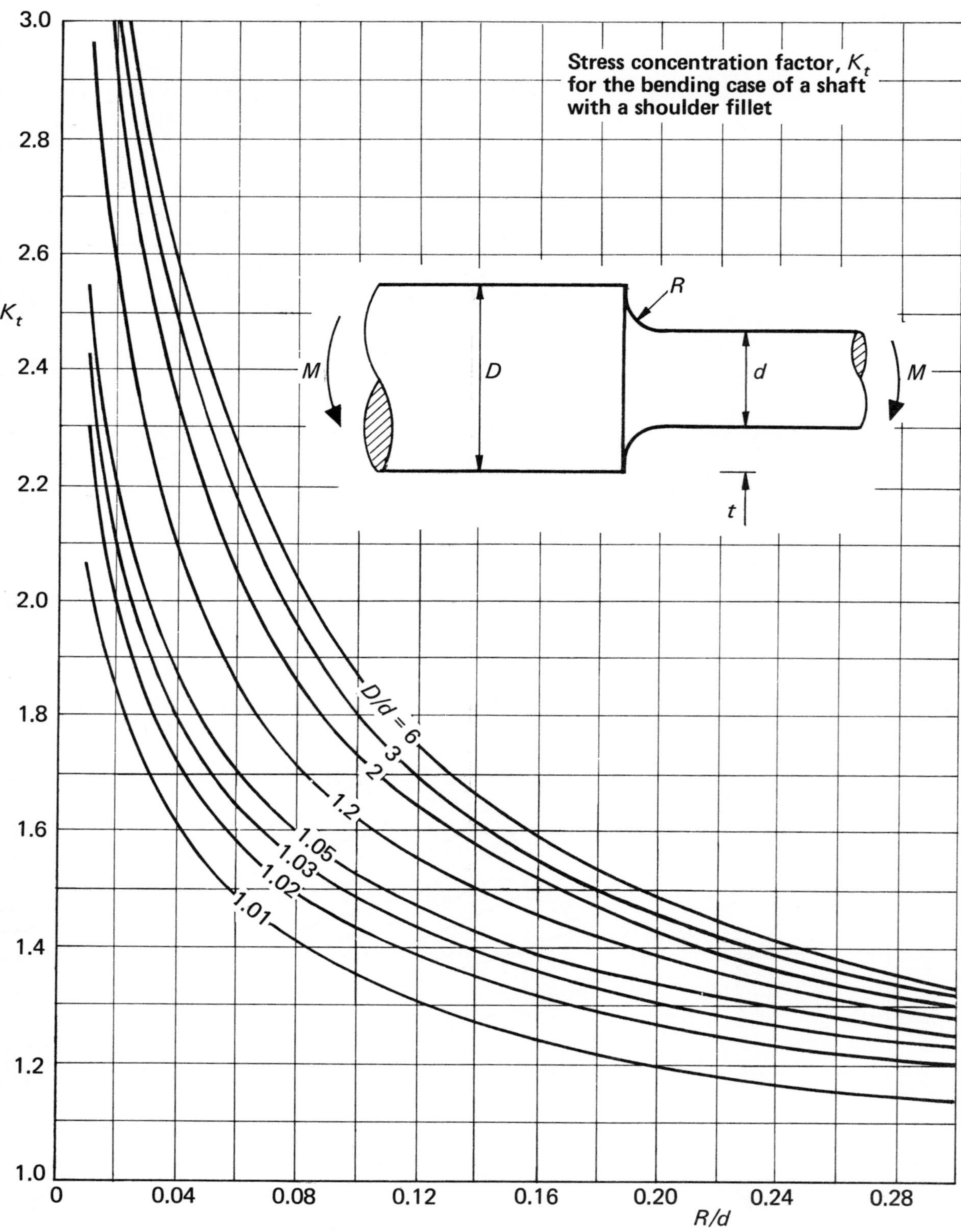
Stress concentration factor, K_t for the bending case of a shaft with a shoulder fillet
3.0
2.8
2.6
2.4
2.2
2.0
1.8
1.6
1.4
1.2
1.0
K_t
0
0.04
0.08
0.12
0.16
0.20
0.24
0.28
R/d
R
M
D
d
M
t
D/d = 6
3
2
1.2
1.05
1.03
1.02
1.01

Fig. A11.7.

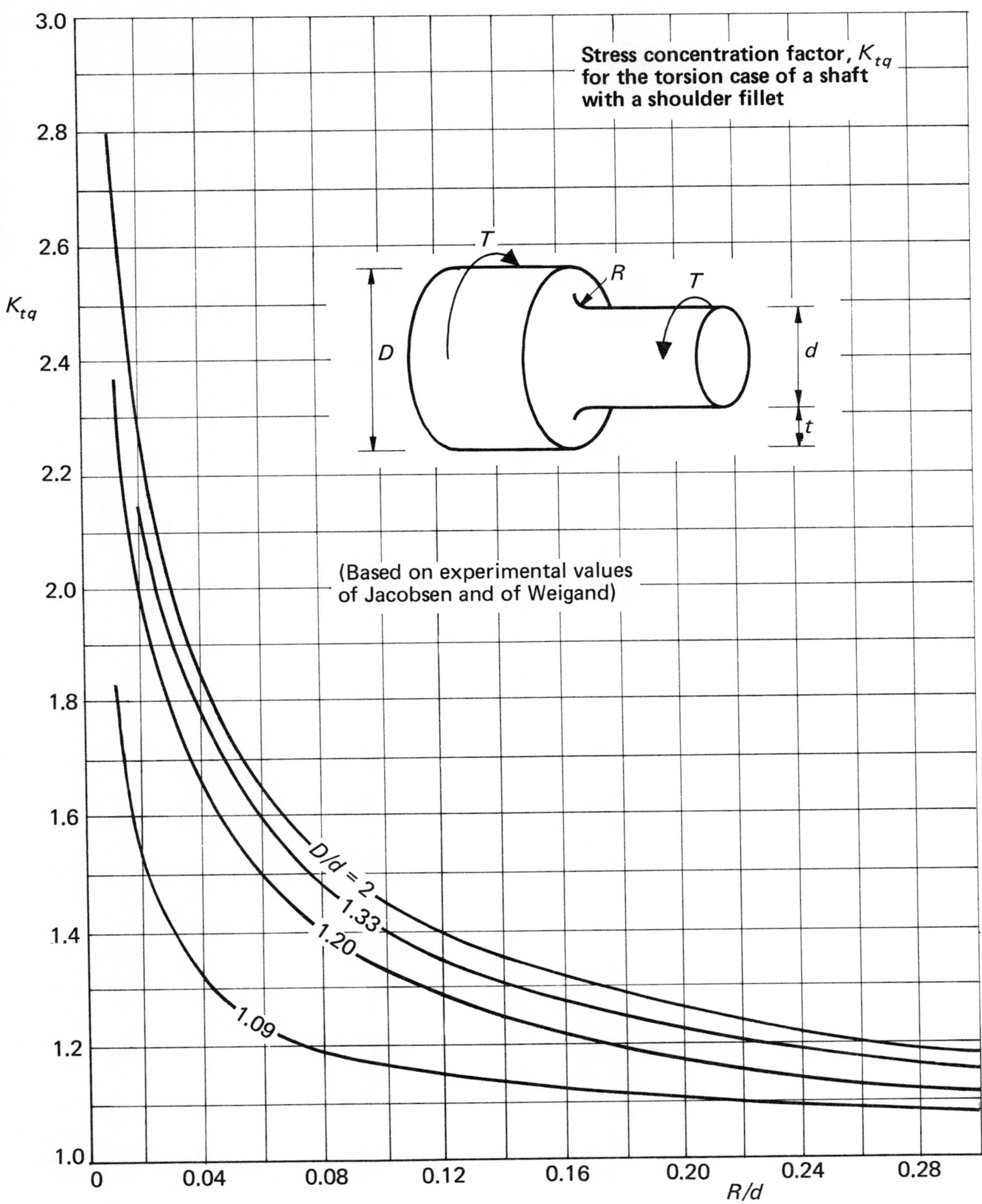

Stress concentration factor, K_{tq} for the torsion case of a shaft with a shoulder fillet
(Based on experimental values of Jacobsen and of Weigand)
K_{tq}
3.0
2.8
2.6
2.4
2.2
2.0
1.8
1.6
1.4
1.2
1.0
0
0.04
0.08
0.12
0.16
0.20
0.24
0.28
R/d
T
R
T
D
d
t
D/d = 2
1.33
1.20
1.09

Fig. A11.8.

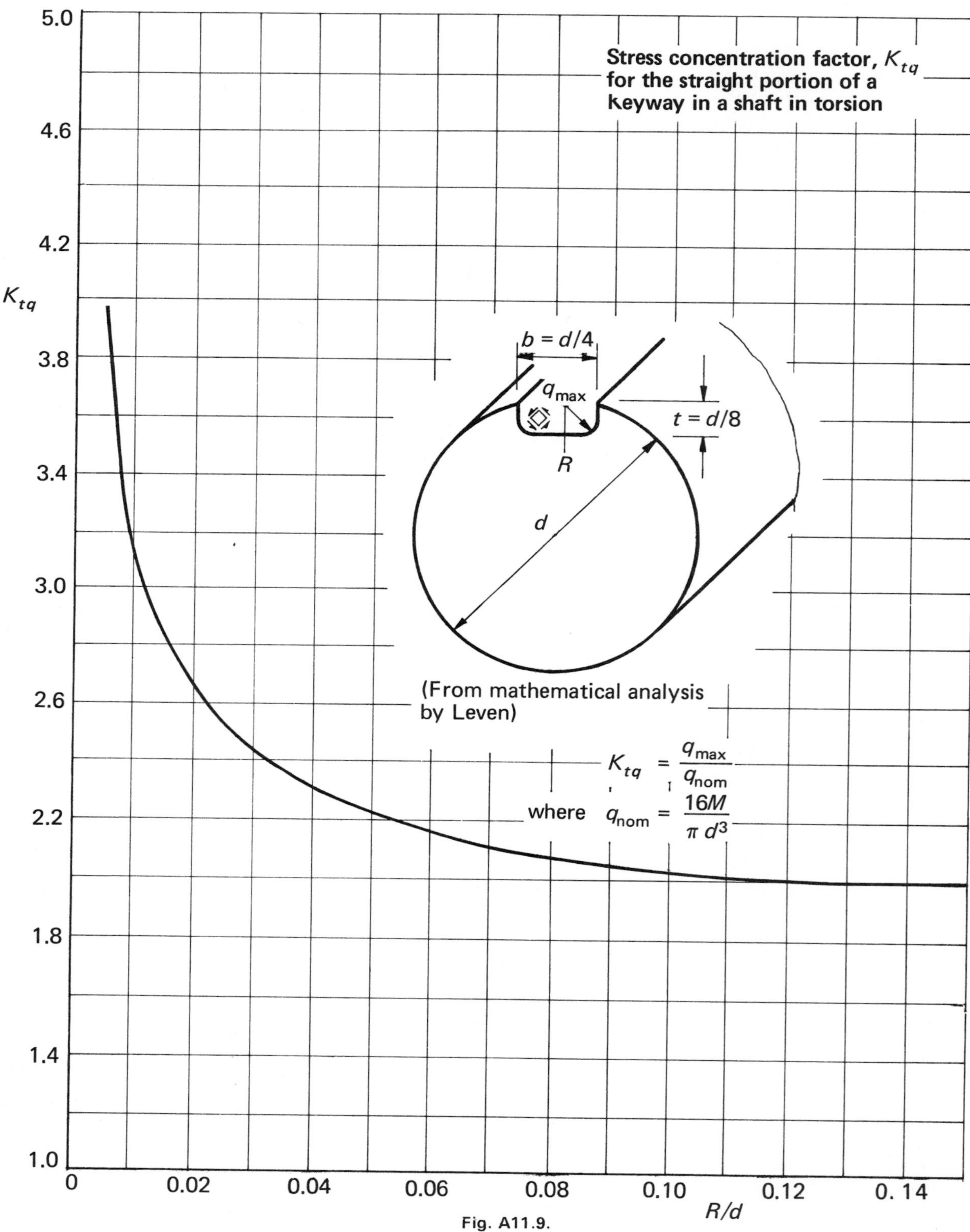

Stress concentration factor, K_{tq} for the straight portion of a keyway in a shaft in torsion
K_{tq}
5.0
4.6
4.2
3.8
3.4
3.0
2.6
2.2
1.8
1.4
1.0
$b = d/4$
q_{max}
$t = d/8$
R
d
(From mathematical analysis by Leven)
$K_{tq} = \frac{q_{max}}{q_{nom}}$
where $q_{nom} = \frac{16M}{\pi d^3}$
0
0.02
0.04
0.06
0.08
0.10
0.12
0.14
R/d

Fig. A11.9.

APPENDIX A12

Prestressing Tendon Details

There are many systems of prestressing tendons and anchorage. The data given here are for one system – the Freyssinet multi-wire system. These data have been adapted from PSC (NZ) Ltd., *Prestressing Tendon Details*, and are published here by permission of Stresscrete (Auckland) Ltd.

Table A12.1 gives details of the Freyssinet multi-wire tendons. Figures A12.1 to A12.4 and Table A12.2 give dimensions of jacking clearances and details of anchorage and anchorage reinforcement.

Table A12.1. Details of Freyssinet multi-wire tendons.

Tendon	Ultimate load*	70% of ultimate load	Weight	Area of cross-section	Sheathing	
					Internal dimensions	
					Beams	Slabs
	(kN)	(kN)	(N/m)	(mm^2)	($mm\,\phi$)	(mm × mm)
12/5 mm	377	264	18.7	244	31.8	38 × 19
12/7 mm	715	500	35.5	463	41.2	44 × 22
12/8 mm	935	655	46.2	605	44.5	51 × 25

*Based on ultimate tensile strength of 1530 MPa.

Table A12.2. Table of measurements.
To be read in conjunction with Figs. A12.1 to A12.4. All dimensions in mm.

Tendon	*A*	*C*	*D*	*E*	*H*	*J*	*L*	*M*	*O*	*P*	*Q**	*R*	*T*	*U*	*V*
12/5 mm	127	51	102	102	89	127	1020	406	685	152	6	152	260	7	6
12/7 mm	152	51	121	127	102	152	1020	406	685	152	8	152	260	7	6
12/8 mm	178	51	152	127	114	178	1020	406	685	152	10	190	235	7	8

**Q* = slip at transfer.

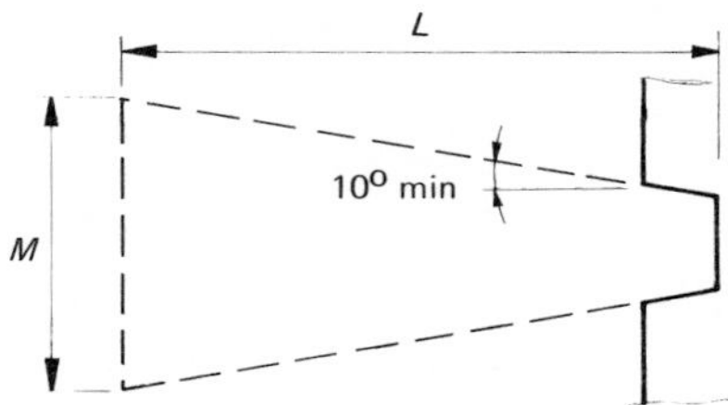

Fig. A12.1. *Jacking clearances behind anchor face. Jacking lengths projecting from face of anchorage—jacking end, O, non-jacking end, P.*

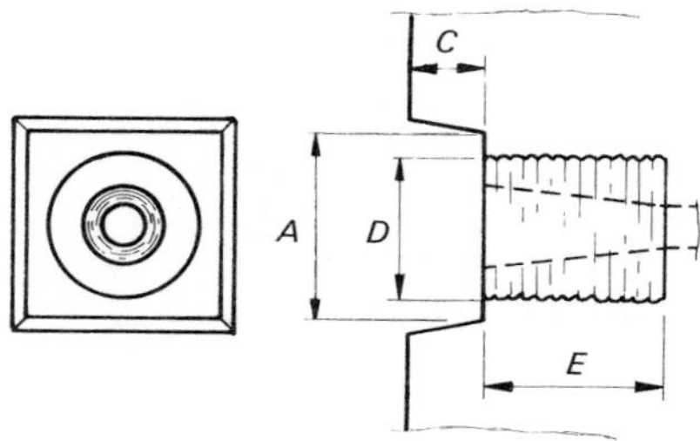

Fig. A12.2. *Anchorage dimensions.*

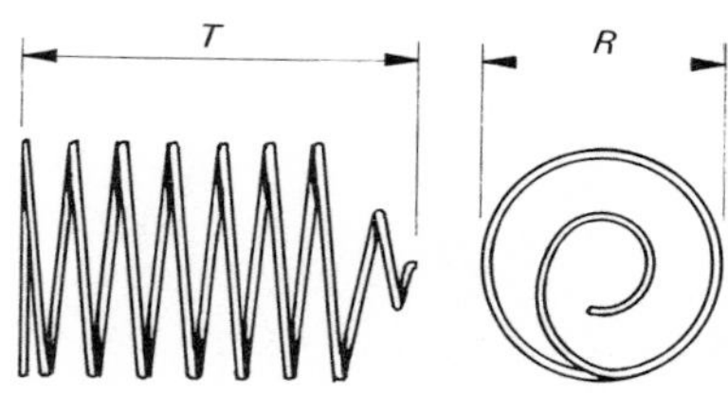

Fig. A12.3. *Anchorage reinforcement. Number of turns = U, material size = V.*

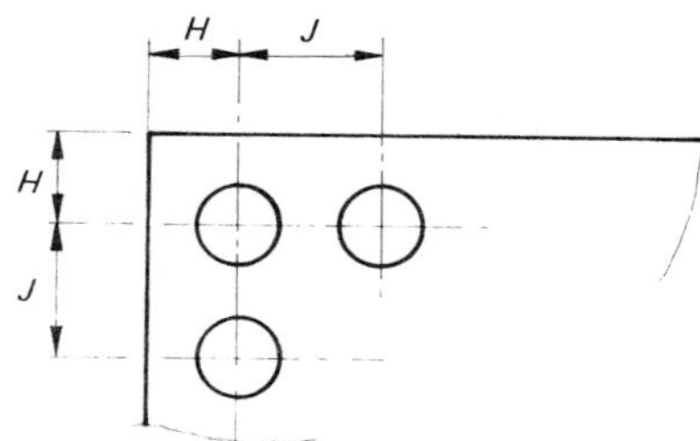

Fig. A12.4. *Anchorage centres and edge distances.*

APPENDIX A13
Dimensions and Tolerances for Square and Rectangular Keys and Keyways

Design information for keys and keyways is given in BS 4235: Part 1: 1972, *Metric Keys and Keyways, Parallel and Taper Keys.* The information in Table A13.1 has been taken from BS 4235 and is reproduced by permission of the British Standards Institution, 2 Park Street, London W1A 2BS. A selection has been made from the sizes and data covered in BS 4235; these will be adequate for most design purposes.

Table A13.1. Dimensions and tolerances for square and rectangular keys and keyways.

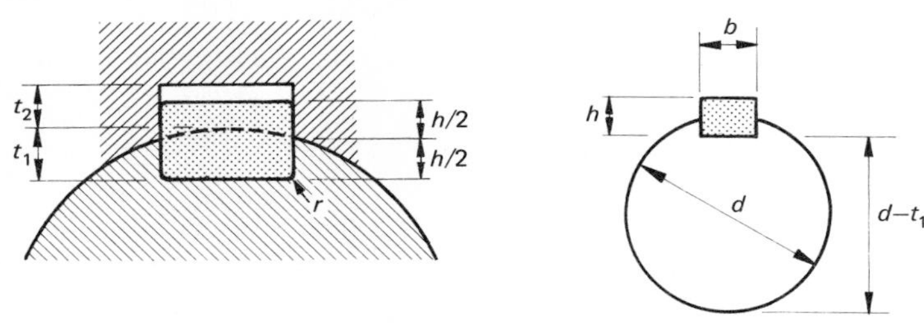

SHAFT diameter d		KEY section $b \times h$	KEYWAY								
			Shaft normal (N9 fit)	Hub (collar) normal (J_s9 fit)	Hub (collar) sliding (D10 fit)	Depth				Radius r	
						Shaft t_1		Hub t_2			
over	to					nom.	tol.	nom.	tol.	max	min
10	17	4 × 4	0 −0.030	±0.015	+0.078 +0.030	2.5	+0.1 0	1.8	+0.1 0	0.16	0.08
17	22	6 × 6	0 −0.030	±0.015	+0.078 +0.030	3.5	+0.1 0	2.8	+0.1 0	0.25	0.16
22	30	8 × 7	0 −0.036	±0.018	+0.098 +0.040	4	+0.2 0	3.3	+0.2 0	0.25	0.16
30	38	10 × 8	0 −0.036	±0.018	+0.098 +0.040	5	↓	3.3	↓	0.40	0.25
38	50	12 × 8	0 −0.043	±0.021	+0.120 +0.050	5	↓	3.3	↓	0.40	0.25
50	65	16 × 10	0 −0.043	±0.021	+0.120 +0.050	6	↓	4.3	↓	0.40	0.25
65	85	20 × 12	0 −0.052	±0.026	+0.149 +0.065	7.5	↓	4.9	↓	0.60	0.40

APPENDIX A14
Allowable Stress in Steel Struts

Table A14.1 shows the allowable compressive stress, p_c, on the gross cross-section of axially loaded steel struts as a function of the slenderness ratio. These stresses are for Grade 43 steel.

Table A14.1 is an extract from BS 449: Part 2: 1969, *The Use of Structural Steel in Building*, "Metric Units", reproduced by permission of the British Standards Institution, 2 Park Street, London, W1A 2BS.

Table A14.1. Allowable stress, p_c, for axially loaded struts of Grade 43 steel.

l/r	p_c (N/mm^2) for Grade 43 steel									
	0	**1**	**2**	**3**	**4**	**5**	**6**	**7**	**8**	**9**
0	155	155	154	154	153	153	153	152	152	151
10	151	151	150	150	149	149	148	148	148	147
20	147	146	146	146	145	145	144	144	144	143
30	143	142	142	142	141	141	141	140	140	139
40	139	138	138	137	137	136	136	136	135	134
50	133	133	132	131	130	130	129	128	127	126
60	126	125	124	123	122	121	120	119	118	117
70	115	114	113	112	111	110	108	107	106	105
80	104	102	101	100	99	97	96	95	94	92
90	91	90	89	87	86	85	84	83	81	80
100	79	78	77	76	75	74	73	72	71	70
110	69	68	67	66	65	64	63	62	61	61
120	60	59	58	57	56	56	55	54	53	53
130	52	51	51	50	49	49	48	48	47	46
140	46	45	45	44	43	43	42	42	41	41
150	40	40	39	39	38	38	38	37	37	36
160	36	35	35	35	34	34	33	33	33	32
170	32	32	31	31	31	30	30	30	29	29
180	29	28	28	28	28	27	27	27	26	26
190	26	26	25	25	25	25	24	24	24	24
200	24	23	23	23	23	22	22	22	22	22
210	21	21	21	21	21	20	20	20	20	20
220	20	19	19	19	19	19	19	18	18	18
230	18	18	18	18	17	17	17	17	17	17
240	17	16	16	16	16	16	16	16	16	15
250	15									
300	11									
350	8									

APPENDIX A15

Timber Connector Joints

Tables A15.1 and A15.2 contain typical data for the design of joints in timber structures. Both tables are extracts from BSCP 112: Part 2: 1971, *The Structural Use of Timber*, "Metric Units", and are reproduced by permission of the British Standards Institution, 2 Park Street, London, W1A 2BS.

Table A15.1 classifies certain species of timber into groups, for the purpose of joint design. Note that the terms "Imported" and "Home-grown" are relevant to the United Kingdom. Similar groupings of species are available in most places. Table A15.2 shows the basic loads for one toothed-plate connector unit.

These tables are included here to facilitate completion of exercises in the book. Designers in practice have access to more extensive collections of local data.

Table A15.1. Group classification of timbers for permissible loads for connectors.

Group	Standard name	
	Imported	**Home-grown**
J1	Afrormosia Greenheart Gurjun/keruing Iroko Opepe Sapele Teak Jarrah Karri	Ash Beech Oak
J2	Abura African mahogany Douglas fir Pitch pine	Douglas fir Larch
J3	Red meranti/red seraya Canadian spruce Parana pine Redwood Western hemlock Whitewood	Scots pine
J4	Western red cedar	European spruce Sitka spruce

Table A15.2. Dry basic loads for one toothed-plate connector unit in softwood.

Nominal size of connector		Diameter of bolt		Thickness of members*		Load parallel to grain			Load perpendicular to grain		
				Connector on one side only	Connector on both sides and on same bolt	Group J2	Group J3	Group J4	Group J2	Group J3	Group J4
mm	in	mm	in	mm	mm	kN	kN	kN	kN	kN	kN
38	$1\frac{1}{2}$	9.5	$\frac{3}{4}$	16	32	2.47	2.20	1.67	1.78	1.56	1.38
round or square				19	38	2.81	2.52	1.92	1.92	1.66	1.46
				22	44	2.92	2.66	2.05	2.05	1.76	1.53
				25	50	2.96	2.72	2.09	2.18	1.86	1.61
51	2	12.7	$\frac{1}{2}$	16	32	3.83	3.27	2.36	2.45	2.11	1.77
round				19	38	4.26	3.61	2.61	2.61	2.23	1 85
				22	44	4.69	3.96	2.86	2.77	2.34	1.94
				25	50	4.89	4.17	3.03	2.92	2.46	2.03
				29	—	5.05	4.38	3.24	3.12	2.61	2.15
				—	60	5.09	4.44	3.30	3.18	2.65	2.18
				—	63	5.16	4.53	3.38	3.26	2.70	2.22
				36	72	5.30	4.72	3.57	3.47	2.89	2.38
				50	100	5.30	4.72	3.57	3.91	3.33	2.73
51	2	12.7	$\frac{1}{2}$	16	32	4.06	3.46	2.50	2.67	2.30	1.90
square				19	38	4.48	3.80	2.74	2.83	2.42	1.99
				22	44	4.91	4.15	2.99	2.99	2.54	2.08
				25	50	5.11	4.36	3.17	3.14	2.65	2.17
				29	—	5.27	4.57	3.38	3.34	2.80	2.29
				—	60	5.31	4.63	3.44	3.40	2.84	2.32
				—	63	5.38	4.72	3.51	3.48	2.90	2.36
				36	72	5.52	4.92	3.71	3.70	3.07	2.50
				50	100	5.52	4.92	3.71	4.13	3.52	2.87
63	$2\frac{1}{2}$	12.7	$\frac{1}{2}$	16	32	4.68	3.98	2.86	3.29	2.82	2.26
round or square				19	38	5.10	4.32	3.10	3.46	2.94	2.35
				22	44	5.53	4.67	3.35	3.61	3.05	2.44
				25	50	5.73	4.88	3.53	3.76	3.17	2.53
				29	—	5.89	5.09	3.74	3.97	3.32	2.65
				—	60	5.93	5.15	3.80	4.03	3.36	2.68
				—	63	6.00	5.24	3.87	4.10	3.41	2.72
				36	72	6.15	5.43	4.07	4.32	3.59	2.86
				50	100	6.15	5.43	4.07	4.75	4.04	3.23
76	3	12.7	$\frac{1}{2}$	16	32	5.06	4.29	3.08	3.67	3.13	2.49
round				19	38	5.48	4.63	3.33	3.84	3.25	2.58
				22	44	5.92	4.98	3.58	3.99	3.37	2.66
				25	50	6.12	5.19	3.75	4.15	3.48	2.75
				29	—	6.28	5.40	3.96	4.35	3.63	2.87
				—	60	6.32	5.47	4.02	4.41	3.67	2.90
				—	63	6.39	5.56	4.10	4.48	3.73	2.94
				36	72	6.53	5.75	4.29	4.70	3.90	3.08
				50 and over	100 and over	6.53	5.75	4.29	5.14	4.35	3.45
76	3	12.7	$\frac{1}{2}$	16	32	6.17	5.22	3.74	4.78	4.06	3.14
square				19	38	6.59	5.56	3.99	4.95	4.18	3.23
				22	44	7.02	5.91	4.23	5.10	4.30	3.32
				25	50	7.22	6.12	4.41	5.25	4.41	3.41
				29	—	7.38	6.33	4.62	5.46	4.56	3.53
				—	60	7.42	6.40	4.68	5.52	4.60	3.56
				—	63	7.49	6.49	4.75	5.59	4.66	3.60
				36	72	7.64	6.68	4.95	5.81	4.83	3.74
				50 and over	100 and over	7.64	6.68	4.95	6.24	5.28	4.11

*Actual thickness. Intermediate thicknesses may be obtained by linear interpolation.

Index